GENOME SCIENCE

A Practical and Conceptual Introduction to Molecular Genetic Analysis in Eukaryotes

ALSO FROM COLD SPRING HARBOR LABORATORY PRESS

RELATED LABORATORY MANUAL

DNA Science: A First Course, Second Edition

GENOME SCIENCE

A Practical and Conceptual Introduction to Molecular Genetic Analysis in Eukaryotes

David A. Micklos
DNA Learning Center
Cold Spring Harbor Laboratory

Bruce Nash
DNA Learning Center
Cold Spring Harbor Laboratory

Uwe Hilgert
BIO5 Institute
The University of Arizona

COLD SPRING HARBOR LABORATORY PRESS
Cold Spring Harbor, New York • www.cshlpress.org

GENOME SCIENCE

A Practical and Conceptual Introduction
to Molecular Genetic Analysis in Eukaryotes

Printed in the United States of America

Publisher and Acquisition Editor	John Inglis
Director of Development, Marketing, & Sales	Jan Argentine
Developmental Editor	Maria Smit
Project Manager	Inez Sialiano
Permissions Coordinator	Carol Brown
Production Editor	Rena Steuer
Desktop Editor	Susan Schaefer
Production Manager	Denise Weiss
Sales Account Manager	Elizabeth Powers
Cover Designer	Susan Lauter

Front cover artwork: Metaphor for the puzzle of understanding the genomes of living things, represented by a base nucleotide sequence overlaid with an assembled set of chromosomes and gene model. Key organisms explored in narratives and laboratories are represented: Hominids, plants (*Arabidopsis* and maize), and the nematode worm *Caenorhabditis elegans*. Speckled corn kernels remind us that transposons are the major component of most advanced genomes. The fossilized Neanderthal skeleton is emblematic of the evolution of our species—and of our capacity to domesticate plants and contemplate the structure of our own genetic material.

Library of Congress Cataloging-in-Publication Data

Micklos, David A.
Genome science : a practical and conceptual introduction to molecular genetic analysis in eukaryotes / David A. Micklos, Uwe Hilgert, Bruce Nash.
p. cm.
Includes bibliographical references and index.
ISBN 978-1-621821-09-0 (hard cover : alk. paper)
1. Genomics. 2. Eukaryotic cells. 3. Molecular genetics. 4. Molecular biology.
I. Hilgert, Uwe, 1960- II. Nash, Bruce, 1968- III. Title.

QH447.M495 2013
572.8'629--dc23

2012007495

Contents

Preface

The DNA Learning Center's effort to develop lab experiments for high school and colleges dates to 1985, when Dave Micklos teamed with Greg Freyer to develop a sequence of experiments to make and analyze a recombinant DNA molecule. The initial testing was done in Rich Roberts' lab, well before he won the Nobel Prize, and incorporated key insights on inserting DNA into bacteria from the dean of transformation, Doug Hanahan. The lab sequence was initially a slim lab manual called "Recombinant DNA for Beginners," whose title was derived from the Graham Nash album "Songs for Beginners." It expressed perfectly our ideal of making complicated gene technology accessible to novices.

Within a year, we were training teachers across the country on summer tours of one, then two, Vector vans. These were customized Ford Econoline vans that packed enough equipment, reagents, and supplies to convert any general science lab into a molecular genetics lab. Along the way, we developed a complementary minitext that presented the concepts behind the labs, as well as extensions of recombinant DNA technology in basic and applied research. The much expanded work was formally published in 1991 as the Cold Spring Harbor Laboratory Press book *DNA Science.* That name, which Jim Watson threw out over lunch one day, seemed to capture the excitement of a new science based on the ability to manipulate and understand the DNA molecule.

DNA Science, now in its second edition, has sold more than 90,000 copies and is credited with helping to catalyze the movement to bring hands-on experiments with DNA into high school and beginning college classrooms. Two experiments found their way into the Advanced Placement biology curriculum, giving these experiments a nationwide audience. Stand-alone kits, developed with Carolina Biological Supply Company, reach well over 100,000 students per year.

The experiments in *DNA Science* are based exclusively on bacterial genetic systems. Now, *Genome Science* aims to take students to a higher level of biological and technological integration—to study the function of eukaryotic genes and genomes. Nineteen laboratories focus on four revolutionary technologies—polymerase chain reaction, DNA sequencing, RNA interference (RNAi), and bioinformatics—across three eukaryotic systems: humans, plants, and *Caenorhabditis elegans.* All labs stress the modern synthesis of molecular biology and computation, integrating in vitro experimentation with in silico bioinformatics. In addition to well-tested biochemical methods, *Genome Science* introduces *DNA Subway*, an intuitive bioinformatics platform that makes easy work of gene and genome analysis.

The four major techonologies are organized into stand-alone chapters with extensive text introductions that place related labs into a common historical and conceptual framework. This modular approach provides options to develop new courses or to integrate labs into existing courses or student research. We especially hope that these protocols will help educators to extend research to classroom settings and distribute experiments in which multiple classes analyze and contribute to common data sets. DNA barcoding is especially amenable to "campaigns" in which many students contribute to understanding diversity within a common biogeographical unit.

Genome Science borrows many user-friendly features from its predecessor, including flow charts, marginal notes, reagent recipes, and extensive instructor information. To ease implementation, most labs are available as ready-to-use kits from Carolina Biological Supply Company. In addition, like its predecessor, *Genome Science* aims to help beginners use modern tools to explore the unseen world of genes and genomes.

In contemplating the cosmos in 1927, the great mathematical geneticist J.B.S. Haldane famously said, "My own suspicion is that the universe is not only queerer than we suppose, but queerer than we can suppose." Had he been alive today, Haldane would almost certainly have the same suspicion about the genomes of higher organisms. In this sense, genome scientists are the new cosmologists of biology, uncovering the strange and beautiful structure of the genetic material that runs through all life.

Happy explorations.

David Micklos
Bruce Nash
Uwe Hilgert

Cold Spring Harbor, New York
March 2012

Acknowledgments

Many people have contributed time, effort, and resources over the years. *Genome Science* would not have been possible without the following individuals.

Jennifer Aizenman, Mark Bloom, Scott Bronson, Tom Bubulya, Greg Chin, Jeanette Collette, Craig Hinkley, Jermel Watkins, and Jason Williams helped develop and test the protocols. Adrian Arva, Matt Christensen, Cornel Ghiban, and Mohammed Khalfan developed our bioinformatics websites and tools. Eun-Sook Jeong, Sue Lauter, and Chun-hua Yang provided art, design, and editing. Stephen Blue turned our ideas into artwork.

Alejandro Sánchez Alvarado and the University of Utah gave permission to use an RNAi targeting plasmid. Nobel Laureate Andy Fire and the Carnegie Institution of Washington graciously provided materials. Ed Lee and Suzy Lewis helped us to implement the Apollo annotation tool and Mark Stoeckle inspired us to make barcoding accessible to students. The iPlant Collaborative took us into the world of cyberinfrastructure and the staff at the DNA Learning Center gave endless support and patience.

Susan Schaefer, Inez Sialiano, Maria Smit, Rena Steuer, and Denise Weiss, and everyone else at Cold Spring Harbor Laboratory Press worked hard and patiently to put this book together and we thank Bruce Stillman, who has kept education at the forefront of Cold Spring Harbor Laboratory's mission. Cold Spring Harbor Laboratory Scientists Greg Hannon, David Jackson, Leemor Joshua-Tor, Rob Martienssen, Dick McCombie, Lincoln Stein, Marja Timmermans, Doreen Ware, and other investigators at Cold Spring Harbor Laboratory shared ideas and materials.

Thanks also to the students and teachers who have tested our labs and guided their improvement, Charles and Helen Dolan who provided us an inspirational place to work, and Jim Watson and Bruce Stillman who developed Cold Spring Harbor Laboratory as a unique village of science.

The biochemical experiments and bioinformatics interfaces used in *Genome Science* evolved over a 20-year period with generous support from the National Science Foundation's Advanced Technological Education, Course Curriculum and Laboratory Improvement, Plant Cyberinfrastructure, Plant Genome Research, and Transforming Undergraduate Education in Science; Applied Biosystems; Biogen Idec; Carolina Biological Supply Company; Eppendorf North America; Roche Molecular Systems; the Alfred P. Sloan Foundation; the Department of Energy Human Genome Project Ethical, Legal, and Social Issues Research Program; the Howard Hughes Medical Institute Precollege Science Education Initiative; and the William A. Haseltine Foundation for Medical Sciences and the Arts.

CHAPTER 1

Genome as Information

Hans Winkler
(Reprinted, with permission, from Brabec F. 1955. Berichte der Deutschen Botanischen Gesellschaft 68: 27, ©Wiley-Blackwell; courtesy of Hunt Institute for Botanical Documentation, Carnegie Mellon University, Pittsburgh.)

J.B.S. Haldane
(Reprinted from Clark RW, 1969. *J.B.S.: The life and work of J.B.S. Haldane*, ©Coward-McCann.)

THE TERM GENOME WAS COINED IN 1920 by the German botanist Hans Winkler. A combination of the words *gene* and chromos*ome*, a genome is the set of genes, located on one or more chromosomes, that defines a living organism. The human genome, for example, is composed of ~25,000 genes that encode proteins needed to carry out the processes of life within the several trillion cells that make up our bodies.

Also working at the beginning of the 20th century, the American geneticist Thomas Hunt Morgan provided an enduring mental picture of the genome as a collection of genes arranged on each chromosome like beads on a string. With the realization that a chromosome is a linear DNA molecule, the concept of "genome" has been expanded to mean the entire sequence of DNA nucleotides or "letters" (A, T, C, and G) that compose the haploid (half set) of chromosomes of an individual. With advances in DNA sequencing technology during the past 30 years, we can now rapidly determine the entire nucleotide sequence of any organism. For humans, this amounts to ~3.2 billion "letters" in the set of chromosomes inherited from one's mother or father.

Although much work is focused on decoding genes that specify proteins, genes also specify several types of RNA molecules that are not translated into proteins. Many unexpected RNA genes have been identified in the past decade, and more unusual sorts of genes may be found in the future. Moreover, almost 99% of the human genome is composed of "spaces" within and between protein-coding genes whose purpose is not fully understood. Included in the non-protein-coding portion of the genome are regulatory sequences that control how genes express their protein products at different times and places. Almost half of the human genome is occupied by so-called "jumping genes" or their remnants, some of which move about using a mechanism that is shared with human immunodeficiency virus (HIV) and other retroviruses. During the evolution of higher living things, genomes have been extensively remodeled by the duplication of individual chromosomes and the exchange of pieces between chromosomes.

Genomes are thus considerably more complicated than originally envisioned by Winkler. We are coming to understand that the genome is both a dynamic structure that changes through evolutionary time and a dynamic concept that changes with our increasing knowledge. For most higher organisms, including humans, Morgan's analogy of a genome should now be envisioned as strings of different sorts spliced together into a very long strand, with many bits of beads scattered here and there.

This book is designed to provide the conceptual and experimental background needed to participate in the new science of genomes. The geneticist J.B.S. Haldane once famously said "The universe is not only queerer than we suppose but queerer than we *can* suppose." With its black holes, curved space, and unaccounted-for dark matter, Haldane's prediction for the universe has certainly come true. The exploration of the human and other genomes is just beginning. They, too, are turning out to be queerer and more exciting than previously imagined and promise to reveal many more surprises in the future.

ESTABLISHING THE PHYSICAL BASIS OF HEREDITY

Living things preserve their own lineages through reproduction—by creating offspring that carry on their inheritance through successive generations. Humans intuitively understand that they pass on some of their physical traits to their children. Thus, men and women endeavor to select vigorous, healthy mates who can give birth to vigorous, healthy children. Over time, the production of vigorous offspring contributed to the development of humans with traits that adapted them to live in a variety of environments. The extension of this concept to other organisms led to the domestication of plants and animals.

During the last 150 years, scientists have sought an increasingly explicit explanation of the hereditary process that allows traits to be passed from one generation to another. Let us start by taking a brief look back to the history of the quest to understand the physical basis of heredity, which is the foundation of genome science.

Charles Darwin (ca. 1859)
(Courtesy of the American Museum of Natural History Library.)

In his 1859 book *On the Origin of Species*, the Englishman Charles Darwin described how heredity operates in populations of organisms, enabling them to adapt to different environmental conditions. In the process of evolution by natural selection, members of the same and different species compete for limited resources needed for survival. The fittest members of a population are more likely to reproduce. On rare occasions, a random physical change in an individual increases its ability to adapt to environmental conditions or exploit new food resources. This "adaptive" change increases the individual's chance to survive and reproduce. Adaptive changes are more likely passed on to offspring, who, in turn, are fitter than their peers; they also have a greater chance of surviving to pass on their physical characteristics to succeeding generations. In this way, beneficial traits accumulate within a population of organisms. Through the process of adaptive radiation, populations expand into new environments and evolve to exploit specialized food resources, thus limiting competition and increasing their chances for survival.

Gregor Mendel (ca. 1860)
(Courtesy of the Austrian Press and Information Service.)

Although Darwin proposed an incorrect mechanism of heredity, termed "pangenesis," he did not know the physical source of individual variation upon which his evolutionary processes acted or how it was passed on to successive generations. In his paper "Experiments in Plant Hybridization," published in 1865, the Moravian monk Gregor Mendel described the hereditary process at the level of the individual organism and provided a mechanism to drive evolution. From the results of controlled crosses of garden peas, he showed that traits are inherited in a predictable manner as "factors," which we now call genes. Mendel related each plant trait to a pair of genes, one of which is inherited from each parent. Although common sense suggests that offspring are a mixture of parental traits, Mendel showed that the parental genes governing each trait do not blend. Instead, each parental gene is maintained as a discrete bit of hereditary information, unchanged through generations.

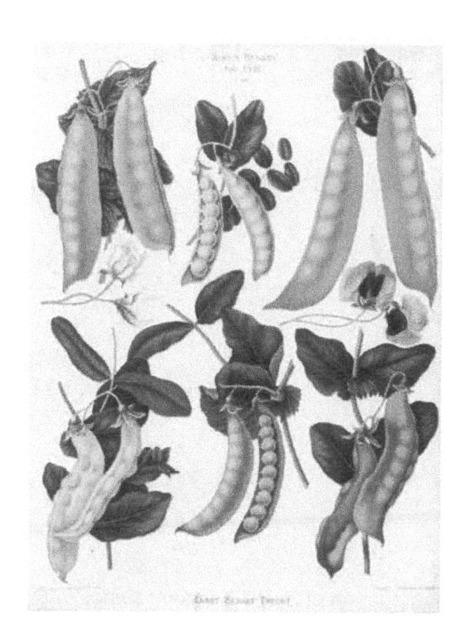

Some pea traits examined by Mendel. Album Bernay (1876–1893) shows some of the pea traits that Mendel examined.

(The John Innes Archives, courtesy of the John Innes Foundation.)

Mendel's notion that a trait is determined by a pair of genes presented a potential problem. If parents pass on both copies of a gene pair, then their offspring would end up with four genes for each trait. This doubling of genetic material would continue in ensuing generations. Mendel deduced that parents contribute only half of their gene set to their offspring. He hypothesized that the gene number is reduced during gametogenesis, so that each gamete (sex cell) receives one copy of each gene pair. During fertilization, the male and female gametes then fuse to restore each pair of genes in the offspring.

Wilhelm Johannsen

(Courtesy of Hunt Institute for Botanical Documentation, Carnegie Mellon University, Pittsburgh, Pennsylvania.)

Mendel's work went essentially unnoticed for 35 years. Then, in 1900, the Dutchman Hugo de Vries, the German Carl Correns, and the Austrian Erich von Tschermak-Seysenegg rediscovered Mendel's paper and published research data that confirmed his earlier work. de Vries realized that Mendel's "factors" were the same entities that he called "pangens," which he had derived from Darwin's "pangenesis." In 1909, Wilhelm Johannsen shortened the term to "gene" and also coined the words "genotype" and "phenotype" to refer to an organism's genetic composition (genes) and its observable characteristics (traits).

In 1902, Theodor Boveri, at the University of Würzburg, and Walter Sutton, a student at Columbia University, were the first to directly relate heredity to chromosome behavior. Boveri found that a sea urchin egg fertilized by two sperm produces daughter cells that divide asymmetrically and have incomplete sets of chromosomes. Sutton found that the genetic material of the grasshopper *Brachystola* consists of 11 pairs of chromosomes and that gametes formed during meiosis receive only one chromosome from each pair. Then, independent work in 1905 by Nettie Stevens and Edmund Wilson (Sutton's mentor at Columbia) showed that sex is determined by separate X and Y chromosomes, with females having two X chromosomes (XX) and males having a single X and Y chromosome (XY). During meiosis, each egg receives a single copy of an X chromosome, whereas each sperm receives either an X or a Y chromosome. These behaviors exactly paralleled the segregation of Mendel's hereditary factors into parental gametes and suggested that genes are physically located on the chromosomes.

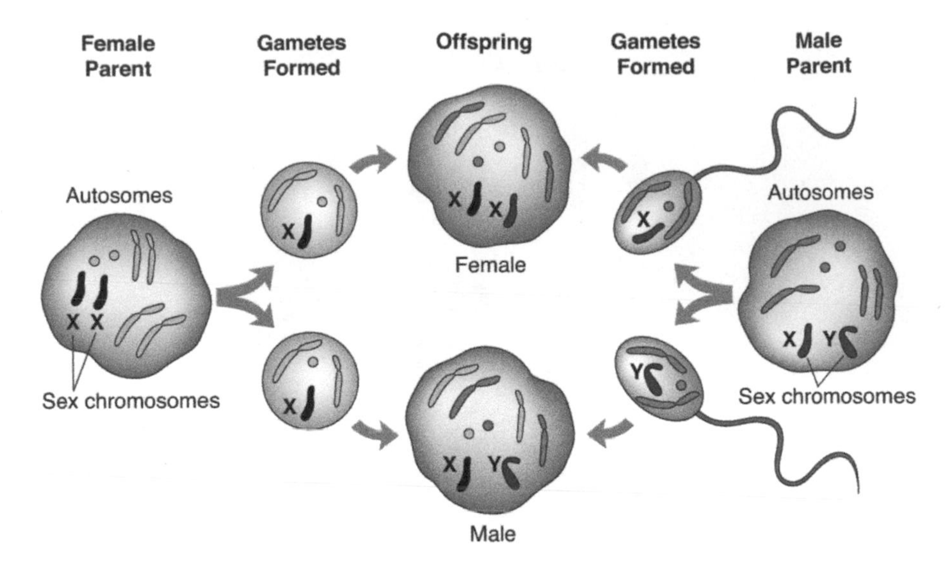

Segregation of X and Y chromosomes in *Drosophila.*

Conclusive evidence that genes are located on chromosomes became available during the second decade of the 20th century. During this period, Thomas Hunt Morgan and his bright cadre of students at Columbia University—Alfred Sturtevant, Calvin Bridges, and Hermann Muller—established the physical basis of heredity. Working with the common fruit fly *Drosophila melanogaster* in 1910, Morgan's group identified a mutation that produces white-colored eyes (as opposed to the normal red color). First, their Mendelian analyses showed that white eyes were confined to males in most crosses, suggesting that white eye color is a sex-linked recessive trait. This meant that the gene for eye color was located on the X chromosome. Next, they identified more than 80 additional mutants and showed that sets of genes are "linked" or inherited together as if they are a single physical unit. All genes sorted into four linkage groups, which corresponded to the number of *Drosophila* chromosomes seen under a microscope.

Thomas Hunt Morgan (ca. 1917), *left*, Courtesy of the American Society of Zoologists; Calvin Bridges in the "fly room" at Columbia University (ca. 1926), *middle*, courtesy of the American Society of Zoologists; and Alfred Sturtevant, *right*, courtesy of the American Philosophical Society.

Frans Alfons Janssens

(Courtesy of the Centre of Microbial and Plant Genetics, K.U. Leuven.)

Working at the Catholic University of Leuven in 1909, Frans Alfons Janssens found that, early in meiosis, homologous chromosomes intertwine and exchange pieces—a process that became known as "crossing-over." Morgan realized that crossing-over could provide a measure of the relative distance between two genes. He reasoned that closely linked genes will rarely be separated by crossing-over, but genes that are far apart will be frequently separated. Therefore, the lower the crossover frequency between two genes, the closer together they should be on the chromosome. Alfred Sturtevant provided support for this concept in his 1913 doctoral thesis, when he made a map of the relative locations of three genes on the *Drosophila* X chromosome.

It was not until 1931, however, that Barbara McClintock and Harriet Creighton, at Cornell University, obtained direct cytological proof of genetic crossing-over. Working in maize, they related the phenotypes caused by gene crossovers to the coinheritance of a visible chromosome "knob." In the same year, Curt Stern, at the University of Berlin, used a similar approach to study the X chromosome of *Drosophila*. Taken together, these experiments conclusively proved that genes reside on chromosomes and are arrayed at specific points along their length.

A clear understanding of the physical basis of heredity came with the discovery that DNA is the genetic material. The first clue came from experiments conducted in

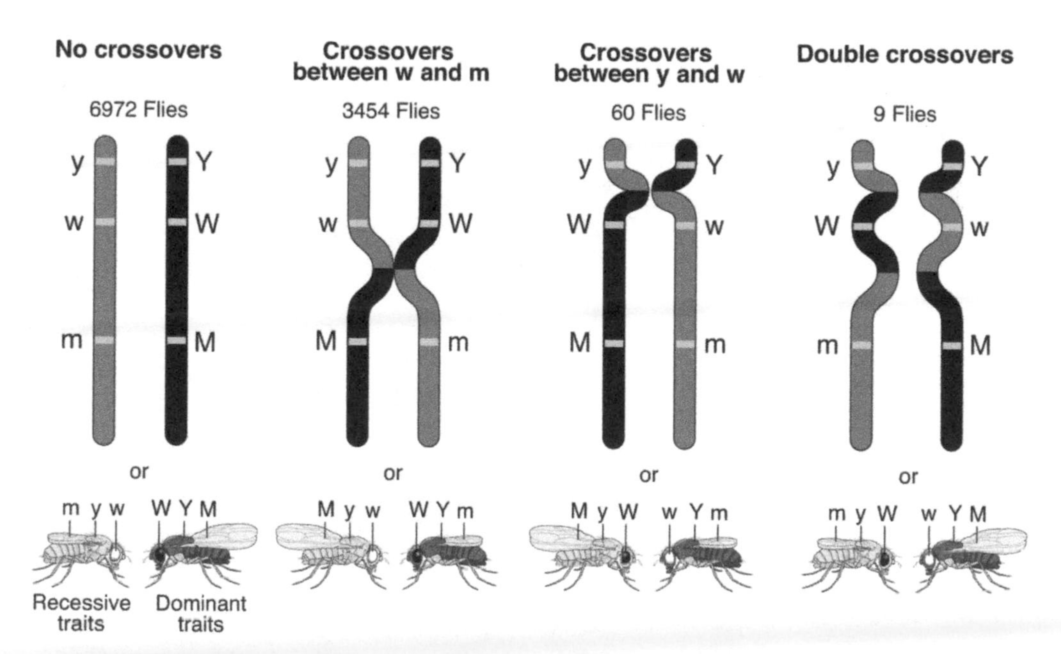

Sturtevant's linkage experiment in *Drosophila*, 1913. Sturtevant examined the X-linked inheritance of three recessive traits: yellow body (*y*), white eyes (*w*), and miniature wings (*m*). He crossed recessive males (y,w,m) with heterozygous females having recessive genes on one X chromosome (*y,w,m*) and dominant genes on the other (*Y,W,M*). Because a male parent can only contribute recessive genes on its single X chromosome, the phenotypes of both male and female offspring are due entirely to the inheritance of the maternal X chromosomes. Mendelian analysis predicts that all of the 10,495 offspring in Sturtevant's experiment would show either a purely dominant phenotype, normal body/eye color/wings, or a purely recessive phenotype, yellow body/*white* eyes/miniature wings. However, offspring inherited various mixtures of dominant and recessive traits. Sturtevant deduced that the mixed phenotypes were caused by genetic exchange between a female's two X chromosomes during gamete formation. The frequency of exchange is a measure of the distance between two genes located on the same chromosome.

Fred Griffith
(From www.wikipedia.org.)

1928 by the English microbiologist Fred Griffith with two strains of pneumococcus bacteria. A virulent smooth (S) strain possesses a smooth polysaccharide capsule that is essential for a pneumonia infection, whereas a nonvirulent, rough (R) strain lacks this outer capsule. Following injection with the S strain, mice succumb in several days to pneumonia. Although neither living R strain nor heat-killed S strain caused illness when injected alone, Griffith found that coinjecting the two produced a lethal infection. Furthermore, he retrieved virulent S strains from mice infected with this mixture of bacteria. He concluded that some principle from the dead S bacteria had "transformed" the innocuous R strain, allowing it to produce the polysaccharide capsule required for virulence.

Although Griffith's experiment hinted at an involvement of metabolism, genes were still known only by their outward manifestation as visible traits. However, in 1941, George Beadle and Edward Tatum at Stanford University finally showed that the job of a gene is to produce a specific enzyme (protein). Their experiment used the simple red bread mold *Neurospora*, which is able to synthesize amino acids and vitamins from simple components (sucrose, salts, and biotin). After exposing *Neurospora* to X rays, they identified strains that grew only when supplemented with a specific amino acid or vitamin. They concluded that, for each deficient strain, irradiation had mutated a single gene that produces an enzyme needed to synthesize one amino acid or vitamin.

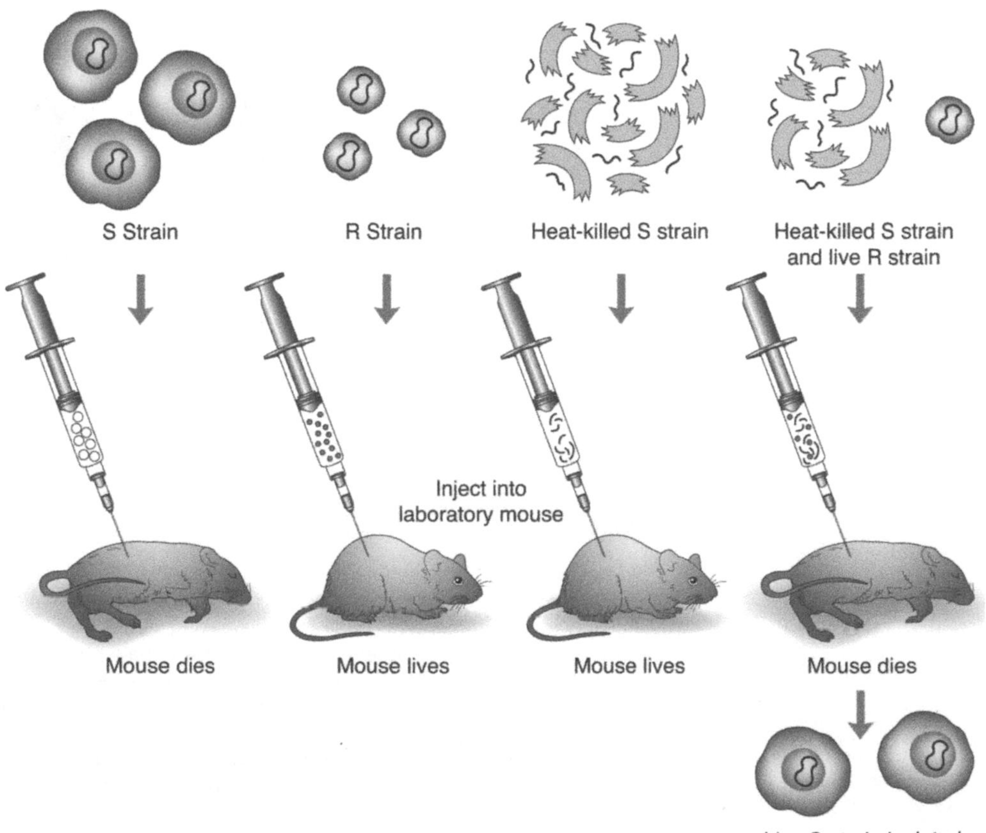

Griffith's transformation experiment with smooth (S) and rough (R) strains of pneumococcus, 1928.

Oswald Avery (center foreground) and associates, 1932. (Seated, *left to right*) Thomas Francis Jr., Avery, and Walther F. Goeble; (standing) Edward E. Terrell, Kenneth Goodner, Rene J. Dubos, and Frank H. Babers. (Courtesy of the Rockefeller Archive Center.)

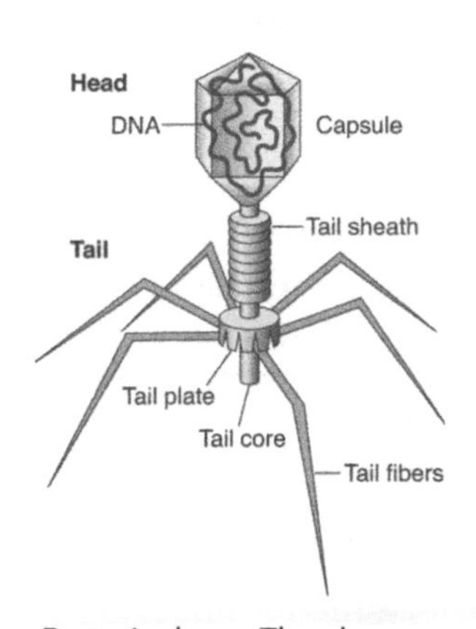

Bacteriophage. The phage particle is essentially a protein capsule surrounding a core of DNA.

In the meantime, Oswald Avery, Collin MacLeod, and Maclyn McCarty followed up on Griffith's transformation experiments at the Rockefeller Institute. They purified the "transforming principle" from killed S bacteria that had readily induced R bacteria to synthesize the outer capsule. Transforming activity was unaffected by treatment with trypsin and chymotrypsin (which digest protein) and ribonuclease (RNase, which digests RNA). However, deoxyribonuclease (DNase, which digests DNA) destroyed all transforming activity, and analysis of molecular composition and weight indicated that the active fraction was primarily DNA. In 1944, they concluded that "The inducing [transforming] substance has been likened to a gene, and the capsular antigen which is produced in response to it has been regarded as a gene product." Thus, the Rockefeller group provided conclusive evidence that a gene is made of DNA.

Lingering dogma that protein was the genetic material prevented most scientists from focusing on DNA until the so-called "blender" experiment was conducted in 1952 by Alfred Hershey and Martha Chase at the Carnegie Department of Genetics at Cold Spring Harbor. They used a bacterial virus, or bacteriophage (phage), which is simply composed of an outer capsule of protein and an inner core of DNA. They attached different radioactive labels to the phage protein and DNA, allowed the phage time to infect bacteria, and then agitated the culture in a Waring blender to detach the phage particles from the bacteria. After centrifuging to separate the detached phages

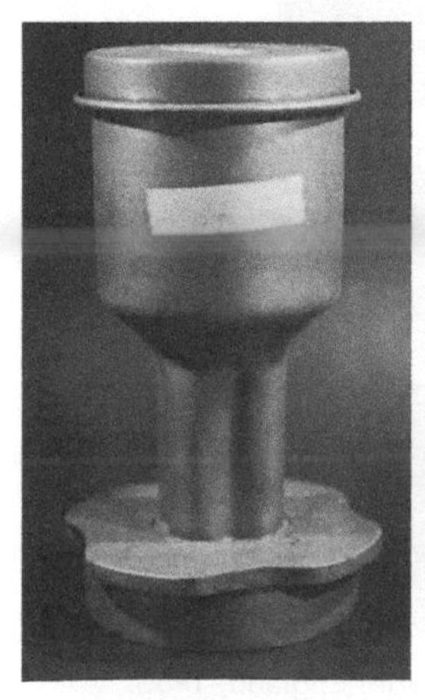

Waring blender used in the Hershey-Chase experiment.

Martha Chase and Alfred Hershey, 1953. (Courtesy of Cold Spring Harbor Laboratory Archives.)

from the bacterial cells, they found that the radioactive DNA that remained with the bacterial fraction was sufficient to produce a new generation of phages. This work further strengthened the concept that DNA is the hereditary material that comprises genes.

DNA AS INFORMATION

Friedrich Miescher
(From www.wikipedia.com.)

Ironically, DNA was discovered in 1869, only 10 years after the publication of Darwin's *On the Origin of Species* and 4 years after Mendel's "Experiments in Plant Hybridization." A Swiss doctor, Friedrich Miescher, isolated a substance he called "nuclein" from the large nuclei of white blood cells. His source of cells was pus from soiled surgical bandages. Building upon Miescher's observation that the substance was rich in phosphorus and nitrogen, by 1900, it had been determined that nuclein was a long molecule composed of three distinct chemical subunits: an acidic phosphate, five types of nitrogen-rich bases (adenine, thymine, guanine, cytosine, and uracil), and a five-carbon sugar. By the 1920s, two forms of nucleic acids were differentiated by virtue of their sugar composition: ribonucleic acid (RNA), based on ribose sugar, and deoxyribonucleic acid (DNA), based on deoxyribose sugar. These forms were also found to differ slightly in base composition; thymine is found exclusively in DNA, whereas uracil is found only in RNA.

The structure of the DNA molecule was solved in 1953 by James Watson and Francis Crick, working at the Cavendish Laboratory in Cambridge, England. They constructed a metal model that showed DNA to resemble a twisting ladder—with the

James Watson and Francis Crick with their DNA model in Cambridge, England, 1953.
(From A. Barrington Brown, Photo Researchers, Inc.)

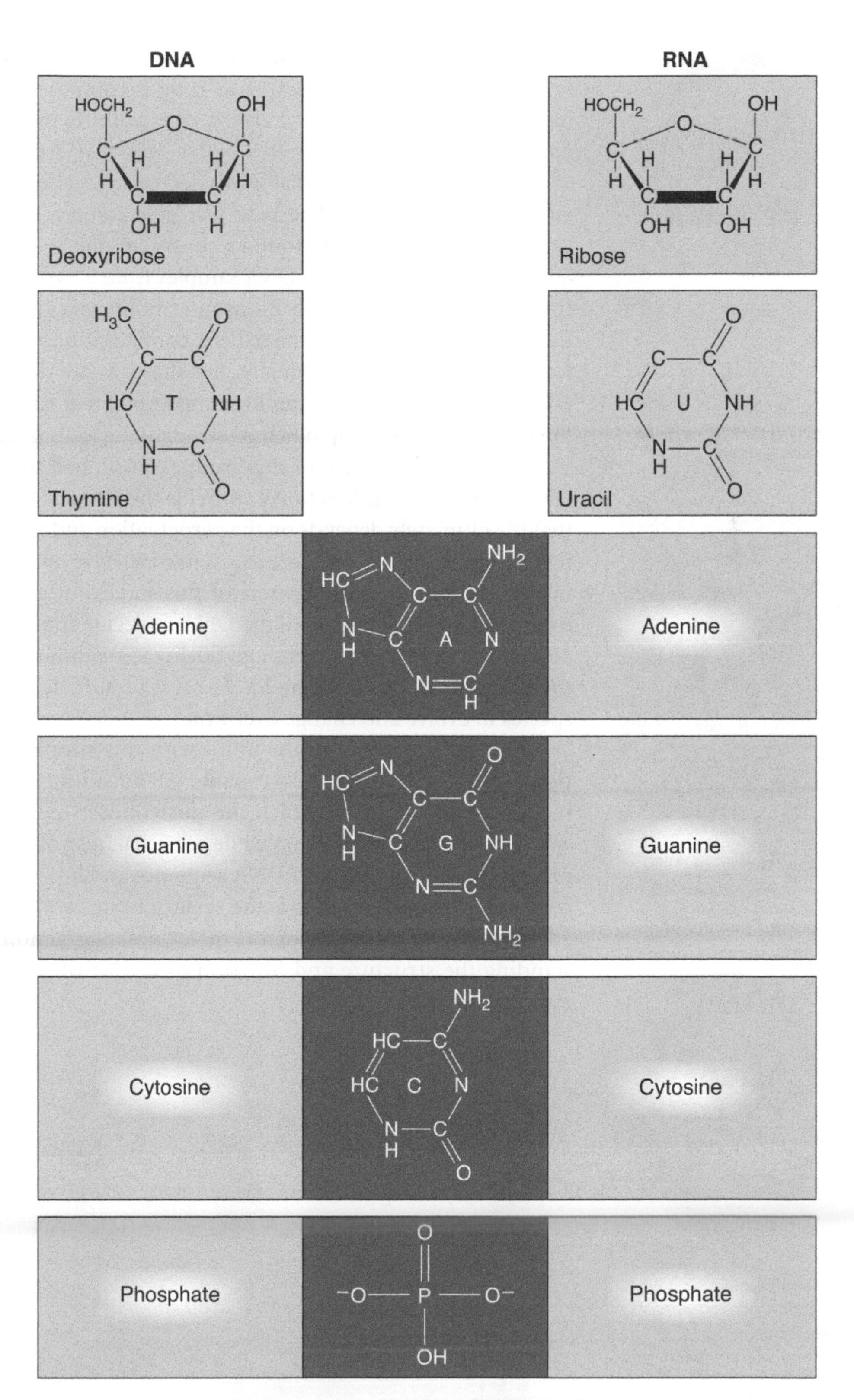

Components of DNA and RNA molecules.
(Art concept developed by Lisa Shoemaker.)

Erwin Chargaff, 1947
(Courtesy of Cold Spring Harbor Laboratory Archives.)

rails formed of alternating units of deoxyribose sugar and phosphate and the rungs formed of nitrogenous bases. Each rung is composed of a two nitrogenous bases, a base pair, where adenine (A) always pairs with thymine (T) and guanine (G) always pairs with cytosine (C).

The Watson-Crick model was based on critical information that had accumulated quickly since 1950. The base-pair rule came from work by Erwin Chargaff of Columbia University, who found a consistent one-to-one ratio of adenine to thymine and guanine to cytosine in DNA samples from a variety of organisms. Linus Pauling, Robert Corey, and Herman Branson at California Institute of Technology provided the atomic dimensions of the α-helix configuration of protein, in which amino acids form a helical structure. Finally, the sharp X-ray diffraction photographs of DNA taken by Maurice Wilkins and Rosalind Franklin at Kings College, London, resembled the patterns of the protein helix—strongly suggesting that DNA is also an α-helix.

Linus Pauling, ca. 1950
(Courtesy of the Archives, California Institute of Technology.)

As the only biologist of this group, Watson had the greatest insight into how the DNA molecule must function to provide the physical basis of heredity. He understood that life ultimately depends on the perpetuation and amplification of a DNA sequence through time. A successful organism must survive and pass on its genome to succeeding generations, and the bearers of this successful genome will increase in number over time. On the one hand, the DNA molecule must be sufficiently stable so that a sequence is inherited with enough fidelity to maintain the identity of each species. On the other hand, the DNA molecule must be sufficiently plastic—mutable—to allow species to evolve and change over time.

Watson later came to the point with this simple definition: "DNA is information." A DNA molecule is capable of encoding information in its nucleotide sequence—the order in which the nucleotides A, T, C, and G follow one another along one strand of the molecule. The balance of this chapter will explore how information is encoded in a DNA sequence and how the DNA sequence of a genome is analyzed. Bioinformatics is the science of understanding the information encoded in DNA and other biological molecules, and genomics is the science of understanding the structure and function of the set of DNA molecules that distinguish each species.

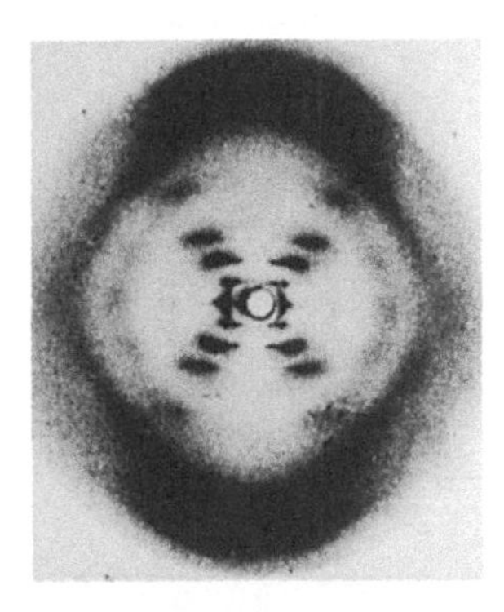
Rosalind Franklin's X-ray diffraction photograph of DNA, 1953
(Reprinted, with permission, from Franklin RE, Gosling RG. 1953. *Nature* 171: 740–741, ©Macmillan; photo courtesy of Cold Spring Harbor Laboratory Archives.)

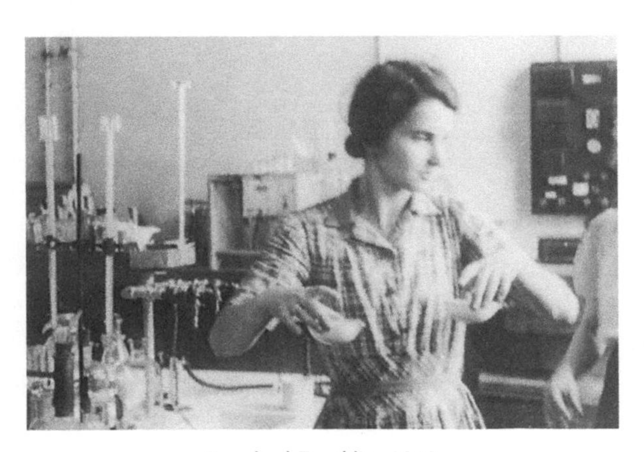
Rosalind Franklin, 1948
(Courtesy of Anne Sayre.)

Maurice Wilkins, ca. 1955
(Courtesy of Cold Spring Harbor Laboratory Archives.)

THE GENETIC CODE

Work in the 1950s and 1960s showed how DNA encodes information and provided the mechanism for Beadle and Tatum's hypothesis that one gene makes one protein. First, Paul Zamecnik, at Massachusetts General Hospital, established that protein synthesis takes place on protein/RNA conglomerates located in the cytoplasm, which we now know as ribosomes. Then, Jerard Hurwitz, at New York University School of Medicine, and Samuel Weiss, at University of Chicago, independently identified RNA polymerase as the enzyme that synthesizes RNA by adding complementary nucleotides to a DNA template. Subsequently, three different RNA polymerases (I, II, and III) were identified in higher organisms. RNA polymerase I synthesizes ribosomal RNA (rRNA), RNA polymerase II synthesizes messenger RNA (mRNA), and RNA polymerase III synthesizes transfer RNA (tRNA) and one small rRNA (5S rRNA).

In addition, Benjamin Hall and Sol Spiegelman, at the University of Illinois, showed that complementary RNA and DNA sequences bind together to form a stable heteroduplex. Collaborators Sydney Brenner (MRC Laboratory), François Jacob (Institut Pasteur), and Matthew Meselson (Harvard University) and a team composed of James Watson, François Gros, and Walter Gilbert (Harvard University) independently showed that immediately after a bacteriophage infects a bacteria, RNA is synthesized and associates with ribosomes. Moreover, the newly synthesized RNA only lasts for several minutes inside the bacterial cells. Taken together, these experiments illuminated the first step of protein production—transcribing the DNA code (a gene) into a complementary RNA code (a messenger RNA).

The next step was to work out how the RNA code is translated into an amino acid code. Whereas RNA is made up of only four different nucleotides (A, C, G, and U), proteins are composed of 20 different amino acids. So it was immediately apparent that a combination of several nucleotides would be required to encode each amino acid. A two-letter code would only have 16 combinations—not enough to specify all 20 amino acids. However, a three-letter code provided more than enough combina-

Sol Spiegelman, ca. 1963
(Courtesy of Cold Spring Harbor Laboratory Archives.)

François Gros and Walter Gilbert, ca. 1970
(Courtesy of Cold Spring Harbor Laboratory Archives.)

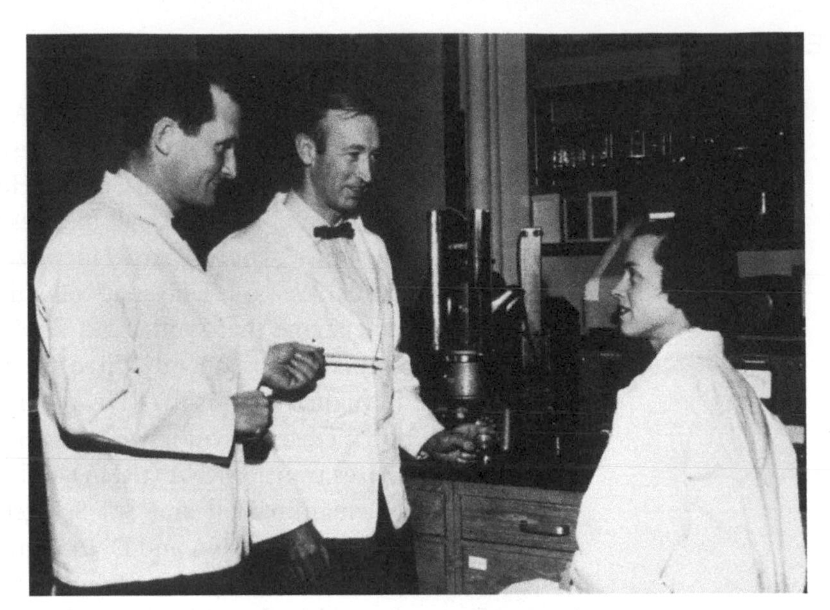

Mahlon B. Hoagland, Paul C. Zamecnik, and Mary L. Stephenson, ca. 1956
(Courtesy of the National Library of Medicine.)

Marshall W. Nirenberg, ca. 1962
(Courtesy of the National Institutes of Health.)

Har Gobind Khorana, ca. 1966
(Courtesy of Cold Spring Harbor Laboratory Archives.)

tions (64), and Francis Crick and Sydney Brenner referred to this nucleotide triplet as a codon. At the same time, Francis Crick realized that some sort of "adaptor" molecule was needed to link a codon to a corresponding amino acid.

Robert Holley, working at Cornell University, discovered that the adaptor molecule is a class of small RNA molecules, 70–80 nucleotides in length, that covalently bind amino acids; these are tRNAs. Then Paul Zamecnik and Mahlon Hoagland, at Massachusetts General Hospital, discovered a class of enzymes called aminoacyl tRNA synthetases that attach a specific amino acid to a specific tRNA. Each tRNA contains a loop structure with a unique three-nucleotide-long sequence—the anticodon—that binds to a complementary codon in mRNA. This aligns the specified amino acid at the ribosome for addition to a polypeptide chain.

By 1966, the laboratories of Marshall Nirenberg (the National Institutes of Health) and Har Gobind Khorana (University of Wisconsin) had broken the genetic code through which mRNA instructs tRNAs to add specific amino acids at the ribosome. Both researchers synthesized RNA molecules composed of repeating units of a single codon, added the synthetic mRNA to a cell-free extract containing all the required tRNAs bound to amino acids, and then monitored the composition of proteins that were synthesized. Initially, Nirenberg found that polyuracil (making codons UUU-UUU-UUU...) produced a protein made up solely of the amino acid phenylalanine. Eventually, all possible codon combinations were tried, yielding a complete genetic "dictionary" for the translation of mRNA into amino acids. Nearly all proteins begin with the amino acid methionine (Met); scientists quickly realized that its codon (AUG) represents the "start" signal for protein synthesis. Three codons for which there are no naturally occurring tRNAs—UAA, UAG, and UGA—are "stop" signals that terminate translation.

Interestingly, only two amino acids, methionine and tryptophan, are specified by a single codon; all other amino acids are specified by two or more different codons.

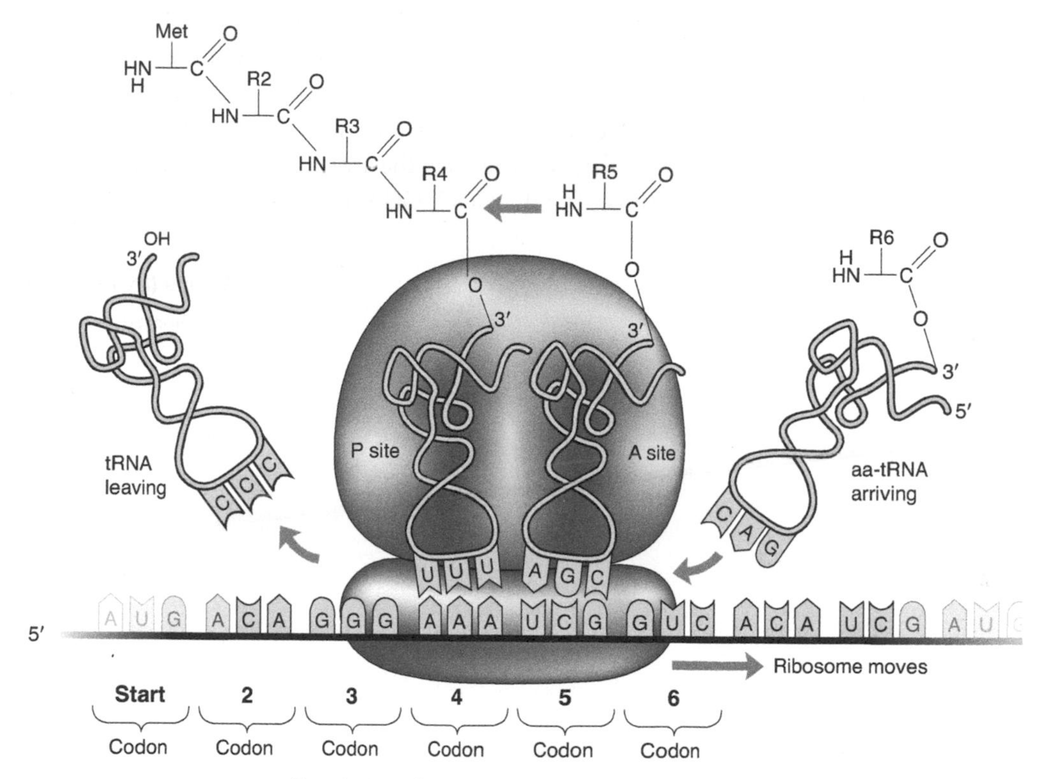

Translation of mRNA codons into amino acids.

2nd position of codon

1st position of codon (5′ terminus)	U	C	A	G	3rd position of codon (3′ terminus)
U	UUU Phe	UCU Ser	UAU Ty	UGU Cys	U
	UUC Phe	UCC Ser	UAC Tyr	UGC Cys	C
	UUA Leu	UCA Ser	UAA Stop	UGA Stop	A
	UUG Leu	UCG Ser	UAG Stop	UGG Trp	G
C	CUU Leu	CCU Pro	CAU His	CGU Arg	U
	CUC Leu	CCC Pro	CAC His	CGC Arg	C
	CUA Leu	CCA Pro	CAA Gln	CGA Arg	A
	CUG Leu	CCG Pro	CAG Gln	CGG Arg	G
A	AUU Ile	ACU Thr	AAU Asn	AGU Ser	U
	AUC Ile	ACC Thr	AAC Asn	AGC Ser	C
	AUA Ile	ACA Thr	AAA Lys	AGA Arg	A
	AUG Met	ACG Thr	AAG Lys	AGG Arg	G
G	GUU Val	GCU Ala	GAU Asp	GGU Gly	U
	GUC Val	GCC Ala	GAC Asp	GGC Gly	C
	GUA Val	GCA Ala	GAA Glu	GGA Gly	A
	GUG Val	GCG Ala	GAG Glu	GGG Gly	G

The genetic code.

Because of this redundancy—also referred to as degeneracy or wobble—single-nucleotide mutations in DNA are often of no functional consequence. Changing a single nucleotide in a degenerate codon to another triplet coding for the same amino acid has no effect on the amino acid sequence of a protein. For example, any codon beginning with GG specifies the amino acid glycine regardless of the nucleotide in the third position (GGU, GGC, GGA, or GGG).

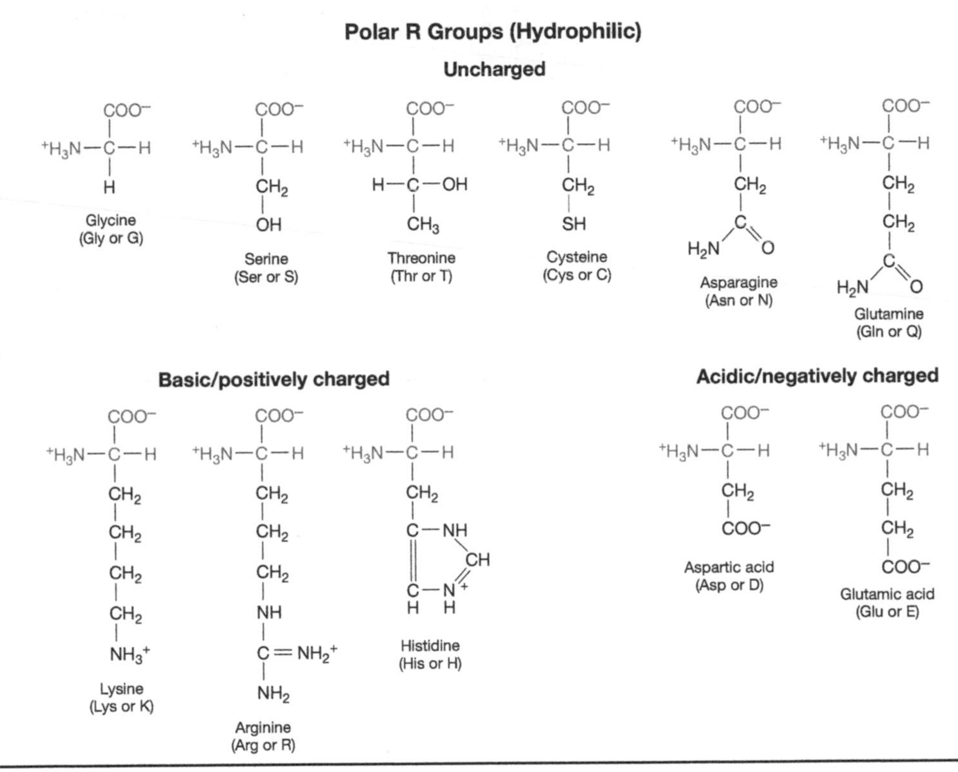

Twenty naturally occurring amino acids grouped by properties. The side chains (gray) determine the characteristic properties of each amino acid.

FINDING SIMPLE PATTERNS IN DNA SEQUENCE

Any sequence of characters (letters or numbers) may be randomly generated or encoded with meaningful information. A receiver might reject a message that is random or, if the sender uses a consistent set of rules to convey meaning, a receiver can decode a message. If we think of hereditary information stored in DNA as a language or code, we can use the English language as a model to introduce some principles of DNA sequence analysis.

It can sometimes be difficult to determine whether a sequence encodes information. Although it is not immediately evident, the meaning of the sequence of 1064 letters (below) becomes clear when we add in the conventions of word spacing and punctuation.

Understanding DNA sequence is not nearly as simple as looking at a language we have learned over a lifetime. English-language speakers may intuitively know that English has meaning, but we can use simple statistics to show, on another level, how a sequence of letters conveys meaning. Frequency analysis evaluates the occurrence of characters to determine if a sequence is random or potentially conveys meaning. First, let's compare

Themainchallengeinbiologywastoundersta
ndgenereplicationandthewayinwhichgenes
controlproteinsynthesisItwasobviousthatth
eseproblemscouldbelogicallyattackedonlyw
henthestructureofthegenebecameknownTh
ismeantsolvingthestructureofDNAThenthis
objectiveseemedoutofreachtotheinterestedg
eneticistsButinourcolddarkCavendishlabwe
thoughtthejobcouldbedonequitepossiblywi
thinafewmonthsOuroptimismwaspartlybas
edonLinusPaulingsfeatindeducingthealpha
helixWealsoknewthatMauriceWilkinshadcr
ystallineXraydiffractionphotographsfromD
NAandsoitmusthaveawelldefinedstructure
Therewasthusananswerforsomebodytoget
Duringthenexteighteenmonthsuntilthedou
blehelicalstructurebecameelucidatedwefre-
quentlydiscussedthenecessitythatthecorrect
structurehavethecapacityforselfreplication
Andinpessimisticmoodsweoftenworriedtha
tthecorrectstructuremightbedullThatisitwo
uldsuggestabsolutelynothingandexciteusno
morethansomethinginertlikecollagenThefi
ndingofthedoublehelixthusbroughtusnoto
nlyjoybutgreatreliefItwasunbelievablyintere
stingandimmediatelyallowedustomakease-
riousproposalforthemechanismofgenedu-
plication

The main challenge in biology was to understand gene replication and the way in which genes control protein synthesis. It was obvious that these problems could be logically attacked only when the structure of the gene became known. This meant solving the structure of DNA. Then this objective seemed out of reach to the interested geneticists. But in our cold, dark Cavendish lab, we thought the job could be done, quite possibly within a few months. Our optimism was partly based on Linus Pauling's feat in deducing the alpha helix... We also knew that Maurice Wilkins had crystalline X-ray diffraction photographs from DNA and so it must have a well-defined structure. There was thus an answer for somebody to get. During the next eighteen months, until the double helical structure became elucidated, we frequently discussed the necessity that the correct structure have the capacity for self-replication. And in pessimistic moods, we often worried that the correct structure might be dull. That is, it would suggest absolutely nothing and excite us no more than something inert like collagen. The finding of the double helix thus brought us not only joy but great relief. It was unbelievably interesting and immediately allowed us to make a serious proposal for the mechanism of gene duplication.

El principal desafio en la biologia fue de comprender replica de gene y la manera en las que genes controlan sintesis de proteina. Fue obvio que estos problemas podrian ser atacados logicamente solo cuando la estructura del gene llego a ser conocida. Este destinado resolviendo la estructura de ADN. Entonces este objetivo parecio fuera de alcance a los genetistas interesados. Pero en nuestro frio, laboratorio oscuro de Cavendish, nosotros pensamos que el trabajo podria ser hecho, bastante posiblemente dentro de unos pocos meses. Nuestro optimismo fue basado en parte en la proeza de Linus Pauling a deducir la helice alfa... Nosotros tambien supimos que Maurice Wilkins tenia fotografias de cristal de difraccion de radiografia de ADN y tan debe tener una estructura bien definida. Habia asi una respuesta para alguien conseguir. Durante los proximos dieciocho meses, hasta que la doble estructura helicoidal llegara a ser aclarada, nosotros discutimos con frecuencia la necesidad que la estructura correcta tiene la capacidad para la auto-replica. Y en humors pesimistas, nosotros a menudo preocupamos que la estructura correcta quizas sea languida. Eso es, sugeriria absolutamente que nada y no nos emociona más que algo inerte quiere colageno. El hallazgo de la doble helice asi nos trajo no solo alegria pero gran alivio. Fue in-creiblemente interesante e inmediatamente nos permitio hacer una propuesta grave para el mecanismo de duplicacion de gene.

Excerpt from James D. Watson's Nobel lecture, December 11, 1962.

the frequency of each letter of the alphabet in the excerpt from Watson's Nobel lecture (see p. 15). If each of the 26 letters of the alphabet occurs equally, we would expect each letter to occur 41 times (1/26 × 1064) or to comprise 3.8% of the text. The graph below clearly shows that the rules used to encode meaning (words) in the English or Spanish language create a bias in the use of characters, such that some letters occur more frequently than expected by chance and others less frequently.

We can potentially uncover additional meaning if we analyze several two-character sequences and their inverses. If each letter pair is equally probable, we would expect 1.36 examples of each combination in the text (1/26 × 1/26 × 1064). However, any fan of crossword puzzles or *Wheel of Fortune* can say that certain letter pairs are much more frequent in the English language. Analysis of individual letters and letter pairs of a Spanish translation of the same text illustrates that different languages encode the same meaning with a different bias in letter use (see the graph at the top of p. 17).

Now, let's turn to DNA sequences composed of the nucleotides A, T, C, and G. Like a language, the genome evolved to convey meaning in DNA sequence—the directions to encode and regulate genes. Because this DNA language is not intuitive to us, merely looking at the three 1064-nucleotide sequences (see p. 18) will not distinguish which was taken from a mammalian protein-coding gene, which was from a mammalian noncoding (intergenic) region, and which was randomly generated by a computer. However, analysis of dinucleotide frequencies clearly shows that the two mammalian genomic sequences have nonrandom distributions—providing evidence of evolutionary selection (see the graph at the bottom of p. 17).

Closer inspection shows a general trend that represents the lowest level of genome organization. The relative abundance of the CG dinucleotide in the genic regions of mammalian genes—and its virtual absence in intergenic regions—can help to focus research efforts on gene-rich regions of the genome. The CG dinucleotide is properly termed CpG to indicate that it is linked by a phosphate on the same strand of the DNA molecule—as opposed to a C ≡ G base pair. The biological explanation for the relationship between "CpG islands" and gene enrichment is that the CpG dinucleotide is

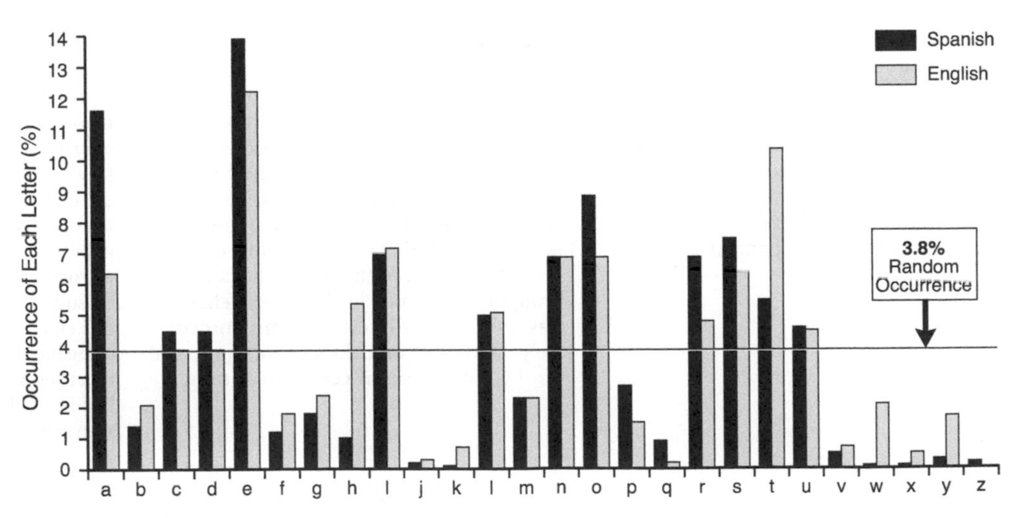

Frequency of letters in excerpt from Watson's Nobel lecture (see p. 15) and Spanish translation.

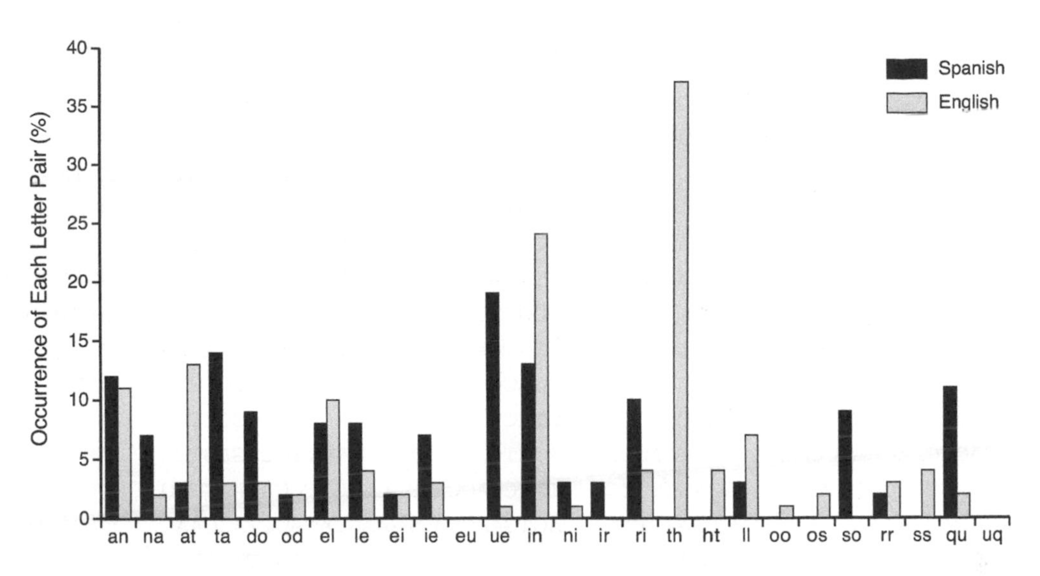

Frequency of some two-letter pairs in excerpt from Watson's Nobel lecture (see p. 15) and Spanish translation.

common in the 5′ promoter sequence that defines the beginning of many plant and animal genes. Promoters are recognized by RNA polymerase and other proteins that bind to the DNA to transcribe a gene sequence into mRNA. Thinking back to the evolution of cells, the cell membrane provided a means to sequester molecules from the environment and increase the concentration of reactants needed to efficiently assemble DNA molecules of sufficient length to encode proteins. At the same time, simple DNA sequences (such as the CpG dinucleotide) may have evolved to aggregate proteins needed to transcribe DNA into mRNA. Thus, the CpG dinucleotide may have been part of the early selection of gene regulators, perhaps analogous to the cosmic background radiation as a remnant of the cosmic dust clouds from which galaxies evolved after the Big Bang.

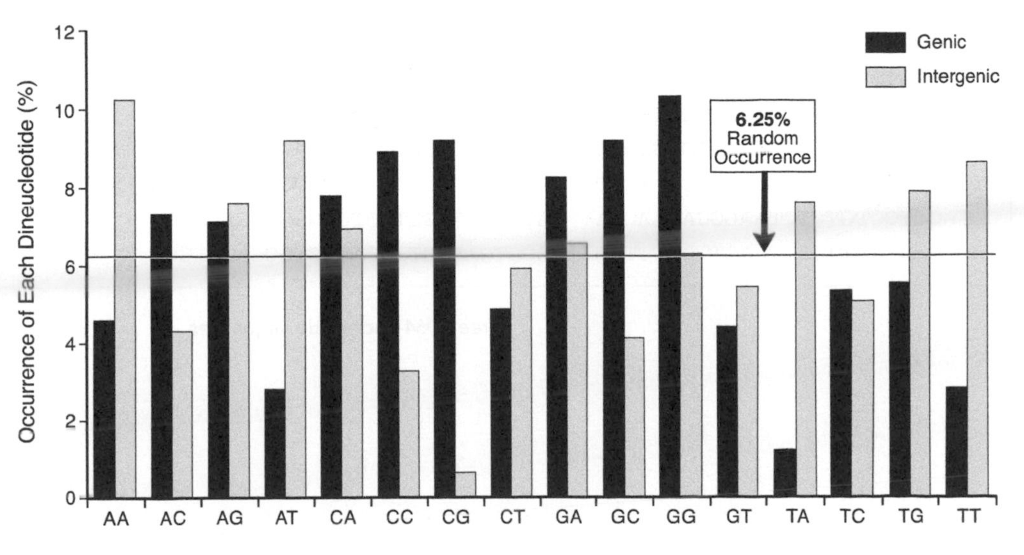

Dinucleotide frequencies in 1064-nucleotide sequences.

Random	Genic	Intergenic
TTGAATGCACTCAGTCGTCGCGGCACC	CCTTTCTGCACTGGGTGGCAACCAGGTC	GAACTGACATGAAAACATGCCTCAGATA
AGTTTCCTCGGACGTTCAAGGAGTTTCC	TTTAGATTAGCCAACTAGAGAAGAGAAG	TATTGTTGAGGGGAAAAGCAAGTTATAA
TCAAGAGCTAGAAGTGTTATATCTCCAA	TAGAATAGCCAATTAGAGAAGTGACATC	ACTAGCATATGCTTTTGATTTTATATAT
GATAAGTAGCACCGATACACTCGTAAGA	ATGTTGACTCTAACTCGCATCCGCACTG	GTATAAAAACATGTGTATACACATATAT
GGGACGGCCAGCGAGCCGTGAACATAAG	TGTCCTATGAAGTCAGGAGTACATTTCT	CCTATATTTAAATGAAAAGACAAATTCC
TTAACAACTTGTGTCAGATTCTAGTAGG	GTTCATTTCAGTCCTGGAGTTTGCAGTG	GCAATGGTAATTCATGAAACTGATAACA
ATACTGATCACTCATTAGTGCCATCTAT	GGGTTTCTGACCAATGCCTTCGTTTTCT	GTCTTTCCCTCTGGGAAACAGCCTAGGC
GTTAATCTTGCGCGCATAGCGGTGAGCG	TGGTGAATTTTTGGGATGTAGTGAAGAG	CTAGGGACATTGATGATCAATGAGAATA
TGTGGAGGGGACGATCGTGTGCACAAAC	GCAGGCACTGAGCAACAGTGATTGTGTG	TTTCTCTATAGGAGATGAATCCTCTTAC
TAAAGAGTGCAGCACTAAATATCCCGTC	CTGCTGTGTCTCAGCATCAGCCGGCTTT	TGCAATAATATATTCATGTTCACAGTTG
ACAGTGAACGATCCAGGACTTTGGACTA	TCCTGCATGGACTGCTGTTCCTGAGTGC	CAAATTGTGGTCTCCTTATCATTAAAGT
TCTAGGAGCGTTTCGGCTCAGAGCGTGC	TATCCAGCTTACCCACTTCCAGAAGTTG	CTTTCATTCCCTGGAAGAATCAGAAAGC
AGAGCGCAAAAGGTTTGAACTTAACTTA	AGTGAACCACTGAACCACAGCTACCAAG	TTGAGTTTATCTTTCAGTAGTTACAGTC
TGGGTGTCAGAACGCTTGTGGATATAT	CCATCATCATGCTATGGATGATTGCAAA	TGTGCTAATGGGGAATATTTTTTATTC
CTCCCACCAGCAGTTGGATCCAATTCGG	CCAAGCCAACCTCTGGCTTGCTGCCTGC	ACTCAAGTATACCAAGATTGACAAAGCG
CACCGGCGACTCTGCTGTCTCACCTTCT	CTCAGCCTGCTTTACTGCTCCAAGCTCA	CATCTAAGGTATCAGGTACGCTAGTAGG
AGCTCTGTGCTCCTCTCAGCCCCCACTG	TCCGTTTCTCTCACACCTTCCTGATCTG	TACCAAGAGAAGTAAATGAAAAGCTCTC
TCACGACAGCCGTGAAAGGTTTAAGAAG	CTTGGCAAGCTGGGTCTCCAGGAAGATC	TTTATTTGGAAGAGCTCACCATCTTGGG
TCAATAGTCGCGTCCCTGTGGTGGTTAC	TCCCAGATGCTCCTGGGTATTATTCTTT	TGGTGGGAGGTAAGACATTTACACAATT
CATCTCTTATCGCCCTACGTAGAGCCTA	GCTCCTGCATCTGCACTGTCCTCTGTGT	AAATAGTTCAATCTATGCAACAAATGCT
CTGTACTGTTCTAACTAGCGTAAGAGCG	TTGGTGCTTTTTTAGCAGACCTCACTTC	ATTATTTCTAGTTTTTCATCCAACAAAT
GACGGTTTGGCCTACGTGGATGCCTGAG	ACAGTCACAACTGTGCTATTCATGAATA	ATTTCCTGAGCACCTGCAGGGCCCAGGC
TATACGCCGCCGTGTTCACTAGTACTGT	ACAATACAAGGCTCAACTGGCAGATTAA	TTTGAGTCATGCACTAAGGATGTGCATG
AAATAACGGGCAGAGGGATGTCAAATCC	AGATCTCAATTTATTTTATTCCTTTCTC	GTTAAATACTTTTCTGCCCTTGAGAAAC
TACTGTTTCCACCTCGTAGCGGCTGCTA	TTCTGCTATCTGTGGTCTGTGCCTCCTT	TCACCTATGTTGCTTGTCTGGTGCATGG
ATAGGTGGAATCGATCTCCGAGAAGTCA	TCCTATTGTTTCTGGTTTCTTCTGGGAT	CCCAGGGCAAAAACCATATCTTACTTAC
ACATTAGCTTGATTAGCTCAGGCGACGG	GCTGACTGTCTCCCTGGGAAGGCACATG	CTCTTTACCCACTGGAGCATCCAGTACC
GACCGTTAAGCCGTGATCTTAGTACAAA	AGGACAATGAAGGTCTATACCAGAAACT	ATGCTTTGTGCATATCAATGGCAGAAGG
GTCTTCGCCCCTGAACAGGCGTACTTGT	CTCGTGACCCCAGCCTGGAGGCCCACAT	TGCACTGCCAGGGTGGGGGTGAATGGAG
GGGCCTGAGACAGATATGGTGTACTCAC	TAAAGCCCTCAAGTCTCTTGTCTCCTTT	GAGGTGAATGAGAGGGAAGAGACACGAA
GATGGTAGAGTTCAGGGTGACCGATGTA	TTCTGCTTCTTTGTGATATCATCCTGTG	GGCGTATAGAATTTCTAACTCGAGTGGC
CAGCCGCCATCAATATTGATCAGGGTGA	CTGCCTTCATCTCTGTGCCCCTACTGAT	TACAGGAAAGTTCAACTTTGTTCATTTT
GAATCGGTTTTACTTATTTTGACTAGGA	TCTGTGGCGCGACAAAATAGGGGTGATG	TAATGTGACGCATGTGCCTGGTAAACAA
TTGCATATTCTGTGCCCGAGGGGTTCGT	GTTTGTGTTGGGATAATGGCAGCTTGTC	GTAGTTAGAAAAATAAGTTTGGGTGTCA
GTAGGGGAAGTCCGTCAAATTGGGTAGT	CCTCTGGGCATGCAGCCATCCTGATCTC	TCTTAGGGCAAGAATTTAGAAATAAAGA
GTGTTTTCAATTTATGCGATGCTCGGAT	AGGCAATGCCAAGTTGAGGAGAGCTGTG	CTGGGGAGTCCTAAGCATGGTGTTGCAG
CGGACGTGCTGTCTCTAAGGCACAAGGA	ATGACCATTCTGCTCTGGGCTCAGAGCA	CCATGGAAGTGAATGTGAATCCTAAGGA
TACTTAACGCTTCACTGAGTTGTCTTGA	GCCTGAAGGTAAGAGCCGACCACAAGGC	GTGAGGGAGAGAAGGGCTATTGATGAAG

Three 1064-nucleotide sequences.

DNA DIRECTIONALITY AND READING FRAMES

Of course, the triplet codons that specify each of the amino acids are the most direct way to search for protein-coding genes. However, to understand the properties of genes that can be discovered by computer algorithms, we first need to consider some fine points of the genetic code and DNA structure.

We previously defined triplet codons as a property of mRNA, translated at the ribosome into amino acids. However, from a bioinformatics standpoint, it is more useful to deal with the genetic code as it exists in the DNA molecule. By convention, the strand of the DNA molecule that encodes a gene is termed the "sense" strand, and the complementary strand is termed the "antisense" strand. The antisense strand is used as the template to produce mRNA, which means that the mRNA then carries the same code as the sense strand, with uracil (U) in mRNA replacing thymine (T) in DNA. In this way, the genetic code carried by a DNA sequence and one that is carried by an mRNA sequence are completely interchangeable.

Each strand of a DNA molecule has a directionality based on the way in which the phosphate–sugar "rails" of the DNA "ladder" are joined when DNA is synthesized (replicated). Each deoxyribose sugar is composed of five carbons, labeled 1′ to 5′. The phosphates that link adjacent nucleotides form covalent bonds between the 3′ and 5′ carbons of adjacent sugars. (These are technically termed phosphodiester linkages, because the hydroxyl groups of phosphoric acid are replaced by oxygen linkages to two deoxyribose molecules.) In every organism ever studied, an incoming nucleotide is always added onto the 3′ carbon during DNA synthesis. Thus, DNA synthesis occurs in a 5′ to 3′ direction, and a DNA strand is referred to as running 5′ to 3′. The sequence of nucleotides forming a gene is also read in a 5′ to 3′ direction. If one considers genetic information as flowing downhill, like a river, 5′ is considered "upstream" and 3′ "downstream."

Importantly, the two strands of any DNA molecule are antiparallel, i.e., they run in opposite 5′ to 3′ orientations. Although the sequence of the paired strands is complementary, each is "read" in the opposite direction. Thus, each strand of a DNA double helix carries a unique coding sequence. Furthermore, the codons on each strand can be "read" in any of three different ways, depending on whether one starts at the first, second, or third nucleotide position of a DNA sequence. Each of these different ways of "spelling out" codons is termed a "reading frame." There are three reading frames on each strand of DNA, making a total of six reading frames per double-stranded DNA molecule. In any given region of the genome, several genes may be present in any of these six reading frames—on opposite strands or even overlapping in different frames of the same strand.

The first details of gene structure and function were analyzed in simple organisms. Bacteria are examples of prokaryotes ("pro," before and "karyon," kernal or nucleus) that lack an organized nucleus. The average prokaryotic protein has ~250 amino acids, so in any given reading frame, there are ~250 codons. However, according to chance alone, each of the three stop codons will occur once in every 64 codons of a given reading frame, so that on average there will be one stop codon in every 21 codons of a reading frame. Any reading frame with frequent stop codons is "closed" to the possibility of a functioning gene, whereas a reading frame with hundreds of contiguous codons that potentially encode amino acids is "open." A prokaryotic gene is a

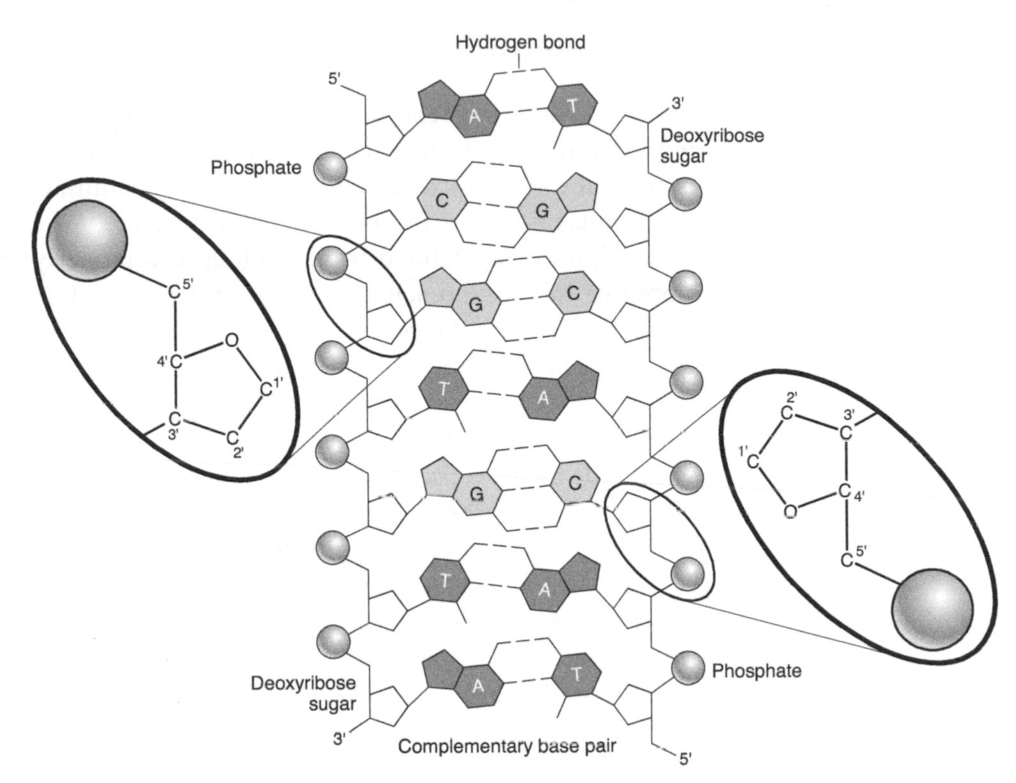

Chemical structure of DNA. Insets show the antiparallel orientation of 5′ and 3′ carbons on the complementary strands.

simple structure—an open reading frame (ORF) beginning with a start codon (ATG), followed by a long string of triplet codons that encode amino acids, and ending with a stop codon (TAA, TAG, or TGA).

ORF prediction programs are straightforward to use, with the main adjustable parameter being the minimum number of consecutive codons that is scored as an

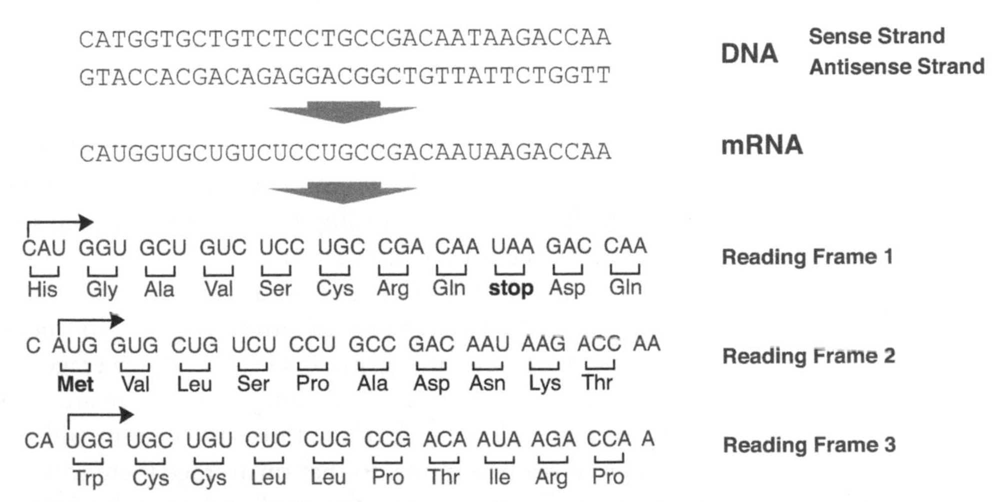

mRNA transcription and reading frames. By convention, mRNA is transcribed from the antisense strand of DNA. The mRNA (coding sequence) has the same sequence as the sense strand, except that thymine in DNA is replaced by uracil in RNA. Each reading frame is a series of triplet codons specifying amino acids. Each of the three reading frames is offset by one nucleotide, yielding a different set of codons.

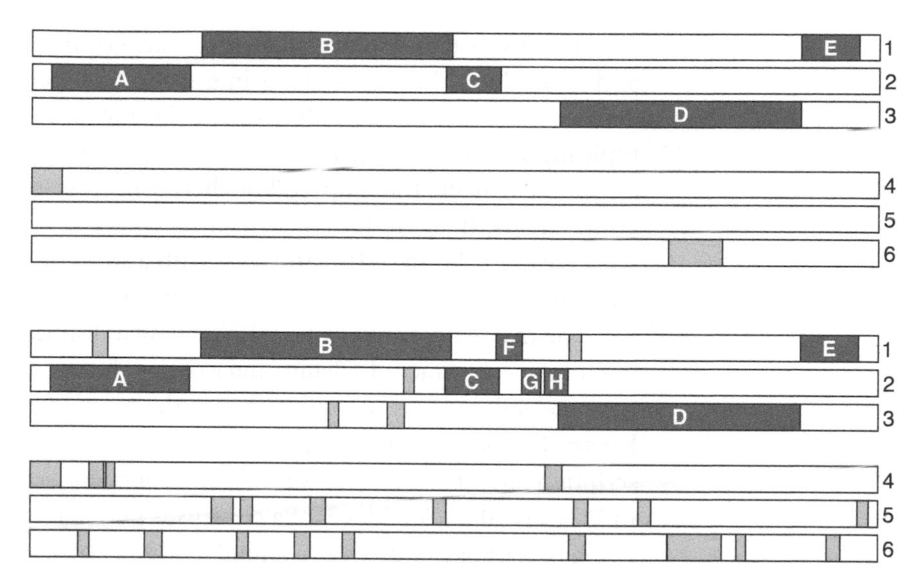

Open reading frame predictions in the HIV genome. (*Top*) Set to a minimum threshold of 300 nucleotides, an ORF prediction program correctly identifies five large genes (*A–E*), as well as two false positives (gray) in six reading frames. (*Bottom*) Lowering the threshold to 100 nucleotides identifies three additional genes (*F–H*), as well as 25 false positives. Shorter sequences—the *rev* gene (91 nucleotides) and part of the *tat* gene (48 nucleotides)—are missed. (A) *gag*, (B) *pol*, (C) *vif*, (D) *env*, (E) *nef*, (F) *vpr*, (G) *tat*, (H) *vpu*.

ORF. This consideration balances sensitivity—failing to detect real genes—and specificity—misidentifying ORFs that are not real genes. If the cutoff is set too high, many smaller genes are not identified as ORFs. If the cutoff is too low, many of the smaller ORFs are not true genes. Although a typical cutoff of 100 codons detects a high percentage of genes in a bacterial genome, it still identifies a substantial number of random ORFs that are not real genes.

3′- AND 5′-UNTRANSLATED REGIONS AND PROMOTERS

A gene is more than a simple open reading frame of codons transcribed into mRNA and translated into protein. During the process of transcription and translation, a number of protein and RNA molecules are recruited to bind directly with DNA and mRNA to regulate the expression of a gene. Because the regulatory sequences carried by mRNA molecules are also encoded in genomic DNA, a computer can search genomic DNA for sequences that regulate both DNA transcription and mRNA translation.

Evolutionary conservation of a common set of regulatory molecules—maintained across a range of organisms—has created a bias toward the use of certain nucleotide combinations in protein-binding sites. However, most protein-binding sites are not identical among species or even among different genes within a species. Rather, each binding site is represented by a consensus sequence of six to ten nucleotides, which is the most frequent combination of nucleotides found at the binding site. Functional binding sites may have combinations that differ by several nucleotides from the consensus. Some consensus sequences have several invariable nucleotide positions in combination with several variable positions. Computer algorithms translate these patterns into a scoring matrix, ratcheting along a DNA sequence in overlapping "windows" and evaluating each against the consensus sequence.

It is important to remember that transcription and translation occur separately and at different times and places in the cell. The start and stop codons that define the ends of a protein do not define the beginning and end of an mRNA. In fact, mRNAs typically include sequences that extend upstream (5′) of the start codon and downstream (3′) from the stop codon. Because these sequences are not translated into amino acids, they are termed untranslated regions (UTRs). Thus, *translation* begins with the start codon (ATG) and ends with a stop codon (TAA, TAG, or TGA), but *transcription* begins with the 5′ UTR and ends with the 3′ UTR.

The majority of eukaryotic mRNAs have a distinctive 3′ feature—a poly(A) tail composed of a long tract of adenine nucleotides. The poly(A) tail stabilizes the mRNA; over time, the tail shortens, and the mRNA is degraded when the poly(A) tail reaches a critical length. The poly(A) tail is not part of the gene sequence, but rather, it is added post-transcriptionally, after the mRNA has been generated. Polyadenylate polymerase cleaves the mRNA 11–30 nucleotides 3′ of a consensus poly(A) signal—A(A/U)UAAA—then adds a string of tens or hundreds of adenine residues. Thus, identifying the poly(A) signal in the DNA—A(A/T)TAAA—helps to define the end of the 3′ UTR in eukaryotes.

Although the poly(A) signal conveniently defines the 3′ end of a gene, the 5′ end is more difficult to define. The 5′ start site can be inferred from promoter sequences that position RNA polymerase at the transcription start site. RNA polymerase II (Pol II), the eukaryotic polymerase that transcribes DNA, cannot initiate transcription on its own but requires the assistance of a number of accessory proteins called transcription factors (TFs), of which TFIIB and TFIID are the most well studied. The core, or basal, promoter provides binding sites for TFs that provide a maintenance level of transcription. Additional elements, termed transcriptional activators, bind promoter sequences with one surface and interact with transcription factors with another surface. In this way, a number of protein molecules work together to recruit Pol II to the transcription start site. Thus, there are a number of binding sites for transcription factors and activators in the 5′ promoter region and in the 5′-coding region. The positions of promoter and activator sequences are denoted by negative numbers to indicate their positions upstream of the transcription start site (position 1).

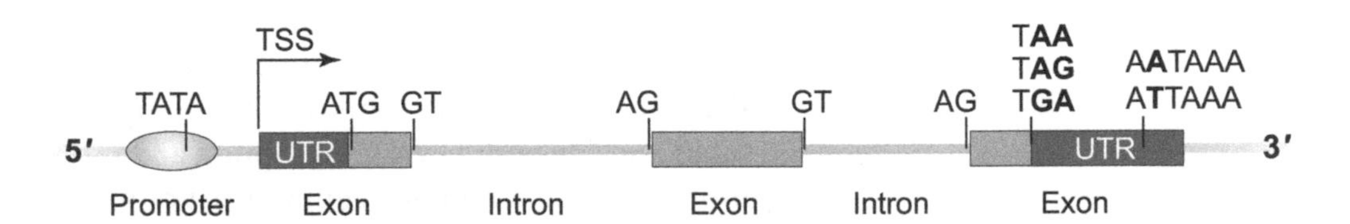

Diagram of a eukaryotic protein-coding gene. The DNA code is "read" from 5′ to 3′, as defined by the orientation of carbon atoms in the deoxyribose backbone. The core, or basal, promoter, located 5′ of the gene, provides binding sites for proteins that position RNA polymerase II at the transcription start site (TSS). Transcription begins approximately 30 nucleotides after the TATA box, one of four major elements that compose the core promoter. Exons (filled boxes) alternate with introns that are spliced out of the pre-mRNA (thinner lines) and whose boundaries are defined by GT and AG sequences. Translation of amino acids begins with the start codon ATG (methionine) and ends with one of three stop codons (TGA, TAG, or TAA). Contrary to popular belief, all exons are not translated into amino acids; instead, the first exon begins and the final exon ends with an untranslated region (UTR; dark box). It is not uncommon for the 3′ UTR to span several exons. The end of the 3′ UTR is defined by the recognition signal for polyadenylate polymerase (A[A/T]TAAA), which cleaves about 20 nucleotides from the end of the transcript and adds a string of adenine residues, the poly(A) "tail."

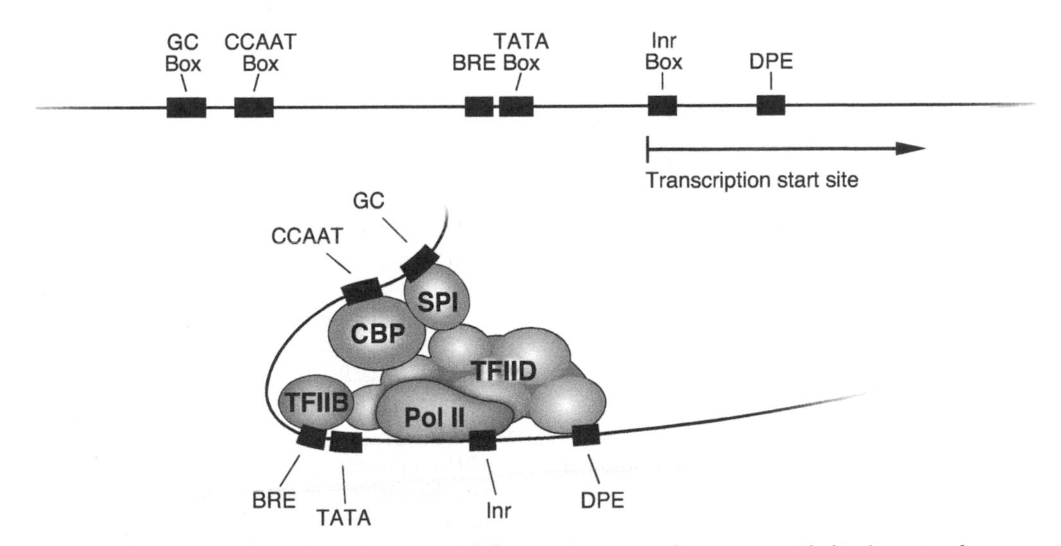

Eukaryotic promoter elements and transcription factors. Promoter elements provide binding sites for transcription factors that help to position RNA polymerase II (Pol II) at the transcription start site. Different genes may have different combinations of promoter sequences that recruit varied sets of transcription factors.

The eukaryotic core promoter is composed of four major sequence elements: TATA box, initiator (Inr) box, downstream promoter element (DPE), and TFIIB recognition element (BRE). Although many genes have a consensus site for only one of these elements, there is strong evidence that additional transcription factors may bind to promoters nonspecifically.

The TATA box is the most well-studied promoter element, and it is found in 20%–50% of eukaryotic genes. Located at positions –23 to –33 before the transcription start site, the eukaryotic consensus sequence, TATAAA, is slightly different from the prokaryotic consensus, illustrating why gene prediction programs must be tuned for different organisms. The TATA box binds to a subunit of TFIID, appropriately called the TATA-binding protein (TBP). The TBP is the first protein to bind DNA to initiate transcription and binds the promoter region even in genes that do not contain a TATA box. TBP binding introduces a kink in the DNA molecule and stresses hydrogen bonds in the region, causing the helix to open slightly. (The double hydrogen bonds in this A = T–rich region denature more easily than G ≡ C triple bonds.) The denatured region allows easier access for RNA polymerase to begin transcription.

The Inr box, which binds TFIID, is the most frequent core element, and it is found in 40%–65% of eukaryotic genes. This pyrimidine (C and T)–rich sequence, (C/T)(C/T)A_{+1}N(A/T)(C/T)(C/T), straddles the transcription start site. DPE also binds TFIID and, in some species, may work with Inr to function as the core promoter when the TATA sequence is absent. DPE is located in the 5′ coding region, +23 to +33, with the consensus (A/G)G(A/T)(C/T)(A/C/G). TFIIB binds to a B-recognition element (BRE) at –42 to –32, with the consensus (C/G)(C/G)(A/G)CGCC.

The CCAAT box, with the consensus GGNCAATCT, is an activator sequence found in about half of vertebrate genes. This sequence binds a number of different CCAAT-box binding proteins (CBPs). Although the CCAAT box is usually located at –40 to –100, it can be located near the Inr box or DPE in promoters without TATA boxes. The transcriptional activator Sp1 binds the GC box located at –40 to –100, which has the consen-

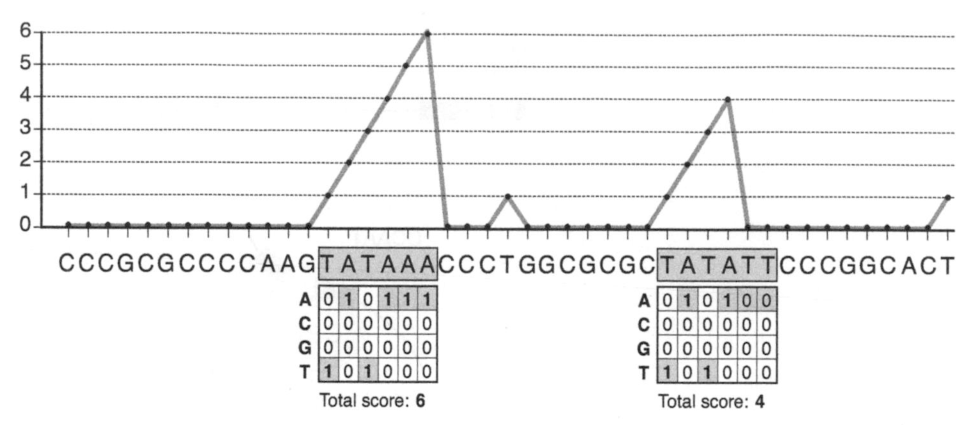

Scoring matrix for TATA box. Computer algorithms use a scoring matrix to search genomic sequence for a close match to a DNA regulatory element, such as the TATA box. The computer ratchets along a DNA sequence, analyzing the positions in each successive "window" of six nucleotides. Each nucleotide that matches its corresponding position in the consensus receives a score of 1; nonmatches score 0. The TATAAA consensus (first gray box) receives a perfect score of 6. Not all TATA boxes perfectly match the consensus sequence; the promoter of the human albumin gene scores 4 with the sequence TATATT.

sus sequence GGGCGG. Sp1 regulates the expression of many "housekeeping" genes that are essential for key cellular functions in vertebrates. The GC box, which may be present in multiple copies in the promoter region, is thought to be the source of the association between elevated G + C levels and CpG dinucleotide enrichment in genic regions.

Enhancers, which further increase transcription, are poorly understood. They may be located hundreds or thousands of nucleotides upstream of the transcription start site, within the transcribed gene or an adjacent gene, or on the opposite DNA strand. Regardless of the enhancer's position, enhancer-binding proteins interact with the enhancer as well as transcription factors assembled at the promoter. The enhancer does not operate from a distance; rather, a loop in the chromosome brings it into proximity with the core promoter.

EXONS AND INTRONS

Richard Roberts
(Courtesy of Cold Spring Harbor Laboratory Archives.)

In 1977, Richard Roberts, at Cold Spring Harbor Laboratory, and Philip Sharp, at the Massachusetts Institute of Technology, independently discovered that most eukaryotic genes are not ORFs composed of a contiguous sequence of codons. Using adenovirus, both groups created heteroduplex molecules by hybridizing an mRNA to single-stranded genomic DNA. Electron microscopy revealed that the mRNA hybridized to discontinuous regions of the DNA, throwing out loops of DNA that were not represented in the mRNA. Their explanation was that adenovirus genes are "split," with protein-coding regions (exons) interrupted by nonprotein-coding regions (introns) that are not represented in mRNA. During transcription, the entire gene is copied into a precursor mRNA (pre-mRNA), which includes exons and introns. Subsequent "RNA processing" removes the intervening introns to form a contiguous coding sequence. This "mature" mRNA passes out of the nucleus and attaches to a ribosome for translation.

Later work explained the mechanism of RNA splicing at the spliceosome, a nuclear complex of numerous proteins and five *s*mall *n*uclear RNAs (snRNAs) aver-

Philip Sharp
(Courtesy of Cold Spring Harbor Laboratory Archives.)

aging ~150 nucleotides in length. The spliceosome assembles on pre-mRNA, aligning 5′ and 3′ splice sites to loop out the intron to form a "lariat." The lariat is then cleaved, and the 5′ and 3′ exon junctions are ligated together. Complementary sequences in snRNAs recognize consensus sequences that define adjacent exon/intron borders, bringing them into proximity for the splicing reactions to occur. The mRNA consensus sequence at the 5′ splice junction is CAG/<u>GU</u>AAGU, whereas the 3′ consensus is UUUUCCCUCC<u>AG</u>/GU. Notably, GU and AG nucleotides define the 5′ and 3′ ends of virtually every eukaryotic intron. At the DNA level, introns invariably begin with GT and end with AG. Using this fact, in combination with other nucleotides in the consensus, a scoring matrix can efficiently identify introns and exons in DNA sequence. After introns are eliminated by splicing, every eukaryotic mRNA is, in fact, an ORF bounded by a start and stop codon.

The simple situation in prokaryotes—where an ORF in the genomic DNA sequence is identical to a contiguous mRNA sequence—is complicated in eukaryotic genes. The "average" human gene has eight exons, each ~135 nucleotides in length (1080 nucleotides in total) and seven introns of ~2200 nucleotides each (15,400 nucleotides in total). It is rare that more than several exons of a gene are "read" by the transcription machinery in the same contiguous reading frame. Rather, the reading frame shifts over the length of a gene to read different exons in different reading frames.

Stop codons are of no consequence inside introns, because these sequences are removed from the pre-mRNA before translation at the ribosome. A shift to one reading frame avoids any stop codons that may be present in others. Frame shifting is actually dictated by intron length. A eukaryotic reading frame shifts after any intron whose length is not divisible by three, compensating +1 or +2 to maintain the triplet codons that define

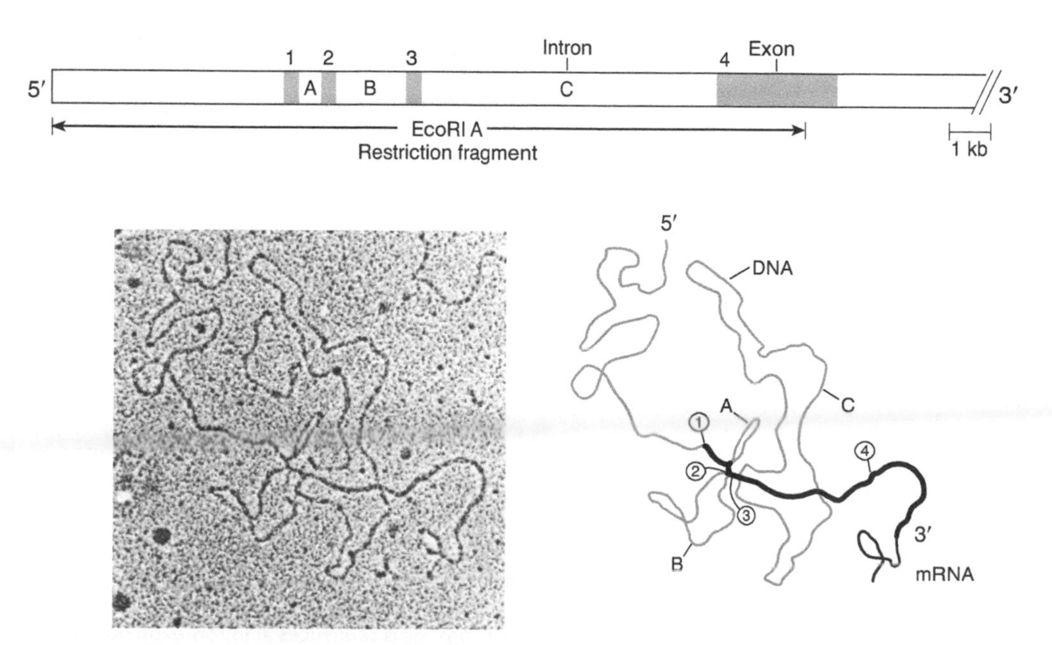

Electron microscopic evidence for RNA splicing. An EcoRI restriction fragment of adenovirus genomic DNA was hybridized to its corresponding mRNA (*bottom left*). In the diagram at right, mRNA (black) and genomic DNA (gray) form a double-stranded molecule in complementary coding regions (1, 2, 3, 4 in gene diagram). Introns A, B, and C are thrown out as loops.
(*Bottom*: Reprinted, with permission, from Berget SM, Moore C, Sharp PA. 1977. *Proc Natl Acad Sci* 74: 3171–3175.)

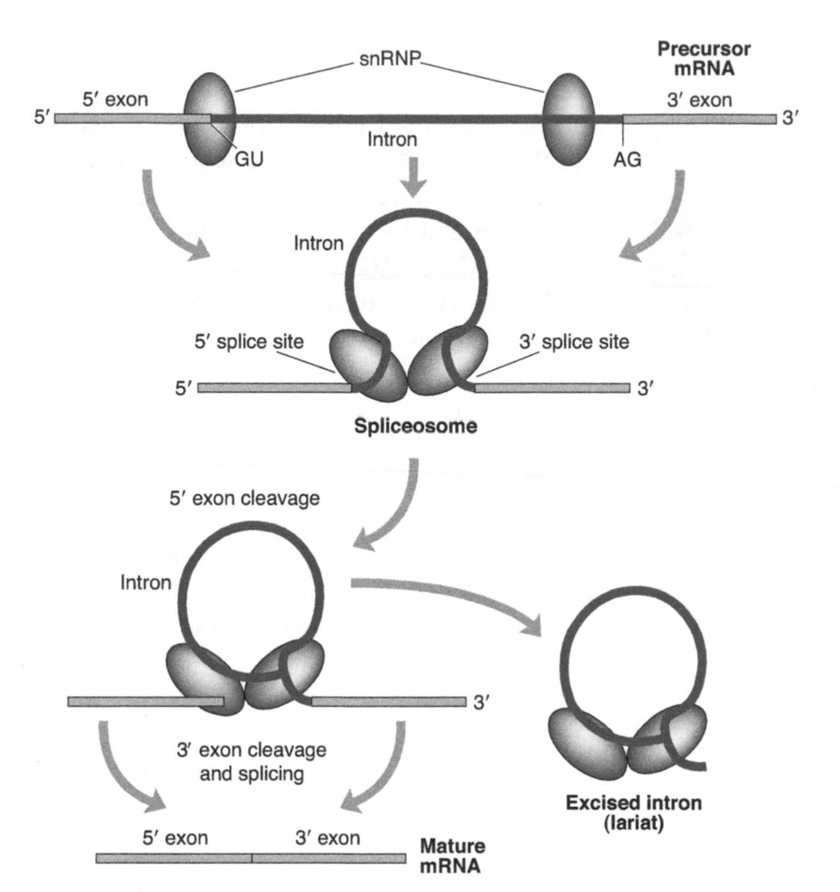

Mechanism of RNA splicing. Conserved GU and AG residues define the intron borders and align a precursor mRNA in the spliceosome. First, the 5′ intron/exon junction is cleaved and an unusual 5′-2′ phosphodiester bond joins the 5′ end of the intron to an adenine residue at the 3′ end of the intron. The resulting loop structure resembles a lariat. The 3′ intron/exon junction is then cleaved to release the lariat, and the two exons are joined.

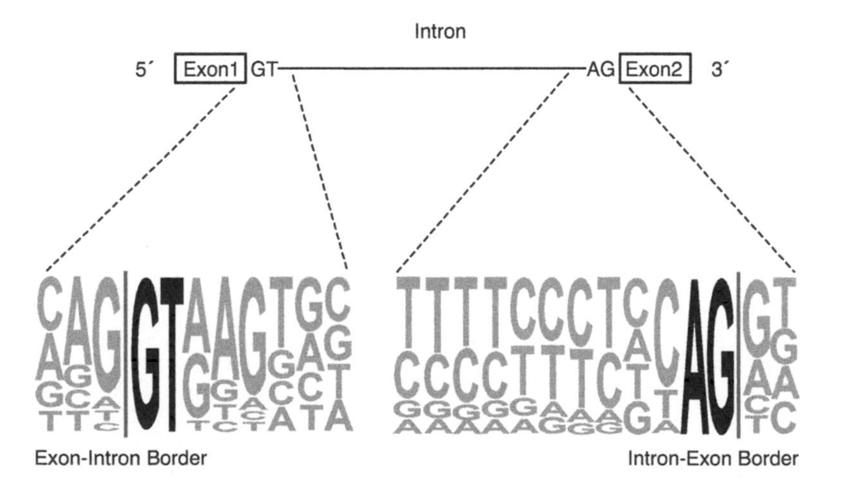

"Sequence logo" of consensus sequences at intron/exon boundaries. Introns are demarcated in genomic sequence by GT and AG. However, because these sequences are common in genic regions, merely searching for GT and AG will turn up many false exons, as well as a few real ones. Computer algorithms use a scoring matrix to evaluate the consensus sequence surrounding the GT and AG dinucleotides. The sequence logo visually summarizes these characteristics, with the height of each letter showing its probability at that position in validated splice junctions.

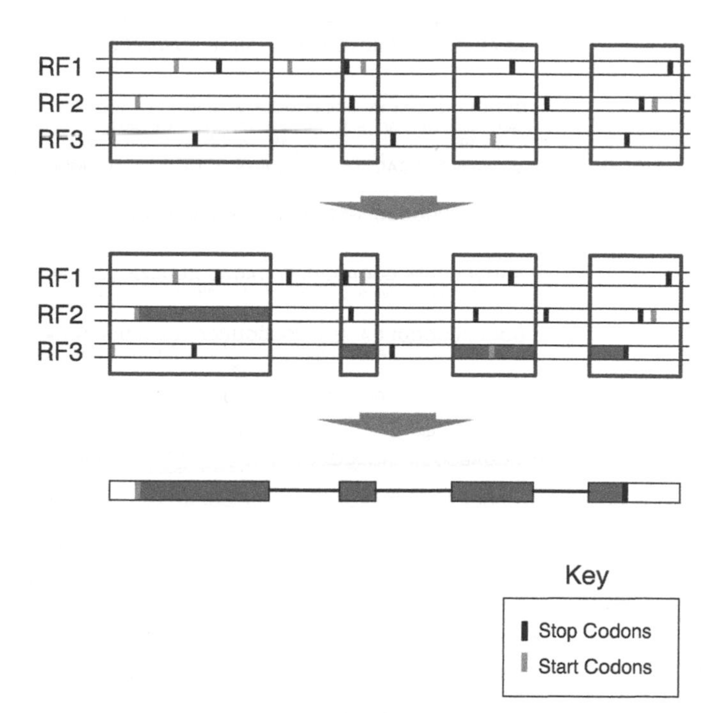

Frame shifting in a eukaryotic gene. Layering start/stop codons in the three reading frames on one DNA strand illustrates the fact that the coding sequences of most eukaryotic genes "shift" between several reading frames. Each of the three reading frames has several stop codons, and none is "open" over the entire coding sequence. The four exons of the gene are boxed. In this case, the coding sequence begins in reading frame 2 for exon 1 and then shifts to reading frame 3 for exons 2, 3, and 4. Stop codons in introns are of no consequence.

the ORF. Thus, the location of splice sites is the most important factor in determining the exon structure of a eukaryotic gene. Put most simply, an ORF is a property of mRNA, whereas intron/exon boundaries are properties of DNA and pre-mRNAs.

METHODS FOR FINDING GENES IN GENOMIC SEQUENCE

The raw DNA sequence is the starting point for understanding a genome. In some cases, the initial search for genes is narrowed by focusing on CpG islands and regions with high G + C content. Whether narrowed in this way or not, two types of computational methods are used to identify genes within DNA sequence. Pattern-based programs use algorithms to search for sequences associated with gene features, including start/stop codons, coding triplets, intron/exon boundaries, promoters, and poly(A) signals. This is termed ab initio (from the beginning) gene finding, because genes are predicted directly from DNA sequence. Comparative programs look for similarities between the genome sequence and independent sequence evidence from the organism under study and from related organisms. The best gene-finding programs now incorporate both pattern-based and comparative strategies, providing increasingly accurate gene models.

Most pattern-based programs are trained on representative genes to develop a "hidden Markov model (HMM)," which identifies the gene patterns "hidden" in DNA sequence. Notably, different organisms have preferences, or biases, among synony-

ACAUUUGCUUCUGACACAACUGUGUUCACUAGCAACCUCAAACAGACACCAUGGUGCACCUGACUCCUGAGGAGAAC
(met) val his leu thr pro glu glu lys
1 5

UCUGCCGUUACUGCCCUGUGGGGCAAGGUGAACGUGGAUGAAGUUGGUGGUGAGGCCCUGGGCAGGUUGGUAUCAAG
ser ala val thr ala leu trp gly lys val asn val asp glu val gly gly glu ala leu gly arg
10 15 20 25 30

GUUACAAGACAGGUUUAAGGAGACCAAUAGAAACUGGGCAUGUGGAGACAGAGAAGACUCUUGGGUUUCUGAUAGGC

ACUGACUCUCUCUGCCUAUUGGUCUAUUUUCCCACCCUUAG**GCUGCUGGUGGUCUACCCUUGGACCCAGAGGUUCUUU**
leu leu val val tyr pro trp thr gln arg phe phe
31 35 40

GAGUCCUUUGGGGAUCUGUCCACUCCUGAUGCUGUUAUGGGCAACCCUAAGGUGAAGGCUCAUGGCAAGAAAGUG
glu ser phe gly asp leu ser thr pro asp ala val met gly asn pro lys val lys ala his gly lys lys val
45 50 55 60 65

CUCGGUGCCUUUAGUGAUGGCCUGGCUCACCUGGACAACCUCAAGGGCACCUUUGCCACACUGAGUGAGCUGCAC
leu gly ala phe ser asp gly leu ala his leu asp asn leu lys gly thr phe ala thr leu ser glu leu his
70 75 80 85 90

UGUGACAAGCUGCACGUGGAUCCUGAGAACUUCAGGGUGAGUCUAUGGGACCCUUGAUGUUUUCUUUCCCCUUCUUU
cys asp lys leu his val asp pro glu asn phe arg
95 100 104

UCUAUGGUUAAGUUCAUGUCAUAGGAAGGGGAGAAGUAACAGGGUACAGUUUAGAAUGGGAAACAGACGAAUGAUUG

CAUCAGUGUGGAAGUCUCAGGAUCGUUUUAGUUUCUUUUAUUUGCUGUUCAUAACAAUUGUGUAUAACAAAAGGAAAU

AUCUCUGAGAUACAUUAAGUAACUAAAAAAAAACUUUACACAGUCUGCCUAGUACAUUACUAUUUGGAAUAUAUGUG

UGCUUAUUUGCAUAUUCAUAAUCUCCCUACUUUAUUUUCUUUUAUUUUAAUUGAUACAUAAUCAUUAUACAUAUUUAUG

GGUUAAAGUGUAAUGUUUUAAUAUGUGUACACAUAUUGACCAAAUCAGGGUAAUUUUGCAUUUGUAAUUUUAAAAAAU

GCUUUCUUCUUUUAAUAUACUUUUUUGUUAUCUUAUUUCUAAUACUUUCCCUAAUCUCUUUCUUUCAGGGCAAUAAUGA

UACAAUGUAUCAUGCCUCUUUGCACCAUUCUAAAGAAUAACAGUGAUAAUUUCUGGGUUAAGGCAAUAGCAAUAUUU

CUGCAUAUAAAUAUUUCUGCAUAUAAAUUGUAACUGAUGUAAGAGGUUUCAUAUUGCUAAUAGCAGCUACAAUCCAG

CUACCAUUCUGCUUUUAUUUUAUGGUUGGGAUAAGGCUGGAUUAUUCUGAGUCCAAGCUAGGCCCUUUUGCUAAUCAU

GUUCAUACCUCUUAUCUUCCUCCCACAG**CUCCUGGGCAACGUGCUGGUCUGUGUGCUGGCCCAUCACUUUGGCAAA**
leu leu gly asn val leu val cys val leu ala his his phe gly lys
105 110 115 120

GAAUUCACCCCACCAGUGCAGGCUGCCUAUCAGAAAGUGGUGGCUGGUGUGGCUAAUGCCCUGGCCCACAAGUAU
glu phe thr pro pro val gln ala ala tyr gln lys val val ala gly val ala asn ala leu ala his lys tyr
125 130 135 140 145

CACUAAGCUCGCUUUCUUGCUGUCCAAUUUCUAUUAAAGGUUCCUUUGUUCCCUAAGUCCAACUACUAAACUGGGGG
his **stop**

AUAUUAUGAAGGGCCUUGAGCAUCUGGAUUCUGCCUAAUAAAAAACAUUUAUUUUCAUUGC

β-Globin gene structure. Three exons with encoded amino acids are in bold, and two introns are in plain font.

mous codons, intron and exon lengths, consensus sequences for promoter elements, intron/exon boundaries, and poly(A) signals. Although these overall sequence characteristics cannot be readily discerned by simply inspecting the DNA sequence, HMMs derive statistical information about these biases from the training set.

In the roundworm *Caenorhabditis elegans* and the fruit fly *Drosophila*, HMMs can correctly identify ~90% of individual exons and every exon in ~40% of genes. However, these figures drop to 70% and 20%, respectively, in human DNA. Gene prediction programs readily identify internal exons, which have two splice junctions (left and right) adjacent to two introns. The first and last exons are frequently missed because they have only half of the sequence information used in prediction. Furthermore, exon prediction generates a large number of false positives. Consensus splice site sequences are very common in introns, making "pseudo-exons" common. Additionally, HHMs cannot predict the correct start codon among several possible ATGs in the 5′ region. Often, definitive information on the translation start site can only be determined directly from protein sequence.

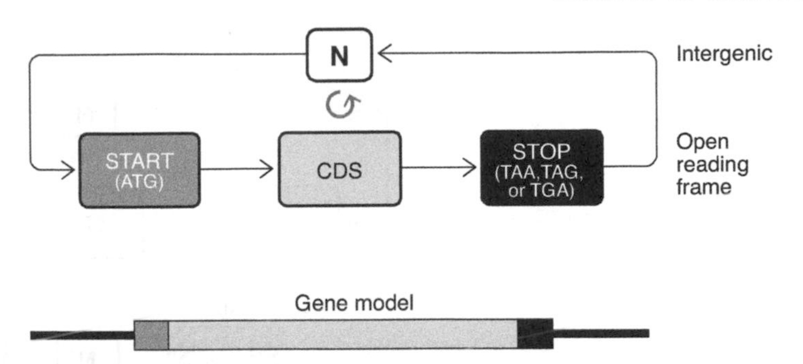

Flow diagram for predicting an ORF gene. A hidden Markov program scans nucleotides (N) until it encounters a potential start codon and then scans for a minimum number of triplet codons that define an ORF. After encountering a stop codon, the program cycles back to the intergenic state (N) and searches for another potential start codon.

Comparative (homology-based) methods find sequence similarities among DNA, RNA, or protein sequences. Homology refers to sequence similarity based on a common origin. Orthologs are similar genes in different species that have arisen due to descent from a common ancestor, whereas paralogs are similar genes within a species that have arisen by the duplication of a single gene. Growing data from parallel sequencing of model species—notably *E. coli*, *Saccharomyces cerevisiae*, *C. elegans*, *Schizosaccharomyces pombe*, fruit fly, mouse, and *Arabidopsis*—provide collections of previously identified genes. A match to a known mRNA sequence from the same species provides direct evidence that a DNA sequence is transcribed—and therefore is a coding sequence—whereas homology with a known gene or protein from another species provides indirect evidence. Homology-based methods identified ~60% of genes in the first draft of the human genome, and 40% of genes were identified ab initio.

Comparative algorithms such as BLAST (Basic Local Alignment Search Tool) or BLAT (BLAST-like Alignment Tool) align raw sequence or gene models to a database of known genes. BLAT is the fastest program for scanning an entire genome, because it stores in RAM memory an index of short DNA or protein sequences (11 or 4 mers) most relevant to the genome under investigation. The BLAT index can fit in the RAM memory of a personal computer, and it can handle long lists of queries simultaneously.

GENE ANNOTATION

The output from a pattern- or homology-based program is termed a gene model, because it may or may not be an accurate representation of the actual gene. Annotation is the process of adding information about the structure and function of a predicted gene. Structural annotations improve the initial gene model by extending 5′- and 3′-noncoding regions, identifying alternative start and stop codons, sorting out exon structure, and finding alternative splice sites. Functional annotations identify conserved amino acid motifs and, therefore, functions that are shared with other organisms. Although early genome-sequencing efforts relied on annotations submitted by human curators, this time-consuming step is now automated. Considering the exponentially increasing rate at which new DNA sequences are being produced, it seems certain that the vast majority of new sequence data will never be carefully examined by human eyes.

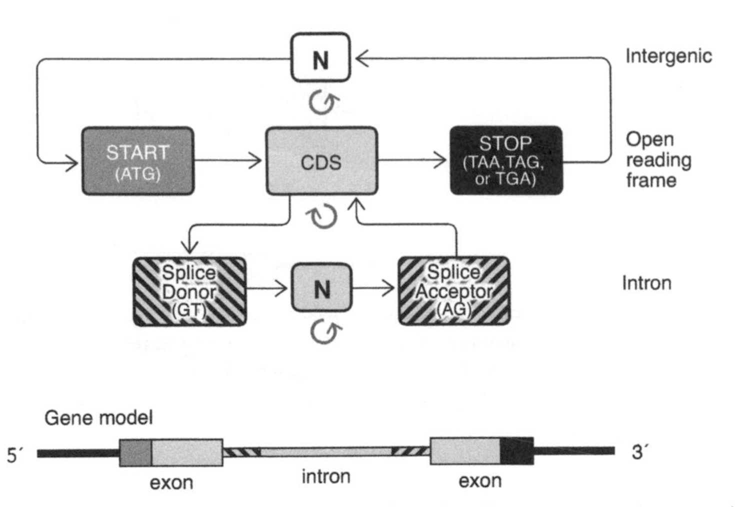

Flow diagram for predicting a gene with an intron. After identifying a potential start codon, a hidden Markov program cycles through triplet codons until it encounters a potential 5′ splice donor (GT) in the context of an appropriate consensus sequence. It then scans intron sequence (N) until it encounters a 3′ splice acceptor (AG) within a consensus sequence. The program then cycles through additional codons until it encounters another 5′ splice donor or a stop codon. When the program encounters a stop codon, it returns to the intergenic (N) state and searches for the next potential start codon.

Confirmation of a predicted gene and annotation of its structure and function must come from independent evidence, usually from homology-based matches to previously annotated genes or mRNA evidence from the species under study and its close relatives. BLASTn uses the predicted gene sequence to search a nucleotide database to discover homologs in closely related species. However, the redundancy of the genetic code usually makes it difficult to uncover more distant relationships at the DNA level. Therefore, BLASTx translates the coding sequence (assembled exons) into an amino acid sequence to search a protein database. BLASTx or BLASTp, which makes protein–protein searches, also highlights any conserved motifs (domains) within the predicted protein, including structures that bind to DNA or other proteins. Programs such as PHYLIP place the predicted protein in a phylogenetic tree that shows its evolutionary relationships to homologs from other organisms.

Much existing evidence for gene annotation has come from complementary DNA (cDNA) libraries, which represent the genes expressed by an organism or by a particular tissue or cell type. To make a cDNA library, mRNA is isolated from living cells, and the enzyme reverse transcriptase is used to convert mRNAs into cDNAs. These are copied into doubled-stranded DNAs, which are ligated into plasmid vectors and transformed into *E. coli*, creating a library of thousands of cDNA clones.

Full-length cDNA sequence is the best source of information about the 5′ and 3′ UTRs, which are not accurately predicted by computer programs. The 5′ UTR is the most difficult part of the gene to annotate. Reverse transcriptase extends from the 3′ end of the mRNA template, so incomplete extension typically produces many cDNAs (and cDNA clones) that are missing sequence at their 5′ ends. Expressed sequence tags (ESTs) are single-read DNA sequences obtained using primers at each end of the cDNA cloning vector. Because they can be generated quickly and inexpensively, ESTs are typically the most abundant biological evidence and are often available from related species. However, EST sequences are short (~500 nucleotides) and highly redundant.

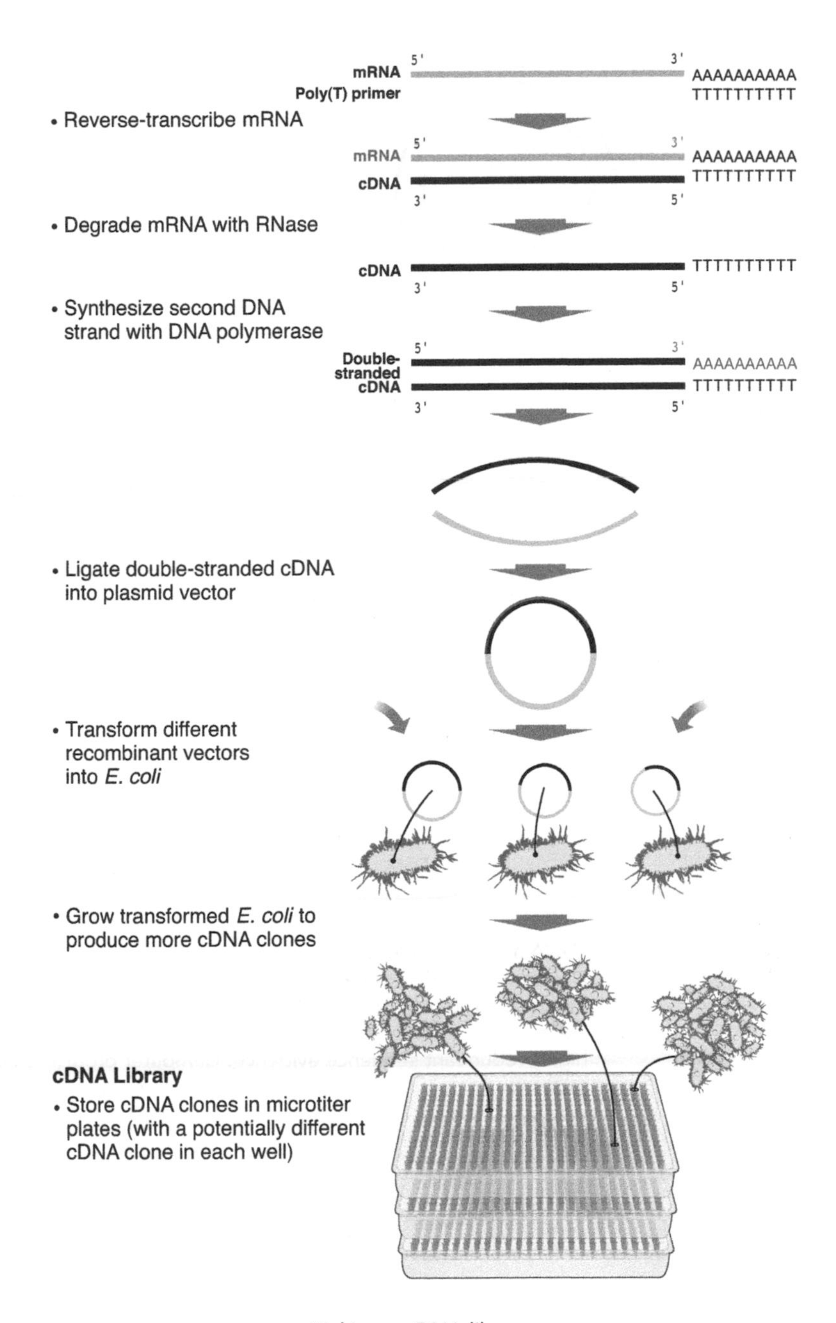

Making a cDNA library.

Annotation programs, such as Apollo, visually align gene models, cDNAs, and ESTs to genome sequence, providing evidence to extend 3′ and 5′ UTRs and confirm exon structure. ESTs and cDNAs with different arrangements of exons define alternatively spliced forms of mRNA that produce different proteins from the same genomic sequence.

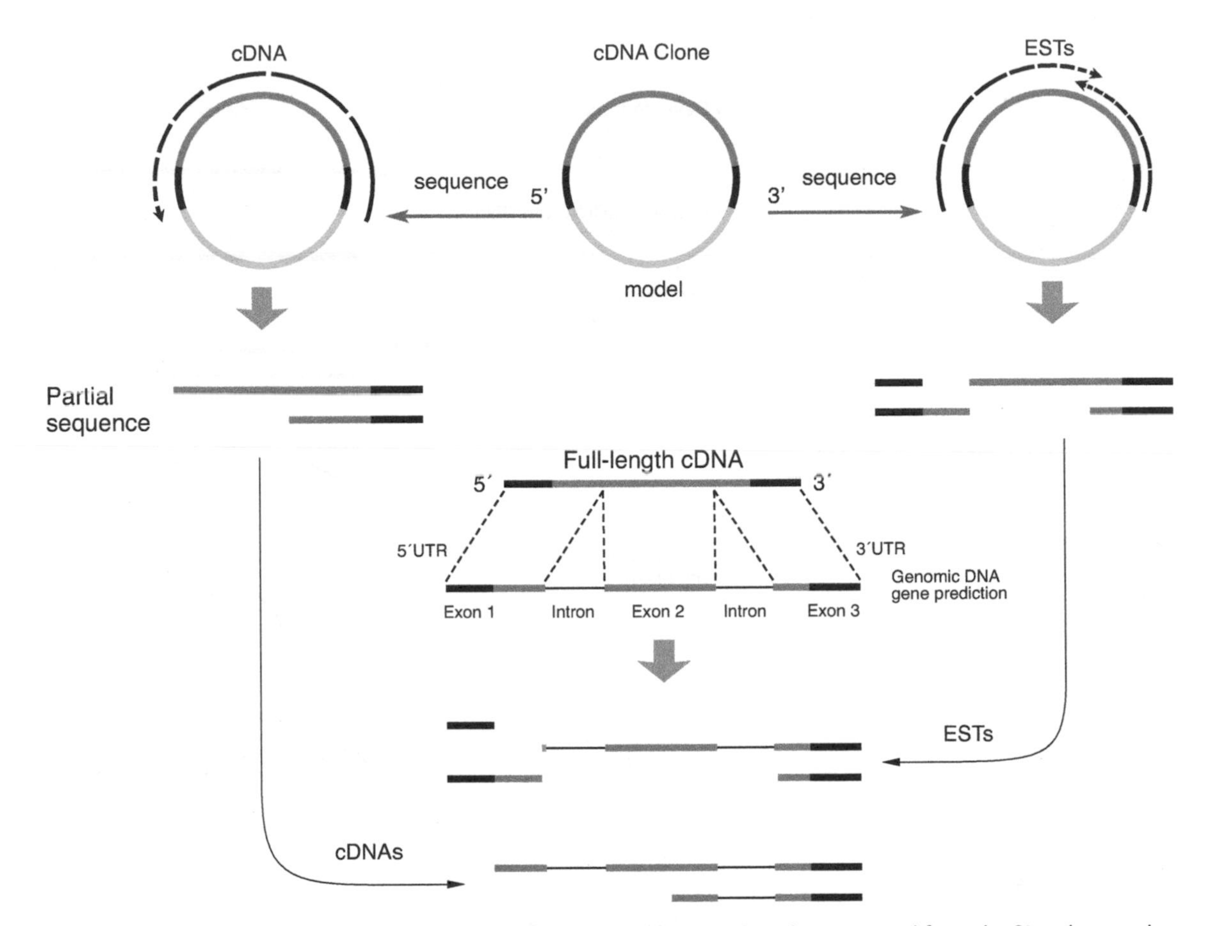

cDNA and EST evidence. A clone from a cDNA library is directly sequenced from the 3′ end to produce high-quality cDNA sequence (*left*). A full-length cDNA sequence offers the best evidence for confirming gene structure. Expressed sequence tags (ESTs) are short reads from the 5′ and 3′ ends of a cDNA clone (*right*). Because they are relatively easy to produce, ESTs typically offer abundant but short and highly redundant sequence evidence. Computer programs align cDNA and EST evidence with genome sequence to confirm the exon/intron structure of predicted genes.

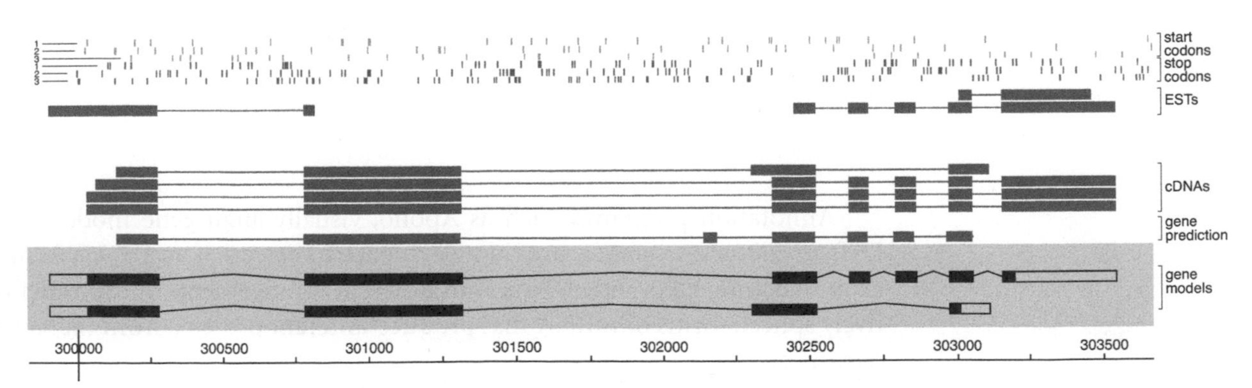

Gene annotation and alternative splicing. This screen shot from the Apollo annotation program shows two alternatively spliced gene models based on a gene prediction plus EST and cDNA evidence.

SANGER DIDEOXY DNA SEQUENCING

Fred Sanger
(Courtesy of Cold Spring Harbor Laboratory Archives.)

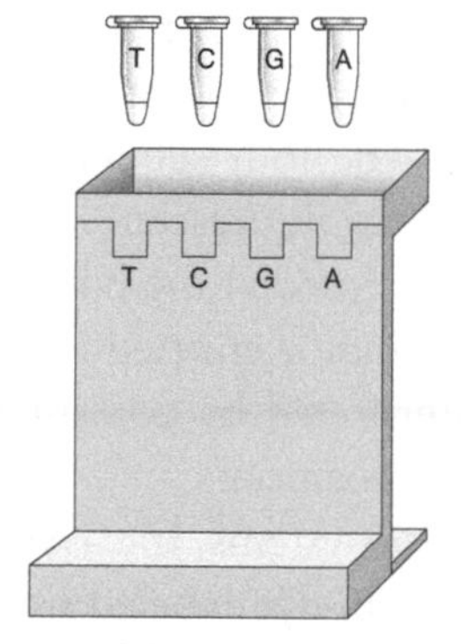

Sanger's dideoxy DNA sequencing. Sequencing reactions are electrophoresed in a vertical polyacrylamide gel.

Modern genome sequencing was launched in 1977 when Fred Sanger at the Medical Research Council's Laboratory of Molecular Biology in Cambridge, England, developed a sequencing method based on enzymatic DNA synthesis. During primer extension by DNA polymerase, discovered in the 1960s, DNA synthesis is "primed" by short single-stranded primers that hybridize to each DNA strand. In the presence of the four deoxynucleotide triphosphates (dNTPs)—dATP, dTTP, dCTP, and dGTP—DNA polymerase will add nucleotides complementary to the single-stranded DNA template and extend the double-stranded region. Sanger found that if dideoxynucleotide triphosphates (ddNTPs) were included in the reaction, DNA elongation stopped when a ddNTP was incorporated. This occurs because ddNTPs lack a 3′ hydroxyl group (–OH), which is needed to form the phosphodiester linkage that joins adjacent nucleotides.

In the original dideoxy sequencing protocol, four reaction tubes (A, T, C, and G) are set up. Each of the reactions contains a DNA template, a primer sequence, DNA polymerase, and the four dNTPs (dATP, dTTP, dCTP, and dGTP), one of which is radioactively labeled. A single type of ddNTP is added to each of the four reactions—ddATP (to tube A), ddTTP (tube T), ddCTP (tube C), or ddGTP (tube G). Working from the primer, DNA polymerase randomly adds dNTPs or ddNTPs that are complementary to the DNA template. The ratio of dNTPs to ddNTPs in the reaction is adjusted so that a ddNTP is incorporated into the elongating DNA chain approximately once every 100 nucleotides. When a ddNTP is incorporated, synthesis stops, and a DNA strand of a discrete size is generated. After replication, there are millions of copies of the DNA sequence, each of which terminated at a different nucleotide position.

When the reactions are complete, the newly synthesized strands are denatured, and each reaction is loaded onto a different lane of a polyacrylamide gel. The synthesized fragments migrate through the gel according to size, and each lane eventually

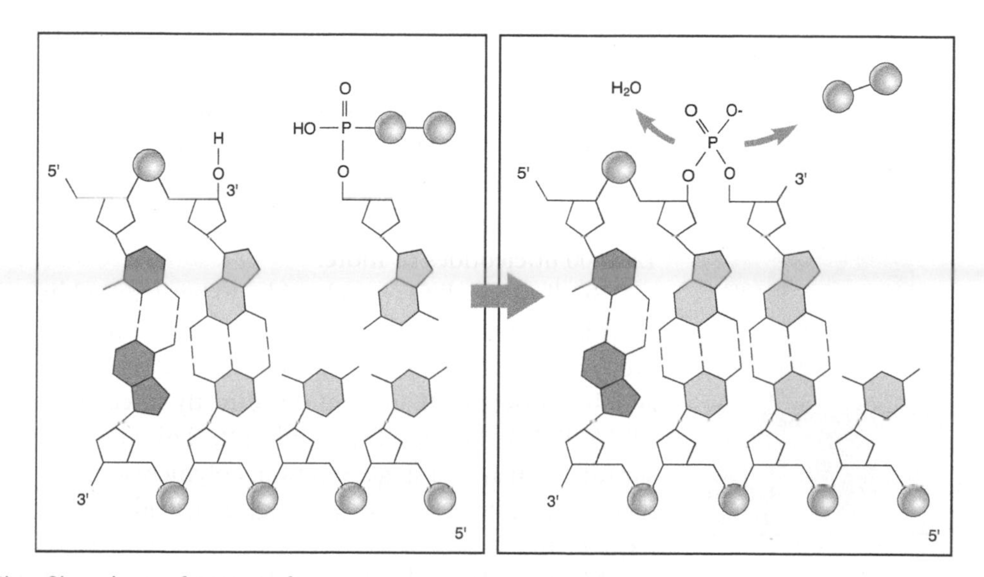

5′ to 3′ synthesis of DNA. *(Left)* An incoming nucleoside triphosphate aligns with the template strand. *(Right)* The 5′ carbon of the incoming nucleotide is joined to the 3′ carbon of the growing strand by way of a phosphodiester linkage. Pyrophosphate (two phosphate balls) and water are liberated.

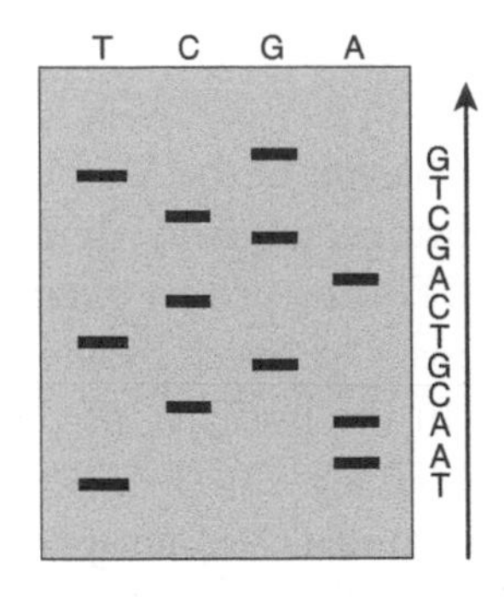

Autoradiogram of dideoxy sequencing gel. Bands are read from bottom to top of gel to generate sequence.

resolves to form a "ladder" of bands. Each band on the gel differs in length by a single nucleotide. Following electrophoresis, the gel is placed in contact with X-ray film. The radioactively labeled nucleotides expose the X-ray film, revealing the series of bands generated in the A, T, C, and G reactions. The gel is then "read" from bottom to top, beginning with the smallest DNA fragment and then scanning across the lanes to identify each successively larger fragment. Optical scanners became the first element of automation in DNA sequencing, producing computer files of the finished sequences. Using this approach, the following first small genomes were sequenced in the 1970s and 1980s:

- The bacterial virus ϕX174, 5386 nucleotides, by Frederick Sanger (1977).
- The mammalian virus SV40, 5224 nucleotides, by Walter Fiers, University of Ghent (1978).
- The human mitochondrion, 16,569 nucleotides, by Stephen Anderson, MRC Laboratory of Molecular Biology (1981).

AUTOMATED DNA SEQUENCING

Leroy Hood
(Courtesy of Leroy Hood.)

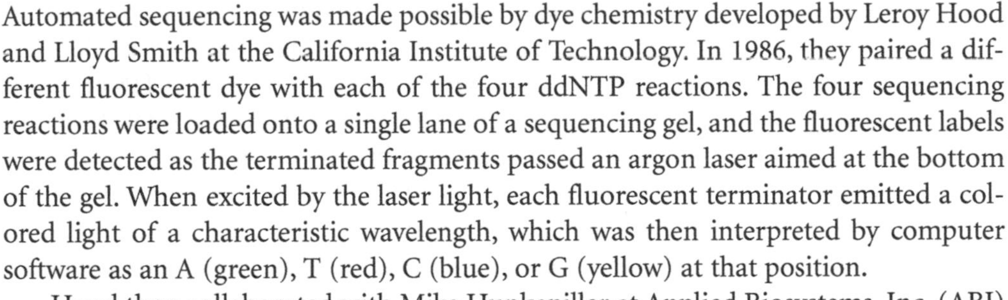

Automated sequencing was made possible by dye chemistry developed by Leroy Hood and Lloyd Smith at the California Institute of Technology. In 1986, they paired a different fluorescent dye with each of the four ddNTP reactions. The four sequencing reactions were loaded onto a single lane of a sequencing gel, and the fluorescent labels were detected as the terminated fragments passed an argon laser aimed at the bottom of the gel. When excited by the laser light, each fluorescent terminator emitted a colored light of a characteristic wavelength, which was then interpreted by computer software as an A (green), T (red), C (blue), or G (yellow) at that position.

Hood then collaborated with Mike Hunkapiller at Applied Biosystems, Inc. (ABI) to produce the first commercial instrument to read sequences from dye-labeled fragments. The ABI Model 370 DNA sequencer, first marketed in 1987, used a polyacrylamide slab gel to resolve ladders of DNA fragments labeled with fluorescent nucleotides. The sequencer incorporated a computer program that built a simulated gel image of fluorescent DNA bands as they were detected by a scanning laser at the bottom of the electrophoresis bed. The final output took the form of an electropherogram, showing colored peaks corresponding to each nucleotide position. The ABI Model 370 DNA sequencer, equipped with a 16-lane polyacrylamide gel, had the capacity to sequence as many as 20,000 nucleotides per day. Increasing the number of lanes to 32, 48, and, ultimately, 96 brought the daily output of each machine to 120,000 nucleotides or more.

Lloyd M. Smith
(Photo by Jim Dahlberg, courtesy of Lloyd M. Smith.)

By allowing all four nucleotides to be analyzed in a single lane, Hood's fluorescent chemistry quadrupled the output of sequencing gels. Parallel improvements in DNA preparation further increased output. DuPont introduced "dye terminators," which attach a different fluorescent dye directly to each of the four terminator nucleotides (didATP, didTTP, didCTP, or didGTP). This allowed all four nucleotides to be labeled simultaneously in a single reaction. Polymerase chain reaction (PCR) was pressed into service to automate dye labeling, a hybrid method that became known as cycle sequencing.

The foundation PCR technology was discovered in 1985 by Kary Mullis at Cetus Corporation and uses enzymatic amplification to increase the copy number of a DNA

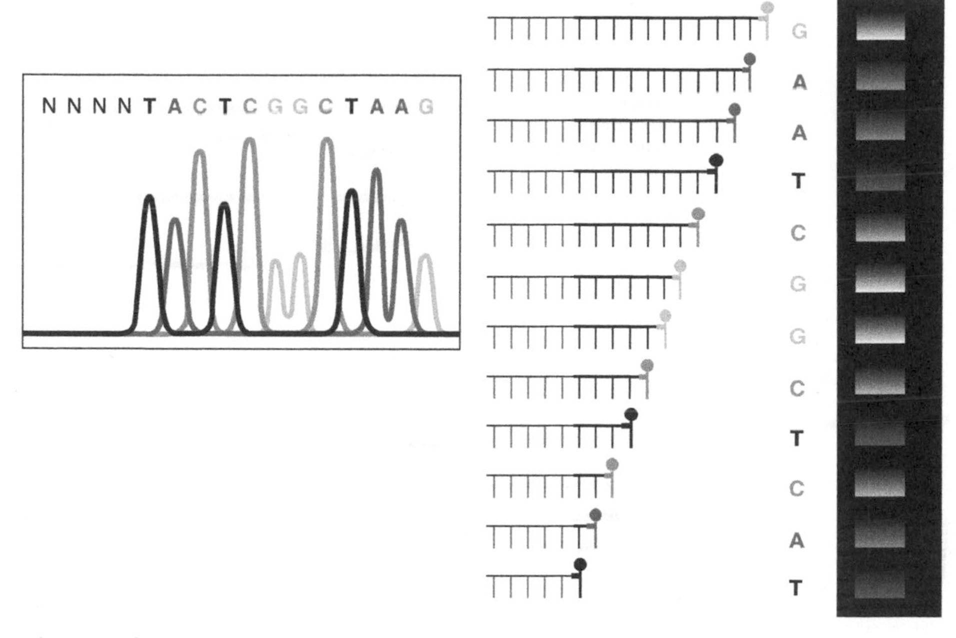

Sample output from an automated DNA sequencer. Electropherogram showing peaks corresponding to each nucleotide position *(left)*. Representation of the dye-terminated sequences that correspond to each peak (*center*). Computer-generated representation of the sequencing gel (*right*).

Mike Hunkapiller
(Courtesy of Mike Hunkapiller.)

Kary Mullis
(Courtesy of Cold Spring Harbor Laboratory Archives.)

fragment. First, a pair of DNA oligonucleotide (oligo, meaning a few) primers of approximately 20 nucleotides in length are synthesized that bracket the "target" region to be amplified. The primers are designed to anneal to complementary DNA sequences at the 5′ end of each strand of the target region. The two primers are mixed in excess with a DNA sample containing the target sequence, a heat-stable polymerase, the cofactor magnesium (Mg^{++}), and the four deoxyribonucleoside triphosphates (dNTPs). *Taq* polymerase from *Thermus aquaticus*, a hot-spring dwelling bacterium, is commonly used.

A thermal cycler then takes the reaction mixture through multiple synthesis cycles, which typically comprise the following three steps:

- *Denaturing.* Heating to near boiling (94°C) denatures the target sequence and creates a set of single-stranded templates. Heating increases the kinetic energy of the DNA molecule to a point at which it is greater than the energy needed to maintain hydrogen bonds between base pairs, and the double-stranded DNA separates into single strands.
- *Annealing.* Cooling to approximately 65°C encourages oligonucleotide primers to anneal to their complementary sequences on the single-stranded templates. The optimum annealing temperature varies according to the proportion of A-T to G-C base pairs in the primer sequence. Because the primers are added in excess and are short, they will anneal to their long target sequences before the two original strands can come back together.
- *Extending.* Heating to 72°C provides the optimum temperature for the DNA polymerase to extend from the oligonucleotide primer. The polymerase synthesizes a second strand complementary to the original template.

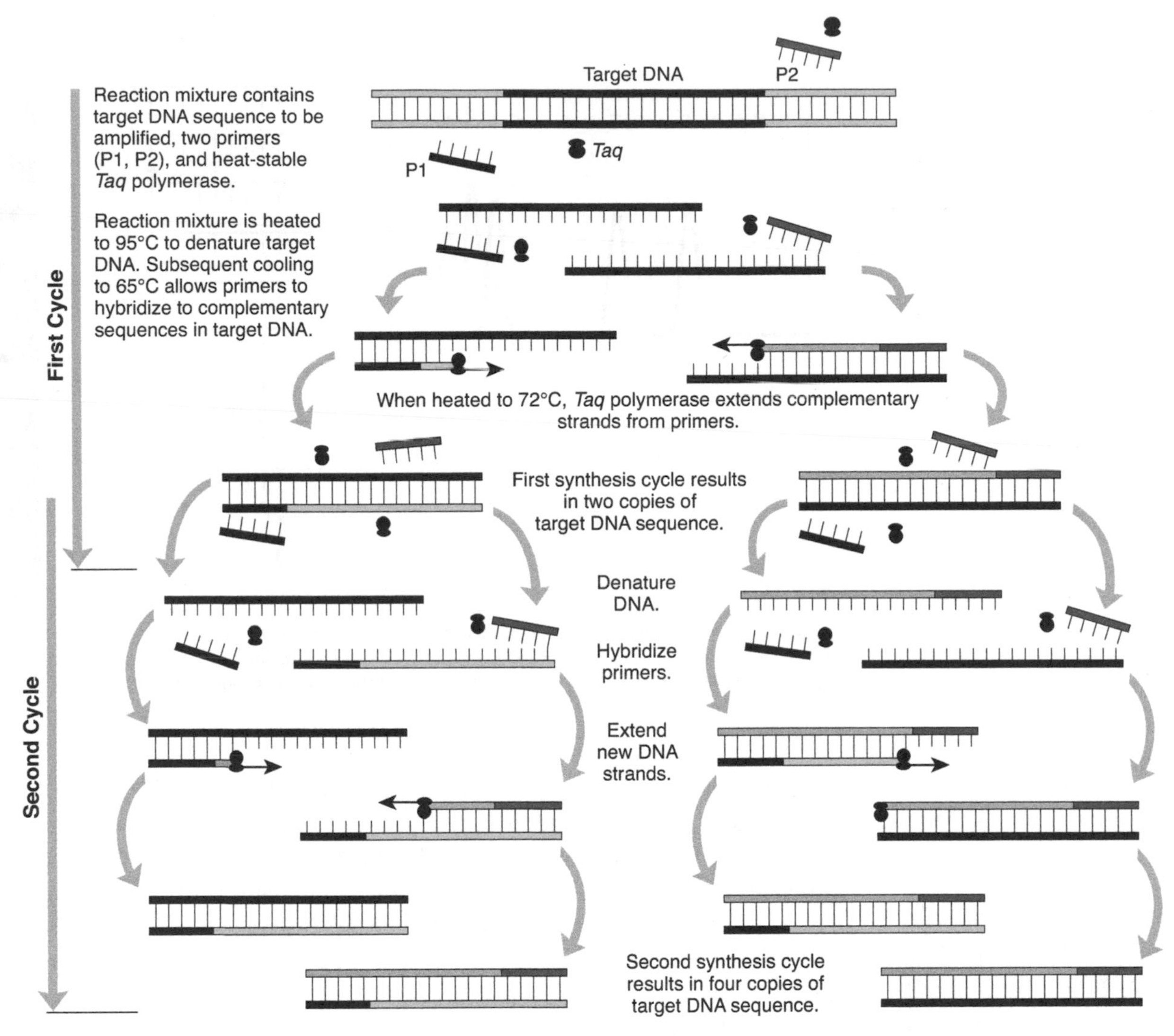

DNA amplification using polymerase chain reaction.

During each synthesis cycle, the number of copies of the target DNA molecule is doubled. Twenty-five rounds of synthesis theoretically produce 1,000,000-fold amplification of the target sequence in as little as 20 minutes (in a two-temperature profile that eliminates a separate annealing temperature).

Cycle sequencing, like PCR, uses multiple rounds of denaturation, annealing, and extension, but uses only one primer and dye terminators. Using this sequencing technology, the Human Genome Project was initiated in 1988 as an international collaboration to determine the entire nucleotide sequence of the haploid human genome. The project ambled along until, in 1998, it received a psychological and technological challenge from J. Craig Venter, a former NIH researcher who had started a biotechnology company, Human Genome Sciences, and a nonprofit organization, The Institute for Genomic Research (TIGR). Venter announced that he had joined with ABI's Hunkapiller to start a new company, Celera, at which he intended to sequence the human genome in 3 years using a new capillary sequencing technology and Venter's shotgun genome assembly (discussed later).

ABI's system replaced each lane of a slab gel with a silica capillary tube, each about the diameter of a human hair and filled with an electrophoresis resin. The capillary reduces heat generated during electrophoresis, allowing higher current and decreasing separation time. A 96-capillary array was linked to a robot mechanism capable of automatically reloading samples from 96-well microtiter plates up to 12 times per day. Throughput was further increased by 384-capillary instruments working from 384-well microtiter plates. This eliminated the time-consuming elements of pouring and loading sequencing gels, reducing human intervention to maintaining reagent levels and loading microtiter plates into the autoloader. Ultimately, Celera and the major centers of the international collaboration became sequencing factories outfitted with 30 or more capillary sequencers, each churning out up to 400,000 nucleotides of sequence per day. This was 400-fold faster than hand-sequencing methods available just prior to the start of the Human Genome Project. Increasing sequence reads to as high as 1000 nucleotides per capillary and limiting human intervention decreased sequencing costs from $1 to $0.01 per nucleotide and increased sequencing accuracy from 99% (one error per 100 nucleotides) to 99.9% (one error per 1000 nucleotides).

NEXT-GENERATION DNA SEQUENCING

DNA sequencing always begins with fragmenting the genome under study using restriction enzymes, sonication (sound waves), nebulization with liquid nitrogen, or mechanical shearing (e.g., passing the DNA through a syringe). Each fragment must be enriched to provide enough DNA from which to detect each nucleotide in the fragment's sequence. In Sanger sequencing, the DNA enrichment is provided by bacterial cloning. Each fragment is ligated into a plasmid vector, which is, in turn, transformed into a bacterium. The bacterium replicates to create a colony of identical clones, each carrying multiple copies of the genome fragment. DNA is extracted from selected clones and forms the basis for individual sequencing reactions. Because some of these steps must be done by human technicians, the entire workflow cannot be fully automated.

Beginning in 2005, "next-generation" sequencers dramatically shortened the sequencing work flow by determining sequence directly from collections of genomic DNA fragments that are amplified by PCR. By 2012, next-generation sequencing had further decreased costs to only $0.10 per megabase (million base pairs). In most next-generation methods, short adapter molecules are first ligated to each end of the genomic DNA fragments. The adapters anchor the fragments to discrete locations on a substrate (a microbead or plate surface) and then act as universal PCR primers to amplify each fragment in situ. Like bacterial colonies on a plate, the PCR colonies, or "polonies," are spatially isolated from one another. In most strategies, polonies are generated by emulsion PCR, in which the adapter-linked templates and water-soluble PCR reagents are emulsified in oil. This creates picoliter-sized reaction vessels in which each PCR is isolated in a tiny water droplet surrounded by oil.

After PCR amplification, the genome fragments are denatured and the single-stranded molecules in each polony serve as templates for sequencing by DNA synthesis. Working from a universal primer within the adapter sequence, sequence is built up by adding nucleotides that are complementary to the genome templates. The reaction

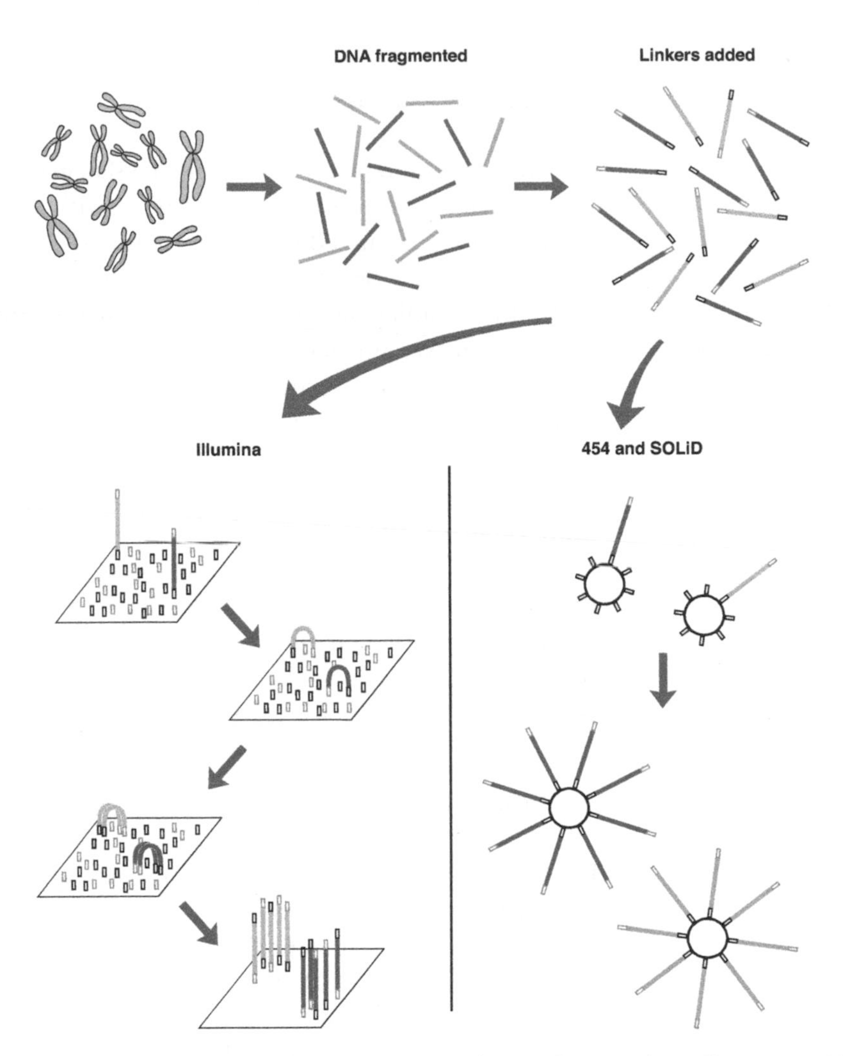

Next-generation sequencing develops PCR colonies (polonies) that are physically separated from one another. Illumina sequencing isolates polonies at discrete locations on solid substrate; 454 and SOLiD sequencing isolate polonies on DNA-capture beads. Each polony, or feature, provides templates for sequencing by DNA synthesis.

vessel in which this occurs is termed a flow cell, because fresh reagents flow past the polony features during each synthesis cycle. The nucleotide added in each synthesis cycle is detected from each feature, building up millions of sequences in parallel. The repeated cycles of DNA synthesis across an array of polonies is referred to as cyclic array sequencing. Collecting sequence information requires a charge-coupled device (CCD) camera, a sensitive imager that can detect fluorescent or visible light "reports" from each tiny polony.

Because each feature can be as small as 1 μm, millions of features can be included in a single array of modest size. This makes it possible to obtain hundreds of millions of nucleotide sequences in a single run on a next-generation sequencer. In fact, next-generation sequencers can generate a genome's worth of sequence in several sequencing runs. The dramatic labor and reagent savings become apparent when one considers that a single run of a cyclic array sequencer replaces more than a million individual sequencing reactions—and the accompanying bacterial cloning, culturing, and DNA preparation. The microliter volumes of each of millions of Sanger sequenc-

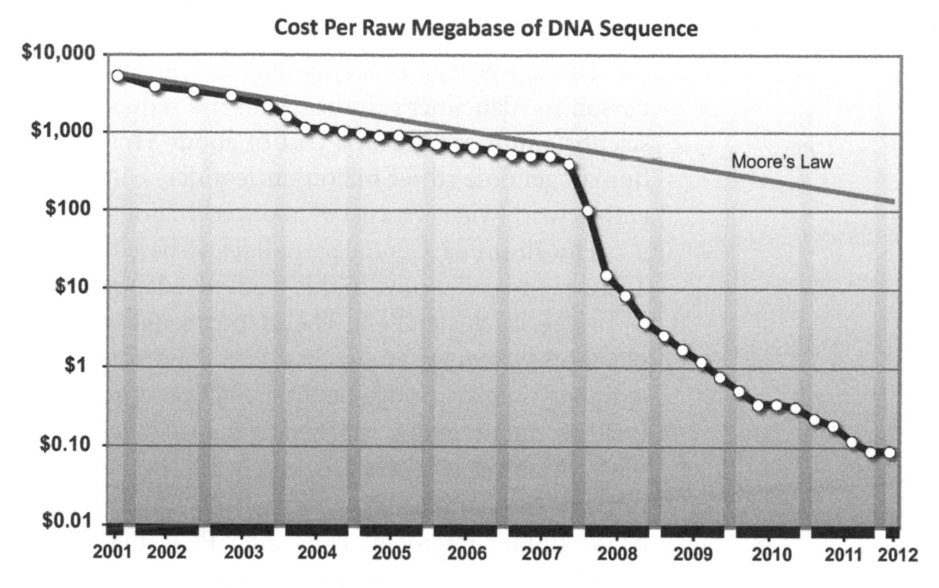

Exponentially declining cost of DNA sequencing. Before the advent of next-generation methods, which develop sequence directly from collections of amplified DNA molecules, the decreasing cost of DNA sequencing generally followed Moore's Law. Named for Intel Corporation founder Gordon Moore, this law predicts that the number of transistors integrated on a computer chip will double approximately every 18 months—with a corresponding decrease in cost. However, since 2008, DNA sequencing costs have decreased 1000 times faster than predicted by Moore's Law.

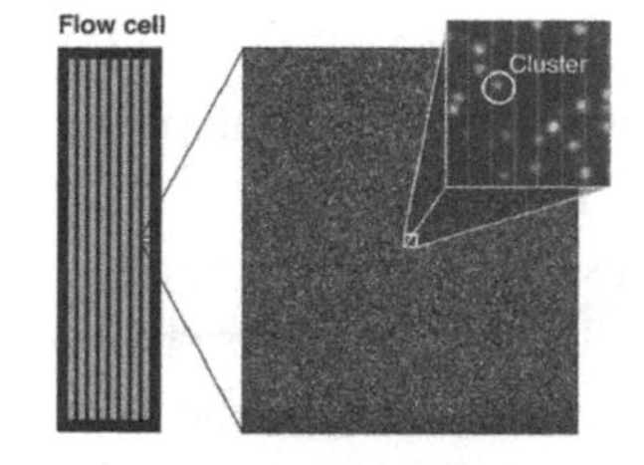

Polony features from Illumina sequencer. Next-generation sequencing produces an array of polonies (PCR colonies). Nucleotide additions in each synthesis round produce light emissions at individual polonies, which are recorded by a charge-coupled device (CCD) camera. Each lane of an Illumina flow cell is imaged as ~100 image "tiles," each of which captures activity from as many as 200,000 features. Each dot in this Illumina random array is an individual polony or cluster of identical templates.

ing reactions are replaced with a single reaction that effectively uses picoliter or femtoliter amounts for each feature.

454 Life Sciences released the first cyclic array sequencer in 2005. In this method, adapter-terminated genome fragments are hybridized to complementary adapter sequences immobilized on DNA-capture beads, such that each bead contains a single library fragment. After emulsion PCR, the beads with attached polonies are arrayed on a picotiter plate—a microfabricated chip containing wells 28 μm in diameter. Each well accommodates a single DNA-capture bead, along with smaller beads containing immobilized enzymes for pyrosequencing (light sequencing). Bst polymerase from *Bacillus stearothermophilus* generates pyrophosphate with each nucleotide addition. The pyrophosphate then reacts with ATP sulfurylase, luciferase, and luciferin (the light-emitting pigment from fireflies) to produce bioluminescence.

During sequencing, a single nucleotide species is added to the array, and any well in which that nucleotide is added to a template produces a burst of light. A CCD camera records light events in each well (channel) during each synthesis cycle. This is an asynchronous system. During each round of synthesis, different features may or may not incorporate the selected nucleotide, and at any point in time, sequences of different lengths are generated from different polonies. Because there is nothing to terminate a sequence, repeated nucleotides in a sequence (such as A-A-A-A) are added during a single synthesis cycle. Hence, the number of repeats must be inferred by a proportional increase in luminescence, with an A-A-A-A repeat producing a light burst about four times the amplitude of a single A nucleotide. The 454 machine produces the longest reads of any of the next-generation synthesizers (700 nucleotides).

Using this technology, the entire diploid genome (six billion nucleotides) of Nobel Prize–winner and DNA structure discoverer James D. Watson was sequenced in 2007. Hoping to downplay the risks of this sort of detailed genetic knowledge, Watson allowed his entire sequence to be made available online (http://jimwatsonsequence.

cshl.edu/cgi-perl/gbrowse/jwsequence/), except for the *ApoE* gene, which predisposes a person to Alzheimer's disease. Watson's sequence was completed by a handful of 454 scientists in 4 months at a cost of about $1.5 million, compared to the first haploid human genome (three billion nucleotides) completed in 15 years by thousands of scientists worldwide at a cost of about $3 billion.

The Illumina Genome Analyzer is based on technology developed by Gerardo Turcatti and colleagues. In this method, adapter-flanked DNA fragments are amplified by bridge, or cluster, PCR. The adapter sequences tether the library fragments to complementary sequences attached to a solid substrate in a flow cell. During PCR, the template forms a bridge between anchored pairs of adapters (primers) so that every PCR product remains attached to the gel. Successive cycles of PCR can be likened to the motion of a "Slinky," in which the motion always proceeds from one anchored point to another. The bridging limits the distance between each amplicon to the length of the library fragment. Thus, after PCR, the ~1000 amplicons of a single genomic template are clustered in a discrete polony.

As with automated Sanger sequencing, the Illumina system uses dye terminators that halt DNA synthesis upon incorporation into an elongating DNA molecule. Whereas traditional dye sequencing uses fluorescent dyes that are irreversibly attached to dideoxy terminators, Illumina sequencing exploits a removable dye. The fluorescent label replaces the 3′ hydroxyl on each nucleotide, blocking further addition and ensuring that only one nucleotide is added per cycle.

In each synthesis cycle, all four fluorescently labeled nucleotides (A, T, C, and G) are added to the flow cell. This is a synchronous sequencing system, because one of the four nucleotides is added to the same nucleotide position, at the same time, in each polony feature. The incorporated nucleotide is excited by a laser, and a CCD camera records the color emitted from each polony. After imaging, the fluorescent dye is chemically removed, regenerating the 3′ hydroxyl and preparing the template for the next cycle of synthesis and imaging. Because each synthesis cycle is a discrete event, the Genome Analyzer accurately detects each nucleotide in a repeated element (such as A-A-A-A). This technology produces paired-end reads of 100 nucleotides per feature.

The Applied Biosystems SOLiD system exploits synthesis by ligation. Like the 454 system, this system generates polonies by emulsion PCR on DNA-capture beads (1 μm). The SOLiD system uses a degenerate collection of short oligonucleotides (eight or nine nucleotides) in which all possible sequence combinations are represented. Each oligonucleotide is labeled with one of four removable fluorescent dyes, which corresponds to the nucleotide at the fifth position. The dye also functions as a blocker, allowing only one oligonucleotide to be ligated in each synthesis cycle.

During the first round of synthesis, a universal primer is annealed to the adapter sequences on each polony feature. An oligonucleotide with a sequence complementary to the template sequence is then ligated to the end of the universal primer. After ligation, the polonies are imaged in four channels to determine the nucleotide in the fifth position. After imaging, the oligonucleotide is cleaved between the fifth and sixth position, releasing the dye and creating a free end for the next ligation. After each successive round of ligation, sequence information is collected on every fifth nucleotide in the template (positions 5, 10, 15, 20, etc.). The universal primer and ligated sequence are then denatured from the template. A new universal primer that has one less nucleotide at the end compared to the first universal primer is then annealed to the template to generate an offset that will allow a different set of positions to be sequenced with the ligated oligonucleotides, e.g., positions 4, 9, 14, 19, etc. After five rounds of synthesis with different universal primers, each with one fewer nucleotide than the previous primer, a complete sequence is generated for each feature. The SOLiD system produces paired-end reads of 60 nucleotides per feature.

After founding 454 Life Sciences, Jonathan Rothberg developed the Ion Torrent. Released in 2010, this instrument uses an entirely different detection system than other next-generation sequencers. Rather than using an optical sensor to detect light emitted from fluorescent dyes, Ion Torrent uses a semiconductor sensor to detect pH changes during DNA synthesis. The semiconductor chip used in the instrument has up to 660 million microwells, each containing a different single-stranded DNA template from the genome under study. During each synthesis cycle, a new deoxynucleotide triphosphate (A, T, C, or G) is flowed through the chip. The addition of the defined nucleotide to its complementary partner on the template strand by DNA polymerase is accompanied by the release of pyrophosphate (P2) and a hydrogen ion (H^+). An ion-sensitive field-effect transistor (ISFET) detects the pH change in the microwell, and an electrical signal from the transistor is directly interpreted by the base-calling software. As with the 454 instrument, Ion Torrent infers the number of nucleotides in a homopolymer tract (such as A-A-A-A) by the relative strength of the signal for that nucleotide addition. Ion Torrent is the fastest of the commercially available sequencers, completing up to 200 bases of sequence per microwell in a 2-hour run.

GENOME ASSEMBLY STRATEGIES

Whether 75 nucleotides are obtained per polony feature or 800 nucleotides are obtained per capillary, each channel of a sequencing instrument represents only a minute fraction of a genome. After sequencing comes the task of assembling millions of sequenced fragments into large contiguous sequences and, ultimately, into whole chromosomes. Two major strategies are used in assembling information obtained from DNA sequencing: whole-genome shotgun and hierarchical cloning. Whole-genome shotgun is a bottom-up method and is the fastest and most economical means to sequence a genome. The name alludes to fragmenting the genome into tiny bits, as with a close-range shotgun blast, and then sequencing the resulting short fragments. The genome under study is typically fragmented by enzyme cleavage, sonication, or mechanical shearing.

Because next-generation sequencing begins with short fragments generated in this way, it is, in fact, the latest evolutionary step in whole-genome shotgun sequencing methods. However, as originally conceived, shotgun sequencing involves ligating genome fragments into bacterial plasmids and transforming the resultant recombinant molecules into *E. coli* bacteria. Each transformed bacterium is cultured in a separate well of a 384-well plate and grows to produce clones of identical bacteria, all of which carry the same insert of genomic DNA. Thus, each of millions of plasmid clones can be identified by a specific position on a master plate. Clones are randomly selected, and 600–800 nucleotides of sequence are generated from each end in a single sequencing run (the middle part of the cloned fragment is generally not sequenced).

In a typical plasmid library, each nucleotide in the genome under study is represented on eight to ten different cloned fragments; this is called 8×–10× coverage. Next-generation sequencing typically provides at least 20× coverage. This level of redundancy is needed for the assembly step, in which computer algorithms sift through the masses

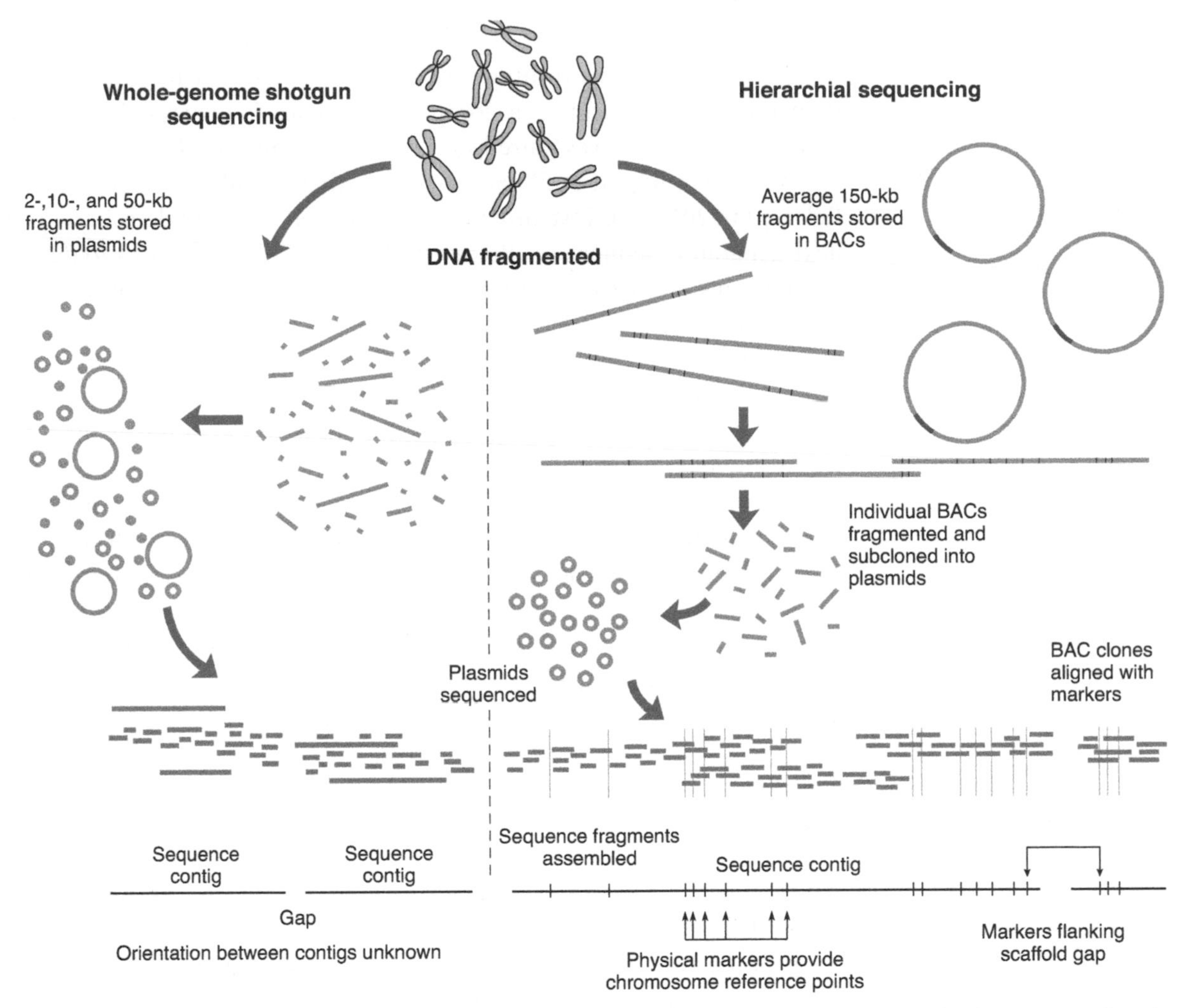

Whole-genome shotgun and hierarchical clone sequencing strategies. Hierarchical sequencing has the advantage of using physical markers to anchor assembled sequences (contigs) to exact chromosome locations.

of DNA data to align the short sequences according to shared regions of overlap. This ultimately produces stretches of contiguous sequence, or contigs.

Assembly of a whole-genome shotgun, however, is confounded by any repetitive DNA whose repeat sequence is longer than the average length of sequence reads: 800 nucleotides in capillary sequencing, 700 nucleotides in 454 sequencing, and 60–200 nucleotides for other next-generation sequencing methods. In addition, many genome regions will not be represented in the library used to generate the sequencing plasmids or features. In the absence of a physical map or existing reference assembly, whole-genome shotgun sequencing produces numerous relatively short contigs whose orientations to one another or to precise chromosome locations are virtually impossible to ascertain. In many cases, assembly of a whole-genome shotgun sequence can be guided by a complete assembly of a close evolutionary ancestor, because chromosome regions display a high degree of synteny (conserved gene order) among related species.

Ultimately, variations of a more laborious method—hierarchical cloning—have been used to generate nearly complete reference genomes for a number of key organisms. Hierarchical cloning is a top-down strategy that creates an ordered library of large DNA molecules, maintained as individual bacterial clones and mapped to specific chromosome locations. Bacterial artificial chromosomes (BACs), developed in the mid 1990s by Mel Simon at the California Institute of Technology, have proven to be the most useful cloning vehicles for large DNA fragments. These circular chromosomes, which can contain up to 250 kb of inserted DNA, are extremely stable and amenable to large-scale automation. Plasmids, in contrast, are much smaller and can hold up to only 2–50 kb of inserted DNA.

To construct a BAC library, genomic DNA is randomly cleaved to produce fragments averaging 150,000 nucleotides in length. Partial digestion can be achieved by using a low concentration of a restriction enzyme, sometimes in combination with a methylase that protects a portion of the cutting sites. Alternately, rare-cutting restriction enzymes such as MluI, NruI, and PvuI, which have recognition sequences that occur infrequently in most eukaryotic genomes, can be used. The resulting restriction fragments are ligated into separate BACs, and these recombinant molecules are transformed into *E. coli*. Each of hundreds of thousands of BAC clones can be identified by a specific position on a master 384-well plate. A typical BAC library achieves 8×–10× coverage of the whole genome.

The BAC libraries are analyzed in several ways to identify sets of BAC clones with shared sequences. For each experiment, samples of the BAC clones are transferred from the master library plates onto replica plates for analysis. In this way, the master library is maintained for future reference. BAC fingerprints are generated by digesting BAC clones with several restriction enzymes and electrophoresing the restriction fragments through a gel. BACs that share a common banding pattern, or fingerprint, must share an overlapping region of sequence. BAC-end sequences are generated by running a sequencing reaction on either end of each BAC clone, and matches are determined by sequence alignment. Finally, genetic markers and "overgo probes" repre-

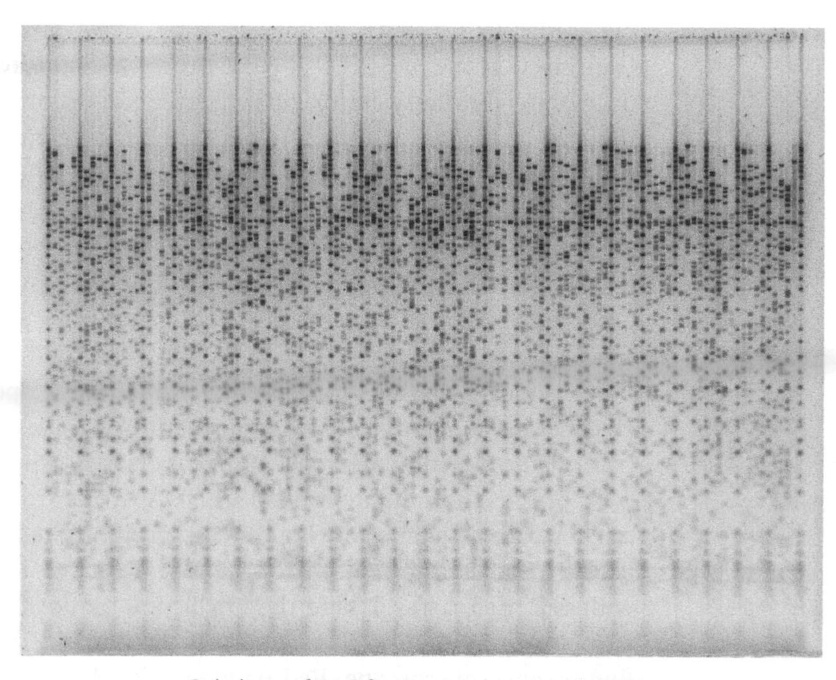

Gel photo of BAC fingerprint clone panel, 2000.
(Courtesy of John McPherson, School of Medicine, Washington University, St. Louis.)

senting known genes are hybridized to the BAC clones. Information from BAC fingerprints, BAC-end sequences, genetic markers, and overgo probes is used to sort BAC clones into "bins" according to their shared sequence information. The genetic markers assign bins to physical locations on chromosomes, whereas the overgo probes provide the relative positions of known genes.

Next, a "tiling path" is selected—an economical set of BACs to sequence that cover the majority of the genome. Each BAC clone along the tiling path is shotgun sequenced. First, each BAC is sheared by forcing the DNA solution through a syringe. This produces fragments several thousand nucleotides in length. The sheared DNA is then separated on a gel, and the fragments are subcloned into plasmids to fill several 384-well plates. "Paired-end reads" are obtained by sequencing 600–800 base pairs of sequence from each end of each subcloned fragment. Finally, a computer aligns overlapping reads to provide the entire sequence of each BAC clone.

The finished BAC clones are then aligned to provide a contiguous sequence, or contig. An assembler program then strings together sequences from local contigs to produce a nearly continuous chromosome sequence. In many cases, unlinked contigs can be oriented with respect to one another using BAC ends, plasmid ends, known mRNAs, and physical/genetic map information to achieve the highest order of large-scale integration called "scaffolds." As the name implies, a scaffold uses map features to affix contigs to chromosome maps, showing the linear relationship of unmerged contigs. Finally, the sequence is improved, or finished, to fill in gaps and extend the contigs. Sequencing primers are used to extend 3′ sequence, whereas 5′ gaps are filled by sequencing PCR products.

DNA MICROARRAYS

DNA sequencing is complemented by DNA microarrays (or DNA arrays), which provided the first means to analyze large numbers of DNA sequences in parallel. Conceived in the mid 1990s by Pat Brown at Stanford University, DNA arrays based on cDNA libraries were originally developed to analyze the expression patterns of thousands of genes at a time. In his method, different cDNAs were spotted at discrete positions on a glass slide coated with polylysine. The spotting can be done with a set of needle-like pins or even with an inkjet printer! The negatively charged DNA molecules form ionic bonds to the positively charged polylysine substrate, holding them firmly in position in the microarray. The finished microarray thus contains immobilized probes representing thousands of genes from a single cell type or from a single species.

Patrick O. Brown
(Courtesy of Pat O. Brown.)

A typical experiment is based on a microarray spotted with cDNAs representing all genes in the genome of an organism. mRNA is isolated from experimental and control cells to be compared, e.g., tumor versus normal cells, mitotic versus quiescent cells, or cells from different tissues. cDNAs made from the mRNA samples from each cell type are then labeled with either a green or red fluorescent dye; the dyes are similar to those used for automated DNA sequencing. The labeled cDNAs are incubated with the microarray, where they hybridize to positions containing complementary sequences. Unbound cDNAs are washed away, and the microarray is imaged under a fluorescence microscope. Red or green signals indicate genes that are differentially expressed in the two populations of cells, and the intensity of the signal indicates the

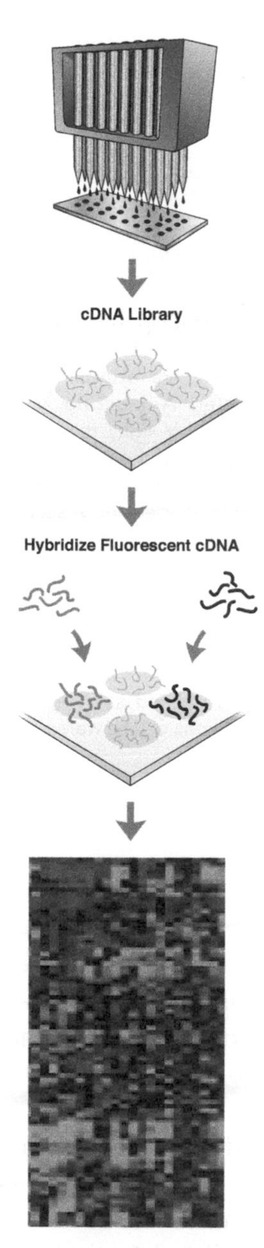

Typical DNA array experiment. A cDNA library is spotted onto a polylysine-coated slide. Differentially labeled cDNA probes are added from two different cell types or experimental treatments, one labeled with a red fluorescent dye (black) and the other labeled with a green fluorescent dye (gray). The probes hybridize to corresponding cDNAs, showing which genes are active under each of the two conditions.

level of expression. In one early experiment, a microarray containing 17,856 cDNAs expressed by the lymph nodes was used to compare diffuse large B-cell lymphomas from different cancer patients. Two different expression patterns, which correlated with different clinical outcomes, were found in tumors that could not be distinguished by microscopic examination.

The company Affymetrix, founded by Steven Fodor, took a different approach to the construction of DNA microarrays. Combining microphotolithography borrowed from computer chip manufacture and combinatorial chemistry from the pharmaceutical industry, Affymetrix patented an industrial method to produce high-quality DNA microarrays. Rather than attaching cDNA probes to the array, oligonucleotides are built anew at individual positions on a quartz wafer using light-directed chemical synthesis. Each wafer may yield 50–400 GeneChips, depending on the number of probes in the microarray.

To make a GeneChip, the wafer is first coated with a linker molecule. Each linker molecule is attached to a single nucleotide with a protecting group that blocks polymerization, in the same manner that a dideoxynucleotide terminates a growing nucleotide chain. The protector group is sensitive to light (photolabile) and is released on exposure to ultraviolet (UV) light. A filter mask is placed between the wafer and the UV light source so that only specific positions are exposed to the light and become deprotected. A new nucleotide is then added to the chain at these deprotected positions. A new protecting group is added to these positions at the end of each synthesis step and the process starts over. A computer program controls the process, and a wafer containing a wide variety of oligonucleotides can be built up in 50–100 synthesis steps.

Oligonucleotide arrays, constructed either by photolithography or by attaching synthesized oligonucleotides to glass, have largely replaced spotted cDNA arrays. Because they control for the copy number and size of each gene probe (typically 50–70 nucleotides), synthetic oligonucleotide arrays are easier to calibrate, and they provide more consistent results in high-throughput applications. Synthetic arrays have another major advantage in that they can be based on annotated genes and complete genome sequences available in public databases. For these reasons, synthetic DNA arrays have supplanted printed arrays for a range of analyses including the following:

- Whole-genome arrays containing probes developed from every annotated gene in a genome sequence are used to study differential gene expression.
- So-called SNPchips containing hundreds of thousands of single-nucleotide polymorphisms (SNPs) are used for generating personal genetic screens or for large-scale population studies. The same is true for copy-number variations (CNVs).
- With multiple probes for each exon of a gene, exon arrays support detailed studies of gene transcription and alternative splicing.
- Tiling arrays investigate 25-nucleotide "tiles" spaced at 10-nucleotide intervals to scan a genome for novel mRNA transcripts or transcription-factor-binding sites.
- High-density resequencing arrays that interrogate each position of an entire genome are used to sequence annotated genomes from multiple individuals to provide information about population variation.

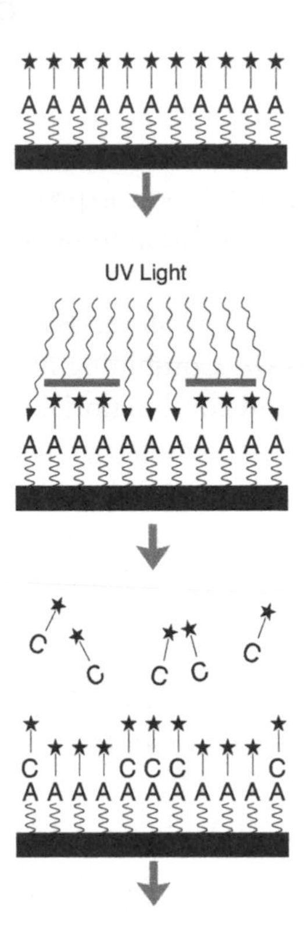

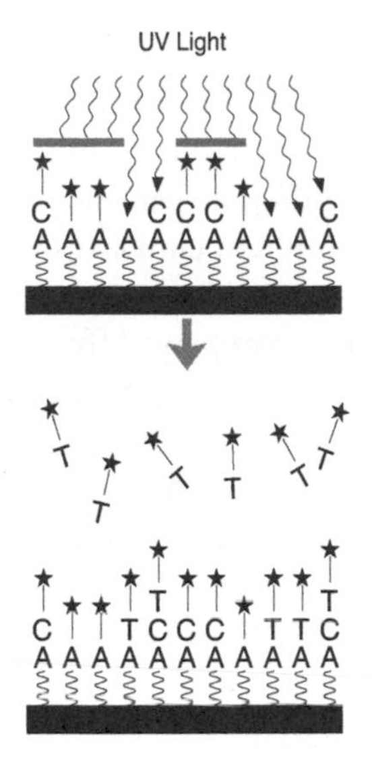

Making a GeneChip. Adenine molecules with photolabile protecting groups (stars) are attached to a quartz wafer. A blocking mask (gray) allows UV light to expose specific regions of the wafer, which removes the protecting group from a population of adenines. A second nucleotide (C) is added to unprotected adenines. Another mask is added, deprotecting a different set of nucleotides, and a new nucleotide (T) is added. The process is repeated to build up olignucleotides with defined sequences at known positions on the wafer.

MINIMAL AND SYNTHETIC GENOMES

Hamilton Smith

(Courtesy of the J. Craig Venter Institute.)

J. Craig Venter

(Courtesy of the J. Craig Venter Institute.)

This consideration of genome structure, function, and sequencing leaves us with several interesting questions. What is the smallest genome—the minimal set of genes—required for life? If a genome is merely biological information encoded in the sequence of a DNA molecule, can scientists synthesize a minimal genome to recreate life in vitro (meaning literally "in glass," but, practically, in a test tube)?

Recently, these questions have been most vigorously pursued by a research team headed by Hamilton Smith and Craig Venter at the J. Craig Venter Institute in Rockville, Maryland. (Smith shared the 1978 Nobel Prize for the discovery of restriction enzymes.) Har Gobind Khorana and coworkers at the National Institutes of Health chemically synthesized the first gene in 1978: the 207-bp gene for tyrosine suppressor tRNA. The advent of automated DNA synthesis in the 1980s, and later the Internet, made it possible for researchers to order small DNA molecules (notably PCR and sequencing primers) online for overnight delivery.

In 2004, Smith and Venter took DNA synthesis into a new realm when they manufactured a biologically active viral genome from scratch. Working from the published sequence of the bacterial virus ϕX174, they designed single-stranded oligonucleotides (42 mers) such that the ends of complementary strand sequences were staggered. When annealed, these formed "sticky" ends that provided templates to align adjacent sequences during a subsequent ligation reaction. Double-stranded ligation products averaging 700 nucleotides were then joined by polymerase cycling assembly that incorporated overlapping templates on each cycle to produce full-length genomes of 5386 bp. The extended products were then electroporated into *E. coli*, where the synthetic molecules directed the production of encapsulated, fully infective viruses. The bacterium *Mycoplasma genitalium* has the smallest genome yet found for any free-living organism. Although it carries 485 protein-coding genes, approximately 100 of these are nonessential when disrupted individually. To determine the smallest gene set that is simultaneously required for life, Smith and Venter proposed to synthesize various reduced genomes and test them inside bacterial cells. As a step toward this goal, in 2008 the team synthesized a correct copy of the 582,970-bp *M. genitalium* genome. Synthetic DNA sequences of 5–7 kb went through several rounds of in vitro ligation to produce overlapping constructs equaling about one-fourth of the *Mycoplasma* genome. These fragments were then joined—by homologous recombination within yeast cells—to create a synthetic *Mycobacterium* chromosome that was stably propagated in yeast.

In 2010, the team produced the first living cell under the control of a synthetic genome. For this experiment, they assembled the entire 1.08 million–bp genome of the fast-growing *Mycoplasma mycoides*. Synthetic 1-kb cassettes went through three rounds of recombination in yeast to produce 10- and 100-kb assemblies—and finally a complete genome, which included several "watermark sequences" that can be decoded into quotes from English literature! Next, they inserted the synthetic genome into a recipient cell of the related species *Mycoplasma capricolom.* These "synthetic cells" reproduced normally and had the expected phenotype of *M. mycoides.* PCR of DNA from synthetic cells identified the watermark sequences, and whole-genome sequencing

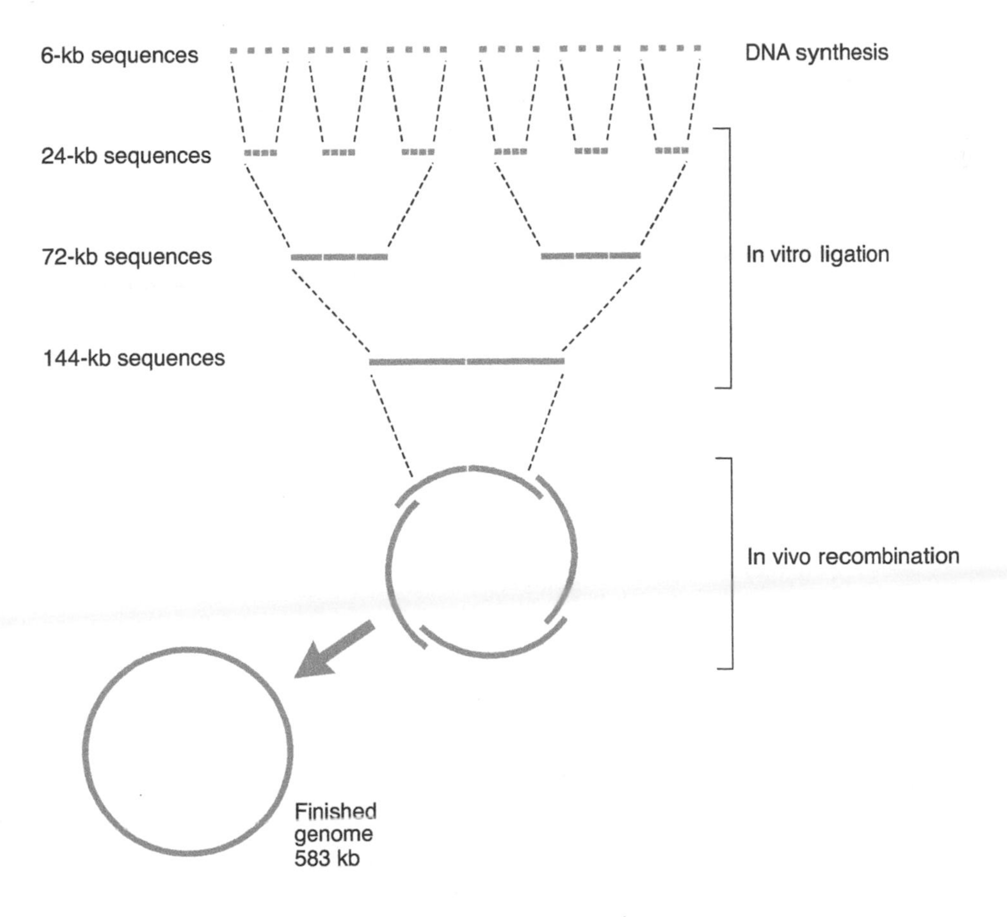

Synthesizing the *Mycoplasma genitalium* genome.

detected several mutations that occurred during assembly but none from the recipient *M. capricolom*.

No vital force is required to "reboot" a living cell from the information inherent in a synthetic DNA molecule. Although organisms have evolved chromosomes to perpetuate their genetic legacy, this information can simply be stored in and retrieved from a digital file. This puts a point on the notion that DNA is information.

LABORATORY 1.1

Annotating Genomic DNA

▼ OBJECTIVES

This laboratory demonstrates several important concepts of modern biology. During the course of this laboratory you will

- Learn that gene annotation adds structural and functional information to a DNA sequence.
- Use gene-finding programs to predict genes in a DNA sequence.
- Use RNA and protein sequences to provide biological evidence for predicted genes.
- Predict gene function using protein domain databases.

In addition, this laboratory uses several bioinformatics methods in modern biological research. You will

- Use RepeatMasker to identify transposons and repeated DNA sequences.
- Use trainable algorithms and hidden Markov models to identify gene features.
- Use the Basic Local Alignment Search Tool (BLAST) to identify sequences in databases.
- Use the genome annotation editor Apollo to build gene models and annotate a DNA sequence.

INTRODUCTION

DNA is an information molecule. Organisms have evolved cellular mechanisms to "read" the DNA code—to convert information stored in their genomes into metabolic processes to sustain and reproduce life.

DNA sequencing and the alignment of sequenced fragments allow scientists to reconstruct the genome of an organism. However, on its own, a genome sequence of A's, T's, C's, and G's tells a biologist nothing about what genes it contains or how they function. Bioinformatics is a relatively new field of science that uses computation and statistics to discover the biological information encoded in DNA and other biomolecules.

Just as one can add notes to written work, gene annotation adds explanations to a DNA sequence. Structural annotation analyzes DNA sequence to determine the physical arrangement of the exons, introns, start/stop codons, and 5′ and 3′ untranslated regions that compose a gene. Predicting gene structure relies on mathematical evidence, provided by gene prediction algorithms, and biological evidence from RNA and protein sequences. Merging mathematical and biological evidence produces a gene model. Functional annotation then uses similarity searches to identify related

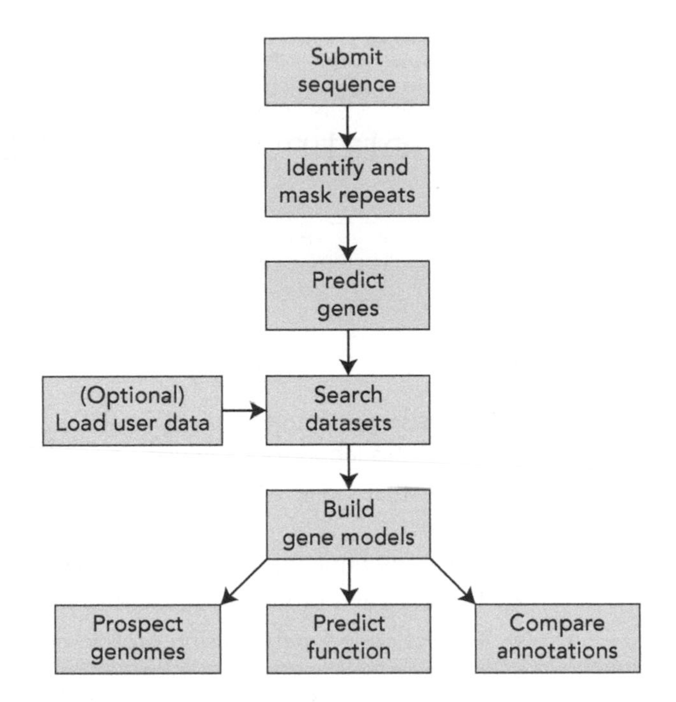

Workflow for annotating protein-coding genes in a genome sequence. Steps include masking repetitive DNA, predicting genes, searching databases for similar sequences, and building gene models.

genes and conserved catalytic and binding motifs; these often reveal roles for the protein product in cell metabolism.

This laboratory uses the simplified bioinformatics workflows in *DNA Subway*, a platform for gene annotation and comparison. The metaphor of a subway map provides an intuitive and appealing interface for working with sophisticated informatics tools. As one "rides" on the Red Line, genes are predicted and annotated in up to 150,000 bp of plant DNA sequence. Each analysis is a "stop" along the way.

Stage A of the laboratory introduces tools used for genome annotation and develops key concepts of gene structure and function. It uses a "synthetic contig," a 16.4-kb sequence that fuses four genes selected from various regions of the *Arabidopsis thaliana* genome. Working with the genes in the order assembled on the contig provides annotation problems that progress from simple to more complex. Stage B supports the independent exploration of a 100-kb sequence from *Arabidopsis* chromosome 5. The two stages build confidence by reinforcing principles of genome annotation.

FURTHER READING

Altschul SF, Gish W, Miller W, Myers EW, Lipman DJ. 1990. Basic local alignment search tool. *J Mol Biol* **215:** 403–410.

Gerstein MB, Bruce C, Rozowsky JS, Zheng D, Du J, Korbel JO, Emanuelsson O, Zhang ZD, Weissman S, Snyder M. 2007. What is a gene, post-ENCODE? History and updated definition. *Genome Res* **17:** 669–681.

Han Y, Burnette JM III, Wessler SR. 2009. TARGeT: A web-based pipeline for retrieving and characterizing gene and transposable element families from genomic sequences. *Nucleic Acids Res* **37:** e78. doi: 10.1093/nar/gkp295.

Snyder M, Gerstein M. 2003. Defining genes in the genomics era. *Science* **300:** 258–260.

Stein L. 2001. Genome annotation: From sequence to biology. *Nat Rev Genet* **2:** 493–503.

PLANNING AND PREPARATION

The following table will help you to plan and integrate the different bioinformatics methods; the workflow can be interrupted at any point and resumed at a later time.

Experiment	Day	Time		Activity
Stage A: Annotate genes in a synthetic contig				
	1	10–60 min	Prelab:	Set up all student computers to run Java and to allow pop-up windows to open. Prepare Apollo by opening the browser to http://gfx.dnalc.org/files/apollo/test.
I. Create a *DNA Subway* project	1	10 min	Lab:	Start a new project on *DNA Subway*'s Red Line with the 100-kb synthetic contig from *Arabidopsis*.
II. Identify and mask repeats	1	10–30 min	Lab:	Analyze repeats in the 100-kb synthetic contig using RepeatMasker and *DNA Subway*'s local browser.
III. Predict genes in DNA sequence	1	10–30 min	Lab:	Predict genes in the 100-kb synthetic contig using Augustus Augustus, FGenesH, SNAP, and tRNAscan. Analyze the results using *DNA Subway*'s local browser.
IV. Name a gene and insert a start codon	2	15 min	Lab:	Use the annotation editor Apollo to create a gene model, name it, and insert a missing start codon.
V. Examine exon/intron borders (splice sites)	2	15–30 min	Lab:	Determine the nucleotide sequences that signify canonical splice sites.
VI. Search databases for biological evidence	2	15 min	Lab:	Search nucleotide (UniGene) and protein (UniProt) databases for matches to the 100-kb synthetic contig using BLASTN and BLASTX.
VII. Edit the ends of a nonspliced gene	3	20 min	Lab:	Adjust the 5′ and 3′ ends of a non-spliced gene using the Exon Detail Editor in Apollo.
VIII. Edit the ends of a spliced gene	3	20 min	Lab:	Adjust the 5′ and 3′ ends of a spliced gene using the Exon Detail Editor in Apollo.
IX. Merge exons and edit the ends of a spliced gene	4	15 min	Lab:	Merge two exons in a gene model using Apollo. Adjust the ends of a spliced gene using the Exon Detail Editor in Apollo.
X. Fix splice sites	4	15 min	Lab:	Adjust the exon borders using Exon Detail Editor in Apollo
XI. Browse your gene models	4	10 min	Lab:	View the gene models in *DNA Subway*'s local browser.
XII. Add expressed sequence tag (EST) evidence to your project	5	10 min	Lab:	Obtain expression data (ESTs), upload the data to *DNA Subway*, and BLAST the 100-kb synthetic contig against this new evidence.
XIII. Examine and fix two gene models	5	10 min	Lab:	Adjust two gene models to match EST evidence using the Exon Detail Editor in Apollo.
XIV. Develop alternatively spliced models for the third gene	5	15–30 min	Lab:	Construct an alternatively spliced gene model that helps make sense of some of the EST evidence using Apollo.
XV. Develop alternatively spliced models for the fourth gene	5	15–30 min	Lab:	Construct an alternatively spliced gene model that helps make sense of some of the EST evidence using Apollo.
XVI. Browse your gene models	5	10 min	Lab:	View the gene models in *DNA Subway*'s local browser.
Stage B: Annotate a 100-kb sequence from *Arabidopsis* chromosome 5				
	6	10–60 min	Prelab:	Set up all student computers to run Java and to allow pop-up windows to open. Prepare Apollo by opening the browser to http://gfx.dnalc.org/files/apollo/test.
	6	60–120 min	Lab:	Annotate genes in the 100-kb *Arabidopsis* chromosome 5 sample from *DNA Subway*.

STAGE A: ANNOTATE GENES IN A SYNTHETIC CONTIG

I. Create a *DNA Subway* Project

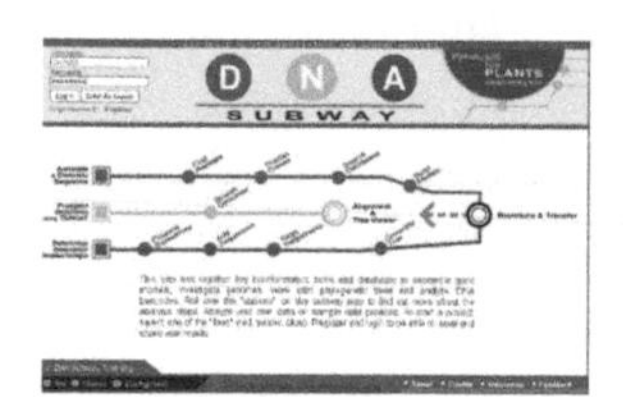

Like many other bioinformatics tools, DNA Subway works best in the Mozilla/Firefox browser.

1. Open *DNA Subway* (http://www.dnasubway.org). If you are a registered user, log in with your username and password (only registered users can save and share work). Alternatively, enter as a guest to gain temporary access.
2. Click on the red square to open a new annotation project.
3. Click on "Select a sample sequence."
4. Click on "*Arabidopsis thaliana* (mouse-ear cress) Synthetic Contig, 16.47" to highlight it.
5. Provide a project title (required), and click on "Continue."

II. Identify and Mask Repeats

RepeatMasker searches for repeated DNA sequences, including transposons, short tandem repeats (STRs), and low-complexity regions (e.g., long sequences composed of a single nucleotide). The program masks these sequences from further analysis.

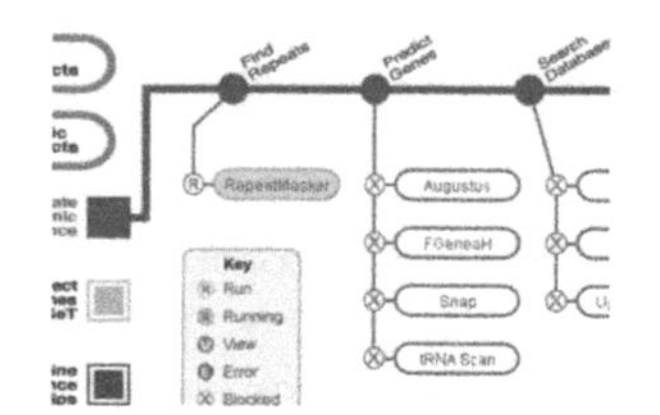

1. Click on "RepeatMasker." The blinking "R" bullet indicates that the program is running. The bullet changes to "V" when results are ready to view.
2. Once the program has finished, click on "RepeatMasker" again to open a new window with a listing of repetitive DNA sequences that have been identified and masked.
3. How many and what types of repetitive DNA were identified? What are their lengths? Is there any relationship between the repeat type and length?
4. Close the table to return to *DNA Subway*.
5. On the "Browsers & Transfer" branch line, click on "Local Browser" to view the results in a graphical interface. Be patient while the local copy of GBrowse loads.
6. Maximize the browser window.
7. Using the drop-down menu, toggle "Scroll/Zoom" to "Show 25 kbp," or adjust the yellow view bar to show "0-16.5k."
8. How many and what types of repetitive DNA does the browser display?
9. Which of the two views, table or browser, do you find easier to work with?
10. Close the local browser screen to return to *DNA Subway*.

GBrowse is a web-based browser that graphically displays a variety of genome data and serves as DNA Subway's local browser.

III. Predict Genes in DNA Sequence

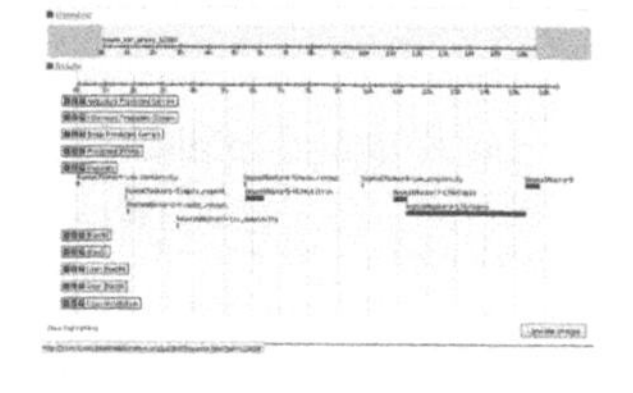

Gene prediction programs search DNA sequence for patterns that are associated with genes. Augustus, FGenesH, and SNAP use hidden Markov models (HMMs), in which gene predictions are the results of "hidden" states (rules that determine genes). They have been "trained" on known genes from specific species. To detect tRNAs, tRNAscan combines a similarity-based approach to detect molecules that have tRNA promoters with a pattern-based approach to identify those molecules that can fold into the three-dimensional (3D) structures of tRNAs.

1. On the "Predict Genes" branch line, click on "Augustus."
2. Once the program has finished, click on "Augustus" again to open a new window with a list of gene features—gene, mRNA, exon, and CDS (coding sequence). Components of a single gene are grouped together as belonging to the same "Parent" and "ID."
3. In turn, click on "FGenesH," then "Snap," and then "tRNA Scan." Examine the tables generated by each. Do any of the programs run significantly longer than any others?
4. Click on "Local Browser" to examine the gene and tRNA predictions, along with the masked repeats.
   i. How many genes did each program predict?
   ii. How are exons, introns, and 5′ and 3′ UTRs represented in the predicted genes?
   iii. Look at each predicted gene. Do all of the programs predict the same structure for each gene? Explain.
5. Close the local browser screen to return to *DNA Subway*.

Apollo is a specialized browser that allows users to view and decorate DNA sequences with graphical and text annotations describing information that they identify in the sequence, such as gene models/functions, homologous regions, etc. To download a manual for Apollo go to DNA Subway and click on "Manual."

Before launching Apollo, make sure that your computer uses a current version of Java and your browser allows pop-up windows. Be patient while Apollo launches via Java Web Start. The initial launch of each editing session typically takes a minute or two and may require you to permit the launch via dialog boxes.

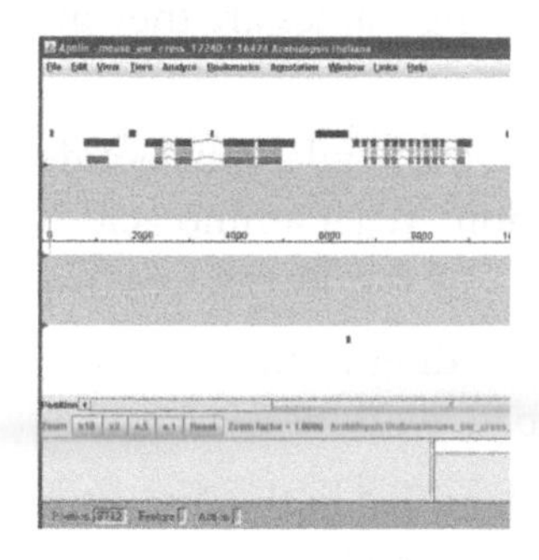

IV. Name a Gene and Insert a Start Codon

1. On the "Build Models" branch line, click on "Apollo." Apollo is a graphical editing system for gene annotation.
2. Familiarize yourself with the Apollo window, which is divided into several functional areas:
   i. A horizontal scale at the center shows the nucleotide coordinates of your sequence.
   ii. Panels above the scale relate to the forward strand, and panels below relate to the reverse strand.
   iii. Blue panels are work spaces for building gene models on the forward (top) and reverse (bottom) strands.
   iv. White panels display evidence on the forward (top) and reverse (bottom) strands.
   v. Windows at the bottom of the page display detailed information about a selected feature.
3. Look at the gene predictions on the forward strand.
   i. Does Apollo use the same conventions for displaying exons and introns as GBrowse?
   ii. Click on any exon to pull up information about the prediction. (This information will be displayed at the bottom of the window.) Click on the "Tiers" tab, and check "Show types panel" for a key to all of the data displayed in Apollo.
   iii. Compare the number and arrangement of exons in the four gene predictions made by Augustus, FGenesH, and SNAP.

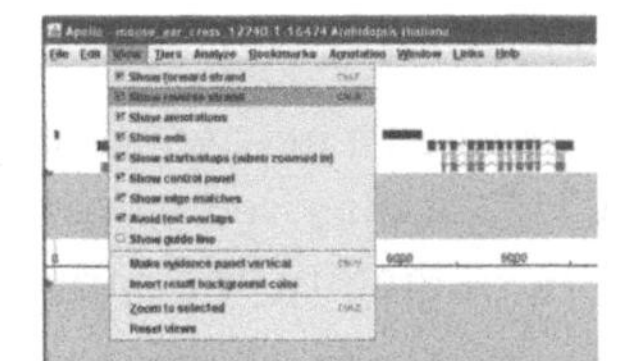

4. Adjust your work space.
   i. Maximize the Apollo window, or stretch it by grabbing and pulling a corner.
   ii. Focus on the forward strand, which contains all of the gene predictions. Click on the "View" tab in the menu bar, and uncheck "Show reverse strand."
   iii. Grab and pull down the divider beneath the "Zoom" buttons to increase the evidence (white) and work space (blue-shaded) panels.
   iv. Grab the red arrow on the left-hand side of the screen, and pull it up/down to adjust the size of the evidence and work space panels.
5. Use the "Zoom" buttons and scroll bar to enlarge and center on the first gene, so that the screen views nucleotide positions 700–1500. The green and red bars at the top of the window are start and stop codons in each of three reading frames on the forward strand.

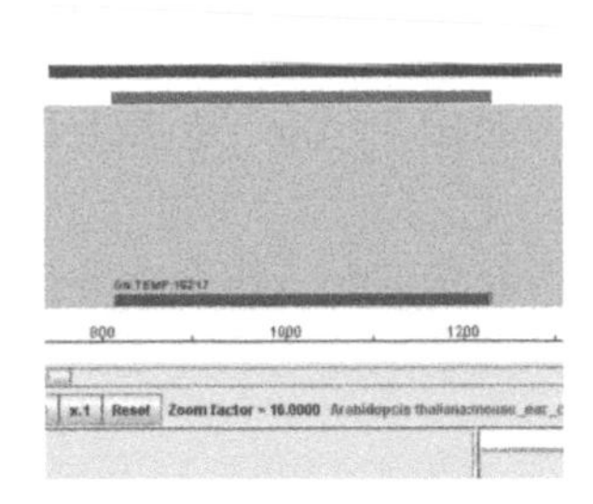

6. Double-click on the FGenesH prediction, and drag it into the blue work space. This creates a "TEMP" gene model with start and stop codons. Note the green and red lines at the ends of the model.
7. Give the model a temporary name to be able to identify it during your work.
   i. Highlight the gene model. Then right-click (PC) or command-click (Mac) and select "Annotation info editor."
   ii. Enter your chosen name in the gene "Symbol" field (top left) and transcript "Symbol" field (top right). You might simply replace the letters *GN* in "GN:TEMP" with the source of the model, so that it reads FGenesH:TEMP. Make sure that you use the same name in both fields. The use of dashes (–) is not permitted.

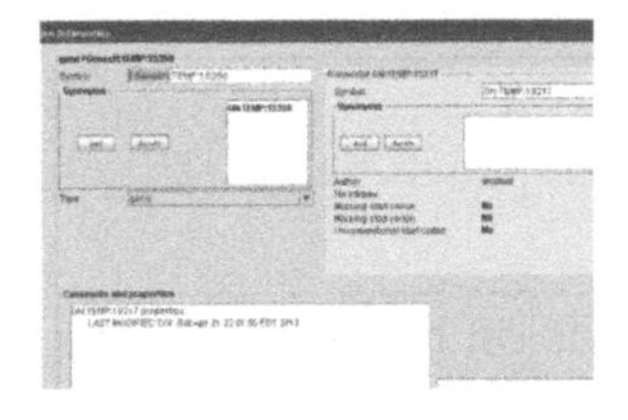

   iii. Delete all other names in the two "Synonyms" fields.
   iv. Click on "Close" and return to the main Apollo screen.
8. Examine the start and stop codons.
   i. Click on the FGenesH temporary gene model to see how the start and stop codons correspond to the panel at the top. Which reading frames are they in? Are they in the same or different reading frames?

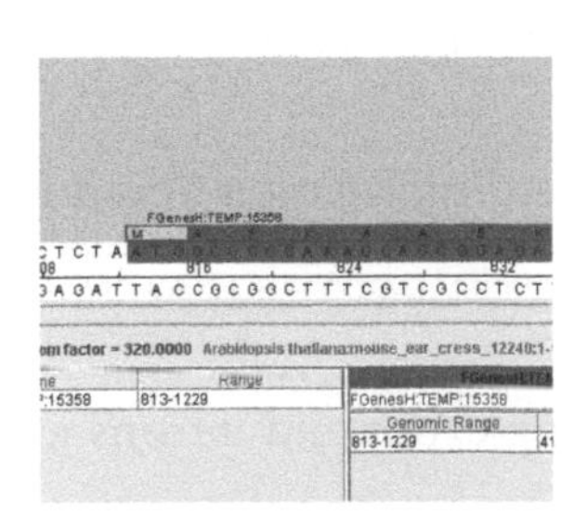

   ii. Zoom in on the start and stop codons to identify the triplet nucleotide codons. What amino acid is encoded by the start codon? What amino acid is encoded by the stop codon?
9. Zoom back out. Double-click on the Augustus prediction, drag it into the work space, and give it a temporary name as described in Step 7.
10. The green arrow means that a start codon has not been identified in the Augustus temporary gene model. One solution for fixing this is to use the first start codon that occurs after the green arrow.

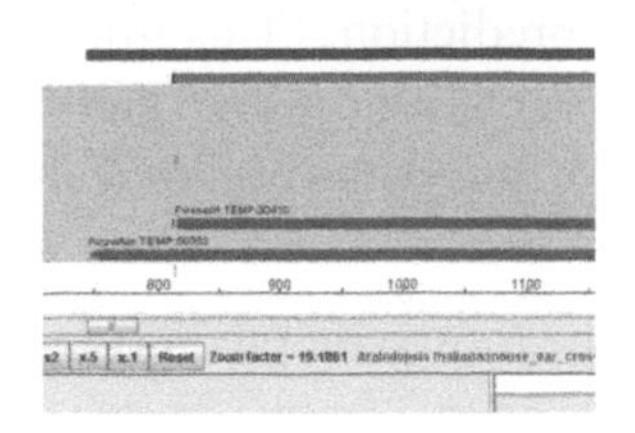

   i. Find this start codon at the top of the window, and use your cursor to drag it into the gene model. What happens?
   ii. What do you notice about the position of the start codon that you have inserted?

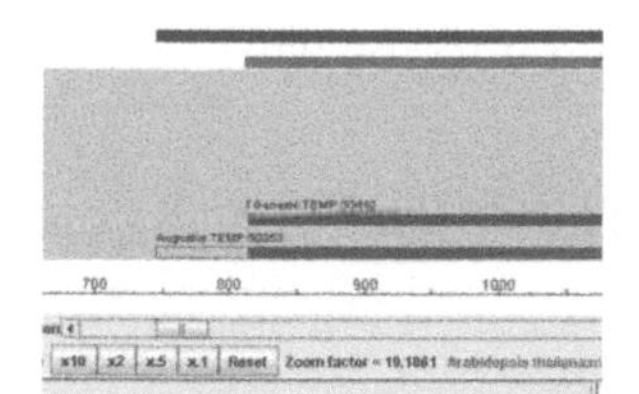

11. Your edit to the start codon of the Augustus gene prediction creates a new gene model.
    i. What do the unfilled segments at the beginning and end of your Augustus temporary gene model represent? (*Hint:* Zoom into the model far enough to find out whether Apollo associates these areas with amino acid sequences.)
    ii. What does this tell you about differences in the way Augustus and FGenesH predict genes?
    iii. How does this relate to the speed of each prediction program? (Refer to your answer in Step 3 of Part III.)

V. Examine Exon/Intron Borders (Splice Sites)

1. Use the "Zoom" buttons and the scroll bar to enlarge and center on the second gene prediction, so that the screen displays nucleotide positions 2000–5200.
2. Double-click on the predictions made by Augustus and FGenesH, drag them into the work space, and rename them as described in Step 7 of Part IV.
    i. How are the two predictions similar or different?
    ii. Could you arrive at a correct gene model based on the gene predictions alone?
3. Double-click on the Augustus gene prediction to highlight exon–intron splice sites. Black lines highlight splice site predictions having the same nucleotide sequence.
4. Determine the conserved (canonical) nucleotide sequence at splice sites.
    i. Zoom in on the gene prediction until you can read the nucleotide sequence on the coordinates bar.
    ii. Move sequentially to each exon/intron and intron/exon boundary, and record the three nucleotides on each side of the splice site as shown in the chart below. (Be sure to use the sequence from the top [forward] strand!)
    iii. Drag the red arrow to position the gene predictions close to the nucleotide coordinates bar. Conduct the same analysis for the Augustus gene prediction at positions 6500–9000, and add the data to the chart.
    iv. What nucleotides are found at every splice site? Make a general rule for the nucleotides that define the beginning and end of each intron.

Splice sites determine the boundaries between introns, which are removed from the primary RNA transcript to produce a mature mRNA, and exons. In Apollo and other genome browsers, splice sites are at the positions where boxes (exons) are met by lines (introns). Box borders that do not connect to a line are generally not splice sites.

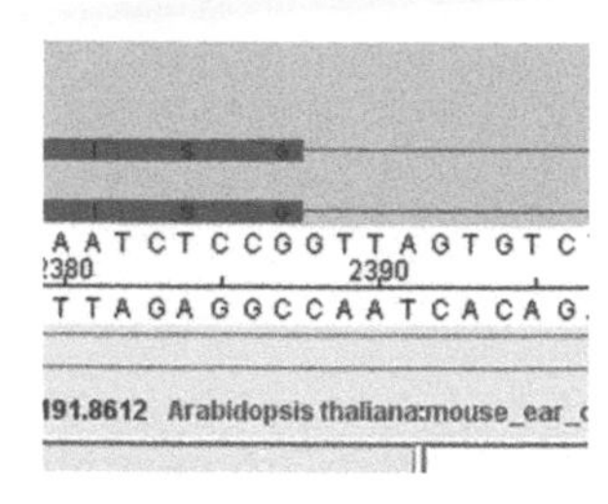

Location	Exon	Intron	Intron	Exon	Location
2387/2388	CCG	GTT	CAG	GTG	2694/2695
3017/3018					3723/3724
4353/4354					4445/4446

5. Zoom out to view the region from positions 2000 to 5200 again.
6. Double-click on the predictions made by SNAP, drag them into the work space, and rename them as described in Step 7 of Part IV.
7. Compare the SNAP models to the Augustus and FGenesH models.
   i. How are the predictions similar or different?
   ii. What additional information is needed to determine the correct structure of the gene?
8. Click on "Reset" to zoom out to view the entire contig.
9. Look at the other two locations that contain predicted genes (positions 6400–9200 and positions 9600–16,400).
   i. How are the predictions similar or different?
   ii. What additional information is needed to determine the correct structures of the genes?
10. Click on the "File" tab in the menu bar, and select "Upload to *DNA Subway*" to save your work. Exit out of Apollo, and return to *DNA Subway*.

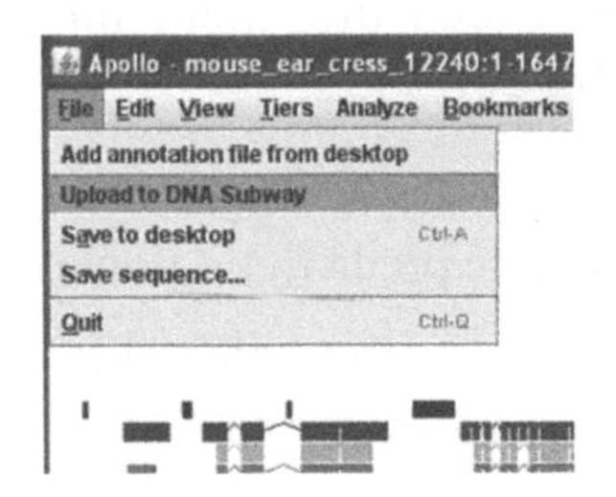

VI. Search Databases for Biological Evidence

1. On the "Search Databases" branch line, click on "BLASTN" to search the UniGene database for genes that match the RepeatMasked *Arabidopsis* sequence.
2. Click on "BLASTX" to search the UniProt database for proteins that match an amino acid translation of the *Arabidopsis* sequence.
3. On the "Browsers & Transfer" branch line, click on "Local Browser" to view the BLASTN and BLASTX matches in the context of the gene predictions. Be patient while the local copy of GBrowse loads.
4. For how many predicted genes did the BLAST searches generate biological evidence?
5. Close the local browser screen to return to *DNA Subway*.

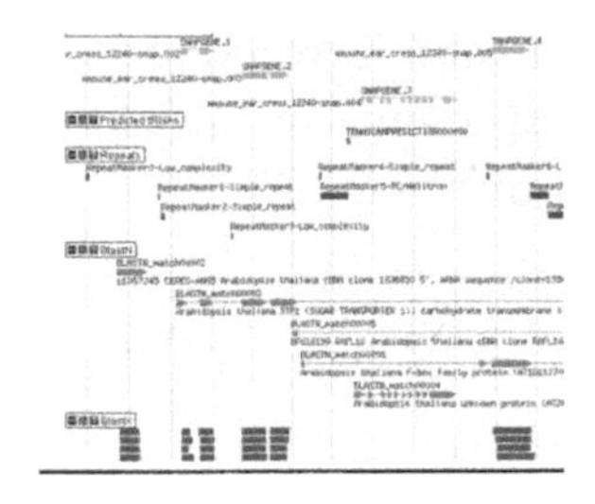

VII. Edit the Ends of a Nonspliced Gene

1. On the "Build Models" branch line, click on "Apollo."
2. Click on the "Tiers" tab in the menu bar, and select "Expand all tiers" to view all of the evidence generated by BLASTN and BLASTX. (Apollo initially collapses all BLASTN and BLASTX evidence onto a single line each. These merged data are often misleading.)
3. Adjust your work space.
   i. Focus on the forward strand, which contains all of the gene predictions. Click on the "View" tab in the menu bar, and uncheck "Show reverse strand."
   ii. Grab and pull down the divider beneath the "Zoom" buttons to increase the evidence (white) panel.
   iii. Scroll up and down with the vertical slider to the right of the evidence (white) panel to view all BLASTN and BLASTX evidence.

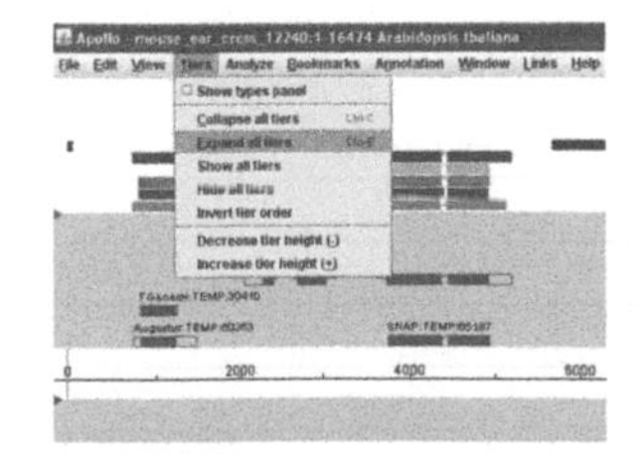

4. Use the “Zoom” buttons and scroll bar to enlarge and center on the first gene, so that the screen shows nucleotide positions 700–1500.
5. Compare the temporary gene models you generated in Part IV to the biological evidence.

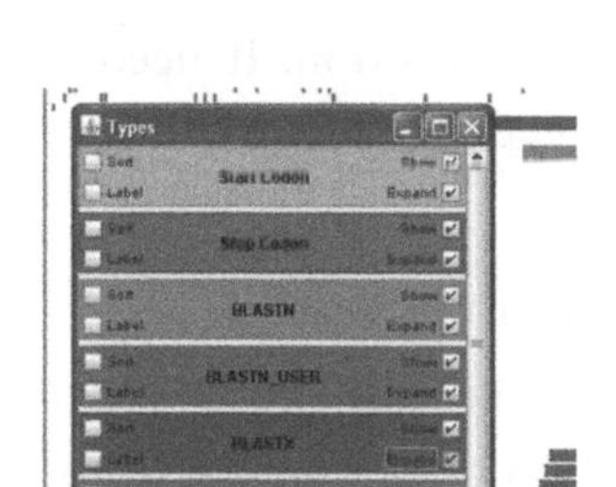

   i. Click on any piece of evidence to pull up information about the prediction. (This information will be displayed at the bottom of the window.) Click on the “Tiers” tab and check “Show types panel” for a key to all of the data displayed in Apollo.
   ii. In this case, the many instances of redundant BLASTX results complicate the analysis. In the “Types” panel, uncheck “Expand” BLASTX to collapse the BLASTX evidence and place key BLASTN evidence close to gene prediction evidence.
   iii. Generally, the best model is the one closest to the longest BLAST evidence. Which of the predictions is best supported by the BLAST results?
6. Delete the shorter FGenesH model. Highlight the model, right-click (PC) or command-click (Mac), and select “Delete selection” from the pop-up menu.
7. Double-click on the BLASTN evidence, drag it into the work space, and give it a temporary name as described in Step 7 of Part IV.
8. Center the screen on the beginning of the two models.
9. Zoom in on the Augustus prediction and the BLASTN evidence until you can read the nucleotide sequence on the Coordinates bar.
10. Compare the sequence at the beginning of the Augustus prediction with the BLASTN evidence by clicking on their temporary gene models. Their sequences appear in blue in the sequence coordinates. What do you notice?
11. Repeat the analysis described in Steps 8–10 for the end of the gene models. What do you notice?
12. Use the Exon Detail Editor to adjust the ends of the Augustus gene model to agree with the BLASTN evidence.

The Exon Detail Editor aligns the gene prediction and the BLASTN evidence sequences on top of each other.

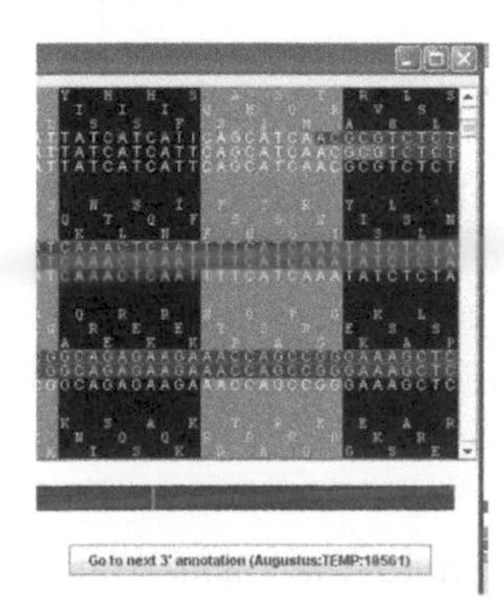

   i. Zoom out until you can see the gene in its entire length.
   ii. Double-click to highlight the Augustus temporary gene model, right-click (PC) or command-click (Mac), and select “Exon detail editor” from the pop-up menu.
   iii. Click on and drag the overhanging nucleotide at the beginning of the Augustus gene model (highlighted with a red frame) until its sequence is flush with the start of the BLASTN evidence.
   iv. Click on the diagram at the bottom to move the sliding window to the end of the gene model. Alternatively, use the slider bar on the right to scroll down.
   v. Click on and drag the overhanging nucleotides of the Augustus gene model until they are flush with the end of the BLASTN evidence.
   vi. Close the Exon Detail Editor.
13. Delete the BLASTN temporary model. Highlight the model, right-click (PC) or command-click (Mac), and select “Delete selection” from the pop-up menu.

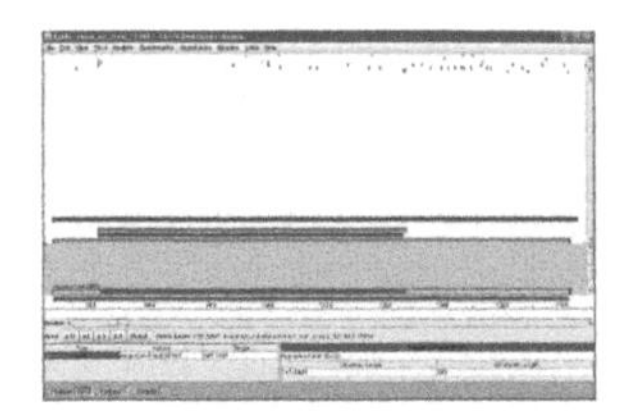

14. Double-click to highlight the Augustus temporary gene model, right-click (PC) or command-click (Mac), and select "Calculate longest ORF." This function uses the start and stop codons and intron/exon boundaries of the model to determine the best open reading frame (ORF).
15. Ensure that the gene model shows a start codon and a stop codon. If needed, insert them at the appropriate positions as described in Step 10 of Part IV.
16. Examine your model.
    i. Does the BLASTN evidence support the extent of your model?
    ii. In the "Types" panel, check "Expand" BLASTX to see all of the BLASTX evidence. Does the BLASTX evidence support the position of the start and stop codons?
    iii. Click several of the BLASTX hits. Can you determine the function of the protein produced by this gene?

Another way to expand or collapse tiers is to use the "Expand all tiers" or "Collapse all tiers" options under the "Tiers" tab in the menu bar.

17. Identify additional functional information through InterProScan.
    i. Highlight the model, right-click (PC) or command-click (Mac), and select "Submit to InterProScan."
    ii. Click on "OK" in a pop-up window.
    iii. Click on "OK" to the prompt that the analysis is complete.
    iv. This will load the results into the Annotation Info Editor and open the InterProScan results in a browser window (make sure your browser is set to allow pop-ups). You can edit the results in the Annotation Info Editor to retain the most useful information.
    v. What information can you extract from the InterProScan results?

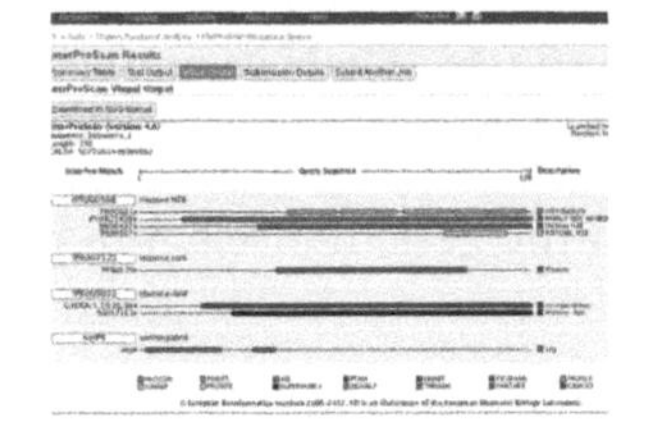

18. Name your model, and add information about your work.
    i. Highlight the gene model. Then right-click (PC) or command-click (Mac) and select "Annotation info editor."
    ii. Enter your chosen name in the gene "Symbol" field (top left) and transcript "Symbol" field (top right). Your gene name may reflect a strong relationship turned up in the BLASTX results (e.g., "Putative Histone H2B"). Make sure that you use the same name in both fields. The use of dashes (–) is not permitted.

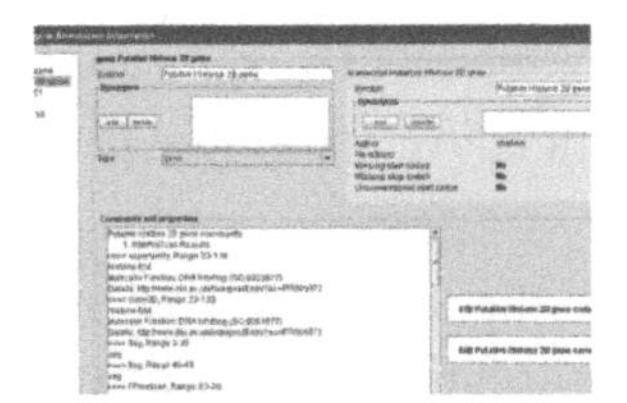

    iii. Delete all other names in the two "Synonyms" fields.
    iv. Click on the respective "Edit comments" button for the name you chose (at the bottom right). Then click on the "Add," and write your comments in the pop-up window. These should include information about protein function obtained from BLAST searches, notes that justify your model, and unresolved annotation problems. Click on "Close" to return to the Annotation Information window.
    v. Close the Annotation Information Editor, and return to the main Apollo screen.
19. In the "File" tab in the menu bar, select "Upload to *DNA Subway*."
20. Zoom out by clicking on "Reset."

VIII. Edit the Ends of a Spliced Gene

1. Zoom and scroll to enlarge the gene models at nucleotide positions 2000–5200.
2. Click on the "Tiers" tab and check "Show types panel" (if it is not already checked) for a key to all of the data displayed in Apollo. In the "Types" panel, uncheck "Expand" BLASTX to collapse BLASTX evidence and place key BLASTN evidence close to gene prediction evidence.
3. Double-click on the BLASTN evidence. Using the BLASTN evidence, determine whether Augustus, FGenesH, or SNAP have predicted the gene(s) in this region most accurately.

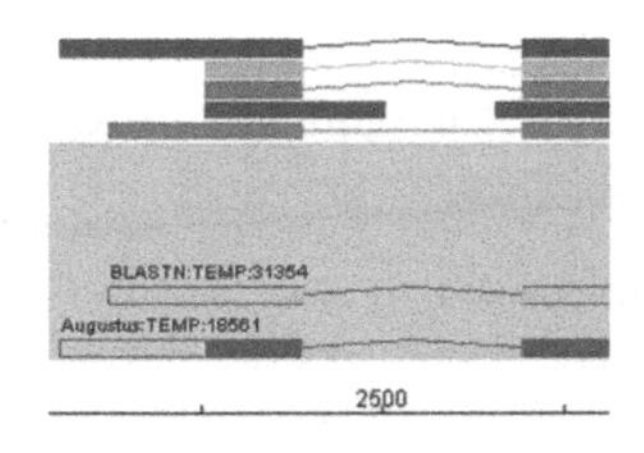

4. Delete the FGenesH and SNAP temporary gene models from the work space as described in Step 6 of Part VII.
5. Double-click on the BLASTN evidence, drag it into the work space, and give it a temporary name as described in Step 7 of Part IV. What do you notice?
6. Use the Exon Detail Editor to adjust the ends of the Augustus gene model to agree with the BLASTN evidence as described in Step 12 of Part VII.
7. Delete the BLASTN temporary model as described in Step 13 of Part VII.
8. Double-click to highlight the Augustus temporary gene model, right-click (PC) or command-click (Mac), and select "Calculate longest ORF." This function uses the start and stop codons and intron/exon boundaries of the model to determine the best ORF.
9. Ensure that the gene model shows a start codon and a stop codon. If needed, insert them in appropriate positions as described in Step 10 of Part IV.
10. Determine domains, functional sites, and putative functions for your gene/protein using InterProScan as described in Steps 16 and 17 of Part VII.
11. Name and add comments to your model, and save it to *DNA Subway*, as described in Steps 18 and 19 of Part VII.
12. Zoom out by clicking on "Reset."

IX. Merge Exons and Edit the Ends of a Spliced Gene

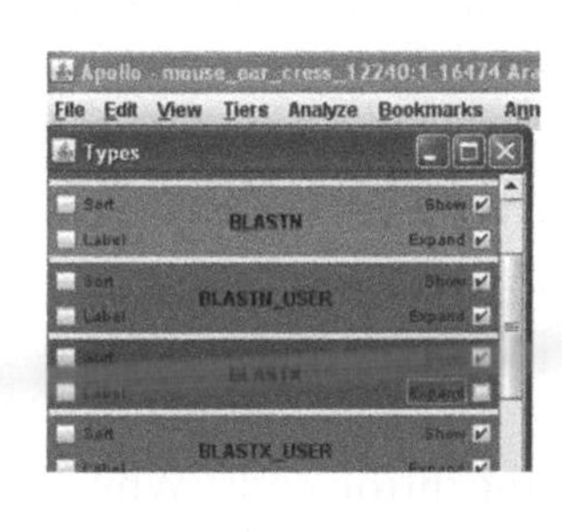

1. Zoom and scroll to enlarge the third gene at nucleotide positions 6500–9000.
2. Click on the "Tiers" tab and check "Show types panel" (if it is not already checked) for a key to all of the data displayed in Apollo. In the "Types" panel, uncheck "Expand" BLASTX to collapse the BLASTX evidence and place key BLASTN evidence close to gene prediction evidence.
3. Which gene prediction most closely matches the BLASTN evidence? Double-click on this prediction, drag it into the work space, and give it a temporary name as described in Step 7 of Part IV. Do the same for the BLASTN evidence.
4. Merge the first two exons of the Augustus prediction to make them closer in length to the first exon of the BLASTN evidence.
   i. Shift-click to highlight the first and second exons of the Augustus prediction.
   ii. Right-click (PC) or command-click (Mac), and select "Merge exons."

5. Repeat Step 4 to merge the last two exons of the Augustus prediction.
6. Use the Exon Detail Editor to adjust the ends of the Augustus gene model to agree with the BLASTN evidence as described in Step 12 of Part VII.
7. Delete the BLASTN temporary model as described in Step 13 of Part VII.
8. Double-click to highlight the Augustus temporary gene model, right-click (PC) or command-click (Mac), and select "Calculate longest ORF." This function uses the start and stop codons and intron/exon boundaries of the model to determine the best ORF.
9. Ensure that the gene model shows a start codon and a stop codon. If needed, insert them in appropriate positions as described in Step 10 of Part IV.
10. Determine domains, functional sites, and putative functions for your gene/protein as described in Steps 16 and 17 of Part VII.
11. Name and add comments to your model, and save it to *DNA Subway*, as described in Steps 18 and 19 of Part VII.
12. Zoom out by clicking on "Reset."

BLASTN has no logic for detecting splice sites, so don't be concerned by all the yellow arrows indicating missing splice sites. Use the Augustus model to identify splice sites, and use BLASTN evidence only to adjust the ends of the gene.

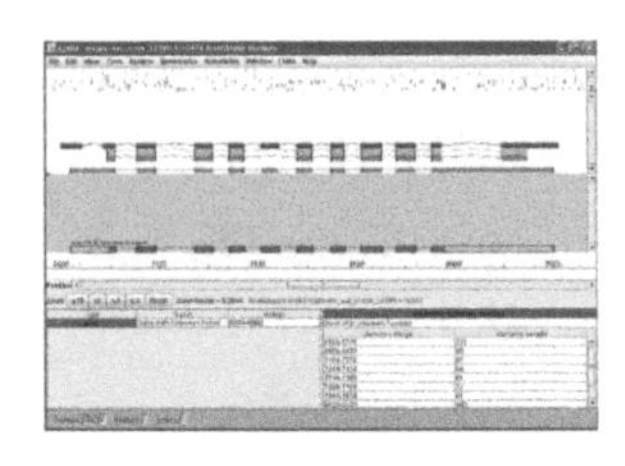

X. Fix Splice Sites

1. Zoom and scroll to enlarge the fourth gene at nucleotide positions 9500–16,400.
2. Click on the "Tiers" tab and check "Show types panel" (if it is not already checked) for a key to all of the data displayed in Apollo. In the "Types" panel, uncheck "Expand" BLASTX to collapse the BLASTX evidence and place key BLASTN evidence close to gene prediction evidence.
3. Compare the gene predictions, the BLASTN evidence, and the RepeatMasker evidence. What interesting situation do you find for this gene?
4. In this case, the BLASTN evidence is more complete than any of the predictions. Double-click on it, drag it into the work space, and give it a temporary name as described in Step 7 of Part IV.
5. Correct the problems with the splice sites indicated by the two yellow arrows.
   i. Double-click to highlight the BLASTN gene model, right-click (PC) or command-click (Mac), and select "Exon detail editor" from the pop-up menu.
   ii. Click on the diagram at the bottom to move the sliding window to the end of the first intron. Alternatively, use the slider bar on the right.
   iii. Click on and drag the intron–exon border of the BLASTN gene model 2 nucleotides to the right until the intron ends with the canonical AG splice site. (Exons are shaded in blue, and introns are designated by a blue line.) How can we tell that this is the correct "AG" at which the intron ends? Why is it not one of the two following "AGs"?
   iv. Move to the end of the second intron, and repeat Step 5.iii to generate a canonical AG splice site.
   v. Close the Exon Detail Editor.

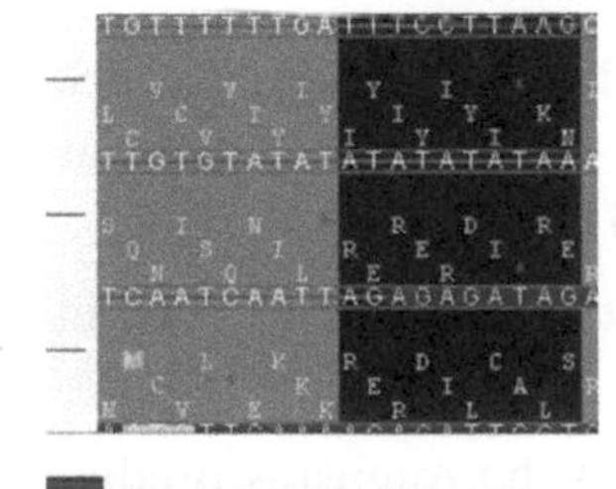

6. Double-click to highlight the BLASTN gene model, right-click (PC) or command-click (Mac), and select "Calculate longest ORF." This function uses the

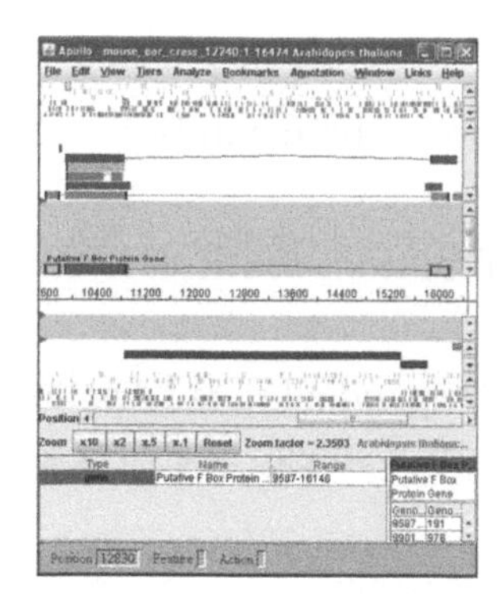

start and stop codons and intron/exon boundaries of the model to determine the best ORF.

7. Ensure that the gene model shows a start codon and a stop codon. If needed, insert them in appropriate positions as described in Step 10 of Part IV.
8. Determine domains, functional sites, and putative functions for your gene/protein as described in Steps 16 and 17 of Part VII. (Remember, names cannot contain dashes.)
9. Name and add comments to your model, and save it to *DNA Subway*, as described in Steps 18 and 19 of Part VII.
10. Exit out of Apollo, and return to *DNA Subway*.

XI. Browse Your Gene Models

1. On the "Browsers & Transfer" branch line, click on "Local Browser" to view your models in a graphical interface. Be patient while the local copy of GBrowse loads.
2. Your models will appear in a track called "User Annotation" at the bottom of the window. If some of your models do not appear, toggle "Scroll/Zoom" to "Show 25 kbp" or adjust the yellow view bar to show "0-16.5k."
3. How do your gene models match the genes predicted by Augustus, FGenesH, and SNAP? Do any of the three prediction algorithms predict all genes accurately?

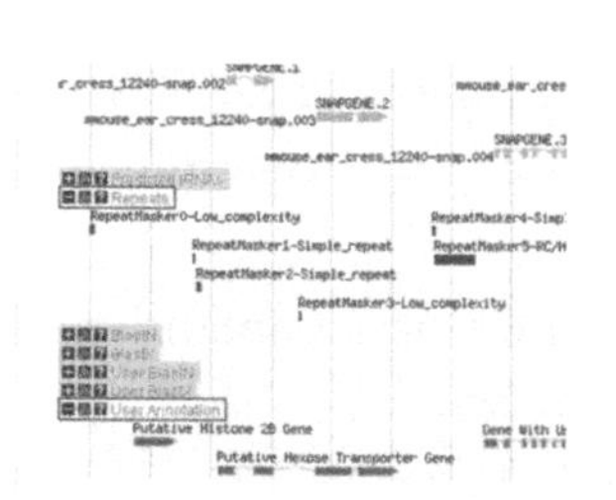

XII. Add Expressed Sequence Tag (EST) Evidence to Your Project

In DNA Subway, click on "Background" at the page bottom, and follow the link for "Evidence" to learn how cDNAs and ESTs are generated and used as evidence for gene models.

Your models currently incorporate limited biological evidence provided by BLAST searches of the UniGene and UniProt databases, which contain only well-validated gene models and protein sequences. The most abundant biological evidence is derived from mRNA transcripts—from complementary DNA (cDNA) sequences and expressed sequence tags (ESTs), and, increasingly, from direct mRNA sequencing (RNA-seq). Here, you will expand your project through a BLAST search of uploaded *Arabidopsis* ESTs.

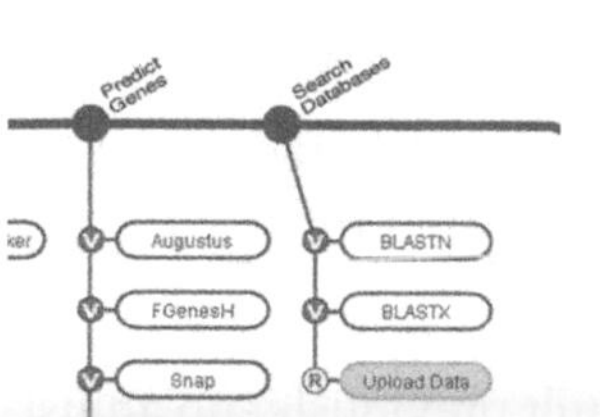

1. Open a new browser window, and go to http://gfx.dnalc.org/files/genomescience/.
2. Right-click (PC) or option-click (Mac) on "at_est_evidence.zip," and save it to your desktop. You will need the file named "at_est_evidence.fasta," which contains the DNA sequences (EST, cDNA) of spliced mRNA transcribed from the regions in the *A. thaliana* genome that were spliced together to form the DNA sequence in the "*Arabidopsis thaliana* (mouse-ear cress) Synthetic Contig, 16.47" contig used for this lab.
3. On the "Search Databases" branch line, click on "Upload Data."

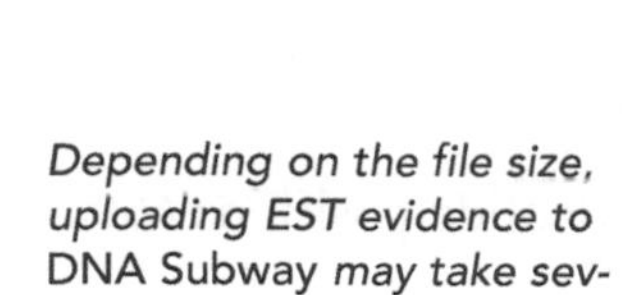

Depending on the file size, uploading EST evidence to DNA Subway may take several minutes.

4. In the pop-up window, click on the top "Browse" button for DNA data, navigate to the EST evidence file ("at_est_evidence.fasta"), and select it. Click "Upload."
5. This generates a new stop on the "Search Databases" branch line, "User BLASTN." Click on this button to BLAST the *Arabidopsis* synthetic contig against the EST sequences in the evidence file.
6. Open "Apollo."

7. Adjust your work space.
   i. Focus on the forward strand. Click on the "View" tab in the menu bar, and uncheck "Show reverse strand."
   ii. Grab and pull down the divider beneath the "Zoom" buttons to increase the evidence (white) panel.
   iii. Click on the "Tiers" tab in the menu bar, and select "Expand all tiers." Apollo initially collapses all BLASTN_USER evidence onto a single line. These merged data are often misleading.
   iv. Click on the "Tiers" tab again, and check "Show types panel" for a key to all of the data displayed in Apollo. In the "Types" panel, uncheck "Expand" BLASTX.

XIII. Examine and Fix Two Gene Models

Here you review the EST evidence for the first two gene models you built previously.

1. Zoom and scroll to enlarge the first gene model at nucleotide positions 700–1500.
2. One EST indicates that the 5′ UTR is 10 nucleotides longer than the BLASTN evidence used to construct the gene model. Double-click that EST evidence and drag it into the work space.
3. Double-click to highlight the gene model, right-click (PC) or command-click (Mac), and select "Exon detail editor" from the pop-up menu.
4. Grab the 5′ end of the gene model and extend it to match the EST evidence.
5. Delete the EST evidence from the work space. Highlight the model, right-click (PC) or command-click (Mac), and select "Delete selection" from the pop-up menu.
6. Double-click to highlight the gene model, right-click (PC) or command-click (Mac), and select "Calculate longest ORF." This function uses the start and stop codons and intron/exon boundaries of the model to determine the best ORF.
7. Ensure that the gene model shows a start codon and a stop codon. If needed, insert them in appropriate positions as described in Step 10 of Part IV.
8. Add comments about the adjustment you made to your model, and save your model to DNA Subway, as described in Steps 18 and 19 of Part VII.
9. Zoom and scroll to enlarge the second model at nucleotide positions 2000–5200.
10. All of the ESTs should support the second gene model built previously; no adjustments are needed.

XIV. Develop Alternatively Spliced Models for the Third Gene

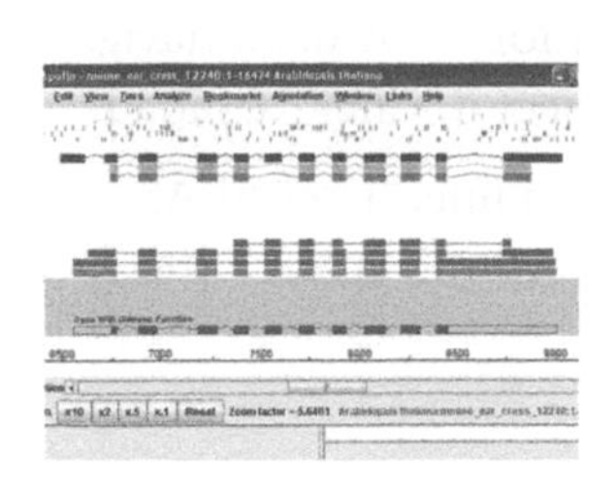

1. Zoom and scroll to enlarge the third gene model at nucleotide positions 6500–9000.
2. Two ESTs "split" the 3′ (last) exon of the gene model into two exons. This is evidence of alternative splicing, which creates different exon arrangements. (Click on them to see their designations: EG433512 and BX820562.) One way to devel-

op an alternatively spliced model is by splitting the final exon as described in Steps 3–5. Another way to make an alternatively spliced model is to merge the last two EST exons with the previous gene model, as described in Steps 6 and 7.

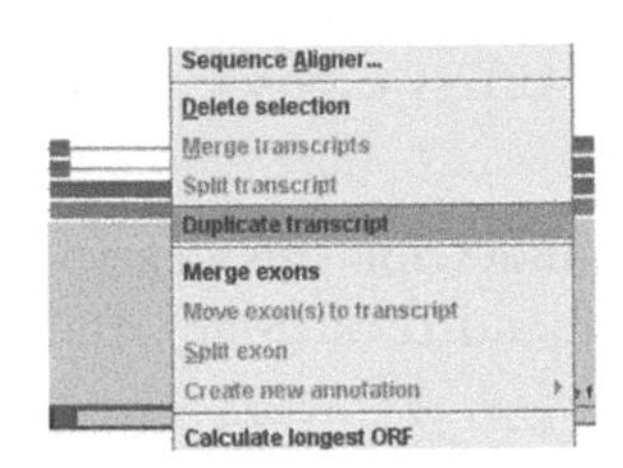

3. To develop an alternatively spliced model by splitting the final exon, the first step is to align the previous gene model with the "split" exon from the EST evidence in the Exon Detail Editor.
   i. Create a duplicate of your gene model. Highlight your gene model, right-click (PC) or command-click (Mac), and select "Duplicate transcript."
   ii. Name the duplicate gene model Alt Splice 3.1 as described in Step 7 of Part IV.
   iii. Double-click on the longer EST (BX820562), drag it into the work space, and name it "EST Evidence."
   iv. Highlight the last exon in the "Alt Splice 3.1" model, right-click (PC) or command-click (Mac), and select "Split Exon." The yellow diamonds indicate missing splice sites.
   v. Double-click to highlight the "Alt Splice 3.1" model, right-click (PC) or command-click (Mac), and select "Exon detail editor" from the pop-up menu.

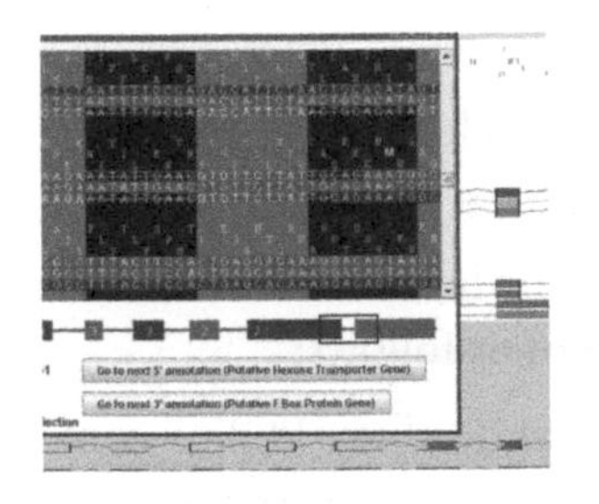

   vi. Click on the diagram at the bottom to move the sliding window to the last exon, over the dark blue/light blue border. The exon split forms two adjacent exons in different reading frames (light and dark blue), separated by a single nucleotide. This single nucleotide is the intron, and it will be lengthened in Steps 4 and 5.
   vii. Enlarge the window to best see the relationship between the exon split and the EST intron. Use the slider bar on the right to make fine adjustments to the view. It will take some manipulation to find a view that includes the split exon and the borders of the EST intron.
4. Create the 5′ (left) intron border.
   i. Click on the "Alt Splice 3.1" model in the Exon Detail Editor, then grab the dark blue nucleotides and drag them to the left until you reach the upper left of the window.
   ii. Move the sliding window to the left to reposition the editor. Continue dragging the dark blue nucleotides until the edge of the exon is flush with "EST Evidence." The intron should now begin with a canonical GT splice site.
   iii. Note that this adjustment removes the yellow arrow from the "Alt Splice 3.1" model in the Apollo window.
5. Create the 3′ (right) intron border.
   i. Move the sliding window to the left to reposition the editor over the edge of the last exon (in light blue).
   ii. Drag the light blue nucleotides to the right until the edge of the exon is flush with "EST Evidence."
   iii. Note that a yellow triangle is still present in the "Alt Splice 3.1" model in the Apollo window, indicating an incorrect splice site.

iv. Continue dragging three more nucleotides, until the intron ends with a canonical AG splice site. This creates a correct splice site in the "Alt Splice 3.1" model in the Apollo window.

v. Close the Exon Detail Editor.

6. Make an alternatively spliced model by merging the last two EST exons with the previous gene model.

i. Create another duplicate of your previous gene model for this region. Highlight your gene model, right-click (PC) or command-click (Mac), and select "Duplicate transcript."

ii. Name the duplicate gene model "Alt Splice 3.2" as described in Step 7 of Part IV.

iii. Delete the last exon of the "Alt Splice 3.2" model. Highlight the exon, right-click (PC) or command-click (Mac), and select "Delete selection" from the pop-up menu.

iv. Shift-click to highlight the last two exons of the longer EST (BX820562), and drag them into the work space.

v. Double-click to highlight the "Alt Splice 3.2" model, and then shift-click to highlight the two EST exons as well.

vi. Right-click (PC) or command-click (Mac), and select "Merge transcripts" from the pop-up menu. Click on "Merge," and then click on "Alt Splice 3.2" to name the merged model.

7. The merged model has yellow arrows at the beginning of the last two exons (exons 10 and 11), indicating improper splice sites. Correct them as follows:

i. Double-click to highlight the "Alt Splice 3.2" model, right-click (PC) or command-click (Mac), and select "Exon detail editor" from the pop-up menu.

ii. Move the sliding window to position the editor at the beginning of the second-to-last exon (exon 10). Drag the light-blue nucleotides to the right so that the intron ends with a canonical AG splice site. This eliminates the yellow arrow in the Apollo window.

iii. Move the sliding window to the beginning of the last exon (exon 11). Drag the dark blue nucleotides to the right so that the intron ends with a canonical AG splice site. This eliminates the second yellow arrow in the Apollo window.

iv. Finally, move the sliding window to the end of the last exon, and extend the 3′ UTR for the last exon to match the length of the original gene model (which was derived from the longest BLASTN evidence).

8. The shortest EST (EG433512) indicates that there is a third alternatively spliced product with a shorter exon 7.

i. Drag EST EG433512 into the work space, and name it "EST Evidence 2."

ii. Open the Exon Detail Editor and position the Edit window at the end of exon 7 of "Alt Splice 3.2."

iii. Drag the nucleotides to the left, until the border aligns with "EST Evidence 2." The intron should now begin with a canonical GT splice site.

9. Double-click to highlight each gene model, right-click (PC) or command-click

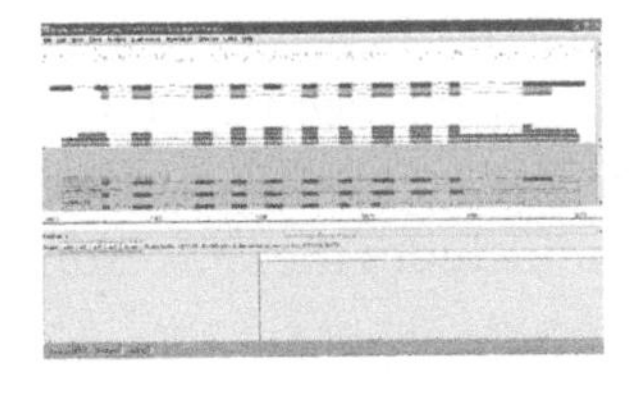

(Mac), and select "Calculate longest ORF." This function uses the start and stop codons and intron/exon boundaries of the model to determine the best ORF.

10. Ensure that each gene model shows a start codon and a stop codon. If needed, insert them in appropriate positions as described in Step 10 of Part IV.
11. Compare gene model "Alt Splice 3.2" with the other gene models. What interesting situation does the shorter exon create?
12. The first exon for EST evidence sequence BX820562 ends before the first exon of the BLASTN evidence and EST evidence BX820662. Why does this not suggest additional alternative splice forms? (Tip: Zoom in to examine the end of the three exons in high magnification.)
13. Add comments about the adjustments you made to your model, and save your model to *DNA Subway*, as described in Steps 18 and 19 of Part VII.

XV. Develop Alternatively Spliced Models for the Fourth Gene

1. Zoom and scroll to enlarge the fourth gene model at nucleotide positions 9500–16,400.
2. Examine the ESTs for the fourth gene model for evidence of alternative splicing. Be careful when interpreting EST data; ESTs usually represent only parts of a gene. Introns can only be deduced from ESTs whose exons are connected by an intron line. When space permits, Apollo will set different ESTs on the same line. Although these may look like exons from the same transcript, they are not connected by an intron line. You can confirm this by clicking on each and seeing that they have different accession numbers. ESTs AK229702.1 and AV535466 indicate two alternatively spliced products.
3. Create an alternatively spliced product based on EST AK229702.1.
   1. Drag EST AK229702.1 into the work space, and name it "Alt Splice 4.1."
   2. Open the Exon Detail Editor, and move the end of the intron 2 nucleotides to the right.
   3. Double-click to highlight the gene model, right-click (PC) or command-click (Mac), and select "Calculate longest ORF." This function uses the start and stop codons and intron/exon boundaries of the model to determine the best ORF.
   4. Ensure that the gene model shows a start codon and a stop codon. If needed, insert them in appropriate positions as described in Step 10 of Part IV.
   5. What interesting situation do you observe in this gene model? Is the protein-coding region interrupted by an intron?
4. Create an alternatively spliced product based on AV535466.
   1. Create a duplicate of your original gene model, and name it "Alt Splice 4.2."
   2. Drag EST AV535466 into the work space.
   3. Highlight the second exon in the "Alt Splice 4.1" model, right-click (PC) or command-click (Mac), and select "Split Exon."
   4. Open the Exon Detail Editor, and center the window on the border between the dark and light blue exons. The exon split forms two adjacent exons in different

reading frames (light and dark blue), separated by a single nucleotide. This single nucleotide is the intron, and it will be lengthened in Steps 4.vi and 4.vii.

v. Enlarge the window to best see the relationship between the exon split and the EST intron. Use the slider bar on the right to make fine adjustments to the view. It will take some manipulation to find a view that includes the split exon and the borders of the EST intron.

vi. Drag the dark blue nucleotides to match the 5′ edge of the EST intron. This forms a canonical GT splice site at the beginning of the exon.

vii. Drag the light-blue nucleotides to match the 3′ edge of the EST intron. Continue dragging two additional nucleotides to create a canonical AG splice site at the 3′ end of the intron.

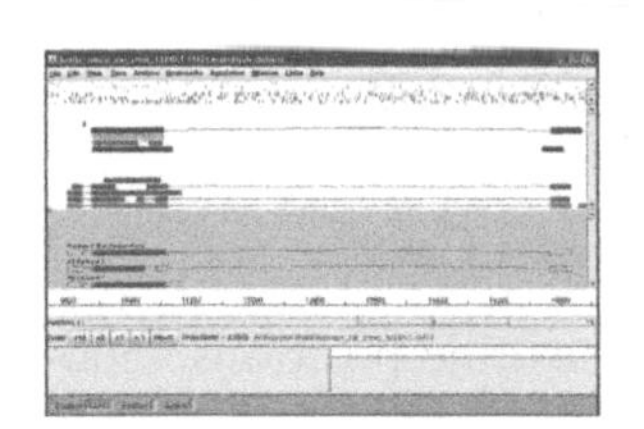

viii. Double-click to highlight the gene model, right-click (PC) or command-click (Mac), and select "Calculate longest ORF." This function uses the start and stop codons and intron/exon boundaries of the model to determine the best ORF.

ix. Ensure that the gene model shows a start codon and a stop codon. If needed, insert them in appropriate positions as described in Step 11 of Part IV.

x. Delete the EST evidence from the work space as described in Step 13 of Part VII.

5. Add comments about the adjustments you made to your model, and save your model to *DNA Subway*, as described in Steps 18 and 19 of Part VII.
6. Click on "Reset" and compare and contrast your gene models. How are the models similar to one another and how are they different?
7. Exit out of Apollo, and return to *DNA Subway*.

XVI. Browse Your Gene Models

1. On the "Browsers & Transfer" branch line, click on "Local Browser" to view your models in a graphical interface. Be patient while the local copy of GBrowse loads.
2. Your models will appear in a track called "User Annotation" at the bottom of the window. If some of your models do not appear, toggle "Scroll/Zoom" to "Show 25 kbp" or adjust the yellow view bar to show "0-16.5k."
3. How do your gene models match the genes predicted by Augustus, FGenesH, and SNAP? Do any of the three prediction algorithms predict all genes accurately? Did any of the evidence provided through the BLASTN protein database search support the gene function identified through InterProScan search? For which of the genes did the evidence indicate alternative splicing?

STAGE B: ANNOTATE A 100-KB SEQUENCE FROM *ARABIDOPSIS* CHROMOSOME 5

Below is a general guide for structurally and functionally annotating gene models with Apollo. Structural annotation improves mathematical gene predictions with biological evidence from cDNAs, ESTs, or RNA-seq obtained by BLASTN searches of appropriate databases or from uploaded experimental data. Functional annotation adds information from BLASTX evidence and InterProScan searches.

Practice the routines detailed below by annotating seven genes in the 100-kb *Arabidopsis* chromosome 5 sample from *DNA Subway*. Download EST evidence for the *Arabidopsis* chromosome 5 sample to your computer (see Part XII of Stage A) and use it to refine your annotations and/or determine the possibility of generating alternatively spliced forms. Note that before opening Apollo, establish a project with the "*Arabidopsis thaliana* (mouse-ear cress) Chr5, 100 kb" contig and run RepeatMasker, the gene prediction programs, and the database searches following the procedures outlined in Laboratory 1.1, Stage A, I, II, and III. In addition, download and add EST evidence to the project following the procedure outlined in Laboratory 1.1, Stage A, XII.

1. Open Apollo following the procedure outlined in Laboratory 1.1., Stage A, IV.1.
2. Zoom and scroll to enlarge a gene or region of interest.
3. Adjust your work space.
   i. Maximize the Apollo window, or stretch it by grabbing and pulling a corner.
   ii. Focus on the forward or reverse strand, whichever contains a gene of interest. Click on the "View" tab in the menu bar, and uncheck "Show [opposite] strand."
   iii. Grab and pull down the divider beneath the "Zoom" buttons to increase the evidence (white) and work space (blue-shaded) panels.
   iv. Grab the red arrow on the left-hand side of the screen, and pull it up/down to adjust the size of evidence and work space panels.
4. Display or hide relevant tiers.

   Apollo initially collapses all BLASTN and BLASTX evidence onto a single line each. These merged data are often misleading.

   i. Click on the "Tiers" tab in the menu bar, and select "Expand all tiers" to view all of the evidence generated by BLASTN and BLASTX, including BLASTN_USER evidence generated against uploaded data.
   ii. Click on the "Tiers" tab, and check "Show types panel" for a key to all of the data displayed in Apollo.
   iii. When redundant BLAST evidence complicates the analysis, uncheck the "Expand" box for that type of evidence to collapse it onto a single line. This will place key evidence closer to the gene predictions, making for easier comparison.
5. Select the best gene model and supporting evidence.
   i. Compare the Augustus, FGenesH, and SNAP gene models with BLASTN results and any uploaded cDNA, EST, or RNA-seq evidence. Generally, the best model is the one that most closely matches the exon–intron configuration of the longest BLASTN or cDNA evidence.
   ii. Click on any gene prediction or evidence to highlight splice sites having the same nucleotide sequence.
   iii. Double-click on the best gene prediction, and drag it into the blue work space.
   iv. Also drag the best supporting BLASTN, cDNA, EST, or RNA-seq evidence into the work space.

Whenever possible, use a gene prediction as the basis for your model. The Augustus, FGenesH, and SNAP algorithms identify start/stop codons and canonical splice sites. BLASTN has no logic for detecting splice sites, thus BLASTN evidence often contains one or more yellow arrows indicating misplaced splice sites. Only Augustus predicts 5′ and 3′ UTRs, thus select it over other predictions having a similar exon–intron configuration.

6. Give your gene model and evidence temporary names to identify them during your work.
   i. Highlight the gene model. Then right-click (PC) or command-click (Mac), and select "Annotation info editor."
   ii. Enter your chosen name, such as Gene Model 1 or Augustus:TEMP, in the gene "Symbol" field (top left) and in the transcript "Symbol" field (top right). Make sure that you use the same name in both fields. The use of dashes (–) is not permitted.
   iii. Delete all other names in the two "Synonyms" fields.
   iv. Repeat Steps 4.i–4.iii to name BLASTN or other evidence.
   v. Close the Annotation Information Editor, and return to the main Apollo screen. Your names will appear along with the models in the work space.
7. Examine the start and stop codons.
   i. A green or red vertical line in a gene model indicates a start or stop codon, respectively.
   ii. A green or red arrow at the beginning or end of a gene model indicates that a start or a stop codon has not been identified. One solution for fixing this is to drag an appropriately positioned start or stop codon from the respective evidence panel into the gene model. (You may have to zoom in to see green and red start and stop codons lined up in rows at the top of the evidence panel [for the forward strand] or at the bottom of the evidence panel [for the reverse strand].)

The Exon Detail Editor aligns the gene prediction and the BLASTN evidence sequences on top of each other. Numbers and different shades of blue indicate different reading frames.

8. Extend or shorten the 5′ and 3′ ends of the model to match the longest BLASTN evidence. (In some cases, different BLASTN results should be used to extend each end of the model.)
   i. Double-click to highlight the gene model, right-click (PC) or command-click (Mac), and select "Exon detail editor" from the pop-up menu.
   ii. Click on the diagram at the bottom to move the sliding window to the end of the gene. Alternatively, use the slider bar on the right to scroll up or down.
   iii. Click on and drag nucleotides of the gene model until they are flush with the 5′ and 3′ ends of the BLASTN evidence.
   iv. Close the Exon Detail Editor.
9. Add any exon from the BLASTN evidence that is not present in the gene model.
   i. Highlight the BLASTN exon, and drag it into the work space.
   ii. Double-click to highlight the entire gene model, and then shift-click to highlight the BLASTN exon.
   iii. Right-click (PC) or command-click (Mac), and select "Merge transcripts."
   iv. Click on "Merge" in the pop-up window, and choose a gene model name in the following window.
10. The merged model may have yellow arrows at either or both ends of the new exon, indicating improper splice sites. Correct them as follows:

i. Double-click to highlight the merged gene model, right-click (PC) or command-click (Mac), and select "Exon detail editor" from the pop-up menu.

ii. Click on the gene model diagram at the bottom to move the sliding window over the exon border with a yellow arrow. Alternatively, use the slider bar on the right to scroll up or down.

iii. Click on and drag one or several nucleotides of the exon to uncover the canonical GT splice site at the beginning of the intron or the AG splice site at the end of the intron.

iv. Close the Exon Detail Editor.

11. Delete any exon from the gene model that is not present in the BLASTN evidence.

i. Highlight the exon in the gene model.

ii. Right-click (PC) or command-click (Mac), and select "Delete selection."

12. Merge an exon in a region where the gene model has two exons but the BLASTN evidence indicates there is only one.

i. Shift-click to highlight the two exons to be merged in the gene model.

ii. Right-click (PC) or command-click (Mac), and select "Merge exons."

13. Split an exon in a region where the gene model has one exon but the BLASTN evidence indicates there are two.

i. Highlight the exon to be split in the gene model, right-click (PC) or command-click (Mac), and select "Split Exon." This will create one or two yellow diamonds that indicate missing splice sites.

ii. Double-click to highlight the gene model, right-click (PC) or command-click (Mac), and select "Exon detail editor" from the pop-up menu.

iii. Click on the gene model diagram at the bottom of the Exon Detail Editor to center the sliding window over the split exon. The exon split forms two adjacent exons in different reading frames (light and dark blue), separated by a single nucleotide. This single nucleotide is the intron that will be lengthened in Steps 14 and 15.

iv. Enlarge the window to best see the relationship between the exon split and the evidence intron. Use the slider bar on the right to make fine adjustments to the view. It will take some manipulation to find a view that includes the split exon and the borders of the evidence intron.

14. Create the 5′ (left) intron border.

i. Click on the gene model in the Exon Detail Editor.

ii. Grab the edge of the nucleotide forming the end of the exon on the left, and drag it to the left to lengthen the intron. Continue dragging until the edge of the exon is flush with the BLASTN evidence.

iii. A yellow triangle may still be present in the gene model in the main Apollo window, indicating an incorrect splice site. If this is the case, change the end of the exon to the left so that the intron begins with a canonical GT splice site.

15. Create the 3′ (right) intron border.
    i. Move the sliding window to reposition the editor over the edge of the exon on the right.
    ii. Grab the edge of the nucleotide forming the beginning of the exon on the right, and drag it to the right to lengthen the intron. Continue dragging until the edge of the exon is flush with the BLASTN evidence.
    iii. If a yellow triangle is present in the gene model in the main Apollo window, move the beginning of the exon on the right so that the intron ends with a canonical AG splice site.
16. Ensure that Apollo displays an appropriate ORF for the model.
    i. Double-click to highlight the gene model, right-click (PC) or command-click (Mac), and select "Calculate longest ORF." This function uses the start and stop codons and intron/exon boundaries of the model to determine the best ORF.
    ii. Ensure that the gene model shows a start codon and a stop codon. If needed, insert them in appropriate positions as described in Step 7.ii.
17. Add functional annotations to your gene model. Layer in additional BLAST evidence as follows:
    i. Highlight the gene model or an individual exon.
    ii. Right-click (PC) or command-click (Mac), and select "Submit to BLAST."
    iii. Verify that the "NCBI-BLAST" tab is selected (in blue), and click on "Run."
    iv. Click on "OK" in the pop-up window giving the expected start time for the analysis.
    v. Click on "OK" to the prompt "Remote analysis complete." This will load the BLAST results into the evidence panel.
    vi. Click on the "Tiers" tab in the menu bar, and select "Expand all tiers" to view the BLAST results.
    vii. Click on BLAST hits to see descriptions in the Information window at the bottom right. Look for a pattern in the names that suggests a function for the protein produced by this gene.
18. Identify additional functional information through InterProScan.
    i. Highlight the gene model, right-click (PC) or command-click (Mac), and select "Submit to InterProScan."
    ii. Click on "OK" in a pop-up window.
    iii. Click on "OK" to the prompt that the analysis is complete.
    iv. This will load the results into the Annotation Info Editor and open the InterProScan results in a browser window (make sure your browser is set to allow pop-ups). You can edit the results in the Annotation Info Editor to retain the most useful information.
19. Name your model, and add information about your work.
    i. Highlight the gene model. Then right-click (PC) or command-click (Mac), and select "Annotation info editor."

ii. Enter your chosen name in the gene "Symbol" field (top left) and transcript "Symbol" field (top right). Your gene name may reflect functional relationships identified in Steps 17 and 18. Make sure that you use the same name in both fields. The use of dashes (–) is not permitted.

iii. Delete all other names in the two "Synonyms" fields.

iv. Click on the respective "Edit comments" button for the name you chose (at the bottom right). Then, click on "Add" and write your comments in the pop-up window. These should include the information about protein function obtained from BLAST searches, notes that justify your model, and unresolved annotation problems. Click on "Close" to return to the Annotation Information window.

v. Close the Annotation Information Editor, and return to the main Apollo screen.

vi. Ensure that the model includes appropriate start and stop codons following Step 16.

20. Check for evidence of alternative splicing, and make additional models.

Basing the alternative model(s) on the finished gene model retains your previous edits, such as 5′ and 3′ UTRs.

i. Look among BLASTN hits (derived from cDNA or EST evidence) for exon structures that are different from those of your gene model (i.e., additional/missing exons and longer/shorter exons). Highlight and drag into the work space any evidence that supports alternative splicing.

ii. Create a duplicate of your finished gene model. Highlight the gene model, right-click (PC) or command-click (Mac), and select "Duplicate transcript."

iii. Name the evidence and duplicate gene model as described in Step 6.

iv. Add, delete, merge, split, and adjust the length of the exons as described in Steps 8, 9, and 11–15.

v. Resolve any yellow-flagged splice sites, as described in Step 10, to reveal the canonical GT splice site at the beginning of introns and the AG splice site at the end of introns.

vi. Ensure that the alternative model includes appropriate start and stop codons following Step 16.

vii. Name and add comments to your alternatively spliced model as described in Step 19.

21. In the "File" tab in the menu bar, select "Upload to *DNA Subway*."

LABORATORY 1.2

Detecting a Lost Chromosome

▼ OBJECTIVES

This laboratory demonstrates several important concepts of modern biology. During the course of this laboratory, you will

- Analyze the genomes of two closely related species.
- Compare the locations of homologous genes in different species.
- Use genes as markers to identify evolutionary genome rearrangements.

In addition, this laboratory uses several bioinformatics methods in modern biological research. You will

- Use the Basic Local Alignment Search Tool (BLAST) and Map Viewer to search for and view the chromosome locations of homologous genes.
- Use the CoGe genome comparison tool to align portions of the human and chimpanzee genomes and to identify syntenic chromosome regions.

INTRODUCTION

A genome is the set of genetic information—carried on chromosomes—that defines a species and provides a blueprint for the inheritance of traits from generation to generation. In eukaryotes, the set of chromosomes contributed by a mother's egg combines with the set from a father's sperm to form a diploid genome. The non-sex chromosomes (autosomes) contributed by the mother and father are essentially duplicates; they have the same genes. Thus, for simplicity's sake, most genome analysis focuses on the haploid genome—a single set of autosomes plus the sex chromosomes (typically X and Y).

Genomes are dynamic and undergo changes over the course of evolution. As two species diverge from a common ancestor, their genomes accumulate changes ranging from point mutations (single-nucleotide changes); to the loss or duplication of genes; to the loss, duplication, or exchange of whole chromosomes or regions of chromosomes. Homologs, genes that are derived from a common ancestor, retain a high degree of sequence similarity despite large-scale chromosome changes in divergent genomes. Furthermore, related species typically retain the same gene order, or *synteny*, over long chromosome regions.

Humans share common ancestry with other Great Apes (members of the family Hominidae). This evolutionary relationship is reflected in many shared anatomical, physiological, and behavioral features—and in genome sequences that are 95%–99.9% similar. Humans are most closely related to chimpanzees, having diverged from a common ancestor 5–7 million years ago. Since that time, genome changes that make humans unique have accumulated.

Humans and chimpanzees have one major difference in their chromosome sets. This laboratory uses bioinformatics tools to identify and explore this difference. A sequence search tool (BLAST) and a graphical browser (Map Viewer) are used to determine the chromosome locations of homologous genes in the human and chimpanzee genomes. The comparative genomics platform CoGe is used to align the two primate genomes and to visually illustrate the major difference between them. The analysis provides insight into how and when chimpanzees and humans diverged from a common ancestor.

FURTHER READING

Avarello R, Pedicini A, Caiulo A, Zuffardi O, Fraccaro M. 1992. Evidence for an ancestral alphoid domain on the long arm of human chromosome 2. *Hum Genet* **89:** 247–249.

Fortna A, Kim Y, MacLaren E, Marshall K, Hahn G, Meltesen L, Brenton M, Hink R, Burgers S, Hernandez-Boussard T, et al. 2004. Lineage-specific gene duplication and loss in human and great ape evolution. *PLoS Biol* **2:** e207. doi: 10.1371/journal.pbio.0020207.

IJdo JW, Baldini A, Ward DC, Reeders ST, Wells RA. 1991. Origin of human chromosome 2: An ancestral telomere-telomere fusion. *Proc Natl Acad Sci* **88:** 9051–9055.

Yunis JJ, Prakash O. 1982. The origin of man: A chromosomal pictorial legacy. *Science* **215:** 1525–1530.

Yunis JJ, Sawyer JR, Dunham K. 1980. The striking resemblance of high-resolution G-banded chromosomes of man and chimpanzee. *Science* **208:** 1145–1148.

PLANNING AND PREPARATION

The following table will help you to plan and integrate the different bioinformatics methods.

Experiment	Day	Time		Activity
I. Compare the human and chimpanzee genomes	1	30–45 min	Lab:	Identify the major difference between the human and chimpanzee genomes using Map Viewer. Develop a hypothesis stating how and when this difference arose during primate evolution.
II. Compare the locations of homologous genes on human and chimpanzee chromosomes	2	30–45 min	Lab:	Use human gene sequences stored in *Sequence Server* to identify the locations of the chimpanzee homologs in Map Viewer. Determine the relationship between human chromosome 2 and chimpanzee chromosomes 2A/2B.
III. Compare the human and chimpanzee genomes by syntenic dotplot	3	30–45 min	Lab:	Identify syntenic regions in the human and chimpanzee genomes using the CoGe comparative genomics tool SynMap.
IV. Determine the origin of the centromere in human chromosome 2	4	30–45 min	Lab:	Use chimpanzee chromosome 2A and 2B sequences stored in *Sequence Server* to identify the locations of the human homologs in Map Viewer. Determine whether the centromere in human chromosome 2 was derived from chimpanzee chromosome 2A or 2B.
V. Extend the analysis of the evolution of human chromosome 2	5	30–120 min	Lab:	Compare the locations of gene homologs from the human, chimpanzee, gorilla, orangutan, and macaque genomes.

BIOINFORMATICS METHODS

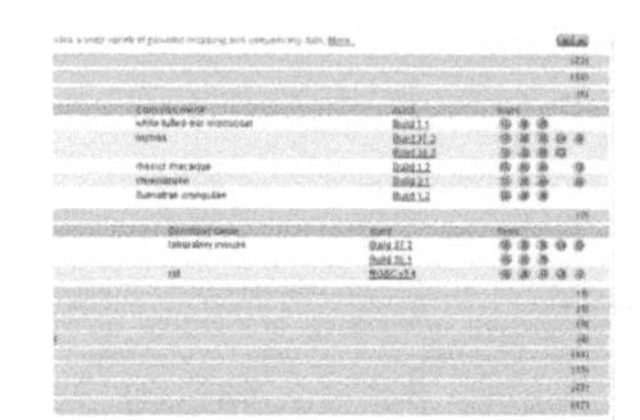

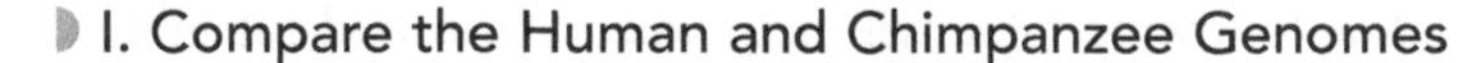

I. Compare the Human and Chimpanzee Genomes

1. Open Map Viewer (http://www.ncbi.nlm.nih.gov/mapview).
2. Find *Homo sapiens* in the table to the right, and click on the most recent version ("Build") of the human genome.
3. Open another Map Viewer in a new browser window. Find *Pan troglodytes* (chimpanzee) in the table to the right, and click on the most recent version ("Build") of the chimpanzee genome.
4. Compare the chromosome configurations of the two genomes. How do they differ?
5. Open another Map Viewer in a new browser window. Find *Pongo abelii* (orangutan) in the table to the right, and click on the most recent version ("Build") of the orangutan genome.
6. Compare the chromosome configurations of the three genomes. How do they differ?
7. Open another Map Viewer in a new browser window. Find another Great Ape in the table to the right, and click on the most recent version ("Build") of the genome.

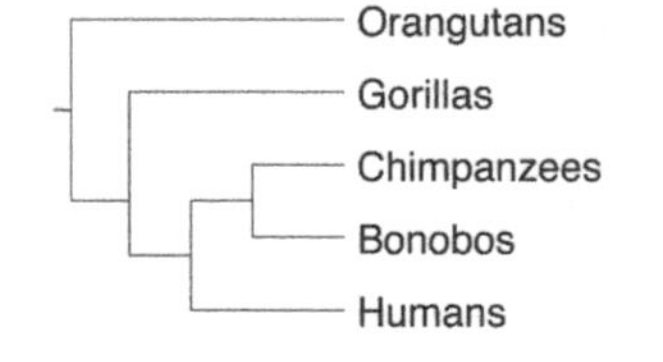

8. Compare the chromosome configurations of the four genomes. How do they differ?
9. The phylogenetic tree at the left shows the relationships among the Great Apes based on genes and genomic features. Based on your analysis in Steps 1–8, what was the likely chromosome configuration of the most recent common ancestor (MRCA) of humans and Great Apes? Explain.

10. In the image to the left, human (H) chromosome 2 is aligned with chromosomes 2A and 2B from the Great Apes (C, chimpanzee; G, gorilla; O, orangutan). Examine the banding patterns, and make a hypothesis regarding how the human and Great Ape chromosomes are related.

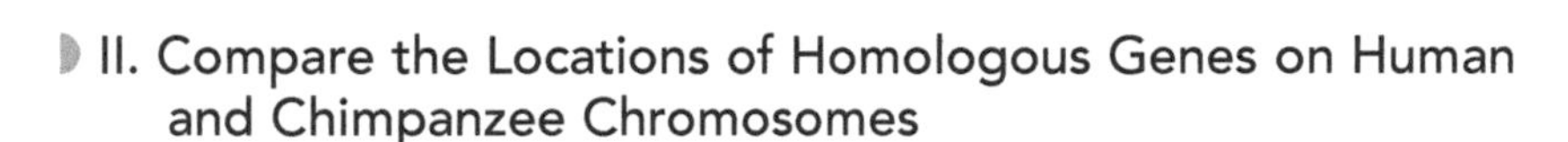

II. Compare the Locations of Homologous Genes on Human and Chimpanzee Chromosomes

1. Obtain a gene sequence from human chromosome 2.
   i. Open the BioServers Internet site at the Dolan DNA Learning Center (http://www.bioservers.org).
   ii. Enter *Sequence Server*. You can register if you want to save your work for future reference, but this is not required.
   iii. The interface is simple to use: Add or obtain data using the top buttons and pull-down menus, and then work with the data in the work space below.
   iv. Click on "MANAGE GROUPS" at the top of the page.
   v. Select "Genome Science Labs" from the "Sequence sources" pull-down menu in the upper right-hand corner.
   vi. Click on the checkbox to select "Chromosome 2 Mystery," and click "OK" to move this group into the work space.

Spastin (SPAST) is an open reading frame (ORF) gene, meaning that its coding sequence begins with a start codon and ends with a stop codon. Unlike most human genes, it has no introns.

vii. Select "Homo sapiens Chr 2 SPAST ORF" from the first pull-down menu in the work space.

viii. Click on "OPEN." Highlight and copy the entire nucleotide sequence. You will use this sequence in Steps 2.iv and 3.i.

2. Locate the *SPAST* gene in the human genome.

i. Open Map Viewer (http://www.ncbi.nlm.nih.gov/mapview).

ii. Find *Homo sapiens* in the table to the right, and click on the most recent version ("Build") of the human genome.

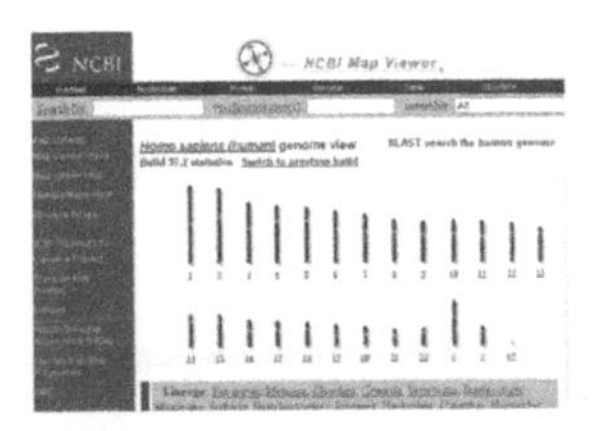

iii. Click on "BLAST search the human genome" at upper right.

iv. Paste the human *SPAST* sequence from Step 1.viii into the search window.

v. Under "Choose Search Set," find "Database" and select the "Genome (reference only)" database from the drop-down menu.

The MegaBLAST algorithm is optimized for comparing closely related sequences that are at least 95% identical.

vi. Under "Program Selection," optimize for "Highly similar sequences (megablast)."

vii. Click on "BLAST." This sends your query sequence to a server at NCBI in Bethesda, Maryland. There, the BLAST algorithm will attempt to match the sequence to the millions of DNA sequences stored in its database. While searching, a page showing the status of your search will be displayed until your results are available. This may take only a few seconds or more than 1 min if many other searches are queued at the server.

viii. In "Other reports," click on the "[Human genome view]" link to see the chromosome location of the BLAST hit. Small horizontal bars on chromosomes indicate the positions of hits. On which chromosome have you landed? Draw a diagram with the approximate location of the hit.

3. Locate the homolog of the human *SPAST* gene in the chimpanzee genome.

i. If the human *SPAST* sequence is not in your buffer memory, obtain it again from *Sequence Server* as described in Step 1.

ii. Search the chimpanzee genome with Map Viewer using the same strategy as described in Step 2 to see the chromosome location. On which chromosome have you landed? Draw a diagram with the approximate location of the hit.

4. Locate additional homologous genes in the human and chimpanzee genomes.

i. Use the strategy from Steps 1–3 to locate the *SUCGL1*, *LCT*, and *BARD1* genes in the human and chimpanzee genomes. Draw a diagram with the approximate location of each hit. When possible, add these to existing diagrams.

ii. Compare the locations of the four genes in the human and chimpanzee genomes. In Step 10 of Part I, you generated a hypothesis regarding the relationship between the human and chimpanzee chromosomes. Does your diagram support your hypothesis? Why or why not?

CoGe uses MegaBLAST to compare two whole-genome sequences to visualize how they have changed during evolution. The CoGe website includes analysis examples and tutorials, including a video tutorial on how to use CoGe to visualize the differences between the human and chimpanzee genomes. To access these items, click the "Help" button at the upper right-hand corner of the screen.

III. Compare the Human and Chimpanzee Genomes by Syntenic Dotplot

1. Open the CoGe genome comparison tool (http://genomevolution.org/CoGe).

2. Under "CoGe's Entrance Tools," click on "SynMap: Whole genome syntenic dotplot analyses."
3. Type Homo sapiens in the "Name" window for "Organism 1 Search." Select "Homo sapiens (human)" to include the entire genome; do not select the mitochondrion.
4. Type Pan troglodytes in the "Name" window for "Organism 2 Search." Select "Pan troglodytes (chimpanzee)" to include the entire genome; do not select the mitochondrion.
5. Ensure that "Genome" is set to "unmasked" and "CDS."
6. Click on the "Display Options" tab.
   i. Set "Sort Chromosomes by" to "Name."
   ii. Set "Master image width" to "500."
7. Click on "Generate SynMap," and wait for the results. In the graph, the entire chimpanzee genome is aligned against the entire human genome. The *x*-axis displays the 22 human autosomes, plus the X and Y chromosomes, head to tail. The *y*-axis shows the 23 chimpanzee autosomes, plus the X and Y chromosomes.

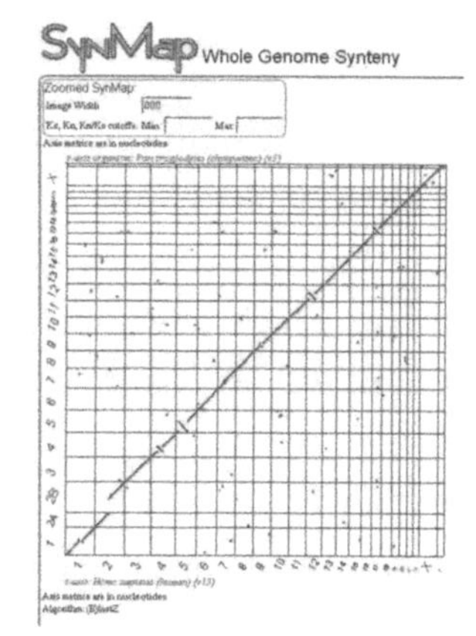

8. In general, homologous chromosomes will align to form a plot running diagonally from the lower left to the upper right. Generally, how would you describe the synteny between the human and chimpanzee chromosomes?
9. Displacements off the main diagonal occur when one chromosome being aligned is significantly shorter than the homologous chromosome from the other species. What do you notice about the alignment for human chromosome 2? Does this alignment support your hypothesis from Step 10 in Part I? Why or why not?
10. Parts of the dotplot on chromosomes 1, 4, 5, 12, and 17 are at 90° from the main diagonal axis. How do you explain this?
11. Zoom in on the dotplot, and analyze gene-to-gene alignments.
    i. Drag your mouse cursor into the dotplot, and center the crosshairs on any part of the dotplot. Click and be patient while the DNA helix spins as a close-up view of the dotplot is generated.
    ii. Now center the crosshairs on any part of the close-up dotplot, and click. The chromosome coordinates of this position are automatically logged into the "Sequence 1" and "Sequence 2" analysis windows on a "GEvo Configuration" page.

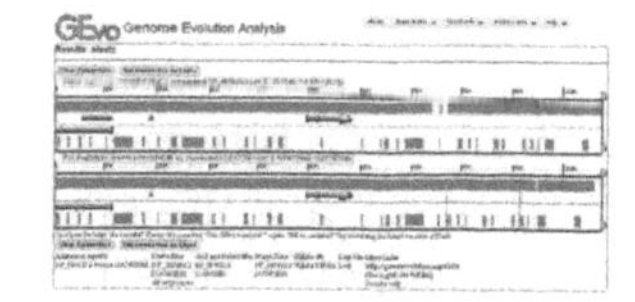

    iii. On the "GEvo Configuration" page, click "Run GEvo Analysis!" Be patient while the DNA helix spins; it will take some time to generate the gene-to-gene alignment.
    iv. Pink blocks are regions of sequence homology. Click for details on specific regions, and to connect homologous regions in the human and chimpanzee sequences.
    v. The thinner gray tubes are introns, whereas the thicker colored tubes are exons. An arrow points to the 3′ end, showing the direction in which the gene is transcribed. Describe the general structures of the homologous human and chimpanzee genes.

vi. Repeat Steps 11.i–11.v for a region of the dotplot where part of the alignment is at 90° from the main diagonal. Does this confirm your answer in Step 10?

IV. Determine the Origin of the Centromere in Human Chromosome 2

1. Obtain two gene sequences that span the centromere of chimpanzee chromosome 2A.

    i. If you are not already in the *Sequence Server* of the BioServers Internet site, open it as described in Steps 1.i and 1.ii in Part II. Then click on "MANAGE GROUPS" at the top of the page.

    ii. Select "Genome Science Labs" from the "Sequence sources" pull-down menu in the upper right-hand corner.

    iii. Click on the checkbox to select "Chromosome 2 Mystery," and click "OK" to move this group into the work space.

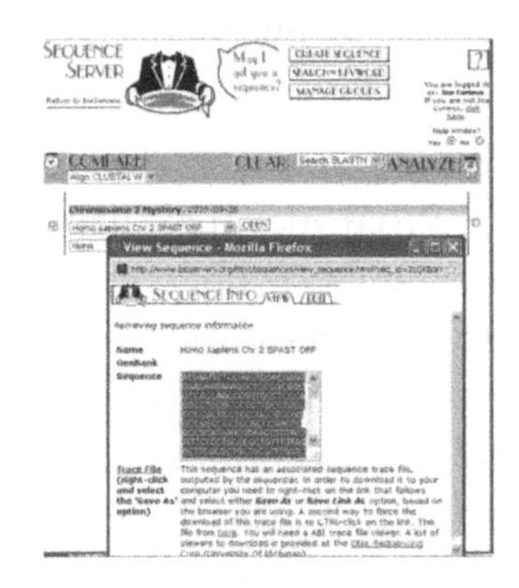

    iv. Select "Pan troglodytes Chr 2A EDAR ORF" from the first pull-down menu in the work space.

    v. Click on "OPEN." Highlight and copy the entire nucleotide sequence. Then paste it into a text document.

    vi. Return to the work space and select "Pan troglodytes Chr 2A TGOLN2 ORF" from the first pull-down menu.

    vii. Click on "OPEN." Highlight and copy the entire nucleotide sequence. Then paste it into the same text document that you created in Step 1.v. Save this text document for use in Step 2.ii.

2. Locate the chimpanzee chromosome 2A gene homologs on human chromosome 2.

    i. Open the human nucleotide BLAST page in Map Viewer as described in Steps 2.i–2.iii in Part II.

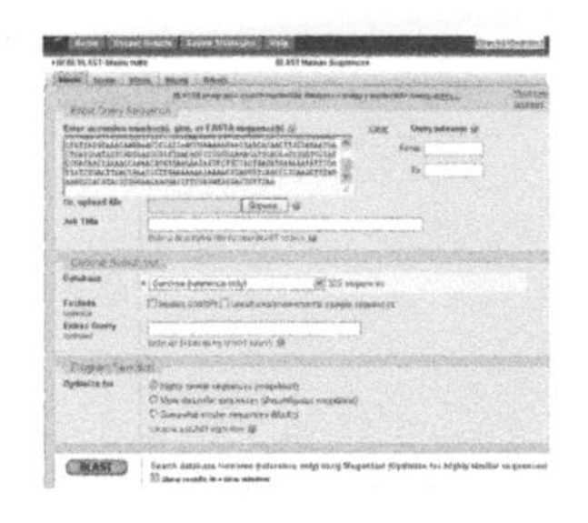

    ii. Paste the chimpanzee *EDAR* and *TGOLN2* ORF sequences from Step 1.vii into the search window, with or without a line break between the two sequences. Omit any nonnucleotide characters from the window because they will not be recognized by the BLAST algorithm.

    iii. Under "Choose Search Set," select the "Genome (reference only)" database from the drop-down menu.

    iv. Under "Program Selection," optimize for "Highly similar sequences (megablast)."

    v. Click on "BLAST" and wait for results.

    vi. In "Other reports," click on the "[Human genome view]" link to see the chromosome location of the BLAST hits. Draw a diagram showing where the two chimpanzee chromosome 2A gene homologs are located in relation to the centromere of human chromosome 2.

3. Obtain two gene sequences that span the centromere of chimpanzee chromosome 2B (*BIN1* and *PKP4*) using the procedures described in Step 1.

4. Locate the chimpanzee chromosome 2B gene homologs on human chromosome 2 using the strategy described in Step 2. Draw a diagram showing where the two

chimpanzee chromosome 2B gene homologs are located in relation to the centromere of human chromosome 2.

5. Where is the human chromosome 2 centromere in relation to the homologs from chimpanzee chromosomes 2A and 2B? What does this say regarding the origin of the human centromere?

V. Extend the Analysis of the Evolution of Human Chromosome 2

1. Use Map Viewer to locate the orangutan (*Pongo abelii*) and gorilla (*Gorilla gorilla*) chromosome 2A and 2B gene homologs in the human genome. Compare your results with those you obtained in Step 5 of Part IV. Do these results support your hypothesis from Step 4 of Part II? Why or why not?

2. Use Map Viewer to locate chimpanzee chromosome 2A and 2B gene homologs in the macaque (*Macaca mulatta*) genome. Comment on the chromosome number and the locations of genes. What do these results indicate regarding the overall organization of primate genomes?

To zoom in on a gene in Map Viewer, click on the number of the chromosome you wish to view. The vertical line labeled "Genes_seq" shows the locations of annotated genes. Use the zoom toggle on the left to zoom in or out for a better perspective of the chromosome. Click the arrows on the "Genes_seq" ruler to scroll up or down along the chromosome.

3. Use Map Viewer to compare the centromere region in chimpanzee chromosome 2B with the homologous region in human chromosome 2. What may have happened to the centromere of chromosome 2B during the fusion of chromosomes 2A and 2B? (*Hint:* Use *BIN1* and *PKP4* from chimpanzee chromosome 2B to locate this region in both the human and chimpanzee genomes. Zoom into these homologous regions to see individual genes, and then scroll to find homologous genes that flank the centromere in chimpanzee chromosome 2B, such as *ARMGEF4* and *GPR39*. Make observations regarding the distance between these homologous genes, and comment on the presence or absence of other genes in these homologous regions.)

LABORATORY 1.3

Comparing Diversity in Eukaryotes

▼ OBJECTIVES

This laboratory demonstrates several important concepts of modern biology. During the course of this laboratory, you will

- Explore genetic diversity and evolutionary history in different plant and animal species.
- Evaluate the nucleotide differences in homologous genes within and between species.

In addition, this laboratory uses a bioinformatics method in modern biological research. You will

- Use CLUSTAL W to align sequences and assess sequence variation.

INTRODUCTION

Genetic diversity is important to the success of a species because it provides a pool of gene variants that allows the species to adapt to different climates and pathogens. Genetic variation is most often equated to mutations in DNA. Because mutations accumulate over time, older populations generally have more genetic diversity than younger ones. Mutations occur more often in large populations than in small ones. A dramatic decrease in the number of individuals in a population creates a "bottleneck" that reduces genetic diversity in the population before it expands again.

Agriculturalists are concerned with preserving germplasm (seed) from wild plants as a continuing source of genetic variation that can be introduced into domestic relatives. For example, several types of teosinte in Mexico are the wild germplasm resources from which genes can be reintroduced into maize (corn), its domestic relative. Plant genetic diversity is becoming more important as scientists look for variants that will allow crops to adapt to rapid changes in temperature, precipitation, and pests brought on by global climate change.

A relatively small proportion of a species' genes produce visible phenotypic differences that characterize different subpopulations. For example, human "races" have been traditionally defined by gene variants that determine hair texture; facial features; and skin, hair, and eye color. Similarly, a handful of gene changes are thought to produce the major stalk, ear, and kernel differences in teosinte and hundreds of domestic corn races and varieties. Conversely, the three to four subspecies of chimpanzee are so similar that even primatologists cannot tell them apart.

This laboratory examines variation in a single gene, alcohol dehydrogenase 1 (*ADH1*), within and between plant and animal species. In animals, the ADH1 enzyme catalyzes the oxidization of ethanol to acetaldehyde. This is the first step in the breakdown of alcohol (which is toxic to cells). ADH1 plays an essentially opposite role in bacteria, yeast, and plants, where it reduces acetaldehyde to ethanol during the final step of fermentation. Increased ADH1 expression in plants is part of a stress response to waterlogging, when fermentation helps plants survive under anaerobic (low oxygen) conditions.

Because the *ADH1* gene has conserved functions, it can be used to analyze genetic variation across the plant and animal kingdoms. It balances having enough sequence conservation to enable comparison between diverse species with having enough sequence variation to gauge diversity within an individual species.

This laboratory uses *Sequence Server*, a simple platform for gene analysis on the BioServers Internet site of the DNA Learning Center (http://www.bioservers.org). Sequence Server uses the CLUSTAL W algorithm to align related DNA (or protein) sequences to produce the fewest gaps.

First, mutations in corn and teosinte *ADH1* sequences are compared. Then, this analysis is repeated for human and chimpanzee *ADH1* genes. Finally, the sequence variation in corn species is compared to that in primate species. Extensions to these analyses assess the sequence variation in *ADH1* genes from rice and barley.

FURTHER READING

Buckler ES, Thornsberry JM, Kresovich S. 2001. Molecular diversity, structure and domestication of grasses. *Genet Res* **77:** 213–218.

Canaran P, Buckler ES, Glaubitz JC, Stein L, Sun Q, Zhao W, Ware D. 2008. Panzea: An update on new content and features. *Nucleic Acids Res* **36:** D1041–D1043.

Diamond J. 1997. *Guns, germs and steel: The fate of human societies.* Norton, New York.

Eyre-Walker A, Gaut RL, Hilton H, Feldman DL, Gaut BS. 1998. Investigation of the bottleneck leading to the domestication of maize. *Proc Natl Acad Sci* **95:** 4441–4446.

Londo JP, Chiang Y-C, Hung K-H, Chiang T-Y, Schaal BA. 2006. Phylogeography of Asian wild rice, *Oryza rufipogon*, reveals multiple independent domestications of cultivated rice, *Oryza sativa. Proc Natl Acad Sci* **103:** 9578–9583.

Tenaillon MI, Sawkins MC, Long AD, Gaut RL, Doebley JF, Gaut BS. 2001. Patterns of DNA sequence polymorphism along chromosome 1 of maize (*Zea mays* ssp. *mays* L.). *Proc Natl Acad Sci* **98:** 9161–9166.

Wright SI, Bi IV, Schroeder SG, Yamasaki M, Doebley JF, McMullen MD, Gaut BS. 2005. The effects of artificial selection on the maize genome. *Science* **308:** 1310–1314.

Yamasaki WM, Tenaillon MI, Bi IV, Schroeder SG, Sanchez-Villeda H, Doebley JF, Gaut BS, McMullena MD. 2005. A large-scale screen for artificial selection in maize identifies candidate agronomic loci for domestication and crop improvement. *Plant Cell* **17:** 2859–2872.

PLANNING AND PREPARATION

The following table will help you to plan and integrate the different bioinformatics methods.

Experiment	Day	Time	Activity
I. Set up a project on *Sequence Server*	1	10 min	Lab: Access *ADH1* gene sequences of corn, teosinte, human, chimpanzee, barley, and rice stored in *Sequence Server*.
II. Align and compare two maize *ADH1* gene sequences	1	15 min	Lab: Align two maize *ADH1* gene sequences using CLUSTAL W. Calculate the percent difference between the two sequences.
III. Align and compare the maize and teosinte *ADH1* gene sequences	1	15 min	Lab: Evaluate the genetic variation in teosinte and maize *ADH1* gene sequences using CLUSTAL W.
IV. Align and compare the human and chimpanzee *ADH1* gene sequences	1	15 min	Lab: Evaluate the genetic variation in human and chimpanzee *ADH1* gene sequences using CLUSTAL W.
V. Extend the analysis of *ADH1* gene sequences	2	30 min	Lab: Evaluate the genetic variation in barley and rice *ADH1* gene sequences using CLUSTAL W. Evaluate the genetic variation in *ADH1* gene sequences of different species using CLUSTAL W.

BIOINFORMATICS METHODS

I. Set Up a Project on *Sequence Server*

1. Open the BioServers Internet site at the Dolan DNA Learning Center (http://www.bioservers.org).
2. Enter *Sequence Server*. You can register if you want to save your work for future reference, but this is not required.
3. The interface is simple to use: Add or obtain data using the top buttons and pull-down menus, and then work with the data in the work space below.
4. Click on "MANAGE GROUPS" at the top of the page.
5. Select "Genome Science Labs" from the "Sequence sources" pull-down menu in the upper right-hand corner.
6. Click on the checkbox to select "Diversity, Adh," and click "OK" to move this group into the work space.

The sequences used in this laboratory are ADH1 gene fragments from corn, teosinte, human, chimpanzee, barley, and rice. The sequences are from genomic DNA and contain both exons and introns.

II. Align and Compare Two Maize *ADH1* Gene Sequences

1. Select "Adh1 sample corn B73" from the first pull-down menu in the work space.
2. Select "Adh1 sample corn Mo17" from the second pull-down menu in the work space.

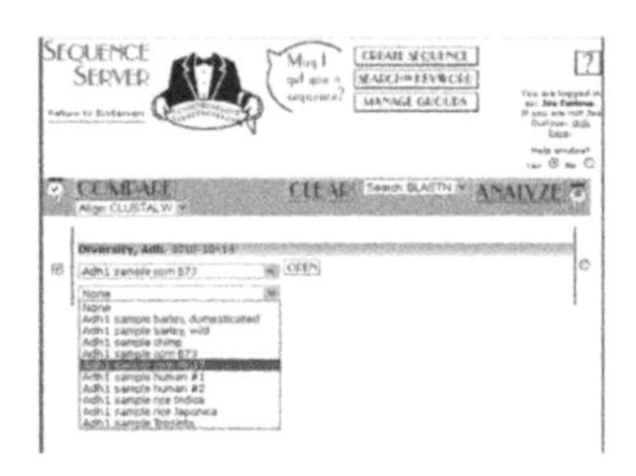

3. Make sure that the boxes to the left of each of the two sequences are checked.
4. Click on "COMPARE" in the gray bar. (The default operation is a sequence alignment using the CLUSTAL W algorithm.) The checked sequences are sent to a server at Cold Spring Harbor Laboratory, where the CLUSTAL W algorithm will attempt to align each nucleotide position. This may take only a few seconds or more than a minute if many other searches are queued at the server.
5. Results will appear in a new window. Maximize this window.
6. By default, only the first 500 bp of the alignment is shown. Enter "**5000**" as the number of nucleotides to display per page, and then click on "Redraw."

7. Examine the sequences, which are displayed in rows of 25 nucleotides. Yellow-shaded dashes (–) indicate nucleotides that are present in one of the sequences but not in the other. Yellow highlighting also denotes mismatches between the sequences. A gray-shaded *N* indicates a sequence error, a position in one or both sequences where a nucleotide could not be determined.
8. Count the number of nucleotide differences between the two sequences as follows:
   i. Count all yellow-shaded nucleotide differences. These are single nucleotide polymorphisms (SNPs).
   ii. Count all insertion/deletions (indels) indicated by internal dashes (–).
   iii. Do not count dashes that precede or follow a nucleotide sequence.
   iv. If you see a string of internal dashes (– – –), this likely arose from a single mutation event and thus should be scored as a single polymorphism.
   v. Do not count any Ns.

9. Record the total number of differences that you counted in Step 8. In addition, record the total length of the sequence analyzed.
10. Calculate the percent difference between the two sequences as follows:

$$\text{Difference (\%)} = \frac{\text{number of nucleotide differences}}{\text{total number of nucleotides counted}} \times 100$$

III. Align and Compare the Maize and Teosinte *ADH1* Gene Sequences

Teosinte (Zea mays ssp. parviglumis) is the wild subspecies of maize from which Native Americans domesticated corn ~10,000 years ago. It is the ancestor of all modern corn varieties.

1. In the *Sequence Server* work space, change one of the two corn sequences used in Part II to "Adh1 sample Teosinte." Click on "COMPARE" in the gray bar.
2. Count the number of nucleotide differences between the two sequences, and calculate the percent difference between the two sequences as described in Steps 8–10 in Part II.
3. Compare the teosinte *ADH1* gene with the second corn *ADH1* sequence by repeating Steps 1 and 2.
4. Is the percent difference between the two corn sequences (calculated in Step 10 in Part II) similar to the percent difference between corn and teosinte?

IV. Align and Compare the Human and Chimpanzee *ADH1* Gene Sequences

1. Use the procedure described in Part II to calculate the percent difference between two human *ADH1* gene sequences.
2. Repeat the procedure two more times to calculate the percent difference between the chimp *ADH1* gene sequence and each of the two human *ADH1* gene sequences.
3. Compare the lengths of the human and chimpanzee alignments with those of corn and teosinte from Parts II and III. What do you notice? How would you explain this?
4. Which has greater genetic diversity: humans or corn? By what factor is one group more diverse than the other? How would you explain this?
5. How does the percent difference between the two corn varieties compare with the percent difference between human and chimpanzee? What does this seem to indicate regarding the ages of these species?

V. Extend the Analysis of *ADH1* Gene Sequences

1. Use the procedure described in Part II to calculate the percent difference between wild and domesticated barley *ADH1* gene sequences.
2. Repeat the procedure to calculate the percent difference between two subspecies of domestic rice: *Oryza sativa* ssp. *japonica* and *Oryza sativa* ssp. *indica*.
3. How does the genetic diversity of barley and rice compare with that of maize? What does this seem to indicate regarding the genetic diversity in the progenitors of modern barley and rice, compared with those of modern corn?
4. Align *ADH1* gene sequences from different species (e.g., corn and human, chimpanzee and rice, or rice and corn). Describe your findings.

[illegible]

$$DD = \frac{\text{number of positions with differences}}{\text{total number of nucleotides counted}} \times 100$$

[illegible]

LABORATORY 1.4

Determining the Transposon Content in Grasses

▼ OBJECTIVES

This laboratory demonstrates several important concepts of modern biology. During the course of this laboratory, you will

- Examine transposons in the genomes of three monocot plants: the model grass *Brachypodium*, rice, and maize (corn).
- Compare the extent and types of transposons in three different plant species.

In addition, this laboratory uses several bioinformatics methods in modern biological research. You will

- Use RepeatMasker to identify repetitive DNA and transposons in genomic DNA sequences.
- Use the annotation editor Apollo to graphically display repetitive DNA and transposons.

INTRODUCTION

DNA variation is the starting point for evolution, providing a source of heritable changes that can improve an organism's fitness in response to natural selection. Some genetic variation arises from errors in DNA replication or from damage caused by chemical mutagens and ionizing radiation. In the 1950s, Barbara McClintock discovered another source of genetic variation—mobile DNA elements that transpose, or "jump," from one chromosome location to another.

McClintock's work challenged the dogma at the time that chromosomes were stable, unchanging carriers of genes. Because they occur in multiple copies, transposons are a type of "repetitive DNA." Repetitive DNA sequences, which also include short tandem repeats (STRs) and variable-number tandem repeats (VNTRs), occupy a large portion of many eukaryotic genomes—from 50% in humans to >95% in lilies. Although transposons were once dismissed by many as mere "junk," the fact that they compose nearly half of the human genome and the vast majority of many cereal genomes has forced scientists to seriously consider their roles in genome evolution. Because McClintock worked only with a light microscope, uncovering the rich details of her world of transposition had to await the advent of modern tools for DNA and genome analysis.

Transposons are divided into two groups, according to their mode of activation. Class I transposons are also called "RNA transposons," "retrotransposons," or "retroposons" because they use the same "copy-and-paste" transposition mechanism as retroviruses. RNA polymerase transcribes the transposon into an mRNA transcript, which, in turn, is converted into mobile DNA by reverse transcriptase. The original transposon is retained at its initial chromosome location (*locus*), and the reverse-transcribed copy inserts elsewhere. Class II transposons, also called "DNA transposons," use a "cut-and-paste" mechanism in which a transposase protein makes double-stranded cuts in DNA to release a transposon from one chromosome locus and insert it at a different one. Because of these different copying mechanisms, RNA transposons accumulate in large numbers and typically comprise the largest component of the genomes of higher organisms, such as mammals and flowering plants. DNA transposons typically constitute only a small fraction of these genomes.

Both classes of transposons contain autonomous and nonautonomous members. Autonomous or complete transposons carry genes that encode enzymes that mobilize a sequence for transposition (e.g., reverse transcriptase in Class I transposons or transposase in Class II transposons). Nonautonomous or defective transposons do not encode these enzymes and, instead, require an autonomous partner to perform transposition. For example, in mammals, the retroposon L1 supplies reverse transcriptase for *Alu*. In maize, the DNA transposon *Ac* provides transposase for *Ds*. Many nonautonomous transposons are derived from autonomous transposons by mutations that inactivate the reverse transcriptase or transposase genes. Autonomous and nonautonomous elements are flanked by direct or inverted repeat sequences that facilitate their integration into a new chromosome locus. At any moment in evolutionary history, only a fraction of transposons are capable of transposition; the vast majority have acquired inactivating mutations in the target sequences, rendering them molecular fossils.

The molecular dissection of the genomes of higher organisms has confirmed that chromosomes are dynamic, changing structures. By remodeling chromosomes, creating new gene combinations, and altering patterns of gene transcription, transposons provide genetic variation upon which evolution can act. Repeated DNA sequences also provide new sites for homologous recombination between chromosomes. There is even evidence to support McClintock's hypothesis that transposition may allow organisms to reorganize the genome rapidly in response to environmental stress. Agriculture—the human-directed evolution of crops—takes advantage of the plastic nature of plant genomes. The grass family has been under intensive human selection for 10,000 years and includes all of the staple crops upon which the world's food supply depends.

This laboratory uses the bioinformatics platform *DNA Subway* to explore the repetitive DNA content of three grass species: *Brachypodium distachyon* (purple false brome), *Oryza sativa* (rice), and *Zea mays* (corn). The highly conserved alcohol dehydrogenase gene *ADH1* serves as a road mark upon which to center the analysis of each genome. The ADH1 enzyme is present in a wide variety of animals, plants, bacteria, and fungi. In animals, the ADH1 enzyme catalyzes the oxidization of ethanol to acetaldehyde. This is the first step in the breakdown of alcohol (which is toxic to cells). ADH1 plays an essentially opposite role in bacteria, yeast, and plants, where it reduces acetaldehyde to ethanol during the final step of fermentation. Increased ADH1 expression is part of a stress response to waterlogging, when fermentation helps the plant survive under anaerobic (low oxygen) conditions. Thus, in both plants and animals, ADH1 is important for surviving stressful conditions.

Analysis routines in *DNA Subway's* Red Line are used to determine and visualize the repetitive DNA and transposon content in the three different grass species. To compare related regions in the three genomes, 100-kb sequence segments were selected that center on each species' *ADH1* gene. RepeatMasker is an algorithm that allows the identification of repetitive DNA and transposons. The annotation editor Apollo is used to visually locate and compare the lengths of different repetitive DNA sequences.

FURTHER READING

Buckler ES, Thornsberry JM, Kresovich S. 2001. Molecular diversity, structure and domestication of grasses. *Genet Res* **77:** 213–218.

Bureau TE, Ronald PC, Wessler SR. 1996. A computer-based systematic survey reveals the predominance of small inverted-repeat elements in wild-type rice genes. *Proc Natl Acad Sci* **93:** 8524–8529.

Feuillet C, Keller B. 2002. Comparative genomics in the grass family: Molecular characterization of grass genome structure and evolution. *Ann Bot* **89:** 3–10.

Han Y, Burnette JM III, Wessler SR. 2009. TARGeT: A web-based pipeline for retrieving and characterizing gene and transposable element families from genomic sequences. *Nucleic Acids Res* **37:** e78. doi: 10.1093/nar/gkp295.

Kellogg EA, Bennetzen JL. 2004. The evolution of nuclear genome structure in seed plants. *Am J Bot* **91:** 1709–1725.

SanMiguel P, Vitte C. 2009. The LTR retrotransposons of maize. In *Handbook of maize: Genetics and genomics* (ed. Bennetzen JL, Hake S), Vol. II, pp. 307–326. Springer, New York.

PLANNING AND PREPARATION

The following table will help you to plan and integrate the different bioinformatics methods.

Experiment	Day	Time		Activity
I. Identify transposable elements in the *ADH1* region of the *Brachypodium* genome	1	10–60 min	Prelab:	Set up all student computers to run Java and to allow pop-up windows to open. Prepare Apollo by opening the browser to http://gfx.dnalc.org/files/apollo/test.
	2	10 min	Lab:	Analyze repeats and transposons in a 100-kd region of the *Brachypodium* genome using RepeatMasker and Apollo.
II. Identify transposable elements in the *ADH1* region of the rice genome	2	10 min	Lab:	Analyze repeats and transposons in a 100-kb region of the rice genome using RepeatMasker and Apollo.
III. Identify transposable elements in the *ADH1* region of the corn genome	2	10 min	Lab:	Analyze repeats and transposons in a 100-kb region of the corn genome using RepeatMasker and Apollo.
IV. Compare the transposable element content in *Brachypodium*, rice, and corn	3	30 min	Lab:	Compare and contrast the amounts and types of transposable elements in three grass species.
V. Extend the analysis of transposons using other tools in *DNA Subway*	4	10 min	Lab:	Evaluate the relationships between genes and transposons. Search other genomes for transposons using *DNA Subway*.

BIOINFORMATICS METHODS

I. Identify Transposable Elements in the *ADH1* Region of the *Brachypodium* Genome

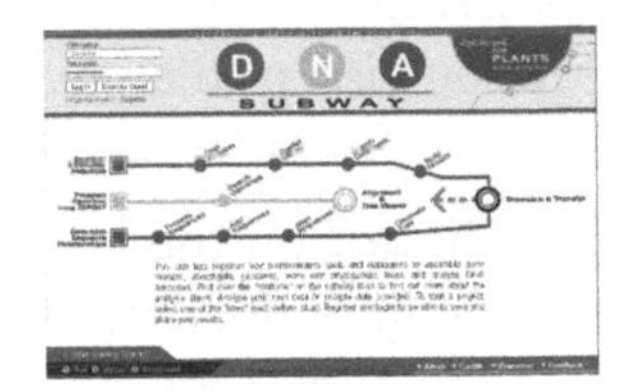

Like many other bioinformatics tools, DNA Subway works best on Mozilla/Firefox and a recent version of Java. For details, follow the "Manual" link at the bottom, left of the home page.

RepeatMasker searches for repeated DNA sequences—including transposons, short tandem repeats (STRs), and low-complexity regions (e.g., long sequences composed of a single nucleotide). The program masks these sequences from further analysis.

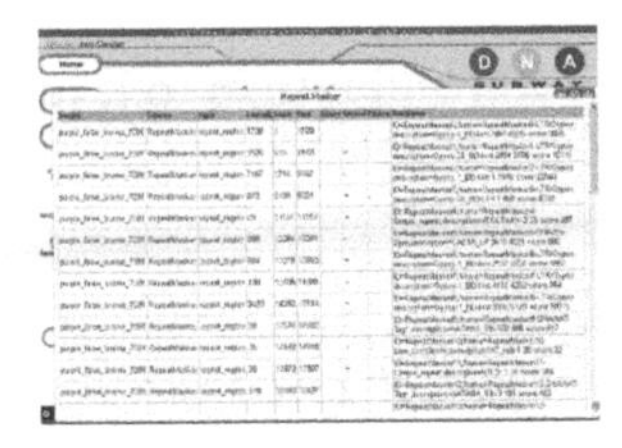

1. Open *DNA Subway* (http://www.dnasubway.org). If you are a registered user, log in with your username and password (only registered users can save and share work). Alternatively, enter as a guest to gain temporary access.
2. Click on the red square to open a new annotation project.
3. Click on "Select a sample sequence."
4. Click on "*Brachypodium distachyon* (purple false brome) Bd4, 100.00 kb" to highlight it.
5. Provide a project title (required), and click on "Continue."
6. Click on "RepeatMasker." The blinking "R" bullet indicates that the program is running. The bullet changes to "V" when results are ready to view.
7. Once the program has finished, click on "RepeatMasker" again to open a new window with a listing of repetitive DNA sequences that have been identified and masked.
   i. The "Attributes" column lists the types of repeats immediately after the dash/hyphen. Specific descriptions are included after the equal sign. "Low_complexity" and "Simple_repeat" refer to specific types of repetitive DNA; descriptions of these repeated sequences are also provided [e.g., "AT_rich" or "(CGG)n"]. All other attributes relate to transposable elements and remnants thereof.
   ii. Types beginning with "DNA" refer to DNA transposons, such as "DNA/TcMar-Stowaway," "DNA/MuDR," "DNA/Tourist," "DNA/En-Spm," and "DNA/hAT-AC." "DNA/hAT-AC" is interesting because it is a member of the same class of transposons that includes *Activator* (*Ac*), the first transposable element discovered.
   iii. Types that don't begin with "DNA" generally signify retroposons, such as "DITTO," "LTR/Copia," "LTR/Gypsy," "LINE/L1," and "SINE."
   iv. Use a search engine to learn more regarding the different repeat types.
8. Gather and analyze your results.
   i. Copy the results table and paste it into a spreadsheet editor.
   ii. How many different repeat types can you identify?
   iii. Group all rows that list repetitive DNAs that are not transposons, and add their "Lengths" to calculate the total amount of DNA occupied by these repeats. Record your result.
   iv. Group all rows that list DNA transposons and retroposons, and add their "Lengths" to calculate the total amount of DNA occupied by transposons. Record your result.
   v. Determine which category of repetitive DNA occupies a larger percentage of DNA in this sample: non-transposon DNA or transposon DNA.

vi. Close the Table view to return to *DNA Subway*.

Before launching Apollo, make sure that your browser allows pop-up windows. Be patient while Apollo launches via Java Web Start. The initial launch of each editing session typically takes a minute or two and may require you to permit the launch via dialogue boxes.

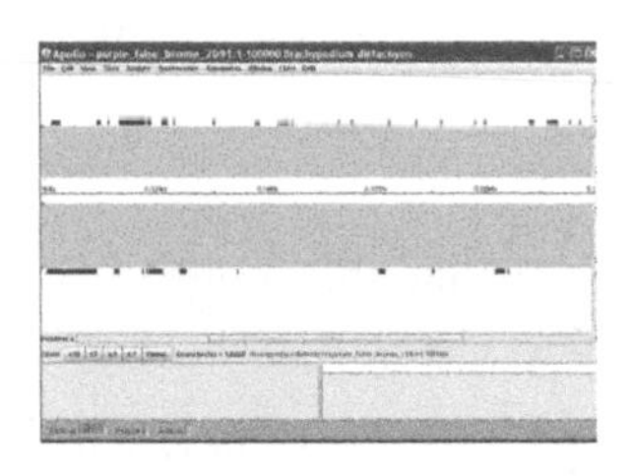

9. Click on "Apollo" to view and analyze the results in a graphical interface.
   i. The tick marks and boxes in the white spaces represent repetitive segments on the chromosome. Roll over and/or click on the tick marks and boxes to identify the nature of the repeats. (This information will be displayed at the bottom of the window.)
   ii. Simple repeats and low-complexity segments are only indicated on the forward strand (upper portion of screen). This saves space because the repetitive sequence on one strand is always complemented by a repetitive sequence on the other strand.
   iii. Transposable elements carry genes that enable them to move. Transposable elements therefore have directionality—and are displayed on the appropriate strand only.
10. Return to *DNA Subway*. (Do not close the Apollo window; you will return to it in Part IV.)

II. Identify Transposable Elements in the *ADH1* Region of the Rice Genome

1. In *DNA Subway*, click on "Home."
2. Repeat Steps 2–6 in Part I above using "*Oryza sativa* (rice) Chr11, 100.00 kb" as the sample sequence.
3. Analyze the RepeatMasker output as described in Steps 7 and 8 of Part I.
4. Browse the results in Apollo as described in Step 9 of Part I.
5. Return to *DNA Subway*. (Do not close the Apollo window; you will return to it in Part IV.)

III. Identify Transposable Elements in the *ADH1* Region of the Corn Genome

1. In *DNA Subway*, click on "Home."
2. Repeat Steps 2–6 in Part I above using "*Zea mays* (corn) 1, 100.00 kb" as the sample sequence.
3. Analyze the RepeatMasker output as described in Steps 7 and 8 of Part I.
4. Browse the results in Apollo as described in Step 9 of Part I.
5. Return to *DNA Subway*. (Do not close the Apollo window; you will return to it in Part IV.)

IV. Compare the Transposable Element Content in *Brachypodium*, Rice, and Corn

1. Arrange the three Apollo windows so that you can see them all at once.
   i. Bring up the Apollo window showing repeats in *Brachypodium* DNA, and resize it to 25% of the height of your monitor.

   ii. Repeat Step 1.i for the other two Apollo windows (rice and corn).

   iii. Position the three Apollo screens in a column, one on top of another.

2. Compare the different types of repeats found in each species by filling in the table below.

		Simple repeats/ low-complexity DNA	Transposons
Bd4	Number		
	Total length (bp)		
	Percentage of DNA occupied		
	Length range (bp)		
	Average length (bp)		
	Median length (bp)		
Os11	Number		
	Total length (bp)		
	Percentage of DNA occupied		
	Length range (bp)		
	Average length (bp)		
	Median length (bp)		
Zm11	Number		
	Total length (bp)		
	Percentage of DNA occupied		
	Length range (bp)		
	Average length (bp)		
	Median length (bp)		

3. Analyze the amounts of transposons in the three species.

   i. Which species contains the greatest number of transposons? First, answer the question by considering the total number of transposons in each sample. Second, answer the question by considering the percentage of DNA occupied by transposable elements in each sample.

   ii. Do you think that higher transposon content would lead to gene loss or would cause genes to be spaced farther apart?

4. Explain why transposable elements, especially those of the same type and with similar descriptions, have a wide range in length.

5. Compare the types of transposable elements across the three species.

   i. Which transposons appear to be present in all three species?

   ii. What would the presence of the same transposon in all three species indicate for the evolutionary history of these transposons: Did they emerge before or after the three species diverged from each other?

   iii. Which transposons appear to be present in only one species?

   iv. Why might you find transposable elements that appear to be species specific?

V. Extend the Analysis of Transposons Using Other Tools in *DNA Subway*

DNA Subway provides three different programs to predict protein-coding genes (Augustus, FGenesH, and SNAP); a fourth program, tRNAscan, searches DNA for the hallmarks of tRNAs. Differences in the ways the three gene prediction algorithms weigh gene evidence often lead to slightly different predictions. *DNA Subway* also uses the Basic Local Alignment Search Tool (BLAST) algorithms, which are used to search sequence databases such as GenBank to identify matches to DNA (BLASTN) and protein (BLASTX) query sequences.

1. In *DNA Subway*, apply the gene prediction algorithms (Augustus, FGenesH, and SNAP) and the tRNAscan algorithm, as well as the BLASTN and BLASTX database search algorithms, to all three plant sequences. Open the local browser, and observe how transposable elements and genes are intertwined. (See Laboratory 1.1 for guidance.)
   i. Can you identify the transposase and reverse transcriptase genes in the DNA transposons and retroposons, respectively? Why or why not?
   ii. Are transposons located within genes? If so, are they more common in exons or in introns?
   iii. Do genes outnumber transposons, or do transposons outnumber genes?
2. Use the "Genome Prospecting" button to transfer a transposon from the Red Line to the Yellow Line, and try to find homologs in the same species as well as in other species. What do you observe?

LABORATORY 1.5

Identifying *GAI* Gene Family Members in Plants

▼ OBJECTIVES

This laboratory demonstrates several important concepts of modern biology. During the course of this laboratory, you will

- Study homologous genes from different plant species and explore their semi-dwarf mutants.
- Compare homologous protein sequences and measure their similarity.
- Align wild-type and mutant sequences to identify mutations.

In addition, this laboratory uses several bioinformatics methods in biological research. You will

- Use the Basic Local Alignment Search Tool (BLAST) to identify members of a gene family in plant genomes.
- Use MUSCLE to align multiple protein or DNA sequences.
- Use PHYLIP to generate phylogenetic trees.
- Use Jalview to view multiple sequence alignments and phylogenetic trees, and to calculate the similarity between sequences.
- Use Tree Analysis of Related Genes and Transposons (TARGeT) to identify gene homologs in other plant genomes.

INTRODUCTION

In the first half of the 20th century, plant breeders generated semi-dwarf varieties of wheat, maize, and rice. Because they don't require nutrients and energy to grow tall, semi-dwarf plants can invest more in seeds, and the stalks of semi-dwarf plants with heavy seed loads break less often than those of taller plants. These varieties therefore spurred the Green Revolution by increasing crop yields. However, although highly desirable for other crops, identifying dwarf varieties by conventional breeding remains a difficult, time-consuming task.

The plant hormone gibberellic acid (GA) is known to stimulate plant growth by activating cell division and cell elongation. The differences in stem length between short and tall varieties are frequently caused by genomic alterations that perturb GA signaling pathways, affecting characteristics such as the timing and amount of endogenous GA production, GA delivery to target cells, and/or the function of GA receptors.

In 1985, Maarten Koornneef and colleagues generated a semi-dwarf strain of *Arabidopsis thaliana*. This mutant could not be rescued by applying GA exogenously, indicating that the mutation did not affect GA synthesis or persistence but had altered GA sensitivity. In 1997, Jin Rong Peng and colleagues isolated the gene that, when mutated, causes insensitivity to GA and stunted growth. This gene was called *gibberellic acid insensitive* (*GAI*); the mutant gene and protein were called *gai* and gai, respectively. The mutated allele appears to be semi-dominant: GA insensitivity occurs in *GAI/gai* heterozygotes, but *gai/gai* homozygotes are more severely stunted.

Arabidopsis GAI was used to isolate homologous genes from wild-type and semi-dwarf wheat, corn, and rice varieties. Subsequent analysis revealed that the early Green Revolution varieties of wheat and corn also carry altered *GAI* genes (*Rht-D1b* and *D8-1*, respectively) that are insensitive to GA. In contrast to these GA-insensitive varieties of wheat and corn, the semi-dwarf rice was shown to grow to regular height when plants were supplemented with GA. Thus, the mutation in rice differed from that in wheat and corn in that it had not affected GA sensitivity but GA biosynthesis and thereby diminished GA production in the rice mutant line.

Therefore, although it was not known at the time, mutations in the corn and wheat *GAI* genes facilitated the Green Revolution. It is now possible to engineer new semi-dwarf crop varieties by producing genetically engineered plants that overexpress the gai protein (e.g., gai overexpression produces dwarf apple trees). This promises to help propel a new phase of the Green Revolution.

In this laboratory, GAI protein sequences from five species—*Arabidopsis*, barley (*Hordeum vulgare*), corn (*Zea mays*), rice (*Oryza sativa*), and wheat (*Triticum aestivum*)—are compared. Then the specific alteration that causes the semi-dwarf phenotype is identified by comparing wild-type GAI and mutant gai protein sequences. This lab uses the MUSCLE analysis routine in *DNA Subway's* Blue Line to determine the differences in GAI DNA and protein sequences from different grass species. The algorithm MUSCLE aligns DNA and amino acid sequences to each other. Jalview is a tool to visualize and examine sequence alignments.

FURTHER READING

Bolle C. 2004. The role of GRAS proteins in plant signal transduction and development. *Planta* **218:** 683–692.

Han Y, Burnette JM III, Wessler SR. 2009. TARGeT: A web-based pipeline for retrieving and characterizing gene and transposable element families from genomic sequences. *Nucleic Acids Res* **37:** e78. doi: 10.1093/nar/gkp295.

Hedden P. 2003. The genes of the Green Revolution. *Trends Genet* **19:** 5–9.

Koornneef M, Elgerma A, Hanhart CJ, van Loenen-Martinet EP, van Rijn F, Zeevaart JAD. 1985. A gibberellin insensitive mutant of *Arabidopsis thaliana. Physiol Plant* **65:** 33–39.

Peng J, Harberd NP. 1993. Derivative alleles of the *Arabidopsis* gibberellin-insensitive (*gai*) mutation confer a wild-type phenotype. *Plant Cell* **5:** 351–360.

Peng J, Carol P, Richards DE, King KE, Cowling RJ, Murphy PM, Harberd NP. 1997. The *Arabidopsis GAI* gene defines a signaling pathway that negatively regulates gibberellin responses. *Gene Dev* **11:** 3193–3205.

Peng J, Richards D, Hartley NM, Murphy GP, Devos KM, Flintham JE, Beales J, Fish LJ, Worland AJ, Pelica F, et al. 1999. 'Green revolution' genes encode mutant gibberellin response modulators. *Nature* **400:** 256–261.

Willige BC, Ghosh S, Nill C, Zourelidou M, Dohmann EMN, Maier A, Schwechheimer C. 2007. The DELLA domain of GA INSENSITIVE mediates the interaction with the GA INSENSITIVE DWARF1A gibberellin receptor of *Arabidopsis. Plant Cell* **19:** 1209–1220.

PLANNING AND PREPARATION

The following table will help you to plan and integrate the different bioinformatics methods.

Experiment	Day	Time		Activity
I. Determine how GAI proteins from different plant species are related	1	10–60 min	Prelab:	Set up all student computers to run Java and allow pop-up windows to open.
	2	45–60 min	Lab:	Analyze the levels of conservation and degrees of relatedness among five GAI protein sequences from different plant species using MUSCLE, PHYLIP, and Jalview.
II. Identify the cause of the gai semi-dwarf phenotype	3	45–60 min	Lab:	Align wild-type GAI and mutant gai proteins from three plant species, and identify the mutations using MUSCLE. Explore what happens if a mutant *Arabidopsis gai* gene is transferred into a wild-type rice plant by working through an animation in *DNA Subway*.
III. Think about the next wave of the green revolution	4	30 min	Lab:	Identify and compare GAI proteins in plant species using TARGeT. Devise a strategy to generate new semi-dwarf crop varieties.
IV. Extend the analysis of GAI sequences	5	45 min	Lab:	Analyze relationships between *GAI* and *gai* genes and their proteins in diverse plant species.

BIOINFORMATICS METHODS

I. Determine How GAI Proteins from Different Plant Species Are Related

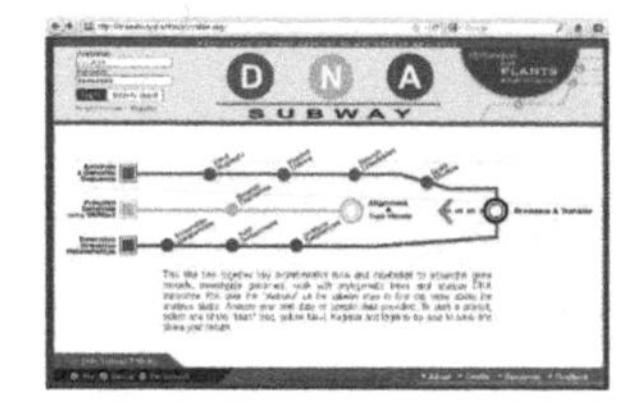

Like many other bioinformatics tools, DNA Subway works best on Mozilla/Firefox and a recent version of Java. For details, follow the "Manual" link at the bottom, left of the homepage.

1. Open *DNA Subway* (http://www.dnasubway.org). If you are a registered user, log in with your username and password (only registered users can save and share work). Alternatively, enter as a guest to gain temporary access.
2. Click on the blue square to generate a new phylogenetic project.
3. Under "Select Project Type" check "Protein."
4. Under "Select a set of sample sequences," click on "GAI Proteins."
5. Provide a project title (required), and click on "Continue."
6. Align the nonmutated protein sequences.
   i. Click on "Select Data."
   ii. Check only sequences whose names contain "GAI" (for wild-type proteins), not "gai."
   iii. Click on "Save Selection" to save your selection.
   iv. Click on "MUSCLE" to align the sequences.

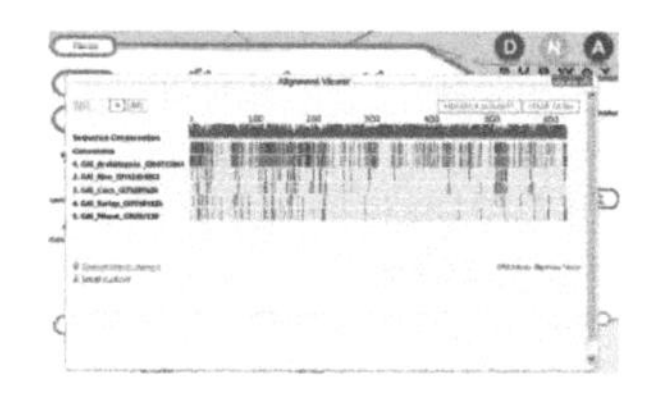

7. Click on "MUSCLE" again to view the alignment.
8. Maximize the screen.
   i. The viewer displays an alignment of the five input sequences, with identical or similar sequences aligned against one another. Dashes have been inserted in places where a sequence does not contain an amino acid that aligns with amino acids in other sequences. The color code indicates the physical or chemical properties of each amino acid; for a key, click on the "Color Codes" button in the upper right-hand corner. To identify the specific amino acids in each position, click on the "SEQ" button in the upper left-hand corner or click sequentially on the "+" button to zoom in until you can discern the single letters that indicate amino acids. (For a key, click on the "COLOR CODES" button in the upper right-hand corner and then click on "View Amino Acids Code.") The "Consensus" row displays the most prevalent amino acid in each position of the multiple sequence alignment. The "Sequence Conservation" bar visually indicates the ratio at which the most prevalent amino acid in each position occurs; a roll-over function reveals the approximate similarity percentage.

   ii. Proteins that share 35% or more amino acid sequence similarity are likely to be related to one another. For a quick estimate of how related the sequences are to one another scroll over the "Sequence Conservation" bar for the entire sequence and observe the similarity values. Do you find evidence that the GAI sequences are related? What is the highest percentage of similarity that you find? What is the lowest percentage of similarity that you find?
   iii. To examine the question of relatedness among the GAI proteins in more detail, click on "Sequence Similarity" in the upper right-hand corner. The pop-up displays the similarity between pairs of GAI proteins as well as

between each of the GAI proteins and the "Consensus" sequence. What is the highest percentage of similarity that you can detect? Between what proteins? What is the lowest percentage of similarity that you can detect? Between what proteins? Use the values for the pairwise similarity between the GAI proteins to discern how closely they are related.

iv. Close the Sequence Similarity matrix by clicking the "SEQUENCE SIMILARITY" button.

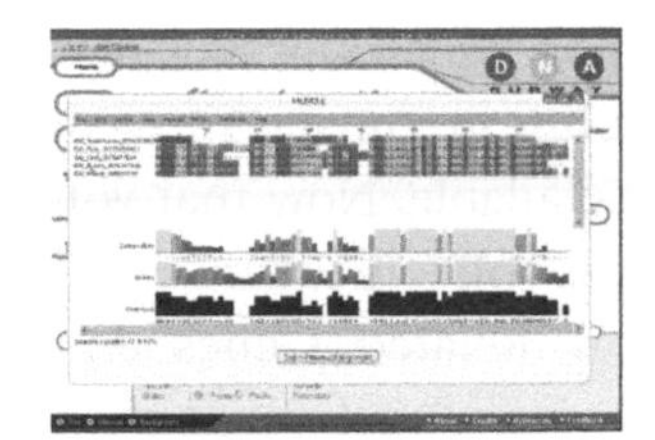

9. GAI proteins have been ascribed to a protein class named "DELLA" proteins. Use the "+" and "–" buttons in the upper left-hand corner to adjust your sequence so that you can read the first 50 amino acids. Scan the alignment from left to right until you reach a region that contains five consecutive amino acids that are identical for each protein. (Identify this region by scrolling along the "Sequence Conservation" bar until it displays "100%" for five consecutive positions.) What is the one-letter "Consensus" amino acid sequence for these five positions?
10. Click on "Launch Jalview" to display the alignment in an alternative alignment viewer. Identify the "DELLA" region in this view.

11. Close Jalview and the *DNA Subway* Alignment Viewer.
12. Click on "PHYLIP NJ" to generate a phylogenetic tree.
13. Once the program has finished, click on "PHYLIP NJ" again to view the tree. Describe the phylogenetic relationships among the GAI proteins of the different plants.
14. Close the phylogenetic tree window.

II. Identify the Cause of the gai Semi-Dwarf Phenotype

1. The Blue Line includes the protein and DNA sequences for wild-type (GAI) and semi-dwarf (gai) mutants for *Arabidopsis*, wheat, and corn. Using the Select Data button and MUSCLE, conduct pairwise alignments between the mutated (gai) and nonmutated (GAI) protein sequences for each of the three species. Once completed, create a new project, select a DNA project, and repeat the pairwise alignments using the "GAI Genes" found in the sample sequences.
2. Examine each alignment and determine the effect of the mutation on the DNA and amino acid sequences. What part of the GAI gene and protein has been affected in each of the mutants? Do the three gai mutants have anything in common?
3. View the following animation.
   i. In *DNA Subway*, click on "Home."
   ii. Click on "Background" on the bottom.
   iii. Select "Evidence."
   iv. Click on "Homology."
   v. Using the red arrows at the bottom of the window, work through the animation to gain an understanding of the relationship between mutant gai and wild-type GAI.
4. Mutations can lead to the loss or gain of functions. The last section in the animation in Step 3 summarizes an experiment whereby scientists transformed the *gai*

gene from *Arabidopsis* into rice. Explain how this experiment showed that the gibberellin-insensitive phenotype is based on a gain of function.

5. Wheat is a hexaploid plant, and corn has more than two copies of each chromosome as well. Therefore, both wheat and corn have multiple copies of wild-type *GAI*. Explain how the gain-of-function character of the mutated *gai* gene has facilitated rapid generation of semi-dwarf corn and wheat varieties, despite the presence of multiple copies of *GAI* in these genomes.

III. Think About the Next Wave of the Green Revolution

The role of DELLA proteins in the generation of semi-dwarf mutations was not understood until long after the first wave of the Green Revolution, during most of which the technology to sequence genes or genomes was unavailable. Now that you know about the protein changes underlying the GAI phenotype, you will identify GAI homologs in other crop genomes using Tree Analysis of Related Genes and Transposons (TARGeT) and then devise a plan to generate semi-dwarf varieties of them that circumvents the long breeding process.

Because genomes are composed of hundreds of millions of nucleotides, the search time increases dramatically if you select several genomes at once.

1. Open *DNA Subway* (http://www.dnasubway.org).
2. Click on the yellow square to generate a genome prospecting project.
3. Click on "Select a sample sequence."
4. Click on "GAI Protein *Arabidopsis thaliana*/Mouse-ear cress" to highlight it.
5. Provide a project title (required), and click on "Continue."
6. Check the genome of a crop plant for which the availability of a semi-dwarf variety might be beneficial (e.g., castor bean, *Ricinus communis*).
7. Click on "Run" to start the search using the TARGeT program.
8. Wait for results to be returned. If an analysis fails the first time, submit it again.
9. Evaluate the matches that TARGeT identified.
   i. Click on "Alignment Viewer" to open the program Jalview.
   ii. Examine the multiple alignment and determine how many matches TARGeT identified.
   iii. Identify any sequences that carry the DELLA motif; these are candidate GAI homologs.
10. Devise a strategy that would use the knowledge regarding potential GAI homologs in your target crop plant to generate a semi-dwarf variety of this plant.

IV. Extend the Analysis of GAI Sequences

1. Use the *Arabidopsis* GAI protein sequence to prospect the rice genome for GAI. How many matches can you identify? How many of these have a DELLA region?
2. How would you expect the results to differ if you used a GAI protein sequence from a species that is more closely related to rice? After predicting what the outcome might be, conduct the experiment using other cereal GAI proteins.

3. How would you expect the results to differ if you would prospect the poplar genome? How about grape? The moss *Physcomitrella*? The green algae *Chlamy-domonas*? (Would you expect green algae to have homologs to a gene that represses stem elongation in higher plants?) Make your predictions, and then conduct the experiments using *DNA Subway*. What might it mean if GAI is present in both trees and mosses—what do trees have in common with mosses, and how are they different?
4. How would you expect the results to differ if you prospected the genomes using the gene sequences instead of the protein sequences? Which would you expect to be more conserved to ensure an equivalent protein function—the genes or the proteins? Think it through, and then conduct the experiments using *DNA Subway*.
5. Examine the *GAI/gai* genes in *DNA Subway*'s Blue Line. Compare the homology among the GAI genes and proteins from the five different species. Why do the gene sequences show different degrees of conservation than those of the protein sequences? Then, use MUSCLE to identify the mutations in the *gai* genes. Compare and contrast the *gai* gene mutation for each species with the protein alteration for the same species.

LABORATORY 1.6 Discovering Evidence for Pseudogene Function

▼ OBJECTIVES

This laboratory demonstrates several important concepts of modern biology. During the course of this laboratory, you will

- Identify the locations of a gene and a related pseudogene in the human genome.
- Compare the structures of a gene and a pseudogene.
- Discover functionality in noncoding gene regions.

In addition, this laboratory uses several bioinformatics methods in biological research. You will

- Use the Basic Local Alignment Search Tool (BLAST) to identify sequences related to a gene in the human genome.
- Use Map Viewer to visualize BLAST results on chromosomes.
- Use *Gene Boy* to evaluate reading frames in sequences.
- Use TargetScan to detect microRNA (miRNA) target sites in genes.
- Use CLUSTAL W to align sequences and assess sequence conservation.

INTRODUCTION

Eukaryotic genomes contain large numbers of gene-like sequences called "pseudogenes," disabled copies of genes that do not produce proteins. They often lack recognizable promoter sequences, include premature stop codons, lack a start codon, or contain frame-shift-causing deletions or insertions. However, the nucleotide sequences in many pseudogenes are highly conserved, indicating that these sequences are important.

Most pseudogenes are either duplicated or retrotransposed versions of their parent genes. Retrotransposed pseudogenes arise when a messenger RNA (mRNA) is reverse-transcribed into DNA, and the DNA copy is then inserted into the genome. Because of this, they are often called processed pseudogenes and do not include the introns from the ancestral gene. In contrast, duplicated or nonprocessed pseudogenes arise via gene duplication and often carry remnants of the ancestral gene's introns.

The number of pseudogenes in the human genome is still being debated but may equal or even surpass the number of functional genes. Pseudogenes were originally thought to be inactive evolutionary fossils and were believed to be transcriptionally silent. However, recent studies have shown that at least 5% of known pseudogenes are transcribed. In some cases, a function for the pseudogene transcript has been identified. For example, it may affect mRNA stability or regulate the transcription or translation of another gene—usually the ancestral gene from which it was derived. But in most cases, the reason why a pseudogene is transcribed is often unclear.

The phosphatase and tensin homolog gene, *PTEN*, is a tumor-suppressor gene. The *PTEN*-encoded protein, a phosphatase, antagonizes two cell signaling pathways involved in tumorigenesis, the mitogen-activated protein kinase (MAPK) pathway and the P13 signaling pathway, thereby blocking cell cycle progression in G_1, the growth phase before DNA synthesis. Cancer cells are frequently found to only have one functional PTEN allele, strongly suggesting that the dose of the *PTEN*-encoded phosphatase is crucial to the normal regulation of cell division. Recently, the identification and validation of numerous *PTEN*-targeting microRNAs (miRNAs), small endogenously expressed regulatory RNAs, confirmed that posttranscriptional regulation plays a pivotal part in determining PTEN abundance in cells. The expression levels of miRNAs that regulate *PTEN* change dramatically in some cancers and affect cancer cell division, migration, and invasion, highlighting their importance.

PTEN has a pseudogene, *PTENP1*, which is more highly transcribed than *PTEN* itself. In this laboratory, the sequences of *PTEN* and its pseudogene *PTENP1* are compared, and a potential function for the pseudogene is identified—all while exploring the newly discovered and surprisingly rich biology of RNA regulation within cells.

This lab uses a variety of bioinformatics tools from the National Center for Biotechnology Information, the DNA Learning Center, and the Massachusetts Institute of Technology. NCBI's Map Viewer is used to locate and visualize *PTEN* and *PTENP1* in the human genome. A simple sequence analysis platform in *Gene Boy* is then used to translate DNA sequence into its corresponding amino acid sequence. The resulting sequences are compared using the Blue Line in *DNA Subway* to explore the function of the pseudogene. Finally, potential regulatory sequences in *PTEN* are identified using TargetScan, a bioinformatics tool that searches for sequences that are targets of short regulatory RNAs called microRNAs (miRNAs). Sequence alignments using the alignment program CLUSTAL W built into *Sequence Server* then refine a function for the pseudogene.

FURTHER READING

Furnari F. 2010. The cancer connection. *Nature* **465:** 1016.

Harrison PM, Zheng D, Zhang Z, Carriero N, Gerstein M. 2005. Transcribed processed pseudogenes in the human genome: An intermediate form of expressed retrosequence lacking protein-coding ability. *Nucleic Acids Res* **33:** 2374–2383.

Poliseno L, Salmean L, Zhang J, Carver B, Haveman W, Pandolfi PP. 2010. A coding-independent function of gene and pseudogene mRNAs regulates tumour biology. *Nature* **465:** 1033–1040.

Rigoutsos I. 2010. Pseudogenes as regulators. *Nature* **465:** 1017.

Zhang Z, Harrison P, Gerstein M. 2003. Millions of years of evolution preserved: A comprehensive catalog of the processed pseudogenes in the human genome. *Genome Res* **13:** 2541–2558.

Zheng D, Frankish A, Baertsch R, Kapranov P, Reymond A, Choo SW, Lu Y, Denoeud F, Antonarakis SE, Snyder M, et al. 2007. Pseudogenes in the ENCODE regions: Consensus annotation, analysis of transcription, and evolution. *Genome Res* **17:** 839–851.

PLANNING AND PREPARATION

The following table will help you to plan and integrate the different bioinformatics methods.

Experiment	Day	Time		Activity
I. Use Map Viewer to locate the *PTEN* gene and its pseudogene, *PTENP1*, in the human genome	1	30 min	Lab:	Locate *PTEN* and *PTENP1* in the human genome using keyword searches in Map Viewer. Determine the size and structure of *PTEN* and *PTENP1*.
II. Use Map Viewer to determine the relationship between *PTEN* and *PTENP1*	1	15 min	Lab:	Use a *PTENP1* sequence stored in *Sequence Server* to perform a BLAST search of the human genome in Map Viewer. Develop a hypothesis stating how *PTENP1* may have been derived from *PTEN*.
III. Use *Gene Boy* to determine the protein-coding capacity of the *PTENP1* pseudogene	2	30 min	Lab:	Translate the *PTEN* and *PTENP1* sequences using *Gene Boy*. Compare the *PTEN* and *PTENP1* amino acid sequences using MUSCLE to determine whether *PTENP1* encodes a protein.
IV. Use TargetScan and CLUSTAL W to identify a potential function for the *PTENP1* pseudogene	3	30–45 min	Lab:	Identify miRNA-binding sites in the *PTEN* 3′ UTR using TargetScan. Determine whether *PTEN* miRNA-binding sites are conserved in *PTENP1* using CLUSTAL W.
V. Extend the analysis of pseudogene sequences	4	45–120 min	Lab:	Evaluate other regions of *PTEN* and *PTENP1*. Compare other gene/pseudogene pairs.

BIOINFORMATICS METHODS

I. Use Map Viewer to Locate the *PTEN* Gene and Its Pseudogene, *PTENP1*, in the Human Genome

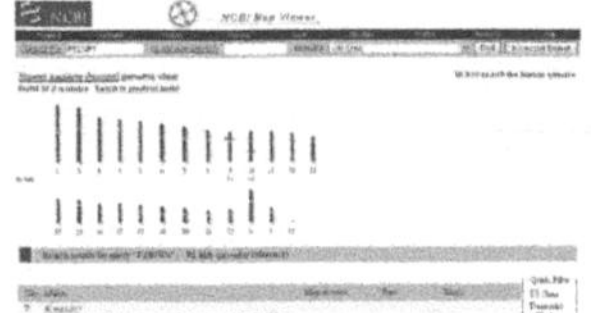
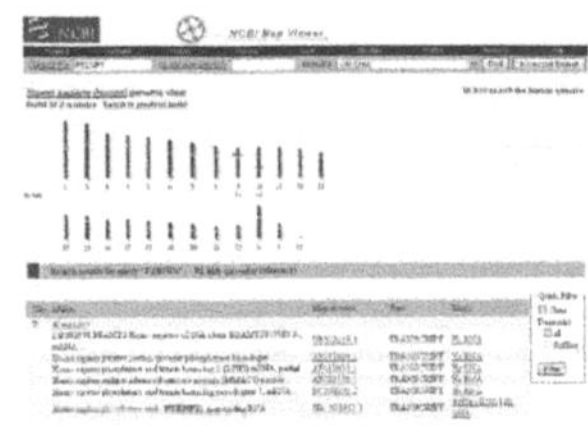

1. Search the human genome for *PTENP1*.

   i. Open Map Viewer (http://www.ncbi.nlm.nih.gov/mapview).

   ii. Find *Homo sapiens* in the table to the right, and click on the most recent version ("Build") of the human genome.

   iii. Enter **PTENP1** into the "Search for" field, change the "assembly" to "reference," and click on "Find." Small horizontal bars on chromosomes indicate the positions of matches.

2. *PTENP1* maps to the short arm of chromosome 9. View and analyze *PTENP1*.

   i. Right-click on the number "9" underneath chromosome 9 to open it in a new window. Do not close the window showing the matches to *PTENP1* in the human genome; you will return to it in Step 3.

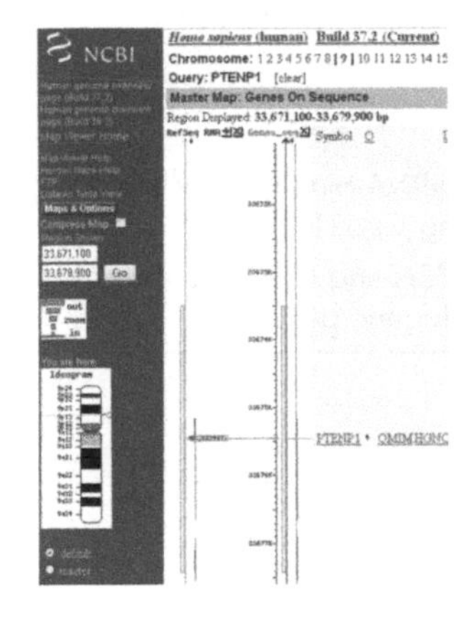

   ii. Click on "Maps & Options" on the left. Under "Tracks Displayed," remove all tracks except "RefSeq Transcripts" and "Gene" by clicking on the gray box with the minus sign to the right of each unwanted track. Click and drag the "Gene" track to the bottom of the list. Click on the gray "R" to the left of the "Gene" track so that it turns blue. Click "OK."

   iii. Zoom to a level that allows you to view *PTENP1* comfortably. First use the zoom toggle on the left, and then refine your position on the chromosome by clicking on the thin vertical line under "Genes_seq" and choosing a specific zoom level. You may need to choose "Recenter" several times as you zoom in.

   iv. Determine the size of *PTENP1* using the map coordinates to the left of the "Genes_seq" map.

   v. Introns and noncoding sequences are denoted by a thin line under "Genes_seq," whereas exons are denoted by thick bars along the line (filled bars are coding sequences; open bars are untranslated regions). Determine the structure of *PTENP1*. How many introns and exons does *PTENP1* have?

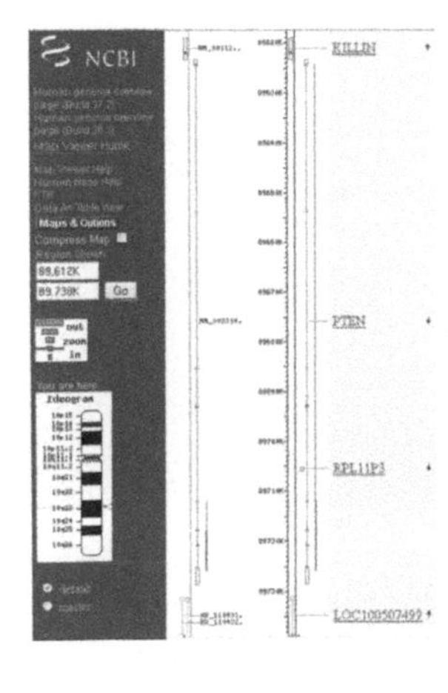

3. Return to the window showing the matches to *PTENP1* in the human genome. *PTEN* maps to chromosome 10. View and analyze the *PTEN* gene.

   i. Click on the number "10" underneath chromosome 10.

   ii. Arrange the window to display only the "RefSeq Transcripts" and "Gene" tracks as described in Step 2.ii.

   iii. Zoom to a level that allows you to view *PTEN* comfortably as described in Step 2.iii.

   iv. Determine the size of *PTEN* as described in Step 2.iv.

   v. Determine the structure of *PTEN* as described in Step 2.v. How many introns and exons does *PTEN* have?

II. Use Map Viewer to Determine the Relationship between *PTEN* and *PTENP1*

1. Obtain the *PTENP1* nucleotide sequence.

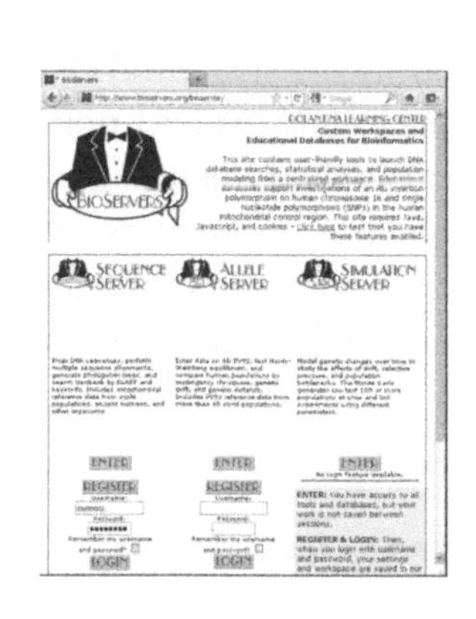

   i. Open the BioServers Internet site at the Dolan DNA Learning Center (http://www.bioservers.org).

   ii. Enter *Sequence Server*. You can register if you want to save your work for future reference, but it is not required.

   iii. The interface is simple to use: Add or obtain data using the top buttons and pull-down menus, and then work with the data in the work space below.

   iv. Click on "MANAGE GROUPS" at the top of the page.

   v. Select "Genome Science Labs" from the "Sequence sources" pull-down menu in the upper right-hand corner.

   vi. Click on the checkbox to select "Pseudogenes, PTENP1," and click "OK" to move this group into the work space.

   vii. Select "PTENP1 gene 3917nt GI:224589821" from the first pull-down menu.

   viii. Click on "OPEN." Highlight and copy the entire nucleotide sequence. Then paste it into a text document. Save this text document for use in Step 2.iv.

2. Align the *PTENP1* and *PTEN* sequences.

   i. Open Map Viewer (http://www.ncbi.nlm.nih.gov/mapview).

   ii. Find *Homo sapiens* in the table to the right, and click on the most recent version ("Build") of the human genome.

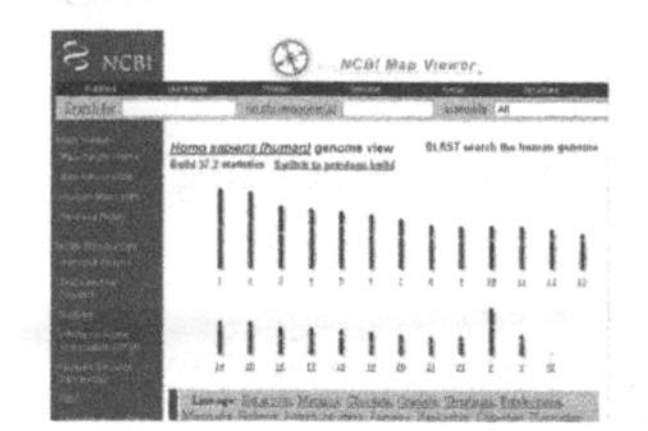

The MegaBLAST algorithm is optimized for comparing closely related sequences that are at least 95% identical.

   iii. Click on "BLAST search the human genome" at the upper right.

   iv. Paste the *PTENP1* sequence from Step 1.viii into the search window.

   v. Under "Choose Search Set," select the "Genome (reference only)" database from the drop-down menu.

   vi. Under "Program Selection," optimize for "Highly similar sequences (megablast)."

   vii. Click on "BLAST." This sends your query sequence to a server at NCBI in Bethesda, Maryland. There, the BLAST algorithm will attempt to match the sequence to the millions of DNA sequences stored in its database. While searching, a page showing the status of your search will be displayed until your results are available. This may take only a few seconds or more than 1 min if many other searches are queued at the server.

   viii. Scroll down to the list of "Sequences producing significant alignments," and identify the two hits with the highest "Total score." Record the numbers of the chromosomes upon which they are located.

   ix. Scroll up again and in "Other reports," click on the "[Human genome view]" link to see the chromosome locations of the BLAST hits.

   x. Click on the number "10" underneath chromosome 10.

   xi. Count the number of hits (in red).

xii. Click on the small blue arrow after the vertical line labeled "Genes_seq" to display the genes. Determine which gene the *PTENP1* query matched.

3. Analyze the relationship between *PTENP1* and *PTEN*.

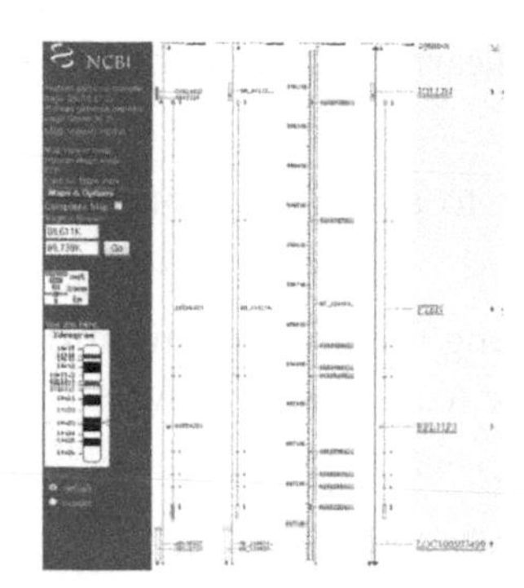

   i. Scroll along the *PTEN* gene in its entire length, and identify the structural elements in *PTEN* that match the *PTENP*1 query sequence. Introns and noncoding sequences are denoted by a thin line, whereas exons are denoted by thick bars along the line (filled bars are coding sequences; open bars are untranslated regions [UTRs]). The black arrow next to the gene symbol indicates the direction of transcription for the gene. Are matches located in exons, introns, or UTRs?

   ii. Is *PTENP1* a processed or a nonprocessed pseudogene? (*Hint:* Does it have any introns?)

   iii. Propose a mechanism by which *PTENP1* may have been derived from *PTEN*.

III. Use *Gene Boy* to Determine the Protein-Coding Capacity of the *PTENP1* Pseudogene

An open reading frame (ORF) is a stretch of triplets that begins with the start codon ATG (AUG in RNA) and ends with a stop codon; it could, therefore, potentially encode a protein. Many pseudogenes lack a start codon and/or contain premature stop codons.

1. Obtain the *PTEN* open reading frame (ORF) sequence.

   i. If you are not already in the *Sequence Server* of the BioServers Internet site, open it as described in Steps 1.i–1.vi of Part II.

   ii. Select "PTEN ORF 1212nt" from the first pull-down menu.

   iii. Click on "OPEN." Highlight and copy the entire nucleotide sequence. You will use this sequence in Step 2.iii.

2. Use *Gene Boy* to translate the *PTEN* ORF.

   i. Open a new browser and point it to *Gene Boy* (http://www.geneboy.org).

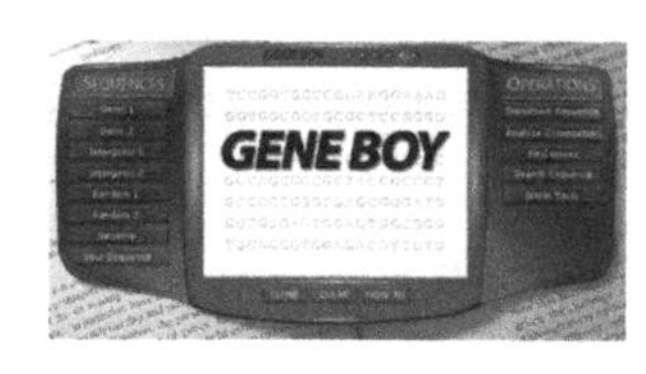

   ii. Click on "Your Sequence" in the "Sequences" panel at the left.

   iii. Paste the *PTEN* ORF sequence into the central "Your Sequence" window.

   iv. Click on "Save Sequence."

   v. Click on "Transform Sequence" in the "Operations" panel at the right.

   vi. Click on "Amino Acids." This will start an algorithm that translates the sequence in all three ORFs, triplet by triplet. The term "reading frame" is used when parsing a DNA sequence into triplets. Reading frame 1 (RF1) starts at the first nucleotide, RF2 at the second nucleotide, and RF3 at the third.

   vii. Examine the amino acid sequences. Determine whether the translated sequences in RF1, RF2, and RF3 begin with a methionine.

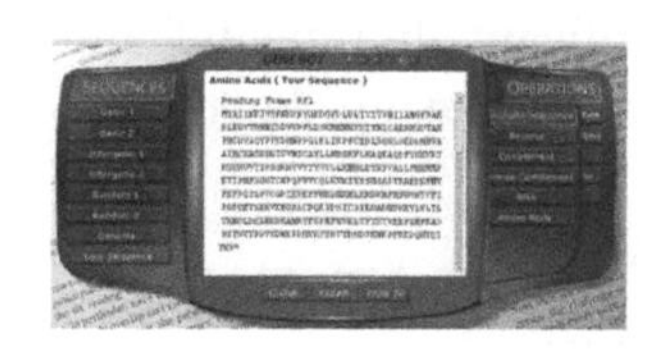

   viii. The program places an asterisk (*) each time a triplet consists of TAA, TAG, or TGA, the three stop codons. Determine whether the translated sequences in RF1, RF2, and RF3 encode a functional protein (not interrupted by a stop codon).

   ix. Highlight and copy the RF1 amino acid sequence (minus its stop codon, symbolized by an asterisk). Then paste it into a text document. Save this text document for use in Step 4.

3. Repeat the analysis described in Steps 1 and 2 with the "PTENP1 ORF 1212nt" sequence from *Sequence Server*. Compare your results with the *PTEN* results generated above. Save the RF1 amino acid sequence from *PTENP1* into the same text document that you created in Step 2.ix.

4. The amino acid sequences must be in FASTA format to be used in *DNA Subway* in Step 5. FASTA is a standard format for displaying sequence. Add one line of text to the beginning of each amino acid sequence to identify the sequence as either ">PTEN" or ">PTENP1." (The ">" at the beginning of the line is required; it tells the program to ignore the contents of this line. There should be no space between the ">" and the identifying name.)

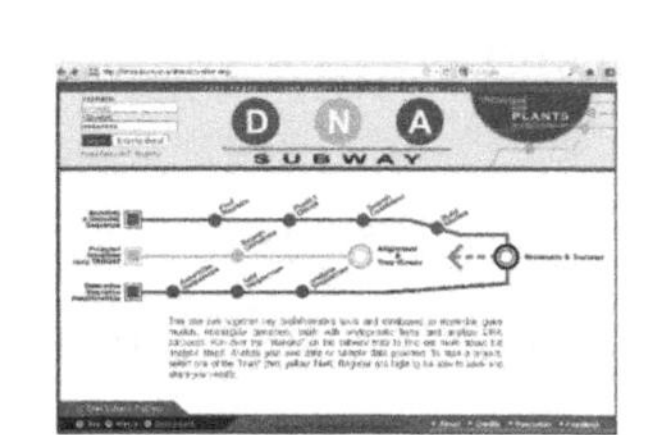

Like many other bioinformatics tools, DNA Subway works best on Mozilla/Firefox and a recent version of Java. For details, follow the "Manual" link at the bottom, left of the home page.

5. Identify differences between the two amino acid sequences using *DNA Subway*.
   i. Open *DNA Subway* (http://www.dnasubway.org). If you are a registered user, log in with your username and password (only registered users can save and share work). Alternatively, enter as a guest to gain temporary access.
   ii. Click on the blue square to start a new sequence alignment project.
   iii. Under "Select Project Type," check "Protein."
   iv. Check "Enter sequences in FASTA format."
   v. Enter the PTEN and PTENP1 amino acid sequences from Step 4.
   vi. Provide a title (required), and click on "Continue."
   vii. Click on "Select Data."
   viii. Check the two sequences.
   ix. Click on "Save Selection" to save your selection.
   x. Click on "MUSCLE" to align the sequences.
   xi. Click on "MUSCLE" again to view the alignment.
   xii. On top, you see an alignment of your input sequences. Identical or similar amino acids have been aligned to each other. Dashes have been inserted in places where a sequence does not contain an amino acid that aligns with an amino acid in the other sequence. The color code indicates the physical or chemical properties of each amino acid. (For a key, click on the "Color Codes" button.)
   xiii. The "Sequence Conservation" bar above the sequence alignment indicates relative levels of conservation at each site. Do the sequences share regions of strong similarity (>90%)?
   xiv. Use the "Consensus" bar to help you to identify amino acid differences in the two sequences. How many amino acid differences can you identify?
   xv. Close the Alignment Viewer window.

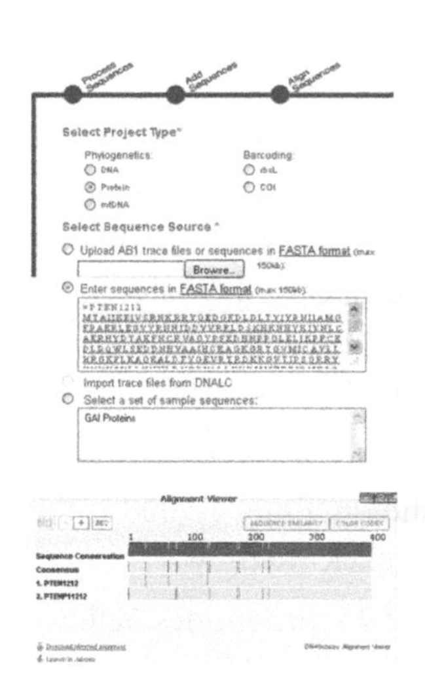

IV. Use TargetScan and CLUSTAL W to Identify a Potential Function for the *PTENP1* Pseudogene

Untranslated regions (UTRs) contain sequences that mediate posttranscriptional regulation by serving as targets for small RNAs that act through RNA interference

(RNAi). Specifically, microRNAs (miRNAs) can flag mRNAs for destruction or translational inhibition by binding to complementary sites in 3′ UTRs. Here, you will use TargetScan, an miRNA target site prediction tool, to identify potential miRNA target sequences in the 3′ UTRs of *PTEN* and *PTENP1*.

1. Identify miRNA-binding sites in *PTEN*.
   - i. Point a browser to TargetScan (http://www.targetscan.org).
   - ii. Under "Select a species," choose "Human."
   - iii. Enter the human gene symbol **PTEN**.
   - iv. Click on "Submit." TargetScan returns a map of the human *PTEN* 3′ UTR that shows binding sites for various miRNAs.
   - v. Count the number of miRNA-binding sites in the *PTEN* 3′ UTR.
2. Determine whether the *PTEN* miRNA-binding sites are conserved in *PTENP1*.

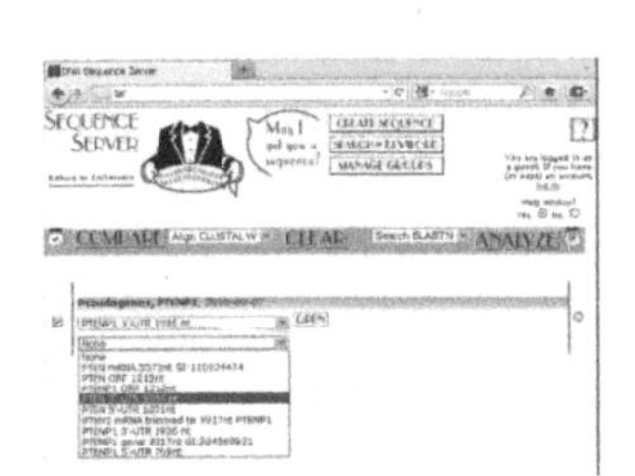

   - i. If you are not already in the *Sequence Server* of the BioServers Internet site, open it as described in Steps 1.i–1.vi of Part II.
   - ii. Select "PTENP1 3′-UTR 1936nt" from the first pull-down menu in the work space.
   - iii. Select "PTEN 3′-UTR 3304nt" from the second pull-down menu in the work space.
   - iv. Make sure that the boxes to the left of each sequence are checked.
   - v. Click on "COMPARE" in the gray bar. (The default operation is a sequence alignment using the CLUSTAL W algorithm.) The checked sequences are sent to a server at Cold Spring Harbor Laboratory, where the CLUSTAL W algorithm will attempt to align each nucleotide position. This may take only a few seconds or more than a minute if many other searches are queued at the server.
   - vi. The results will appear in a new window. Maximize this window.
   - vii. By default, only the first 500 bp of the alignment is shown. Enter **10,000** as the number of nucleotides to display per page, and then click "Redraw."
   - viii. Examine the sequences, which are displayed in rows of 25 nucleotides. Yellow-shaded dashes (–) indicate nucleotides that are present in one of the sequences but not in the other. Yellow highlighting also denotes mismatches between the sequences. A gray-shaded "N" indicates a sequence error, a position in one or both sequences where a nucleotide could not be determined.

   - ix. Arrange the TargetScan and the CLUSTAL W alignment windows side by side.
   - x. Click on the name of an miRNA in TargetScan to highlight its position and sequence in the gray-shaded box below the map.
   - xi. Navigate to the same position in the CLUSTAL W alignment.
   - xii. Determine whether *PTENP1* contains the same binding sequence as the *PTEN* sequence highlighted in the TargetScan box.
   - xiii. Repeat Steps 2.x–2.xii for all miRNA-binding sites shown on the TargetScan map. Record the names of the miRNAs for which the *PTEN* and *PTENP1* 3′-UTR sequences are conserved.

3. Assess the potential for *PTENP1* to serve a function. How does the level of sequence conservation in the 3′ UTR (as evaluated in Step 2) compare with the level of sequence conservation in the ORF (as evaluated in Step 5.xiii of Part III)? Do your findings indicate selective pressure toward maintaining a purpose for the pseudogene? Explain.

V. Extend the Analysis of Pseudogene Sequences

All sequences needed to conduct these analyses are stored in Sequence Server *and can be accessed as described in Steps 1.i–1.v of Part II.*

1. Use *Sequence Server* and/or *DNA Subway* to compare in detail various structural features of *PTEN* and *PTENP1* (e.g., 5′ UTRs, 3′ UTRs, and mRNA sequences).
2. Use Map Viewer to determine whether other gene/pseudogene combinations (*KRAS/KRAS1P, E2F3/E2F3P1, BRCA1/BRCA1P*, and *HBZ/HBZP*) are closely related.
3. Use *Gene Boy* to determine what renders the pseudogenes *KRAS1P, E2F3P1, BRCA1P*, and *HBZP* nonfunctional.
4. Use TargetScan and CLUSTAL W to identify any conserved miRNA-binding sites in *KRAS1P, E2F3P1*, and *BRCA1P*.
5. Align the *HBZP* and the *HBZ* ORFs in *Sequence Server*, and determine how *HBZP* has been generated from *HBZ*. How does this mechanism of pseudogene genesis compare with that proposed for *PTENP1* in Step 3.iii of Part II?

Laboratory Planning and Preparation

Laboratories 1.1–1.6

Bioinformatics laboratories demonstrate important concepts of modern biology and genetics. Engaging students in these laboratories allows them to make their own discoveries using the same data and tools that are used by research scientists, albeit often through more intuitive interfaces than scientists use, such as *DNA Subway*. As biology changes from a data-limited to a data-unlimited discipline, it is becoming increasingly important for students to understand that the toolkit of biomedical researchers is expanding to include sophisticated computational analysis tools—and that it is instrumental for biologists and medical specialists to understand how to use these tools.

The bioinformatics resources listed in this section include two substantially different entities: tools and platforms. Tools, such as the sequence search and alignment tool BLAST or the gene predictor Augustus, are based on a defined algorithm and conduct one basic task. Platforms, such as the National Center of Biotechnology Information's (NCBI) Internet suite, *Gene Boy*, *Sequence Server*, or *DNA Subway*, on the other hand, provide access to multiple bioinformatics tools and data—either presenting them as lists of tools (NCBI, *Gene Boy*, *Sequence Server*) or organizing them into logical workflows that are optimized for specific tasks (*DNA Subway*).

OVERVIEW

All tools and platforms necessary to complete the bioinformatics laboratories are accessed through graphical user interfaces; none requires specialized computer equipment, programming knowledge, or use of the command line. However, at the time of writing this book (summer 2011), most tools are optimized for use with a recent version of Mozilla/Firefox on any computer operating system. Many of the tools also require a recently updated version of Java (downloadable from java.com), and some require the browser to open pop-up windows.

If tools appear to be working incorrectly, remove cached information, close and restart the browser and/or restart the computer, and this is likely to resolve the issue. Ultimately, switching to another computer may solve the problem. For some issues, an e-mail to the webmaster of the respective site should provide additional help.

All tools can be run through wireless or wired Internet connections. However, the amount of data being analyzed and the complexity of the tool being used have a strong impact on how quickly analysis routines will run. Among the more taxing tools are the annotation editor Apollo (see below), which can run very slowly if accessed by multiple students at once through the same wireless hub—especially if the hub connects to the Internet through a low-bandwidth connection such as a DSL line. Take this into consideration when selecting a lab and test your systems before instruction, because insufficient Internet connection speed can create significant delays. Where necessary, the different tool sections include set-up and troubleshooting instructions.

ANNOTATION EDITOR APOLLO (LABS 1.1A, 1.1B, 1.4)

Apollo is a specialized browser that allows users to view and decorate DNA sequences with graphical and text annotations describing information that they identify in the sequence, such as gene models/functions, homologous regions, etc.

Apollo was developed by members of the *Drosophila* genome annotation group at the University of Berkeley, California, to facilitate the distributed annotation of the *Drosophila* genome. Apollo is written by experts for experts and may appear quirky and complicated to the novice user. However, working on guided exercises allows novices to quickly aquire the skills needed to access this very powerful and versatile tool.

Several resources are available to help learn Apollo basics:

- The instructions in Laboratories 1.1.A and 1.1.B in this textbook;
- A *DNA Subway* manual that can be downloaded by clicking on "Manual" at the lower left-hand corner of http://www.dnasubway.org;
- An extensive manual that can be downloaded from the Apollo website at http://apollo.berkeleybop.org/current/index.html.

As with many other tools, Apollo has menu shortcuts and special key combinations that increase ease of use. The following three tables provide overviews of the function of certain mouse clicks as well as Apollo menu and pop-up summaries.

Different computers/operating systems may require different mouse combinations to yield the effects described in Table 3. To simulate the center wheel/button, use of the Alt key with the left mouse button is sometimes required. With Apple computers, you can simulate a right mouse click by holding down the control or Alt key while clicking the mouse, and you can simulate a middle mouse click by holding down the apple key while clicking the mouse. If you are trying to copy text from

Table 1. Important Mouse Functions in Apollo

Click mouse key	Action
Left	Select feature (or deselect if you are not over any features)
Double left	Select all features in one track (a prediction or whole piece of evidence)
Shift left	Add feature to current selection (or remove if already selected)
Left drag	Drag selection into the annotation workspace
Left drag onto annotation	Add a feature to an annotation (e.g., new start codon)
Center button/wheel	Center display on clicked location
Center button/wheel drag	Select multiple features
Shift-center button/wheel drag	Rubberband multiple features, add to currently selected
Right	Open pop-up menu
Shift-right drag	Tier drag—move currently selected tier

Table 2. Important Menu Functions in Apollo

Direction	Action
Click on "File" > "Upload to DNA Subway"	Save annotations to *DNA Subway*
Click on "Edit" > "Undo"	Undo last operation (can be used repeatedly)
Click on "View" > Select/deselect "Show… strand"	Show or hide forward or reverse strand
Go to View > Select/deselect "Show reverse strand"	Show reverse complement
Go to Analysis > Select "Show GC plot"	Show GC content plot
Click on "Tiers" > Select "Show Tiers panel"	Show tiers panel (close when already open)
Click on "Tiers" > Select "Collapse all tiers"/"Expand all tiers"	Collapse or expand all tiers
Click on "Links" > "Submit to Blast"	Submit sequence to NCBI Blast to search GenBank for homologs

Table 3. Important Pop-Up Menu Functions[a] in Apollo

Direction	Action
Click on "Sequence…"	Display sequence for selection
Click on "Annotation info editor"	Add or edit annotation information
Click on "Exon detail editor"	Edit extent of exons
Click on "Delete selection"	Delete selection
Click on "Merge transcripts" or "Merge Exons"	Merge selection
Click on "Duplicate transcript"	Generate copy of selection
Click on "BLAST region"	Submit selection to NCBI BLAST to search Gen Bank for homologs
Click on "Submit to InterProScan"	Submit selection to InterProScan to search protein and domain databases for functional homologs
Click on "Calculate longest ORF"	Calculate longest ORF (this may have to be followed by inserting a start codon if Apollo displays a green arrow at the beginning of the coding sequence)

[a]Functions are available by right-clicking on annotation.

Apollo (for example, sequence residues from a Sequence window) to paste into another application, use ctrl-c (for a PC) to copy the text and apple-v (for a Mac) to paste it (or, if you are trying to paste into a Web browser, you can use the "Paste" command from the browser's Edit menu). Please note that if you are running an old version of Mac OS X (Version 10.2.2 or earlier), you may find strange mouse-button behavior. With a three-button mouse, you may find that the behavior of the right and middle buttons is switched. If you have an old Mac and a five-button mouse, both the left button and the middle (wheel) button are treated as the left button, whereas the two buttons on the far left side of the mouse are treated as the middle button.

COMPARATIVE GENOMICS TOOL COGE (LAB 1.2)

CoGe is a comparative genomics platform to help researchers visualize how genomes have changed. This online system makes the retrieval and comparison of genomic information and sequence quick and easy; the name, CoGe, stands for comparative genomics. CoGe gives users access to all publicly available genomes that are sequenced or being sequenced. Users can also submit genomes. CoGe's analysis tools include BLAST searches, dot plots, and other graphic display options to find, compare, and analyze homologous sequences from sets of organisms, including analysis routines that allow the detection and visualization of syntenic regions—genomic regions with conserved ancestry.

CoGe runs may can take a long time. However, CoGe stores results from previous analyses and these can be accessed quickly. In general, accessing previously run analyses will be easier. However, users who wish to generate a new set of results can do so. Each tool is somewhat different, but, in general, new analyses can be accessed through the "Display Options."

DNA SUBWAY (LABS 1.1A, 1.1B, 1.4, 1.5, 1.6)

DNA Subway is a platform that provides bioinformatics analysis routines, sample sequences, and utilities to input and store user data. Genomic data analysis usually requires the use of a number of different databases, software packages, and/or Internet sites that are frequently hampered by a lack of clarity and/or interoperability. *DNA Subway* bundles tools necessary for genome analysis into one simplified bioinformatics platform. The analogy of a subway was chosen to represent the interconnected workflows that enable computer-based genome analysis. Each of three lines in *DNA Subway* represents a specific workflow to (1) annotate genomic DNA of up to 150,000 bp (Red Line), (2) prospect entire genomes for members of gene or transposon families (Yellow Line), and (3) conduct phylogenetic analyses (Blue Line). *DNA Subway* can be used as a guest or after registering, with registered users being able to save and share their projects and data with other *DNA Subway* users.

Computer and Network Requirements

To ensure smooth use of *DNA Subway*, users need to

- Use a recent version of a browser (preferably, Firefox 3.5+ or Safari 10+);
- Use a recent version of Java (test and update the version installed on your computer at Java.com. Mac users must ensure that the most recent version is installed and enabled.);
- Enable browser pop-ups;
- Avoid accessing *DNA Subway* over a busy wireless hub. If you work wirelessly and notice that *DNA Subway* routines become sluggish, switch to a wired connection, if possible.

DNA Subway Accounts

Registration is free and available by following the link at the top left-hand corner of the *DNA Subway* main window.

DNA Subway Workflows

DNA Subway includes workflows that are tailored to different types of genome analyses including DNA sequence annotation, genome prospecting, and phylogenetic analysis. Each workflow lists tools in a logical order at "stops" with related functions and includes options for extended analysis on other lines, so-called "transfers." Skipping "stops" is often possible but sometimes prevents or limits further analysis.

Time Requirements

Most stops in *DNA Subway* take less than 1 minute to complete. However, depending on the Internet connection and complexity of the analysis, times can vary widely. If an analysis is taking a long time, you can work on a different project or leave—once started, *DNA Subway* will complete any analysis. After you return to the project, the completed analyses will be accessible.

DNA Subway Lines

Each line has different purposes and capabilities; the following table provides a few examples.

If you have	And wish to	Use the	Example
Genomic DNA	generate models for genes or transposons	Red Line	Find and annotate genes or transposons
A gene or protein	find similar genes	Yellow Line	Identify members of a gene family in a genome
A gene or protein	determine the relationship to similar genes	Blue Line	Generate a phylogenetic tree
A DNA barcode sequence	determine the species from which it originates	Blue Line	DNA barcoding

Sequence Preparation

```
SB: Sequence in FASTA format:

>GN:TEMP:9671 (amino acid sequence): 119 residues.
SGISVNEFKIFTGGVTSMPSFLKRFFPSVYRKQQEDAST-
NQYCQYDSPTLTMFTSSLYLAALISSLVASTVTRKFGR-
RLSMLFGGILFCAGALINGFAKHVWMLIVGRILLGFGIG-
FAN
```

Bioinformatics programs generally recognize (and often require) input sequences to be presented in FASTA format. FASTA is organized in two segments. A first segment, usually the first line, contains a definition or title for the sequence. This starts with a ">" symbol and ends at a manual line break (use the "Enter" key). The second segment contains the sequence such as nucleotides or single-letter amino acid symbols. The sequence ends at the end of the file or at the next "hat" symbol (>) if the file contains more than one sequence.

Common *DNA Subway* Routines

Create a project. The first step in using any line is to create a new project (click the square to the left of the line), work on a previous personal project (click "My Projects" and select the project), or work on a project made public by another user (click "Public Projects").

Run an analysis. Click the buttons to run analyses. Sometimes analyses are dependent on previous routines and will only become available and display an "R" (for "Run") once those routines are finished.

View or browse analysis results. To view analysis results, you often must click the button after the analysis is completed and a "V" is displayed, or it may be necessary to click buttons to launch browsers or viewers. Wherever feasible, *DNA Subway* provides results in tabular as well as graphical formats. As a general guideline, it is best to view your results after running each routine.

Save results and annotations. DNA Subway saves your results automatically with one exception: Annotations generated in the annotation editor Apollo require an explicit operation to upload the results to *DNA Subway* (see Table 2.)

Transfer data. Where indicated, *DNA Subway* allows you to transfer results and annotations between lines to extend an analysis.

Export data. Where indicated, *DNA Subway* allows you to export results and annotations to genome hubs or other external sites to view your results in the context of other scientific data.

Troubleshoot. Most problems with *DNA Subway* are the result of using outdated browsers, an outdated version of Java, or a browser that has not been enabled to open pop-up windows. In general, *DNA Subway* and any of the associated programs are optimized to be used by a recent version of Mozilla/Firefox that has been enabled to allow opening pop-ups and that uses a recent version of Java. If it is unclear whether your computer uses recent versions of the programs, update the version you have before opening *DNA Subway*. Recent versions of Internet Explorer, Chrome, and Safari should also run *DNA Subway,* but these have not been tested as extensively.

Additional troubleshooting advice is listed at the end of the *DNA Subway* manual that can be downloaded by clicking on "Manual" in the lower left-hand corner of the *DNA Subway* website (http://www.dnasubway.org).

GENE BOY (LAB 1.6)

Gene Boy is a simple, multifunctional DNA sequence analysis tool. Users can draw sequence data from buttons on the left side of the console and perform bioinformatics analyses with buttons on the right: transform/format DNA sequence, analyze nucleotide composition, search for restriction sites or other short motifs, and find open reading frames. With the "clone screen" feature, users can duplicate a result in a new screen for comparison with a second result. Users can enter and analyze DNA sequences of their own and personalize *Gene Boy* by changing its "skin" color. *Gene Boy* was developed for the DNA Interactive Internet site at http://www.dnai.org, where it can be accessed in the context of the "Genome" section.

GENE PREDICTION TOOLS AUGUSTUS, FGenesH, AND SNAP (LABS 1.1A, 1.1B, 1.4)

Genes are fragments of nucleotide sequences that carry the information to synthesize transcripts. Transcripts can be divided into two major groups: those that get translated into polypeptides and those that function as transcripts. Examples for the latter include rRNA, tRNA, and miRNA.

Genes can be identified using two fundamentally different methods: by searching genomes for regions similar to known genes or transcripts (extrinsic, evidence-based gene finding) or by searching genomes for the hallmarks of genes (ab initio gene finding or gene prediction). Extrinsic approaches require the availability of sufficiently homologous genes or proteins from other organisms or the sequencing of large amounts of RNA from the organism under analysis. High-throughput RNA sequencing (RNAseq), has become increasingly common as the cost of sequencing has been decreasing dramatically, which makes gathering evidence of genes and gene structure more widely accessible. Nevertheless, obtaining and analyzing large amounts of sequencing data and the existence of genes with low-abundance transcripts continues to limit extrinsic approaches.

For ab initio gene finding, gene prediction programs such as Augustus, FGenesH, and SNAP attempt to detect patterns or signals in nucleotide sequence that are associated with protein-coding genes, such as start codons, stop codons, coding regions, and splice sites. Different gene prediction programs often produce different gene predictions with rates for false positives (predicted genes that are probably not real) and false negatives (missed genes) in the range of 15%–40%, depending on the program and the organism examined.

The patterns associated with genes are functional and related directly to how cells express genetic information. The principal steps for using the information in protein-coding genes are transcription, posttranscriptional RNA processing, and translation. Genes consist of three major structural parts: a transcribed region, a promoter, and transcription factor-binding sites (TBSs). Although the promoter and transcribed sequences are usually closely associated, TBSs can be positioned at large and variable distances from the coding region. TBSs usually make a gene accessible for transcription by opening the DNA when associated with transcription factors. Promoters bind proteins and contain nucleotides that can be methylated; these in turn regulate the synthesis of transcripts. Transcripts are then subject to 5′ capping, 3′ polyadenylation, splicing, differential transport, regulation by localization or degradation, and ribosomal attachment and translation. All of these processes require tightly coordinated interactions between proteins and nucleic acids that require specific nucleotide patterns.

Advanced gene finders such as Augustus, FGenesH, and SNAP typically use complex probabilistic models, such as hidden Markov models (HMMs), to combine information from a variety of different signal and content measurements. To optimize gene prediction algorithms for different genomes, the models are initially trained on representative genes in a specific genome to develop a statistical model that identifies the signals associated with genes for this organism. This model is then applied to the genome to create gene predictions. These gene models consist of an open reading frame (nonspliced gene) or exons and introns (spliced gene), a coding sequence (cds), and, in the case of Augustus, 5′ and 3′ untranslated regions (UTRs).

GENOME BROWSERS MAP VIEWER (LABS 1.2, 1.6) AND GBROWSE (LABS 1.1A, 1.1B)

Genome Browsers allow users to view genomic data in a graphical, interactive interface. They derive their data from a database and display genomic data that include genes, gene predictions, gene structures, proteins, and other data that pertain to expression, regulation, variation, and comparative analyses data. This interactive display of multiple data sets makes the browsers superior to static views or tables of the data. Most genome browsers have zoom, pan, scroll, and search capabilities and allow users to select the tracks that they wish to view. Map Viewer, the genome browser of the National Center for Biotechnology Information (NCBI), provides special browsing capabilities for a subset of organisms in the Entrez Genomes database. GBrowse is a powerful generic genome browser that is available free of charge and has been adapted by many genome sequencing initiatives.

INTERPROSCAN (LABS 1.1A, 1.1.B)

InterProScan is a tool that supports the prediction of protein function and can be launched from inside Apollo. It scans a given protein sequence for the signatures of known protein families, domains, and functional sites. InterProScan combines a number of different protein recognition methods by searching the databases of the members of the InterPro collaborative project such as PROSITE, Pfam, ProDom, SMART, and TIGRFAMMs. Searching against the different secondary protein and functional site databases has become a vital resource for the prediction of protein function, because each tool uses different approaches that combine to optimize predictions. For example, PROSITE is ideal for finding short motifs but is not appropriate for identifying members of highly divergent protein families. In contrast, the PRINTS database is excellent for determining specific subfamily membership but does not find short motifs. Methods based on HMMs and profiles are best for identifying members of divergent superfamilies but they do not perform well in determining specific subfamily membership. To ensure the best results, InterProScan searches all of the secondary databases simultaneously. The results can then be viewed in the context of the InterProScan website and are also integrated into Apollo's Annotation Info Editor.

MULTIPLE SEQUENCE ALIGNMENT AND PHYLOGENETIC TREE EDITOR, JALVIEW (LAB 1.5)

Sequences of DNA, RNA, and proteins are the fundamental currency of modern biological research and link different levels of the biological hierarchy, from gene to three-dimensional (3D) structure. Multiple sequence alignments permit the identification of common features among species and/or of functionally important sequence residues. Multiple sequence alignments constitute the first step in studying molecular phylogeny and identifying genomic rearrangements, and they provide a convenient framework for displaying common features and complex annotations relating to sequences and their functions. It is therefore important to obtain the best alignment possible.

Many multiple alignment techniques exist, but no single method is perfect for all situations. As a consequence, all alignments require inspection, interpretation, and, often, adjustment by hand to produce an alignment that best represents the biological context of the sequences. Editing tools are essential for this task, not least because they provide visual feedback on an alignment's quality in the light of all known and computationally predicted annotations. Jalview is an alignment editor first developed in 1996 as an improvement to static alignment visualization tools. In addition to alignment editing, coloring, and generation of figures as postscript or HTML, Jalview includes methods for alignment conservation analysis, phylogenetic tree construction, sequence annotation, secondary structure prediction, a 3D structure viewer, and links to external bioinformatics web services and data bases (from Waterhouse et al. 2009. Jalview version 2: A multiple sequence alignment and analysis workbench. *Bioinformatics* **25:** 1189–1191, doi: 10.1093/bioinformatics/btp033).

PHYLOGENETIC TREE TOOL, PHYLIP (LAB 1.5)

Phylogenetic tree tools estimate the relationships among organisms, genes, or proteins and display them in a branched graph or "phylogenetic tree." These graphs consist of nodes and branches whose topology (the way in which the nodes are connected) represents the relationships, inferred from the differences in DNA, protein sequences, or physical characteristics, and the branch lengths often display the degree of difference. Sequences on connected branches share a node, which indicates that they share a common ancestor. Nodes or common ancestors are frequently inferred, because there may not be data available for them, but their existence is postulated to explain the branches and subsequent nodes for which data exist. Branch tips (also called leaves or terminal nodes) each correspond to genes, proteins, or organisms for which data were analyzed. In scaled trees, the length of each branch is a measure of evolutionary distance from the (inferred) ancestral sequence at the node; shorter branches indicate that a sequence is more similar to the ancestral sequence than sequences with longer branches. If the branch lengths for both sequences are zero, the sequences are usually identical or too similar to resolve.

Differences between sequences, such as substitutions, insertions, and deletions, accumulate over evolutionary time. The rate at which mutations become fixed represents a "molecular clock." The elapsed time from when two sequences diverged can be estimated by measuring the number of differences between the sequences and calibrating the rate of mutation with external data such as the fossil record. For instance, if two sequences from organisms known to have diverged 6 million years ago have 12 mutations every 100 base pairs, the rate of mutation is approximately 2% per million years. If two other sequences had six mutations every 100 base pairs, an estimated 3 million years has passed since they diverged. However, mutation rates differ among species, over evolutionary time in one lineage, and even among different sequences from one organism. This means that the concept of a molecular clock must be applied with care and is most useful when analyzing closely related lineages.

REPEATMASKER (LABS 1.1A, 1.1B, 1.4)

The genomes of higher eukaryotes such as flowering plants or mammals generally contain significant amounts of recurring DNA sequence. These include transposons but also low-complexity DNA (example: TTTTAAATATAAATAATTTT), simple repeats (example: AAAAAAA), and tandem repeats (example: AACCCTAACCCTAACCCT). In general, these DNA repeats contain as much useful information as a string of repetitions of the utterance "uhh" in a report. Taken together, transposons and other repeats can occupy large portions of a genome; the human genome consists of more than 50% repetitive DNA, and that of lilies more than 95%. Repetitive DNA has long been thought to include little information relevant to an organisms' genetic makeup. However, transposon researchers suggest that transposons may have led to spliced genes and enable species to quickly rearrange their genomes to adapt to and survive environmental challenges; many important roles for repeats and repetitive regions of genomes have been described. The large number of repeats makes finding other genes difficult, so most genomic analysis workflows include a step to identify and mask repetitive DNA. This also helps in minimizing the time to complete an analysis. RepeatMasker combines two types of algorithms to identify repeats: a mathematical algorithm to detect low-complexity DNA and simple repeats, and a similarity search to identify transposons. Repetitive DNA is masked and "hidden" from successive analysis tools either by converting repeat nucleotides from upper into lower case or replacing them with Ns.

SEQUENCE ALIGNMENT TOOLS CLUSTAL W (LAB 1.3) AND MUSCLE (LABS 1.5, 1.6)

Identifying mutations, genome alterations, and evolutionary relationships usually involves the comparison of two or more sequences by means of pairwise or multiple sequence alignments. Database search algorithms such as those provided in BLAST use pairwise alignments to search sequence databases for matches to query sequences. Alignment tools arrange sequences in a way that reflects the level of similarity among them. The final alignment represents a hierarchy of similarities among all input sequences, usually by aligning the most similar sequences first and the more divergent sequences next. Assuming evolutionary relationships among input sequences, alignments can help to determine the degree of relatedness among them.

Pairwise alignments. A sequence alignment usually arranges one sequence on top of another where the residues at each position are best aligned in the context of the overall alignment. If the same letter representing a nucleotide or amino acid occurs in both sequences, this position is likely to have been conserved in evolution (or mutations have given rise to the same letter twice). If the letters differ, it is likely that the two derive from an ancestral form. Length differences in homologous sequences are due to deletions or insertions that either removed or added letters from one of the sequences. Often, it is hard to determine which has happened, and which sequence, so scientists often call them "indels." An alignment may then pair a letter or a stretch of letters in one sequence with dashes in the other to indicate gaps created by indels. To determine the optimal alignment between two sequences, each position is assigned a value, with matches being scored most favorably, mismatches less favorably, and gaps least favorably. By adding these values along an alignment, the most likely alignment between two sequences is determined.

Multiple alignments. Frequently, database searches with nucleotide or protein sequences will yield multiple matches for which to discern the evolutionary relationships and sequence conservation. Pairwise alignment methods can miss subtle similarities that become more obvious after several sequences are compared. As in pairwise alignments, multiple sequence alignment tools arrange sequences one above the other after computing the sites that are likely to be homologous, with dashes to account for differences in sequence lengths.

SEQUENCE SEARCH TOOL BLAST (LABS 1.1A, 1.1B, 1.2, 1.5, 1.6)

BLAST (Basic Local Alignment Search Tool) is a sequence search tool that identifies matching sequences by aligning the characters of a query sequence to the characters of sequences in databases. Matches are scored according to the number of similar characters, the number and extent of gaps, and the overall lengths of query and match. BLAST is not restricted to finding full matches to a query and also returns partial matches. Besides a score, BLAST also provides an E value, which estimates the chance that a match could have occurred simply by chance. Small E values indicate "better" matches than larger ones. An E value of 6.88E-11 is 6.88×10^{-11} or 0.0000000000688, a very small number indeed; this suggests that the hit is very likely to be real.

SEQUENCE SERVER (LABS 1.2, 1.3, 1.6)

Sequence Server provides a user-friendly platform to store and access DNA sequences, conduct CLUSTAL W multiple sequence alignments, construct phylogenetic trees, and launch BLAST sequence searches.

TARGeT (TREE ANALYSIS OF RELATED GENES AND TRANSPOSONS (LAB 1.5)

DNA Subway uses the TARGeT workflow to prospect genomes for matches to genes or transposons. TARGeT bundles similarity searches with a multiple alignment and a phylogenetic tree tool. Using DNA for a query may identify closely related members of a gene (or transposon) family; using proteins may reveal more distantly related members due to the conservation critical amino acid functions in proteins that can be masked by the redundancy of the genetic code. TARGeT searches genomes and can facilitate the discovery of hitherto unknown gene family members.

TARGETSCAN (LAB 1.6)

RNA interference (RNAi) is a system within living cells that mediates antiviral defenses and regulates gene expression. Very small RNA molecules (20–22 nucleotides) are central to this system because they can bind to DNA and RNA and influence their function, localization, or stability. Small RNAs can affect transcription, direct mRNA cleavage, and affect translation. RNAi is an ancient and conserved phenomenon in eukaryotes and is important for normal development, disease, and viral and transposon silencing. RNAi has also become a valuable tool for research scientists who wish to study the function of genes without having to generate mutants by silencing them. Finally, RNAi promises to become an important therapeutic, possibly treating many different diseases.

Several classes of small RNA molecules have been found to mediate RNAi, including miRNA, shRNA, and siRNA. Well over one-third of human genes contain conserved targets for miRNAs and are likely to be subject to posttranscriptional regulation by them. TargetScan predicts biological targets of miRNA by searching for the presence of conserved 8- and 7-mer sites that match the seed regions of each known miRNA. As an option, nonconserved sites can also be predicted. In mammals, these sites are ranked based on the predicted efficacy of miRNA targeting. TargetScanHuman also considers matches to annotated human UTRs and their orthologs.

tRNAscan-SE (Labs 1.1.A., 1.1.B)

tRNAscan-SE uses two steps to arrive quickly at highly accurate tRNA gene predictions, first generating a selection of likely tRNA genes and then examining these genes to eliminate false positive results. The first stage involves algorithms that use similarity searches to detect eukaryotic intragenic regions and RNA polymerase III promotors and terminators, and an algorithm that can detect whether molecules have the ability to form the base pairs present in tRNA stem-loop structures. The second stage screens these results by applying a covariance model that scores the likelihood for the transcript of a presumptive tRNA gene candidate to fold into the three-dimensional structure of a tRNA molecule.

Answers to Questions
Laboratories 1.1–1.6

LABORATORY 1.1: ANNOTATING GENOMIC DNA

Answers to Questions in Stage A: Annotate Genes in a Synthetic Contig

II. Identify and Mask Repeats

3. How many and what types of repetitive DNA were identified? What are their lengths? Is there any relationship between the repeat type and length?

RepeatMasker identified 10 repetitive DNA segments in the synthetic contig. Low-complexity DNA and simple repeats range from 25 nt to 64 nt in length; *Helitron* and *Copia* transposons range from 440 nt to 4077 nt in length. Low-complexity DNA and simple repeats are generally shorter, whereas transposons and their remnants can span up to several thousand nucleotides.

8. How many and what types of repetitive DNA does the browser display?

The local browser displays 10 repetitive DNA segments, all of which were also listed in the RepeatMasker result table.

9. Which of the two views, table or browser, do you find easier to work with?

Your answer to this question may depend somewhat on the purpose of your work. The browser view is useful to get a quick overview; exact position data and lengths of specific DNA segments are easier to extract from the table.

III. Predict Genes in DNA Sequence

3. Do any of the programs run significantly longer than any others?

Augustus runs usually take longer than SNAP or FGenesH runs.

4.i. How many genes did each program predict?

All gene prediction programs predicted four protein-coding genes; tRNAscan predicted one tRNA gene.

4.ii. How are exons, introns, and 5′ and 3′ UTRs represented in the predicted genes?

Genes are represented as boxes and connecting lines. Boxes represent exons (DNA segments that are transcribed and processed into the final transcript, the mRNA), and lines represent introns (DNA segments that are transcribed but removed from the transcript via splicing).

4.iii. Do all of the programs predict the same structure for each gene? Explain.

All three gene predictors predict four genes, but the individual predictions deviate significantly. For example, the first gene on the 5′ end (to the left) was predicted by Augustus and FGenesH only, but not by SNAP. At other locations, the programs predict different gene structures (e.g., the second gene, predicted by Augustus and FGenesH as one gene, is "seen" by SNAP as two, shorter genes). Furthermore, Augustus predictions often include 5′ and 3′ UTRs (unshaded box segments that flank the shaded, protein-coding box segments), thus they are longer than FGenesH or SNAP predictions, which begin and end at predicted start and stop codons. Conflicting gene predictions generally cannot be resolved by looking at gene predictions alone, and require additional evidence.

IV. Name a Gene and Insert a Start Codon

3.i. Does Apollo use the same conventions for displaying exons and introns as GBrowse?

Apollo uses the same elements as the local browser: Boxes represent exons (DNA segments that are transcribed and processed into the final transcript, the mRNA), and lines represent introns (DNA segments that are transcribed but removed from the transcript via splicing).

8.i. Which reading frames are they in? Are they in the same or different reading frames?

A DNA sequence can be parsed into triplets in three different ways, depending on whether one starts with the first, second, or third nucleotide from the beginning. Accordingly, DNA sequences can be read in six different reading frames, three on the forward strand and three on the reverse strand. Potential start codons and stop codons are in rows at the top of the upper white evidence panel (for the forward strand) and at the bottom of the lower white evidence panel (for the reverse strand). There are three rows of start codons and three rows of stop codons for each of the two strands. The start for the FGenesH prediction is located in the third row from the top, as is the stop codon, indicating that they are in the same reading frame.

8.ii. What amino acid is encoded by the start codon? What amino acid is encoded by the stop codon?

The start codon (ATG) encodes the amino acid M, which is the one-letter code for methionine. The stop codon (TAG for this gene) does not encode an amino acid but instead determines the position at which the ribosome would terminate translating the mRNA.

10.i. What happens?

The start codon gets inserted into the gene model.

10.ii. What do you notice about the position of the start codon that you have inserted?

The position of the start codon in the Augustus model is the same position at which FGenesH predicted the start codon for this gene.

11.i. What do the unfilled segments at the beginning and end of your Augustus temporary gene model represent?

These segments represent the 5′- and 3′-untranslated regions (UTRs) at the beginning and end of the transcribed region of the gene.

11.ii. What does this tell you about differences in the way Augustus and FGenesH predict genes?

FGenesH and SNAP do not predict 5′ and 3′ UTRs; their gene predictions begin with a start codon and end at a stop codon.

11.iii. How does this relate to the speed of each prediction program?

Augustus, which attempts to predict UTRs, runs longer than FGenesH or SNAP, because predicting UTRs is computationally intensive.

V. Examine Exon/Intron Borders (Splice Sites)

2.i. How are the two predictions similar or different?

Augustus and FGenesH predict a spliced gene with four exons. The splice sites for these exons are in the same position. Because Augustus attempts to predict the gene's UTRs, its first and last exons extend further outward than those of the FGenesH-predicted model.

2.ii. Could you arrive at a correct gene model based on the gene predictions alone?

Although this is possible, it would not be appropriate to edit the gene's extent by using the gene predictions only. Instead, defining the beginning and the end of the gene's transcript requires material or "biological" evidence such as transcript (mRNA) sequences that map to this location.

4.ii. Move sequentially to each exon/intron and intron/exon boundary, and record the three nucleotides on each side of the splice site as shown in the chart below.

The splice sites in the Augustus-predicted models are surrounded by the following nucleotides:

Location	Exon	Intron	Intron	Exon	Location
2387/2388	CCG	GTT	CAG	GTG	2694/2695
3017/3018	CAG	GTC	CAG	GCT	3723/3724
4353/4354	CAG	GTA	TAG	GCT	4445/4446
6602/6603	CAG	GTG	CAG	AGA	6709/6710
6770/67701	AAG	GTT	AAG	CTA	6887/6888
6977/6978	CAG	GTT	CAG	TAC	7183/7184
7270/7271	CCA	GTT	CAG	GCA	7357/7358
7426/7427	TTG	GTA	CAG	TGG	7518/7519
7599/7600	GCT	GTA	CAG	GCT	7697/7698
7769/7770	GAA	GTA	TAG	GGC	7868/7869
7829/7830	ACT	GTT	TAG	TTC	8022/8023
8125/8126	AAG	GTA	CAG	GTG	8205/8206
8302/8303	ATG	GTG	CAG	GTA	8388/8389
8435/8436	CAG	GTA	CAG	AAA	8735/8736

4.iv. What nucleotides are found at every splice site? Make a general rule for the nucleotides that define the beginning and end of each intron.

The sequences at the ends of exons do not show a pattern. All introns, however, begin with GT and end in AG. If you were to include a sufficiently large number of splice sites in this analysis, you would discover that this is true for the vast majority of splice sites in eukaryotic genes; they have therefore been termed "canonical" splice sites, although rare exceptions do occur.

7.i. How are the predictions similar or different?

Instead of one gene with four exons, SNAP predicts two genes with two exons each. The beginning of the first gene and the end of the second gene align with the first and last exons, respectively, of the FGenesH prediction as well as with the positions of the start and stop codons in the Augustus prediction.

7.ii. What additional information is needed to determine the correct structure of the gene?

Deciding which of the two models is correct—one four-exon gene or two two-exon genes—requires material evidence such as transcript (mRNA) sequences that map to this location.

9.i. How are the predictions similar or different?

Augustus, FGenesH, and SNAP predict genes in approximately the same locations. At position 6400–9200, all predict a multi-exon gene, with Augustus predicting a 12-exon gene and SNAP and FGenesH predicting a 10-exon gene. The splice sites of these 10 exons align with the splice sites of the same exons in the Augustus-predicted model. In position 9600–16,400, each of the three prediction programs predicts a different model, with Augustus and FGenesH predicting two-exon genes of widely different extent, and SNAP predicting a nonspliced gene. Additional information to guide edits is needed to arrive at valid gene models for both locations.

9.ii. What additional information is needed to determine the correct structures of the genes?

Deciding which models are correct requires material evidence such as transcript (mRNA) sequences that map to these locations.

VI. Search Databases for Biological Evidence

4. For how many predicted genes did the BLAST searches generate biological evidence?

The BLASTN search yielded matches for each of the four gene-bearing locations; BLASTX yielded matches for only three of these. The BLASTN results represent transcript/mRNA evidence for genes (cDNAs and ESTs), whereas BLASTX results represent protein evidence. BLAST is not programmed to discover splice sites and may align sequences that have already been aligned to a previous exon; this often leads to faulty splice site indication. BLASTX results may not always synch well with gene structures indicated by the gene predictors and/or by BLASTN evidence. The major reasons are that (a) protein matches from other organisms may deviate in their amino acid sequence from the sequence you analyze; and (b) protein matches may be from paralogs of the gene elsewhere in the genome and therefore deviate from the gene being annotated.

VII. Edit the Ends of a Nonspliced Gene

After following the steps in Part VII, you will have generated a model for the first gene in the synthetic contig:

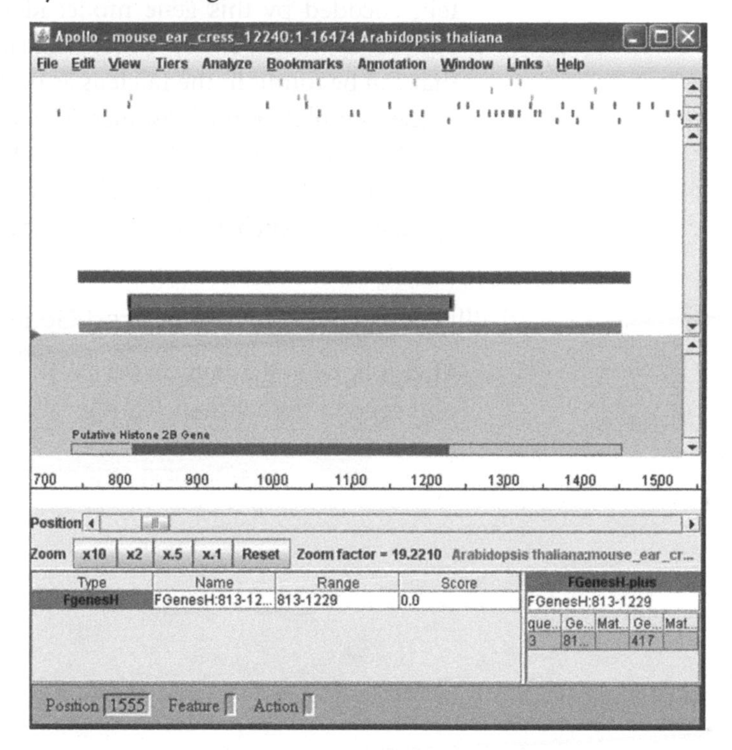

5.iii. Which of the predictions is best supported by the BLAST results?

The longest piece of BLAST evidence is a BLASTN hit that represents a cDNA. It best supports the Augustus prediction in that it extends beyond the FGenesH prediction to include the gene's UTRs.

10. What do you notice?

The Augustus model is one nucleotide shorter than the BLASTN model.

11. What do you notice?

The Augustus model is 10 nucleotides shorter than the BLASTN model.

16.i. Does the BLASTN evidence support the extent of your model?

The BLASTN evidence supports the extent of the gene model.

16.ii. Does the BLASTX evidence support the position of the start and stop codons?

The BLASTX evidence, which is protein evidence and can be used to help determine the start and stop of translation, supports the location of the start and stop codons for this gene model.

16.iii. Can you determine the function of the protein produced by this gene?

BLASTX matches from different species indicate that protein encoded by this gene shares homology with histone 2B, a protein involved in securing the supercoiled structure of DNA in chromosomes.

17.v. What information can you extract from the InterProScan results?

Searching the databases of the members of the InterPro collaborative project such as PROSITE, Panther, Pfam, ProDom, SMART, and TIGRFAMMs with the protein encoded by this gene model identifies matches to several known protein regions, domains, and/or functional sites. These domains are typical of proteins that can be found in the nucleus as part of the nucleosome, bind DNA, and serve in the assembly of nucleosomes. Several of the results specify the protein as putative histone 2B. Gene Ontology (GO) numbers associated with features allow users to quickly identify proteins with related domains in gene or protein ontology databases such as http://www.geneontology.org.

VIII. Edit the Ends of a Spliced Gene

After following the steps in Part VIII, you will have generated a model for the second gene in the synthetic contig:

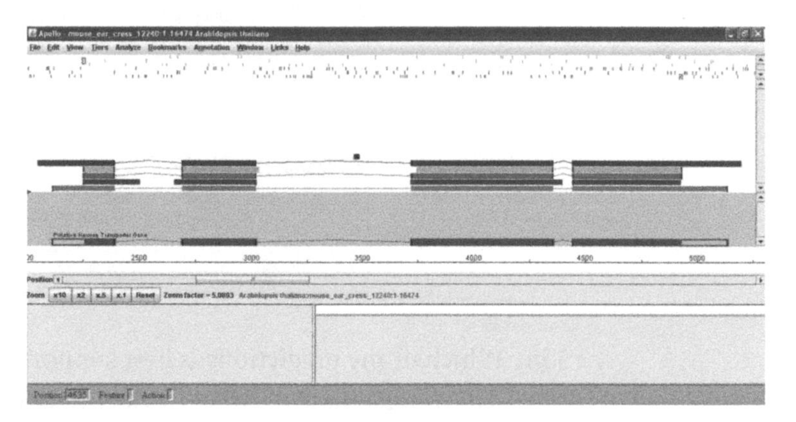

3. **Using the BLASTN evidence, determine whether Augustus, FGenesH, or SNAP have predicted the gene(s) in this region most accurately.**

   The Augustus gene prediction matches the BLASTN evidence best because it shows one four-exon gene and includes supported UTRs.

5. **What do you notice?**

   The model shows several yellow arrows. Apollo uses arrows to indicate positions that require further attention: Green and red arrows indicate missing start and stop codons, and yellow arrows indicate splice sites that do not follow the canonical pattern (i.e., introns that do not start with GT and/or end with AG). BLASTN matches transcript evidence to the fullest extent possible and often uses nucleotides twice if they match at the beginning and at the end of an intron. Using BLASTN evidence therefore often leads to faulty splice sites that need adjustment. One can use gene predictions to correct splice sites because gene predictors are programmed to adhere to the canonical splice site pattern.

IX. Merge Exons and Edit the Ends of a Spliced Gene

After following the steps in Part IX, you will have generated a model for the third gene in the synthetic contig at the top of the next page.

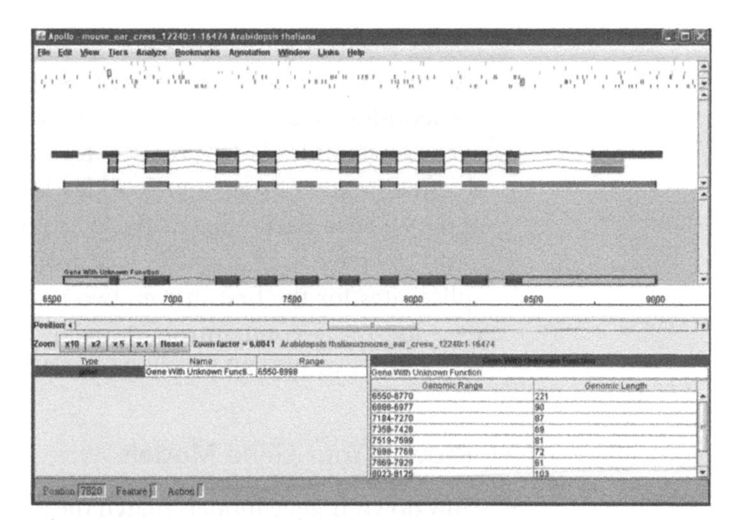

3. **Which gene prediction most closely matches the BLASTN evidence?**

   The gene predicted by Augustus most closely resembles the BLASTN evidence.

X. Fix Splice Sites

After following the steps in Part X, you will have generated a model for the fourth gene in the synthetic contig:

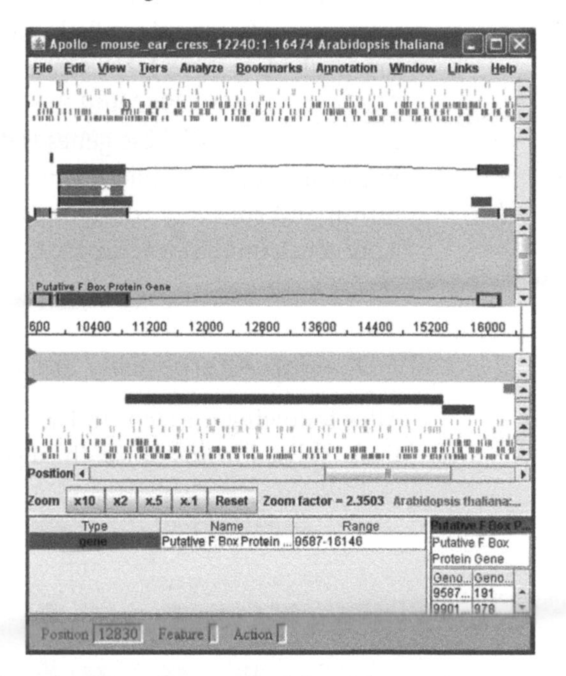

3. **What interesting situation do you find for this gene?**

   RepeatMasker locates a *Copia* transposon in the long intron indicated by the Augustus prediction and the BLASTN evidence.

5.iii. **How can we tell that the intron ends at the correct "AG"? Why is it not one of the two following "AGs"?**

   To determine how many nucleotides a faulty splice site needs to be moved, highlight the BLASTN evidence on the white work space panel. This will display details regarding this evidence piece in the bottom part of the Apollo screen. Examining the column "Match Range," you will find that BLASTN aligns nucleotides twice: the match range of the first segment that aligns to the contig spans nucleotides 1–191; the second segment includes nucleotides 190–1169; and the final segment spans nucleotides 1168–1438. In other words, nucleotides 190 and 191 have been aligned twice around the first intron, and nucleotides 1168 and 1169 have been aligned twice around the second intron. Therefore, to fix the splice sites for the first intron, two nucleotides need to be removed. The same is true for the second intron. Moving the end of each of the two introns two nucleotides to the right fixes the model and establishes canonical splice sites.

XI. Browse Your Gene Models

3. **How do your gene models match the genes predicted by Augustus, FGenesH, and SNAP? Do any of the three prediction algorithms predict all genes accurately?**

   Augustus, FGenesH, and SNAP predict a total of seven different genes, some of which overlap among the different prediction algorithms. After verifying and annotating the genes in this synthetic contig, it is apparent that it contains four genes that all match the predicted genes in some way. The Augustus algorithm comes closest to predicting these four genes accurately. The gene in position 747–1451 is not interrupted by an intron. Genes that are not interrupted by introns are frequently called one-exon genes (despite the fact that exons were originally defined as DNA sequences that flank an intron). The genes in positions 2123–5132, 6550–8982 and 9587–16,146 are spliced genes. The coding sequence for the first two of these genes is distributed over several exons, whereas the coding sequence in the spliced gene in position 9587–16,146 is not interrupted by an intron and runs as a contiguous open-reading frame (ORF) in the second exon. This gene contains a tranposon inserted into its second intron.

XIV. Develop Alternatively Spliced Models for the Third Gene

After following the steps in Part XIV, you will have generated models for alternatively spliced forms of the third gene in the synthetic contig:

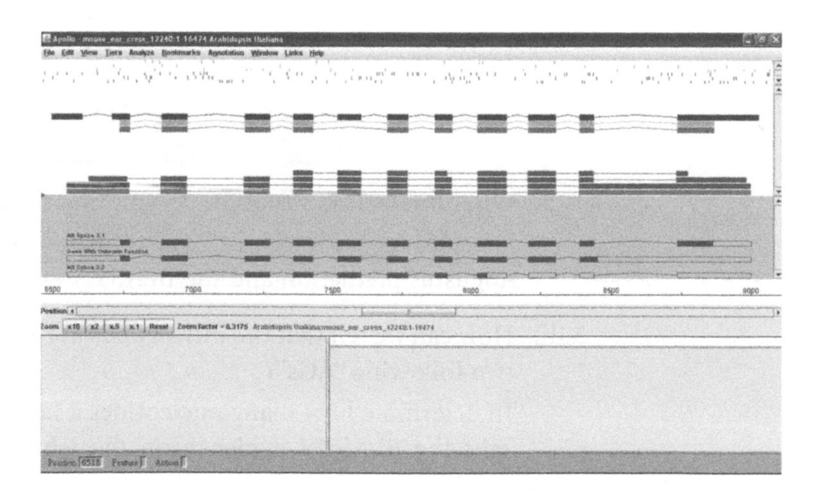

11. **What interesting situation does the shorter exon create?**

    The shorter exon causes the 3′ UTR for this model to be spliced three times. The reason for this is that the shorter exon is 45 nucleotides long, whereas the previous, longer exon was 61 nucleotides long. Although this difference of 16 nucleotides creates a frame shift that shortens the open reading frame (ORF) for this gene, the model still includes the exons downstream from the new stop codon.

12. **The first exon for EST evidence sequence BX820562 ends before the first exon of the BLASTN evidence and EST evidence BX820662. Why does this not suggest to additional alternative splice forms? (Tip: Zoom in to examine the end of the three exons in high magnification.)**

    The first exon of EST BX820562 does not end at GT and, therefore, does not indicate a correct splice site. The EST evidence provided for this gene does not support more than three alternative splice forms for this gene.

XV. Develop Alternatively Spliced Models for the Fourth Gene

After following the steps in Part XV, you will have generated a model for an alternatively spliced form of the fourth gene in the synthetic contig:

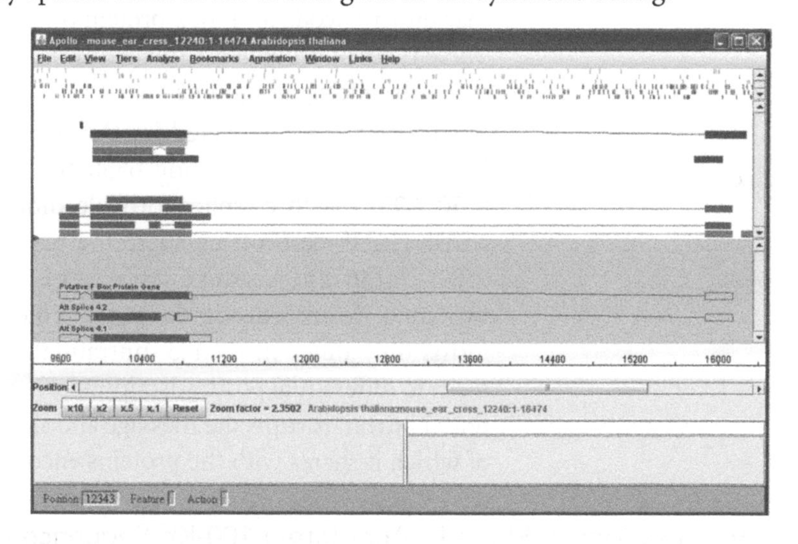

3.v. **What interesting situation do you observe in this gene model? Is the protein-coding region interrupted by an intron?**

Whereas this alternatively spliced form consists of a spliced transcript, the protein-coding region consists of one contiguous ORF that is not interrupted by an intron.

6. **Click on "Reset" and compare and contrast your gene models. How are the models similar to one another and how are they different?**

   All four gene models are supported by evidence for gene expression such as BLASTN and EST data. Three of the four genes are spliced genes and one is a non-spliced gene. Introns interrupt the coding sequence (cd) in only two of the three spliced genes; in two of the three splice forms of the fourth gene, the cd is contained as an ORF in one exon. Two genes show indication of alternative splicing, which affects the cds in these genes as well as the structure and extent of the 3′ UTR.

XVI. Browse Your Gene Models

3. **How do your gene models match the genes predicted by Augustus, FGenesH, and SNAP? Do any of the three prediction algorithms predict all genes accurately? Did any of the evidence provided through the BLASTN protein database search support the gene function identified through InterProScan search? For which of the genes did the evidence indicate alternative splicing?**

   Augustus, FGenesH, and SNAP predict a total of seven different genes, some of which overlap in the different prediction algorithms. After verifying and annotating the genes in this synthetic contig, it is apparent that it contains four genes that all match the predicted genes in some way. The Augustus algorithm comes closest to predicting these four genes accurately. Both BLASTX and InterProScan search results indicate that the gene in position 737–1451 may encode a histone 2B protein. For the gene in position 2123–5132, both searches indicate that it may encode a carbohydrate transporter protein, specifically a hexose/inositol transporter protein. For the gene in position 6550–8982, the BLASTX search did not return results at publication date. InterProScan returned sparse results from indicating that the gene may encode a protein with catalytic activity involved in lipid metabolism. For the gene in position 9587–16,146, both the BLASTX evidence and InterProScan results indicate that it may either encode an F-box protein or a protein that can bind to an F-box protein. These types of proteins are frequently involved in marking a target protein for degradation by adding ubiquitins.

   The EST evidence uploaded to *DNA Subway* in Part XII indicated alternative splicing for the genes in position 6550–8982 and 9587–16,146. The gene in position 6550–8982 has three splice forms leading to proteins of 170, 251, and 289 amino acids. The three proteins share the first 159 amino acids. The gene in position 9587–16,146 also has three splice forms, with two, three, and four exons. The first two splice forms encode the same protein of 312 amino acids; these two splice forms only differ in their 3′-untranslated regions (UTRs), potentially subjecting them to differential posttranscriptional regulation. In the third splice form, an additional intron disrupts the message and shortens the protein to 223 amino acids, 219 of which it shares with the proteins encoded by the other two splice forms.

Answers to Questions in Stage B: Annotate a 100-Kb Sequence from *Arabidopsis* Chromosome 5

The following table displays the main work routines required to correctly annotate seven gene examples in this contig.

Gene example	1	5	2	6	3	7	4
Position	14,000–18,500	26,000–24,500	27,800–29,800	54,500–52,000	67,500–69,000	78,000–75,000	89,500–92,500
Strand	forward	reverse	forward	reverse	forward	reverse	forward
Type	spliced	nonspliced	spliced	spliced	spliced	spliced	spliced
Insert start or stop codon	X				X		
Insert exon							X
Merge exons			X		X		X
Delete exon				X		X	X
Split exon							X
Adjust ends	X		X	X	X	X	X
Alternative splicing					X		X

The seven gene examples are illustrated below.

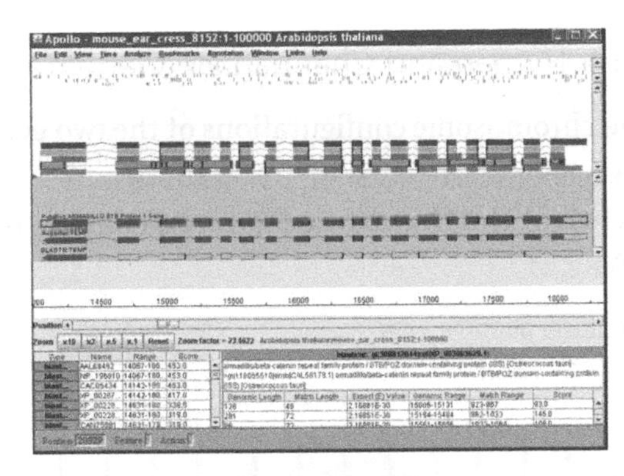

Gene example 1

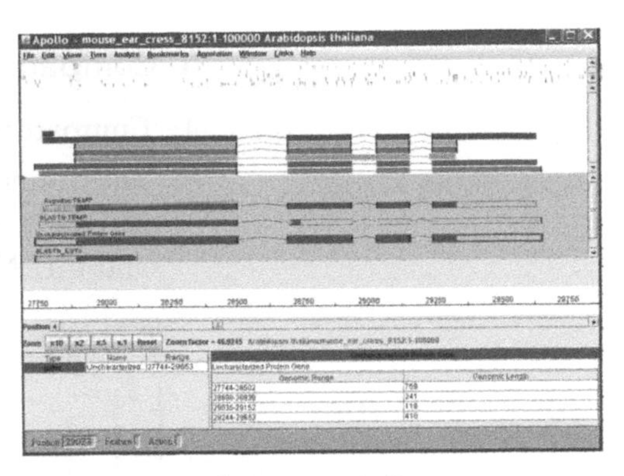

Gene example 2

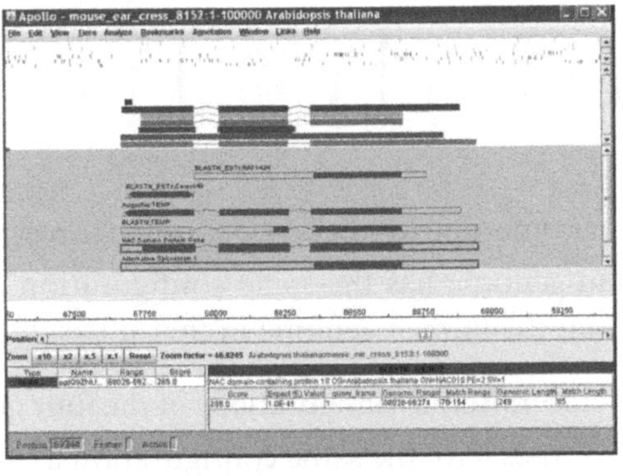

Gene example 3

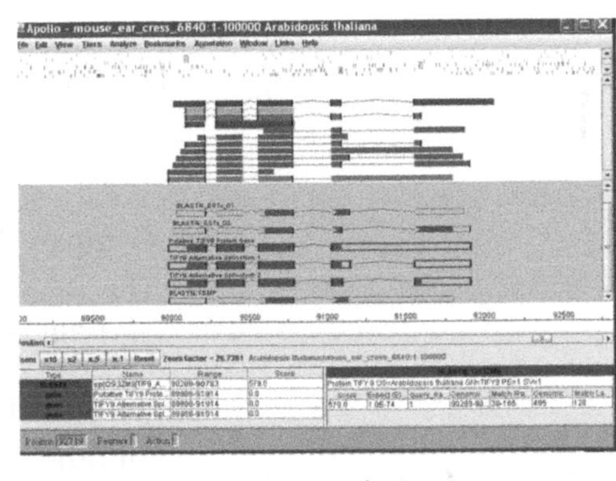

Gene example 4

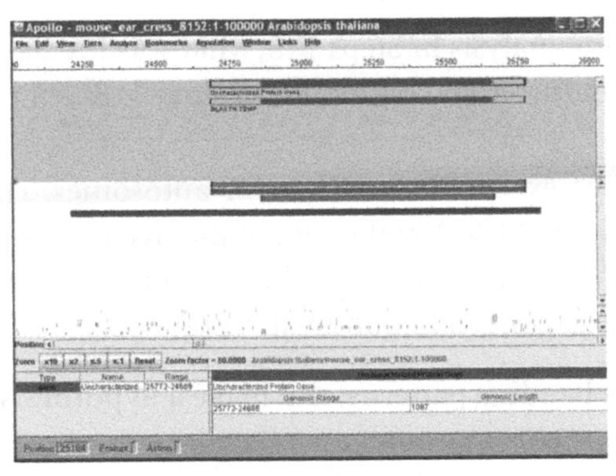

Gene example 5

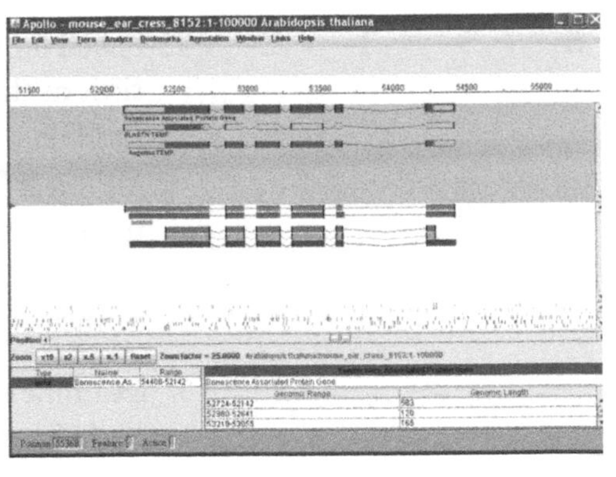

Gene example 6

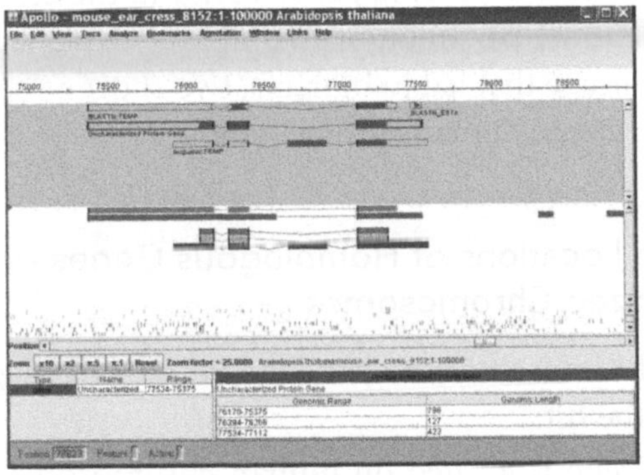

Gene example 7

LABORATORY 1.2: DETECTING A LOST CHROMOSOME

I. Compare the Human and Chimpanzee Genomes

4. **Compare the chromosome configurations of the two genomes. How do they differ?**

   The human genome consists of 25 chromosomes (22 autosomes, X, Y, and mtDNA), whereas the chimpanzee genome consists of 26 (23 autosomes, X, Y, and mtDNA).

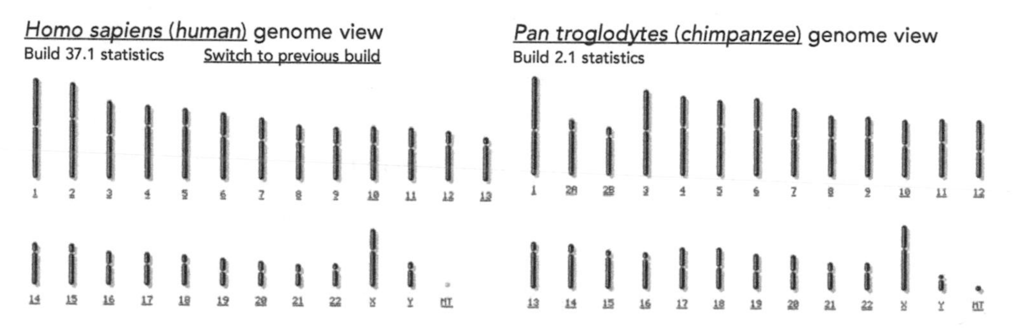

6. **Compare the chromosome configurations of the three genomes. How do they differ?**

   The orangutan genome has the same configuration as the chimpanzee genome and differs from the human genome in that it has 23 autosomes instead of 22.

8. **Compare the chromosome configurations of the four genomes. How do they differ?**

   All other Great Apes have the same configuration as the chimpanzee genome and differ from the human genome in that they have 23 autosomes instead of 22.

9. **Based on your analysis in Steps 1–8, what was the likely chromosome configuration of the most recent common ancestor (MRCA) of humans and Great Apes? Explain.**

   All Great Apes, except humans, have 23 autosomes. The relationships depicted in the tree show the MRCA at the root of the tree in a position that makes sense only if the MRCA had 23 autosomes. These 23 autosomes were passed on through the branching lineages to all nonhuman Great Apes.

10. **Examine the banding patterns, and make a hypothesis regarding how the human and Great Ape chromosomes are related.**

    The banding patterns at the ends of the short arms of Great Ape chromosomes 2A and 2B resemble the banding pattern around the centromere of human chromosome 2. The most parsimonious (simplest) explanation for these similarities is that in the lineage that led to humans, two short ancestral chromosomes (2A and 2B) fused to create human chromosome 2.

II. Compare the Locations of Homologous Genes on Human and Chimpanzee Chromosomes

2.viii. **On which chromosome have you landed? Draw a diagram with the approximate location of the hit.**

The *SPAST* gene is located on human chromosome 2. (See the diagram in the answer to Step 4.ii below.)

3.ii. On which chromosome have you landed? Draw a diagram with the approximate location of the hit.

The *SPAST* gene is located on chimpanzee chromosome 2A. (See the diagram in the answer to Step 4.ii below.)

4.i. Draw a diagram with the approximate location of each hit. When possible, add these to existing diagrams.

SUCLG1 is located on human chromosome 2 and on chimpanzee chromosome 2A. *LCT* and *BARD1* are located on human chromosome 2 and on chimpanzee chromosome 2B. (See the diagram in the answer to Step 4.ii below.)

4.ii. Compare the locations of the four genes in the human and chimpanzee genomes. In Step 10 of Part I, you generated a hypothesis regarding the relationship between the human and chimpanzee chromosomes. Does your diagram support your hypothesis? Why or why not?

The locations of the human gene homologs in chimpanzee confirm that human chromosome 2 is a fusion of chimpanzee chromosomes 2A and 2B, the ancestral chromosome configuration in the Great Apes (Sp, *SPAST*; Su, *SUCLG1*; Lct, *LCT*; Ba, *BARD1*).

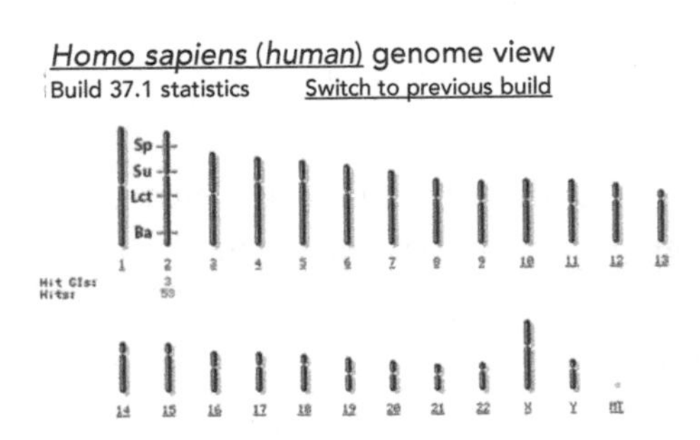

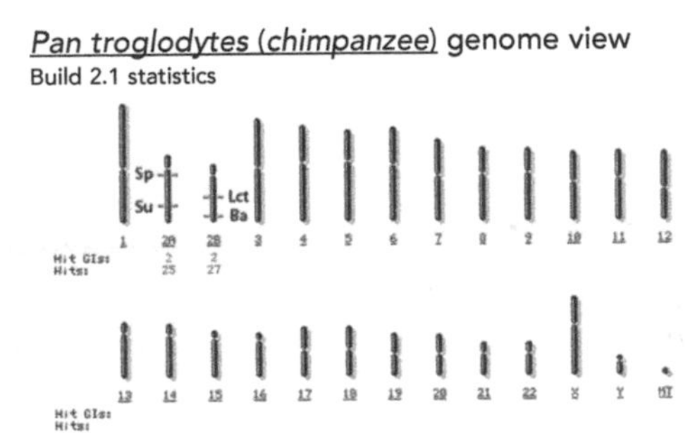

III. Compare the Human and Chimpanzee Genomes by Syntenic Dotplot

8. Generally, how would you describe the synteny between the human and chimpanzee chromosomes?

The strong, uninterrupted diagonal dotplot shows that long stretches of the human and chimpanzee chromosomes have virtually perfect synteny (identical gene order). This degree of synteny indicates that there is strong conservation of genome structure between the two species.

9. What do you notice regarding the alignment for human chromosome 2? Does this alignment support your hypothesis from Step 10 in Part I? Why or why not?

Human chromosome 2 is syntenic with two different, smaller chimpanzee chromosomes: 2A and 2B. This agrees with previous conclusions.

10. How do you explain this?

Each of these segments is an inversion, where a block of chromosome sequence recombined in the reverse orientation in either the human lineage or the chimpanzee lineage. These regions are still syntenic—the gene order is conserved, but reversed.

11.v. Describe the general structures of the homologous human and chimpanzee genes.

Genes that are homologous between human and chimpanzee have, in general, identical numbers of exons (and introns).

11.vi. Does this confirm your answer in Step 10?

Yes, the gene-to-gene alignment confirms that homologous genes in genomic regions that are displaced by 90° against each other are still syntenic.

IV. Determine the Origin of the Centromere in Human Chromosome 2

2.vi. Draw a diagram showing where the two chimpanzee chromosome 2A gene homologs are located in relation to the centromere of human chromosome 2.

Chimpanzee chromosome 2A genes *EDAR* (Ed) and *TGOLN2* (Tg) flank the human chromosome 2 centromere.

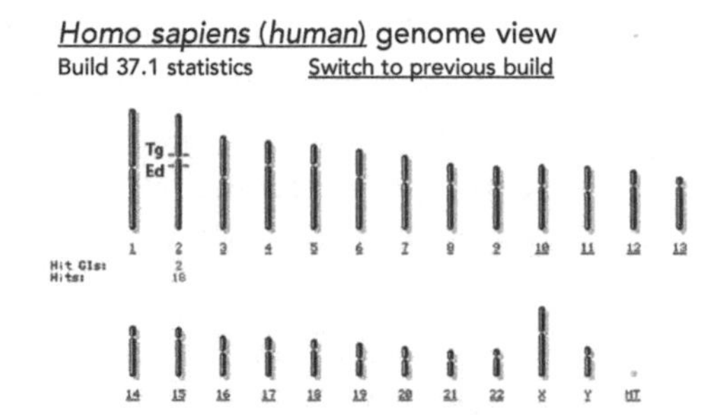

4. Draw a diagram showing where the two chimpanzee chromosome 2B gene homologs are located in relation to the centromere of human chromosome 2.

Genes *BIN1* (Bi) and *PKP4* (Pk), which flank the chimpanzee chromosome 2B centromere, localize to the long arm of human chromosome 2.

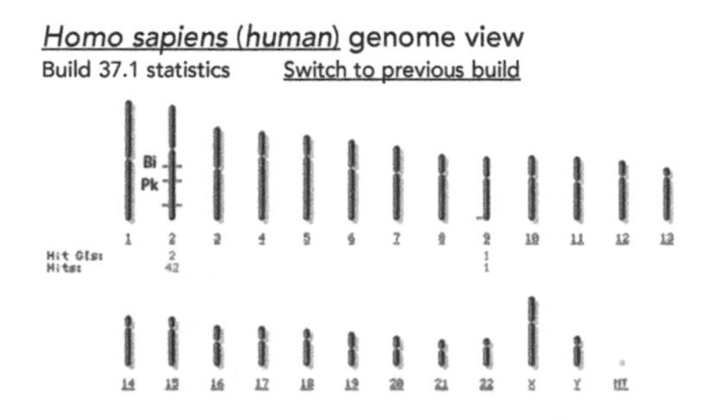

5. Where is the human chromosome 2 centromere in relation to the homologs from chimpanzee chromosomes 2A and 2B? What does this say regarding the origin of the human centromere?

Chimpanzee chromosome 2A genes *EDAR* and *TGOLN2* flank the human chromosome 2 centromere, indicating that the ancestral Great Ape chromosome 2A provided the centromere in the fusion that created human chromosome 2. Genes *BIN1* and *PKP4* flank the chimpanzee chromosome 2B centromere but localize to the long arm of human chromosome 2, more than 30 Mb away from the human centromere. Therefore, the ancestral Great Ape chromosome 2B is unlikely to have provided the centromere for human chromosome 2. This interpretation does not change when taking into account an additional match on chromosome 2 (a *PKP4* pseudogene) and a short match on human chromosome 9 (only 75 bp of the 3579-bp *PKP4* gene probe). (See the diagram in the answer to Step 4 above.)

V. Extend the Analysis of the Evolution of Human Chromosome 2

1. **Compare your results with those you obtained in Step 5 of Part IV. Do these results support your hypothesis from Step 4 of Part II? Why or why not?**

   The gene probes for chromosomes 2A and 2B in gorilla and orangutan follow the same pattern as that in chimpanzee—localizing to the same regions of human chromosome 2. This supports the contention that two ancestral chromosomes of the Great Apes (2A and 2B) fused in the human lineage to form a single chromosome (2).

2. **Comment on the chromosome number and the locations of genes. What do these results indicate regarding the overall organization of primate genomes?**

   The primate lineages leading to human and chimpanzee, on the one hand, and to macaque, on the other, diverged ~25 million years ago. The macaque genome consists of 20 autosomes, as opposed to 23 in chimpanzees (and 22 in humans). Chimpanzee chromosome 2A genes *SPAST* (Sp) and *SUCLG1* (Su), as well as 2B gene *LCT* (Lct), localize to macaque chromosome 13. The chimpanzee 2B gene, *BARD1* (Ba), localizes to macaque chromosome 12. Resolving the relationships between the macaque and Great Ape genomes in more detail may be possible by extending the analysis to include many more gene sequences as probes. However, locating just these four genes in the macaque genome indicates that, although primate genomes share a large number of homologous genes, the overall structures of their genomes and the distribution of these genes across their chromosomes vary widely.

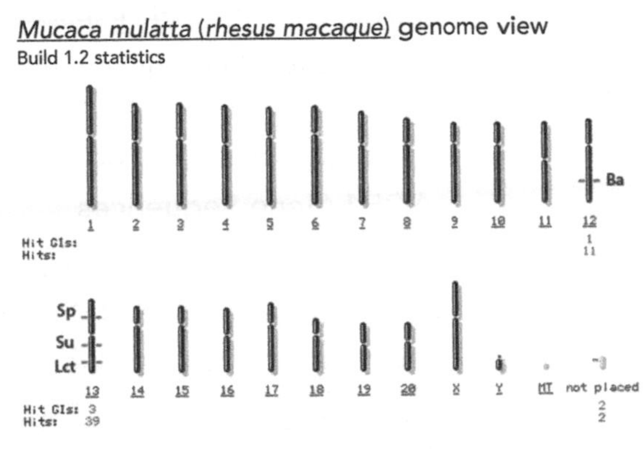

3. **What may have happened to the centromere of chromosome 2B during the fusion of chromosomes 2A and 2B?**

   Having two centromeres in one chromosome would be very unstable during mitosis or meiosis, thus selection presumably favored the loss of the centromere from ancestral chromosome 2B after the fusion to produce human chromosome 2. This becomes clear by comparing the locations of genes that most closely flank the centromere in chimpanzees with the locations of their human homologs. The chimpanzee genes *ARMGEF4* and *GPR39* are separated by a centromere region of >4 Mb on chromosome 2B, whereas the human homologs of these two genes are located much closer together on chromosome 2 (~1.5 Mb from each other). However, human chromosome 2 may contain genes between *ARMGEF4* and *GPR39* that have not been identified in chimpanzees, indicating that the absence of the centromere may be due to a more complex rearrangement that may have occurred at the same time as the fusion of 2A and 2B.

LABORATORY 1.3: COMPARING DIVERSITY IN EUKARYOTES

II. Align and Compare Two Maize *ADH1* Gene Sequences

9. **Record the total number of differences that you counted in Step 8. In addition, record the total length of the sequence aligned.**

   There are 31 differences, four of which consist of two neighboring gaps. The total number of differences is therefore 29; the total length of the alignment is 1386 nucleotides.

10. **Calculate the percent difference between the two sequences.**

   The percent difference between the two samples is 29/1386 x 100 = 2.1%.

III. Align and Compare the Maize and Teosinte *ADH1* Gene Sequences

2./3. **Count the number of nucleotide differences between the two sequences, and calculate the percent difference between the two sequences as described in Steps 8–10 in Part II. Compare the teosinte *ADH1* gene with the second corn *ADH1* sequence by repeating Steps 1 and 2.**

   The number of nucleotide differences between B73 and teosinte is 32, and the number of nucleotide differences between Mo17 and teosinte is 28. These differences amount to 2.3% and 2.0%, respectively.

4. **Is the percent difference between the two corn sequences (calculated in Step 10 in Part II) similar to the percent difference between corn and teosinte?**

   The sequence diversity between the corn sequences and teosinte is approximately the same as the sequence diversity between two corn lines.

IV. Align and Compare the Human and Chimpanzee *ADH1* Gene Sequences

1. **Use the procedure described in Part II to calculate the percent difference between two human *ADH1* gene sequences.**

   The percent difference is 0.1%.

2. **Repeat the procedure two more times to calculate the percent difference between the chimp *ADH1* gene sequence and each of the two human *ADH1* gene sequences.**

   The percent differences are 1.5% and 1.4%.

3. **Compare the lengths of the human and chimpanzee alignments with those of corn and teosinte from Parts II and III. What do you notice? How would you explain this?**

   The alignments between human sequences or between human and chimpanzee sequences (animals) are >5000 nucleotides long, whereas those between corn and teosinte (plants) are ∼1400 nucleotides long. This experiment uses *ADH1* sample sequences that include exons and introns, and plant introns are, on average, significantly shorter than introns in spliced genes in animals.

4. **Which has greater genetic diversity: humans or corn? By what factor is one group more diverse than the other? How would you explain this?**

The data indicate that the nucleotide diversity among humans is much smaller than that among corn cultivars. In fact, corn is considered 100 times more diverse than humans.

5. **How does the percent difference between the two corn varieties compare with the percent difference between human and chimpanzee? What does this seem to indicate regarding the ages of these species?**

   The nucleotide diversity between two corn cultivars exceeds that between two humans and even that between human and chimpanzee. This seems to indicate that the lineages between corn cultivars diverged much longer ago than the lineages that led to humans and chimpanzees. How is that possible, given that paleontological and genetic evidence indicates that humans and chimpanzees shared a common ancestor several million years ago, whereas we have no evidence that humans started breeding corn from teosinte any earlier than 10,000 years ago? In fact, Mo17 and B73 are likely to have shared common ancestors as recently as ~100 years ago, when breeders began to generate inbred lines. The resolution to this apparent contradiction is that teosinte itself has a large degree of variation. Given this highly diverse progenitor, a domestication process based on just a handful of heterozygous individuals could result in a high level of sequence diversity in current corn cultivars. One must also consider that recombination and transposition have contributed to genetic diversity within maize after its domestication and continue to do so today. In addition, ancestral human populations seem to have gone through a series of bottlenecks in their evolution that have limited the genomic diversity of modern humans. There is strong genomic evidence that the human lineage may have been so severely affected by catastrophic events in the past that all modern humans descended from a few thousand (or fewer) ancestral human individuals. Therefore, the degree of nucleotide variation between humans and between corn cultivars does not accurately reflect the age of the respective species.

   Another consideration is domestication. As pointed out by Jared Diamond in his book *Guns, Germs, and Steel: The Fates of Human Societies* (Norton, 1999), the domestication of crops in the Old World was followed by east-to-west movement into similar climate zones. That meant that people carried and planted seeds in places where they knew their crops would thrive. In contrast, in the New World, domestication was followed by north-to-south migrations. People moved into entirely different climate zones (highland mountains, deserts, seaside locations, savannahs, etc.), planted crops, and propagated the few individual plants that thrived. The rare successful plants may have had alleles that were adaptive and could be maintained through selection. A very important factor for plant growth and flowering is the length of the day (i.e., how many hours of light they experience). Day length is maintained when moving from east to west. In contrast, when moving from the tropics, with 12 h of light, to either the north or the south means that there are much longer summer days; under these conditions, only "mutant" tropical plants will flower. This selection on corn very likely contributed to its current diversity. Just a few thousand years after domestication, people had moved corn from the highlands of Mexico to as far south as Patagonia and as far north as Canada. Developing new dog breeds, another domesticated companion of people, illustrates just how rapidly selection can yield distinctive morphology, size, and behavior.

V. Extend the Analysis of *ADH1* Gene Sequences

1. **Use the procedure described in Part II to calculate the percent difference between wild and domesticated barley *ADH1* gene sequences.**

   The percent difference is 0.3%.

2. **Repeat the procedure to calculate the percent difference between two subspecies of domestic rice: *Oryza sativa* ssp. *japonica* and *Oryza sativa* ssp. *indica*.**

   The percent difference is 0.3%.

3. **How does the genetic diversity of barley and rice compare with that of maize? What does this seem to indicate regarding the genetic diversity in the progenitors of modern barley and rice, compared with those of modern corn?**

   The variation between rice cultivars and between domesticated and wild barley is much smaller than that between modern corn cultivars; they resemble the nucleotide variation between modern humans. The progenitors of current rice and barley cultivars were significantly less diverse than the teosinte plants that founded today's corn lineages. Additionally, until recently, both barley and rice were moved mainly east or west.

   The experiments in this bioinformatics laboratory were conducted on a portion of the *ADH1* gene that includes coding sequences and introns. Studies on other genes and intergenic genomic loci confirm these results and show that corn is, indeed, unusually diverse when compared with other cereals or humans. On the practical side, this means that there are many distinctive alleles for most corn genes; this is "raw material" for generating new lines of corn that will be well suited for new uses and locations.

4. **Align *ADH1* gene sequences from different species (e.g., corn and human, chimpanzee and rice, or rice and corn). Describe your findings.**

   The diversity of *ADH1* genes, even just between different cultivars of the same plant species, is very high. One reason for this high degree of diversity is that genomic DNA (i.e., containing introns) is being compared. However, the *ADH1* sequences are also well suited for comparing the diversity within one species with that in another species.

LABORATORY 1.4: DETERMINING THE TRANSPOSON CONTENT IN GRASSES

I. Identify Transposable Elements in the *ADH1* Region of the *Brachypodium* Genome

8.ii. **How many different repeat types can you identify?**

There are seven types: simple repeats, low-complexity DNA, and five different types of transposons.

8.v. **Determine which category of repetitive DNA occupies a larger percentage of DNA in this sample: nontransposon DNA or transposon DNA.**

In this *Brachypodium* DNA sample, 0.5% consists of simple repeats and low-complexity DNA, and 12% consists of transposons.

II. Identify Transposable Elements in the *ADH1* Region of the Rice Genome

3. **Analyze the RepeatMasker output as described in Steps 7 and 8 of Part I.**

   There are 14 types: simple repeats, low-complexity DNA, and 12 different types of transposons. In this *Oryza sativa* DNA sample, 1.4% consists of simple repeats and low-complexity DNA, and 36% consists of transposons.

III. Identify Transposable Elements in the *ADH1* Region of the Corn Genome

3. **Analyze the RepeatMasker output as described in Steps 7 and 8 of Part I.**

   There are five types: simple repeats, low-complexity DNA, and three different types of transposons. In this *Zea mays* DNA sample, 0.1% consists of simple repeats and low-complexity DNA, and 81% consists of transposons.

IV. Compare the Transposable Element Content in *Brachypodium*, Rice, and Corn

2. Compare the different types of repeats found in each species by filling in the table at the top of the next page.

3.i. **Which species contains the greatest number of transposons? First, answer the question by considering the total number of transposons in each sample. Second, answer the question by considering the percentage of DNA occupied by transposable elements in each sample.**

Rice has the largest number of transposons, but in corn, transposons account for the largest percentage of sequence.

3.ii. **Do you think that higher transposon content would lead to gene loss or would cause genes to be spaced further apart?**

Higher transposon content seems to be correlated with larger intergenic spaces.

4. **Explain why transposable elements, especially those of the same type and with similar descriptions, have a wide range in length.**

   Transposons mutate, and their sequences will differ from the sequences of the transposons with which they share a common ancestor.

5.i. **Which transposons appear to be present in all three species?**

The transposons common in all species are DNA/En-Spm, DNA/MuDR, LINE, LTR/*Copia*, and LTR/*Gypsy*.

5.ii. **What would the presence of the same transposon in all three species indicate for the evolutionary history of these transposons: Did they emerge before or after the three species diverged from each other?**

The simplest (and most probable) explanation would be that they emerged before the ancestors of the three species diverged from each other.

5.iii. **Which transposons appear to be present in only one species?**

The transposons limited to one species (all only found in rice) include DNA/hAT, DNA/TcMar-*Stowaway*, DNA/*Tourist*, and SINE.

		Simple repeats/ low-complexity DNA	Transposons
Bd4	Number	12	24
	Total length (bp)	509	11,780
	Percentage of DNA occupied	0.5%	12%
	Length range (bp)	22–151	44–6772
	Average length (bp)	42	1563
	Median length (bp)	32	1184
Os11	Number	25	62
	Total length (bp)	1375	36,005
	Percentage of DNA occupied	1.4%	36%
	Length range (bp)	21–124	36–2140
	Average length (bp)	55	491
	Median length (bp)	37	304
Zm11	Number	4	52
	Total length (bp)	108	81,281
	Percentage of DNA occupied	0.1%	81%
	Length range (bp)	23–31	36–9278
	Average length (bp)	27	581
	Median length (bp)	27	192

5.iv. Why might you find transposable elements that appear to be species specific?

It is possible that new transposons emerge after a species has separated. On the other hand, sample sizes of 100 kb are not sufficient to judge the entire diversity of the transposable elements in a species, and transposons lacking from a 100-kb contig may be present elsewhere in the genome. Currently, with only a few plant genomes sequenced, we may only be seeing the tip of the iceberg of diversity among plants, including transposon diversity.

V. Extend the Analysis of Transposons Using Other Tools in *DNA Subway*

1.i. Can you identify the transposase and reverse transcriptase genes in the DNA transposons and retroposons, respectively? Why or why not?

The gene prediction algorithms do not mark genes in transposons because RepeatMasker masks these genes out before gene prediction and BLAST search algorithms are applied.

1.ii. Are transposons located within genes? If so, are they more common in exons or in introns?

Yes; transposons are frequently found in introns. (See Lab 1.1, Stage A, Section X.7.)

1.iii. Do genes outnumber transposons or do transposons outnumber genes?

That depends on the species. However, the RepeatMasker results confirm that a large proportion of the genomes of higher eukaryotes contains repetitive DNA, predominantly transposons. More than 50% of the human genome, for example, consists of repetitive DNA, whereas protein-coding genes only account for ~25% of its DNA with only ~1.5% of the human genome coding for amino acid sequences (the rest are UTRs and introns). This situation is likely to be more pronounced in plants, which, in general, have shorter introns than humans and animals.

2. **What do you observe?**

   The outcome for this exercise will depend on the transposon transferred from the Red Line to the Yellow Line. However, because DNA sequences are less well conserved across species than are protein sequences, transferring transposon DNA may only yield matches in the same organism. To try this out, transfer the DNA of the 2140-bp DANIELA LTR/*Gypsy* transposon at the leftmost position in the *Brachypodium* chromosome 4 sample to the Yellow Line, and prospect the *Brachypodium* genome for similar transposons. Then prospect the rice genome with the same sequence. Searching *Brachypodium* with DANIELA yields more than 100 hits; searching rice yields none.

LABORATORY 1.5: IDENTIFYING *GAI* GENE FAMILY MEMBERS IN PLANTS

I. Determine How GAI Proteins from Different Plant Species Are Related

8.ii. Do you find evidence that the GAI sequences are related? What is the highest percentage of similarity that you find? What is the lowest percentage of similarity that you find?

Yes, the sequence conservation is 40% or higher across the entire protein alignment. The percentage of similarity varies between 40% and 100%.

8.iii. What is the highest percentage of similarity that you can detect? Between what proteins? What is the lowest percentage of similarity that you can detect? Between what proteins? Use the values for the pairwise similarity between the GAI proteins to discern how closely they are related.

The similarity between the GAI proteins varies between 96.75% for barley and wheat and 62.10% for barley and *Arabidopsis*. Because all pairwise alignments show a similarity far above 35% the GAI protein is highly conserved, not just among the grass species barley, corn, rice, and wheat—but also between the grasses and *Arabidopsis*.

9. **What is the one-letter "Consensus" amino acid sequence for these five positions?**

   The five amino acids in all five GAI proteins read "DELLA."

13. **Describe the phylogenetic relationships among the GAI proteins of the different plants.**

    Barley and wheat *GAI* are most closely related to each other followed by corn and then by rice. *Arabidopsis GAI* is most distantly related, as could be expected given the fact that *Arabidopsis* is a dicotyledonous plant while the others are monocotyledonous grasses.

II. Identify the Cause of the gai Semi-Dwarf Phenotype

2. What part of the GAI gene and protein has been affected in each of the mutants? Do the three gai mutants have anything in common?

Each of the mutated proteins is affected within its first 65 amino acids. The *Arabidopsis gai* allele carries a 51-bp deletion, leading to a deletion of 17 amino acids in the gai protein, rendering the mutant insensitive to GA. The mutation in wheat *Rht-D1b* consists of a point mutation G-T in the 5′ coding region of the gene, leading to a premature stop codon GAG-TAG that causes the deletion of the first 64 amino acids from the protein. Notice that the premature stop codon in position 181–183 does not abolish the gai protein altogether. Instead, transcription picks up the ATG in position 193–195, leading to a truncated protein that causes semi-dwarfism in wheat *Rht-D1b*. The semi-dwarf phenotype in corn D8-1 is caused by a deletion in the 5′ region of the gene, which changes the amino acid in position 55 from aspartic acid to glycine and causes an adjacent deletion of four amino acids.

The gai proteins encoded by *Arabidopsis gai* and wheat *Rht-D1b* lack the DELLA region altogether. In contrast, the gai protein encoded by *D8-1* has retained the DELLA amino acid sequence. Although the corn *D8-1* mutation does not affect the DELLA sequence, it is part of the DELLA region and is likely to explain the semi-dwarf appearance.

4. Explain how this experiment showed that the gibberellin-insensitive phenotype is based on a gain of function.

Transforming rice with a mutated gai gene from *Arabidopsis* produced semi-dwarf transformants that could not be complemented by supplying gibberellic acid exogenously. This result was achieved despite the fact that the transformants continued to harbor their endogenous, unaltered rice *GAI* genes. The semi-dwarf phenotype is therefore based on a dominant gain-of-function mutation that is able to mask the function of the endogenous wild-type *GAI* gene complement of a plant.

5. Explain how the gain-of-function character of the mutated *gai* gene has facilitated rapid generation of semi-dwarf corn and wheat varieties, despite the presence of multiple copies of *GAI* in these genomes.

The dominant gain-of-function character of the *gai* mutation allows the mutant phenotype to be expressed regardless of the presence of many copies of the wild-type *GAI* genes in polyploid genomes.

III. Think About the Next Wave of the Green Revolution

10. Devise a strategy that would use the knowledge regarding potential GAI homologs in your target crop plant to generate a semi-dwarf variety of this plant.

Many crop plants have *GAI* gene homologs that encode the DELLA region. The presence of these genes indicates a potential lever for generating semi-dwarf mutants by either altering the endogenous *GAI* gene or by transforming the plant with a mutated *gai* gene from another species. The generation of semi-dwarf, high-yielding Basmati rice by transforming it with *Arabidopsis gai* was proof for the principle that Green Revolution varieties of cereals and other crop plants can be produced by transforming tall-growing varieties with a mutated *gai* gene—even if the gene might be derived from a different plant species.

IV. Extend the Analysis of GAI Sequences

1. **How many matches can you identify? How many of these have a DELLA region?**

   Using the *Arabidopsis* GAI protein to prospect the rice genome yields three matches, one of which appears to carry a DELLA region. This particular protein is ~58% identical to *Arabidopsis* GAI; the other two proteins share 34% identity with *Arabidopsis* GAI.

2. **How would you expect the results to differ if you used a GAI protein sequence from a species that is more closely related to rice?**

   Using a GAI protein from a closer relative should yield more matches and/or a higher degree of similarity. Using other cereal GAI proteins to prospect the rice genome yields the DELLA protein with degrees of similarity between the DELLA proteins of 86% for barley (*H. vulgare*) and for wheat (*T. aestivum*) and 87% for corn (*Z. mays*). Using the wheat GAI protein as a probe yields an additional match in rice that was not identified with the *Arabidopsis* GAI. This match, however, is not a DELLA protein but a member of an over-arching gene family, the GRAS proteins (named after GAI, RGA, and SCR, the first three members that were identified in the family). GRAS proteins have two leucine-rich areas flanking a VHIID motif; try to find these in the alignments in *DNA Subway*. GRAS proteins are important regulatory components in several different cellular processes ranging from meristem maintenance to hormone signaling.

3. **How would you expect the results to differ if you would prospect the poplar genome? How about grape? The moss *Physcomitrella*? The green algae *Chlamydomonas*? (Would you expect green algae to have homologs to a gene that represses stem elongation in higher plants?) Make your predictions, and then conduct the experiments using *DNA Subway*. What might it mean if GAI is present in both trees and mosses—what do trees have in common with mosses, and how are they different?**

   As in Step 2, it is difficult to predict the potential to identify GAI in other species. Using the *Arabidopsis* GAI protein as a probe yields the following matches:

   - Poplar: 22 matches, four of which contain a DELLA region.
   - Grape: nine matches, two of which contain a DELLA region.
   - *Physcomitrella:* three matches, none of which contains a DELLA region.
   - *Chlamydomonas:* no matches.

   The lack of *Chlamydomonas* proteins matching the *Arabidopsis* GAI protein could reflect the fact that the alga is too distantly related to find a match. Prospecting the *Chlamydomonas* genome with *Arabidopsis* ADH1, however, yields *Chlamydomonas* ADH1. Therefore, the absence of a GAI match from *Chlamydomonas* may indicate that the alga does not have a GAI homolog. *Chlamydomonas* is unicellular and may not need a gene that controls height.

   Poplar, on the other hand, is a tree, one that is known for its rapid stem growth and is farmed for paper and biofuel production. The rapid stem elongation of poplar may be ensured by multiple *GAI* genes that may enhance GA sensitivity, although this hypothesis has not been tested.

4. **How would you expect the results to differ if you prospected the genomes using the gene sequences instead of the protein sequences? Which would you expect to be more conserved to ensure an equivalent protein function—the genes or the proteins?**

The following table presents the number of matches obtained following prospecting various plant genomes with the *Arabidopsis* GAI protein and gene sequences. The first number indicates the total number of matches; the second indicates the number of DELLA-containing matches.

	Query sequence	
TARGeT genome	*A. thaliana* GAI protein	*A. thaliana GAI* gene
Arabidopsis	7/2	2/1
Grape	9/2	0/0
Poplar	22/4	0/0
Corn	9/2	0/0
Rice	3/1	0/0
Sorghum	9/0	0/0
Physcomitrella	3/0	0/0
Chlamydomonas	0/0	0/0

Using protein queries yields significantly more matches than using nucleotide queries. This is because functionally or structurally important amino acids are conserved in proteins; the redundancy of the genetic code, however, allows nucleotide changes without amino acid changes. Therefore, genes may have changed to a much higher degree than the resulting proteins, allowing protein searches to identify more hits than nucleotide searches. As the results for the search of the *Arabidopsis* genome show, this is true not only for members of gene families in different organisms but also for members of gene families in the same organism: The search with the GAI protein yielded seven matches in *Arabidopsis*, two of which contain the DELLA region. Querying *Arabidopsis* with the *Arabidopsis GAI* gene yielded two matches, one of which contained the DELLA region.

The tables below show additional search results for various genomes using GAI proteins and genes from three different cereals as queries. In all cases, the first number indicates the total number of matches; the second indicates the number of DELLA-containing matches.

	Query sequence	
TARGeT genome	Rice GAI protein	Rice *GAI* gene
Arabidopsis	5/2	0/0
Grape	4/2	0/0
Poplar	5/4	0/0
Corn	4/2	2/2
Rice	1/1	1/1
Sorghum	1/0	1/0
Physcomitrella	3/0	0/0
Chlamydomonas	0/0	0/0

	Query sequence	
TARGeT genome	Wheat GAI protein	Wheat *GAI* gene
Arabidopsis	6/2	0/0
Grape	4/2	0/0
Poplar	6/4	0/0
Corn	4/2	2/2
Rice	2/1	1/1
Sorghum	1/0	1/0
Physcomitrella	5/0	0/0
Chlamydomonas	0/0	0/0

	Query sequence	
TARGeT genome	Corn GAI protein	Corn *GAI* gene
Arabidopsis	5/2	0/0
Grape	5/2	0/0
Poplar	5/4	0/0
Corn	2/2	2/2
Rice	1/1	1/1
Sorghum	1/0	1/0
Physcomitrella	3/0	0/0
Chlamydomonas	0/0	0/0

5. **Examine the *GAI/gai* genes in *DNA Subway*'s Blue Line. Compare the homology among the GAI genes and proteins from the five different species. Why do the gene sequences show different degrees of conservation than those of the protein sequences? Then, use MUSCLE to identify the mutations in the *gai* genes. Compare and contrast the *gai* gene mutation for each species with the protein alteration for the same species.**

   The *GAI* genes show much more diversity than GAI proteins. However, the redundancy of the genetic code by which the same amino acid can be placed by different triplets ensures that proteins that result from diverse DNA sequences can contain the same amino acids in key positions responsible for the function of the protein. Therefore, changes in genes do not necessarily lead to changes in proteins, and related amino acid sequences frequently remain more conserved than the underlying coding DNA sequences may suggest.

LABORATORY 1.6: DISCOVERING EVIDENCE FOR PSEUDOGENE FUNCTION

I. Use Map Viewer to Locate the *PTEN* Gene and Its Pseudogene, *PTENP1*, in the Human Genome

2.iv. **Determine the size of *PTENP1* using the map coordinates to the left of the "Genes_seq" map.**

PTENP1 consists of one nonspliced sequence of 3917 bp.

2.v. Determine the structure of *PTENP1*. How many introns and exons does *PTENP1* have?

There are no introns and, therefore, no exons.

3.iv. Determine the size of *PTEN* as described in Step 2.iv.

The *PTEN* gene is 105,337 bp long.

3.v. Determine the structure of *PTEN* as described in Step 2.v. How many introns and exons does *PTEN* have?

The *PTEN* gene consists of nine exons and eight introns.

II. Use Map Viewer to Determine the Relationship between *PTEN* and *PTENP1*

2.viii. Record the numbers of the chromosomes upon which they are located.

Chromosomes 9 and 10.

2.xi. Count the number of hits (in red).

Nine hits, all located in *PTEN* exons.

2.xii. Determine which gene the *PTENP1* query matched.

PTEN.

3.i. Are matches located in exons, introns, or UTRs?

PTENP1 matches the *PTEN* exons as well as the proximal segments of the *PTEN* 5′ and 3′ UTRs.

3.ii. Is *PTENP1* a processed or a nonprocessed pseudogene?

Nonprocessed pseudogenes arose via gene (DNA) duplications; processed pseudogenes arose via retrotransposition of mRNAs. Because *PTENP1* resembles the *PTEN* mRNA, it is likely to be a processed (or retrotransposed) pseudogene.

3.iii. Propose a mechanism by which *PTENP1* may have been derived from *PTEN*.

PTENP1 could be derived from *PTEN* mRNA. *PTENP1* is most likely derived from the reverse transcription of *PTEN* mRNA and insertion of this DNA into a new location in the genome. Pseudogenes that are generated by this mechanism are called "processed pseudogenes."

III. Use *Gene Boy* to Determine the Protein-Coding Capacity of the *PTENP1* Pseudogene

2.vii. Examine the amino acid sequences. Determine whether the translated sequences in RF1, RF2, and RF3 begin with a methionine.

The PTEN protein encoded by RF1 carries an M for the amino acid methionine. Neither of the translated sequences for RF2 or RF3 begins with an M.

2.viii. Determine whether the translated sequences in RF1, RF2, and RF3 encode a functional protein (not interrupted by a stop codon).

In RF1, *PTENP1* is not interrupted by stop codons, with the exception of the stop codon at the end of the ORF. In RF2 and RF3, *PTENP1* is interrupted by multiple stop codons that are dispersed across the sequence.

3. **Compare your results with the *PTEN* results generated above.**

   Translating the *PTEN* and *PTENP1* ORF sequences into amino acids shows that, in RF1, neither *PTEN* nor *PTENP1* is interrupted by stop codons, with the exception of the stop codon at the end of each ORF. In RF2 and RF3, both genes are interrupted by multiple stop codons dispersed across the sequences. Therefore, like *PTEN* the *PTENP1* ORF can be translated into a protein of 403 amino acids and does not contain a premature stop codon. However, none of the three *PTENP1* reading frames begins with methionine. No evidence has been found that *PTENP1* mRNA is translated into a protein in vivo, most likely because of the lack of the canonical start codon (ATG). This makes it a pseudogene.

5.xiii. **Do the sequences share regions of strong similarity (>90%)?**

   Yes, the two translated DNA sequences are more than 95% similar, indicating a high degree of conservation. This is somewhat curious because the *PTENP1* gene is considered to be a pseudogene, the amino acid sequence of which should be free to mutate without consequences for the organism. The pseudogene character of the PTENP1 sequence is confirmed by the absence of a functional start codon and, therefore, the absence of a leading M (methionine) in the amino acid sequence.

5.xiv. **Count the amino acid differences. How many amino acid differences can you identify?**

   The sequences differ in 11 amino acids, the most significant of which appears to be the absence of a leading methionine in PTENP1.

IV. Use TargetScan and CLUSTAL W to Identify a Potential Function for the *PTENP1* Pseudogene

1.v. **Count the number of miRNA-binding sites in the *PTEN* 3′ UTR.**

   The PTEN 3′ UTR contains 23 predicted miRNA-binding sites.

2.xiii. **Record the names of the miRNAs for which the *PTEN* and *PTENP1* 3′-UTR sequences are conserved.**

   The binding sites for all miRNAs in the first 1000 bp of the *PTEN* 3′ UTR are conserved in *PTENP1*: miR26ab, miR17-5p, miR542, miR19, miR130, mir29abc, miR22, miR486, and miR205.

3. **Assess the potential for *PTENP1* to serve a function. How does the level of sequence conservation in the 3′ UTR (as evaluated in Step 2) compare with the level of sequence conservation in the ORF (as evaluated in Step 5.xiii of Part III)? Do your findings indicate selective pressure toward maintaining a purpose for the pseudogene? Explain.**

   The sequence similarity between *PTENP1* and PTEN is far greater within the ORF and flanking regions of the UTRs than in the more distant UTR regions. This suggests selection for the interior part of the *PTENP1* pseudogene, implying a potential function for these regions. The pseudogene has lost its start codon and is unlikely to yield a functional protein unless a different ATG takes on that role (such as the ATG in position 103–105 of the *PTENP1* ORF, which would be in-frame with the *PTEN* ORF sequence). Because it appears that *PTENP1* can bind to some of the same miRNAs as *PTEN*, the PTENP1 transcript may function by reducing the effect of miRNAs on the regulation of PTEN. Bioinformatics and experimental studies on *PTEN/PTENP1* indicate that *PTENP1* might serve as a decoy for miRNAs targeting PTEN. This, in turn, would maximize the abundance of the tumor-suppressor PTEN and minimize the chance for tumor formation. Therefore, even though it is called a pseudogene, *PTENP1* may function as an auxiliary tumor-suppressor gene.

V. Extend the Analysis of Pseudogene Sequences

1. **Use *Sequence Server* and/or *DNA Subway* to compare in detail various structural features of *PTEN* and *PTENP1* (e.g., 5′ UTRs, 3′ UTRs, and mRNA sequences).**

   The *PTENP1* gene shares a high degree of homology with the mRNA of the *PTENP1* gene. This similarity begins ~675 nucleotides upstream of the start codon in the PTEN 5′ UTR, spans the entire coding region (ORF) and extends by about 1086 nucleotides into the *PTEN* 3′ UTR.

   The last 675 nucleotides of the 5′ UTR are 96.7% identical. The ORFs for the two genes are 98.5% identical. The first 1086 nucleotides of the 3′ UTRs are 97.6% identical. The high degree of conservation for this pseudogene, again, is an indicator that it continues to have an important function for the cell and organism.

2. **Use Map Viewer to determine whether other gene/pseudogene combinations (*KRAS/KRAS1P, E2F3/E2F3P1, BRCA1/BRCA1P*, and *HBZ/HBZP*) are closely related.**

   The coding sequence and 3′ UTRs in *KRAS1P* and E2F3P1 are 89%–95% conserved with those of *KRAS* and *E2F3*. *BRCA1P* shows similarity (90%) to BRCA1 within the first 1500 bp of the gene, including the 5′ UTR, first introns, and beginning of the coding sequence; the similarity also extends 1200 bp upstream into the promoter. The *HBZP* sequence is similar to the entire *HBZ* gene sequence and contains some duplicated intronic sequences. The 5′ UTR, coding sequence, and the first 40 bp of the 3′ UTR are highly conserved (99%). This level of similarity extends ~130 bp upstream into the promoter region, to include identical TATA-box sequences.

3. **Use *Gene Boy* to determine what renders the pseudogenes *KRAS1P, E2F3P1, BRCA1P*, and *HBZP* nonfunctional.**

   KRAS1P, E2F3P1, and *BRCA1P* have no start codons and are interrupted by stop codons in all three reading frames. *HBZP* has a start codon but is interrupted by a nonsense mutation (stop codon) 19 bp into the ORF. However, RF3 of *HBZP* has no stop codon and could encode part of a new protein if the gene were transcribed and spliced appropriately.

4. **Use TargetScan and CLUSTAL W to identify any conserved miRNA-binding sites in *KRAS1P, E2F3P1*, and *BRCA1P*.**

   At the time of this writing, miRNA bioinformatics is still in its infancy, and you might find better tools with which to perform this analysis at the time when you work through these laboratories. The conserved 3′ UTRs in *KRAS1P* and *E2F3P1* may attenuate RNA interference, targeting the functional counterparts of these pseudogenes (*KRAS1* and *E2F3*, respectively). The *BRCA1P* 3′ UTR, on the other hand, has no sequence similarity to that of *BRCA1*. *BRCA1P* and *BRCA1* share similar 5′ regions, including 1200 bp of the *BRCA1* promoter. The same is true for the 5′ regions of the *HBZ*/*HBZP* gene pair, but the reason for these conserved regions remains unclear.

5. **Align the *HBZP* and the *HBZ* ORFs in *Sequence Server*, and determine how *HBZP* have been generated from *HBZ*. How does this mechanism of pseudogene genesis compare with that proposed for *PTENP1* in Step 3.iii of Part II?**

   HBZP has introns and is unlikely to have been derived by reverse transcription of *HBZ* mRNA. *HBZP* appears to be a nonprocessed pseudogene that, most likely, arose by duplication. However, *HBZP* and *HBZ* show a high degree of sequence conservation; this could point toward a recent duplication event or indicate that the *HBZP* pseudogene serves a function. One way to try to answer this question would be to determine the age of the duplication. For instance, if chimpanzees have this gene/pseudogene combination, then the pseudogene would predate the divergence of humans and chimps (which occurred 5–6 million years ago) and strongly indicate that this pseudogene has a function.

CHAPTER 2

The Human Genome

THE HUMAN GENOME IS AT ONCE A PUBLIC and intensely private record. Written in each person's DNA is a shared history of the evolution of our species and a personal portent of both the health and disabilities we may encounter as individuals. The publication of the finished reference human genome sequence in April 2003 was the culmination of a biological revolution that began with Watson and Crick's discovery of the DNA structure half a century earlier. Now, scans of the genomes of thousands of individuals are uncovering an exponentially increasing number of genes implicated in human diseases. Pharmaceutical companies are racing to convert this new knowledge into new drugs, while the medical profession ponders how it will keep up with a flood of new diagnostic and treatment options. And the implications extend far beyond medicine, as scientists seek genes behind the behaviors—and misbehaviors—that make us uniquely human.

An amazing amount of personal DNA information is becoming affordable to the interested person willing to provide a saliva sample by mail. With the tagline "genetics just got personal," the company 23andMe, for less than $300, offers a sophisticated genome scan of DNA extracted from saliva. The company uses a research-grade microarray to analyze more than 600,000 DNA variations (mainly single-nucleotide polymorphisms, or SNPs) and then derive risk information about 150 diseases and health-related traits—as well as genetic ancestry—for that person. Several other companies will do a similar analysis at a competitive price.

However, direct DNA sequencing methods are becoming so inexpensive that examination of whole-genome sequences is replacing microarray methods. To date, whole-genome sequences have been published for about 20 individuals, but within the next year, the 1000 Genomes Project will increase that number 50-fold and provide a large-scale survey of structural variation of human chromosomes. We are only a year or two away from the day when an entire human genome sequence can be generated for a cost of $1000. Perhaps not everyone will be able to afford to carry a complete copy of their genome in a medical alert locket, but rapid scans of thousands of medically and behaviorally important genes will almost certainly become part of standard medical care in developed countries.

On a personal level, will a genome-wide scan resemble a genetic tarot, predicting the future course of our lives? What will it be like when we have a precise catalog of all the good, bad, and middling gene variants—and the wherewithal to determine who has which? In the face of such knowledge, will society continue to acquiesce to those

James D. Watson
(Courtesy of Cold Spring Harbor Laboratory Archives.)

who prefer to let nature take its course or will we gravitate toward a prescribed definition of the "right" genetic stuff?

Although ethicists have debated the consequences of such genetic knowledge, it may be more banal than originally thought. The finished reference human genome sequence was a composite of the genomes of several anonymous individuals, for which an extensive informed consent protocol had been followed. However, on becoming the first known individual to have his genome sequenced, James D. Watson released his entire sequence online—with the exception of the *APOE* gene, which estimates the risk of Alzheimer's disease. Watson's sequence has been available for public examination since 2007, with no negative consequences for him.

THE STRUCTURE OF THE HUMAN GENOME

The nucleus of each human cell contains 23 pairs of chromosomes. There are 22 pairs of autosomes and 1 pair of sex chromosomes (X and Y). With the exception of the X/Y pair in males, each member of a homologous chromosome pair has the same linear arrangement of genes. One tiny circular chromosome is also contained in the mitochondrion. For the sake of simplicity, biologists are concerned only with the haploid genome composed of a single set of 25 chromosomes: the 22 autosomes, the X and Y chromosomes, and the mitochondrial chromosome. Each chromosome is a complex

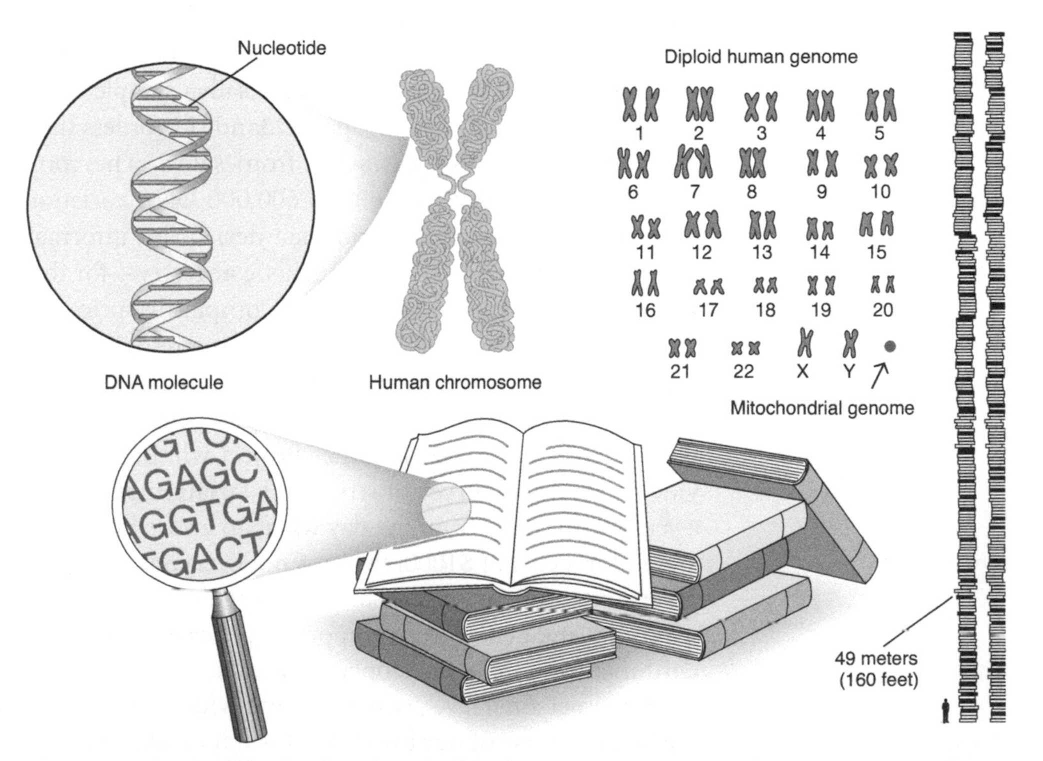

The genomic book of life. If each nucleotide of the human genome is equated to a single letter in this textbook, and the nucleotide sequence is written without spaces, punctuation, illustrations, or tables, then 3675 nucleotides would fill one page of type. At this rate, the DNA sequence of the haploid human genome (the 22 autosomes, the X and Y chromosomes, and the mitochondrial chromosome) would fill 843,500 pages. If contained in volumes the size of this textbook, the genomic book of life would comprise a stack 49.21 meters tall. Twice this amount would be needed to represent the genetic information of the full diploid genome.

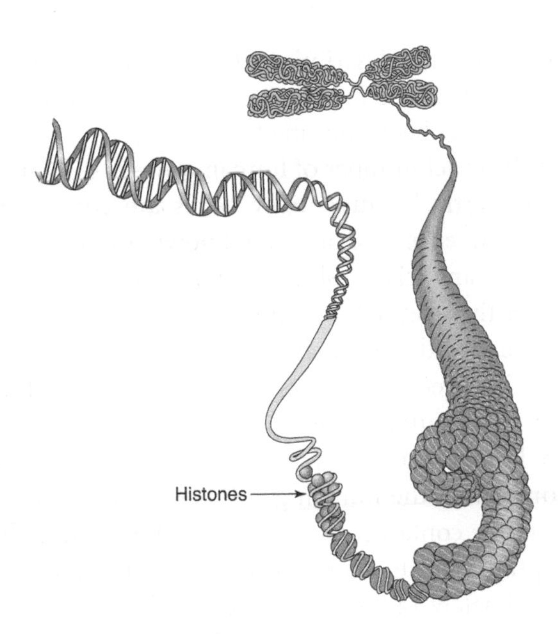

DNA packing in a chromosome. Complex folding around histone proteins allows the approximately 6 feet of DNA in each human cell to be compacted into the 46 nuclear chromosomes.

Electron micrograph of a histone-depleted chromosome. The entire protein scaffold takes the shape of a chromosome with DNA spilling all around.

(Reprinted from Paulson J, Laemmli UK. 1977. *Cell* 12: 817–828 with permission from Elsevier.)

package containing a single, unbroken strand of DNA. The bulk of the chromosome structure is contributed by protein molecules (histones) that form a scaffold around which the DNA strand is wound, like coils of rope around a capstan. Thus, it is useful to think of the human genome as a set of very long DNA molecules, each one corresponding to a chromosome. The 25 DNA molecules of the haploid human genome contain approximately 3.1 billion (3.1×10^9) nucleotides.

The average human gene is ~35,000 nucleotides long and has seven exons (containing the protein-coding sequence) and six introns. At 6400 nucleotides in length, the average intron is 20 times as long as the average exon (310 nucleotides). Genes span ~25% of the human chromosomes, with protein-coding sequences accounting for a little more than 1% and intragenic noncoding sequences (introns) accounting for ~24%. Intergenic noncoding regions, located between genes, comprise ~75% of the genome. Approximately 20%–30% of genes are clustered in "CpG islands" that comprise <10% of the genome. Gene "deserts," tracts of 500,000 or more nucleotides without genes, comprise ~20% of the human genome.

The most recent build of the reference human genome contains ~22,400 protein-coding genes. According to Online Mendelian Inheritance in Man (OMIM), an authoritative catalog of human genes and genetic disorders, only ~3000 of these have detailed

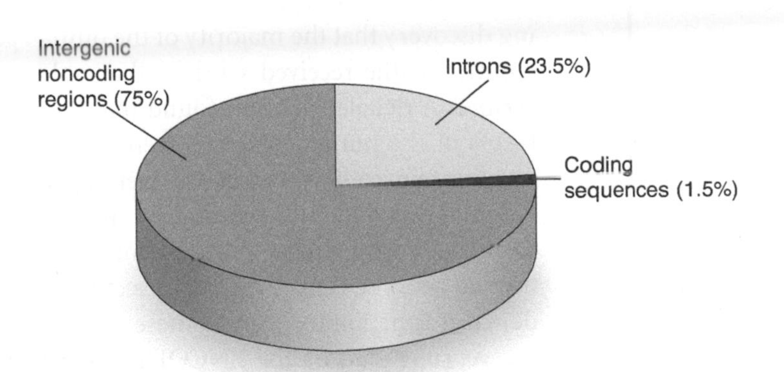

Composition of the human genome.

phenotype and molecular descriptions. OMIM still contains 1600 traits or disease susceptibilities that have been linked to specific chromosome regions but whose gene sequences have not yet been identified.

Ironically, the exact number of human genes is still unknown—and may never be known. This is partly because many genes are computer-based predictions from genomic DNA sequence and are not known to contribute to any phenotype in humans or to have any relationship to genes that have been identified in other organisms. It is also partly because the definition of a gene is evolving. A decade ago, the definition of a gene would have included only sequences that encode proteins and the RNA molecules involved in protein synthesis (transfer and ribosomal RNA). However, now the collection of human genes includes an increasing number of small RNA molecules that are known to have important roles in gene regulation.

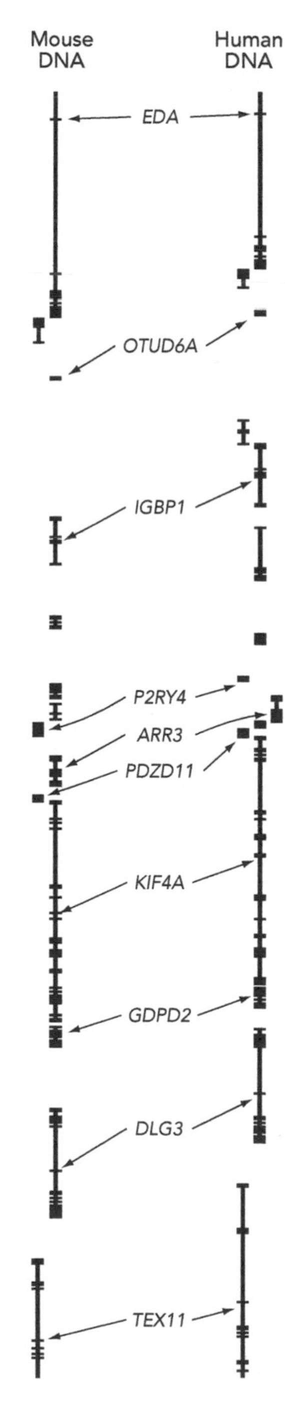

Synteny of human and mouse chromosomes. The order of shared genes is perfectly conserved over a 650,000-nucleotide region of the long arm of the X chromosome (Xq13).

Before completing the human genome sequence, most biologists estimated that the human genome contained at least 100,000 protein-coding genes. This was based on the assumption that the complex human body would require several times more than the approximately 14,000 genes in the fruit fly *Drosophila* or the 20,500 genes in the roundworm *Caenorhabditis elegans*. The revelation that humans have about the same number of genes as *C. elegans* indicates that gene number is not a very accurate measure of organism complexity. Since the discovery of RNA splicing in 1977, biologists have realized that exons from one gene can be assembled in different configurations to encode multiple proteins. However, the extent of this alternative splicing was not known. There is now compelling evidence that 95% of human genes are alternately spliced. On average, each human gene produces approximately seven alternatively spliced products. This posttranscriptional modification effectively expands the human protein repertoire many-fold beyond the actual number of genes. Most genes seem to preferentially express one form, with others in minor concentrations; however, the biological importance of the minor forms is often not known.

In 2002, a whole-genome comparison in humans and mice revealed that ~90% of chromosomal regions have maintained the same order of genes (synteny) in the two species since they diverged from a common ancestor ~75 million years ago. Furthermore, ~5% of 50–100-bp sequences in each of the two genomes are virtually identical—indicating that they have been highly conserved by purifying selection, which eliminates deleterious mutations in key functional units. This is about four times greater than the amount of protein-coding sequence in the genome, suggesting that noncoding regulatory elements within introns and in intergenic regions, chromatin structural elements, and RNA genes create transcriptional diversity that makes each mammal unique.

In 2007, the Encyclopedia of DNA Elements (ENCODE) project made the surprising discovery that the majority of the human genome is transcribed into RNA. This ran counter to the received wisdom that only protein-coding and RNA genes are transcribed. A detailed analysis found that the majority of nucleotides in a 30-Mb sample (~1% of the human genome) were present in at least one primary transcript. Many non-protein-coding transcripts overlapped known protein-coding regions, and many novel transcripts were found in regions thought to be transcriptionally silent. Consistent with earlier mouse comparative data, the project found that 5% of the human genome is under strong selection and provided independent experimental evidence of function for 60% of these conserved nucleotides in the sample studied.

As suggested by the ENCODE project, the non-protein-coding regions between and within genes are far from vacant or featureless. In addition to regulatory elements,

Pre-mRNA

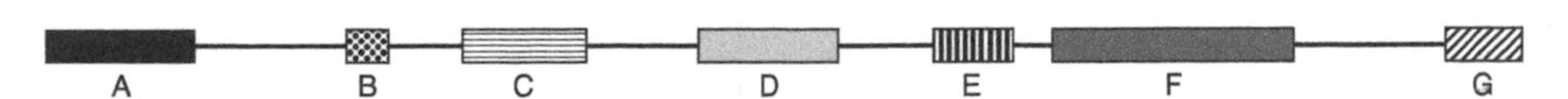

Mature mRNA

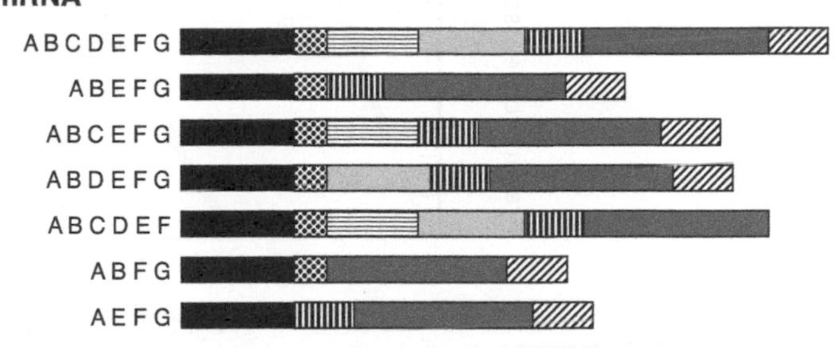

Alternative splicing of an average human gene. The average human gene contains seven exons (A–G) that produce an average of seven alternatively spliced products. Introns are not drawn to scale, as the average human intron is 20 times the length of an average exon.

such as promoters and enhancers, noncoding regions are chock full of several types of repeated DNA sequences. Ranging from dinucleotide repeats to functional and non-functional genes to duplicated chromosome segments, repeat DNA comprises at least two-thirds of the human genome.

REPETITIVE DNA

Roy Britten, 1970s
(Courtesy of Roy Britten, California Institute of Technology.)

David Kohne
(Courtesy of Connie Kohne.)

The extent of repetitive DNA in genomes first became apparent in hybridization studies conducted by Roy Britten and David Kohne in the 1950s and 1960s at the Carnegie Institution of Washington. They sheared DNA, denatured it using high temperatures, and then measured the rate at which the single-stranded fragments reassociated (renatured) with complementary sequences to form duplex DNA. Reassociation was measured by passing the DNA through a hydroxyapatite column and comparing the amount of double-stranded DNA bound by the column with the amount of single-stranded DNA in the flowthrough. The fraction of reassociated DNA was then plotted versus the log of the product of DNA concentration and time (C_0t).

Britten and Kohne reasoned that reassociation is a function of DNA complexity. The more complex the DNA—the greater the number of unique sequences—the longer it takes for one single-stranded DNA molecule to "find" its complementary match. Bacterial DNA showed a single reassociation curve, indicating that it was composed mainly of unique, open reading frame genes. However, eukaryotic DNA showed three distinct components. Two components of eukaryotic DNA reassociated more quickly than bacterial DNA; the fastest component renatured nearly as quickly as a control in which poly(A) reassociated with poly(U).

The fastest renaturing component was composed of highly repetitive DNA, which, like poly(A), finds complementary matches very quickly. The second component was composed of "middle" repetitive sequences. The third component, the slowest renaturing fraction containing unique DNA sequences, hybridized strongly with radioactive mRNA, indicating that only a minority of eukaryotic DNA encodes protein. This was the beginning of our understanding that the genomes of many higher organisms are composed primarily of repetitive DNA sequences.

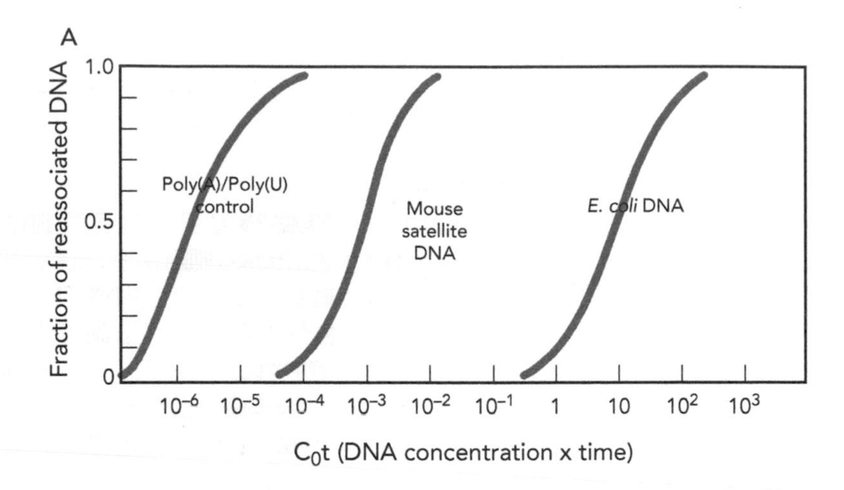

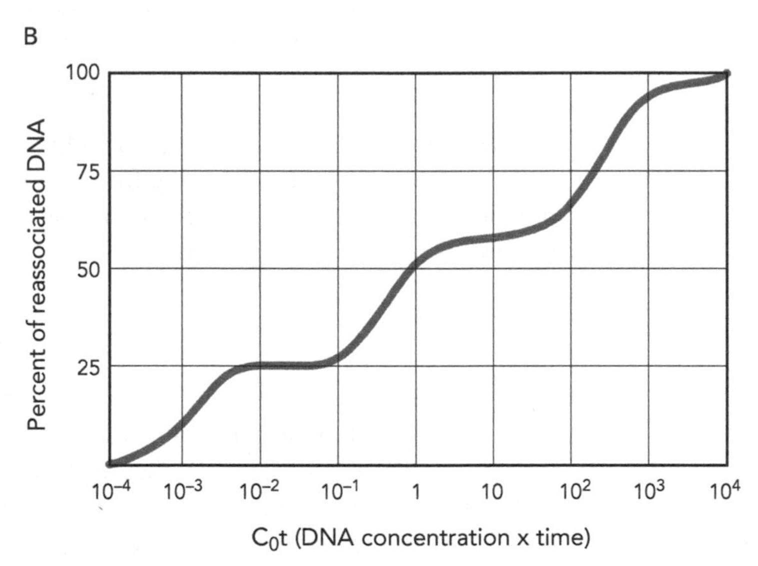

C_0t graphs of DNA reassociation. (*A*) Strands of poly(A) and poly(U) were synthesized and used as a control. These strands reassociate very quickly. DNA from *E. coli* takes much longer to reassociate because repetitive sequences are not present. Mouse satellite DNA contains many repeated sequences, and thus it reassociates faster than that of *E. coli.* (*B*) The average eukaryotic genome has a three-step reassociation curve. The first part of the curve is the fast component and represents highly repetitive DNA. The second part of the curve is the intermediate component where moderately repetitive DNA reassociates. The third part of the curve is the slow component, containing mainly single-copy genes.

Highly repetitive DNA is composed primarily of short tandem repeats (STRs), in which several to thousands of repeated units line up head to tail, like the cars of a train. Tandem repeats are classified by the size of the repeated unit: microsatellites (2–5 nucleotides per repeat), minisatellites (6–50 nucleotides), and satellites (up to several hundred nucleotides). Tandem repeats are thought to arise when DNA polymerase "slips" during replication, mispairs with a previous repeat unit, and adds a new unit. The number of tandem repeats at a given location in the genome often varies from person to person; these sequences are called variable-number tandem repeats (VNTRs). A high proportion of tandemly repeated DNA is located in centromere and telomere regions. Repeat DNA in the centromere region may aid in chromatid pairing during cell division, whereas the telomere repeat TTAGGG protects the free chromosome ends from degradation.

Middle repetitive DNA sequences are found outside the centromeres and telomeres and are not due to errors of replication. The vast majority of middle repetitive DNA is derived from transposable elements (so-called jumping genes) that have copied themselves and moved from chromosome to chromosome. Transposable elements comprise more than half of the human genome. Short interspersed nucleotide elements (SINEs) are 75–500 base pairs (bp) in length, whereas long interspersed nucleotide elements (LINEs) are up to 6.5 kb long. Also included in the group of middle repetitive DNA sequences are ribosomal DNA (rDNA) genes that encode ribosomal RNA; several hundred copies of these genes exist in the human genome. An additional 10% of the human genome contains large-scale duplications >5 kb. Many of these duplications are pseudogenes, nonfunctional copies of bona fide genes that may be located on entirely different chromosomes. Pseudogenes, along with other types of repetitive DNA, greatly confound sequence assembly.

Two transposable elements—a LINE called L1 and a SINE called *Alu*—are the most frequent gene-size sequences in the human genome. Approximately 1.1 million copies of *Alu* and 500,000 copies of L1 comprise ~27% of human DNA by weight. Each of these hundreds of thousands of copies arose from an individual "jump" at some point in human evolution. L1 transposons came into the genome ~150 million years ago, so we share these sequences with many vertebrate animals. *Alu* elements originated ~65 million years ago and are found exclusively in primates, the "monkey" branch of the evolutionary tree. Several thousand *Alu*s are found only in humans, so they have made the jump in the past 6–7 million years, after humans diverged from a common ancestor with chimps.

L1 is a "complete" transposon, in that it carries a gene for the enzyme reverse transcriptase (RT), which converts L1 RNA into a mobile DNA copy; that DNA copy can then insert into a new chromosome location. This same enzyme enables retroviruses, such as human immunodeficiency virus (HIV), to insert into positions on the human

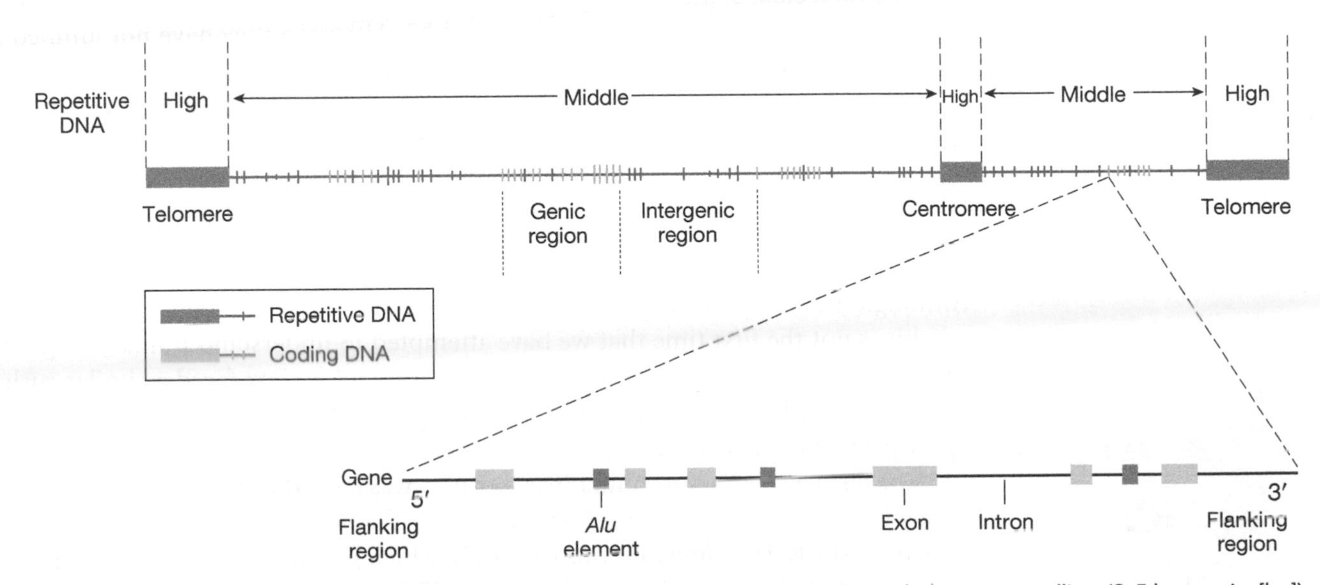

Structure of a human chromosome. Highly repetitive DNA includes microsatellites (2–5 base pairs [bp]) and minisatellites (6–50 bp). The telomere repeat is an example of a minisatellite. Middle repetitive DNA includes short interspersed nucleotide elements (SINEs) (75–500 bp), long interspersed nucleotide elements (LINEs) (up to 6.5 kb), and ribosomal RNA genes. *Alu* is an example of a SINE.

chromosomes from which they cause infection. Current thinking holds that the retroviruses borrowed the RT gene from LINE transposons in the host genome. As the largest source of RT in the genome, L1 is largely responsible for the creation of pseudogenes. These gene-like sequences arise when an mRNA transcript is reverse-transcribed into a complementary DNA (cDNA) molecule that is integrated into a new chromosome position and subsequently mutated to become nonfunctional.

With no functional genes, *Alu* is a defective transposon. It must be mobilized for transposition using the RT enzyme produced by L1; therefore, *Alu* and L1 exist in a sort of molecular symbiosis. Some scientists believe that transposons have been successful because they promote the evolution of the genomes that they inhabit. *Alu* reached a jumping peak ~40 million years ago, with a rate of perhaps one new jump in each newborn. This period roughly coincided with the evolutionary "radiation" that led to the development of the modern branches of the primate family.

Alu accumulates in gene-rich regions, and transcription is activated under stressful conditions. *Alu*s located in introns provide splice sites in ~5% of alternatively spliced genes. There is some evidence that L1 is active in nerve cell progenitors, and increased numbers of L1 copies have been detected in several regions of the brain, as compared to heart and liver tissue. These observations have led some scientists to believe that transposons were positively selected as evolutionary agents because they provided variants that created a selective advantage for evolving primates.

Other biologists believe that any advantageous changes are merely random byproducts of transposition and that the evolutionary success of *Alu* and other transposons rests entirely in their own reproductive ability. *Alu* makes no protein and may exist solely for its own replication—thus fitting Richard Dawkins' description of "selfish DNA." Continuing on this line of thought, one might craft a definition of life as *the perpetuation and amplification of a DNA sequence through time*. By this view, *Alu* is a supremely successful life form, with 1 million copies of itself perpetuated in each of the billions of humans and primates alive today.

The molecular symbiosis of *Alu* and L1 replaced an earlier and virtually identical symbiosis between sequences called *Mer* and L2. Although they have not jumped for at least 100 million years, *Mer* and L2 "fossils" still litter ~5% of the human genome. One might wonder what made the *Mer*/L2 symbiosis successful in its own time and why it died out.

THE BIRTH OF EUGENICS

Francis Galton
(©The Galton Collection, University College London; http://www.eugenicsarchive.org.)

Although the level of detail provided by the human genome sequence is unprecedented, this is not the first time that we have attempted to understand humans as genetic organisms. At the turn of the last century, science and society faced a similar rush to understand and exploit human genes. Eugenics was the name of this early effort to apply principles of Mendelian genetics to improve the human species.

The term eugenics—meaning "well born"—was coined in 1883 by Francis Galton, a scientist at University College London. Galton's conception of eugenics arose from his earlier study, *Hereditary Genius* (1869), in which he concluded that superior intelligence and abilities were inherited with an efficiency of ~20% among primary relatives in noteworthy British families. Galton's emphasis on the voluntary improvement of a family's genetic endowment became known as "positive eugenics" and remained the focus of this British movement.

"Fitter Families" contest winner at the 1920 Kansas State Free Fair, flanked by nurses and examining staff.
(Courtesy of the American Philosophical Society; http://www.eugenicsarchive.org.)

During the first decade of the 20th century, eugenics was organized as a scientific field by the confluence of ideas from evolutionary biology, Mendelian genetics, and experimental breeding. In the United States, Charles Davenport embodied this synthesis. In 1898, Davenport became the director of the Biological Laboratory at Cold Spring Harbor, New York, a field station to study evolution in the natural world. On adjacent property, in 1904, he founded the Station for Experimental Evolution, whose researchers were among the very first adherents of Mendelian genetics.

Davenport became interested in eugenics through his association with the American Breeders Association (ABA), the first scientific body in the United States to actively support eugenics research. The ABA members—including Luther Burbank, a pioneer of the American seed business—aimed to directly apply principles of agricultural breeding to human beings. This is aptly illustrated by Davenport's book *Eugenics: The Science of Human Improvement by Better Breeding* as well as by the "Fitter Family" contests held at state fairs throughout the United States during the 1920s. These competitions judged families in the same context as the fastest racehorses, the fattest pigs, and the largest pumpkins.

In 1910, Davenport obtained funding to establish a third scientific organization at Cold Spring Harbor, the Eugenics Record Office (ERO). A series of ERO bulletins, including Davenport's *Trait Book* and *How to Make a Eugenical Family Study*, helped to standardize methods and nomenclature for constructing pedigrees to track traits through successive generations. Constructing a pedigree entailed three important elements: (1) finding extended families that express the trait under study, (2) "scoring" each family member for the presence or absence of the trait, and (3) attempting to discern one of three basic modes of Mendelian inheritance: dominant, recessive, or sex limited (X linked).

Davenport effectively used pedigrees to contribute to the first genetic description of *Homo sapiens*. In 1907, he published the classical (although still incomplete) description of the inheritance of human eye color. He contributed early studies on the genetics of skin pigmentation, epilepsy, Huntington's disease, and neurofibromatosis. The circus performers on midways of nearby Coney Island offered eugenics researchers a trove of unusual physical trait differences, and Davenport's correspondence with an albino circus family resulted in the first Mendelian study of albinism. Other researchers described Mendelian inheritance in alkaptonuria, brachydactyly, hemophilia, and color blindness.

However, scoring traits was a difficult problem, especially when eugenicists attempted to measure complex traits (such as intelligence or musical ability) and mental illnesses (such as schizophrenia or manic depression). In general, eugenicists were lax in defining the criteria for measuring many of the "traits" that they studied. This led them to conclude that many real and imagined traits—including alcoholism, feeblemindedness, pauperism, social dependency, shiftlessness, nomadism, and lack of moral control—were single-gene defects inherited in a simple Mendelian fashion.

Davenport's study of naval officers amusingly illustrates the extent to which eugenicists sought genetic explanations of human behavior to the exclusion of environmental influences. After analyzing the pedigrees of notable seamen—including those of Admiral Lord Nelson, John Paul Jones, and David Farragut—he concluded that they shared several heritable traits. Among these traits was thalassophilia, "love of the sea," which Davenport determined was a sex-limited trait because it was found only in men. Davenport failed to consider the equally likely explanations that sons of naval officers often grew up in environments dominated by boats and tales of the sea or that women were prohibited from seafaring occupations throughout the 19th and early 20th centuries.

Although Galton had focused on the positive aspects of human inheritance in Britain, the American movement increasingly focused on a "negative eugenics" program to prevent the contamination of the American germplasm with supposedly unfit traits. Although the concept of genetically inferior groups dates back to the Bible, in the 18th century, degeneracy theory supplied the "scientific" explanation that unfit people arose from bad environments that damaged heredity and perpetuated degenerate offspring. Richard Dugdale, of the New York Prison Association, brought the concept of degenerate inheritance to eugenics in *The Jukes* (1877), a pedigree study of a clan of 700 petty criminals, prostitutes, and paupers living in the Hudson River Valley north of New York City.

Dugdale held the Lamarckian view that the environment induces heritable changes in human traits. He compassionately concluded that the Jukes' situation could be corrected by providing them with improved living conditions, schools, and job opportunities. However, this interpretation was discredited by later eugenicists, who embraced Gregor Mendel's genetics and August Weismann's theory of the germplasm. Together, Mendel's and Weismann's findings formed an interpretation that human traits are determined by genes that are passed from generation to generation without any interaction with the environment. Thus, when the ERO's field worker Arthur Estabrook reevaluated the Jukes family in 1915, he found continued degeneration and placed the blame squarely on bad genes.

In addition to family studies, eugenicists collected data on the "unfit" from insane asylums, prisons, orphanages, and homes for the blind. Surveys filled out by superintendents provided the ethnic makeup of societal "dependents" and the costs of main-

August Weismann
(Courtesy of the National Library of Medicine.)

Arthur Estabrook, ca. 1921
(Courtesy of University at Albany, State University of New York; http://www.eugenicsarchive. org.)

Robert Yerkes, ca. 1917
(Source unknown.)

taining them in public institutions. With the mobilization for World War I, tens of thousands of men inducted for the draft provided a ready source of anthropometric and intelligence data. Notably, the Army Alpha and Beta Intelligence Tests, developed by Robert Yerkes of Harvard University, supposedly measured the innate intelligence of army recruits. African-American and foreign-born recruits were much more likely to do poorly on the Yerkes tests, because they mostly measured knowledge of white American culture and language.

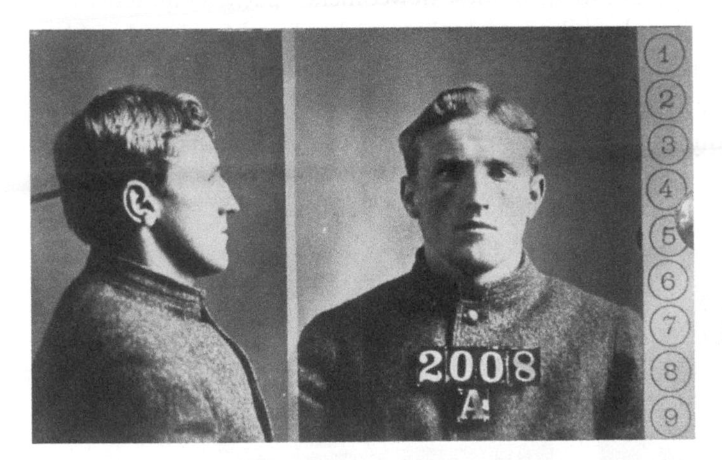

"Jukes" family member, ca. 1910. This young man was described in *The Jukes* in 1915: "At 10 he was sent to the Children's Home for truancy...At 11 incorrigibility sent him to the State Industrial School for Boys...In 1902, aged 17, he was paroled from this institution to his stepfather. In the following year he was fined $3 for breach of the peace, and for burglary was in the county jail 2 months. At 19 he was sent to the county jail for a year for burglary. At 21 he served 2 months in the county jail for burglary. At 22 he was convicted of an assault with intent to kill and a sentence of from 10 to 15 years in State prison was given him...He died of tuberculosis in State prison at the age of 27."
(Courtesy of the University at Albany, State University of New York; http://www.eugenicsarchive.org.)

Ulster County, New York, ca. 1910. A field photo by Arthur Estabrook illustrates the poor living conditions he found in rural New York during his reevaluation of the "Jukes" family.
(Courtesy of the University at Albany, State University of New York; http://www.eugenicsarchive.org.)

EUGENIC SOCIAL ENGINEERING

Eugenics arose in the wake of the Industrial Revolution, when the fruits of science were improving public and private life. A growing middle class of professional managers believed that scientific progress offered the possibility of rational cures for social problems. Placing the blame for social ills on bad genes—and the people who carried them—raised the question, "Why bother to build more insane asylums, poorhouses, and prisons when the problems that necessitated them could be eliminated at their source?" Thus, negative eugenics seemed to offer a rational solution to age-old social problems. American eugenicists successfully lobbied for social legislation on three fronts: restricting European immigration, preventing "race" mixing (miscegenation), and sterilizing the "genetically unfit."

As the eugenics movement was gathering strength, a phenomenal tide of immigration was rolling into the United States. During the first two decades of the 20th century, 600,000 to 1,250,000 immigrants per year entered the country through the facility on Ellis Island in New York Harbor (except during World War I). It is estimated that 100 million Americans alive today can trace their ancestry to an immigrant who arrived at Ellis Island. In addition, during this period, the nativity of the majority of immigrants shifted away from the northern and western European countries that had contributed most immigrants during the Colonial, Federal, and Victorian eras. Increasingly, the immigrant stream was dominated by southern and eastern Europeans, including large numbers of displaced Jews.

American eugenicists—overwhelmingly white, of northern and western European extraction, and members of the educated middle and upper classes—looked with disdain on the new immigrants. The plight of many of these newcomers—packed into tenements in lower Manhattan, plagued by tuberculosis and crime, and reduced to virtual serfdom in sweat shops—was sympathetically chronicled by Jacob Riis and the "muckraking" journalists. But to eugenicists, the immigrants' lot had little to do with poverty or lack of opportunity and had everything to do with their poor genes that eugenicists feared would

Doctor examining immigrants at Ellis Island, 1904.
(Courtesy of National Park Service: Statue of Liberty Monument; http://www.eugenicsarchive.org.)

Immigrants bound for New York on an Atlantic liner.
(Courtesy of National Park Service: Statue of Liberty Monument; http://www.eugenicsarchive.org.)

Harry Laughlin and Charles Davenport outside the Eugenics Record Office, 1912.
(Courtesy of the Harry H. Laughlin Papers, Special Collections Department, Pickler Memorial Library, Truman State University.)

quickly "pollute" the national germplasm. The eugenics movement provided a scientific rationale for growing anti-immigration sentiments in American society. Labor organizations fed on fears that working class Americans would be displaced from their jobs by an oversupply of cheap immigrant labor, while anti-Communist factions stirred up fears of the "red tide" entering the United States from Russia and eastern Europe.

As "expert agent" for the Committee on Immigration and Naturalization of the U.S. House of Representatives, ERO Superintendent Harry Laughlin became the anti-immigration movement's most persuasive lobbyist in the early 1920s. During three separate testimonies, he presented data that purported to show that southern and eastern European countries were "exporting" to the United States genetic defectives that had disproportionately high rates of mental illness, crime, and social dependency. The resulting Immigration Restriction Act of 1924 cut immigration to 165,000 per year and restricted immigrants from each country according to their proportion in the U.S. population in 1890—a time before the major waves of immigration from southern and eastern Europe. This had the desired effect of reducing southern and eastern European immigrants to <15,000 per year. Immigration did not regain prerestriction levels again until the late 1980s.

Of all the legislation enacted during the first four decades of the 20th century, sterilization laws adopted by 30 states most clearly bear the stamp of the eugenics lobby. This owed much to Henry Goddard's influential book, *The Kallikaks*, published in 1912. (The fictitious family name is a contraction of *kalli* from the Latin *kallos* for "goodness and beauty" and *kakos* for "bad.") This moral story of the ancestors of a feebleminded woman, Deborah Kallikak, purported to describe a controlled experiment in positive and negative eugenics. The experiment began with Martin Kallikak, who as a young militiaman in the Revolutionary War had an illicit union with an attractive but feebleminded barmaid. This "bad" lineage produced 262 feebleminded individuals. After the war, Martin "straightened up and married a respectable girl of good family," siring a second lineage of respectable and prominent individuals. Thus, the pri-

Henry Goddard
(Reprinted, with permission, from Archives of the History of American Psychology, The University of Akron.)

mary intent of eugenic sterilization was to curb the silent threat of the feebleminded, who threatened to contaminate good lineages as surely as the case of Martin Kallikak.

Many of the early sterilization laws were legally flawed and did not meet the challenges of state court tests. To address this problem, Laughlin designed a model eugenics law that was reviewed by legal experts. Virginia's use of the model law was tested in Buck v. Bell, heard before the Supreme Court in 1927. Carrie Buck, her mother Emma, and daughter Vivian were the subjects of the case. After having Vivian out of wedlock, Carrie was judged to be promiscuous and "feebleminded" and joined Emma in incarceration at

DEBORAH KALLIKAK, AS SHE APPEARS TO-DAY AT THE TRAINING SCHOOL.

"Deborah Kallikak," 1912. Henry Goddard based his moral story on the ancestors of "Deborah Kallikak," who he found at The Training School at Vineland, New Jersey, a mental institution for the feebleminded.
(Courtesy of the University at Albany, State University of New York; http://www.eugenicsarchive.org.)

Carrie's Baby, Vivian, 1924. This photo, taken the day before the Virginia trial, is believed to capture the "standard mental test" used by Arthur Estabrook to determine that Vivian Buck was feebleminded. Six-month-old Vivian appears uninterested as foster mother Mrs. Dobbs attempts to catch her attention with a coin.

(Courtesy of the University of Albany, State University of New York; http://www.ugenicsarchive.org.)

Carrie and Emma Buck, 1924. Daughter and mother on a bench at the Virginia Colony for the Epileptic and the Feebleminded, in Lynchburg, the day before the start of the Virginia trial that would lead all the way to the U.S. Supreme Court.

(Courtesy of the University at Albany, State University of New York; http://www.eugenicsarchive.org.)

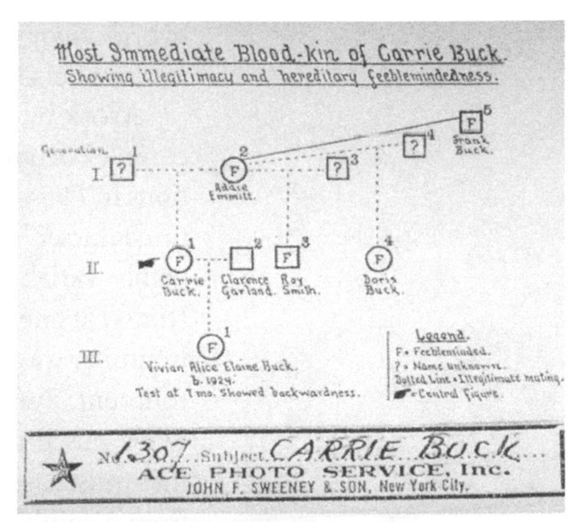

Pedigree of the Buck Family, 1924. An exhibit from the Virginia trial clearly shows three supposed generations of feebleminded females in the Buck family: Emma (Addie Emmitt), Carrie, and Vivian. Vivian's father, Clarence Garland, was the nephew of Carrie's foster parents, Mr. and Mrs. John Dobbs. Clarence had promised to marry Carrie, but he disappeared before her trial in 1924.

(Courtesy of Paul Lombardo, Ph.D., J.D., and the American Philosophical Society; http://www.eugenicsarchive.org.)

the Virginia Colony for the Epileptic and the Feebleminded. Estabrook examined Vivian and found her to be "not quite normal," establishing a pedigree of mental deficiency. Oliver Wendell Holmes, Jr., delivered the court's decision upholding the legality of eugenic sterilization, which included the infamous phrase, "Three generations of imbeciles are enough!"

Carrie Buck was the first person sterilized under Virginia's law, and 30 states eventually passed eugenic sterilization statutes. Buck v. Bell was never overturned, and sterilization of the mentally ill continued into the 1970s, by which time ~60,000 Americans had been sterilized—most without their consent or the consent of a legal guardian. Psychologists no longer recognize feeblemindedness as a class of mental illness. Carrie left the Virginia Colony and was married for 25 years to one man. Although she died at a young age, Vivian also was no imbecile. Her first-grade report card showed her to be a solid "B" student and she received an "A" in deportment (manners). On May 2, 2002, the 75th anniversary of the Buck v. Bell decision, Virginia Governor Mark Warner publicly apologized for Virginia's past involvement in eugenics. A roadside marker was erected in memory of Carrie in her hometown of Charlottesville.

OPPOSITION AND THE END OF EUGENICS

Scientific opposition to eugenics was present on many fronts and began even as it was being organized as a scientific discipline. In 1909, George Shull, at the Carnegie Station for Experimental Evolution, showed that the hybrid offspring produced by crossing two inbred strains of corn are more vigorous than their inbred parents. The phenomenon of hybrid vigor also held true in mongrel animals, refuting eugenicists' notion that racial purity offers biological advantage or that race mixing destroys "good" racial types.

Work by a number of scientists countered the eugenicists' simplistic assertions that complex behavioral traits are governed by single genes. Hermann Muller's survey of mutations in *Drosophila* and other organisms from 1914 to 1923 showed variation in the "gene-to-character" relationship that defied simple Mendelian explanations. Many genes are highly variable in their expression, and a single gene may affect several characteristics (traits) at one time. Conversely, mutations in many different genes can affect the same trait in similar ways. Moreover, the expression of a gene can be altered significantly by the environment. Twin studies conducted by Horatio Hackett Newman showed that identical twins raised apart after birth averaged a 15-point I.Q. difference. Lionel Penrose found that most cases of mental illness at a state-run institution in Colchester, England, resulted from a combination of genetic, environmental, and pathological causes.

Horatio Hackett Newman
(Reprinted, with permission, from the University of Chicago, Special Collections.)

Mathematical models of population genetics provided evidence against the simplistic claim that degenerate families were increasing the societal load of dysgenic genes. The equilibrium model of Godfrey Hardy and Wilhelm Weinberg showed that, although the absolute number of dysgenic family members might increase over time, the frequency of any "negative" trait does not increase relative to the normal population. Feeblemindedness, thought to be a recessive disorder, presented a particular quandary. The Hardy–Weinberg equation showed that sterilization of affected individuals would never appreciably reduce the incidence of the disorder. Despite this, feeblemindedness was thought to be so rampant that many geneticists believed reproductive control could still prevent the birth of tens of thousands of affected individuals per generation.

Although he was a founding member of the board of the ERO, Thomas Hunt Morgan resigned after several years. He criticized the eugenics movement in the 1925 edition of his popular textbook *Genetics and Evolution*, warning against the wholesale application of genetics to mental traits and comparing whole races as superior or inferior. In 1928, Johns Hopkins geneticist Raymond Pearl charged that most eugenics preaching was "contrary to the best established facts of genetical science." A visiting committee of the Carnegie Institution in 1935 concluded that the ERO's work was without scientific merit and recommended that it cease sponsorship of programs in sterilization, race betterment, and immigration restriction. Thus, the negative emphasis of American eugenics was completely discredited among scientists by the mid 1930s.

Godfrey Hardy
(Reproduced, with permission, from Trinity College, Cambridge, England.)

Wilhelm Weinberg
(Reprinted from Stern C. 1962. Wilhelm Weinberg, 1862–1937. *Genetics* 47: 1–5; © Genetics Society of America.)

Raymond Pearl, ca. 1913
(From www.wikipedia.org.)

Eugen Fischer, ca. 1938. Fischer's text *The Principles of Human Heredity and Race Hygiene* provided a "scientific" basis for Hitler's concept of a master Aryan race.

(Courtesy of the Max Planck Institute for Medical Research.)

In the meantime, eugenics gathered steam in Germany—with American help. In 1927, the Rockefeller Foundation provided funds for the construction of the Kaiser Wilhelm Institute of Anthropology, Human Genetics, and Eugenics in Berlin. Under the direction of Eugen Fischer, the Institute was at the forefront of the movement of "race hygiene" to cleanse the German race. Laughlin's model sterilization law was the basis for the Nazis' own law in 1933, and his contributions to German eugenics were recognized with an honorary degree from the University of Heidelberg in 1936. During the next several years, some 400,000 people—mainly in mental institutions—were sterilized. In 1939, euthanasia replaced sterilization as a solution for mental illness, and the lives of nearly 100,000 patients were ended "mercifully" by carbon monoxide gas—or simply by starvation or induced pneumonia. Overt euthanasia of mental patients ceased in 1941, when physicians with experience in euthanasia were reassigned to concentration camps in Poland, where they were needed to apply the "final solution" for Nazi racial purity. Growing public knowledge of Germany's radical program of race hygiene led to a wholesale abandonment of popular eugenics in the United States. The ERO was closed in December 1939.

THE CHROMOSOMAL AND MOLECULAR BASIS OF HUMAN DISEASE

Following the shocking revelations of euthanasia and human experimentation that took place in Nazi concentration camps, it is easy to understand why human genetics research was largely avoided during the years immediately following World War II. It is worth remembering that during the entire reign of eugenics, nothing was known about the physical basis of mutation and gene variation. DNA had not yet been shown to be the molecule of heredity and even the correct number of human chromosomes was not known. (Humans were thought to have 48 chromosomes, the correct number that had been previously determined for chimps.)

The development of new cytological methods, beginning in the 1950s, helped to bring human genetics out of its "dark ages." T.-C. Hsu's hypotonic (low salt) solution

Tao-Chiuh Hsu
(Reprinted, with permission, from The University of Texas, MD Anderson Cancer Center.)

caused cells to swell, separating the chromosomes and making them easier to count. Wright stain, and later Giemsa and quinacrine stains, made it possible to identify each chromosome based on size and distinctive staining patterns. In 1956, J.H. Tjio and A. Leven, of the National Institutes of Health, used these techniques to conclusively show that humans have 46 chromosomes.

Joe Hin Tjio
(Reprinted from Hsu TC. 1979. *Human and Mammalian Cytogenetics.* ©1979, with permission of Springer Science + Business Media.)

Within several years, cytologists established a direct relationship between human genetic disorders and abnormal chromosome number, or aneuploidy. Trisomy—the condition in which an extra copy of one chromosome is present—was found for Down's syndrome (chromosome 21), Patau's syndrome (13), and Edward's syndrome (18). Abnormalities in sex chromosome number were described for Turner's syndrome (XO) and Klinefelter's syndrome (XXY). In the early 1960s, translocations were identified in some cases of Down's syndrome. Studies in the 1970s showed that chronic myeloid leukemia, Burkitt's lymphoma, and several other blood cancers are characterized by specific chromosome translocations.

The British physician Sir Archibald Garrod proposed in 1908 that alkaptonuria and, by extension, other inherited disorders are caused by "inborn errors in metabolism." A deficiency results from the lack of a specific enzyme needed to perform a biochemical reaction. In 1941, George Beadle and Edward Tatum proved that this is exactly the case in *Neurospora*. In 1956, Veron Ingram and John Hunt, at the University of Cambridge, discovered a molecular mechanism for sickle cell anemia that proved that Garrod's thesis was correct for humans.

Albert Levan
(Reprinted from Hsu TC. 1979. *Human and Mammalian Cytogenetics.* ©1979, with permission of Springer Science + Business Media.)

Sickle cell disease was first described in 1910 by Chicago physician James Herrick, whose patient had anemia characterized by unusual sickle-shaped red cells. Over the years, evidence accumulated confirming the disease to be a recessive disorder. In the mid 1940s, Linus Pauling and Harvey Itano, at the California Institute of Technology, isolated hemoglobin, the red pigment of blood, from normal individuals and patients with sickle cell disease. When separated by electrophoresis, sickle cell hemoglobin (Hb_S) migrated more slowly than normal hemoglobin (Hb), showing that it was less negatively charged. This was consistent with work by Vernon Ingram, of the Cavendish Laboratory in Cambridge, England, showing that Hb contains more glutamic acid (a negatively charged amino acid) and Hb_S contains more valine (a neutral amino acid).

Archibald Garrod, ca. 1910

(From *Genetics: A periodical record of investigations bearing on heredity and variation.* 1967. 56: Frontispiece.)

Vernon Ingram, 1950s

(Reprinted from Perutz V. 2009. *What a time I am having: Selected letters of Max Perutz.* Cold Spring Harbor Laboratory Press, Cold Spring Harbor, NY. Photo courtesy of Beth Ingram.)

Yuet Wai Kan

(Courtesy of University of California, San Francisco.)

Andrea-Marie Dozy

(Courtesy of Y.W. Kan, University of California, San Francisco.)

In 1956, Vernon Ingram and John Hunt independently sequenced the Hb and Hb_S proteins, finding that a glutamic acid at position 6 in Hb is replaced with valine in Hb_S. From this information, they used a genetic code table (showing that glutamic acid is encoded by GAG and valine by GTG) to predict that the A-to-T point mutation in the sixth codon is responsible for sickle cell disease. This made possible the first molecular diagnosis of a human genetic disorder in 1978 by Yuet Wai Kan and Andrea-Marie Dozy at the University of California, San Francisco.

Human genetics made striking advances in the 1970s and 1980s, when recombinant DNA techniques were used to isolate genes for important therapeutic proteins—notably, insulin, clotting factors VIII and IX, human growth hormone (HGH), tissue plasminogen activator (TPA), erythropoietin, and interferon. Although these proteins are produced in minute amounts in the body, cloning their genes made it possible to produce abundant quantities in cultured cells. Previously, these proteins were laboriously isolated from pooled tissues. Treating a single patient for 1 year required insulin purified from 10 pounds of pancreas from 70 pigs or 14 cows, clotting factor from 8000 pints of human blood, or HGH from pituitary glands from 80 human cadavers. Producing these therapeutics in cultured cells also ended the risk for viruses and prion diseases, which could be coisolated from human or animal tissues. These first products of the recombinant DNA revolution are now widely available for the treatment of diabetes, hemophilia, short stature, stroke, anemia, and infectious diseases.

Biologists, however, then had to come to grips with the difficult work of identifying a medically important gene in the absence of knowledge about its protein product. For this, disease genes needed to be mapped to specific positions by proximity, or linkage, to known chromosome locations (loci).

MAPPING THE HUMAN GENOME

Genetic linkage maps first constructed for *Drosophila* by Thomas Hunt Morgan and his colleagues in the early 1900s gave rise to the classical measure of genetic distance, the centiMorgan (cM). During meiosis, paired chromosomes align and homologous regions are exchanged when chromatids "cross over" with one another. The further apart two loci are on a chromosome, the greater the possibility they will become separated during a crossover event. The frequency of recombination is a measure of the

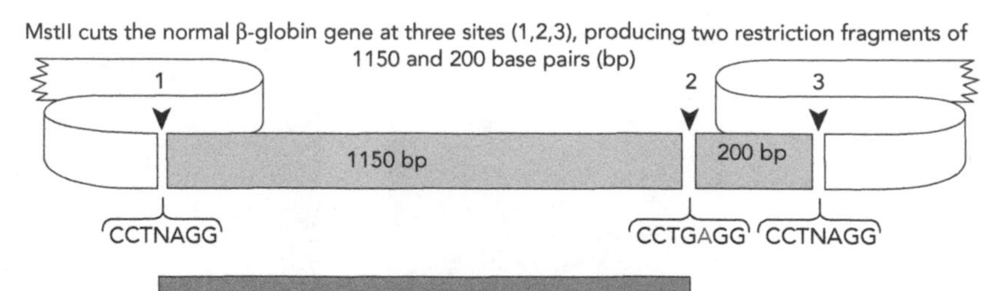

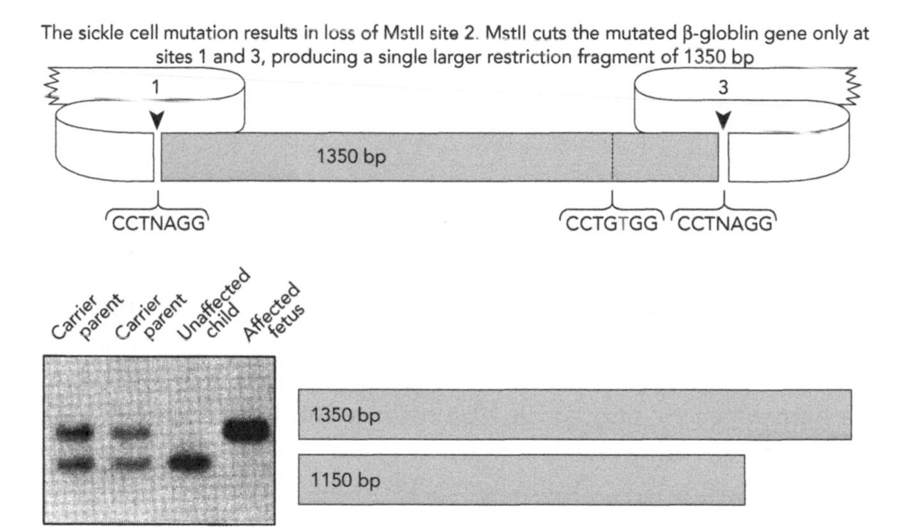

Prenatal molecular diagnosis of sickle cell anemia, 1982. The Southern blot shows the RFLP patterns of two carrier parents, an unaffected offspring, and amniotic fluid from an affected fetus. The carrier parents show a single copy of each RFLP: the 1350-bp fragment associated with the disease allele and the 1150-bp fragment associated with the normal allele (the 200-bp fragment is not detected by the probe). The unaffected child shows the 1150-bp fragment, and the affected fetus shows the 1350-bp fragment. Both offspring are homozygous and thus show a single relatively thick band, denoting two chromosomal copies of the normal (Hb) or mutated (Hb$_S$) gene.

(Reprinted, with permission, from Chang JC, Kan YW. 1982. *N Engl J Med* 307: 30–32; Massachusetts Medical Society.)

Some approved drugs produced from cloned human genes

Recombinant protein	Example of product name (and developer)	Year of first FDA approval	Use
Insulin	Humulin (Genentech, Inc.)	1982	diabetes mellitus
Human growth hormone (HGH)	Protropin (Genentech, Inc.)	1985	growth hormone deficiency in children
Interferon alfa-2b	Intron A (Biogen)	1986	hairy cell leukemia
		1988	genital warts
		1988	Kaposi's sarcoma
		1991	hepatitis C
		1992	hepatitis B
Hepatitis B surface antigen	Recombivax HB (Chiron)	1986	component of hepatitis B vaccine
Erythropoietin	EPOGEN (Amgen Ltd.)	1989	anemia
Interferon-γ	Actimmune (Genentech)	1990	chronic granulomatous disease
Colony-stimulating factor (CSF)	Leukine (Immunex Corp.)	1991	bone marrow transplantation
Clotting factor VIII	Recombinate (Genetics Institute)	1992	hemophilia A
DNase I	Pulmozyme (Genentech, Inc.)	1993	cystic fibrosis
Tissue plasminogen activator (TPA)	Retavase (Boehringer Mannheim)	1996	heart attack
Clotting factor IX	AlphaNine SD (Alpha Therapeutic Corp.)	1996	hemophilia B
Interleuken-11	Neumega (Genetics Institute)	1997	platelet depletion during chemotherapy
Chorionic gonadotropin	Ovidrel (Serono Laboratories)	2000	infertility
Antithrombin	Atryn (GTC BioTherapeutics)	2009	embolism

genetic distance between two loci on a chromosome. If two loci have a recombination frequency of 1%, they will become separated during one in every 100 meiotic recombination events. A recombination frequency of 1% is referred to as 1 cM. In the human genome, 1 cM = ~1 million nucleotides in physical distance.

For any two loci on the same chromosome, there is equilibrium between two states: crossover and linkage. For distant loci, the equilibrium shifts toward a high recombination frequency. At a distance of 50 cM, the recombination frequency is 50%, with an equal chance that the loci will stay together or become separated during meiosis; this is the maximum possible recombination frequency. Loci separated by 50 cM or more are said to be unlinked. Linkage increases as distance and recombination frequency decrease, reaching 0% when the loci are very close to each other. At 0 cM, the only possible state is linkage, so equilibrium with crossover is formally impossible. Thus, two loci that never cross over are said to be in linkage *dis*equilibrium.

In 1911, Edmund B. Wilson, of Columbia University, had, by default, mapped the first human gene to the X chromosome when he discovered that the inheritance of color blindness is "sex limited." The presence of a testis-determining factor was inferred in 1959, making it the first Y-linked "gene." By 1980, 135 genes had been mapped to specific locations on *Drosophila* chromosomes, and 120 genes had been mapped to human chromosomes. Considering that *Drosophila* has four pairs of chromosomes and humans have 23, this meant that 34 genes had been mapped per *Drosophila* chromosome, whereas only five genes were mapped per human chromosome.

Chromosome mapping was revolutionized in 1978, when David Botstein, Ronald Davis, and Mark Skolnick proposed that the human chromosome map could be populated with physical variations in the DNA molecule itself. These DNA polymorphisms ("poly" for many; "morph" for forms) extended linkage analysis far beyond the phenotypic or biochemical variants that were previously used.

Based on earlier work with small viral genomes, the first physical maps of human chromosomes were based on cutting patterns of restriction enzymes that cleave DNA at specific sequences. Point mutations can create or delete restriction enzyme recognition sites along the length of a chromosome, and the addition of repeated DNA sequences (tandem repeats) can increase the distance between enzyme cutting sites. Each mutation or tandem repeat addition creates a different allele of the DNA molecule that can vary from chromosome to chromosome or person to person. Different alleles produce different-sized restriction fragments, or restriction-fragment-length polymorphisms (RFLPs), that can be identified using gel electrophoresis.

Classical genetics defined alleles as alternative forms of a trait or gene. RFLPs extended the definition of alleles to mean alternative sequences of a region of DNA (a locus)—whether or not it is a gene. Measuring the crossover frequency between RFLP markers provided the first means to relate genetic distance (measured in cM) to physical distance (measured in nucleotides). The first linkage map of the human genome, incorporating 403 RFLP markers, was published in 1987 by Helen Donis-Keller and colleagues at Collaborative Research, Inc., in Boston, Massachusetts.

In 1989, Maynard Olson (Washington University), Leroy Hood (California Institute of Technology), Charles Cantor (Lawrence Berkeley Laboratory), and David Botstein (Genentech, Inc.) showed that a common language of DNA markers could allow scientists to easily exchange sequence information and obtain probes for known regions of the genome. In retrospect, what they proposed seemed obvious—that a DNA sequence itself can act as a chromosome marker! Any single-copy sequence of 200–500 nucleotides—a

Helen Donis-Keller
(Courtesy of Helen Donis-Keller.)

Maynard Olson
(Courtesy of Cold Spring Harbor Laboratory Archives.)

Charles Cantor
(Courtesy of Cold Spring Harbor Laboratory Archives.)

David Botstein
(Courtesy of Cold Spring Harbor Laboratory Archives.)

sequence-tagged site (STS)—could be used as physical landmark on a chromosome. By the time the rough draft of the human genome sequence was published in 2001, the resolution of the human genome map included 30,000 RFLP and STS markers at intervals of ~0.1 cM (100,000 bp).

DNA POLYMORPHISMS AND HUMAN IDENTITY

Although fingerprints and thumbprints have been used as personal identifiers since ancient times, only in the 20th century did they come into use in criminal cases. Francis Galton studied the fingerprints of thousands of schoolchildren from around the British Isles. His 1892 treatise *Finger Prints* showed how to analyze the various patterns of whorls and loops, noted their relative frequencies, and suggested how fingerprints could be rigorously used in criminal cases. This book formed the basis of forensic DNA fingerprinting still in use today.

Alec Jeffreys, at the University of Leicester, realized that DNA polymorphisms can be used to establish human identity. He coined the term "DNA fingerprinting" and was the first to use DNA polymorphisms in forensic (legal) science. Whereas classic fingerprinting analyzes a phenotypic trait, DNA fingerprinting, or DNA profiling, directly analyzes genotypic information. In 1984, Jeffreys identified minisatellite VNTRs of 9–80 bp at multiple loci near the human myoglobin gene; the length of each of these VNTRs varies from person to person. Visualized with a set of radioactive "Jeffreys' probes," the DNA fingerprint was a composite of many VNTRs near the myoglobin gene, producing 20–30 interpretable bands on a Southern blot of a polyacrylamide gel.

VNTR polymorphisms are inherited in a Mendelian fashion and can be used to track heredity. The first practical test of the DNA fingerprinting method was the "Ghana Immigration Case" (1985), where Jeffreys proved a genetic relationship between a woman in England and her son, who wished to emigrate from Ghana to rejoin her. A year later, Jeffreys' method was used to screen blood samples from more than 5000 adult male suspects in a serial rape murder case in the British Midlands, which was the subject of Joseph Wambaugh's popular book *The Blooding* (1989).

Jeffreys' multilocus probes proved difficult to analyze and were supplanted by single-locus probes that identify a polymorphism at one location in the genome. Beginning in 1987, Yusuke Nakamura, Ray White, and other investigators, at the University of Utah, began a systematic search for single-locus VNTRs, ultimately providing more than

Alec Jeffreys, 1989
(Courtesy of Alec Jeffreys, University of Leicester.)

100 useful polymorphic loci scattered throughout the genome. Each probe hybridized to a unique hypervariable region of the genome and generated a pattern consisting of one or two bands from an individual's DNA, depending on whether they were homozygous or heterozygous at that locus.

Yusuke Nakamura
(Courtesy of Yusuke Nakamura.)

Because each probe identified the alleles at a discrete locus, the frequency of each allele could be determined for a population of individuals. The Hardy–Weinberg equation could be used to calculate the occurrence probability of each genotype in that population. The VNTR loci first used in forensics were those with 10 or more alleles and a high degree of heterozygosity in a population, thus maximizing the ability to discriminate between two individuals. A good example is D1S80, a VNTR on chromosome 1 in which a 16-nucleotide unit can be repeated from 14 to more than 41 times, creating 29 different alleles.

By studying the occurrence of alleles in human populations, one can calculate the probability of an individual's DNA fingerprint having this or that combination of DNA bands or the probability of two DNA samples matching each other. Examining more and more VNTRs further differentiates the DNA profiles and improves the ability to distinguish among individuals. Provided that each VNTR is taken from a different chromosome, and is therefore unlinked from the others, the probability of any combination of VNTR genotypes being inherited together is the product of their individual occurrences. By the mid 1990s, as polymerase chain reaction (PCR) supplanted Southern blotting in DNA profiling, most forensic laboratories were producing types with five to eight unlinked markers.

Ray White
(Reprinted, with permission, from the University of Utah.)

The use of microsatellites—or short tandem repeats (STRs)—in DNA profiling was first suggested in 1992 by Thomas Caskey, of the Baylor College of Medicine. With repeat units of 2–5 nucleotides, STRs have smaller alleles, providing a greater chance of "rescuing" a DNA profile from a degraded sample and allowing the STRs to be effectively analyzed on automated DNA sequencers. Because each STR polymorphism produces a tight range of allele sizes, three to four STRs with differing ranges in allele size can be labeled with the same fluorescent dye (red, green, blue, or yellow) and detected in a single channel.

Virtually all crime laboratories now use the Combined DNA Index System (CODIS), developed by the Federal Bureau of Investigation (FBI) in 1997. The CODIS

1	2 Type	3 Allele	4 Frequency	5 Hardy–Weinberg	6 Calculation	7 D1S80 prob.	8 Locus 2 prob.	9 Combined prob.
C	18/31	18 31	0.263 0.058	2pq	2 (0.263 x 0.058)	0.0305	0.0050	0.000153
1	24/37	24 37	0.318 0.003	2pq	2 (0.318 x 0.003)	0.0002	0.0035	0.0000007
2	18/18	18 18	0.263 0.263	p^2	(0.263 x 0.263)	0.0692	0.0075	0.000519
3	28/31	28 31	0.050 0.058	2pq	2 (0.050 x 0.058)	0.0058	0.0025	0.0000145
4	18/25	18 25	0.263 0.055	2pq	2 (0.263 x 0.055)	0.0289	0.0045	0.000013
5	17/24	17 24	0.013 0.318	2pq	2 (0.013 x 0.318)	0.0008	0.0065	0.0000052

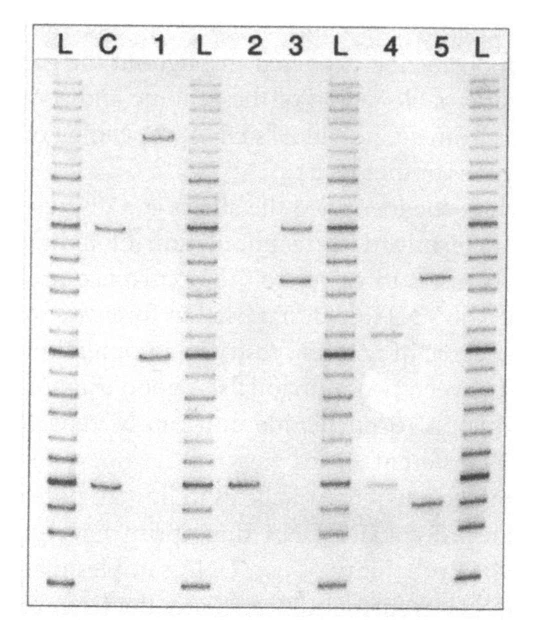

PCR amplification of the D1S80 locus, 1991. Shown in Column 1 of the table are a control (C) and five types (1–5). In the gel, the type lanes show the major alleles of the system, composed of 14–41 repeats of a 16-nucleotide unit (L lanes show DNA size marker ladder). The table shows how to "score" the alleles (columns 2–3), with allele frequencies for a Hispanic-American population (column 4). Assuming Hardy–Weinberg equilibrium, the frequency of each DNA type is calculated in columns 5–7. Hypothetical frequencies for DNA types at a second locus are given in column 8. Provided the two loci are unlinked, the combined frequency of their coinheritance is the product of their individual occurrences (column 9). With each additional marker, the DNA types become increasingly diversified, with some types being orders of magnitude rarer than others.
(Image courtesy of Life Technologies, Carlsbad, CA.)

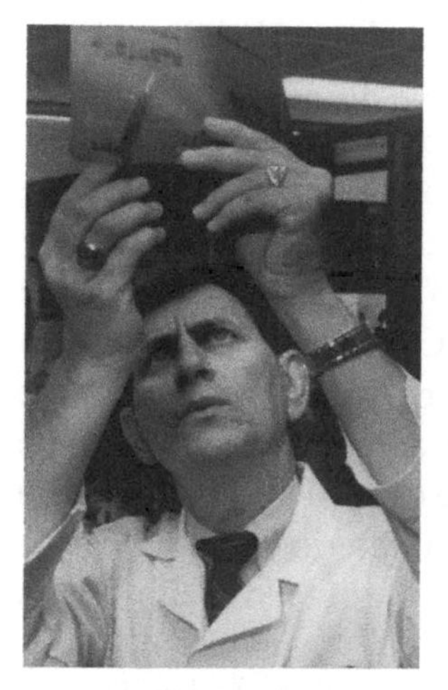

Thomas Caskey
(Courtesy of Jan Witkowski.)

system uses a 13-marker panel of STRs and an XY marker that are typically amplified together in a single "megaplex" PCR reaction. With this number of independently inherited polymorphisms, the probability of even the most common combination is one in 10s of billions. Thus, modern DNA testing has the capability of uniquely identifying each and every person alive today. As of August 2012, the FBI's National DNA Index contained 11.1 million profiles of convicted offenders and 447,300 forensic profiles collected at crime scenes. DNA matches to the CODIS database have assisted in more than 180,000 investigations.

All that is required for DNA fingerprinting is a small tissue sample from which DNA can be extracted. This can be blood or cheek cell samples in a paternity case, a semen sample from a rape victim, dried blood from fabric, skin fragments from under the fingernails of a victim after a struggle, or even several hairs (with the attached roots) combed from a crime scene. Ted Kaczynski, the "Unabomber," was definitively linked to the case when his DNA type matched the one obtained from cells left when he licked a stamp used on a letter. Using the best available techniques, a DNA type can be obtained from cells in

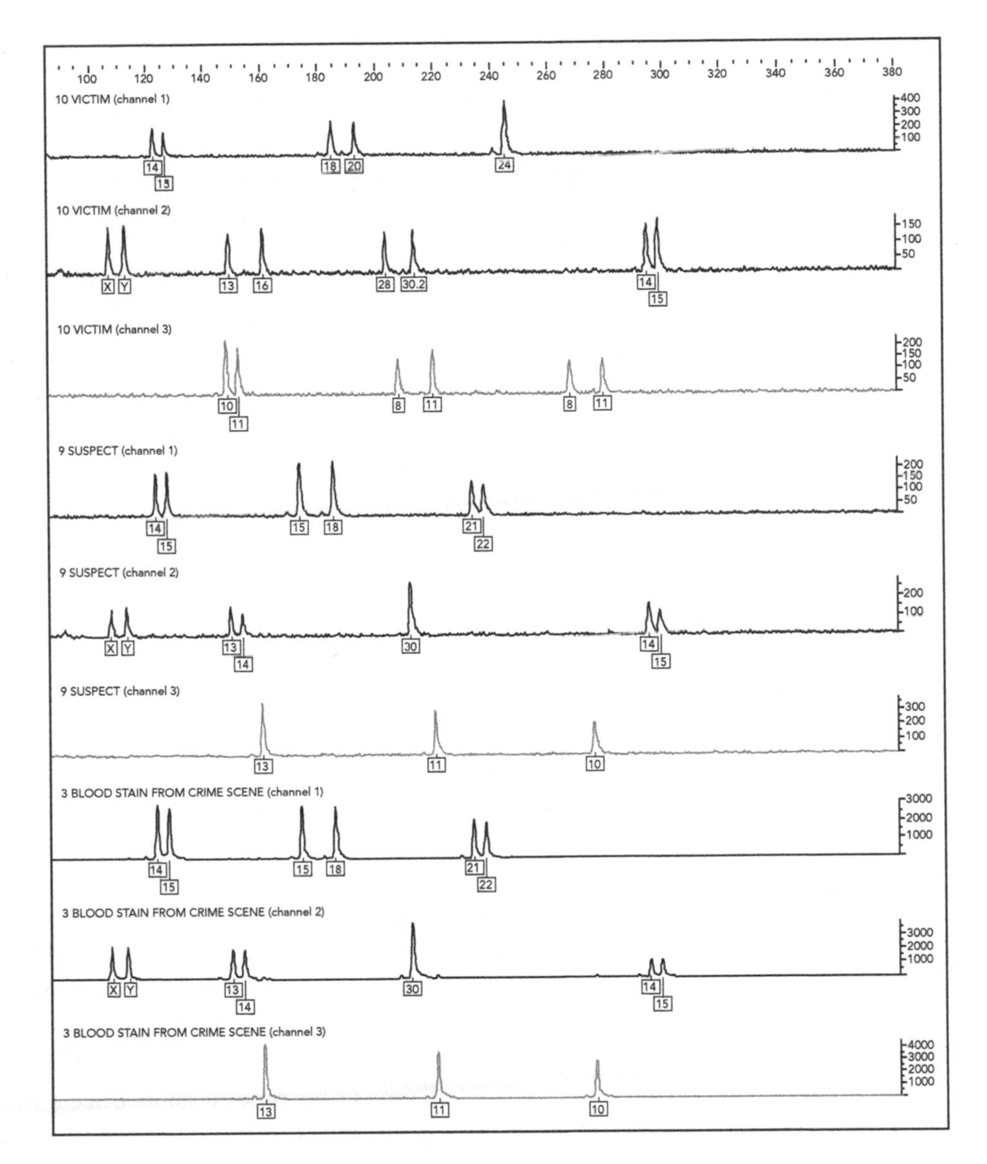

Alleles Detected

Sample	Amelogenin	D3S1358	vWA	FGA	D8S1179	D21S11	D18S51
Victim	XY	14, 15	18, 20	24	13, 16	28, 30.2	14, 15
Suspect	XY	14, 15	15, 18	21, 22	13, 14	30	14, 15
Blood Stain From Crime Scene	XY	14, 15	15, 18	21, 22	13, 14	30	14, 15

Sample	D5S818	D13S317	D7S820	D16S539	THO1	TPOX	CSF1PO
Victim	10,11	8, 11	8, 11	9, 11	7, 9	9, 11	10, 12
Suspect	13	11	10	9, 12	6, 9	8, 11	9, 12
Blood Stain From Crime Scene	13	11	10	9, 12	6, 9	8, 11	9, 12

CODIS panel from a criminal case, Suffolk County, New York, 2000. A typical criminal case from a 1997 homicide in which a blood stain from the crime scene was tested against blood samples from the victim and from the suspect, who was wounded during a struggle. Shown are sequencing results for nine STR loci, plus an XY marker, using three color channels. Four additional loci were run in another channel, which is not shown. The frequency of the suspect/blood stain type in different populations is Caucasian 0.000000000000000372 (3.73×10^{-16}), African-American 0.000000000000000103 (1.03×10^{-16}), and Hispanic 0.0000000000000000267 (2.67×10^{-17}).

(Courtesy of the Suffolk County Crime Laboratory.)

fingerprints on a glass or other hard surface. The time is approaching when a criminal will not be able to afford to leave even a single cell at a crime scene.

In addition to linking suspects to crime scenes, as of 2012, DNA fingerprints have exonerated 290 people serving prison time for crimes they did not commit. These people spent on average 13 years in prison before their release. The Innocence Project, founded in 1992 by Barry Scheck and Peter Neufeld of the Benjamin Cardoza School of Law, has led the way in DNA exonerations and calling attention to the causes of wrongful convictions. Three-fourths involved eyewitness misidentifications in improperly conducted photo or live lineups, and half involved unreliable or invalid forensic evidence—including hair microscopy, firearm tool marks, shoe prints, and bite marks. Suspects identified by local law enforcement agencies were excluded in about 25% of cases where the FBI conducted pre-trial DNA testing. The potential extent of wrongful conviction is staggering—the Innocence Project has about 300 active cases and is evaluating approximately 7000 potential cases.

FINDING GENES BEHIND SIMPLE GENETIC DISORDERS

In the 1980s, armed with increasingly detailed physical maps, biologists combined linkage analysis and cloning strategies to identify disease genes solely by their chromosome positions—in the absence of any information about their protein products. This method became known as positional cloning. Linkage analysis begins by screening hundreds or thousands of polymorphic loci (markers) to find one or more that are associated (coinherited) with the disease phenotype. Because the alleles present at a polymorphic locus may differ from family to family, the alleles are typically traced through each family pedigree under study. This establishes the allele associated with the disease state in each family. It is important to identify heterozygous carriers of the disease gene in which one polymorphic allele segregates with the disease gene and a different polymorphism segregates with the normal gene.

Accurate DNA diagnosis becomes feasible once a marker has been located within 5 cM of the disease gene. At this distance, the recombination frequency (and probability that the marker and gene become unlinked) is 5%. Conversely, the coinheritance of the marker and the disease gene, as well as the accuracy of diagnosis, is 95%. The addition of a "flanking" marker within 5 cM on the other side of the gene theoretically increases the accuracy of predictions to 99.75%. (The chance of both markers becoming unlinked is $0.05 \times 0.05 = 0.0025$.)

Flanking markers within 1 cM are typically the starting point for a strategy of "chromosome walking" to generate a detailed map of the disease locus. A genomic library of large DNA fragments (20,000–40,000 bp) is constructed that encompasses the disease region. The closest linked marker is used as a probe to isolate its corresponding genomic clone from the library. Following restriction mapping of the clone, a restriction fragment is isolated from the end of the clone closest to the disease locus. This fragment is used to reprobe the library to identify an overlapping clone. The endmost fragment of this clone is then used to reprobe the library, and another overlapping clone is isolated. Through such a succession of overlapping clones, one "walks" along the chromosome region spanning the disease locus, eventually reaching the flanking marker on the other side. The overlapping fragments are then assembled to produce a map of the disease locus. Candidate genes within the region are then scrutinized for mutations or functions that relate to the disorder under study.

Some human disease genes isolated by positional cloning

Disease	Genes
Achondroplasia	*FGFR3*
Alzheimer's disease	*APOE*
Crohn's disease	*NOD2*
Cystic fibrosis	*CFTR*
Duchenne muscular dystrophy	*DMD*
Early-onset breast/ovarian cancer	*BRCA1, BRCA2*
Hereditary nonpolyposis colon cancer	*MSH2, MLH1, PMS1, PMS2*
Huntington's disease	*HTT*
Neurofibromatosis types 1 and 2	*NF1, NF2*
Polycystic kidney disease (autosomal dominant)	*PKD1, PKD2*
Retinitis pigmentosa	*RHO*
Retinoblastoma	*RB1*
Spinal muscular atrophy	*SMN2*

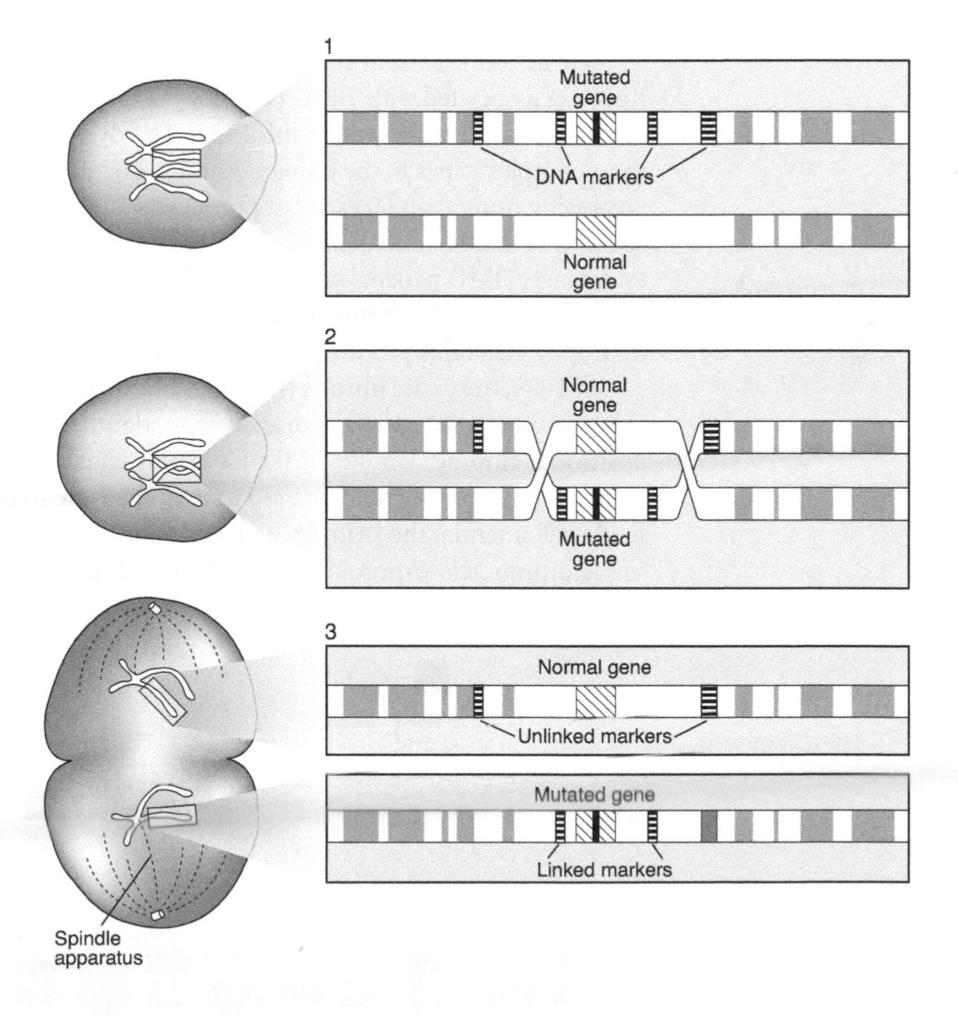

Fate of linked and unlinked markers during meiotic recombination. (1) Homologous chromosomes, each composed of two sister chromatids, pair (synapse) during prophase of the first meiotic division. (2) Recombination occurs when chromatids cross over, exchanging DNA fragments. Linked markers remain with the original chromatid, whereas unlinked markers become separated from it. (3) During anaphase, the recombined chromatids separate into two different daughter cells (in a subsequent meiotic division, the sister chromatids will segregate into separate haploid sex cells).

The cloning and analysis of the genes for Huntington's disease (HD), Duchenne muscular dystrophy (DMD), and cystic fibrosis (CF) illustrate different challenges and strategies of positional cloning—and different disease mechanisms. In 1983, HD was the first major disease locus mapped by RFLP linkage analysis. However, it took 10 years to clone the huntingtin (*HTT*) gene because its location very near to the end of chromosome 4 made it difficult to find a flanking marker on the telomere side. The causative mutation proved to be an expanded region of nucleotide triplets. The normal huntingtin gene has 6–35 CAG repeats; the mutated version in HD patients has 36–180 repeats. The number of repeats correlates with the age when symptoms appear: Individuals with 36–41 repeats may never have symptoms, whereas those with more than 50 repeats develop symptoms before age 20. Because the repeat occurs in a coding exon, each additional repeat adds another unit of the amino acid glutamine to the expressed huntingtin protein. Inherently unstable, triplet repeat expansions such as this are also the causative mutations in Fragile X mental retardation and several neuromuscular ataxias.

Cloning the dystrophin (*DMD*) gene in 1987 was simplified by the knowledge that a number of DMD patients have large deletions clustered in a region of the X chromosome known as Xp21. In addition, females having the disease were found to have a break in their active X chromosome at position Xp21. This suggested that the deletions at Xp21 are associated with DMD pathology and that they may have resulted from the loss of part of the normal gene at this locus. The dystrophin gene is one of the largest and most complex genes in the human genome. Encompassing more than 2,000,000 bp and possessing more than 60 exons, it produces a 14,000-bp mRNA that codes for a protein containing 4000 amino acids. As large as it is, the dystrophin gene is particularly prone to damage. DMD patients show many different deletions of exons that effectively knock out production of any functional dystrophin. In the milder form of Becker muscular dystrophy, deletions produce a semifunctional dystrophin protein.

In 1989, the cystic fibrosis transmembrane conductance regulator (*CFTR*) gene on chromosome 7 was the first disease gene identified entirely by using the methods of positional cloning. Unlike DMD, CF is not characterized by large-scale deletions or rearrangements that could be used to map the gene to its chromosomal location. As in sickle cell anemia, the primary genetic lesion in CF is a specific mutation affecting a single amino acid. Approximately 70% of CF patients show a 3-bp deletion, named

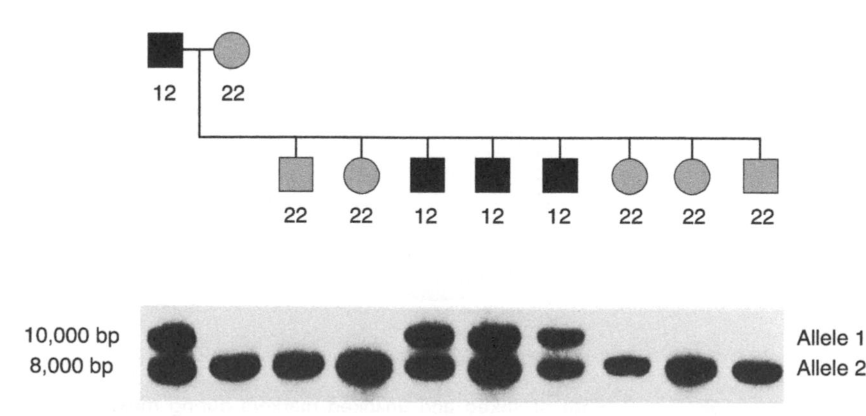

DNA diagnosis of Huntington's disease, 1987. This pedigree shows the coinheritance of a polymorphic allele and Huntington's disease. Allele 1 is linked to the Huntington's disease gene, and allele 2 is linked to the normal gene. The affected father and three affected sons (black) all carry one copy of each allele. The unaffected mother and five of the offspring (gray) have copies of the normal allele.
(Courtesy of T.C. Gilliam, Columbia University.)

Triplet repeat disorders

Disease	Gene	Repeat	Normal repeat number	Disease repeat number
Dentatorubropallidoluysian atrophy (DRPLA)	*ATN1* or *DRPLA*	CAG	6–35	49–88
Fragile X				
Mental retardation (FRAXA)	*FMR1*	CGG	6–53	230+
Associated tremor/ataxia syndrome (FXTAS)	*FMR1*	CGG	6–53	55–200
XE mental retardation (FRAXE)	*AFF2* or *FMR2*	GCC	6–35	200+
Freidreich's ataxia (FRDA)	*FXN* or *X25*	GAA	7–34	100+
Huntington's disease (HD)	*HTT*	CAG	10–35	36+
Myotonic dystrophy (DM)	*DMPK*	CTG	5–37	50+
Spinobulbar muscular atrophy (SBMA)	androgen receptor	CAG	9–36	38–62
Spinocerebellar ataxia				
SCA1 (type 1)	*ATXN1*	CAG	6–35	49–88
SCA2 (type 2)	*ATXN2*	CAG	14–32	33–77
SCA3 (type 3)	*ATXN3*	CAG	12–40	55–86
SCA6 (type 6)	*CACNA1A*	CAG	4–18	21–30
SCA7 (type 7)	*ATXN7*	CAG	7–17	38–120
SCA8 (type 8)	*OSCA* or *SCA8*	CTG	16–37	50+
SCA12 (type 12)	*PPP2R2B* or *SCA12*	NNN (on 5′ end)	7–28	66–78
SCA17 (type 17)	*TBP*	CAG	25–42	47–63

ΔF508, that results in the loss of a single phenylalanine residue at amino acid position 508 of the CFTR polypeptide.

USING SNPs TO FIND GENES BEHIND COMPLEX DISORDERS

Positional cloning was effective in finding the causes of many "simple" disorders that are caused by single genes. Straightforward diagnosis and high heritability (the component of disease attributed to hereditary as opposed to environment) made it possible to trace linked markers through family pedigrees. Unfortunately, it has proved much more difficult to find genes involved in common disorders that cause the greatest amount of human sickness and loss of productivity—including heart disease, asthma, non-insulin-dependent diabetes, depression, schizophrenia, autism, and bipolar disorder.

Although these diseases are common, each is thought to arise from the complex interaction of a number of genes. Each gene likely contributes a relatively small effect, and different sets of genes may cause the disease in different populations. These common disorders are also significantly influenced by environmental factors such as air quality, diet, and stress. To further complicate matters, psychiatric disorders have a range of severity and expression that makes it difficult to standardize diagnosis.

Because of this heterogeneity, it became clear that finding genes for common disorders might require studies of large populations in which a range of genotypes and phenotypes would be at play. Population studies, like family studies, aim to associate particular DNA polymorphisms with a disease phenotype. However, imagine the complexity of searching for potential genes involved in a heterogeneous disorder in which different genes or gene combinations may produce similar phenotypes in different populations. Any of several contributing genes would likely be associated with different markers in different population groups.

The LOD (logarithm of the odds) score is the key statistical method used to establish linkage in family- and population-based association studies. On the basis of an observed recombination frequency between a marker and a putative disease locus, the LOD score is a ratio of the probability (odds) of a pedigree occurring at a given linkage divided by the probability (odds) of no linkage. A LOD score of 3, which is generally considered the threshold for possible linkage in a complex disorder, means that linkage is 1000 times more likely than no linkage. As a logarithmic function, like the Richter earthquake scale, each LOD score increases by a factor of 10. Thus, a LOD score of 3 represents a 10 times closer association between a marker and a locus than does a LOD score of 2.

The LOD score is extremely sensitive to changes in data analysis and laboratory errors. A change in diagnosis of one or several people in a study may be enough to lessen the association between the marker and a phenotype and, thus, weaken statistical linkage. Furthermore, relatively low LOD scores are expected for any of several genes contributing to a complex disorder.

The search for genes behind complex disorders has been significantly bolstered by the development of a high-density map of SNPs by the International Haplotype Map (HapMap) Project. This project catalogs the simplest and most frequent source of DNA variation in the human genome—point mutations that change a nucleotide at a single position. The project looks at the unique set of polymorphisms on each of the paired chromosomes (maternal and paternal); each set is a haplotype or "half type." This involves genotyping the chromosomes from hundreds of individuals from different population groups, identifying nucleotide positions that vary between any two chromosomes (within one person or between people), and measuring the frequency of each SNP allele. As of March 2010, more than 4 million unique SNPs had been identified in the human genome, with at least 18,000 SNPs on the smallest autosome (chr 22) and 112,000 on the largest (chr 1).

The density of these SNP markers makes it possible to look for linkage to disease genes in heterogeneous populations of unrelated individuals. Genome-wide association studies (GWAS) compare hundreds of thousands of SNPs across the genomes of diseased individuals and healthy controls. Then, computer algorithms search for a consensus haplotype—a set of SNP markers—that is associated with the disease locus. Because haplotypes may encompass tens or hundreds of SNPs, this type of association analysis is much more complex than determining linkage with one marker at a time. Because genome-wide association is not very sensitive, very large samples—on the order of tens of thousands of individuals—are needed to reach statistically significant LOD scores.

GWAS have identified haplotypes or gene variants associated with increased risk for a variety of disorders—including asthma, obesity, type 2 diabetes, coronary heart disease, prostate cancer, Crohn's disease, schizophrenia, and bipolar disorder. GWAS have often confirmed linkage with loci previously identified in family or population studies. Although the linkage reported in any single study is modest, the same regions

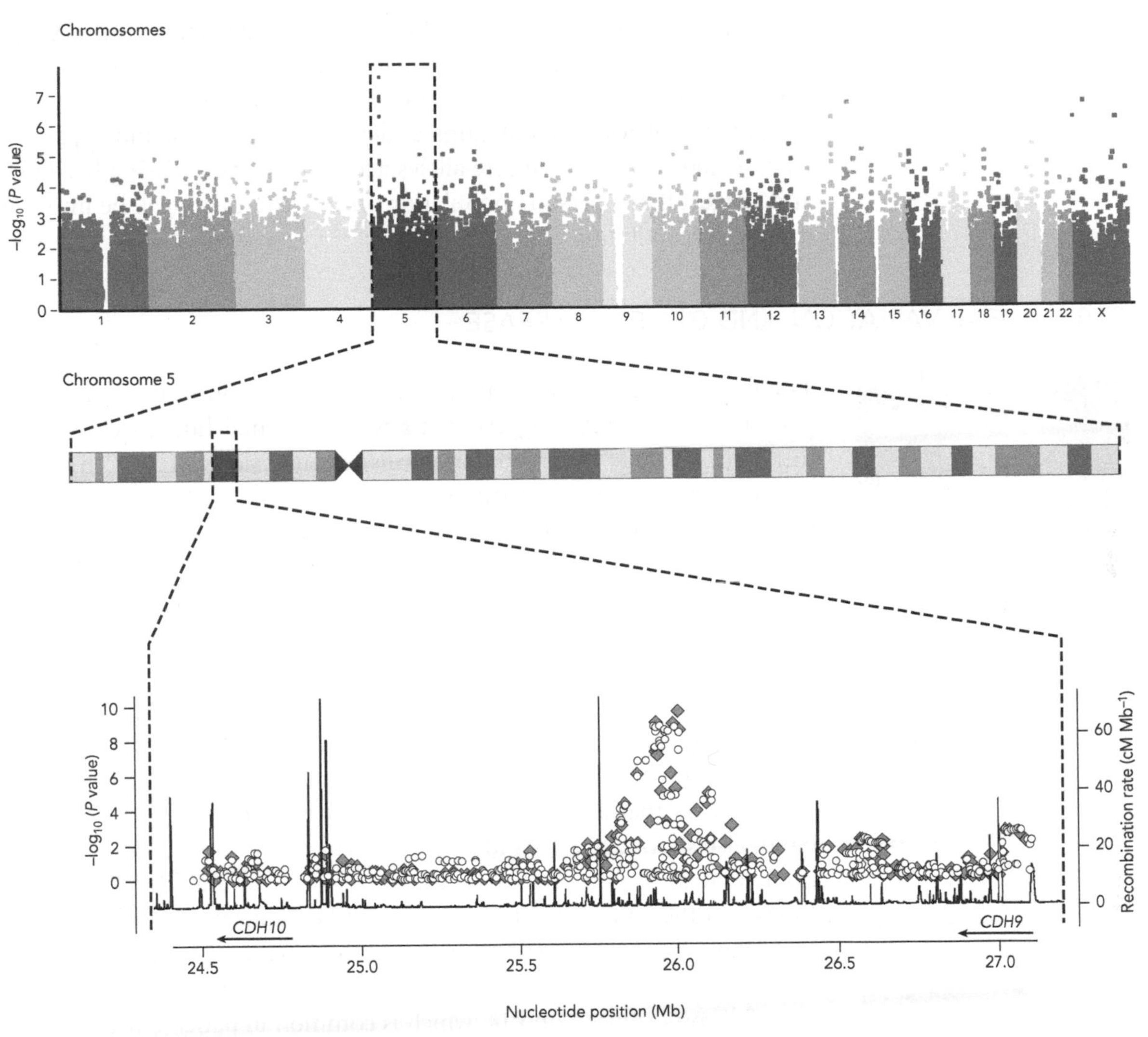

Genome-wide association study of autism spectrum disorder (ASD). (*Top*) A meta-analysis combined results from two studies that looked at approximately 500,000 common SNPs across 10,796 individuals (ASD patients and normal controls). With a *P* value below the generally accepted threshold of 5×10^{-8} (0.00000005), one SNP on the short arm of chromosome 5 was significantly associated with ASD; several other less-strongly associated SNPs were also detected at the same locus. (The lower the *P* value, the smaller the probability that an association is due to chance—and the greater the probability of a true association.) (*Bottom*) The addition of two more data sets identified a number of significantly associated SNPs. (Grey diamonds) SNPs that were genotyped in one or more of the studies, (open circles) additional SNPs that were imputed (inferred) from individual haplotypes. The very low recombination rate across the locus provided evidence that the SNPs are in linkage disequilibrium with ASD; i.e., they are never separated by chromatid exchange (crossover).
(Adapted, with permission, from Wang K, et al. 2009. *Nature* 459: 528–533; ©Macmillan.)

have turned up again and again in different pedigrees and populations. This is consistent with polygenic inheritance for many common disorders, with multiple susceptibility genes contributing to an individual's overall risk. Unfortunately, the significant genes and haplotypes ("hits") uncovered by GWAS typically explain only a small fraction of phenotypic variation or heritability—the proportion of phenotypic variance that is inherited. At the extreme, a 2010 GWAS of 183,727 adults identified 180 loci associated with height, but that explained only ~10% of phenotypic variance.

Thus far, genome sequencing has focused on detecting common SNPs and alleles with population frequencies >5%. However, there is growing consensus that most common diseases result from combinations of rare mutations that were concentrated as "private" alleles as populations underwent bottlenecks and expansions during the past 100,000 years. These rare alleles account for the "missing heritability" in GWAS and suggest that a person's population affinity (heritage) will be important in gauging disease susceptibility.

STRUCTURAL VARIATION AND GENETIC DISEASE

Janet Rowley
(Courtesy of the Albert and Mary Lasker Foundation.)

Lore Zech, ca. 2004
(Courtesy of Christine Scholz, Munich.)

The focus on SNPs initially underestimated variation in the human genome. Recent genome-wide surveys suggest that any two normal human chromosomes differ by hundreds of large insertions, deletions, or inversions of 1000 nucleotides or more. These copy-number variations (CNVs) are now thought to contribute more nucleotide variation in the human genome than do SNPs. In 2008, a detailed analysis of structural variants greater than 8000 nucleotides across eight human genomes showed that at least half of 1700 large insertions, deletions, and inversions were seen in more than one individual.

More than 1000 genes map to regions of known structural variation, so is not surprising that large-scale chromosome changes are involved in human disease. In 1890, David von Hansemann, a pathology student of Rudolf Virchow's in Berlin, first described abnormal chromosomes in a variety of tumor cells. The staining techniques of the 1950s that allowed human chromosomes to be identified by their distinctive bands also showed that translocations—exchanges of chromosome pieces between nonhomologous chromosomes—are characteristic of several blood cancers. In 1972, Janet Rowley, at the University of Chicago, identified a diagnostic translocation between the long arms of chromosomes 9 and 22 in 95% of patients with chronic myeloid leukemia (CML). In 1976, Lore Zech, at the Karolinksa Institute, identified a translocation of the tips of chromosomes 8 and 14, which is common in patients with Burkitt's lymphoma.

Although translocations occur most frequently in blood tumors, many are associated with solid tumors. About 50 common translocations have been identified, with most associated with disruptions in genes involved in the control of cell division. Detailed molecular analysis showed that the CML and Burkitt's translocations cause qualitative and quantitative changes in gene expression that contribute to oncogenesis. The CML translocation occurs in the breakpoint cluster region (BCR) of chromosome 22—fusing the 5′ portion of the resident *BCR* gene with the 3′ portion of the *c-ABL* gene on chromosome 9. Although the ABL kinase normally transmits weak signals about cell adhesion and replication, the BCR-ABL fusion protein has greatly increased kinase activity and lacks normal regulation. The altered BCR-ABL protein is essentially a growth factor stuck in a hyperactive "on" position. The translocation in Burkitt's lymphoma causes a quantitative change in protein expression. The breakpoint on chromosome 14 coincides with a major immunoglobulin locus, stimulating transcription of the *MYC* oncogene under the control of a strong immunoglobulin enhancer.

Expression of an oncogene is also amplified if it is present in multiple copies—as visibly occurs in a homogeneously staining region (HSR) or in duplicated chromosome fragments called double minutes (DMs). In situ hybridization, where a radioactively labeled probe is hybridized to intact chromosomes, has confirmed that numer-

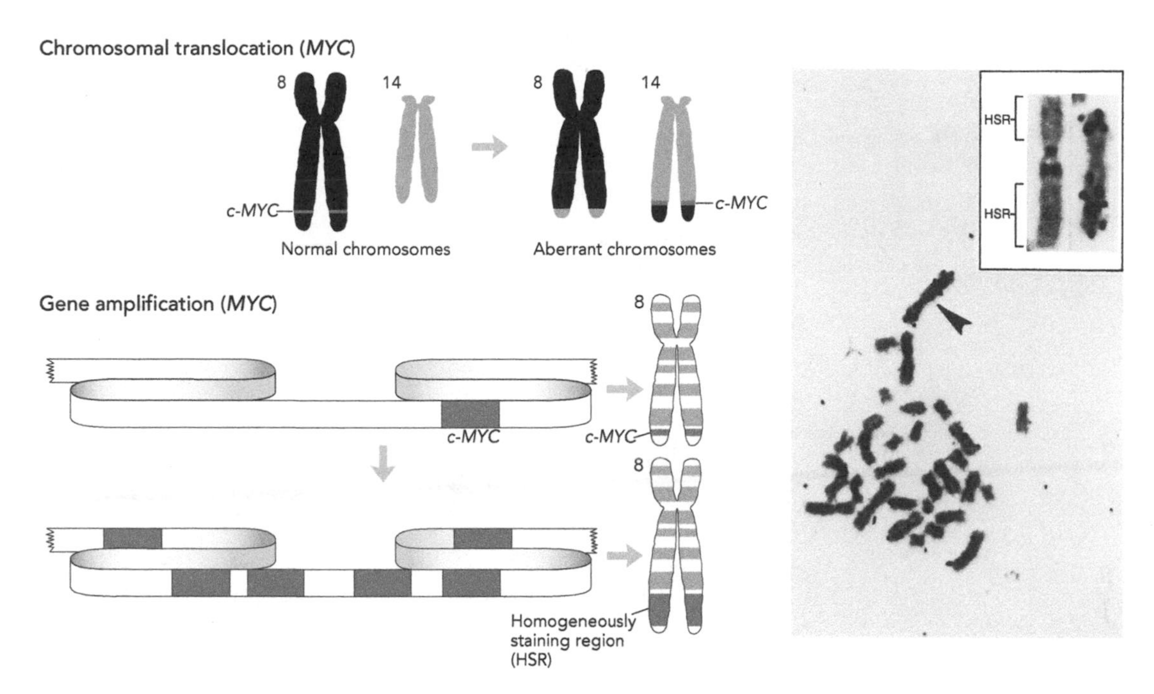

MYC oncogene translocation and amplification. (*Left*) Two examples of molecular mechanisms of oncogenesis: chromosomal translocation and gene amplification. (*Right*) Dark grains indicate regions where a radiolabeled probe for the *MYC* oncogene hybridizes to amplified copies in homogeneously staining regions (HSRs) of chromosomes from neuroendocrine tumor cells. (*Inset*) Comparison of labeled and unlabeled chromosomes.

(Courtesy of J. Bishop and H. Varmus, University of California, San Francisco. From Alitalo K, et al. 1983. *Proc Natl Acad Sci* 80: 1707.)

Eric Fearon
(Courtesy of Eric Fearon.)

Bert Vogelstein
(Courtesy of Bert Vogelstein.)

ous copies of the *MYC* oncogene are present in HSRs and DMs from neuroendocrine tumor cells. Because *MYC* gene copy number tends to correlate with cancer progression, gene amplification may be a secondary result of chromosome "crisis" during tumorigenesis.

Deletions are an important mechanism in disabling tumor-suppressor genes, whose proteins limit cell division. A tumor-suppressor gene is said to be "recessive acting" because both copies must be mutated to contribute to the cancer phenotype. For example, patients with the childhood tumor retinoblastoma typically show chromosome deletions and rearrangements of the *RB* gene on chromosome 13. A first deletion of chromosome 13—inherited in "familial" cases or occurring randomly in "sporadic" cases—eliminates one copy of the *RB* gene. Heterozygous cells, with one functioning copy of RB and one mutated copy of *RB*, behave normally. However, deletion of the second copy of the *RB* gene—referred to as a loss of heterozygosity (LOH)—eliminates RB's tumor-protective function and results in a clone of tumor cells.

Work in the 1990s by Eric Fearon and Bert Vogelstein, at Johns Hopkins Medical School, convincingly showed a multistep model for colon cancer. The deletion of several tumor suppressors combines with oncogene activation in the progression from a benign polyp to a malignant carcinoma. The first step in carcinogenesis is typically the loss of both copies of the *APC* (adenomatous polyposis coli) gene on chromosome 5. As in retinoblastoma, patients with familial colon cancer are born with a single functional copy of the *APC* tumor suppressor, greatly increasing the chance of LOH. Then, mutational activation of one or more oncogenes is typically followed by deletions that inactivate tumor-suppressor genes.

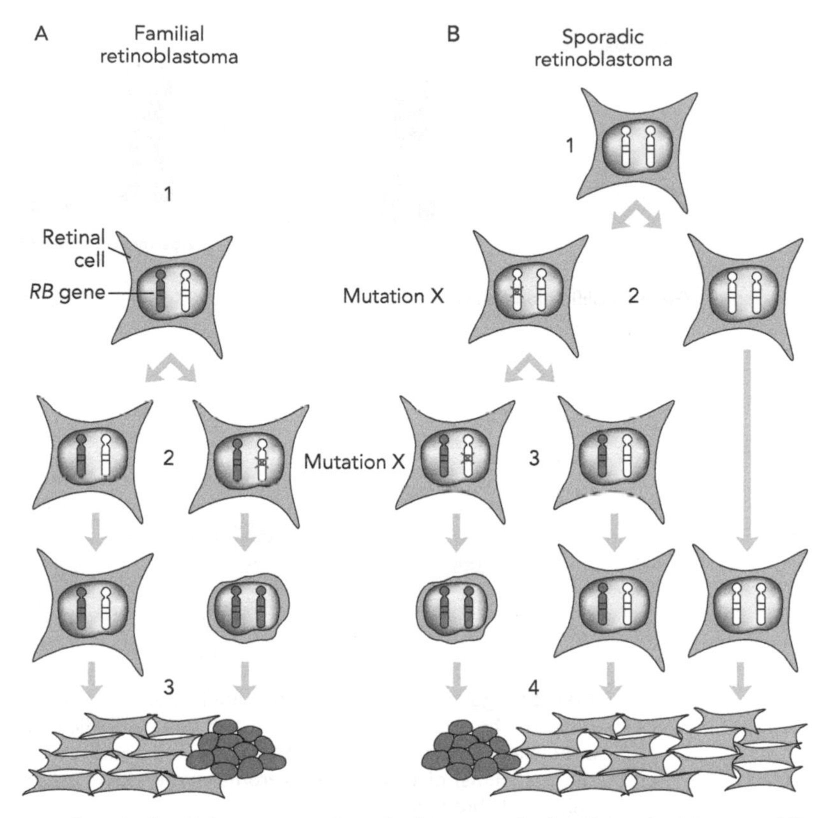

RB gene mutations in familial and sporadic retinoblastoma. In familial retinoblastoma (*A*), one normal (white) and one mutated *RB* gene (gray) are inherited on chromosome 13 (1). Subsequent mutation (mutation X) in any retinal cell inactivates the remaining *RB* gene (2), leading to loss of growth control in a clone of tumor cells (3). In sporadic retinoblastoma (*B*), two normal *RB* genes are inherited (1). The first mutation inactivates one copy of the *RB* gene (2); subsequent mutation within the same retinal cell inactivates the remaining copy of the *RB* gene (3), leading to loss of growth control in a clone of tumor cells (4).

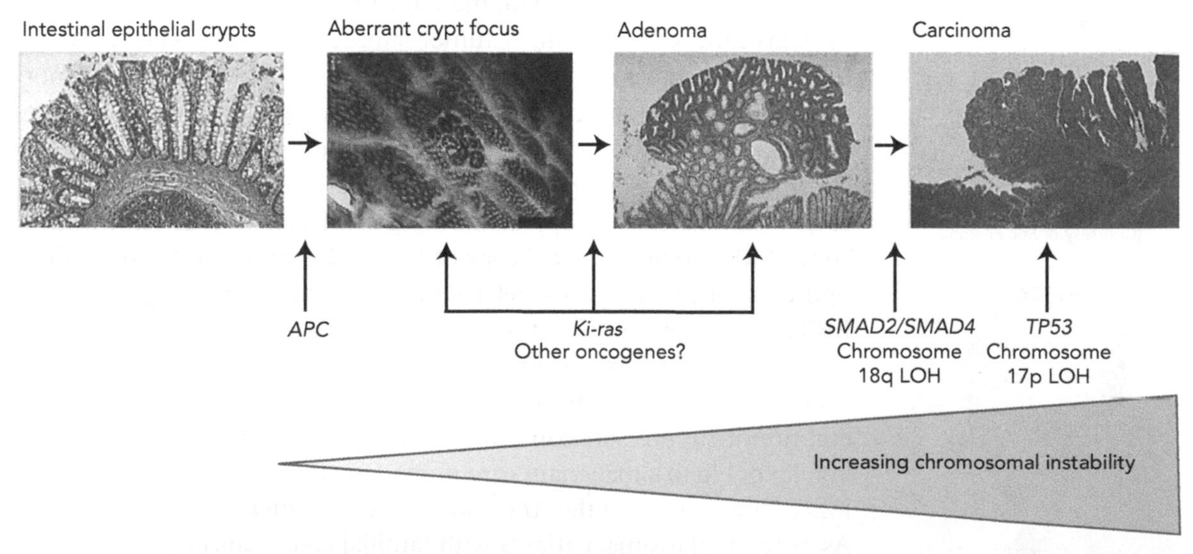

Multiple steps in colon cancer. Patients with familial colon cancer are born with a single functional copy of the *APC* tumor suppressor, greatly increasing the chance of LOH. Then, mutational activation of the *Ki-ras* oncogene is typically followed by deletions in chromosomes 17 and 18 that inactivate tumor suppressors *TP53* and *SMAD2/SMAD4*.

(Adapted, with permission, from Takayama T, et al. 1998. *N Engl J Med* 339: 1277–1284;)

Robert Lucito

(Courtesy of Cold Spring Harbor Laboratory Archives.)

Jonathan Sebat and Michael Wigler

(Courtesy of Cold Spring Harbor Laboratory Archives.)

Extending this earlier work on tumor cells, in 2004 Robert Lucito, Jonathan Sebat, and Michael Wigler of Cold Spring Harbor Laboratory first demonstrated that CNVs are a major source of genetic variation in normal cells. Conducting pairwise hybridizations of DNA from 20 human genomes, they identified more than 200 very large CNVs, with any two individuals differing by an average of 11 CNVs of about 500,000 nucleotides in length. Their subsequent 2007 study showed that de novo CNVs are a significant risk factor for autism spectrum disorder (ASD). CNVs—mainly deletions of about 100,000 to 6 million nucleotides—were found in 10% of sporadic cases (children without an affected sibling) but in only 3% of patients with an affected sibling and 1% of controls. Of 12 deletions in autism patients, one region contained the candidate gene oxytocin, which is involved in social recognition. Other deletions affected genes that are

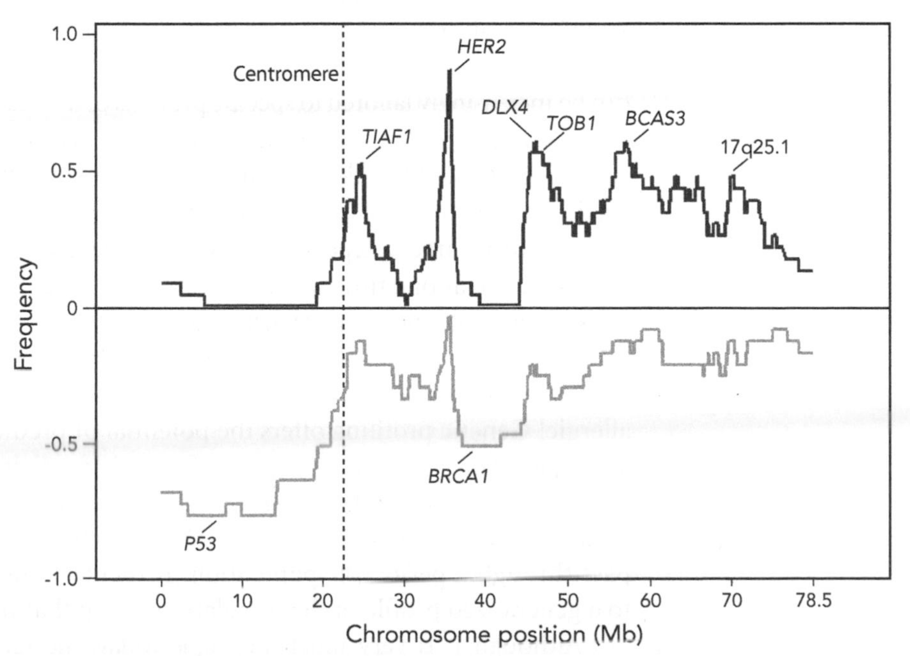

Copy-number variations in human breast cancer, 2006. Representational oligonucleotide microarray analysis (ROMA) detects a "firestorm" of amplifications and deletions on chromosome 17 in 243 breast tumors. Included are deletions in the breast cancer susceptibility gene *BRCA1* and the gene for the tumor suppressor p53. Among the amplified genes, *HER2* (human epidermal growth factor receptor) is the target of the antibody treatment Herceptin.

(Adapted, with permission, from Hicks J, et al. 2006. *Genome Res* 16: 1465–1479; ©Cold Spring Harbor Laboratory Press.)

among the largest in the human genome. This is consistent with other large genes that are involved in spontaneous genetic diseases, such as retinoblastoma, neurofibromatosis, and DMD.

PHARMACOGENOMICS AND PHARMACOGENETICS

Throughout the second half of the 20th century, major pharmaceutical companies amassed "libraries" containing hundreds of thousands of chemical compounds. These numbers have increased dramatically with the advent of combinatorial chemistry, which builds up compounds from simple chemical components—analogous to the way in which DNA probes are built up on a photolithographic DNA chip. Each of these compounds is a potential pharmaceutical that can fight disease by altering the activity of a gene or its corresponding protein. For example, a number of compounds have been developed against histamines and other molecules involved in allergic reactions.

Pharmaceutical development, however, has been hampered by a relative lack of metabolic "targets" against which companies can test their huge compound libraries. The availability of the human genome sequence promises to solve this problem by presenting drug developers with a trove of new targets. Using the human genome sequence to inform drug discovery is termed pharmacogenomics. Each gene that is definitively linked to a disease becomes a validated target for drug discovery. Knowledge of mutations in disease-causing genes, and the corresponding changes in the three-dimensional structures of the encoded proteins, allows one to develop strategies for screening compound libraries. Rational drug design carries this concept a step further by using the target protein's structure to predict the properties of small molecules that can bind to an active site or otherwise modulate the protein's activity. This was the triumph of Gleevec, the first anticancer drug developed using detailed knowledge of protein kinase receptors.

As more genes, and therefore proteins, are identified in a disease pathway, treatments can be increasingly tailored to specific proteins of a metabolic or signal transduction pathway. Defects in different proteins in the same pathway may cause the same disease or symptoms. (Recall Beadle and Tatum's experiment, where mutations in different genes produced the same metabolic phenotype.) Thus, the same apparent disease may present different drug targets, depending on which gene in a pathway is mutated.

Everyone at one time must have taken pause at the paradox of a physician asking us if we are allergic to a particular drug. After all, the doctor should be the one to inform us of a potential problem. Unfortunately, trial and error is the only way to determine a response to most drugs: It takes an allergic reaction to know if we are allergic! Genetic profiling offers the potential of predicting a negative response before a drug is taken. Thus, the endgame of genetic medicine is pharmacogenetics, predicting drug response and tailoring treatment to each person's genetic makeup. However, before we enter this era of personalized medicine, experts today believe that we must pass through a period of "population medicine," where drugs are targeted according to a generalized profile of the population group that most closely matches the patient.

Although it is very much in vogue today, the term "pharmacogenetics" was first coined in 1959 by Friedrich Vogel based on earlier evidence that drug responses are inherited and vary among population groups. Notably, African-American soldiers serving in Italy during World War II suffered adverse effects, including hemolysis, from the anti-malaria drug primaquine. This was correlated with glucose-6-phosphate dehydrogenase (G6PD) deficiency, which, ironically, provides some protection against malaria.

Drug response is largely mediated by so-called metabolic enzymes in the liver—the cytochrome P450 mono-oxidases (CPY450s)—that detoxify compounds and metabolize many drugs into their bioactive forms. People who are "extensive metabolizers" efficiently convert a given drug to its active form and/or metabolize it at a rate that provides the desired therapeutic effect. "Poor metabolizers" fail to convert enough of the drug to its active form or metabolize it at a rate that fails to produce a therapeutic effect. "Toxic metabolizers" convert the drug into a toxic product or metabolize it so slowly that it accumulates to toxic levels.

In the late 1970s, Robert Smith of St. Mary's Hospital, London, noticed an unusually high incidence of side effects, including an unusual fainting response, among patients prescribed the antihypertension drug debrisoquine. He found that about 8% of Caucasians (but less than 2% of African and Asian populations) are poor metabolizers, handling debrisoquine 10–200 times less efficiently than extensive metabolizers. Michel Eichelbaum, of the University of Bonn, found similar disparities in the metabolism of sparteine, used for treating heart arrhythmias. This led to the realization that both drugs are metabolized by the CPY2D6 enzyme and that poor metabolizers inherit a defective CPY2D6 enzyme. Subsequent work revealed that CPY2D6 is involved in deficient responses to at least 40 common drugs, including codeine, dextromethorphan, beta-blockers, monoamine oxidase inhibitors, tricyclic antidepressants, warfarin (blood thinner), antipsychotics (neuroleptics), and fluoxetine (Prozac). Cloning and sequencing of the *CPY2D6* gene in 1988 showed that poor metabolizers have polymorphisms that produce splicing errors or amino acid substitutions.

Affymetrix markets a gene chip that assays for 1936 pharmacogenetic markers in *CPY2D6* and 230 other genes involved in drug metabolism and transport. However, this and other molecular screens for personalized drug response have not yet become a "standard of care" in general medicine. Cigna and other major insurance companies do not cover *CPY2D6* screening, because they consider the tests experimental or lacking sufficient evidence to guide therapy.

Because of its long emphasis on analyzing mutations and understanding gene pathways, cancer treatment remains the best example of pharmacogenetics. In use since the 1970s, Tamoxifen is one of the earliest examples of individualized cancer therapy. A majority of breast tumors have estrogen receptors and require estrogen to grow. Tamoxifen blocks estrogen uptake, so stained breast cancer cells are examined for estrogen receptors to determine those patients who could benefit from the drug. Tumor genotyping detects key mutations that predict drug response and is increasingly used to guide treatment. The tyrosine kinase inhibitors Iressa and Tarceva have proven to be very effective in non-small-cell lung cancer patients with specific mutations in the tyrosine kinase domain of the epidermal growth factor receptor, which ironically are found most commonly in nonsmokers. Although melanomas have traditionally been classified by histology and morphology, rapid genotyping can now detect mutations in a number of key oncogenes, some of which are sensitive to targeted therapy. Similarly, genotyping of stomach tumors identifies five distinct subtypes—two of which are sensitive to available therapeutics. Tumor genotyping can reveal the signatures of highly malignant tumors that require aggressive therapy but also identifies low-grade tumors for which patients can be spared overtreatment.

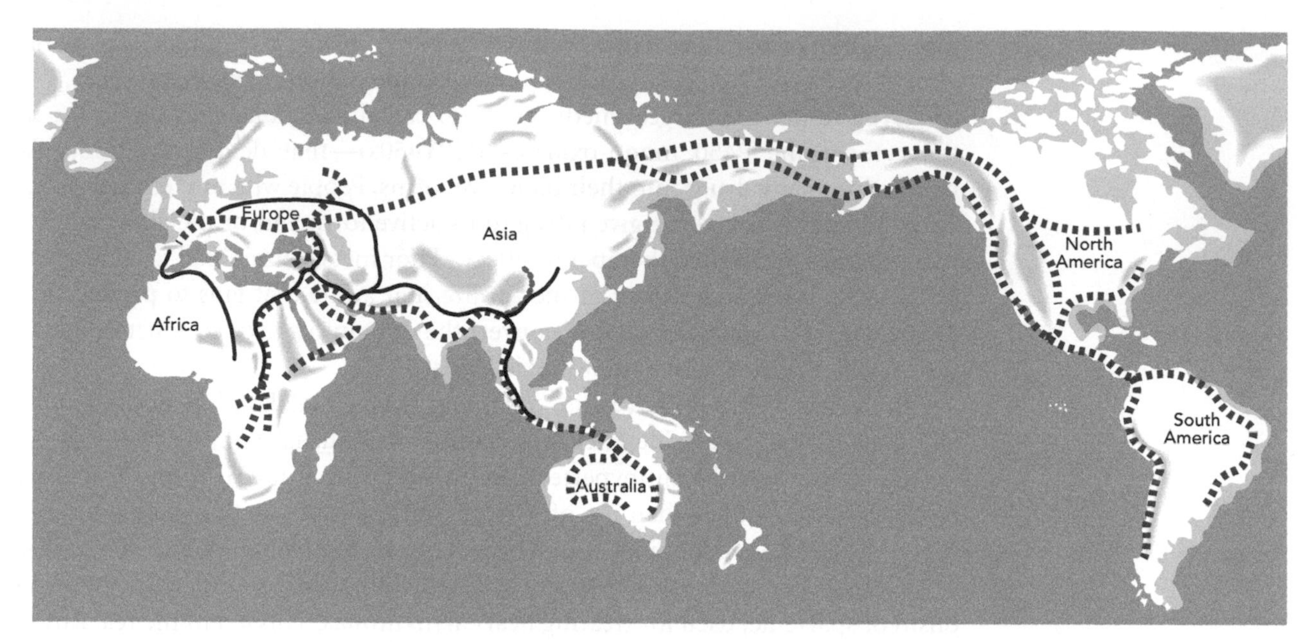

Theories of the origin of modern humans. Fossil evidence shows that *Homo erectus* left Africa ~1.8 million BP, rapidly reaching Europe and Asia (solid line). The multiregional theory holds that modern humans arose independently in Africa, Europe, and Asia from local populations of *H. erectus*. The displacement theory holds that ancestors of modern human arose from *H. erectus* populations in Africa, which colonized Europe and Asia within the last 60,000 years (dotted line).

WHAT THE FOSSIL RECORD TELLS US ABOUT HUMAN EVOLUTION

The fossil record shows that the human species arose in Africa, and all people alive today share a common ancestor from there. Anthropologists estimate that the human lineage diverged from that of other primates ~6–7 million BP (years before present), with chimps being our closest living relative. Among the most primitive human ancestors are members of the genus *Ardipithecus*, which lived 5.6–4.4 million BP, and members of the genus *Australopithecus*, which lived ~3.5 million BP. Evidence of these ancient hominids has been discovered primarily in the Rift Valley of East Africa. Early members of our own genus, *Homo erectus*, arose in the same region ~2.5 million BP. These "archaic" hominids migrated out of Africa ~1.8 million BP to found populations in Europe, the Middle East, and southern Asia.

The earliest fossils of modern humans, or artifacts made by them, have been found in southern and eastern Africa, dating to about 140,000 BP. Remains of modern humans dating to 100,000 BP have been found in the Middle East, to 60,000 years in Asia and Australia, and to 45,000 years in Europe. By modern humans, we mean members of our own species, *Homo sapiens*, who share with us important anatomical features (skull shape and size) and behavioral attributes (use of blades, bone tools, pigments, burial goods, representational art, long-distance trade, and varied environmental resources). These humans subsequently spread to Micronesia, Polynesia, and the "New World" (North and South America).

Analysis of the fossil record gave rise to two opposing theories of human origins. The multiregional theory contends that modern human populations developed independently from archaic hominid (*H. erectus*) populations in Africa, Europe, and Asia. Early modern groups evolved in parallel with one another and exchanged members to give rise to modern population groups. The displacement theory, also known as the "out of Africa" theory, holds that modern humans arose from a single *H. erectus* stock that left Africa beginning ~80,000 BP. This founding group migrated throughout the Old World, displacing any surviving archaic hominids. Thus, scientists all agree that our earliest hominid relatives arose in Africa, but they disagree on when the direct ancestors of living humans left Africa to populate the globe. In the 1980s and 1990s, DNA evidence seemed to settle the debate soundly in favor of the displacement theory. However, whole-genome sequencing in 2010 confirmed that several ancient hominids made recent genetic contributions to the human genome in Europe and Asia.

THE DNA MOLECULAR CLOCK

At first thought, there does not seem to be any way in which DNA could provide us with information about the origin of modern humans. The oldest hominid DNA isolated thus far dates back only ~60,000 years, well after our emergence as a species. But, in fact, we can study our evolutionary past by looking at the DNA variation of humans and primates alive today. Although this might seem to be a contradiction, remember that the DNA of any individual bears the accumulated genetic history of its species.

When two groups split off from a common ancestor, each accumulates a unique set of random DNA mutations. Provided mutations accumulate at a constant rate and occur sequentially (one at a time), and the number of mutations is proportional to the length of time that two groups have been separated. This relatively constant accumulation of mutations in the DNA molecule over time is called the "molecular clock." An event that has been independently established by anatomical, anthropological, or geochronological data is used to attach a timescale to the clock. For example, the human molecular clock is typically set using fossil and anatomical evidence suggesting that humans and chimps diverged 6–7 million BP.

Because of its high mutation rate, the mitochondrial (mt) control region evolves more quickly than other chromosome regions—it has a faster molecular clock. The fast mutation rate means that lineages diversify rapidly, amplifying differences among populations. However, rapid mutation also introduces the confounding problem of "back mutation," where the same nucleotide mutates more than once, returning it to its original state. Multiple mutations at the same position also cause an underestimation of the total number of mutation events. Thus, the number of observed differences between human and chimp sequences are less than one would expect to have occurred in the 6–7 million years since the lineages diverged. However, the chance of back or multiple mutations is much smaller over the period during which modern humans have arisen. So, the number of observed mutations among living humans is very close to the actual number that has accumulated since we arose as a species.

Mitochondrial DNA (mtDNA) offers another important advantage in reconstructing human evolution: With very few documented exceptions, the mitochondrial chromosome is inherited exclusively from the mother. This is because mitochondria are inherited from the cytoplasm of the mother's large egg cell. Any paternal mitochondria that may enter the ovum at the moment of conception are marked with ubiquitin and destroyed by proteases. The lack of paternal chromosomes with which to recombine greatly simplifies the analysis of mitochondrial inheritance. The mitochondrial genome is inherited intact over thousands of generations, without the confounding effect of crossover with a paternal chromosome. Because the mitochondrial genome is haploid, having only a contribution from the mother, mtDNA types are termed haplotypes ("half types").

WHAT DNA TELLS US ABOUT HUMAN EVOLUTION

Allan Wilson
(Courtesy of Cold Spring Harbor Laboratory Archives.)

Throughout the 20th century, fossils provided the only tools for reconstructing human origins, and the field remained virtually the sole province of anthropologists. In 1987, Allan Wilson and coworkers at the University of California at Berkeley moved anthropology into the molecular age when they made mtDNA haplotypes for 145 living humans. Using a molecular clock like that described above, they constructed a tree that extrapolated back to a common ancestor who lived about 200,000 BP.

It is important to note that the tips of the branches of Wilson's evolutionary tree were the 145 individual humans whose mtDNA types he had determined. Although most individuals generally came out on branches with others from their regional population group, some individuals fell on branches with other groups. This illustrated a high degree of mixing among human population groups. Importantly, Africans turned up on several non-African branches, but only African individuals were found on the branch closest to the root of the tree. Thus, Wilson concluded that the so-called mitochondrial "Eve" most likely lived in Africa, providing DNA evidence for the recent dispersion of modern humans "out of Africa."

1. Six allelic sequences

Sequence (1) agctggctgaatgctatctgcgtcgcgcgaaataacgtcagcaattcgttacat**c**tctctagggc
Sequence (2) agctggctgaatgctatctgc**c**tcgcgcgaaataacgtcagcaattcgttacatttctctagggc
Sequence (3) agctggctgaatgctatctgcgtcgcgcgaaa**c**aacgtcagcaattcgttacatttctctagggc
Sequence (4) agctggctgaatgctatctgcgtcgcgcgaaataacgtcagcaattcgttacatttctctagggc
Sequence (5) agctggctga**g**tgctatctgc**c**tcgcgcgaaataacgtcagcaattcgttacatttctctagggc
Sequence (6) agctggctgaatgctatctgcgtcgcgcgaaataacgtcagca**g**ttcgttacat**c**tctctagggc

2. The six haplotypes for the five SNPs in the allelic sequences above

Position	11	22	33	44	55
Sequence (1)	a	g	t	a	c
Sequence (2)	a	c	t	a	t
Sequence (3)	a	g	c	a	t
Sequence (4)	a	g	t	a	t
Sequence (5)	g	c	t	a	t
Sequence (6)	a	g	t	g	c

3. Two possible trees for the evolutionary relationships among the haplotypes above, assuming that mutations occur sequentially

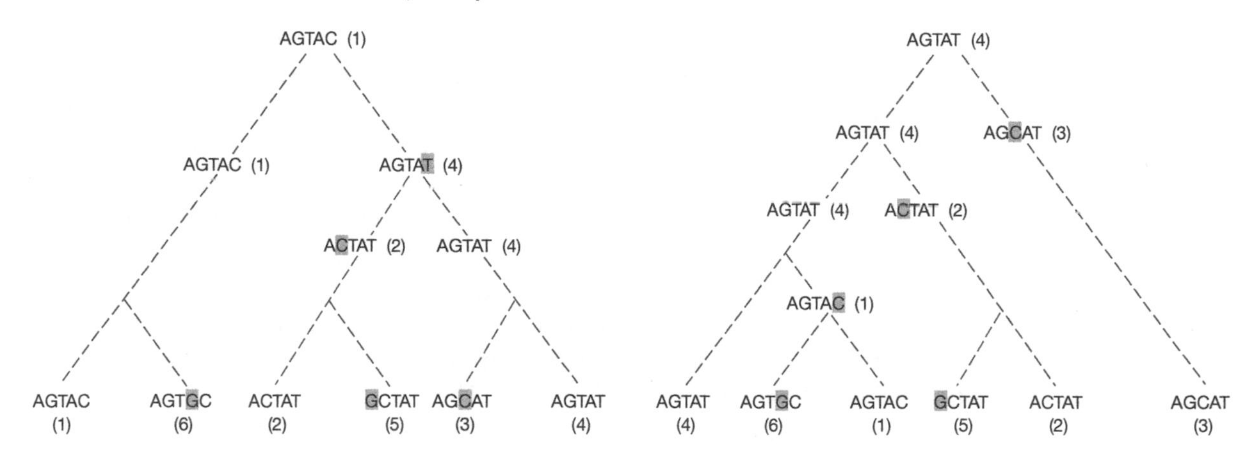

Using mutations in DNA to reconstruct evolutionary relationships.

During the next decade, molecular reconstructions of human lineages were conducted with autosomal and Y-chromosome SNP polymorphisms. Unlike mitochondrial SNPs, which have a high rate of back mutation, each Y-chromosome SNP is believed to represent a unique mutation event that occurred once in evolutionary history. Whereas the mitochondrial chromosome is inherited only from the mother, the Y chromosome is passed only from father to son. The accumulated DNA evidence has confirmed Wilson's original story. Several lines of evidence point to the emergence of modern humans in Africa ~150,000 BP.

- The greatest amount of DNA variation occurs in Africa, suggesting that African populations have been accumulating mutations for the longest period of time. (For example, Yoruban African chromosomes show 15% more SNPs than European or Asian chromosomes.)
- Most Asian and European variations are a subset of variations found in African populations, suggesting that Asian and European populations are derived from an African source.
- The deepest roots of a tree diagram of human variation contain only Africans. Ancient alleles have not been found in non-African populations.

Interestingly, comparisons of mitochondrial and Y-chromosome polymorphisms suggest that men and women have had different roles in the peopling of the planet and in the mixing of genes among population groups. Generally, mtDNA types show a gradation (or cline) of allele frequencies from one geographic region to another. This is the signature of "gene flow," the slow and steady exchange of genes between adjacent populations, which occurs in many cultures when women leave their families to live in their husbands' villages. Y polymorphisms, on the other hand, show discontinuities between adjacent regions, suggesting that men have not moved freely between local groups. However, related Y-chromosome types do leave the signature of migrations and war campaigns that abruptly transplant genes over long distances. Thus, members of the Black South African tribe, the Lemba, have the telltale signature of Cohanin Jewry displaced from the Middle East. The most common Y chromosomes in cosmopolitan southern Japan clearly were transplanted from Korea in the last several hundred years, but the Ainu of the northern islands of Japan have an ancient affinity with Tibetans.

GENETIC LEGACIES OF NEANDERTAL AND DENISOVAN

Since their initial discovery in the Neander Valley of Germany in 1856, the heavy-set bones of Neandertal have fascinated scientists as well as the general public. Neandertal had a brain capacity within the range of modern humans and was certainly an archaic member of the genus *Homo*. Neandertal ranged throughout Europe, the Middle East, and Western Russia beginning ~300,000 BP and became extinct ~30,000 BP. Clearly, during part of its span on earth, Neandertal shared its habitat with modern humans. Thus, there has been a long-standing controversy about whether Neandertal was the direct ancestor of modern humans.

According to the multiregional model, modern humans developed concurrently from several distinct archaic populations living in different parts of the world. Under this model, Neandertal must be the intermediate ancestor of modern Europeans. Other archaic hominid fossils, Java and Peking man (*H. erectus*), were the ancestors of modern Asians. According to the "out of Africa" model, Neandertal was displaced by modern *H. sapiens*, who arrived in Europe ~40,000 BP.

Svante Pääbo, 1997
(Photo by Margot Bennett, Cold Spring Harbor Laboratory.)

In 1997, at the University of Munich, Svante Pääbo, then a student of Allan Wilson, further revolutionized human molecular anthropology when he added a 40,000-year-old DNA sample to the reconstructions of hominid evolution. Pääbo extracted DNA from the humerus of the original Neandertal-type specimen, amplified the sample by PCR, and cloned the resulting products in *Escherichia coli*. The cloned fragments were then used to reconstruct a 379-bp stretch of the mt control region. Pääbo drew ~1000 human mt control region sequences from the GenBank database and compared them in pairs. He found an average of seven mutations between these pairs of modern humans, representing the average variation accumulated since the divergence from a common ancestor. However, he found an average of 27 mutations when he compared each of the 1000 modern human sequences against the reconstructed Neandertal sequence. This placed Neandertal outside the range of variation of modern humans.

On average, one new mutation accumulates in the mt control region of a human lineage during the course of 20,000 years, in accordance with the earliest fossil record of modern humans. The 27-mutation difference between living humans and Neandertal suggests that our lineage diverged from a common ancestor ~550,000 BP. This provided strong evidence in favor of the "out of Africa" model.

However, because the mitochondrial genome is inherited only through female lineage, this analysis was mute on whether our ancestors might have "regionally" interbred with Neandertal. The answer to that question had to await the 2010 publication of the draft sequence of the Neandertal nuclear genome by Pääbo's group, which had moved to the Max Planck Institute for Evolutionary Anthropology in Leipzig.

The Neandertal and chimpanzee genomes, and genomes of contemporary humans, were compared to identify "fixed" nucleotide substitutions—locations where all humans share the same nucleotide but where Neandertal and chimp share a different (ancestral) nucleotide. This genome-wide analysis turned up only 78 nucleotide substitutions that cause amino acid differences in proteins between humans and Neandertal. Pair-wise comparison of SNPs between Neandertal and contemporary humans showed a significantly greater number of matches with Europeans and Asians than with Africans. Neandertal matched long-range SNP patterns (haplotypes) in 10 of 12 Asian and European chromosome regions showing high SNP diversity. This provided strong evidence that Neandertal contributed 1%–4% of the DNA of modern Europeans and Asians. The "gene flow" from Neandertal into the human genome likely occurred 50,000–80,000 BP in the Middle East, after modern humans migrated out of Africa but before their dispersion into Europe and Asia.

The evolution of modern humans became even more complicated when, in 2010, Pääbo's group sequenced the entire mitochondrial and low-coverage autosomal genome from a single finger bone from Denisova Cave in the Altai mountains of southern Siberia. Dated to ~50,000 BP, the bone provided DNA sequence that was unlike either Neandertal or modern humans. Rather, it was from an unknown hominid group, dubbed Denisovan, which, by the mtDNA clock, diverged from a common ancestor with Neandertal and humans about 1 million BP. Autosomal analysis showed that the Denisovan were a sister group of Neandertal. Most surprising, a comparison of SNP variation with humans alive today showed that Denisovan contributed 4%–6% of the DNA of southeast Asian populations. This admixing suggests that the Denisovan may have once ranged throughout a large region of Asia.

The human leucocyte antigens (HLAs) have a vital role in immune defense and are highly variable among human populations. A 2011 study by Peter Parham and Laurent

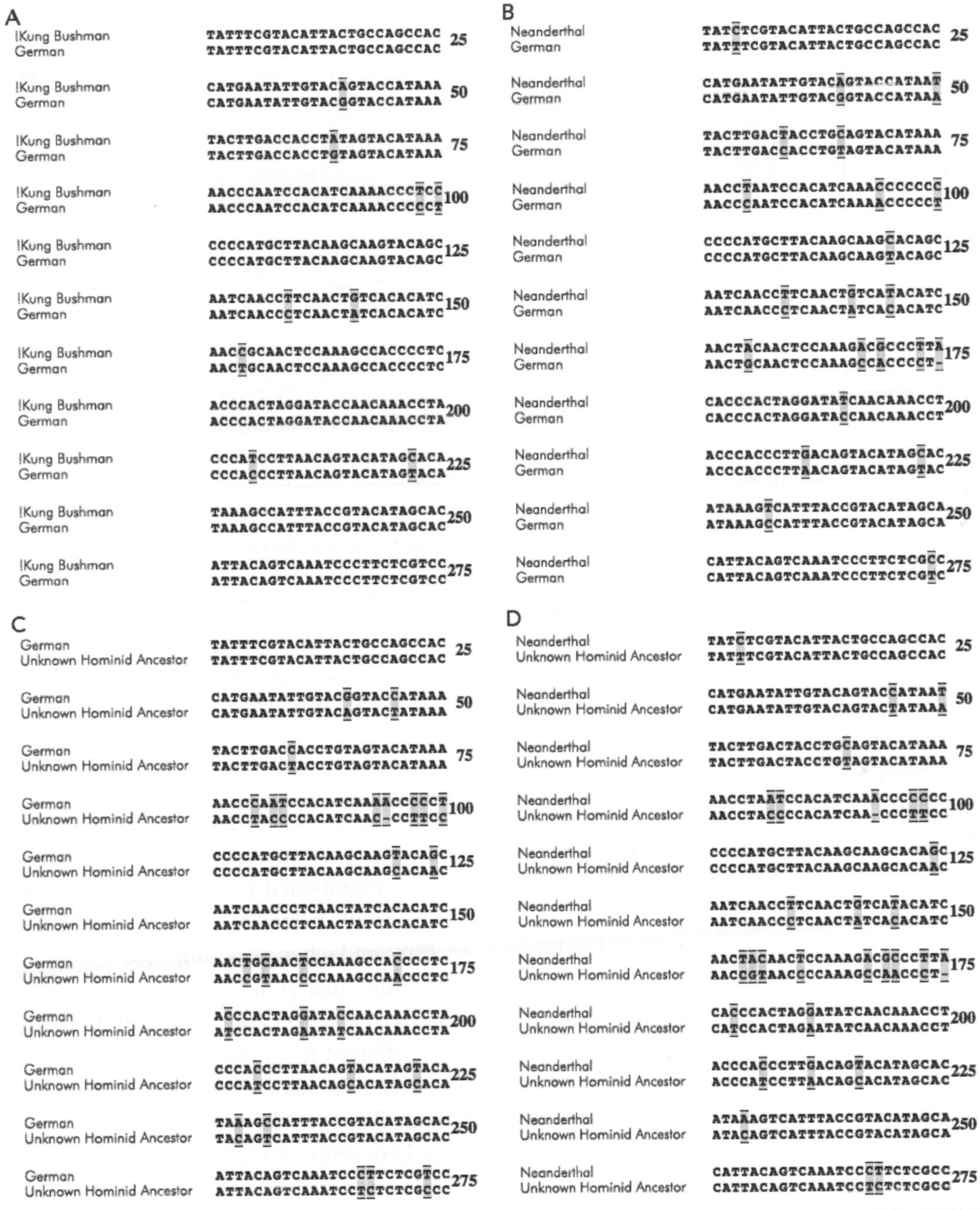

A

!Kung Bushman	TATTTCGTACATTACTGCCAGCCAC	25
German	TATTTCGTACATTACTGCCAGCCAC	
!Kung Bushman	CATGAATATTGTACAGTACCATAAA	50
German	CATGAATATTGTACGGTACCATAAA	
!Kung Bushman	TACTTGACCACCTATAGTACATAAA	75
German	TACTTGACCACCTGTAGTACATAAA	
!Kung Bushman	AACCCAATCCACATCAAAACCCTCC	100
German	AACCCAATCCACATCAAAACCCCCT	
!Kung Bushman	CCCCATGCTTACAAGCAAGTACAGC	125
German	CCCCATGCTTACAAGCAAGTACAGC	
!Kung Bushman	AATCAACCTTCAACTGTCACACATC	150
German	AATCAACCCTCAACTATCACACATC	
!Kung Bushman	AACCGCAACTCCAAAGCCACCCCTC	175
German	AACTGCAACTCCAAAGCCACCCCTC	
!Kung Bushman	ACCCACTAGGATACCAACAAACCTA	200
German	ACCCACTAGGATACCAACAAACCTA	
!Kung Bushman	CCCATCCTTAACAGTACATAGCACA	225
German	CCCACCCTTAACAGTACATAGTACA	
!Kung Bushman	TAAAGCCATTTACCGTACATAGCAC	250
German	TAAAGCCATTTACCGTACATAGCAC	
!Kung Bushman	ATTACAGTCAAATCCCTTCTCGTCC	275
German	ATTACAGTCAAATCCCTTCTCGTCC	

B

Neanderthal	TATCTCGTACATTACTGCCAGCCAC	25
German	TATTTCGTACATTACTGCCAGCCAC	
Neanderthal	CATGAATATTGTACAGTACCATAAT	50
German	CATGAATATTGTACGGTACCATAAA	
Neanderthal	TACTTGACTACCTGCAGTACATAAA	75
German	TACTTGACCACCTGTAGTACATAAA	
Neanderthal	AACCTAATCCACATCAAACCCCCCC	100
German	AACCCAATCCACATCAAAACCCCCT	
Neanderthal	CCCCATGCTTACAAGCAAGCACAGC	125
German	CCCCATGCTTACAAGCAAGTACAGC	
Neanderthal	AATCAACCTTCAACTGTCATACATC	150
German	AATCAACCCTCAACTATCACACATC	
Neanderthal	AACTACAACTCCAAAGACGCCCTTA	175
German	AACTGCAACTCCAAAGCCACCCCT-	
Neanderthal	CACCCACTAGGATATCAACAAACCT	200
German	CACCCACTAGGATACCAACAAACCT	
Neanderthal	ACCCACCCTTGACAGTACATAGCAC	225
German	ACCCACCCTTAACAGTACATAGTAC	
Neanderthal	ATAAAGTCATTTACCGTACATAGCA	250
German	ATAAAGCCATTTACCGTACATAGCA	
Neanderthal	CATTACAGTCAAATCCCTTCTCGCC	275
German	CATTACAGTCAAATCCCTTCTCGTC	

C

German	TATTTCGTACATTACTGCCAGCCAC	25
Unknown Hominid Ancestor	TATTTCGTACATTACTGCCAGCCAC	
German	CATGAATATTGTACGGTACCATAAA	50
Unknown Hominid Ancestor	CATGAATATTGTACAGTACTATAAA	
German	TACTTGACCACCTGTAGTACATAAA	75
Unknown Hominid Ancestor	TACTTGACTACCTGTAGTACATAAA	
German	AACCCAATCCACATCAAAACCCCCT	100
Unknown Hominid Ancestor	AACCTACCCCACATCAAC-CCTTCC	
German	CCCCATGCTTACAAGCAAGTACAGC	125
Unknown Hominid Ancestor	CCCCATGCTTACAAGCAAGCACAAC	
German	AATCAACCCTCAACTATCACACATC	150
Unknown Hominid Ancestor	AATCAACCCTCAACTATCACACATC	
German	AACTGCAACTCCAAAGCCACCCCTC	175
Unknown Hominid Ancestor	AACCGTAACCCCAAAGCCAACCCTC	
German	ACCCACTAGGATACCAACAAACCTA	200
Unknown Hominid Ancestor	ATCCACTAGAATATCAACAAACCTA	
German	CCCACCCTTAACAGTACATAGTACA	225
Unknown Hominid Ancestor	CCCATCCTTAACAGCACATAGCACA	
German	TAAAGCCATTTACCGTACATAGCAC	250
Unknown Hominid Ancestor	TACAGTCATTTACCGTACATAGCAC	
German	ATTACAGTCAAATCCCTTCTCGTCC	275
Unknown Hominid Ancestor	ATTACAGTCAAATCCTCTCTCGCCC	

D

Neanderthal	TATCTCGTACATTACTGCCAGCCAC	25
Unknown Hominid Ancestor	TATTTCGTACATTACTGCCAGCCAC	
Neanderthal	CATGAATATTGTACAGTACCATAAT	50
Unknown Hominid Ancestor	CATGAATATTGTACAGTACTATAAA	
Neanderthal	TACTTGACTACCTGCAGTACATAAA	75
Unknown Hominid Ancestor	TACTTGACTACCTGTAGTACATAAA	
Neanderthal	AACCTAATCCACATCAAACCCCCCC	100
Unknown Hominid Ancestor	AACCTACCCCACATCAA-CCCTTCC	
Neanderthal	CCCCATGCTTACAAGCAAGCACAGC	125
Unknown Hominid Ancestor	CCCCATGCTTACAAGCAAGCACAAC	
Neanderthal	AATCAACCTTCAACTGTCATACATC	150
Unknown Hominid Ancestor	AATCAACCCTCAACTATCACACATC	
Neanderthal	AACTACAACTCCAAAGACGCCCTTA	175
Unknown Hominid Ancestor	AACCGTAACCCCAAAGCCAACCCT-	
Neanderthal	CACCCACTAGGATATCAACAAACCT	200
Unknown Hominid Ancestor	CATCCACTAGAATATCAACAAACCT	
Neanderthal	ACCCACCCTTGACAGTACATAGCAC	225
Unknown Hominid Ancestor	ACCCATCCTTAACAGCACATAGCAC	
Neanderthal	ATAAAGTCATTTACCGTACATAGCA	250
Unknown Hominid Ancestor	ATACAGTCATTTACCGTACATAGCA	
Neanderthal	CATTACAGTCAAATCCCTTCTCGCC	275
Unknown Hominid Ancestor	CATTACAGTCAAATCCTCTCTCGCC	

Mitochondrial haplotype comparisons show evolutionary relationships. Two-way comparisons of mt control region sequences show relationships among modern humans, Neandertal, and an unknown hominid ancestor (the mt sequence is from an anatomically untyped hominid bone from Denisova cave in southern Siberia). Sequence differences are shaded. (*A*) Modern !Kung Bushman and German have nine differences over 275 nucleotides. (*B*) Neandertal and modern German have 22 differences. (*C*) German and unknown hominid ancestor have 28 differences. (*D*) Neandertal and unknown hominid ancestor have 30 differences.

Abi-Rached of Stanford University showed that the Denisovan and Neandertal genomes share some HLA alleles with either Asians or Europeans but not with Africans—suggesting introgression into the genome after humans migrated out of Africa. Several Denisovan HLA alleles are common in Asian populations, and Neandertal alleles constitute half of the HLA diversity of modern Europeans. HLA al-

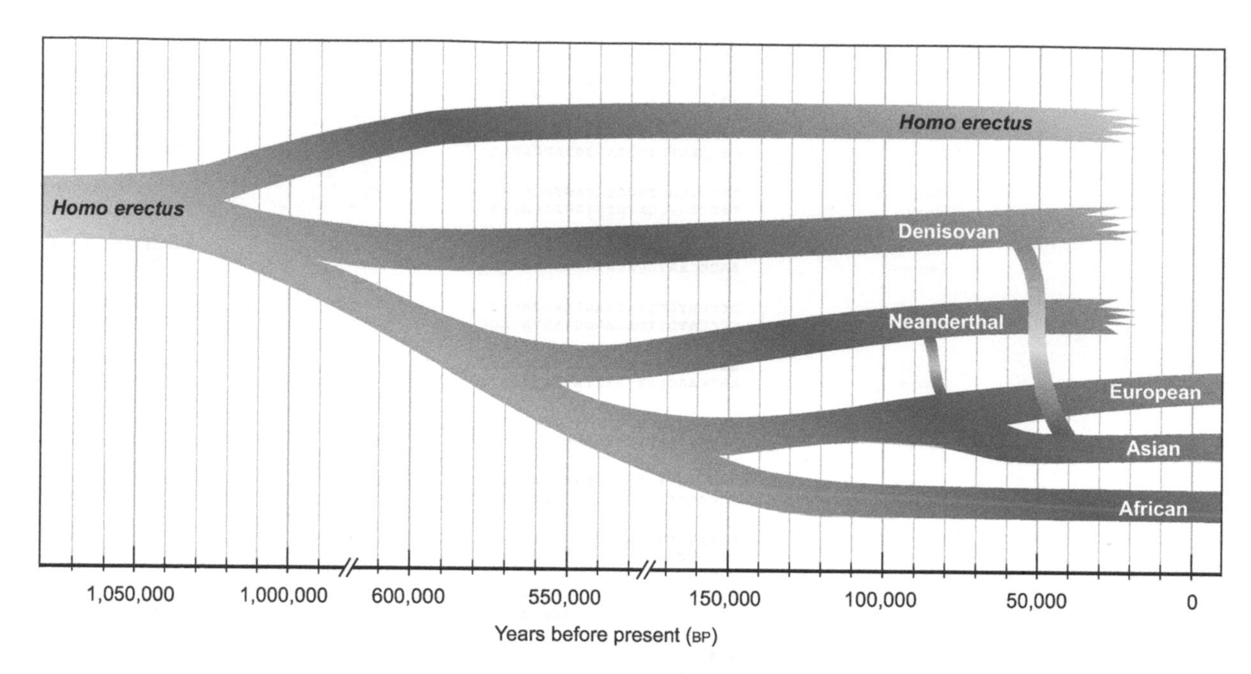

DNA mixing in the hominid family tree. *Homo erectus* left Africa ~1.8 million BP, giving rise to local populations of Denisovans in Asia and Neandertals in Europe. The three "old world" groups of modern humans arose from *H. erectus* in Africa ~150,000 BP. Neanderthal contributed 1%–4% of DNA (vertical branch) to modern Europeans and Asians, suggesting that interbreeding occurred in the Middle East at the time that humans emerged from Africa, ~80,000 BP. Denisovan DNA is restricted to Asians (4%–6%), suggesting that interbreeding occurred in Southeast Asia, ~50,000 BP.

leles are under strong selection in the human genome, and some of the ancient alleles are unique or strong binders of natural killer cell receptors. The fact that Neandertal and Denisovan alleles were maintained in the human genome suggests that regional gene flow from archaic hominids helped to rapidly adapt the immune system to challenges that humans encountered when they migrated out of Africa.

Detailed analysis of individual genes paints a picture of a decidedly "modern" Neandertal. Recent analysis of three loci demonstrates that Neandertal may have been far more similar to than different from modern humans. Neandertal has the human form of the FOXP2 transcription factor, which is encoded by the first gene implicated in language development. The *FOXP2* gene was positionally cloned in 2001 by Anthony Monaco, of the University of Oxford, through an analysis of the KE family, in which severe speech impediments and difficulty with grammatical language were associated with a point mutation in one copy of the *FOXP2* gene. Pääbo then showed that *FOXP2* is highly conserved in mammals, but with humans and Neandertals sharing two amino acid differences from chimps. These changes are "fixed" in every copy of *FOXP2* examined across human populations, suggesting its evolutionary importance. However, it is impossible to tell whether Neandertals had speech or even when this ability developed in humans.

Carles Lalueza-Fox, of the University of Barcelona, and Holger Römpler, of the University of Leipzig, found, in 2007, that Neandertal has a point mutation in the gene for melanocortin 1 receptor (MC1R), which regulates the balance between red-yellow pigment (phaeomelanin) and brown-black pigment (eumelanin). Mutations in this receptor strongly contribute to the characteristic Irish phenotype of fair skin and red hair. Although it was not found in any genes checked from 3700 human samples, the

Neandertal receptor variant had lowered activity when expressed in cultured human cells—consistent with reduced dark pigmentation in homozygotes. This appears to be a case of parallel evolution, with modern humans and Neandertal taking different genetic pathways to skin and hair diversity. A 2009 study found that Neandertal also shares with modern humans a key mutation of the *Tas2R38* taste receptor that diminishes perception of bitter taste. This presumably freed hominids to include more bitter foods in their diet.

CLIMATE CHANGES AND POPULATION BOTTLENECKS

Several lines of evidence suggest that the lineage shared by all humans alive today came perilously close to extinction at one or more points during the last 100,000 years. On the surface, humans appear to be very diverse. Different populations have acquired distinctive morphological adaptations, including skin color, body shape, and pulmonary capacity, that allowed humans to inhabit virtually every biogeographical region of the earth. Despite these morphological differences, the human species as a whole has surprisingly little genetic diversity. Differences among populations account for only ~10% of human genetic variance, and there is no evidence for separate human subspecies.

This contrasts sharply with chimpanzees, who are restricted to similar habitats in equatorial Africa and have few morphological differences. Despite this seeming homogeneity, scientists recognize as many as four distinct subspecies of chimps living in eastern, central, and western Africa and Nigeria. There is substantial genetic diversity

Distribution of chimpanzee subspecies.

among, and within, chimp subspecies. Notably, there is a greater diversity of mtDNA types among members of a single troop of western chimps than among all humans alive today. This striking lack of genetic diversity in the human species supports the contention that we are a young species that has gone through several "bottlenecks" that drastically reduced the human population—and genetic diversity. During these periods, the entire human population may have shrunk to as few as 1000 individuals, clinging to life in scattered refuges.

The fossil record shows that modern humans first left Africa ~100,000 BP, traveling via the Sinai Peninsula into the Middle East. However, this group seems to have stalled, never reaching Asia or Europe. The track of modern human migrations grows cold worldwide until ~60,000 BP, when we find evidence of a second movement out of Africa, hugging the coast of the Sinai Peninsula and across the narrow neck of the Red Sea into Asia and Australia. There is no sign of *H. sapiens* in Europe until ~40,000 BP, when they appear to burst on the scene at a number of sites almost simultaneously.

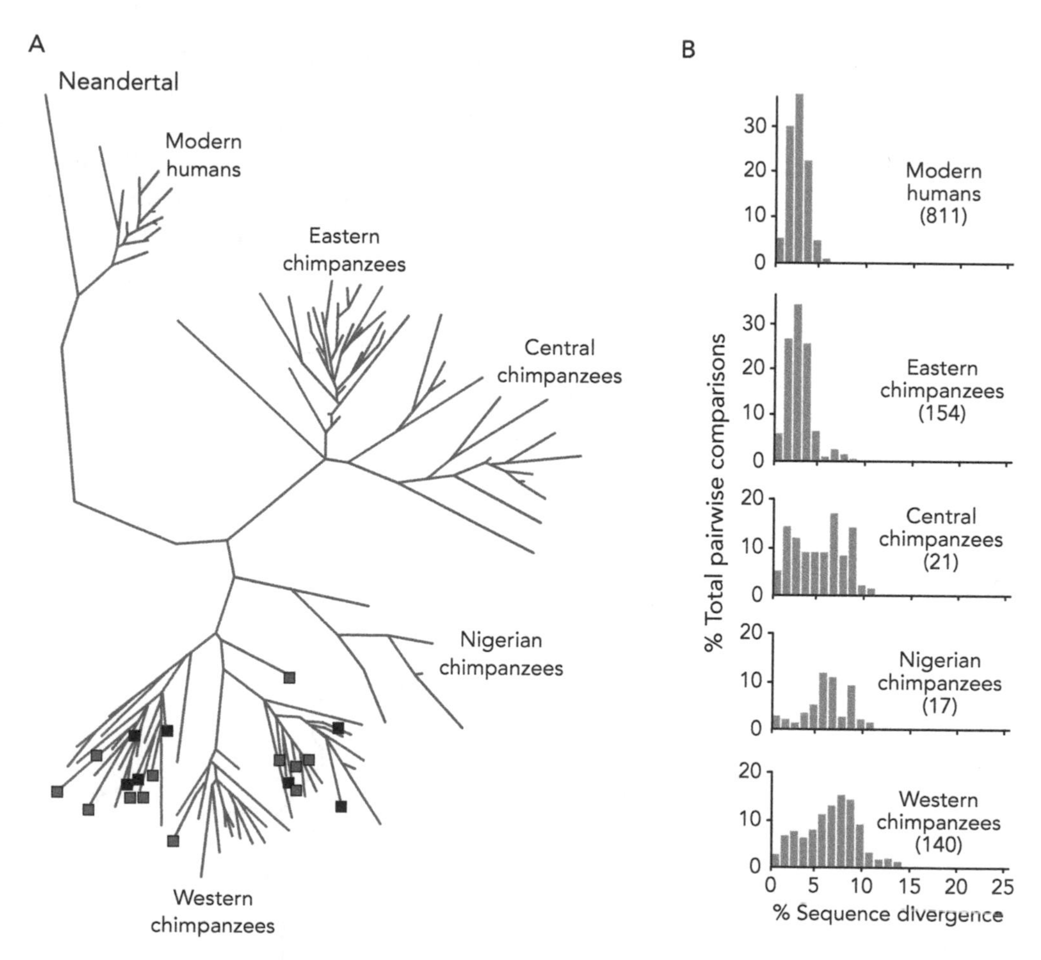

Comparison of genetic diversity in humans and chimpanzees. (*A*) The family tree compares mitochondrial control region (MCR) sequences from 811 humans and 332 chimps. Note the extensive branching of chimp groups, especially the western subspecies, indicating a high degree of diversity. (Black/gray boxes) Chimps from two social groups, each of which exhibits greater diversity than the entire human population. (*B*) Bar graphs show pairwise sequence differences. Note the tight clustering of humans with <5% sequence divergence, compared to the broader distribution and greater sequence divergence among central, Nigerian, and western chimps.

(Reprinted, with permission, from Ambrose SH. 1998. *J Hum Evol* 34: 623–651; ©Elsevier.)

Toba Caldera, Sumatra, Indonesia. Landsat Geocover image of the caldera formed by the Toba Supervolcano. (Image composed by geology.com using Landsat data from NASA.)

Stanley Ambrose, of the University of Illinois, has provided a plausible explanation for the aborted first venture of *H. sapiens* out of Africa. This was the volcanic eruption of Mt. Toba in Sumatra, ~71,000 BP. Toba's 30 × 100-km caldera (more than 30 times the size of Crater Lake in Oregon) was produced by the largest known eruption of the Quaternary Period, the most recent 1.8 million years of the earth's history. The eruption produced as much as 2000 km^3 of ash, leaving ash beds across the Indian Ocean and into mainland India. By comparison, the 1984 eruption of Mt. St. Helens in Oregon produced 0.2 km^3 of ash.

Although the airborne ash would have settled after several months, Mt. Toba also injected huge amounts of sulfur into the atmosphere, where it combined with water vapor to form sulfuric acid. Greenland ice cores show heavy sulfur deposition for 6 years following the eruption, indicating a lingering, sun-obscuring haze worldwide. The reduced solar radiation reaching the earth's surface would have lowered sea surface temperatures by about 3°C for several years. Pollen records suggest that much of Southeast Asia was deforested following the eruption, and significant changes are also recorded in the pollen profile of Grand Pile, in France. Greenland ice cores show that the eruption of Toba was followed by 1000 years of the lowest oxygen isotope ratios of the last glacial period, indicating the lowest temperatures of the last 100,000 years. Thus, it is not difficult to believe that the eruption of Toba produced several years of volcanic winter, followed by 1000 years of unrelenting cold. This surely would have decimated human populations outside of the scattered refuges in tropical Africa.

The fossil record shows that the modern humans who first reached the Middle East were replaced after the Toba eruption by cold-tolerant Neandertals, illustrating that adaptations are relative to environmental factors. Interestingly, Neandertals seem to have gone through similar population bottlenecks during their 250,000 or so years on earth. In 2009, Pääbo's group used next-generation sequencing methods to compare the completed mt genomes of Neandertal specimens representing most of Neandertal's geographic range—Germany, Croatia, Spain, and western Russia. He found that these individuals, who lived 38,000–70,000 BP, had only about one-third as much genetic diversity as modern human populations, despite the fact that Neandertal existed more than twice as long as *H. sapiens*.

THINKING ABOUT HUMAN HISTORY, POPULATIONS, AND RACE

Written in each person's DNA is a record of our shared ancestry and our species' struggle to populate the earth. Our ancient ancestors moved in small groups, following river valleys, coastlines, and bridges that appeared during recurring Ice Ages. As these early people wandered, their DNA accumulated mutations. Some provided advantages that allowed these pioneers to adapt to new homes and ways of living. Most were nonessential. Mutations are the grist of evolution, producing gene and protein variations that have allowed humans to adapt to a variety of environments—and to become the most far-ranging mammal on the planet. But the same mutational processes that generated human diversity—point mutations, insertions/deletions, transpositions, and chromosome rearrangements—also generated disease.

It may be hard to see from our current vantage point, but the entire industrial revolution has occupied only ~0.1% of our 150,000-year history as a species. The cradles of western civilization—classical Greece and Rome—take us back into only 2% of our history. The earliest city-states of Mesopotamia, Babylonia, Assyria, and China take us back only 4% into our past. At 6%, we reach the watershed of agriculture, which changed forever the way humans live and work. Reaching further back to the remaining 94% of our history, to the dawn of the human species, we lived only as hunter-gatherers.

The fastest evolving part of our genome, the mitochondrial (mt) control region, accumulates about one new mutation every 20,000 years. Mutations are an estimated 100-fold less frequent in most regions of the nuclear chromosomes. Thus, virtually every gene in our genome is, on average, only one event away from our hunter-gatherer heritage. This leads to two far-reaching conclusions that substantially broaden our understanding of evolutionary processes and the origin of human disease:

- Throughout most of human history, the hunter-gatherer group was a basic population unit upon which evolution acted.
- Our basic anatomy and physiology, and many aspects of behavior, are essentially identical to the hunter-gatherers who ranged through the ancient landscapes of Africa, Europe, Asia, Australia, and the Americas.

Today, it is difficult to conceive of what is meant, in a genetic sense, by a human population. During the past half century, people have become extremely mobile. Airplanes and four-wheel vehicles have made it possible to travel virtually anywhere in the inhabited world within 1 or 2 days. Major urban centers have become cosmopolitan, with mixes of people representing many ethnic groups and cultures. Even so, today there are still regions of the world where people are born, reproduce, and die all in the same village. This essentially defines the "classical" definition of a human population: a group of people who, by reason of geography, language, or culture, preferentially mate with one another.

Unique human populations—e.g., the Saami of Finland, Ainu of Japan, Nanuit of Alaska, Yanomami of Brazil, Pygmies of Central Africa, and Bushmen of Southern Africa—have preserved unique cultures and languages. Their genomes preserve the genetic residue of a time when all human beings lived in smaller and more cohesive groups. Small populations are subject to the founder effect, "inbreeding," and genetic drift (a random fluctuation of nonessential alleles). Over millennia, these effects join with selection to concentrate particular gene variations within different population groups. Gene variations come into Hardy–Weinberg equilibrium when a population grows to several thousand individuals.

Most people can readily define characteristics that make them different from others. The most obvious difference among people is the color of their skin, followed by hair and eye color, hair texture, and shapes of body and facial features. These physical characteristics, in combination with cultural and religious practices, have been generalized into the related concept of ethnicity. Unfortunately, ethnic prejudices have fueled many of the worst events in human history.

The physical characteristics we associate with ethnicity likely are controlled by a mere handful of genes in the human genome. Variation in human skin color is determined by levels of two different forms of a pigment produced by melanocytes in the dermis layer of the skin. Eumelanin is brown-black and pheomelanin is red-yellow. However, the genetic basis of pigmentation is not well understood, and only one gene involved in human pigment variation has been located. But why did different population groups develop different skin colors in the first place?

Biologists assume that early human ancestors had light skin covered by dense hair—like our nearest primate relative, the chimpanzee. Australopithecines and other early human ancestors probably looked and acted like tall chimpanzees, but with the ability to walk upright for longer periods of time. *H. erectus*, with its striding gate, spent more time tracking prey and foraging in the open African savannas. Increased activity in the open sun, and a larger brain to be protected from overheating, necessitated efficient evaporative cooling. This would have selected for individuals with larger numbers of sweat glands. (Chimps have very few sweat glands.) However, wet hair hinders evaporation, so a trend toward evaporative cooling also favored a reduction in body hair. This is a plausible explanation of how humans came to have nearly hairless bodies.

In the absence of protective hair, it is generally assumed that darkly pigmented skin developed in hominid populations in Africa as protection against the damaging effects of UV radiation. Most skin cancers develop later in life, well after reproductive age. Thus, it seems unlikely that melanin's anticancer effect, alone, could have provided enough selective advantage for dark skin.

In 1967, W. Farnsworth Loomis, of Brandeis University, offered an explanation as to why lighter skin evolved among populations living at higher latitudes. Short-wavelength UV (UVB) radiation in sunlight triggers a reaction in the skin to produce vitamin D, which is important in skeletal formation and immune function. Vitamin D synthesis by the skin is especially important for people without diets rich in this vitamin, as would have been the case for most early hominids. Thus, Loomis hypothesized that lighter skin offered a selective advantage as people migrated out of Africa, allowing them to absorb more of the reduced UV light that penetrates the atmosphere in the higher latitudes.

Recent modeling of worldwide UVB radiation, based on satellite mapping of the earth's ozone layer, shows a correlation between levels of UVB that are sufficient for vitamin synthesis and skin pigmentation. Thus, darkly pigmented skin is found in the tropics where there is sufficient UVB to synthesize vitamin D year-round. Lighter skin, but with the ability to tan, is found in subtropical and temperate regions, which have at least 1 month of insufficient UVB radiation. Very light skin that burns easily is found north of 45 degrees, where there is insufficient UVB year-round.

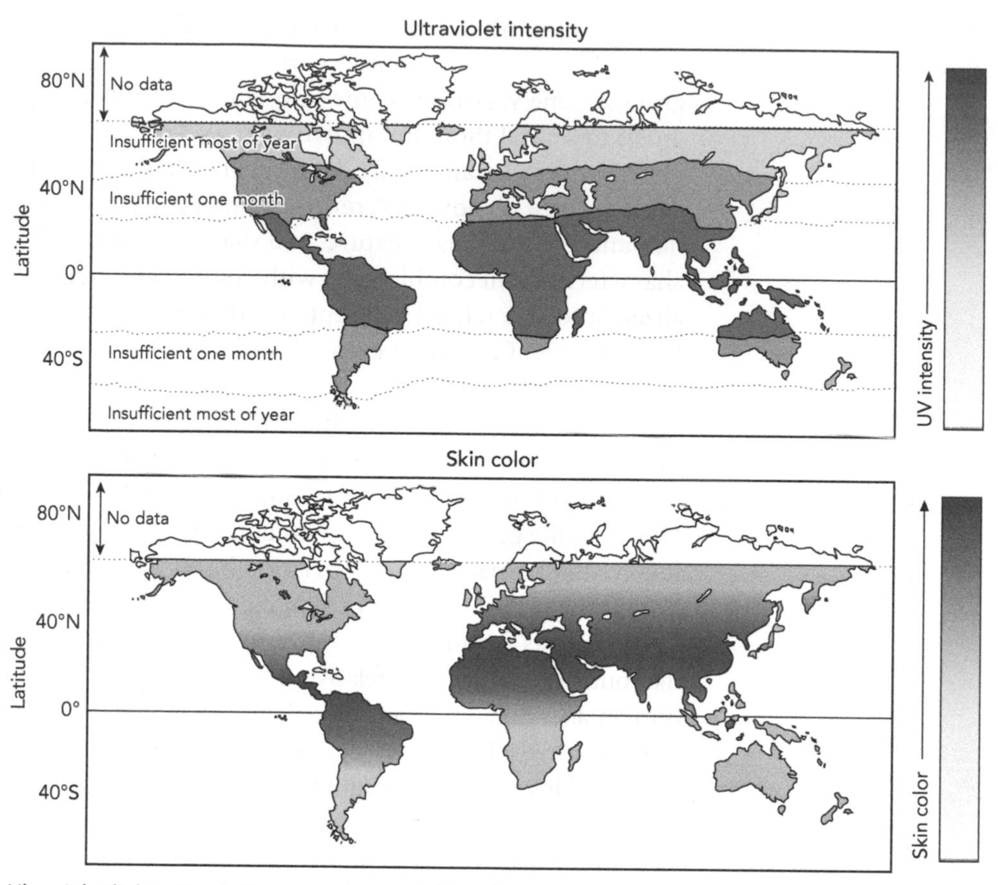

Ultraviolet light, vitamin D production, and skin color. (*Top*) Populations that live in the tropics near the equator receive enough UV light from the sun to synthesize vitamin D all year long. In temperate zones, people lack sufficient UV to make vitamin D at least 1 month of the year. Those nearer to the poles do not get enough UV light most months for vitamin D synthesis. (*Bottom*) Predicted skin colors for humans based on UV light levels. In the Old World, the skin color of indigenous peoples closely matches predictions. In the New World, however, the skin color of long-term residents is generally lighter than expected, probably because of their recent migration and factors such as diet.
(Adapted, with permission, from Jablonski NG, Chaplin G. 2002. *Sci Am* 287: 74–81.)

In 2000, Nina Jablonski and George Chaplin, of the California Academy of Sciences, offered a more complete explanation for the evolution of darkly pigmented skin among early hominids in Africa. They proposed that melanin protects the body's stores of the B vitamin folate, which is essential for reproduction and embryonic development. This conclusion came from the synthesis of several lines of research: (1) Exposure to sunlight rapidly reduces folate levels in the blood. (2) Treating male rodents with folate inhibitors impairs sperm development and induces infertility. (3) Folate deficiency during pregnancy, including the reduction apparently induced by overuse of tanning beds, increases risk of neural cord defects in infants. Thus, as early hominids spent more time hunting and gathering on the open savanna, those with darker skin would have had greater reproductive success and produced more healthy offspring.

Douglas Wallace
(Courtesy of University of California, Irvine.)

Clearly, moving out of the tropics presented challenges to human metabolism. At the most basic level, the oxidative-phosphorylation reactions of mitochondria produce ATP, the energy currency of cells, as well as heat to maintain constant temperature. As humans evolved in Africa, the tropical climate selected for mitochondria that maximize energy and limit heat production. Douglas Wallace, of the University of

Nina Jablonski and George Chaplin at a cave site in China.
(Courtesy of California Academy of Sciences.)

California, found evidence in 2003 that migration to colder climates selected for mutations in the mitochondrial genome that retuned the mitochondria to produce more heat and less energy. He found marked differences in three key proteins involved in oxidative phosphorylation among people living in tropical, temperate, and arctic climates. Functionally significant amino acid changes in ATP6 synthase were most frequent in the Arctic, cytochrome b in the temperate zone, and cytochrome oxidase I in the tropics.

EVIDENCE OF RECENT SELECTION IN THE HUMAN GENOME

In 2005, comparison of whole-genome sequences confirmed that chimps and humans are remarkably similar at the genetic level. At the sequence level, only ~1% of nucleotides differ between the species, with insertions/deletions accounting for 1.5% of sequence differences. Orthologs are equivalent genes in different species that are related by descent from a common ancestor. Approximately 30% of human and chimp orthologs encode identical amino acid sequences.

Likenesses aside, many biologists are interested in finding examples of genes that have evolved since we split from our common ancestor with chimpanzees 6–7 million BP and that were presumably important in making us "human." This involves looking for gene or genome regions that differ from our nearest relatives—chimps alive today and extinct Neandertals and Denisovans.

Olfactory receptors (ORs) are the largest vertebrate gene family, one that has shown rapid evolution in the human lineage. Mammals have a repertoire of ~1000 olfactory genes that encode receptors lining the nasal cavity and detect odor molecules. If one has ever watched a dog or mouse absorbed in sniffing its environment, it is not hard to believe that most mammals smell an entirely different world than do humans. This is because as one progresses "up" the tree of mammal evolution, an increasing proportion of ORs has been disabled by point mutations—primarily by generating premature stop codons that truncate the receptor protein. These nonfunctional copies are termed pseudogenes.

In mice and probably dogs, ~800 of OR genes are functional, but this number drops to ~600 in chimps—a loss of about 200 receptors in the 80 million years since mice and chimps shared a common ancestor. However, a similar number has been lost since humans and chimps diverged, leaving humans with only ~400 functional receptors. This means that humans have only half as many ORs as mice or dogs! Approximately 60 of these are literally caught in the act of disappearing. Termed segregating pseudogenes, functional (+) and nonfunctional (−) alleles of these genes are found in the human population—with three possible genotypes (++, +−, and −−). Without a selective advantage, the allele frequency of many of these pseudogenes will reach 100%, at which point they are "fixed" in the human population. The exact reason for this rapid loss of ORs is unknown. Presumably, many olfactory functions in identifying food sources, marking territory, and finding mates were replaced by cultural conventions as the human brain developed and humans fanned out around the globe.

Indeed, there is growing consensus that human culture—the act of being human—has influenced gene evolution. Here, biologists focus on very recent genome changes, which have the footprint of strong selection. When an advantageous mutation occurs in the human lineage, this gene version and its surrounding chromosome region is passed on intact through generations as a haplotype, supplanting less adaptive genes and their attending haplotypes. Thus, the preponderance of the same haplotype—with its identical suite of SNPs—across population groups is evidence of purifying selection or a "selective sweep." Under strong selection, a region of the genome is purified or swept clean of alternative alleles. Loci that have undergone selective sweeps can be identified first by scanning the genome for regions of diminished SNP variation and then identifying the minimum conserved haplotype. Finally, candidate genes are screened for mutations that are fixed throughout human populations.

Lactase is one gene that shows evidence of strong selection in the last 7000 years and that has also been influenced by culture. Mammalian genomes contain a lactase

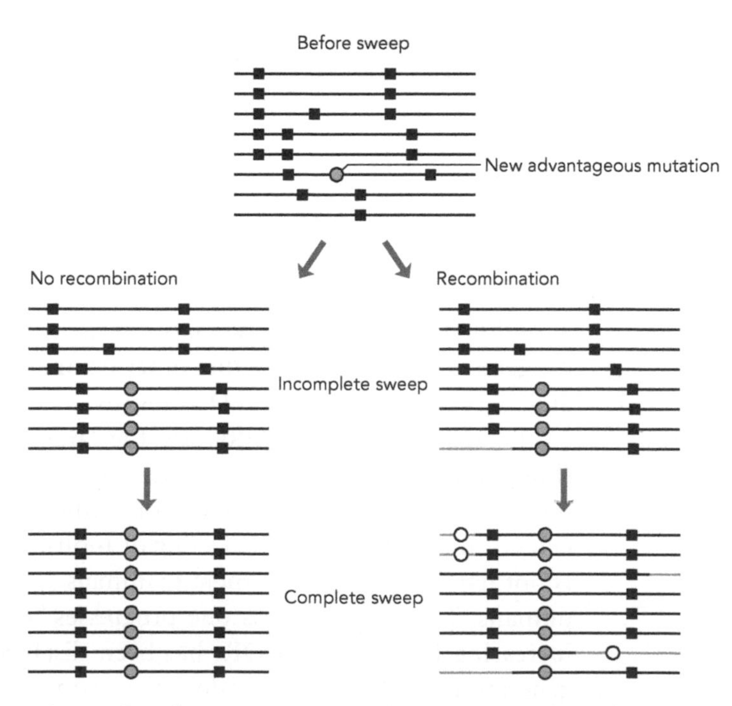

Haplotype evidence of a selective sweep. Most chromosomes in a population show a variety of SNPs at a specific locus. A selective sweep fixes an advantageous mutation (gray circle) along with a set of homozygous SNPs within a nonrecombining region of the chromosome.

(Redrawn, with permission, from Nielsen R, et al. 2007. *Nat Rev Genet* 8: 857–868; © Macmillan.)

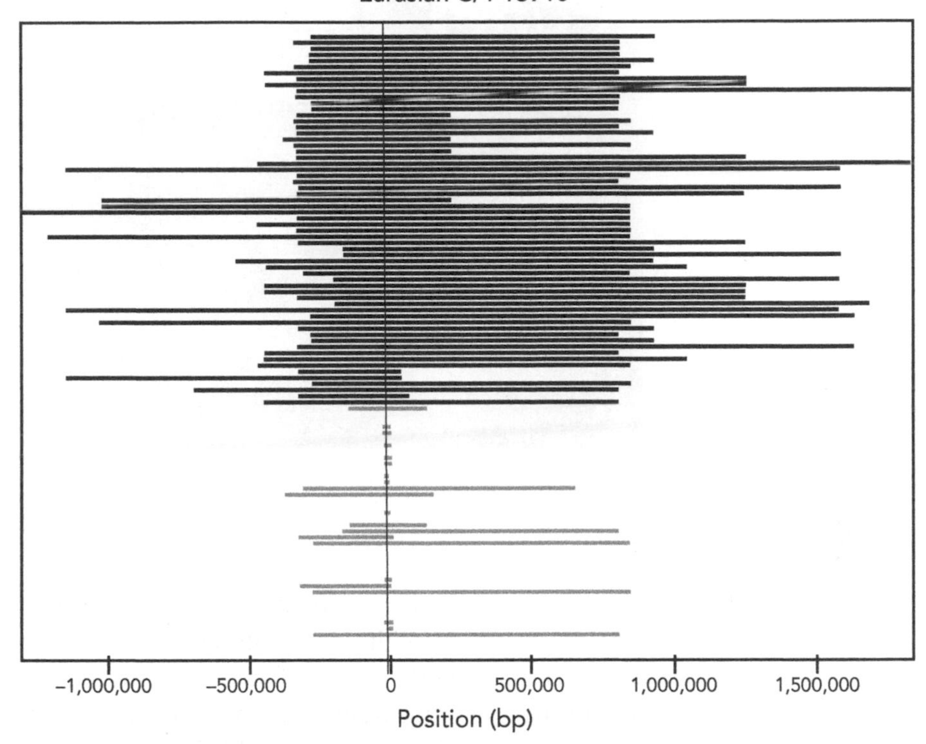

Selective sweep at the lactase locus, 2006. Each line represents the haplotype of a Eurasian individual over a region of ~315,000 base pairs (bp). The lactase gene begins at the 0 position and extends to ~50,000 bp. The vertical line to the left of the 0 position (at approximately −14,000) is the location of the lactose-tolerance mutation in an enhancer. Lactose-tolerant individuals at the top (black bars) share long tracks of homozygous haplotypes. Intolerant individuals, at the bottom (gray bars), show shorter tracks, with some too short to plot.

(Adapted, with permission, from Tishkoff SA, et al. 2007. *Nat Genet* 39: 31–40; ©Macmillan.)

gene that allows infants to digest lactose in its mother's milk into glucose and galactose. Levels of lactase decrease after weaning, and adult primates cannot digest whole milk. In humans, undigestable lactose causes stomach upset. This "natural" lactose intolerance affects nearly 80%–100% of adults in Asian, Native American, and West African populations, whose diets traditionally had little or no milk. Conversely, up to 50%–90% of adults in European and African populations who have traditionally raised cows are lactose tolerant and can readily digest milk.

Milk is a good source of protein, so the cultural activity of dairying provided selective pressure for DNA variations that changed gene regulation to allow lactase activity to persist into adulthood. Lactose tolerance is inherited as a dominant trait, with SNPs on chromosome 2 correlating with lactase persistence in adults—as measured by lactase activity in intestinal tissue or blood glucose levels after drinking milk. Initial work in 2002 by Nabil Enattah, of the University of Helsinki, identified an SNP in Finnish families that is strongly associated with lactase persistence in European populations. In 2007, Sarah Tishkoff, of the University of Maryland, identified three different SNPs that correlate with lactase persistence in East African pastoralist populations, showing the independent (convergent) evolution of this trait in different human populations. Interestingly, these SNPs are located ~14,000 base pairs upstream of the lactase gene—within intron 13 of an adjacent gene, *MCM6*. In vitro assays confirmed that these SNPs act as enhancers, increasing expression of the lactase gene.

Sarah Tishkoff conducts a lactose tolerance test among the Maasai in Tanzania. Tishkoff used this test and genetic studies to identify novel genetic variants that regulate lactose tolerance in East African pastoralists.
(Courtesy of Sarah Tishkoff.)

THE SPREAD OF AGRICULTURE

The advent of farming at ~10,000 BP marks the boundary between the Paleolithic (Old Stone Age) and Neolithic (New Stone Age). In 1978, Luigi Luca Cavalli-Sforza, Paolo Menozzi, and Alberto Piazza, of Stanford University, provided the first genetic description of the "Neolithic transition" to farming. Before the advent of restriction fragment length polymorphism (RFLP) and SNP data, disease and protein polymorphisms of the blood system (ABO and Rh groups, human leukocyte antigens, and globin variants) provided the only means to study human population variation quantitatively. Using these data, Sforza and his colleagues developed principle component analyses to compare genetic variation in multiple allele systems across European population groups. Their first principal component showed the highest allele frequencies in the Middle East, diminishing through central Europe to reach the lowest frequencies in northwestern Europe. They interpreted this gradient, or cline, as genes that followed

Luigi Luca Cavalli-Sforza, Alberto Piazza, and Paolo Menozzi
(Courtesy of Paolo Menozzi.)

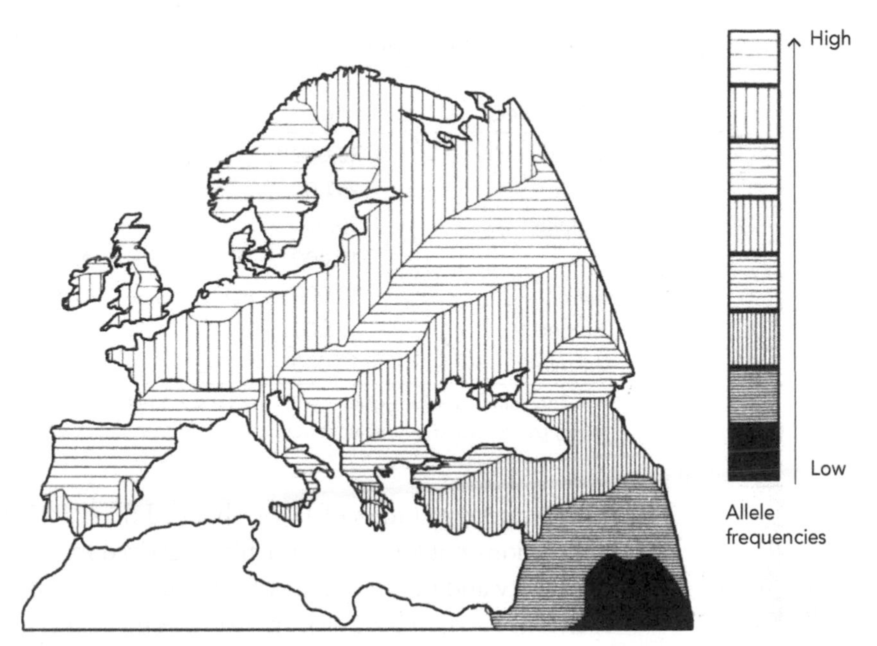

Spread of agriculture in Europe. Principal component analysis of genetic markers yielded this map of allele frequencies in a southeast-to-northwest cline. Cavalli-Sforza, Menozzi, and Piazza interpreted this as the genetic residue of the migration of farmers out of the fertile crescent and through Europe, beginning ~10,000 BP.

(Adapted, with permission, from Cavalli-Sforza LL, Menozzi P, Piazza A. 1994. *The history and geography of human genes.* ©Princeton University Press, Princeton, NJ.)

the spread of agriculture, from the origin of wheat and barley domestication in southeastern Turkey northward through Europe, at a rate of about 1 km per year. Local farming populations (demes) increased, expanding into adjacent agricultural lands and reaching England and Scandinavia by ~5000 BP.

Brian Sykes
(©Rob Judges.)

DNA typing of living Europeans and ancient remains has leant increasing support to the "demic diffusion" model, in which Neolithic farmers spread throughout Europe. Analysis of mtDNA from European populations by Brian Sykes, of Oxford University, and Antonio Torroni, of University of Rome, identified seven major mitochondrial haplogroups. Sykes popularized the founders of these lineages as "the seven daughters of Eve." Six of these lineages, representing ~80% of Europeans alive today, are derived from Paleolithic stocks dating back before the advent of agriculture. Only ~20% of European haplotypes are young enough to represent the new genes of Neolithic farmers. This is not far different from the 28% of variance described by Cavalli-Sforza's first principle component.

The first ancient data came from "Ötzi the Iceman," a 5300-BP mummy found frozen in the Tyrolean Alps in 1991. Kernels of domesticated Einkorn wheat and barley showed that he was a member of an agrarian society. Initial mtDNA sequencing in 1994 showed that Ötzi was fully modern, with a haplotype identical to or nearly identical to ~8% of living people of European ancestry. This provided practical confirmation that 5000 years is less than a single mutation from the present, even by the fast mitochondrial clock. Analysis of Ötzi's whole-genome sequence, in 2012, showed a southern and eastern European ancestry, consistent with the demic diffusion model. Principle component analysis comparing tens of thousands of autosomal SNPs across 1400 European samples placed him closest to present-day Sardinians. Ötzi's Y chromosome haplogroup, G2a3, is most frequent in the Caucasus.

"Seven Daughters of Eve"

"Daughters"	Age	Origin	Percent of modern Europeans
Ursula	45,000	Greece	11
Xenia	25,000	Southern Russia	6
Helena	20,000	Southern France	46
Velda	17,000	Northern Spain	5
Tara	17,000	Central Italy	9
Katrine	15,000	Northeastern Italy	6
Jasmine	10,000	Middle East	17

The fictitious names given by Bryan Sykes to the founders of the seven major European mitochondrial haplogroups are based on the alphabetic classification system of Antonio Torroni.

Two recent studies of DNA obtained from ancient cemeteries provide snapshots of populations not long after agriculture arrived in northern Europe. A team from Uppsala University and the University of Copenhagen compared partial genome sequences from skeletons dated to 5000 BP from two sites in southern Sweden. Three skeletons were from a hunter-gatherer village and one was from a farming village, separated by <400 km. Principle component analysis and extent of shared SNPs placed the hunter-gathers closest to living populations in northern Europe, whereas the farmer was closest to southern European populations. Mitochondrial DNA types from 22 skeletons in a cemetery in Derenberg, Germany, dated to 7100 BP, showed strongest associations with modern populations from central Europe but also from Turkey, Syria, and Iraq. One of three Y haplotypes obtained at Derenberg, G2a3, matched that of Ötzi, also suggesting a Middle Eastern origin.

Taken together, these studies suggest that farmers from southern Europe initially remained distinct from local hunter-gathers for about 1000 years after their arrival in northern regions. Gene flow over several thousand ensuing years brought allele frequencies to intermediate levels between the two Neolithic groups, as now seen in most European populations. Similar movements of people, genes, and the agricultural way of life transformed Asia and the New World.

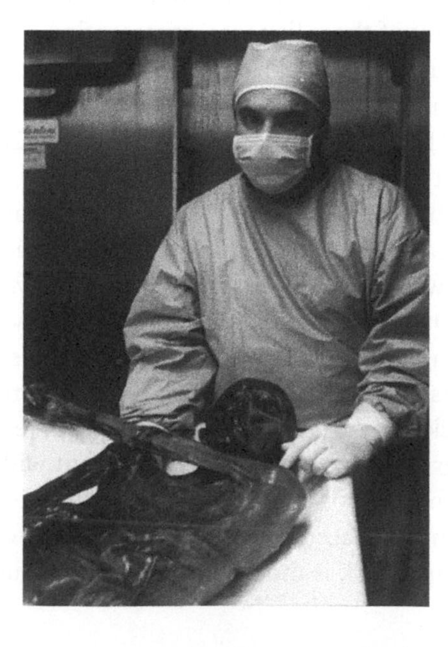

Ötzi the Iceman

(Reprinted, with permission, from Schiermeir Q, Stehle K. 2000. *Nature* 407: 550; ©Macmillan.)

LABORATORY 2.1

Using Mitochondrial DNA Polymorphisms in Evolutionary Biology

▼ OBJECTIVES

This laboratory demonstrates several important concepts of modern biology. During the course of this laboratory, you will

- Learn how mitochondrial DNA (mtDNA) differs from nuclear DNA and why mtDNA is useful for forensic analyses, archaeological investigations, and population genetic studies.
- Collect and analyze genetic information in human populations.
- Use single-nucleotide polymorphisms (SNPs) to study human evolution.
- Analyze population genetic data to trace identity by descent from a common ancestor.
- Move between in vitro experimentation and in silico computation.

In addition, this laboratory utilizes several experimental and bioinformatics methods in modern biological research. You will

- Extract and purify DNA from your own cells.
- Amplify a specific region of your mitochondrial genome by polymerase chain reaction (PCR) and analyze PCR products by gel electrophoresis.
- Align mtDNA sequences to identify mutations and examine evolutionary patterns in ancient and modern humans.
- Use the Basic Local Alignment Search Tool (BLAST) to identify sequences in databases.
- Use MITOMAP to explore the role of mitochondrial mutations in human health and aging.

INTRODUCTION

In addition to the 46 chromosomes found in the nucleus of each human cell, each mitochondrion in the cell cytoplasm has several copies of its own genome. The mitochondrial genome contains only 37 genes, many of which are involved in the process of oxidative phosphorylation—the production of energy and its storage in ATP.

There is strong evidence that mitochondria once existed as free-living bacteria that were taken up by a primitive ancestor of eukaryotic cells. In this endosymbiosis, the host cell provided a ready source of energy-rich nutrients, and the mitochondria provided a means to extract energy using oxygen. This attribute was key to survival as oxygen accumulated in the primitive atmosphere. Mitochondria are in the same size range as bacteria, and the mitochondrial genome retains several bacterial features. Like bacterial chromosomes and plasmids, the mitochondrial genome is a circular molecule and mitochondrial genes are not interrupted by introns. These features are different from those of nuclear genes in eukaryotes, where chromosomes are linear and genes contain numerous introns.

The entire DNA sequence of the human mitochondrial genome (16,569 nucleotides) was determined in 1981, well in advance of the Human Genome Project. Genes make up the majority of the mitochondrial genome. However, a noncoding region of approximately 1200 nucleotides contains signals that control replication of the chromosome and transcription of the mitochondrial genes. The DNA sequence of this control region is termed "hypervariable" because it accumulates point mutations at approximately 10 times the rate of nuclear DNA. This high mutation rate results in distinctive patterns of SNPs.

The combination of SNPs inherited by each person is termed a haplotype (or "half type"). The mitochondrial genome provides only a half set of genes because it is inherited exclusively from the mother with no paternal contribution. The female egg is a huge cell, with on the order of 100,000 mitochondria; in contrast, the tiny sperm cell is powered by fewer than 100 mitochondria at the base of the flagellum. Any male mitochondria that may enter the egg cell at conception are identified by ubiquitin surface proteins as "foreign" and actively destroyed by enzymes in the egg cytoplasm.

In the 1980s, Allan Wilson and coworkers at the University of California at Berkeley used mtDNA polymorphisms to create a "family tree" showing ancestral relationships among modern populations. Reasoning that all human populations arose from a common ancestor in the distant evolutionary past, Wilson's group calculated how long it would take to accumulate the pattern of mitochondrial mutations observed in modern populations. They concluded that the ancestor of all modern humans arose in Africa about 200,000 BP. This common ancestor was widely reported as the "mitochondrial Eve."

mtDNA sequences from ancient humans are quite rare, but several key ones aid in the study of human evolution. One sequence is from "Ötzi the Iceman," a 5300-year-old human discovered frozen in the Ötztal Alps between Austria and Italy. Although Ötzi had the tools and weapons of a hunter, he carried the domesticated wheat of an agricultural society.

An archaic member of the genus *Homo*, Neandertal lived in what is now Europe and central Asia—beginning about 300,000 BP and becoming extinct ~30,000 BP. Clearly, during part of its span on earth, Neandertal shared the European habitat with modern humans (*Homo sapiens*). According to the multiregional model, modern humans developed concurrently from different archaic populations in Africa, Europe, and Asia—and Neandertal was the direct ancestor of modern Europeans. According to the displacement model ("out of Africa"), the ancestors of modern humans left Africa only 50,000–70,000 BP and so share with Neandertal an ancient ancestor in Africa. In 1997, an international research team headed by Svante Pääbo settled the debate about the relationship of Neandertals to modern humans. They extracted DNA from the humerus of the original Neandertal specimen, amplified the sample by PCR, and cloned the resulting products in *Escherichia coli*. The cloned fragments were then used to reconstruct a 379-bp stretch of the mitochondrial control region.

Although each cell contains only two copies of a given nuclear DNA sequence—one on each of the paired chromosomes—there are hundreds to thousands of copies of a given mtDNA sequence in each cell. This increases the chances that enough mtDNA can be obtained for forensic analysis when tissue samples are old or badly degraded and offers a simple means to visualize a discrete region of one's own genetic material. Polymorphisms in the mitochondrial hypervariable region have been used to

- Identify remains from wars and natural disasters.
- Identify the remains of the Romanov royal family assassinated during the Russian Revolution.
- Determine the relationships of Ötzi, the Tyrolean iceman, and the ancient hominid Neandertal to modern humans.

This experiment examines a sequence within the hypervariable region of the mitochondrial genome. A sample of human cells is obtained by saline mouthwash (alternatively, DNA may be isolated from hair sheaths). DNA is extracted by boiling with Chelex resin, which binds contaminating metal ions, and the control region sequence is amplified by PCR. The PCR products (amplicons) are then visualized on agarose gels. However, because SNPs do not change the size of the amplicon, gel electrophoresis shows no differences among student samples. To analyze SNPs that vary from person to person, the nucleotide sequence of each student amplicon must be determined.

The laboratory includes bioinformatics exercises that complement the experimental methods. The individual nucleotide sequences from students in the class are evaluated and compared with mitochondrial sequences from other modern human populations. Tools for comparing student sequences, generating mitochondrial haplotypes, and studying human evolution are found at the *BioServers* Internet site of the Dolan DNA Learning Center (www.bioservers.org). BLAST is used to determine the exact size of the fragment amplified by the primer set and to identify sequences in biological databases, and the features of a complete human mitochondrial chromosome sequence are examined. MITOMAP is used to learn about the impact of mitochondrial gene mutations on human health and aging.

FURTHER READING

Biesecker LG, Bailey-Wilson JE, Ballantyne J, Baum H, Bieber FR, Brenner C, Budowle B, Butler JM, Carmody G, Conneally PM, et al. 2005. Epidemiology. DNA identifications after the 9/11 World Trade Center attack. *Science* **310:** 1122–1123.

Cann RL, Stonekin M, Wilson AC. 1987. Mitochondrial DNA and human evolution. *Nature* **325:** 31–36.

Gill P, Iavanov PL, Kimpton C, Piercy R, Benson N, Tully G, Evett I, Haqelberg E, Sullivan K. 1994. Identification of the remains of the Romanov family by DNA analysis. *Nat Genet* **6:** 130–135.

Handt O, Richards M, Trommsdorff M, Kilger C, Simanainen J, Georgiev O, Bauer K, Stone A, Hedges R, Schaffner W, et al. 1994. Molecular genetic analyses of the Tyrolean Ice Man. *Science* **264:** 1775–1778.

Krings M, Stone A, Schmitz RW, Krainitzki H, Stoneking M, Pääbo S. 1997. Neandertal DNA sequences and the origin of modern humans. *Cell* **90:** 19–30.

Mullis K. 1990. The unusual origin of the polymerase chain reaction. *Sci Am* **262:** 56–65.

PLANNING AND PREPARATION

The following table will help you to plan and integrate the different experimental methods.

Experiment	Day	Time		Activity
I. Isolate DNA from cheek cells	1	60 min	Prelab:	Prepare and aliquot saline solution. Prepare and aliquot 10% Chelex. Make centrifuge adapters. Set up student stations.
		30 min	Lab:	Isolate student DNA.
I. (Alternate) Isolate DNA from hair sheaths	1	30 min	Prelab:	Aliquot proteinase K Set up student stations.
		30 min	Lab:	Isolate student DNA.
II. Amplify DNA by PCR	2	15 min	Prelab:	Aliquot mt primer/loading dye mix. Set up student stations.
		15 min	Lab:	Set up PCRs.
		60–150 min	Postlab:	Amplify DNA in thermal cycler.
III. Analyze PCR products by gel electrophoresis	3	30 min	Prelab:	Dilute TBE electrophoresis buffer Prepare agarose gel solution Set up student stations.
		30 min	Lab:	Cast gels.
	4	45+ min	Lab:	Load DNA samples into gel. Electrophorese samples. Photograph gels.

OVERVIEW OF EXPERIMENTAL METHODS

I. ISOLATE DNA FROM CHEEK CELLS

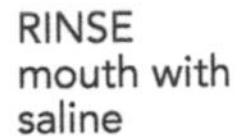

RINSE mouth with saline

TRANSFER saline

CENTRIFUGE

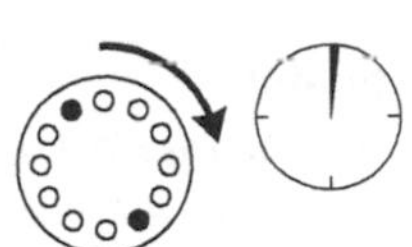

POUR OFF supernatant

RESUSPEND

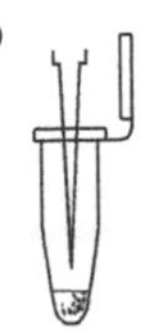

TRANSFER cell suspension into Chelex

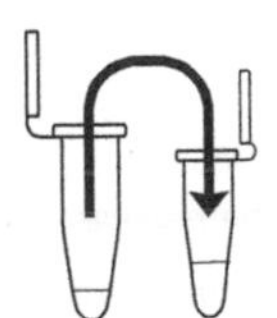

BOIL in thermal cycler

SHAKE vigorously

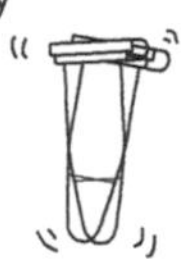

CENTRIFUGE

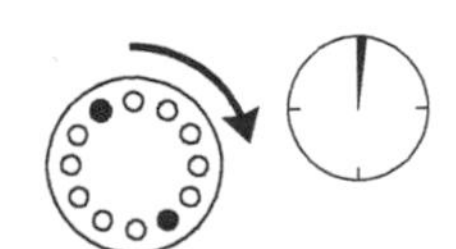

TRANSFER supernatant

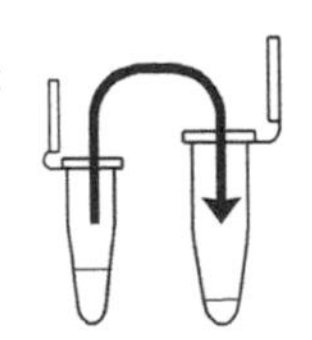

STORE on ice

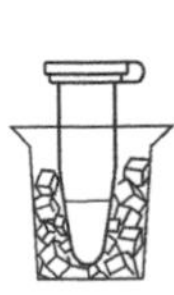

I. (ALTERNATE) ISOLATE DNA FROM HAIR SHEATHS

CUT hairs

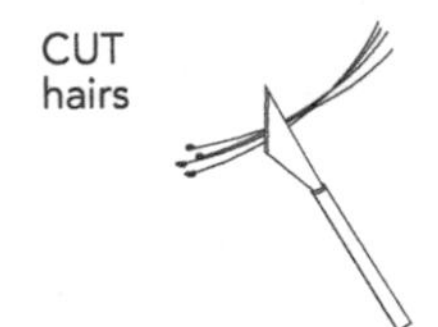

ADD hairs to proteinase K

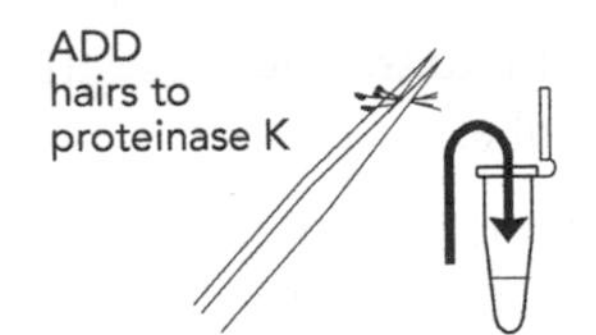

INCUBATE in thermal cycler

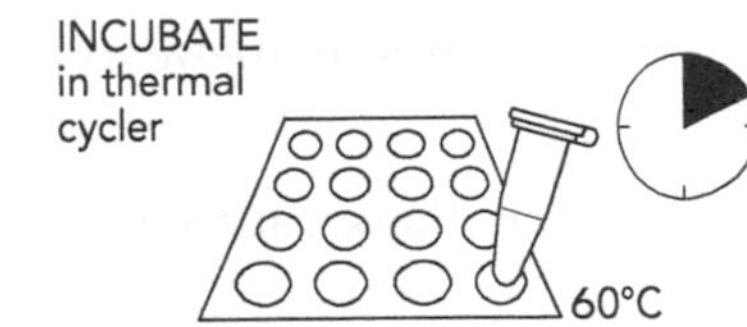

VORTEX

BOIL in thermal cycler

99°C

MIX by pipetting in and out

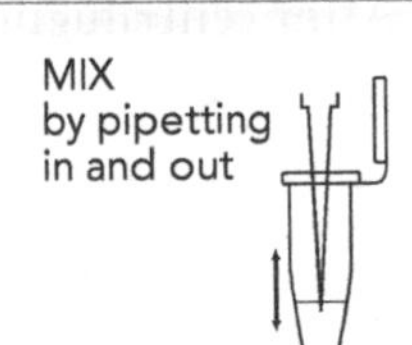

STORE on ice

II. AMPLIFY DNA BY PCR

ADD primer/loading dye mix

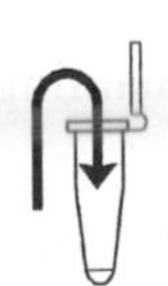

ADD DNA

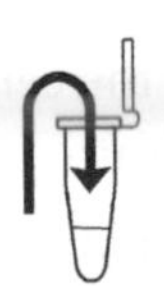

AMPLIFY in thermal cycler

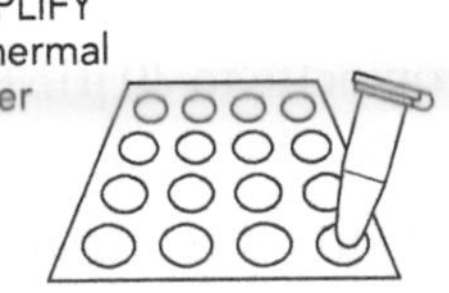

III. ANALYZE PCR PRODUCTS BY GEL ELECTROPHORESIS

POUR gel

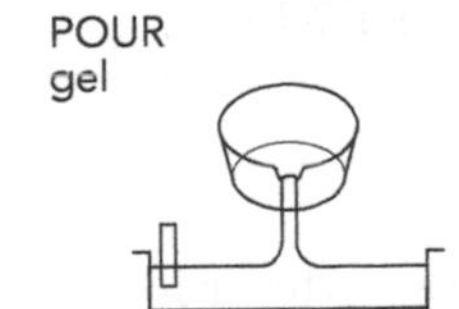

SET 20 min

LOAD gel

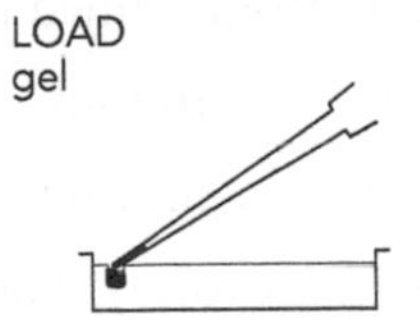

ELECTROPHORESE 130 V

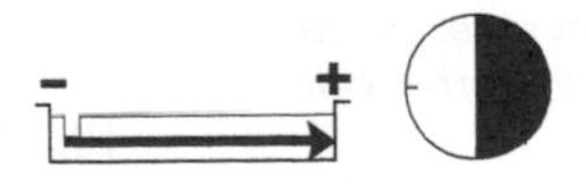

EXPERIMENTAL METHODS

I. Isolate DNA from Cheek Cells

REAGENTS, SUPPLIES, & EQUIPMENT

For each student

10% Chelex (100 μL) in a 0.2- or 0.5-mL PCR tube (or in a microcentrifuge tube)
2 Microcentrifuge tubes (1.5 mL)
Paper cup
Saline solution (10 mL) in a 15-mL tube

For each group

Container with cracked or crushed ice
Microcentrifuge adapters for 0.2- or 0.5-mL PCR tubes
Microcentrifuge tube rack
Micropipettes and tips (10–1000 μL)
Permanent marker
Vortexer (optional)

To share

Microcentrifuge
Thermal cycler (or water bath or heat block)

1. Use a permanent marker to label a 1.5-mL microcentrifuge tube and paper cup with your assigned number.
2. Pour all 10 mL of saline solution into your mouth and vigorously rinse your cheek pockets for 30 sec.
3. Expel the saline solution into the paper cup.
4. Swirl the cup gently to mix the cells that may have settled to the bottom. Use a micropipette with a fresh tip to transfer 1000 μL of the solution into your labeled 1.5-mL microcentrifuge tube.
5. Place your sample tube, along with those from other students, in a balanced configuration in a microcentrifuge. Centrifuge the tubes at full speed for 90 sec.
6. After centrifuging, check to see if your pellet is firmly attached to the bottom of the tube.
   - If the pellet is firmly attached, carefully pour off the supernatant into the paper cup. Gently tap the inverted tube against the cup to remove most of the supernatant, but be careful not to disturb the cell pellet at the bottom of the tube.
   - If the pellet is loose or unconsolidated, centrifuge for 90 sec. Then, carefully use a micropipette with a fresh tip to remove as much saline as possible.

If your microcentrifuge tube is graduated, the volume remaining after Step 6 will approximately reach the 0.1-mL mark.

7. Set a micropipette to 30 μL. Resuspend the cells in the remaining saline by pipetting in and out. Work carefully to minimize bubbles.

Food particles will not resuspend. Your teacher may instruct you to examine a sample of the cell suspension under a microscope.

8. Withdraw 30 μL of cell suspension and add it to a PCR tube containing 100 μL of 10% Chelex. Label the cap and side of the tube with your assigned number.

As an alternative to Steps 8 and 9, you may add the cell suspension to Chelex in a 1.5-mL tube and incubate for 10 min in a boiling water bath or 99°C heat block.

9. Place your PCR tube, along with those from other students, in a thermal cycler that has been programmed for one cycle of the following profile:

   Boiling step: 10 min 99°C

   The profile may be linked to a 4°C hold program.

The near-boiling temperature lyses the cell and nuclear membranes, releasing DNA and other cell contents.

10. After boiling, vigorously shake your tube for 5 sec. Use a vortexer if available.

Skip Step 11 if your sample is in a microcentrifuge tube (not in a PCR tube).

11. To prepare your PCR tube for centrifugation, "nest" it within adapter tubes as follows:
    - If your sample is in a 0.5-mL PCR tube, "nest" it within a capless 1.5-mL tube.
    - If your sample is in a 0.2-mL PCR tube, "nest" it within a capless 0.5-mL tube and then place both tubes into a capless 1.5-mL tube.
12. Place your tube, along with those from other students, in a balanced configuration in a microcentrifuge. Centrifuge the tubes at full speed for 90 sec.
13. After centrifugation, use a micropipette with a fresh tip to transfer 30 µL of the clear supernatant into a clean 1.5-mL microcentrifuge tube. Be careful to avoid pipetting any cell debris or Chelex beads.
14. Use a permanent marker to label the cap and side of the tube with your assigned number. This tube contains your DNA and will be used for setting up the PCRs in Part II.
15. Store your sample on ice or at –20°C until you are ready to continue with Part II.

I. (Alternate) Isolate DNA from Hair Sheaths

REAGENTS, SUPPLIES, & EQUIPMENT

For each student

100-mg/mL proteinase K (100 µL)* <!> in a 0.2- or 0.5-mL PCR tube (or in a microcentrifuge tube)

Micropipettes and tips (100 µL)
Permanent marker
Scalpel or razor blade
Vortexer (optional)

For each group

Container with cracked or crushed ice
Forceps or tweezers
Hand lens or dissecting microscope
Microcentrifuge tube rack

To share

Thermal cycler (or water bath or heat block)

*Store on ice.
See Cautions Appendix for appropriate handling of materials marked with <!>.

HAIR WITH SHEATH
HAIR ROOT
BROKEN HAIR

1. Pull out several hairs. Use a hand lens or dissecting microscope to inspect them for the presence of a sheath. The sheath is a barrel-shaped structure surrounding the base of the hair and is most easily observed on dark hair. By holding the hair up to a light source, you may observe the glistening sheath with your naked eye.
2. Select one to several hairs with good sheaths. Alternatively, select hairs with the largest roots. Broken hairs, without roots or sheaths, will not yield enough DNA for amplification.
3. Use a fresh razor blade or scalpel to cut off each hair shaft just above the sheath.

Your teacher may instruct you to prepare a hair sheath to observe under a compound light microscope.

4. Use forceps or tweezers to transfer the hairs to a PCR tube containing 100 µL of proteinase K. Make sure that the sheath is submerged in the solution and not stuck to the wall of the tube.
5. Use a permanent marker to label the cap and the side of your PCR tube with your assigned number.
6. Place your PCR tube, along with those of other students, in a thermal cycler that has been programmed for one cycle of the following profile:

As an alternative to Steps 4–6, add the hairs to proteinase K in a 1.5-mL microcentrifuge tube and incubate in a water bath or heat block for 10 min at 60°C. Then, in Step 8, incubate the tube for 10 min in a boiling water bath or 99°C heat block.

Incubation step:	10 min	60°C

7. Place your sample tube at room temperature. To dislodge the cells from the hair shaft, vortex your tube by machine or vigorously with a finger for 15 sec.
8. Place your PCR tube, along with those of other students, in a thermal cycler that has been programmed for one cycle of the following profile:

Boiling step:	10 min	99°C

   The profile may be linked to a 4°C hold program.
9. Place your sample tube at room temperature. Use a micropipette and fresh tip to mix your sample by pipetting in and out for 15 sec. This sample contains your DNA and will be used for setting up the PCRs in Part II.
10. Store your sample on ice or at –20°C until you are ready to begin Part II.

II. Amplify DNA by PCR

REAGENTS, SUPPLIES, & EQUIPMENT

For each student

Cheek cell or hair sheath DNA (from Part I)*
mt primer/loading dye mix (25 µL)*
Ready-To-Go PCR Bead in a 0.2- or 0.5-mL PCR tube

For each group

Container with cracked or crushed ice

Microcentrifuge tube rack
Micropipette and tips (1–100 µL)
Permanent marker

To share

Thermal cycler

*Store on ice.

1. Obtain a PCR tube containing a Ready-To-Go PCR Bead. Label the tube with your assigned number.
2. Use a micropipette with a fresh tip to add 22.5 µL of mt primer/loading dye mix to the tube. Allow the bead to dissolve for approximately 1 min.

The primer/loading dye mix will turn purple as the PCR bead dissolves.

3. Use a micropipette with a fresh tip to add 2.5 µL of your DNA (from Part I) directly into the primer/loading dye mix. Ensure that no DNA remains in the tip after pipetting.
4. Store your sample on ice until your class is ready to begin thermal cycling.

If the reagents become splattered on the wall of the tube, pool them by pulsing the sample in a microcentrifuge or by sharply tapping the tube bottom on the lab bench.

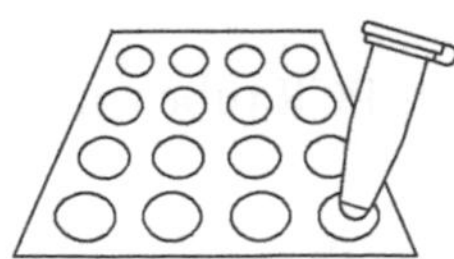

5. Place your PCR tube, along with those of other students, in a thermal cycler that has been programmed for 30 cycles of the following profile:

Denaturing step:	30 sec	94°C
Annealing step:	30 sec	58°C
Extending step:	30 sec	72°C

   The profile may be linked to a 4°C hold program after the 30 cycles are completed.
6. After thermal cycling, store the amplified DNA on ice or at –20°C until you are ready to continue with Part III.

III. Analyze PCR Products by Gel Electrophoresis

REAGENTS, SUPPLIES, & EQUIPMENT

For each student

Latex gloves
Microcentrifuge tube (1.5 mL)
PCR product* from Part II
SYBR Green <!> DNA stain (5 μL)

For each group

2% Agarose in 1x TBE (hold at 60°C) (50 mL per gel)
Container with cracked or crushed ice
Gel-casting tray and comb
Gel electrophoresis chamber and power supply
Masking tape
Microcentrifuge tube rack
Micropipette and tips (1–100 μL)
pBR322/BstNI marker (20 μL per gel)*
1x TBE buffer (300 mL per gel)

To share

Digital camera or photodocumentary system
UV transilluminator <!> and eye protection
Water bath for agarose solution (60°C)

*Store on ice.
See Cautions Appendix for appropriate handling of materials marked with <!>.

1. Seal the ends of the gel-casting tray with masking tape and insert a well-forming comb.
2. Pour the 2% agarose solution into the tray to a depth that covers about one-third the height of the open teeth of the comb.
3. Allow the gel to completely solidify; this takes approximately 20 min.

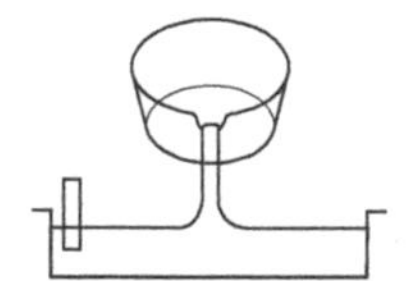

Avoid pouring an overly thick gel, which will be more difficult to visualize.

The gel will become cloudy as it solidifies.

4. Remove the masking tape, place the gel into the electrophoresis chamber, and add enough 1x TBE buffer to cover the surface of the gel.
5. Carefully remove the comb and add additional 1x TBE buffer to fill in the wells and just cover the gel, creating a smooth buffer surface.

Do not add more buffer than necessary. Too much buffer above the gel channels electrical current over the gel, increasing the running time.

6. Orient the gel according to the diagram in Step 8, so that the wells are along the top of the gel.
7. Use a micropipette with a fresh tip to transfer 15 μL of your PCR product (from Part II) to a fresh 1.5-mL microcentrifuge tube. Add 2 μL of SYBR Green DNA stain to the 15 μL of PCR product. In addition, add 2 μL of SYBR Green DNA stain to 20 μL of pBR322/BstNI marker.

A 100-bp ladder may also be used as a marker.

8. Use a micropipette with a fresh tip to load each sample from Step 7 into your assigned well, according to the following diagram:

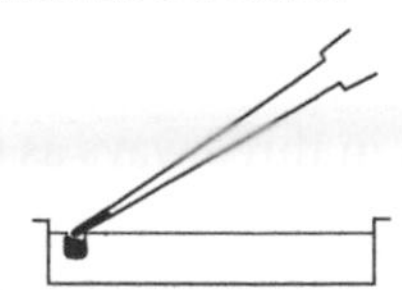

Expel any air from the tip before loading. Be careful not to push the tip of the pipette through the bottom of the sample well.

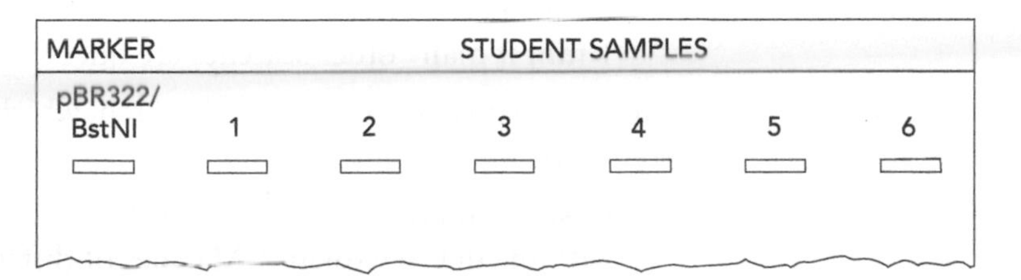

Store the remaining 10 μL of your PCR product on ice for subsequent use in DNA sequencing.

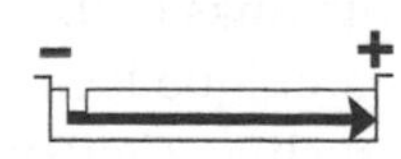

9. Run the gel for approximately 30 min at 130 V. Adequate separation will have occurred when the cresol red dye front has moved at least 50 mm from the wells.

Transillumination, where the light source is below the gel, increases brightness and contrast.

10. View the gel using UV transillumination. Photograph the gel using a digital camera or photodocumentary system.

BIOINFORMATICS METHODS

For a better understanding of the experiment, perform the following bioinformatics exercises before you analyze your results.

I. Use BLAST to Find DNA Sequences in Databases (Electronic PCR)

1. Perform a BLAST search as follows:

   i. Do an Internet search for "ncbi blast."

   ii. Click on the link for the result "BLAST: Basic Local Alignment Search Tool." This will take you to the Internet site of the National Center for Biotechnology Information (NCBI).

   iii. Click on the link "nucleotide blast" (blastn) under the heading "Basic BLAST."

   iv. Enter both primer sequences into the search window. These are the query sequences.

   > The following primer set was used in the experiment:
   >
   > 5′-TTAACTCCACCATTAGCACC-3′ (forward primer)
   > 5′-GAGGATGGTGGTCAAGGGAC-3′ (reverse primer)

   v. Omit any nonnucleotide characters from the window because they will not be recognized by the BLAST algorithm.

   vi. Under "Choose Search Set," select the "Nucleotide collection (nr/nt)" database from the drop-down menu.

   vii. Under "Program Selection," optimize for "Somewhat similar sequences (blastn)."

   viii. Click on "BLAST." This sends your query sequences to a server at NCBI in Bethesda, Maryland. There, the BLAST algorithm will attempt to match the primer sequences to the millions of DNA sequences stored in its database. While searching, a page showing the status of your search will be displayed until your results are available. This may take only a few seconds or more than 1 min if many other searches are queued at the server.

2. Analyze the results of the BLAST search, which are displayed in three ways as you scroll down the page:

   i. First, a graphical overview illustrates how significant matches (hits) align with the query sequence. Matches of differing lengths are indicated by color-coded bars. What do you notice about the lengths (and colors) of the matches (bars) as you look from the top to the bottom?

   ii. This is followed by a list of significant alignments (hits) with links to the corresponding accession numbers. (An accession number is a unique identifier given to a sequence when it is submitted to a database such as GenBank.)

Note the scores in the "E value" column on the right. The Expectation or E value is the number of alignments with the query sequence that would be expected to occur by chance in the database. The lower the E value, the higher the probability that the hit is related to the query. For example, an E value of 1 means that a search with your sequence would be expected to turn up one match by chance. Longer query sequences generally yield lower E values. An alignment is considered significant if it has an E value of less than 0.1. What is the E value of the most significant hit and what does it mean? Note the names of any significant alignments that have E values of less than 0.1. Do they make sense? What do they have in common?

iii. Third is a detailed view of each primer (query) sequence aligned to the nucleotide sequence of the search hit (subject, abbreviated "Sbjct"). Note that the forward primer (nucleotides 1–20) and the reverse primer (nucleotides 21–40) often align within the same accession.

3. Click on the accession number link to open any hit that is labeled "complete genome."

   i. At the top of the report, note basic information about the sequence, including its length (in base pairs, or bp), database accession number, source, and references to papers in which the sequence is published. What is the source and size of the sequence in which your BLAST hit is located?

   ii. In the middle section of the report, the sequence features are annotated, with their beginning and ending nucleotide positions ("xx..xx"). These features may include genes, coding sequences (CDS), ribosomal RNA (rRNA), and transfer RNA (tRNA). You will examine these features more closely in Part II.

   iii. Scroll to the bottom of the data sheet. This is the nucleotide sequence to which the term "Sbjct" refers.

   iv. Use your browser's "Back" button to return to the BLAST results page.

4. Predict the length of the product that the primer set would amplify in a PCR (in vitro) as follows:

   i. Scroll down to the alignments section (third section) of the BLAST results page. Select a hit that is labeled "complete genome." Both primers should align to the sequence.

   ii. To which positions do the primers match in the subject sequence?

   iii. The lowest and highest nucleotide positions in the subject sequence indicate the borders of the amplified sequence. Subtract the lowest nucleotide position in the subject sequence from the highest nucleotide position in the subject sequence. What is the difference between the coordinates?

   iv. Note that the actual length of the amplified fragment includes both ends, so add 1 nucleotide to the result that you obtained in Step 4.iii to obtain the exact length of the PCR product amplified by the two primers.

5. Obtain the nucleotide sequence of the amplicon that the primer set would amplify in a PCR (in vitro) as follows:

   i. Open the sequence data sheet for the hit that you identified in Step 4.i by clicking on the accession number link. Scroll to the bottom of the data sheet.

In Step 5.iii, you can retain the nucleotide coordinates and spacers to ease readability. Reduce the point size of the font so that each row of 60 nucleotides sits on one line. Then, set the font to Courier or another nonproportional font to align the blocks of sequence.

ii. The bottom section of the data sheet lists the entire nucleotide sequence that contains the PCR product. Highlight all of the nucleotides between the coordinates that you identified in Step 4.iii, from the beginning of the forward primer to the end of the reverse primer.

iii. Copy and paste the highlighted sequence into a text document. Then, delete all nonnucleotide characters and spaces. This is the amplicon, or amplified PCR product. Save this text document.

II. Examine the Structure of the Mitochondrial Chromosome and Its Genes

1. Continue working with the same hit that you used in Steps 4 and 5 of Part I. Open the data sheet for that hit. To help you answer the questions below, print the report or copy the report into a text document.
2. Look at the features section in the middle of the report: Regulatory region, gene, coding sequence (CDS), ribosomal RNA (rRNA), and transfer RNA (tRNA) features may be annotated. Next to each feature are its nucleotide coordinates on the mitochondrial chromosome ("xx..xx").
3. Use the information in the features section and the nucleotide positions that match the primers (from Step 4.ii in Part I above) to identify the features that the amplicon spans.

   i. How many features does the amplicon span? What are they called? What is the size of each feature?

   ii. The largest feature that you identified in Step 3.i has several names. Do some research to learn and understand these names.

4. Study the general features of the mitochondrial genome.

   i. Look at the features list for your "complete genome" hit. Remember that genes can be located on either the forward or reverse strand; in the features section, the word "complement" indicates that a feature is transcribed from the reverse strand. List the genes that are transcribed from the forward strand and those that are transcribed from the reverse (complement) strand. On which strand are most genes found?

   ii. What kind of product do most of the genes produce? For what biological function are these products used?

   iii. What is the spacing like between genes? Is there much intergenic DNA between two adjacent genes?

   iv. Focus on one of the protein-coding genes. Compare the nucleotide coordinates for the gene and the CDS. What do you notice? What does this tell you about the structure and origin of mitochondrial genes?

III. Use MITOMAP to Explore the Role of Mitochondrial Mutations in Human Health and Aging

1. MITOMAP is a database of polymorphisms in the human mitochondrial genome. Open the MITOMAP Internet site (http://www.mitomap.org).
2. Scroll down the page to the section entitled "MtDNA Mutations with Reports of Disease-Associations."

3. Follow the links under "Organized by mtDNA location" ("rRNA/tRNA Mutations" and "Coding & Control Region Mutations") to explore several diseases. Jot down notes on several disorders and share your findings with your classmates.
   i. Use "Nucleotide Position" to correlate the location of the causative mutation with the locations of RNA and protein genes in the features section of the GenBank report from Part II above.
   ii. Follow the "Disease" link to a list of links to OMIM (Online Mendialian Inheritance in Man). Click on the link for the disease and select the link that appears on the OMIM page (usually the first one).
   iii. The OMIM report will provide a detailed description of the disease.
4. What types of mitochondrial mutations generally cause diseases? What do many of these diseases have in common?
5. Research the mitochondrial theory of aging.

RESULTS AND DISCUSSION

I. Interpret Your Gel and Think About the Mitochondrial Genome

1. Observe the photograph of the stained gel containing your PCR samples and those from other students. Orient the photograph with the sample wells at the top. Use the sample gel shown below to help to interpret the band(s) in each lane of the gel.

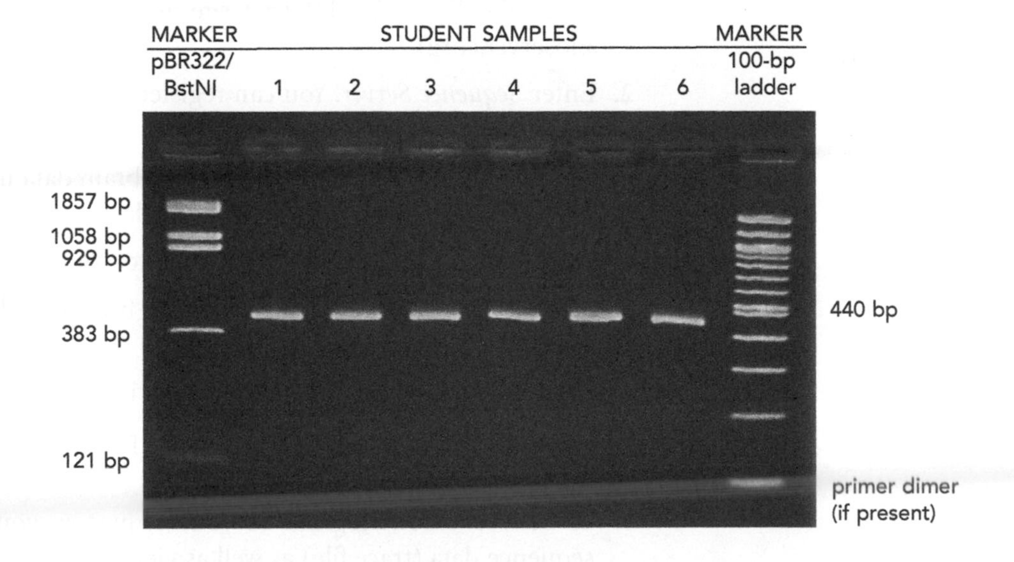

If a 100-bp ladder was used, as shown on the right-hand side of the sample gel, the bands increase in size in 100-bp increments starting with the fastest-migrating band of 100 bp.

2. Locate the lane containing the pBR322/BstNI markers on the left side of the gel. Working from the well, locate the bands corresponding to each restriction fragment: 1857, 1058, 929, 383, and 121 bp. The 1058- and 929-bp fragments will be very close together or may appear as a single large band. The 121-bp band may be very faint or not visible.
3. Scan across the row of student results. You should notice that virtually all student lanes contain one prominent band. The amplification product of 440 bp should roughly align with the 383-bp marker.

4. It is common to see a diffuse (fuzzy) band that runs ahead of the 121-bp marker. This is "primer dimer," an artifact of the PCR that results from the primers overlapping one another and amplifying themselves. How would you interpret a lane in which you observe primer dimer but no bands as described in Step 3?
5. The mitochondrial control region mutates at approximately 10 times the rate of nuclear DNA. Propose a biological reason for the high mutation rate of mtDNA.
6. The high mutability of the mitochondrial genome means that it evolves more quickly than the nuclear genome. This makes the mitochondrial control region a laboratory for the study of DNA evolution. However, can you think of any drawbacks to this high mutation rate when studing evolution?
7. There are numerous insertions of mtDNA in nuclear chromosomes. Most of these are very ancient and are the result of a mitochondrial genome reduction process that transferred functional mitochondrial genes to the nuclear genome. A 540-bp insertion of the mitochondrial control region on chromosome 11 is an example of a relatively recent event that occurred about 350,000 years ago. Would you expect any difference in the mutation rates of the control region sequence in the mitochondrial genome versus the chromosome 11 insertion? What implication does this have for the study of human evolution?

II. Visualize and Assess Your DNA Sequence

If your DNA has not been sequenced, you can follow these directions to assess another student's mtDNA sequence.

1. Open the *BioServers* Internet site at the Dolan DNA Learning Center (www.bioservers.org).
2. Enter *Sequence Server*. You can register if you want to save your work for future reference, but this is not required.
3. The interface is simple to use: Add or obtain data using the top buttons and pull-down menus and then work with the data in the work space below.
4. Click on "MANAGE GROUPS" at the top of the page.
5. Scroll down the default "Classes" list to find your class and click on the check box to select it.
6. Click on "OK" to move your class data into the work space.
7. Select your sample number from the pull-down menu. Click "OPEN" to view your sequence.
8. To view a chromatogram of your sequence, you will need to download the sequence data (trace file) as well as viewer software.
   i. Follow the "Trace File" link and save the .abi file to your desktop.
   ii. Follow the "DNA Sequencing Core (University of Michigan)" link and download a chromatogram viewer. (Note that some of the programs are specific to Mac or PC platforms. FinchTV is a nice viewer that works on both platforms.)
9. Double-click on the trace file to launch your sequence in the chromatogram viewer. (If this does not work, first open the chromatogram viewer, click on "File" and then on "Open," and navigate to select your .abi trace file.)

10. Inspect your chromatogram. Remember that the primers amplify a 440-nucleotide sequence, so it is physically impossible to generate a sequence (read) longer than this.
    i. Each peak represents the fluorescence measured at that nucleotide position. Whenever possible, the software "calls" each peak as an A, T, C, or G. What does the amplitude (height) of each peak represent? What do you notice about the amplitude of the peaks at the beginning of the sequence?
    ii. Every sequence will begin with nucleotides (A, T, C, G) interspersed with Ns. In "clean" sequences, where experimental conditions were near optimal, the initial Ns will end within the first 25 nucleotides. The remaining sequence will have very few, if any, internal Ns. Large numbers of Ns scattered throughout the sequence indicate poor-quality sequence. Then, at the end of the read, the sequence will abruptly change over to Ns. In what part(s) of your sequence read are the most Ns found? What do you notice about the trace at N positions?
11. Load two different trace files into different viewers and attempt to match (align) the DNA sequences. Compare the two reads. Do the nucleotide numbers (positions) correspond to the same nucleotide sequence in both reads? What trick did you use to align them?
12. In several percent of cases, a clean sequence ends abruptly about midway through the read and then gives way to a poor-quality sequence with many Ns. Check to see if this situation describes any of your classmates' sequences; if so, manually align their sequence with a clean sequence. What sequence feature coincides with the abrupt transition from a clean to a poor-quality sequence?

III. Assess the Extent of Mitochondrial Variation in Modern Humans

1. If you are not already in the *Sequence Server* of the *BioServers* Internet site, open it as described in Steps 1 and 2 of Part II. Then click on "MANAGE GROUPS" at the top of the page.
2. Select "Modern Human mtDNA" from the "Sequence sources" pull-down menu in the upper right-hand corner. Click in the check boxes to select the African, Asian, and European mitochondrial DNA samples and click "OK" to move these groups into the work space.
3. If your class data are not already in the work space, follow Steps 4–6 from Part II to add them. (If your DNA was not sequenced, you can follow these instructions to add data from another class.)
4. Use the pull-down menus and check boxes to select two sequences to compare. If you determined in Step 10 of Part II that your sequence read was clean, make sure that the check box next to your sequence is marked. (If your sequence was not clean, choose a sequence from another student that was clean.) Then choose one African, Asian, or European reference sequence and click on the check box to select it. Uncheck all other boxes; make sure that you end up with only two sequences checked.
5. Click on "COMPARE" in the gray bar. (The default operation is multiple sequence alignment using the CLUSTAL W algorithm.) The checked sequences are sent to a server at Cold Spring Harbor Laboratory, where the CLUSTAL W algorithm will

attempt to align each nucleotide position. This may take only a few seconds or more than a minute if a lot of other searches are queued at the server.

6. The results will appear in a new window. Examine the sequences, which are displayed in rows of 25 nucleotides. The alignment typically begins with yellow-shaded dashes (–), indicating an initial stretch of nucleotides that is present in one of the samples but not in the other. This occurs because the sequence read usually begins at different points in different samples. Yellow highlighting also denotes mismatches between the sequences. A gray-shaded "N" indicates a sequence error, a position in one or both sequences where a nucleotide could not be determined.
7. Scan into the sequence beyond all sequence errors and mismatches until you encounter a relatively long, unbroken string of perfect matches between the two sequences. This typically occurs at position 25–40. Once you are into clean sequence, count the number of polymorphisms (differences) between the two individuals as follows:
   i. Count all yellow-shaded nucleotide differences. These are single-nucleotide polymorphisms (SNPs).
   ii. Count all deletions (–). If you see a string of internal dashes (– – – –), this likely arose from a single mutation event and thus should be scored as a single polymorphism.
   iii. Do not count any Ns. Sequences with more than one or two internal Ns or with multiple nucleotide differences on every line are not reliable. If you detect this, select other sequences with which to work.
   iv. Stop counting when you again encounter frequent Ns and/or sequence mismatches at the end of the alignment. This should occur at about position 375–400.
8. Record the names of your samples and the total number of polymorphisms that you counted in Step 7. In addition, record the number of nucleotides of clean sequence you counted (typically 325–400).
9. Use the pull-down menus and check boxes in your work space to select a different pair of individuals and repeat Steps 5–8. Continue until you have made at least five comparisons that sample a variety of populations and recorded your results. (You do not need to use the sequences from your class for all comparisions; the idea is to compare many pairs of modern human mtDNA sequences.)
10. For each pair of sequences, calculate the percent difference:

$$\text{difference (\%)} = \frac{\text{number of nucleotide differences}}{\text{total number of nucleotides counted}} \times 100.$$

11. Share class data to determine the largest possible number of unique pairs. Exclude any "outliers" with suspiciously large numbers of SNPs; these comparisons likely included unreliable sequences with too many sequence errors. Use these data to determine the average number of mutations, as well as the average percent difference, between any two pairs. What does this number represent?
12. Mutations accumulate one at a time. If one assumes a constant mutation rate, then we can estimate how often a new mutation arises. However, this "mutation-

al clock" needs to be set by some external event. In this case, we can use anthropological data that suggest that modern humans arose about 200,000 years ago. If this is so, what is the approximate mutation rate for the mitochondrial control region? Why is your calculation only approximate?

IV. Assess Mitochondrial Variation in Ancient Hominids

A. Ötzi the Iceman

1. If you are not already in the *Sequence Server* of the *BioServers* Internet site, open it as described in Steps 1 and 2 of Part II. Then click on "MANAGE GROUPS" at the top of the page.
2. Select "Ancient Human mtDNA" from the "Sequence sources" pull-down menu in the upper right-hand corner.
3. Click in the check box to select "'Ötzi' the Iceman mtDNA," and click "OK" to move this sequence onto the work space.
4. Use the pull-down menus and check boxes to select a modern human from your class, Africa, Asia, or Europe to compare with Ötzi. (If these sequences are not already in your work space, follow Steps 4–6 from Part II and Step 2 of Part III to add them.)
5. Perform Steps 5–8 and Step 10 of Part III to compare the number of differences and percent difference between Ötzi and the chosen sequence.
6. Use the pull-down menus and check boxes in your work space to select a different individual to compare to Ötzi and repeat Steps 5–8 and 10 of Part III. Continue until you have compared Ötzi to each of at least five modern humans and recorded your results.
7. Pool the class data to determine the average number of differences and average percent difference between Ötzi and modern humans. What does this tell you about Ötzi and the DNA mutational clock?

B. Neandertal

1. If you are not already in the *Sequence Server* of the *BioServers* Internet site, open it as described in Steps 1 and 2 of Part II. Then click on "MANAGE GROUPS" at the top of the page.
2. Select "Neanderthal mtDNA" from the "Sequence sources" pull-down menu in the upper right-hand corner.
3. Click in the check boxes to select several samples of Neandertal mtDNA and click "OK" to move these sequences onto the work space.
4. Using the pull-down menus and check boxes, select a modern human from your class, Africa, Asia, or Europe to compare with one of the Neandertal samples. (If these sequences are not already in your work space, follow Steps 4–6 from Part II and Step 2 of Part III to add them.)
5. Perform Steps 5–8 and 10 of Part III to compare the number of differences and percent difference between the Neandertal sequence and the modern human sequence.

6. Use the pull-down menus and check boxes in your work space to select a different Neandertal–modern human pair and repeat Steps 5–8 and 10 of Part III. Continue until you have compared at least five Neandertal–modern human pairs and recorded your results.
7. Compare each possible pair of Neandertals to assess the extent of variation in Neandertals across their habitat range.
8. Pool the class data to determine the average number of differences and average percent difference between Neandertals and modern humans—and between Neandertals. What does this tell you about the relationship between Neandertals and modern humans?
9. If modern humans arose 200,000 BP, approximately how long ago did humans and Neandertals share a common mitochondrial ancestor? (Assume that the mutational clock is running at the same rate as in Step 12 of Part III.)
10. Scientists estimate that Neandertals lived on Earth for approximately 300,000 years. What does the level of genetic variation in Neandertals suggest about Neandertal population changes during the course of their existence on Earth?

V. Discover What DNA Says About Human Evolution

Use the chart below to record your answers to the questions that follow.

African 1	African 2	European or Asian	African different	European or Asian different

1. If you are not already in the *Sequence Server* of the *BioServers* Internet site, open it as described in Steps 1 and 2 of Part II. Then follow Step 2 of Part III to retrieve the African, Asian, and European mtDNA samples if they are not already in your work space.
2. Use the pull-down menus and check boxes to select two African samples and one European or Asian sample. Record the names of the sample populations in the chart.
3. Click on "COMPARE" (Align: CLUSTAL W) to perform a multiple sequence alignment.
4. Evaluate each yellow-shaded mismatch, where one sample is different from the other two (see Steps 6 and 7 of Part III). For each position, determine whether one of the Africans is different or the European/Asian is different. Record tick marks for each mismatch in the chart.
5. Use the pull-down menus and check boxes in your work space to select a different set of individuals and repeat Steps 2–4. Continue until you have made a total of five comparisons and recorded your results in the chart.

6. What do your results show? Is this what you expected?
7. Pool the class results to confirm or refute your own results.
8. What does your analysis say about African mtDNA diversity? What population factors could account for this?
9. What does your analysis say about European and Asian mtDNA diversity? What population factors could account for this?

LABORATORY
2.2

Using an *Alu* Insertion Polymorphism to Study Human Populations

▼ OBJECTIVES

This laboratory demonstrates several important concepts of modern biology. During the course of this laboratory, you will

- Collect and analyze genetic information in human populations.
- Use allele and genotype frequencies to test for Hardy–Weinberg equilibrium.
- Use DNA polymorphisms to study human evolution.
- Analyze population genetic data to trace identity by descent from a common ancestor.
- Move between in vitro experimentation and in silico computation.

In addition, this laboratory utilizes several experimental and bioinformatics methods in modern biological research. You will

- Extract and purify DNA from your own cells.
- Amplify a specific region of your genome by polymerase chain reaction (PCR).
- Analyze PCR products by gel electrophoresis.
- Use the Basic Local Alignment Search Tool (BLAST) to identify sequences in databases.
- Use the Map Viewer tool to visualize genes on chromosomes.

INTRODUCTION

Although DNA from any two people is more alike than different, many chromosomal regions exhibit sequence differences among individuals. Such variable sequences are termed "polymorphic" (meaning "many forms") and are used in the study of human evolution as well as for disease and identity testing. Many polymorphisms are located in the estimated 98% of the human genome that does not encode protein.

This experiment examines a polymorphism in the human genome that is caused by the insertion of an *Alu* transposon, which is a mobile element in the genome. Although *Alu* is sometimes called a "jumping gene," it is not properly a gene, because it does not produce a protein product. *Alu* is a member of the family of short interspersed nucleotide elements (SINEs) and is approximately 300 nucleotides in length. *Alu* owes its name to a recognition site for the endonuclease AluI that is located in the middle of its sequence.

Alu transposons are found only in primate genomes and have accumulated in large numbers since primates diverged from other mammals. Human chromosomes contain more than one million *Alu* copies, comprising about 10% of the genome. This accumulation was made possible by a transposition mechanism that reverse-transcribes *Alu* messenger RNAs (mRNAs) into mobile DNA copies. Another transposon, the long interspersed nucleotide element (LINE) L1, supplies a specialized reverse transcriptase enzyme needed for *Alu* to jump. Hence, *Alu* and L1 exist in a sort of molecular symbiosis.

At any point in evolutionary time, only one or several *Alu* "masters" were capable of transposing. Although the rate of transposition was once much higher, a new *Alu* jump is estimated to now occur once per 200 live human births.

There is lively debate about whether *Alu* serves some larger purpose in primate genomes or is merely "selfish DNA" that has been successful at replicating. *Alu* insertions in coding exons are implicated in a number of human diseases, including neurofibromatosis, thalassemia, cancer, and heart attack. However, the majority of *Alu*s are located in introns or intergenic regions, where they appear to have no phenotypic effect. *Alu*s in introns have had a potentially important impact on protein evolution: They provide alternative splice sites in approximately 5% of genes that produce multiple protein products.

Each *Alu* is the "fossil" of a unique transposition event that occurred once in primate history. After the initial jump, an *Alu* was inherited from parents by offspring in a Mendelian fashion. The vast majority of *Alu* insertions occurred millions of years ago and are "fixed." This means that, for a particular locus, all primates have *Alu*s inserted on both copies of the locus.

Several thousand *Alu*s, however, have inserted in our genome since humans branched from other primates. Some of these are not fixed, meaning that the *Alu* insertion may be present or absent on each of the chromosomes, thus creating two possible alleles (+ and –). These "dimorphic" *Alu*s jumped within the last several hundred thousand years and have reached different allele frequencies in different human populations. Thus, *Alu* insertion polymorphisms are useful tools for reconstructing human evolution and migration.

This experiment examines a human *Alu* dimorphism at the PV92 locus in the heart cadherin gene. A sample of human cells is obtained by saline mouthwash (alternatively, DNA may be isolated from hair sheaths). DNA is extracted by boiling with

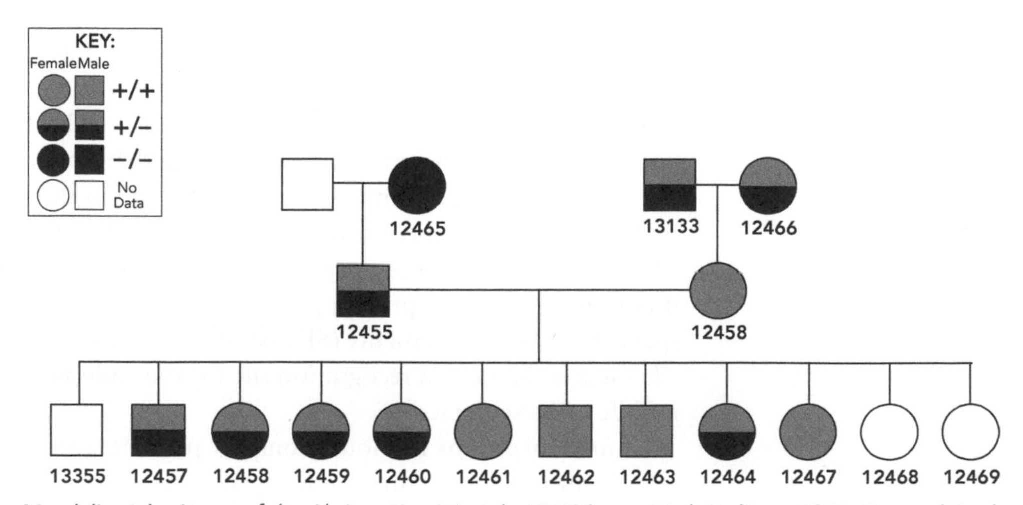

Mendelian inheritance of the *Alu* insertion (+) at the PV92 locus. (Utah Pedigree 1356, Centre d'Etude du Polymorphisme Humain; genotyping by Renato Robledo, University of Cagliari.)

Chelex resin, which binds contaminating metal ions. PCR is then used to amplify a chromosome region that contains the PV92 *Alu* dimorphism. The *Alu* insertion allele (+) is about 300 nucleotides longer than the noninsertion allele (–); thus, the two alleles are readily separated by agarose gel electrophoresis.

The laboratory includes bioinformatics exercises that complement the experimental methods. Each student scores his or her genotype, and the compiled class results are used as a case study in human population genetics. Tools for testing Hardy–Weinberg equilibrium, comparing the PV92 insertion in world populations, and simulating the inheritance of a new *Alu* insertion are found at the *BioServers* Internet site of the Dolan DNA Learning Center (www.bioservers.org). BLAST is used to identify sequences in biological databases, make predictions about the outcome of the experiments, and identify additional alleles at the PV92 locus. Another bioinformatics tool, Map Viewer, is used to discover the chromosome location of the PV92 insertion.

FURTHER READING

Batzer MA, Stoneking M, Alegria-Hartman M, Barzan H, Kass DH, Shaikh TH, Novick GE, Iannou PA, Scheer WD, Herrera RJ, et al. 1994. African origin of human-specific polymorphic *Alu* insertions. *Proc Natl Acad Sci* **91:** 12288–12292.

Comas D, Plaza S, Calafell F, Sajantila A, Bertranpetit J. 2001. Recent insertion of an *Alu* element within a polymorphic human-specific *Alu* insertion. *Mol Biol Evol* **18:** 85–88.

Deininger PL, Batzer MA. 1999. *Alu* repeats and human disease. *Mol Genet Metab* **67:** 183–193.

Mullis K. 1990. The unusual origin of the polymerase chain reaction. *Sci Am* **262:** 56–65.

Prak ETL, Kazazian HH. 2000. Mobile elements and the human genome. *Nat Rev Genet* **1:** 134–144.

PLANNING AND PREPARATION

The following table will help you to plan and integrate the different experimental methods.

Experiment	Day	Time		Activity
I. Isolate DNA from cheek cells	1	60 min	Prelab:	Prepare and aliquot saline solution. Prepare and aliquot 10% Chelex. Make centrifuge adapters. Set up student stations.
		30 min	Lab:	Isolate student DNA.
I. (Alternate) Isolate DNA from hair sheaths	1	30 min	Prelab:	Aliquot proteinase K. Set up student stations.
		30 min	Lab:	Isolate student DNA.
II. Amplify DNA by PCR	2	15 min	Prelab:	Aliquot PV92 primer/loading dye mix. Set up student stations.
		15 min	Lab:	Set up PCRs.
		60–150 min	Postlab:	Amplify DNA in thermal cycler.
III. Analyze PCR products by gel electrophoresis	3	30 min	Prelab:	Dilute TBE electrophoresis buffer. Set up student stations. Prepare agarose gel solution.
		30 min	Lab:	Cast gels.
	4	45+ min	Lab:	Load DNA samples into gel. Electrophorese samples. Photograph gels.

OVERVIEW OF EXPERIMENTAL METHODS

I. ISOLATE DNA FROM CHEEK CELLS

RINSE mouth with saline

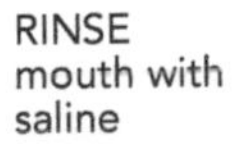

TRANSFER saline

CENTRIFUGE

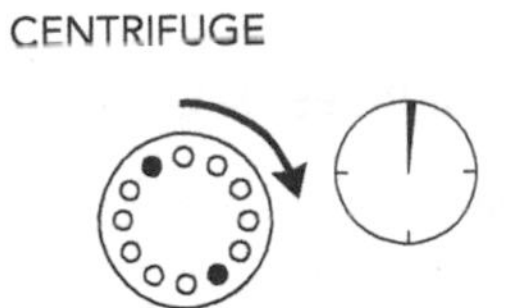

POUR OFF supernatant

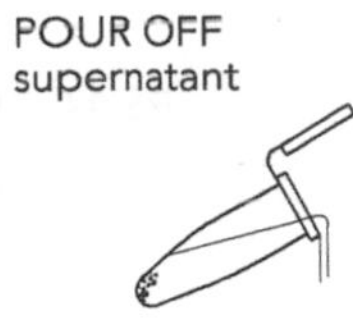

RESUSPEND

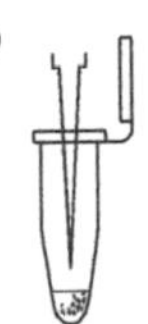

TRANSFER cell suspension into Chelex

BOIL in thermal cycler

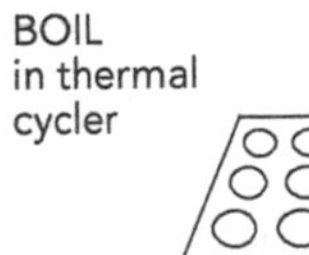

SHAKE vigorously

CENTRIFUGE

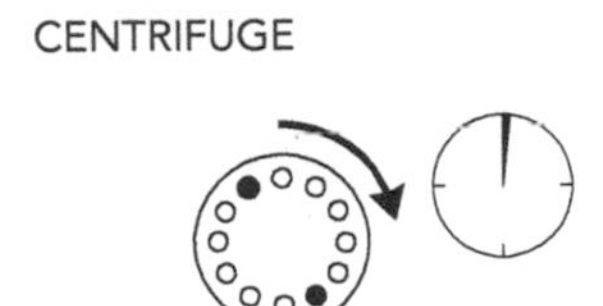

TRANSFER supernatant

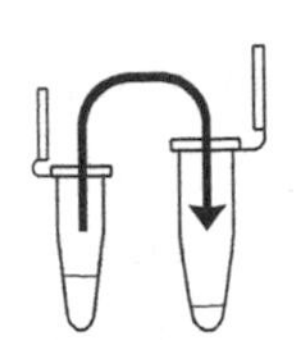

STORE on ice

I. (ALTERNATE) ISOLATE DNA FROM HAIR SHEATHS

CUT hairs

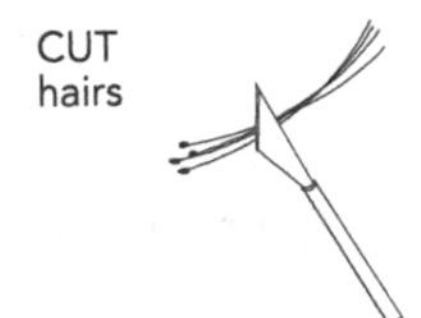

ADD hairs to proteinase K

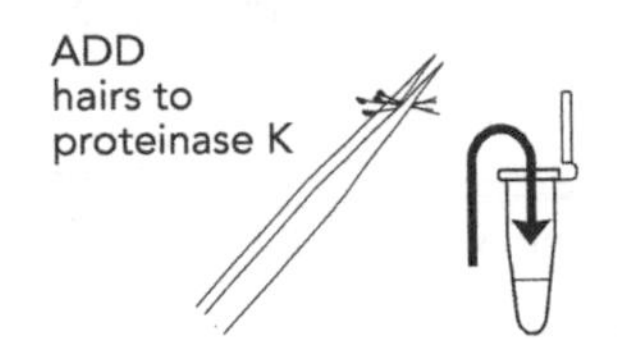

INCUBATE in thermal cycler

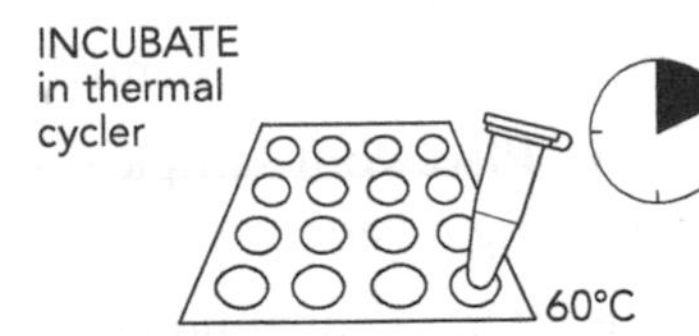

VORTEX

BOIL in thermal cycler

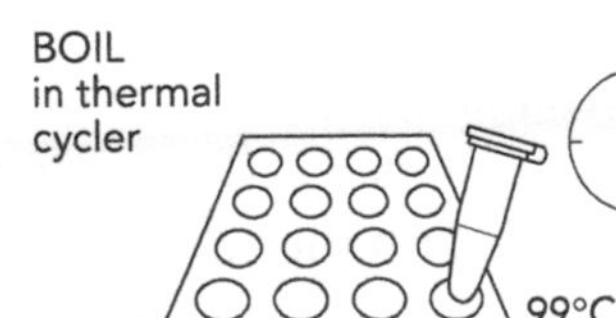

MIX by pipetting in and out

STORE on ice

II. AMPLIFY DNA BY PCR

ADD primer/ loading dye mix

ADD DNA

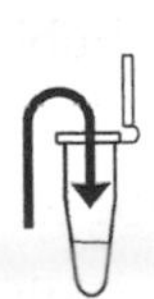

AMPLIFY in thermal cycler

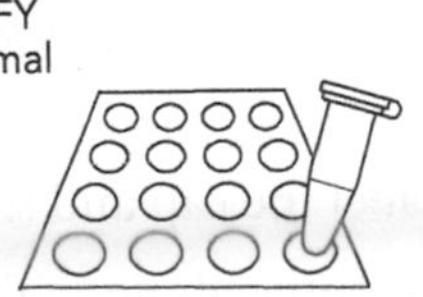

III. ANALYZE PCR PRODUCTS BY GEL ELECTROPHORESIS

POUR gel

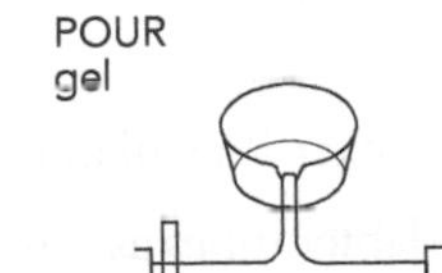

SET 20 min

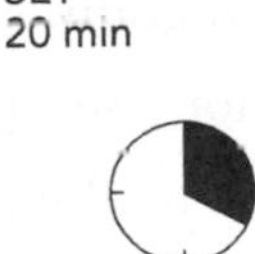

LOAD gel

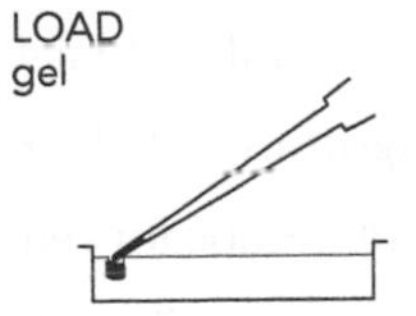

ELECTROPHORESE 130 V

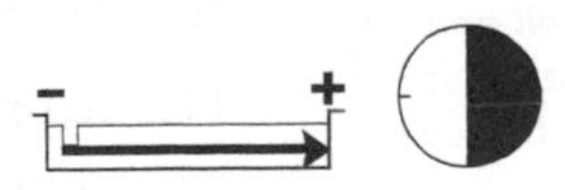

EXPERIMENTAL METHODS

I. Isolate DNA from Cheek Cells

REAGENTS, SUPPLIES, & EQUIPMENT

For each student

10% Chelex (100 μL) in a 0.2- or 0.5-mL PCR tube (or in a microcentrifuge tube)
2 Microcentrifuge tubes (1.5 mL)
Paper cup
Saline solution (10 mL) in a 15-mL tube

For each group

Container with cracked or crushed ice

Microcentrifuge adapters for 0.2- or 0.5-mL PCR tubes
Microcentrifuge tube rack
Micropipettes and tips (10–1000 μL)
Permanent marker
Vortexer (optional)

To share

Microcentrifuge
Thermal cycler (or water bath or heat block)

1. Use a permanent marker to label a 1.5-mL microcentrifuge tube and paper cup with your assigned number.
2. Pour all 10 mL of saline solution into your mouth and vigorously rinse your cheek pockets for 30 sec.
3. Expel the saline solution into the paper cup.
4. Swirl the cup gently to mix the cells that may have settled to the bottom. Use a micropipette with a fresh tip to transfer 1000 μL of the solution into your labeled 1.5-mL microcentrifuge tube.
5. Place your sample tube, along with those from other students, in a balanced configuration in a microcentrifuge. Centrifuge the tubes at full speed for 90 sec.
6. After centrifuging, check to see if your pellet is firmly attached to the bottom of the tube.
   - If the pellet is firmly attached, carefully pour off the supernatant into the paper cup. Gently tap the inverted tube against the cup to remove most of the supernatant, but be careful not to disturb the cell pellet at the bottom of the tube.
   - If the pellet is loose or unconsolidated, centrifuge again for 90 sec. Then, carefully use a micropipette with a fresh tip to remove as much saline as possible.

If your microcentrifuge tube is graduated, the volume remaining after Step 6 will approximately reach the 0.1-mL mark.

7. Set a micropipette to 30 μL. Resuspend the cells in the remaining saline by pipetting in and out. Work carefully to minimize bubbles.

Food particles will not resuspend. Your teacher may instruct you to examine a sample of the cell suspension under a microscope.

8. Withdraw 30 μL of cell suspension and add it to a PCR tube containing 100 μL of 10% Chelex. Label the cap and side of the tube with your assigned number.
9. Place your PCR tube, along with those from other students, in a thermal cycler that has been programmed for one cycle of the following profile:

   Boiling step: 10 min 99°C

   The profile may be linked to a 4°C hold program.

As an alternative to Steps 8 and 9, you may add the cell suspension to Chelex in a 1.5-mL tube and incubate for 10 min in a boiling water bath or 99°C heat block.

The near-boiling temperature lyses the cell and nuclear membranes, releasing DNA and other cell contents.

10. After boiling, vigorously shake your tube for 5 sec. Use a vortexer if available.
11. To prepare your PCR tube for centrifugation, "nest" it within adapter tubes as follows:
    - If your sample is in a 0.5-mL PCR tube, "nest" it within a capless 1.5-mL tube.
    - If your sample is in a 0.2-mL PCR tube, "nest" it within a capless 0.5-mL tube and then place both tubes into a capless 1.5-mL tube.

Skip Step 11 if your sample is in a microcentrifuge tube (not in a PCR tube).

12. Place your tube, along with those from other students, in a balanced configuration in a microcentrifuge. Centrifuge the tubes at full speed for 90 sec.
13. After centrifugation, use a micropipette with a fresh tip to transfer 30 μL of the clear supernatant into a clean 1.5-mL microcentrifuge tube. Be careful to avoid pipetting any cell debris or Chelex beads.
14. Use a permanent marker to label the cap and side of the tube with your assigned number. This tube contains your DNA and will be used for setting up the PCRs in Part II.
15. Store your sample on ice or at –20°C until you are ready to continue with Part II.

I. (Alternate) Isolate DNA from Hair Sheaths

REAGENTS, SUPPLIES, & EQUIPMENT

For each student

100-mg/ml proteinase K (100 μL) <!> in a 0.2- or 0.5-mL PCR tube (or in a microcentrifuge tube)*

For each group

Container with cracked or crushed ice
Forceps or tweezers
Hand lens or dissecting microscope
Microcentrifuge tube rack
Micropipettes and tips (100 μL)
Permanent marker
Scalpel or razor blade
Vortexer (optional)

To share

Thermal cycler (or water bath or heat block)

*Store on ice.
See Cautions Appendix for appropriate handling of materials marked with <!>.

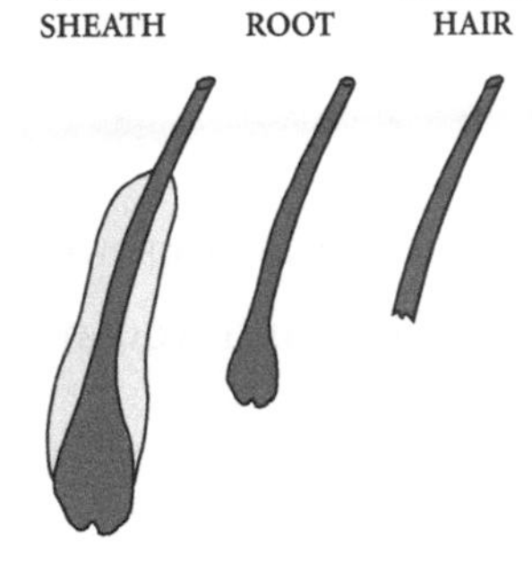

1. Pull out several hairs. Use a hand lens or dissecting microscope to inspect them for the presence of a sheath. The sheath is a barrel-shaped structure surrounding the base of the hair and is most easily observed on dark hair. By holding the hair up to a light source, you may observe the glistening sheath with your naked eye.
2. Select one to several hairs with good sheaths. Alternatively, select hairs with the largest roots. Broken hairs, without roots or sheaths, will not yield enough DNA for amplification.
3. Use a fresh razor blade or scalpel to cut off each hair shaft just above the sheath.
4. Use forceps or tweezers to transfer the hairs to a PCR tube containing 100 μL of proteinase K. Make sure that the sheath is submerged in the solution and not stuck to the wall of the tube.
5. Use a permanent marker to label the cap and the side of your PCR tube with your assigned number.

Your teacher may instruct you to prepare a hair sheath to observe under a compound light microscope.

As an alternative to Steps 4–6, add the hairs to proteinase K in a 1.5-mL microcentrifuge tube and incubate in a water bath or heat block for 10 min at 60°C. Then, in Step 8, incubate the tube for 10 min in a boiling water bath or 99°C heat block.

6. Place your PCR tube, along with those of other students, in a thermal cycler that has been programmed for one cycle of the following profile:

Incubation step:	10 min	60°C

7. Place your sample tube at room temperature. To dislodge the cells from the hair shaft, vortex your tube by machine or vigorously with a finger for 15 sec.
8. Place your PCR tube, along with those of other students, in a thermal cycler that has been programmed for one cycle of the following profile:

Boiling step:	10 min	99°C

   The profile may be linked to a 4°C hold program.

9. Place your sample tube at room temperature. Use a micropipette and fresh tip to mix your sample by pipetting in and out for 15 sec. This sample contains your DNA and will be used for setting up the PCRs in Part II.
10. Store your sample on ice or at –20°C until you are ready to begin Part II.

II. Amplify DNA by PCR

REAGENTS, SUPPLIES, & EQUIPMENT

For each student

Cheek cell or hair sheath DNA* from Part I
PV92 primer/loading dye mix (25 µL)*
Ready-To-Go PCR Bead in a 0.2- or 0.5-mL PCR tube

For each group

Container with cracked or crushed ice

Microcentrifuge tube rack
Micropipette and tips (1–100 µL)
Permanent marker

To share

Thermal cycler

*Store on ice.

1. Obtain a PCR tube containing a Ready-To-Go PCR Bead. Label the tube with your assigned number.
2. Use a micropipette with a fresh tip to add 22.5 µL of PV92 primer/loading dye mix to the tube. Allow the bead to dissolve for approximately 1 min.
3. Use a micropipette with a fresh tip to add 2.5 µL of your DNA (from Part I) directly into the primer/loading dye mix. Ensure that no DNA remains in the tip after pipetting.
4. Store your sample on ice until your class is ready to begin thermal cycling.
5. Place your PCR tube, along with those of other students, in a thermal cycler that has been programmed for 30 cycles of the following profile:

Denaturing step:	30 sec	94°C
Annealing step:	30 sec	68°C
Extending step:	30 sec	72°C

The profile may be linked to a 4°C hold program after the 30 cycles are completed.

6. After thermal cycling, store the amplified DNA on ice or at –20°C until you are ready to continue with Part III.

The primer/loading dye mix will turn purple as the PCR bead dissolves.

If the reagents become splattered on the wall of the tube, pool them by pulsing the sample in a microcentrifuge or by sharply tapping the tube bottom on the lab bench.

III. Analyze PCR Products by Gel Electrophoresis

REAGENTS, SUPPLIES, & EQUIPMENT

For each student

Latex gloves
PCR product* from Part II
SYBR Green <!> DNA stain (5 µL)

For each group

2% Agarose in 1x TBE (hold at 60°C) (50 mL per gel)
Container with cracked or crushed ice
Gel-casting tray and comb
Gel electrophoresis chamber and power supply
Masking tape

Microcentrifuge tube rack
Micropipette and tips (1–100 µL)
pBR322/BstNI marker (20 µL per gel)*
1x TBE buffer (300 mL per gel)

To share

Digital camera or photodocumentary system
UV transilluminator <!> and eye protection
Water bath for agarose solution (60°C)

*Store on ice.
See Cautions Appendix for appropriate handling of materials marked with <!>.

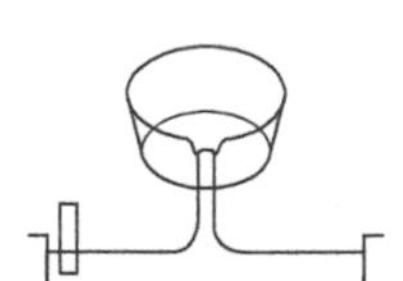

1. Seal the ends of the gel-casting tray with masking tape and insert a well-forming comb.
2. Pour the 2% agarose solution into the tray to a depth that covers about one-third the height of the open teeth of the comb.

Avoid pouring an overly thick gel, which will be more difficult to visualize.

3. Allow the gel to completely solidify; this takes approximately 20 min.

The gel will become cloudy as it solidifies.

4. Remove the masking tape, place the gel into the electrophoresis chamber, and add enough 1x TBE buffer to cover the surface of the gel.
5. Carefully remove the comb and add additional 1x TBE buffer to fill in the wells and just cover the gel, creating a smooth buffer surface.

Do not add more buffer than necessary. Too much buffer above the gel channels electrical current over the gel, increasing the running time.

6. Orient the gel according to the diagram in Step 8, so that the wells are along the top of the gel.
7. Add 2 µL of SYBR Green DNA stain to your PCR product (from Part II). In addition, add 2 µL of SYBR Green DNA stain to 20 µL of pBR322/BstNI marker.

A 100-bp ladder may also be used as a marker.

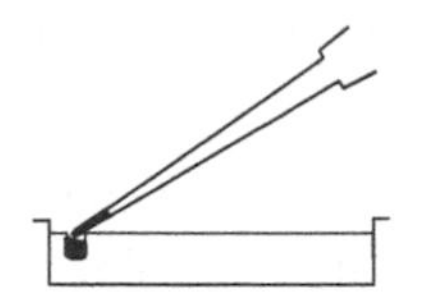

8. Use a micropipette with a fresh tip to load each sample from Step 7 into your assigned well, according to the following diagram:

MARKER	STUDENT SAMPLES					
pBR322/ BstNI	1	2	3	4	5	6

Expel any air from the tip before loading. Be careful not to push the tip of the pipette through the bottom of the sample well.

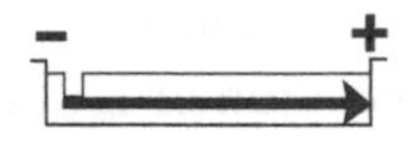

9. Run the gel for approximately 30 min at 130 V. Adequate separation will have occurred when the cresol red dye front has moved at least 50 mm from the wells.
10. View the gel using UV transillumination. Photograph the gel using a digital camera or photodocumentary system.

Transillumination, where the light source is below the gel, increases brightness and contrast.

BIOINFORMATICS METHODS

For a better understanding of the experiment, perform the following bioinformatics exercises before you analyze your results.

I. Use BLAST to Find DNA Sequences in Databases (Electronic PCR)

1. Perform a BLAST search as follows:
   - i. Do an Internet search for "ncbi blast."
   - ii. Click on the link for the result "BLAST: Basic Local Alignment Search Tool." This will take you to the Internet site of the National Center for Biotechnology Information (NCBI).
   - iii. Click on the link "nucleotide blast" (blastn) under the heading "Basic BLAST."
   - iv. Enter both primer sequences into the search window. These are the query sequences.
   - v. Omit any nonnucleotide characters from the window because they will not be recognized by the BLAST algorithm.

The following primer set was used in the experiment:
5′-GGATCTCAGGGTGGGTGGCAATGCT-3′ (forward primer) 5′-GAAAGGCAAGCTACCAGAAGCCCCAA-3′ (reverse primer)

vi. Under "Choose Search Set," select the "Nucleotide collection (nr/nt)" database from the drop-down menu.

vii. Under "Program Selection," optimize for "Somewhat similar sequences (blastn)."

viii. Click on "BLAST." This sends your query sequences to a server at NCBI in Bethesda, Maryland. There, the BLAST algorithm will attempt to match the primer sequences to the millions of DNA sequences stored in its database. While searching, a page showing the status of your search will be displayed until your results are available. This may take only a few seconds or more than 1 min if many other searches are queued at the server.

2. Analyze the results of the BLAST search, which are displayed in three ways as you scroll down the page:

   i. First, a graphical overview illustrates how significant matches (hits) align with the query sequence. Matches of differing lengths are indicated by color-coded bars. What do you notice about the lengths (and colors) of the matches (bars) as you look from the top to the bottom?

   ii. This is followed by a list of significant alignments (hits) with links to the corresponding accession numbers. (An accession number is a unique identifier given to a sequence when it is submitted to a database such as GenBank.) Note the scores in the "E value" column on the right. The Expectation or E value is the number of alignments with the query sequence that would be expected to occur by chance in the database. The lower the E value, the higher the probability that the hit is related to the query. For example, an E value of 1 means that a search with your sequence would be expected to turn up one match by chance. Longer query sequences generally yield lower E values. An alignment is considered significant if it has an E value of less than 0.1. What is the E value of the most significant hit and what does it mean? Note the names of any significant alignments that have E values of less than 0.1. Do they make sense?

   iii. Third is a detailed view of each primer (query) sequence aligned to the nucleotide sequence of the search hit (subject, abbreviated "Sbjct"). Note that the first match to the forward primer (nucleotides 1–25) and to the reverse primer (nucleotides 26–51) are within the same subject (accession number).

3. Click on the accession number link to open the data sheet for the first hit (the first subject sequence).

   i. At the top of the report, note basic information about the sequence, including its length (in base pairs, or bp), database accession number, source, and references to papers in which the sequence is published. What is the source and size of the sequence in which your BLAST hit is located?

   ii. In the middle section of the report, the sequence features are annotated, with their beginning and ending nucleotide positions ("xx..xx"). These features may include genes, coding sequences (CDS), regulatory regions, ribosomal RNA (rRNA), and transfer RNA (tRNA).

iii. Scroll to the bottom of the data sheet. This is the nucleotide sequence to which the term "Sbjct" refers.

iv. Use your browser's "Back" button to return to the BLAST results page.

4. Predict the length of the product that the primer set would amplify in a PCR (in vitro) as follows:

   i. Scroll down to the alignments section (third section) of the BLAST results page. Examine the alignments with the lowest E values and identify a human (*Homo sapiens*) sequence to which both primers align.

   ii. To which positions do the primers match in the subject sequence?

   iii. The lowest and highest nucleotide positions in the subject sequence indicate the borders of the amplified sequence. Subtract the lowest nucleotide position in the subject sequence from the highest nucleotide position in the subject sequence. What is the difference between the coordinates?

   iv. Note that the actual length of the amplified fragment includes both ends, so add 1 nucleotide to the result that you obtained in Step 4.iii to obtain the exact length of the PCR product amplified by the two primers.

   v. At this stage, do you have enough information to determine whether this is the + or the – allele?

5. Obtain the nucleotide sequence of the amplicon that the primer set would amplify in a PCR (in vitro) as follows:

   i. Open the sequence data sheet for the hit that you identified in Step 4.i by clicking on the accession number link. Scroll to the bottom of the data sheet.

   ii. The bottom section of the data sheet lists the entire nucleotide sequence that contains the PCR product. Highlight all of the nucleotides between the coordinates that you identified in Step 4.iii, from the beginning of the forward primer to the end of the reverse primer.

   iii. Copy and paste the highlighted sequence into a text document. Then, delete all nonnucleotide characters and spaces. This is the amplicon, or amplified PCR product. Save this text document for use in Step 2 of Part II and/or in Step 3 of Part III.

In Step 5.iii, you can retain the nucleotide coordinates and spacers to ease readability. Reduce the point size of the font so that each row of 60 nucleotides sits on one line. Then, set the font to Courier or another nonproportional font to align the blocks of sequence.

II. Use BLAST to Identify Additional Alleles at the PV92 Locus

1. Open the nucleotide BLAST page by following Steps 1.i–1.iii in Part I.
2. Paste the amplicon from Step 5.iii of Part I into the search window.
3. Ensure that "Nucleotide collection (nr/nt)" and "Somewhat similar sequences (blastn)" are selected, click on "BLAST," and wait until the BLAST results are displayed.
4. Compare the E values obtained by this search with those values that you obtained in Step 2.ii of Part I. What do you notice? Why is this so?
5. Why does the first hit have an E value of 0?
6. Now focus on the hit named "Human Alu repeat"; this is the *Alu* insertion at PV92.

   i. Follow the link for the accession number and then, in the "Features" section, click on the second "repeat_unit" (the *Alu* repeat unit). Look at the top of the report. What is the length of the *Alu* repeat (insert) at PV92?

ii. Scroll down to the bottom of the report. What do you notice about the 3′ end of the *Alu* repeat sequence?

iii. Click on your Internet browser's "Back" button to return to the full report. In the "Features" section, click on the first and third "repeat_unit" links for the insertion target sequences on either side of the *Alu* repeat. Compare these sequences. What appears to have happened to this sequence when the *Alu* repeat inserted itself into the PV92 locus?

iv. If you assume that the amplicon that you identified in Part I is the – allele, what is the length of the + allele? When answering this question, keep in mind that, in addition to the *Alu* repeat that you identified in Step 6.i, the insertion target sequence (that you examined in Step 6.iii) increases the total length of the + allele.

7. Now return to the BLAST results page and open accession number AF302689.1 (one of the lowest E-value hits). Carefully examine the "Features" and follow the links. What is going on here? How are the three repeated sequences related to one another?

III. Use Map Viewer to Determine the Chromosomal Location of the PV92 Insertion

1. Open Map Viewer (http://www.ncbi.nlm.nih.gov/mapview).
2. Find *Homo sapiens* in the table to the right and click on the "B" icon under the "Tools" header. If more than one build is displayed, select the one with the highest number; this will be the most recent version.
3. Paste the amplicon from Step 5.iii of Part I into the search window. (Usually, primers are not long enough to produce a result using map BLAST.) Omit any nonnucleotide characters from the window because they will not be recognized by the BLAST algorithm.
4. Use the default settings and click on "Begin Search."
5. Click on "View report" to retrieve the results.
6. In "Other reports," click on the "[Human genome view]" link to see the chromosome location of the BLAST hit. Small horizontal bars on chromosomes indicate the positions of hits. On which chromosome have you landed?
7. Click on the marked chromosome number to move to the PV92 locus.
8. The chromosome is represented by one or more vertical lines. Click on the small blue arrow after the vertical line labeled "Genes_seq" to display the genes. The amplicon (vertical red bar) occupies the whole field of the default view. Which gene contains the amplicon? What is the function of that gene? Click on the name of the gene under the "Symbol" track and follow the links to find out.
9. Use the zoom toggle on the left to zoom out for a better perspective of the gene. You may refine your position on the chromosome by clicking on the thin vertical line under "Genes_seq" and choosing a specific zoom level.
10. Introns and noncoding sequences of the gene are denoted by a thin line, whereas exons are denoted by thick bars along the line.

    i. Determine the size of the gene using the map coordinates to the left of the "Contig" map.

ii. How many introns and exons does the gene have?

iii. Where in the gene has the PV92 *Alu* inserted: an exon or an intron? How does this explain the fact that the PV92 insertion is believed to be neutral, i.e., to have no phenotypic effect?

RESULTS AND DISCUSSION

The following diagram shows how PCR amplification identifies the *Alu* insertion polymorphism at the PV92 locus.

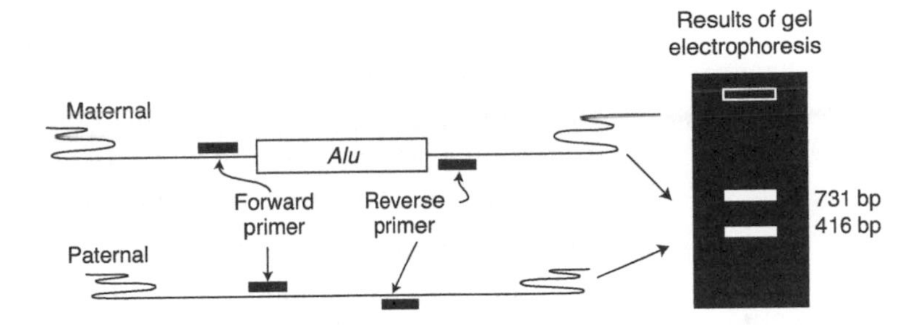

I. Determine Your PV92 Genotype

1. Observe the photograph of the stained gel containing your PCR samples and those from other students. Orient the photograph with the sample wells at the top. Use the sample gel shown below to help interpret the band(s) in each lane of the gel.

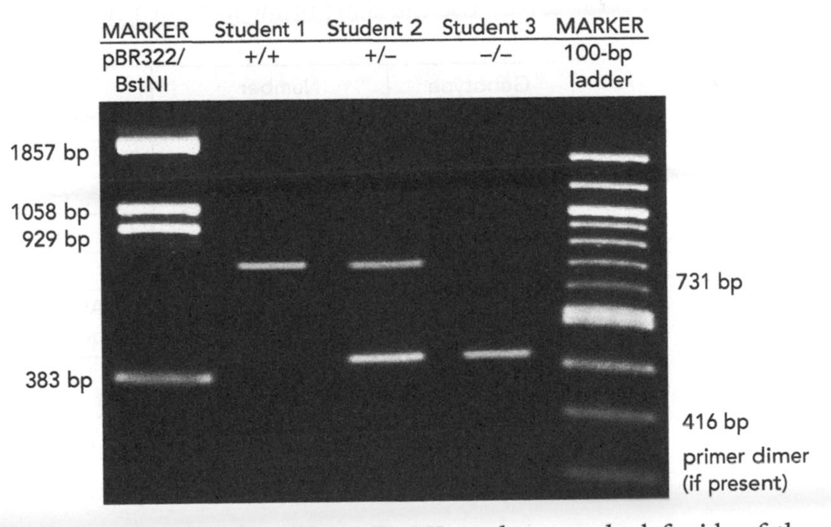

If a 100-bp ladder was used, as shown on the right-hand side of the sample gel, the bands increase in size in 100-bp increments starting with the fastest-migrating band of 100 bp.

2. Locate the lane containing the pBR322/BstNI markers on the left side of the gel. Working from the well, locate the bands corresponding to each restriction fragment: 1857, 1058, 929, 383, and 121 bp. The 1058- and 929-bp fragments will be very close together or may appear as a single large band. The 121-bp band may be very faint or not visible.

3. Scan across the row of student results. You should notice that virtually all student lanes contain one or two prominent bands.

   i. A heterozygous +/– genotype shows two prominent bands. The + allele (731 bp) should be slightly ahead of the 929-bp marker. The – allele (416 bp) should be about even with the 383-bp marker.

ii. A homozygous +/+ genotype shows a single prominent band slightly ahead of the 929-bp marker.

iii. A homozygous –/– genotype shows a single prominent band about even with the 383-bp marker.

Additional faint bands at other positions occur when the primers bind to chromosome loci other than the exact locus and give rise to "nonspecific" amplification products.

4. To "score" your genotype, compare your PCR product with the markers and the other student results in the row. The analysis will be simple if your row contains a heterozygous genotype (+/–) that shows the positions of both alleles. Homozygotes of each genotype (+/+ and –/–) will also help. If your row contains only a single homozygous genotype, you will need to rely entirely on markers to determine the allele.

5. It is common to see a diffuse (fuzzy) band that runs ahead of the 121-bp marker. This is "primer dimer," an artifact of the PCR that results from the primers overlapping one another and amplifying themselves. How would you interpret a lane in which you observe primer dimer but no bands as described in Step 3?

6. An *Alu* insertion has only two states: + and –. How does this relate to the most basic unit of information stored in digital form by a computer? What equivalent in digital information is provided by an *Alu* genotype?

II. Determine the Observed Genotype and Allele Frequencies for Your Class

Use the chart below to record your answers to the questions that follow.

Genotype frequency	Number of students	Genotype	Number of + alleles	Number of – alleles
		+/+		
		+/–		
		–/–		
Totals				
		Allele frequencies		

1. Count the number of students of each genotype: +/+, +/–, and –/–. Exclude from the analysis any students whose genotypes could not be determined.

2. Calculate the frequency of each genotype, where

$$\text{frequency of genotype } x\ (\%) = \frac{\text{number of students of } x \text{ genotype}}{\text{total number of students}} \times 100.$$

3. Determine the total number of + and – alleles in the class as follows:

   i. Multiply the number of students of each genotype by the number of + or – alleles in that genotype. Remember that each +/+ or –/– student contributes two copies of that allele, whereas each +/– student contributes one of each allele.

   ii. Add up the total number of copies of each allele. The total number of + and – alleles in the class is twice the number of students.

4. Calculate the frequency of each allele, where

$$\text{frequency of allele } x\ (\%) = \frac{\text{number of } x \text{ alleles}}{\text{total alleles in class}} \times 100.$$

5. All students with at least one + allele should raise their hands. Is the + allele confined to any particular racial or ethnic group in your class? What can you say about individuals in the class who have at least one + allele?

III. TeOst for Hardy–Weinberg Equilibrium in Your Class

INSTRUCTORS

To complete Part III, you must first register at the BioServers Internet site (www.bioservers.org). This will allow you to enter data from your students into the online Allele Server database. After you've registered, enter the Allele Server, click on "Manage Groups," and wait for the existing data to load. This may take a moment. Select "Your Groups" from the "Select Group Type" pull-down menu. Click "Add Group." Provide the requested information and be sure to make the group "Public." Then create a password and enter the number of students who will submit data. Click "OK." Your class now appears in the list of "Your Groups," which can now be accessed by class members.

Under certain conditions, a population comes into genetic equilibrium, where the genotype frequencies at a single locus remain constant over time. The Hardy–Weinberg equation describes the genotype frequencies that are expected in a population at equilibrium:

$$p^2 + 2pq + q^2 = 1,$$

where p and q represent the allele frequencies, p^2 and q^2 are the homozygote frequencies, and $2pq$ is the heterozygote frequency.

A Chi-square test is used to compare observed genotype frequencies with those predicted by the Hardy–Weinberg equation. The Chi-square statistic tests the "null hypothesis"—that there is no significant difference between observed and expected genotype frequencies. The Chi-square result is associated with a p value or probability that observed and expected frequencies are substantially alike and that frequency differences are merely due to chance. Scientists generally accept that the results are statistically significant at a p value of 0.05 or less. This technically means that there is only a 5% chance such results could be obtained by chance or, more to the point, the observed differences in genotype frequencies are likely real.

Population statistics are tedious to calculate by hand, but they are easily determined using algorithms at the *BioServers* Internet site. In this exercise, you will enter your data into a class file that has been set up by your teacher on the *Allele Server* database at the *BioServers* site. Then, using the tools available at the site, you will determine whether your class is in Hardy–Weinberg equilibrium at the PV92 locus.

1. Calculate the genotype frequencies expected for your class under Hardy–Weinberg equilibrium as follows:
   i. Use the allele frequencies calculated for your class in Step 4 of Part II to determine the genotype frequencies expected under Hardy–Weinberg equilibrium. Make + = p and – = q in the equation.
   ii. How do the genotype frequencies that you observed in your experiment compare with those expected by the Hardy–Weinberg equation? Would you say that they are very similar or very different?
2. Open the *BioServers* Internet site at the DNA Learning Center (www.bioservers.org).
3. Enter *Allele Server*. You can register if you want to save your work for future reference, but this is not required.

4. The interface is simple to use: Add or obtain data using the top buttons and pull-down menus and then work with the data in the work space below.
5. Click on "ADD DATA" located at the top of the page and find your group in the pull-down menu. Enter the password supplied by your teacher as well as your sample number. Then click "OK."
6. Use the pull-down menus to add your sex, descent, and genotype. Then click "OK." Your data have been added to your group.
7. Once your class has finished entering data, click on "MANAGE GROUPS," then wait while the existing data load. This may take a moment.
8. Find your class in the list and click on the check box to select it.
9. Click "OK" to move your class data into the work space.
10. Click "OPEN" to get basic information on your population: number in the sample, frequencies of the + and – alleles, and frequencies of the three genotypes +/+, +/–, and –/–.
11. In your work space, mark the dot to the right of your group name and click "ANALYZE."
12. The pie chart provides a visual comparison of your observed versus expected results. When you ask yourself if the sections of the two pies are substantially similar or rather different, you are performing an informal Chi-square analysis.
13. The Chi-square result is at the top of the page. Is your p value greater or less than the 0.05 cutoff? What does this mean?
14. What conditions are required for a population to come into genetic equilibrium? Does your class satisfy these requirements?

IV. Compare Genotype Frequencies in World Populations

The Chi-square statistic is also used to compare the genotype frequencies of two populations. A p value of 0.05 or less indicates that two populations have significantly different genetic structures.

1. If you are not already in the *Allele Server* of the *BioServers* Internet site, open it as described in Steps 2 and 3 of Part III. Click on "MANAGE GROUPS," then wait for the existing data to load.
2. Under the "Select group type" pull-down menu, select "Reference" to get a list of PV92 experiments that have been conducted by scientists with individuals from a number of relatively distinct populations around the world.
3. Browse the list and click on the check boxes of a number of populations that interest you. Take samples that represent different continents and regions of the world.
4. Press "OK" to move the populations into the work space.
5. Test Hardy–Weinberg equilibrium in any population by marking the dot in the right-hand column and clicking "ANALYZE." (Only one population can be tested at a time.)
6. Next, compare your class to one of the world populations by checking the appropriate boxes in the left-hand column and clicking "COMPARE." (Only two populations can be compared at a time.)

7. Do the pie charts look similar or different? Does the Chi-square statistic and associated *p* value support your visual impression?
8. Continue to compare your class to other world populations. Also, compare any two reference populations. Uncheck populations that you have completed.
9. Which groups have significantly different genotype frequencies? What is the most frequent genotype in each group?

V. Compare Allele Frequencies in World Populations

Genetic distance is a relatively simple statistic that uses differences in allele frequency to gauge the relative distance that separates two populations in genetic space, with 0 being the least distance and 1 the greatest. Perform this exercise immediately after Part IV.

1. Click on the check boxes to select any two populations you selected in Part IV above.
2. Select "Fst Genetic Distance" from the pull-down window next to the "COMPARE" button.
3. Click "COMPARE."
4. Compare the pie charts with the calculated genetic distance.
5. Continue to compare populations that you selected in Part IV above and note the + allele frequency for each. (You can also obtain the + allele frequency by clicking the "OPEN" button next to each population in the work space.)
6. Now, plot the + allele frequency for each group on the map of world populations (see next page).
7. Do you notice any pattern in the allele frequencies?
8. Suggest a hypothesis about the origin and dispersal of the + allele that accounts for your observation.
9. Calculations suggest that the original *Alu* insertion at the PV92 locus occurred about 200,000 BP. If this is so, in what sort of hominid did the jump occur, and what implications does this have for your suggested hypothesis of Step 8 above?

VI. Simulate a New *Alu* Jump in an Ancient Hominid Population

In this exercise, you will simulate the sort of populations in which the PV92 insertion occurred about 200,000 BP. A Hardy–Weinberg simulator will allow you to model population changes over time. In each generation, parents are chosen at random and offspring are generated using an approach similar to a Punnett square analysis. The survival rate of a particular genotype (+/+, +/–, or –/–) determines the probability that an individual will reproduce in his/her generation. This process is repeated in each generation, producing enough offspring to maintain the population at a constant size.

1. Enter *Simulation Server* from the *BioServers* home page (www.bioservers.org). Wait while the Java applet loads on your computer.
2. Create a node (#1) by clicking in the white work space. This node represents a human population.

3. The red circle indicates that the parameters for Node #1 are available for editing in the right-hand control panel. Think about how to represent this population at the start of the simulation.
   i. How did hominids live 200,000 BP and what size population group would be supported? Enter this number into the "Starting Pop." window at the top right.
   ii. What would be the allele frequency if a single new *Alu* jump occurred in one child born into a group of this size? Enter this number into the "Starting % '+'" window.
   iii. Leave the "# Generations" at 100.
   iv. Assume that this *Alu* jump is neutral and has no effect on gene expression. Accordingly, leave the "Survival %" for each genotype at 100%. This means that individuals with each of the three genotypes have an equal chance of surviving to reproduce.
   v. At the top of the window, set the "# Runs" to 100. The computer will perform 100 experiments with these parameters. You can think of this as 100 different population groups in which a new *Alu* jump occurs. These 100 groups would be equivalent to estimates of the size of the entire hominid population in Africa during several bottlenecks before the advent of agriculture.
   vi. Click "Enter Values" to program the node.
4. Click on "Begin Run" at the top left. Do not touch or move the screen until the calculations are complete or the application may freeze. The progress of the run is indicated in "% Complete" at the top of the window.
5. Scroll down to see the results of the simulation. The histogram is difficult to interpret, so click on the "Graph" tab at the upper left. Then, check "Node #1" and click on "Press here to graph."
6. Allele frequency is on the *y* axis and generations are on the *x* axis. Each blue line traces one population over 100 generations.
7. What happens to the new *Alu* insertion in the 100 populations?
8. Follow the allele frequency in one population over 100 generations. What happens to the allele frequency and what causes this?
9. Try another experiment with the same parameters. Scroll to the top of the page and click the "Restart" and "Begin Run" buttons.

VII. Simulate Population Expansion

Next, determine what happens to an *Alu* insertion when a small population dramatically expands. This simulates what happened to neutral alleles when hunter-gatherer groups became agriculturalists and settled down to form the first urban centers. It also illustrates the so-called "founder effect," the effect on an allele frequency when a large population is derived from a small group of original settlers.

1. Immediately after completing Step 9 of Part VI, click Restart, and then add Node #2 on the work space.

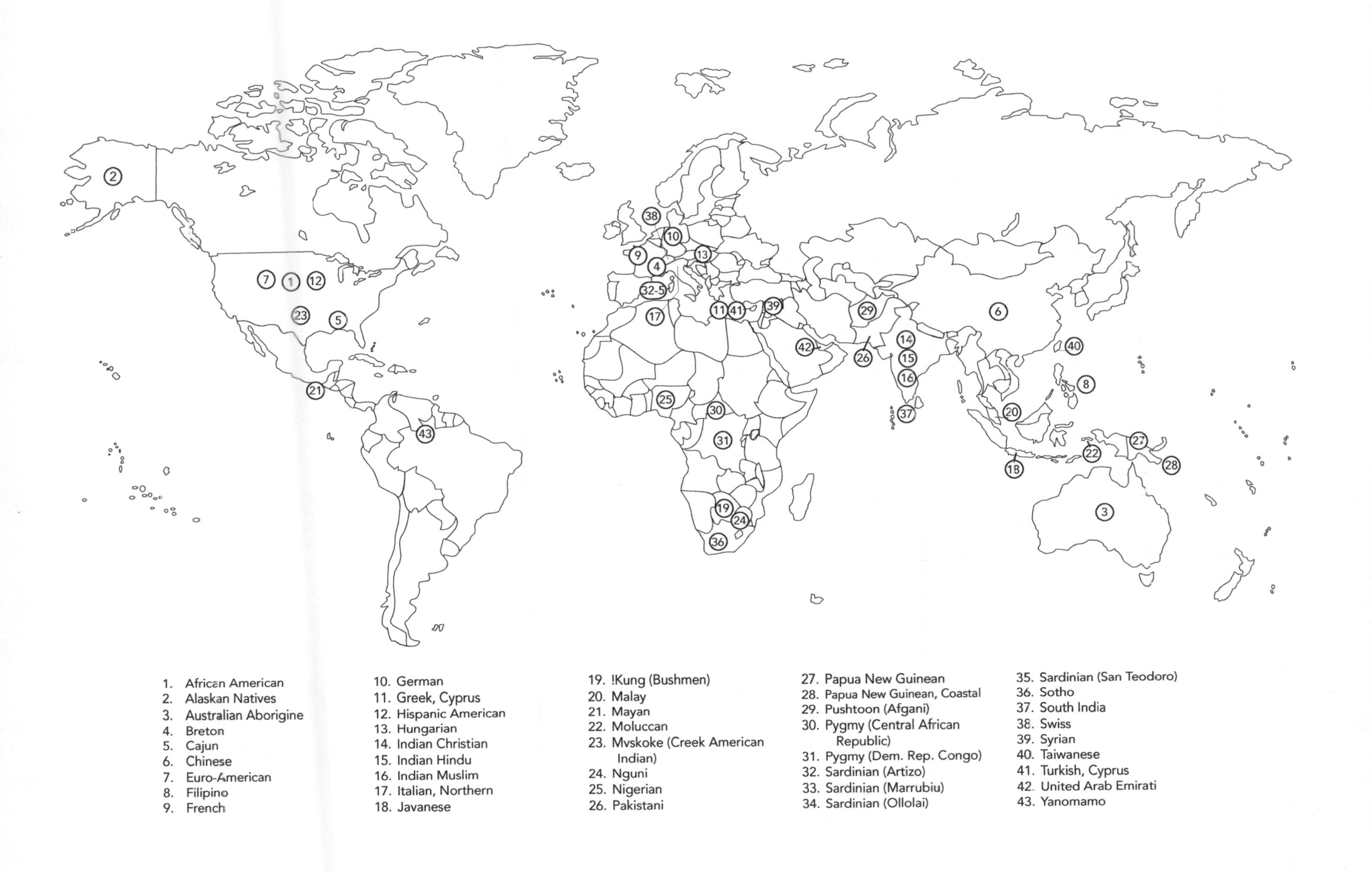
1. African American
2. Alaskan Natives
3. Australian Aborigine
4. Breton
5. Cajun
6. Chinese
7. Euro-American
8. Filipino
9. French
10. German
11. Greek, Cyprus
12. Hispanic American
13. Hungarian
14. Indian Christian
15. Indian Hindu
16. Indian Muslim
17. Italian, Northern
18. Javanese
19. !Kung (Bushmen)
20. Malay
21. Mayan
22. Moluccan
23. Mvskoke (Creek American Indian)
24. Nguni
25. Nigerian
26. Pakistani
27. Papua New Guinean
28. Papua New Guinean, Coastal
29. Pushtoon (Afgani)
30. Pygmy (Central African Republic)
31. Pygmy (Dem. Rep. Congo)
32. Sardinian (Artizo)
33. Sardinian (Marrubiu)
34. Sardinian (Ollolai)
35. Sardinian (San Teodoro)
36. Sotho
37. South India
38. Swiss
39. Syrian
40. Taiwanese
41. Turkish, Cyprus
42. United Arab Emirati
43. Yanomamo

2. With Node #2 active, change one parameter in the right-hand column. Enter 5000 in the "Starting Pop." window. Then click "Enter Values" to program the node.
3. Change the second window in the lower-right corner to read "Link 1 to 2." Click on the "Link" button and a red line will appear between Nodes #1 and #2. In the link mode, Node #1 feeds its results into Node #2. The initial population thus mates randomly for 100 generations and then feeds the resulting + allele frequency into an expanded population, which mates for an additional 100 generations at Node #2. (This is why the "Starting % '+'" is inactivated in Node #2.)
4. Click "Begin Run" at the top left. The calculations take longer with the larger population, so be patient.
5. When the calculations are complete, scroll down to see the results.
6. In the "Graph" mode, check "Node #1," "Node #2," and "Graph Linked." Then click "Press here to graph."
7. The left-hand side of the graph shows the first 100 generations of the small population, and the right-hand side shows the next 100 generations as a larger population.
8. What do you notice about the allele frequency in those populations that maintain the + allele over 200 generations?
9. Click on the "Restart" and "Begin Run" button to see another set of experiments with the same parameters.
10. Add additional nodes to simulate other effects, such as population bottlenecks, or create scenarios in which the + allele confers some survival advantage or disadvantage.

LABORATORY 2.3

Using a Single-Nucleotide Polymorphism to Predict Bitter-Taste Ability

▼ OBJECTIVES

This laboratory demonstrates several important concepts of modern biology. During the course of this laboratory, you will

- Learn about the relationship between genotype and phenotype.
- Discover how single-nucleotide polymorphisms (SNPs) can be used to predict drug response (pharmacogenetics).
- Observe how a number of SNPs can be inherited together as a haplotype.
- Move between in vitro experimentation and in silico computation.

In addition, this laboratory utilizes several experimental and bioinformatics methods in modern biological research. You will

- Extract and purify DNA from your own cells.
- Amplify a specific region of your genome by polymerase chain reaction (PCR).
- Digest PCR products with restriction enzymes.
- Analyze PCR products by gel electrophoresis.
- Use the Basic Local Alignment Search Tool (BLAST) to identify sequences in databases.
- Use the Map Viewer tool to visualize genes on chromosomes.
- Perform multiple sequence alignments to identify mutations and examine evolutionary patterns.

INTRODUCTION

Mammals are believed to distinguish only five basic tastes: sweet, sour, bitter, salty, and umami (the taste of monosodium glutamate). Taste recognition is mediated by specialized taste cells that communicate with several brain regions through direct connections to sensory neurons. Taste perception is a two-step process. First, a taste molecule binds to a specific receptor on the surface of a taste cell. Then, the taste cell generates a nervous impulse that is interpreted by the brain. For example, stimulation of "sweet cells" generates a perception of sweetness in the brain. Recent research has shown that taste sensation is ultimately determined by the wiring of the taste cell to the brain, rather than by the type of receptor on the surface of the cell or by the molecule that is bound by that receptor. So, for example, if a bitter-taste receptor is expressed on the surface of a "sweet cell," a bitter molecule is perceived as tasting sweet.

A serendipitous observation at DuPont in the early 1930s first showed a genetic basis for taste. Arthur Fox had synthesized some phenylthiocarbamide (PTC), and some

Albert Blakeslee uses a voting machine to tabulate results of taste tests at the American Association for the Advancement of Science Convention in 1938.
(Courtesy of Cold Spring Harbor Laboratory Archives.)

of the PTC dust escaped into the air as he was transferring it into a bottle. Lab-mate C.R. Noller complained that the dust had a bitter taste, but Fox tasted nothing—even when he directly sampled the crystals. Subsequent studies by Albert Blakeslee at the Carnegie Department of Genetics (the forerunner of Cold Spring Harbor Laboratory) showed that the inability to taste PTC is a recessive trait that varies in the human population.

Bitter-tasting compounds are recognized by specific receptor proteins on the surface of taste cells. Mammals have approximately 30 genes for different bitter-taste receptors. The gene for the PTC taste receptor, *TAS2R38*, was identified in 2003. Sequencing analysis revealed that three nucleotide positions within the *TAS2R38* gene vary within the human population—each variable position is termed a SNP. One specific combination of the three SNPs, termed a haplotype, correlates most strongly with PTC-tasting ability.

Analogous variation in other cell surface molecules influences the activity of many drugs. For example, SNPs in serotonin transporter and receptor genes predict adverse responses to antidepression drugs, including PROZAC and Paxil.

In this experiment, a sample of human cells is obtained by saline mouthwash (alternatively, DNA may be isolated from hair sheaths). DNA is extracted by boiling with Chelex resin, which binds contaminating metal ions. PCR is then used to amplify a short region of the *TAS2R38* gene. The amplified PCR product is digested with the restriction enzyme HaeIII, whose recognition sequence includes one of the PTC-tasting SNPs. One allele is cut by the enzyme and one is not—producing restriction-fragment-length polymorphisms (RFLPs) that are readily separated by gel electrophoresis.

Students score their genotypes, predict their tasting ability, and then taste PTC paper. Class results show how well PTC-tasting actually conforms to classical Mendelian inheritance and illustrates the modern concept of pharmacogenetics—where a SNP genotype is used to predict drug response.

The laboratory includes bioinformatics exercises that complement the experimental methods. BLAST is used to identify sequences in biological databases, including the human PTC-taster and -nontaster alleles, and the *TAS2R38* gene from several primate species. Another bioinformatics tool, Map Viewer, is used to discover the chromosome location of the *TAS2R38* gene. Finally, to identify mutations and examine evolutionary patterns in the *TAS2R38* gene, multiple sequence alignments are performed using the CLUSTAL W algorithm.

FURTHER READING

Blakeslee AF. 1932. Genetics of sensory thresholds: Taste for phenyl thio carbamide. *Proc Natl Acad Sci* **18:** 120–130.

Fox AL. 1932. The relationship between chemical constitution and taste. *Proc Natl Acad Sci* **18:** 115–120.

Kim U, Jorgenson E, Coon H, Leppert M, Risch N, Drayna D. 2003. Positional cloning of the human quantitative trait locus underlying taste sensitivity to phenylthiocarbamide. *Science* **299:** 1221–1225.

Mueller KL, Hoon MA, Erlenbach I, Chandrashekar J, Zuker CS, Ryba NJP. 2005. The receptors and coding logic for bitter taste. *Nature* **434:** 225–229.

Scott K. 2004. The sweet and the bitter of mammalian taste. *Curr Opin Neurobiol* **14:** 423–427.

PLANNING AND PREPARATION

The following table will help you to plan and integrate the different experimental methods.

Experiment	Day	Time		Activity
I. Isolate DNA from cheek cells	1	60 min	Prelab:	Prepare and aliquot saline solution. Prepare and aliquot 10% Chelex. Make centrifuge adapters. Set up student stations.
		30 min	Lab:	Isolate student DNA.
I. (Alternate) Isolate DNA from hair sheaths	1	30 min	Prelab:	Aliquot proteinase K. Set up student stations.
		30 min	Lab:	Isolate student DNA.
II. Amplify DNA by PCR	2	15 min	Prelab:	Aliquot PTC primer/loading dye mix. Set up student stations.
		15 min	Lab:	Set up PCRs.
		60–150 min	Postlab:	Amplify DNA in thermal cycler.
III. Digest PCR products with HaeIII	3	30–60 min	Prelab:	Aliquot HaeIII restriction enzyme. Set up student stations.
		45 min	Lab:	Set up HaeIII restriction digests.
		30 min	Postlab:	Incubate restriction digests at 37°C.
IV. Analyze PCR products by gel electrophoresis	4	30 min	Prelab:	Dilute TBE electrophoresis buffer. Prepare agarose gel solution. Set up student stations.
		30 min	Lab:	Cast gels.
	5	45+ min	Lab:	Load DNA samples into gel. Electrophorese samples. Photograph gels.

OVERVIEW OF EXPERIMENTAL METHODS

I. ISOLATE DNA FROM CHEEK CELLS

RINSE mouth with saline

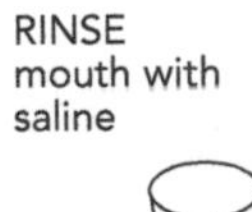

TRANSFER saline

CENTRIFUGE

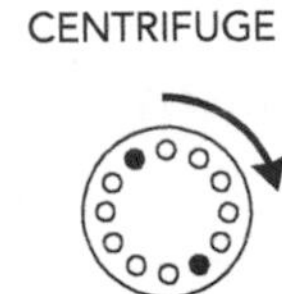

POUR OFF supernatant

RESUSPEND

TRANSFER cell suspension into Chelex

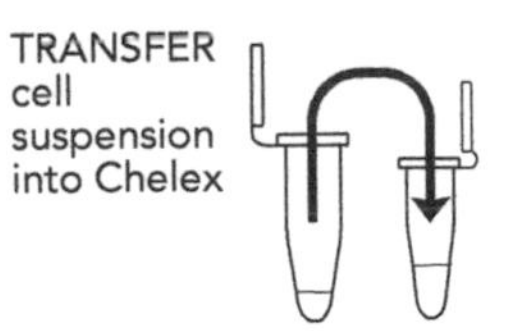

BOIL in thermal cycler

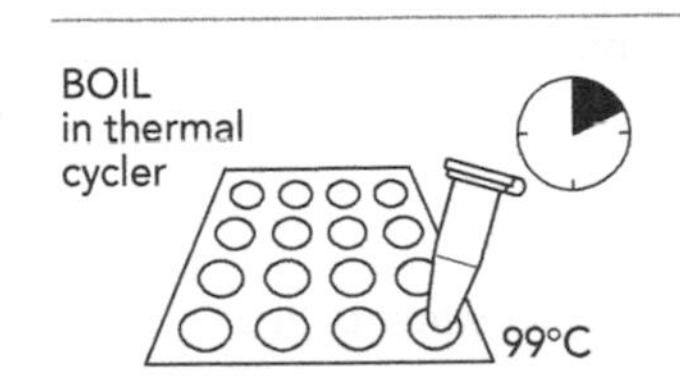

SHAKE vigorously

CENTRIFUGE

TRANSFER supernatant

STORE on ice

I. (ALTERNATE) ISOLATE DNA FROM HAIR SHEATHS

CUT hairs

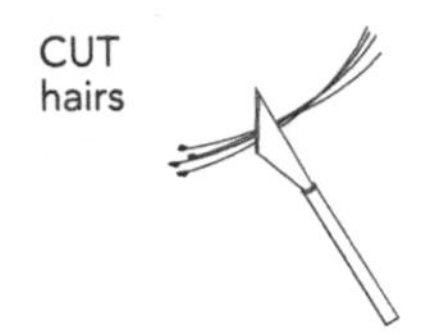

ADD hairs to proteinase K

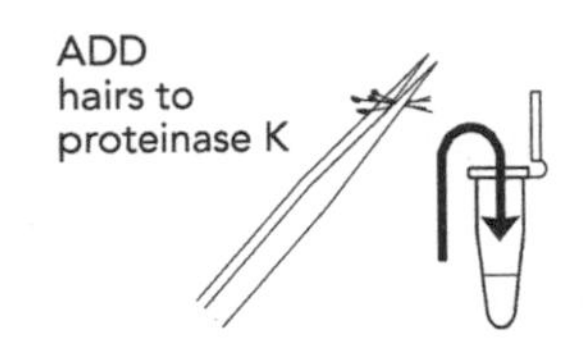

INCUBATE in thermal cycler

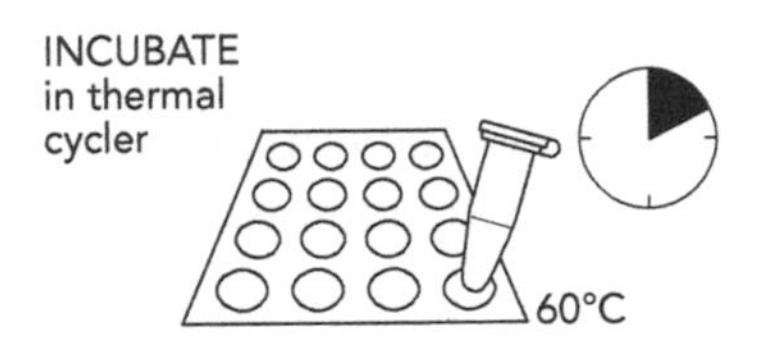

VORTEX

BOIL in thermal cycler

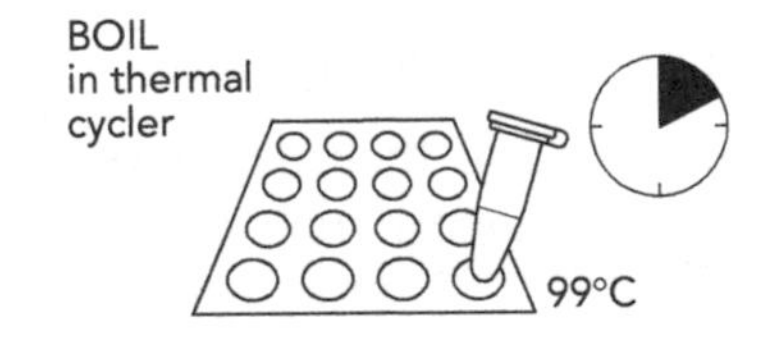

MIX by pipetting in and out

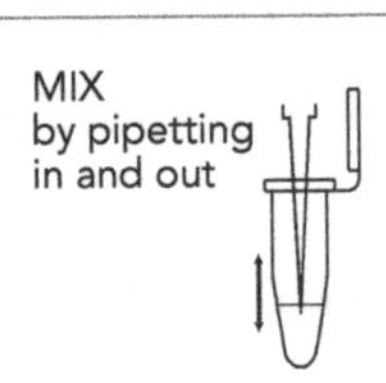

STORE on ice

II. AMPLIFY DNA BY PCR

ADD primer/loading dye mix

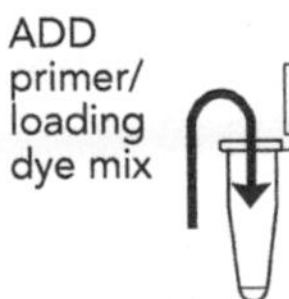

ADD DNA

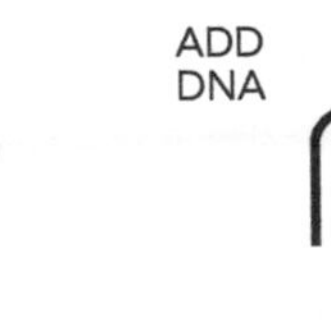

AMPLIFY in thermal cycler

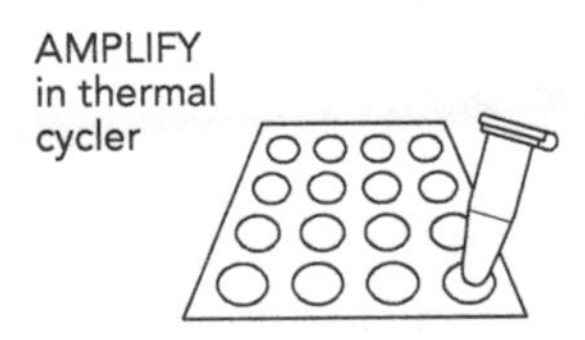

III. DIGEST PCR PRODUCTS WITH HAEIII

TRANSFER PCR product

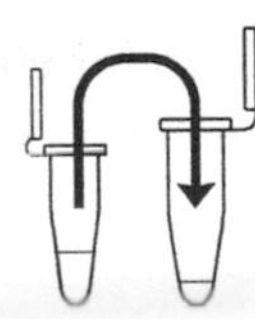

ADD HaeIII

MIX

INCUBATE in thermal cycler

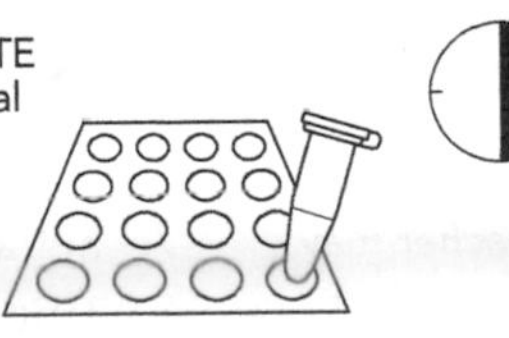

IV. ANALYZE PCR PRODUCTS BY GEL ELECTROPHORESIS

POUR gel

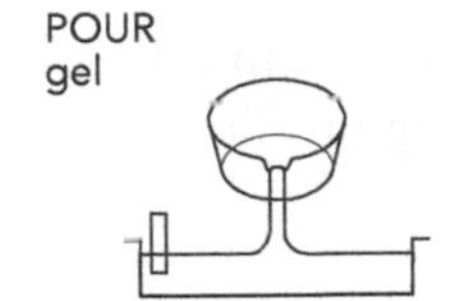

SET 20 min

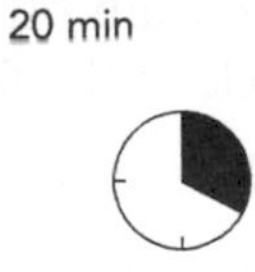

LOAD gel

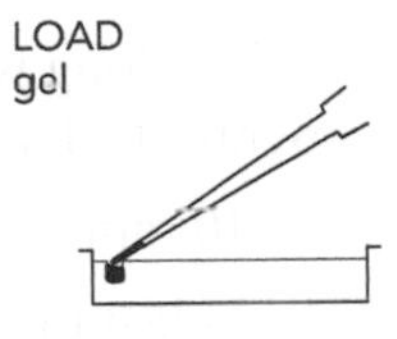

ELECTROPHORESE 130V

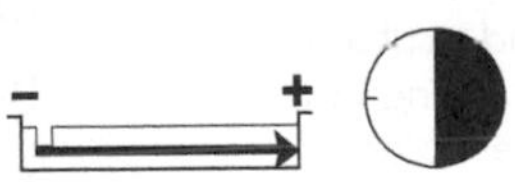

EXPERIMENTAL METHODS

I. Isolate DNA from Cheek Cells

REAGENTS, SUPPLIES, & EQUIPMENT

For each student

10% Chelex (100 μL) in a 0.2- or 0.5-mL PCR tube (or in a microcentrifuge tube)
2 Microcentrifuge tubes (1.5 mL)
Paper cup
Saline solution (10 mL) in a 15-mL tube

For each group

Container with cracked or crushed ice
Microcentrifuge adapters for 0.2- or 0.5-mL PCR tubes
Microcentrifuge tube rack
Micropipettes and tips (10–1000 μL)
Permanent marker
Vortexer (optional)

To share

Microcentrifuge
Thermal cycler (or water bath or heat block)

1. Use a permanent marker to label a 1.5-mL microcentrifuge tube and paper cup with your assigned number.
2. Pour all 10 mL of saline solution into your mouth and vigorously rinse your cheek pockets for 30 sec.
3. Expel the saline solution into the paper cup.
4. Swirl the cup gently to mix the cells that may have settled to the bottom. Use a micropipette with a fresh tip to transfer 1000 μL of the solution into your labeled 1.5-mL microcentrifuge tube.
5. Place your sample tube, along with those from other students, in a balanced configuration in a microcentrifuge. Centrifuge the tubes at full speed for 90 sec.
6. After centrifuging, check to see if your pellet is firmly attached to the bottom of the tube.
   - If the pellet is firmly attached, carefully pour off the supernatant into the paper cup. Gently tap the inverted tube against the cup to remove most of the supernatant, but be careful not to disturb the cell pellet at the bottom of the tube.
   - If the pellet is loose or unconsolidated, centrifuge again for 90 sec. Then, carefully use a micropipette with a fresh tip to remove as much saline as possible.

If your microcentrifuge tube is graduated, the volume remaining after Step 6 will approximately reach the 0.1-mL mark.

7. Set a micropipette to 30 μL. Resuspend the cells in the remaining saline by pipetting in and out. Work carefully to minimize bubbles.

Food particles will not resuspend. Your teacher may instruct you to examine a sample of the cell suspension under a microscope.

8. Withdraw 30 μL of cell suspension and add it to a PCR tube containing 100 μL of 10% Chelex. Label the cap and side of the tube with your assigned number.
9. Place your PCR tube, along with those from other students, in a thermal cycler that has been programmed for one cycle of the following profile:

   Boiling step: 10 min 99°C

   The profile may be linked to a 4°C hold program.

As an alternative to Steps 8 and 9, you may add the cell suspension to Chelex in a 1.5-mL tube and incubate for 10 min in a boiling water bath or 99°C heat block.

10. After boiling, vigorously shake your tube for 5 sec. Use a vortexer if available.

The near-boiling temperature lyses the cell and nu-

clear membranes, releasing DNA and other cell contents.

Skip Step 11 if your sample is in a microcentrifuge tube (not in a PCR tube).

11. To prepare your PCR tube for centrifugation, "nest" it within adapter tubes as follows:
    - If your sample is in a 0.5-mL PCR tube, "nest" it within a capless 1.5-mL tube.
    - If your sample is in a 0.2-mL PCR tube, "nest" it within a capless 0.5-mL tube and then place both tubes into a capless 1.5-mL tube.
12. Place your tube, along with those from other students, in a balanced configuration in a microcentrifuge. Centrifuge the tubes at full speed for 90 sec.
13. After centrifugation, use a micropipette with a fresh tip to transfer 30 µL of the clear supernatant into a clean 1.5-mL microcentrifuge tube. Be careful to avoid pipetting any cell debris or Chelex beads.
14. Use a permanent marker to label the cap and side of the tube with your assigned number. This tube contains your DNA and will be used for setting up the PCRs in Part II.
15. Store your sample on ice or at –20°C until you are ready to continue with Part II.

I. (Alternate) Isolate DNA from Hair Sheaths

REAGENTS, SUPPLIES, & EQUIPMENT

For each student

100-mg/ml proteinase K (100 µL) <!> in a 0.2- or 0.5-mL PCR tube (or in a microcentrifuge tube)*

For each group

Container with cracked or crushed ice
Forceps or tweezers
Hand lens or dissecting microscope
Microcentrifuge tube rack

Micropipettes and tips (100 µL)
Permanent marker
Scalpel or razor blade
Vortexer (optional)

To share

Thermal cycler (or water bath or heat block)

*Store on ice.
See Cautions Appendix for appropriate handling of materials marked with <!>.

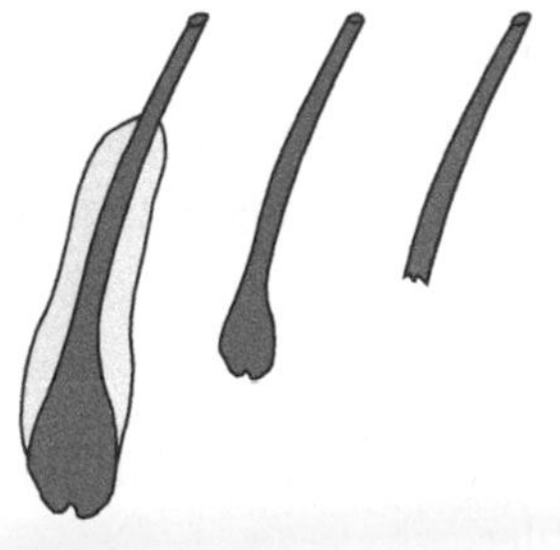

Your teacher may instruct you to prepare a hair sheath to observe under a compound light microscope.

As an alternative to Steps 4–6, add the hairs to proteinase K in a 1.5-mL microcentrifuge tube and incubate in a water bath or

1. Pull out several hairs. Use a hand lens or dissecting microscope to inspect them for the presence of a sheath. The sheath is a barrel-shaped structure surrounding the base of the hair and is most easily observed on dark hair. By holding the hair up to a light source, you may observe the glistening sheath with your naked eye.
2. Select one to several hairs with good sheaths. Alternatively, select hairs with the largest roots. Broken hairs, without roots or sheaths, will not yield enough DNA for amplification.
3. Use a fresh razor blade or scalpel to cut off each hair shaft just above the sheath.
4. Use forceps or tweezers to transfer the hairs to a PCR tube containing 100 µL of proteinase K. Make sure that the sheath is submerged in the solution and not stuck to the wall of the tube.
5. Use a permanent marker to label the cap and the side of your PCR tube with your assigned number.

heat block for 10 min at 60°C. Then, in Step 8, incubate the tube for 10 min in a boiling water bath or 99°C heat block.

6. Place your PCR tube, along with those of other students, in a thermal cycler that has been programmed for one cycle of the following profile:

   Incubation step: 10 min 60°C

7. Place your sample tube at room temperature. To dislodge the cells from the hair shaft, vortex your tube by machine or vigorously with a finger for 15 sec.
8. Place your PCR tube, along with those of other students, in a thermal cycler that has been programmed for one cycle of the following profile:

   Boiling step: 10 min 99°C

   The profile may be linked to a 4°C hold program.

9. Place your sample tube at room temperature. Use a micropipette and fresh tip to mix your sample by pipetting in and out for 15 sec. This sample contains your DNA and will be used for setting up the PCRs in Part II.
10. Store your sample on ice or at –20°C until you are ready to begin Part II.

II. Amplify DNA by PCR

REAGENTS, SUPPLIES, & EQUIPMENT

For each student

Cheek cell or hair sheath DNA* from Part I
PTC primer/loading dye mix (25 µL)*
Ready-To-Go PCR Bead in a 0.2- or 0.5-mL PCR tube

For each group

Container with cracked or crushed ice

Microcentrifuge tube rack
Micropipette and tips (1–100 µL)
Permanent marker

To share

Thermal cycler

*Store on ice.

1. Obtain a PCR tube containing a Ready-To-Go PCR Bead. Label the tube with your assigned number.
2. Use a micropipette with a fresh tip to add 22.5 µL of PTC primer/loading dye mix to the tube. Allow the bead to dissolve for approximately 1 min.

The primer/loading dye mix will turn purple as the PCR bead dissolves.

3. Use a micropipette with a fresh tip to add 2.5 µL of your DNA (from Part I) directly into the primer/loading dye mix. Ensure that no DNA remains in the tip after pipetting.

If the reagents become splattered on the wall of the tube, pool them by pulsing the sample in a microcentrifuge or by sharply tapping the tube bottom on the lab bench.

4. Store your sample on ice until your class is ready to begin thermal cycling.
5. Place your PCR tube, along with those of other students, in a thermal cycler that has been programmed for 30 cycles of the following profile:

   Denaturing step: 30 sec 94°C
   Annealing step: 45 sec 64°C
   Extending step: 45 sec 72°C

   The profile may be linked to a 4°C hold program after the 30 cycles are completed.

6. After thermal cycling, store the amplified DNA on ice or at –20°C until you are ready to continue with Part III.

III. Digest PCR Products with HaeIII

REAGENTS, SUPPLIES, & EQUIPMENT

For each student

Microcentrifuge tube (1.5 mL)
PCR product* from Part II
Micropipette and tips (1–20 µL)
Permanent marker

For each group

Container with cracked or crushed ice
HaeIII restriction enzyme (10 µL)*
Microcentrifuge tube rack

To share

Microcentrifuge (optional)
Thermal cycler (or water bath or heat block)

*Store on ice.

1. Label a 1.5-mL microcentrifuge tube with your assigned number and with a "U" (for "undigested").
2. Use a micropipette with a fresh tip to transfer 10 µL of your PCR product (from Part II) to the "U" tube. Store this sample on ice or at –20°C until you are ready to begin Part IV.

The DNA in the tube labeled "U" will not be digested with the restriction enzyme HaeIII.

3. Use a micropipette with a fresh tip to add 1 µL of the restriction enzyme HaeIII directly into the PCR product remaining in the PCR tube. Label this tube with a "D" (for "digested").
4. Mix and pool the reagents in the tube by pulsing in a microcentrifuge or by sharply tapping the tube bottom on the lab bench.
5. Place your PCR tube, along with those from other students, in a thermal cycler that has been programmed for one cycle of the following profile:

   Digesting step: 30 min 37°C

   The profile may be linked to a 4°C hold program.

As an alternative to Step 5, you may incubate the reaction in a 37°C water bath or heat block. Thirty minutes is the minimum time needed for complete digestion. If time permits, incubate the reaction for 1 or more hours. A lengthy incubation helps to ensure that the restriction enzymes digest the PCR products completely. A partial digest can make homozygous individuals appear heterozygous, confounding the analysis.

6. Store your sample on ice or at –20°C until you are ready to begin Part IV.

IV. Analyze PCR Products by Gel Electrophoresis

REAGENTS, SUPPLIES, & EQUIPMENT

For each student

HaeIII-digested (D) PCR product* from Part III
Latex gloves
SYBR Green <!> DNA stain (10 µl)
Undigested (U) PCR product* from Part II
Microcentrifuge tube rack
Micropipette and tips (1–100 µL)
pBR322/BstNI marker (20 µL per gel)*
1x TBE buffer (300 mL per gel)

For each group

2% Agarose in 1x TBE (hold at 60°C) (50 mL per gel)
Container with cracked or crushed ice
Gel-casting tray and comb
Gel electrophoresis chamber and power supply
Masking tape

To share

Digital camera or photodocumentary system
UV transilluminator <!> and eye protection
Water bath for agarose solution (60°C)

*Store on ice.
See Cautions Appendix for appropriate handling of materials marked with <!>.

1. Seal the ends of the gel-casting tray with masking tape and insert a well-forming comb.

Avoid pouring an overly thick gel, which will be more difficult to visualize.

2. Pour the 2% agarose solution into the tray to a depth that covers about one-third the height of the open teeth of the comb.

The gel will become cloudy as it solidifies.

3. Allow the gel to completely solidify; this takes approximately 20 min.

Do not add more buffer than necessary. Too much buffer above the gel channels electrical current over the gel, increasing the running time.

4. Remove the masking tape, place the gel into the electrophoresis chamber, and add enough 1x TBE buffer to cover the surface of the gel.
5. Carefully remove the comb and add additional 1x TBE buffer to fill in the wells and just cover the gel, creating a smooth buffer surface.
6. Orient the gel according to the diagram in Step 8, so that the wells are along the top of the gel.

A 100-bp ladder may also be used as a marker.

7. Add 2 µl of SYBR Green DNA stain to the undigested (U) PCR product and add 2 µl of SYBR Green DNA stain to the digested (D) PCR product. In addition, add 2 µl of SYBR Green DNA stain to 20 µl of pBR322/BstNI marker.

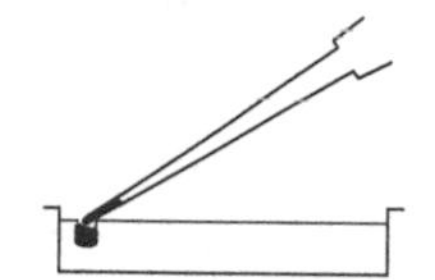

8. Use a micropipette with a fresh tip to load each sample from Step 7 into your assigned wells, according to the following diagram:

Expel any air from the tip before loading. Be careful not to push the tip of the pipette through the bottom of the sample well.

MARKER	STUDENT 1		STUDENT 2		STUDENT 3	
pBR322/ BstNI	U	D	U	D	U	D

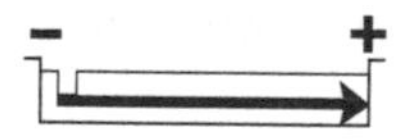

Transillumination, where the light source is below the gel, increases brightness and contrast.

9. Run the gel for approximately 30 min at 130 V. Adequate separation will have occurred when the cresol red dye front has moved at least 50 mm from the wells.
10. View the gel using UV transillumination. Photograph the gel using a digital camera or photodocumentary system.

BIOINFORMATICS METHODS

For a better understanding of the experiment, perform the following bioinformatics exercises before you analyze your results.

I. Use BLAST to Find DNA Sequences in Databases (Electronic PCR)

1. Perform a BLAST search as follows:
   i. Do an Internet search for "ncbi blast."
   ii. Click on the link for the result "BLAST: Basic Local Alignment Search Tool." This will take you to the Internet site of the National Center for Biotechnology Information (NCBI).
   iii. Click on the link "nucleotide blast" (blastn) under the heading "Basic BLAST."
   iv. Enter both primer sequences into the search window. These are the query sequences.

> The following primer set was used in the experiment:
>
> 5′-CCTTCGTTTTCTTGGTGAATTTTTGGGATGTAGTGAAGAGG CGG-3′ (forward primer)
>
> 5′-AGGTTGGCTTGGTTTGCAATCATC-3′ (reverse primer)

v. Omit any nonnucleotide characters from the window because they will not be recognized by the BLAST algorithm.

vi. Under "Choose Search Set," select the "Nucleotide collection (nr/nt)" database from the drop-down menu.

vii. Under "Program Selection," optimize for "Somewhat similar sequences (blastn)."

viii. Click on "BLAST." This sends your query sequences to a server at NCBI in Bethesda, Maryland. There, the BLAST algorithm will attempt to match the primer sequences to the millions of DNA sequences stored in its database. While searching, a page showing the status of your search will be displayed until your results are available. This may take only a few seconds or more than 1 min if many other searches are queued at the server.

2. Analyze the results of the BLAST search, which are displayed in three ways as you scroll down the page:

   i. First, a graphical overview illustrates how significant matches (hits) align with the query sequence. Matches of differing lengths are indicated by color-coded bars. What do you notice about the lengths (and colors) of the matches (bars) as you look from the top to the bottom?

   ii. This is followed by a list of significant alignments (hits) with links to the corresponding accession numbers. (An accession number is a unique identifier given to a sequence when it is submitted to a database such as GenBank.) Note the scores in the "E value" column on the right. The Expectation or E value is the number of alignments with the query sequence that would be expected to occur by chance in the database. The lower the E value, the higher the probability that the hit is related to the query. For example, an E value of 1 means that a search with your sequence would be expected to turn up one match by chance. Longer query sequences generally yield lower E values. An alignment is considered significant if it has an E value of less than 0.1. What is the E value of the most significant hit and what does it mean? Note the names of any significant alignments that have E values of less than 0.1. Do they make sense? What do they have in common?

   iii. Third is a detailed view of each primer (query) sequence aligned to the nucleotide sequence of the search hit (subject, abbreviated "Sbjct"). Note that the first match to the forward primer (nucleotides 1–42) and to the reverse primer (nucleotides 44–68) are within the same subject (accession number).

3. Click on the accession number link to open the data sheet for the first hit (the first subject sequence).

   i. At the top of the report, note basic information about the sequence, including its length (in base pairs, or bp), database accession number, source, and references to papers in which the sequence is published. What is the source and size of the sequence in which your BLAST hit is located?

   ii. In the middle section of the report, the sequence features are annotated, with their beginning and ending nucleotide positions ("xx..xx"). These features may include genes, coding sequences (CDS), regulatory regions, ribosomal RNA (rRNA), and transfer RNA (tRNA).

iii. Scroll to the bottom of the data sheet. This is the nucleotide sequence to which the term "Sbjct" refers.

iv. Use your browser's "Back" button to return to the BLAST results page.

4. Predict the length of the product that the primer set would amplify in a PCR (in vitro) as follows:

   i. Scroll down to the alignments section (third section) of the BLAST results page. Examine the alignments with the lowest E values and identify a human (*Homo sapiens*) sequence to which both primers align.

   ii. To which positions do the primers match in the subject sequence? Note that position 43 of the forward primer is missing. What does this mean?

   iii. The lowest and highest nucleotide positions in the subject sequence indicate the borders of the amplified sequence. Subtract the lowest nucleotide position in the subject sequence from the highest nucleotide position in the subject sequence. What is the difference between the coordinates?

   iv. Note that the actual length of the amplified fragment includes both ends, so add 1 nucleotide to the result that you obtained in Step 4.iii to obtain the exact length of the PCR product amplified by the two primers.

5. Obtain the nucleotide sequence of the amplicon that the primer set would amplify in a PCR (in vitro) as follows:

   i. Open the sequence data sheet for the hit that you identified in Step 4.i by clicking on the accession number link. Scroll to the bottom of the data sheet.

   ii. The bottom section of the data sheet lists the entire nucleotide sequence that contains the PCR product. Highlight all of the nucleotides between the coordinates that you identified in Step 4.iii, from the beginning of the forward primer to the end of the reverse primer.

   iii. Copy and paste the highlighted sequence into a text document. Then, delete all nonnucleotide characters and spaces. This is the amplicon, or amplified PCR product. Save this text document for use in Step 2.iii of Part III.

In Step 5.iii, you can retain the nucleotide coordinates and spacers to ease readability. Reduce the point size of the font so that each row of 60 nucleotides sits on one line. Then, set the font to Courier or another nonproportional font to align the blocks of sequence.

II. Use Map Viewer to Determine the Chromosomal Location of the *TAS2R38* Gene

1. Open Map Viewer (http://www.ncbi.nlm.nih.gov/mapview).
2. Find *Homo sapiens* in the table to the right and click on the "B" icon under the "Tools" header. If more than one build is displayed, select the one with the highest number; this will be the most recent version.
3. Paste the primer sequences from Step 1.iv of Part I into the search window. Omit any nonnucleotide characters from the window because they will not be recognized by the BLAST algorithm.
4. Select "BLASTN: Compare nucleotide sequences" from the drop-down menu under "Program" and click on "Begin Search."
5. Click on "View report" to retrieve the results.
6. In "Other reports," click on the "[Human genome view]" link to see the chromo-

some location of the BLAST hit. Small horizontal bars on chromosomes indicate the positions of hits. On which chromosome have you landed?

7. Click on the marked chromosome number to move to the *TAS2R38* locus.
8. The chromosome is represented by one or more vertical lines. Click on the small blue arrow after the vertical line labeled "Genes_seq" to display the genes. The *TAS2R38* gene occupies the whole field of the default view. What is the function of the *TAS2R38* gene? Click on the name of the gene under the "Symbol" track and then follow the links to find out.
9. Use the zoom toggle on the left to "show 1/1,000th of chromosome." Now you can see the chromosomal region surrounding *TAS2R38* and its nearest gene "neighbors." What genes are found on either side of *TAS2R38*?
10. Introns and noncoding sequences of the genes are denoted by thin lines, and exons are denoted by thick bars along the line.
    i. Determine the size of the *TAS2R38* gene using the map coordinates to the left of the "Contig" map.
    ii. How many introns and exons does the *TAS2R38* gene have?
    iii. How do the structures of the neighboring genes differ from the structure of the *TAS2R38* gene?
11. What are the functions of the neighboring genes? Click on their names under the "Symbol" track and then follow the links to find out.
12. Click on the blue arrow at the top of the chromosome image to scroll up the chromosome. Look at the names of each of the genes. Then scroll up one more screen and look at the names of those genes. What do some of these genes have in common with *TAS2R38*? What can you conclude about the chromosomal locations of genes with similar functions?

III. Use Multiple Sequence Alignment to Explore the Evolution of the *TAS2R38* Gene

In this exercise, you will compare the following *TAS2R38* nucleotide sequences:

- PTC amplicon from Step 5.iii of Part I
- Human (*Homo sapiens*) PTC-taster allele
- Human (*Homo sapiens*) PTC-nontaster allele
- Chimpanzee (*Pan troglodytes*)
- Bonobo (*Pan paniscus*)
- Gorilla (*Gorilla gorilla*)

1. Return to your original BLAST results from Part I or repeat Step 1 of Part I to obtain a list of significant alignments.
2. Obtain the last five *TAS2R38* nucleotide sequences listed above by performing the following steps for each sequence:
    i. Examine the list of significant alignments with the lowest E-values and select the appropriate hit. Use only entries listed as "complete CDS" ("CDS" stands for "coding sequence").

ii. Click on the accession number link to open the sequence data sheet for the appropriate hit. The bottom section of the report lists the entire nucleotide sequence. Highlight the entire sequence.

iii. Copy and paste the highlighted sequence into the same text document that you created in Step 5.iii of Part I above. Make sure to identify each sequence in the text document so you remember which species or allele it represents. Then, delete all nonnucleotide characters and spaces.

iv. Repeat Steps 2.i–2.iii for the remaining sequences. After you have gathered all five sequences, save the text document for use in Step 5.

3. Open the *BioServers* Internet site at the Dolan DNA Learning Center (www.bioservers.org).

4. Enter *Sequence Server*. You can register if you want to save your work for future reference, but it is not required.

5. The interface is simple to use: Add or obtain data using the top buttons and pull-down menus, then work with the data in the work space below. Enter your nucleotide sequences for comparison as follows:

   i. Click on "CREATE SEQUENCE" at the top of the page.

   ii. Copy the first *TAS2R38* sequence from Step 2.iv, and paste it into the "Sequence" window. Enter a name for the sequence and click "OK." Your new sequence will appear in the work space at the bottom half of the page.

   iii. Repeat Steps 5.i and 5.ii for each of the human and primate sequences from Step 2.iv. In addition, enter the sequence of the forward primer used in your PCR amplification (from Step 1.iv of Part I) and the sequence of the PTC amplicon (from Step 5.iii of Part I).

6. In Steps 7–10 below, you will compare the following sets of sequences:

   - Human PTC-taster allele, human PTC-nontaster allele, and PTC amplicon (Step 7).
   - Human PTC-taster allele and human PTC-nontaster allele (Step 8).
   - Human PTC-taster allele, human PTC-nontaster allele, chimpanzee, bonobo, and gorilla (Step 9).
   - Forward primer, human PTC-taster allele, and human PTC-nontaster allele (Step 10).

   To compare the sequences, perform the following steps:

   i. Click on the check boxes to the left of the sequences that you wish to compare.

   ii. Click on "COMPARE" in the gray bar. (The default operation is a multiple sequence alignment using the CLUSTAL W algorithm.) The checked sequences are sent to a server at Cold Spring Harbor Laboratory, where the CLUSTAL W algorithm will attempt to align each nucleotide position. This may take only a few seconds or more than a minute if a lot of other searches are queued at the server.

   iii. The results will appear in a new window. Examine the sequences, which are displayed in rows of 25 nucleotides. The alignment may begin with yellow-shaded dashes (–), indicating an initial stretch of nucleotides that is present in one of the samples but not in the other. This occurs because the sequence read usually begins at different points in different samples. Yellow highlighting also denotes mismatches between the sequences. A gray-shaded "N" indicates a sequence error, a position in one or both sequences where a nucleotide could not be determined.

iv. By default, only the first 500 bp of the alignment are shown. To view the entire alignment, enter "1100" as the number of nucleotides to display per page and then click "Redraw."

7. Compare the human PTC-taster allele, the human PTC-nontaster allele, and the PCR amplicon by performing Steps 6.i–6.iv.

   i. Why is the initial stretch of the alignment (as well as the last stretch of the alignment) highlighted in yellow?

   ii. At what positions does the amplicon track along with the two human alleles?

   iii. At what position in the sequence is the SNP that you examined in this experiment? What is the nucleotide difference between the taster and nontaster alleles?

8. Compare the human PTC-taster allele to the human PTC-nontaster allele by performing Steps 6.i–6.iv.

   i. In the left column of the table below, list the nucleotide position(s) of all SNP(s) that you see in the sequence, including the SNP that you identified in Step 7.iii. Then, in the columns labeled "Nucleotide," record the nucleotides that are present for each allele. What is one combination of three nucleotides called?

   ii. Print the alignment. Then, using the printed copy, count triplets of nucleotides from the initial ATG start codon. In the table below (under "Codon"), list the codon(s) that are affected by the SNP(s).

   iii. Use a standard genetic code chart to determine which amino acids use the codons that you identified in Step 8.ii. Record your findings in the table below (under "Amino acid").

SNP position	PTC-taster allele			PTC-nontaster allele		
	Nucleotide	Codon	Amino acid	Nucleotide	Codon	Amino acid

9. Compare the human PTC-taster, human PTC-nontaster, chimpanzee, bonobo, and gorilla sequences by performing Steps 6.i–6.iv.

   i. In the sequence alignment, find the SNP positions that you identified in the table above. What nucleotide is present at each of these positions in the chimpanzee, bonobo, and gorilla sequences? These nucleotides represent the ancestral (original) state of this gene.

   ii. Based on your findings in Step 9.i, are chimpanzees, bonobos, and gorillas tasters or nontasters? What does this indicate about the functional significance of bitter taste receptors over evolutionary time?

   iii. What patterns do you notice in SNPs at other locations in the gene?

10. Compare the forward primer, the human PTC-taster allele, and the human PTC-nontaster allele by performing Steps 6.i–6.iv.
    i. At what nucleotide positions does the primer bind?
    ii. What discrepancy do you notice between the primer sequence and the human *TAS2R38* sequences? At what position is this discrepancy located? Print and save a copy of the alignment; you will learn how this discrepancy is important for the experiment when you work through Part III of Results and Discussion below.

RESULTS AND DISCUSSION

The diagram below shows how PCR amplification and restriction enzyme digestion identifies the C > G polymorphism in the *TAS2R38* gene. The taster allele, with the C

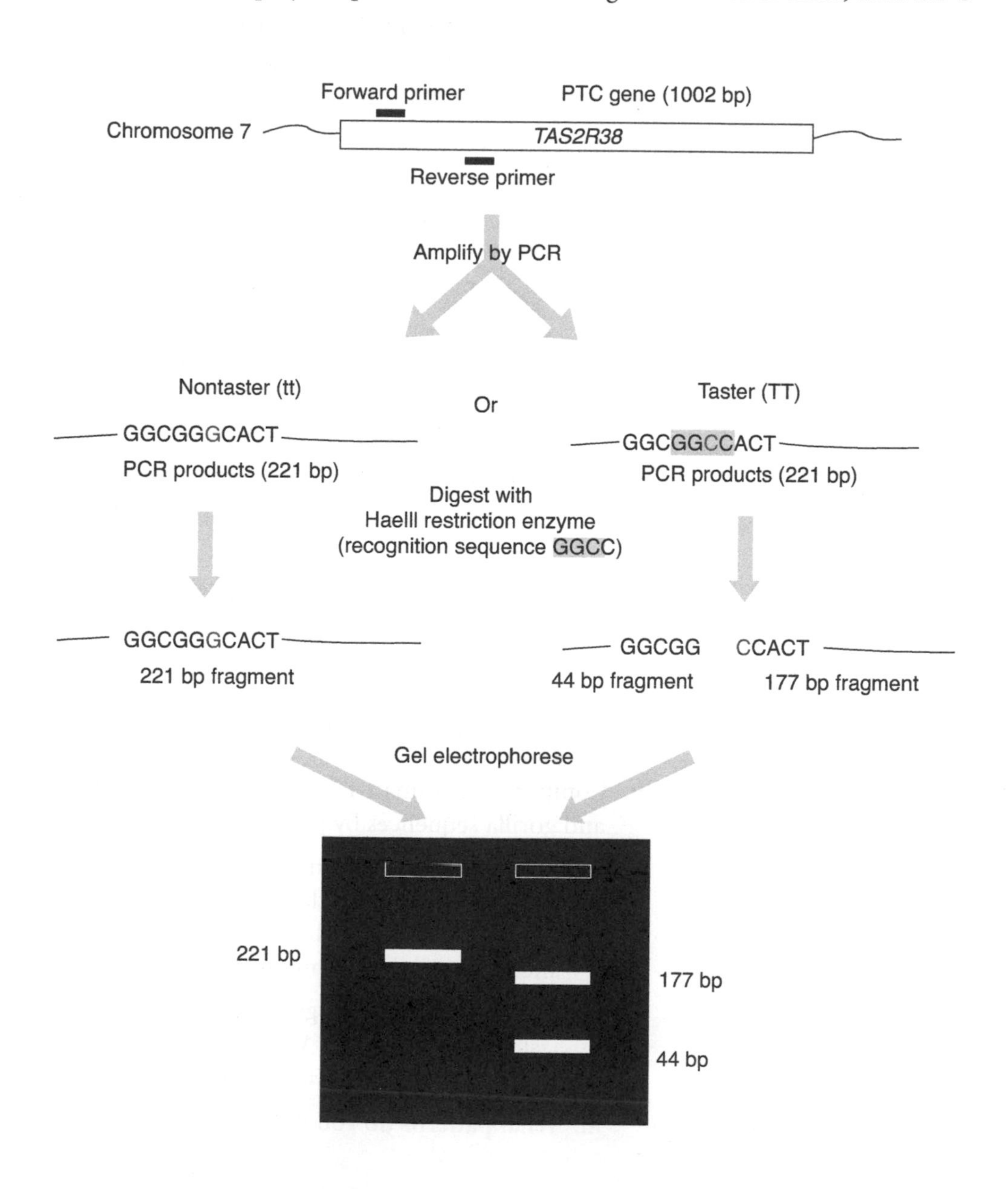

nucleotide, is digested by HaeIII (as shown on the right), whereas the nontaster allele with the G nucleotide is not digested by HaeIII (as shown on the left).

I. Determine Your PTC Genotype

1. Observe the photograph of the stained gel containing your PCR digest and those from other students. Orient the photograph with the sample wells at the top. Use the sample gel shown below to help to interpret the band(s) in each lane of the gel.

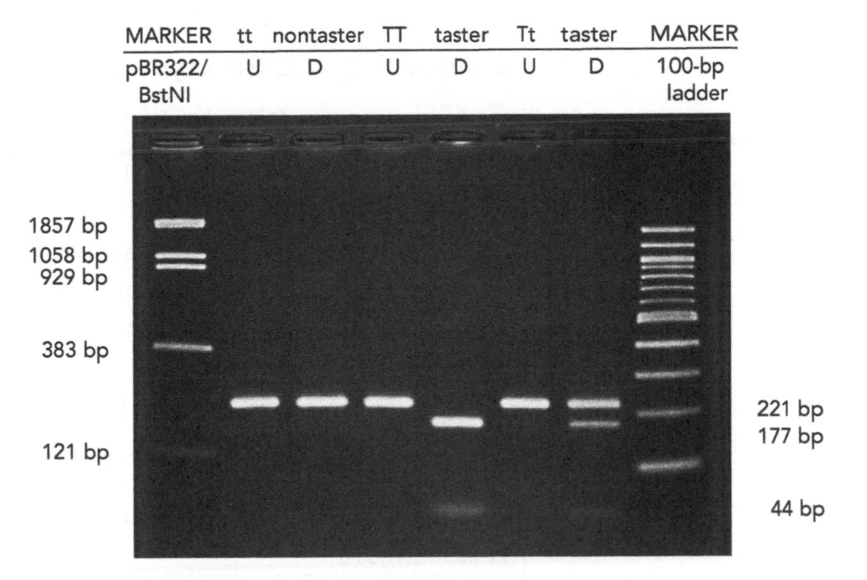

If a 100-bp ladder was used, as shown on the right-hand side of the sample gel, the bands increase in size in 100-bp increments starting with the fastest-migrating band of 100 bp.

2. Locate the lane containing the pBR322/BstNI markers on the left side of the gel. Working from the well, locate the bands corresponding to each restriction fragment: 1857, 1058, 929, 383, and 121 bp. The 1058- and 929-bp fragments will be very close together or may appear as a single large band. The 121-bp band may be very faint or not visible.
3. Locate the lane containing the undigested PCR product (U). There should be one prominent band in this lane. Compare the migration of this band with the migration of the 383- and 121-bp bands in the pBR322/BstNI lane. Confirm that the undigested PCR product corresponds to a size of about 221 bp.
4. Scan across the row of student results. Note that virtually all student lanes contain one to three prominent bands.
   i. A tt nontaster (homozygous recessive) shows a single band in the same position as the uncut control.
   ii. A TT taster (homozygous dominant) shows two bands: 177 and 44 bp. The 177-bp band migrates just ahead of the uncut control; the 44-bp band may be faint. (Incomplete digestion may leave a small amount of uncut product at the 221-bp position, but this band should be clearly fainter than the 177-bp band.)
   iii. A Tt taster (heterozygous) shows three bands that represent both alleles: 221, 177, and 44 bp. The 221-bp band must be stronger than the 177-bp band. (If the 221-bp band is fainter, it is an incomplete digest of TT.)

5. To "score" your genotype, compare your digested PCR product (D) with the markers, the other student results in the row, and the uncut control. The analysis will be simple if your row contains a heterozygous genotype (Tt) that shows the positions of both alleles. Homozygotes of each genotype (TT and tt) will also help.

Additional faint bands at other positions occur when the primers bind to chromosomal loci other than the exact locus and give rise to "nonspecific" amplification products.

6. It is common to see a diffuse (fuzzy) band that runs just ahead of the 44-bp fragment. This is "primer dimer," an artifact of the PCR that results from the primers overlapping one another and amplifying themselves. How would you interpret a lane in which you observe primer dimer, but no bands as described in Steps 3 and 4?

II. Correlate Your PTC Genotype with Your PTC Phenotype

1. First, determine your PTC phenotype as follows:
   i. Place one strip of control taste paper in the center of your tongue for several seconds. Note the taste.
   ii. Remove the control paper and place one strip of PTC taste paper in the center of your tongue for several seconds. How would you describe the taste of the PTC paper as compared to the control: strongly bitter, weakly bitter, or no taste other than paper?
2. Record the class results in the table below.

	Phenotype		
Genotype	Strong taster	Weak taster	Nontaster
TT (homozygous)			
Tt (heterozygous)			
tt (homozygous)			

3. According to your class results, how well does the *TAS2R38* genotype predict the PTC-tasting phenotype? If the genotypes were not predictive of the phenotypes, explain why this might be the case.

III. Discover How the HaeIII Enzyme Discriminates the PTC-taster and -nontaster Alleles

1. HaeIII recognizes and cuts the sequence 5′-GGCC-3′. The forward primer used in this experiment incorporates the first three nucleotides of the HaeIII recognition sequence. Examine the printout of the multiple sequence alignment that you generated in Part III (Step 10.ii) of Bioinformatics Methods. How is the forward primer sequence different from the human *TAS2R38* sequence?
2. What characteristic of the PCR allows the primer sequence to "override" the natural gene sequence? Draw a diagram to support your contention.
3. The forward primer used in this experiment did not incorporate the fourth nucleotide of the HaeIII recognition sequence. Why was this important for the experiment? How did this enable the HaeIII enzyme to discriminate between the taster and nontaster alleles?

IV. Apply and Expand Your Knowledge

1. Define the terms synonymous and nonsynonymous mutation. Is the C > G polymorphism in the *TAS2R38* gene a synonymous or nonsynonymous mutation? By what mechanism does this influence bitter-taste perception?
2. Research other polymorphisms in the *TAS2R38* gene and how they may influence bitter-taste perception.
3. The frequency of PTC nontasting is higher than would be expected if bitter-tasting ability were the only trait upon which natural selection had acted. In 1939, the geneticist R.A. Fisher suggested that the gene controlling PTC tasting is under "balancing" selection, meaning that a possible negative effect of losing this tasting ability is balanced by some positive effect. Under some circumstances, balancing selection can produce heterozygote advantage, where heterozygotes are fitter than homozygous dominant or recessive individuals. What advantage might this be in this case?
4. Research how the methods of DNA typing used in this experiment differ from those used in forensic crime labs. Focus on (a) types of polymorphism used, (b) methods for separating alleles, and (c) methods for ensuring that samples are not mixed up.
5. What ethical issues are raised by human DNA typing experiments?

Laboratory Planning and Preparation

Laboratories 2.1–2.3

STUDENT PARTICIPATION IN EXPERIMENTS INVOLVING HUMAN DNA raises real-life questions about the use of personal genetic data: For what is my DNA sample being used? Does my DNA type tell me anything about my life or health? Can my data be linked personally to me? There is consensus that a human DNA sample be obtained only with the willing consent of a donor, who understands the purpose for which it is being collected. Thus, these experiments should be explained to the donor ahead of time and students should be given the option to refrain from participating. (Some teachers may wish to have parents sign a consent form, such as the one on page 258.) There is also consensus that a DNA sample be used only for the express purpose for which it is collected and that student DNA samples be thrown away after completing the experiment.

The DNA polymorphisms used in these laboratory exercises were specifically selected because they have no known relationship to any disease state or to sex determination. However, the alleles are inherited in a Mendelian fashion and can give indications about family relationships. To avoid the possibility of suggesting inconsistent inheritance, it is best to avoid generating genotypes from parent-child pairs.

The chance that student samples will be mixed up when isolating DNA, setting up polymerase chain reactions (PCRs), and loading electrophoresis gels provides no certainty to any of the genotypes obtained in these experiments. (A forensic laboratory would use approved methods for maintaining "chain of custody" and for tracking samples.) To minimize sample mix-up, each student should be assigned a unique number before starting each laboratory exercise. This will make it easier to mark and identify the tubes used in each experiment.

ISOLATING HUMAN DNA FROM CHEEK CELLS

Saline mouthwash is the most reproducible of the simple methods used to obtain human DNA for PCR. The mouthwash gently loosens a large number of single cells and small clusters of cheek cells. This maximizes the surface area of cells, allowing for virtually complete lysis during boiling. Cheek brushes and swabs generally yield larger clumps of cells, which are less effectively lysed by boiling.

Surprisingly, food particles rinsed out with the mouthwash have little effect on PCR amplification. Still, it is best to avoid eating before the experiment because food particles, especially from fruits, may block the pipette tip and make pipetting difficult.

Parental Consent Form

As part of this activity, your child will have the opportunity to conduct several experiments with his/her own DNA. DNA samples are collected from cells that are normally present in saliva. The student will simply swish his/her mouth with a saline solution and spit the sample into a cup. The DNA samples that are extracted from these cells are amplifed by a process called polymerase chain reaction (PCR) and examined for specific DNA markers, which vary from person to person. The DNA markers we will examine play no role in an individual's health. Student samples will be discarded after completing the experiments in this workshop and will not be used for any other purposes.

There is a consensus that human DNA experiments should not be conducted without the willing consent of the donor, who understands the purpose for which his/her DNA is being used. Thus, these experiments will be explained clearly beforehand, and students will be given the option to not participate.

Please sign below indicating authorization for your child's participation in these experiments.

____

Name of Student
please print

____

Name of Parent/Legal Guardian
please print

____

Parent or Guardian Signature

____

Date

Although saline solution is simple to make, remember that this will be put in the students' mouths. As a failsafe against the possibility of a different (distasteful or even dangerous) salt solution being prepared by an inattentive lab aid, consider using commercially prepared saline.

With careful lab management, up to 90% of students should ultimately be able to score their genotypes using the mouthwash method. Be especially watchful after the initial centrifugation step (Step 5). Most students will have compact pellets, and the supernatant can be poured off. However, about 10% of students will have diffuse or slimy masses that do not pellet well. In these cases, remove as much supernatant as possible using a micropipette and fresh tip. It may be necessary to centrifuge these samples again.

DNA is liberated from cheek cells by boiling in 10% Chelex, which binds the contaminating metal ions that are the major inhibitors of PCR. The boiling step is most easily accomplished using the same thermal cycler used for PCR. To do this, provide each student with 100 μL of 10% Chelex suspension in a PCR tube that is compatible with the thermal cycler you will be using: either 0.2 mL or 0.5 mL. It is not necessary to use a "thin-walled" tube. Alternatively, use 1.5-mL tubes in a heat block or a boiling water bath. To make a simple water bath, maintain a beaker of water at a low boil on a hot plate. Place the 1.5-mL tubes in a floating rack or tightly cover the beaker with a double layer of aluminum foil and use a pencil to punch holes in the aluminum foil to hold the tubes. If using aluminum foil, ensure that the tubes are immersed and add hot water as necessary to maintain the water level. Watch out for lids opening as the tubes heat.

It is a worthwhile diversion to allow students to view their own squamous epithelial cells under a compound microscope. Add several microliters of cell suspension remaining after Step 8 to a microscope slide, add a drop of 1% methylene blue (or other stain), and add a coverslip.

ISOLATING HUMAN DNA FROM HAIR SHEATHS

Hair roots provide the simplest source of DNA for PCR amplification; no special equipment is required for extraction. Hairs are also an extremely safe source of cells. The risk of spreading an infectious agent is minimized by "dry" collection, which does not involve any body fluid nor generate any supernatant. This method also stresses the power of PCR in forensic cases—even one growing hair root provides enough DNA for excellent amplification. However, forensic biologists generally rate hair as a poor source of DNA for analysis, for the same reason that it can prove to be difficult in the classroom: Most plucked or shed hairs are broken off from the root, which is the source of cells for DNA extraction.

The success of this method is entirely dependent on finding large roots from growing hairs. This can be tricky and time-consuming—and often hilarious. With vigilance, up to 80% of students may find hairs with good roots from which to isolate DNA. However, it is more likely that only about 60%–70% of students will ultimately be able to score their genotypes using this method.

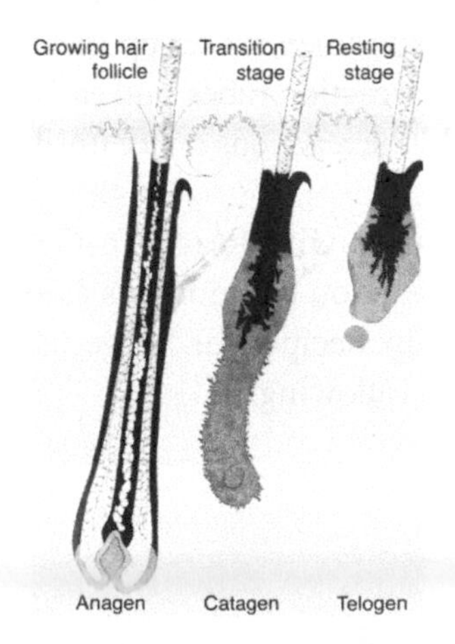

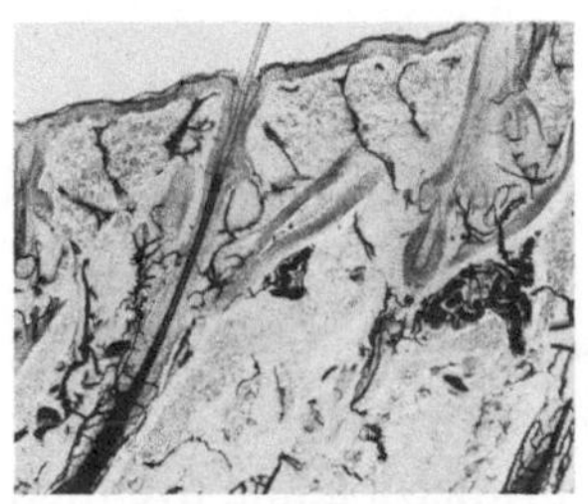

Photo credit: Joel Ito, ONPRC.

Each hair is anchored to the skin by a follicle, or "root," whose growing cells produce the hair shaft. Hair goes through a growth cycle with alternating periods of growth and quiescence, during which the follicle increases and decreases in size. During the growth phase, the follicle extends up the hair shaft in a structure called the sheath, which is a rich source of cells. The sheath membrane is easily digested by treatment with proteinase K, releasing squamous cells singly or in small clusters. A high percentage of these cells is lysed by boiling and releases DNA.

The sheath decreases in size as the hair follicle enters a resting stage (see the drawing and the micrograph of growing and resting follicles). The withered bulb of a resting follicle is, in fact, what most people would consider a "root." Resting follicles usually yield little DNA for analysis. First, there are fewer cells. Second, proteinase K treatment does not effectively digest the shriveled root mass, and only cells at the edge are lysed by boiling.

Successful amplification of a locus, which is available in only two copies per cell, is closely correlated with the presence of a sheath on the hair shaft. One or two hairs with long sheaths will provide plenty of DNA for PCR amplification. Three or four good-sized roots will usually work, especially if they have at least small sheaths.

A good sheath is unmistakable. Especially contrasted on a dark hair, it glistens when held up to the light and extends several millimeters up the hair shaft. Make sure to show off the first several good sheaths that turn up so that other students will know what to look for. Because of the hair growth cycle, most people find sheaths only on some hairs. Students whose hair grows slowly may have difficulty finding sheaths, and thin or brittle hair is likely to break off before the root. If students are having difficulty finding sheaths on hairs pulled from their scalps, have them try hairs from the eyebrow or arm.

Sheaths are the most underrated source of squamous cells for microscopic examination. Give them a try! Simply place a sheath on a microscope slide and add a drop of proteinase K (100 mg/mL). Let it stand for several minutes to allow the proteinase K to digest the sheath membrane. Then add a drop of 1% methylene blue (or other stain) and a coverslip, and gently press to disrupt the sheath membrane. Observe the cells under medium power and at several time points to see the effect of enzyme digestion. If you gently press the coverslip while the slide is on the microscope stage, you should be able to observe squamous cells squirting out of tears in the sheath membrane.

AMPLIFYING DNA BY PCR

Each Ready-To-Go PCR Bead contains reagents so that a final reaction volume of 25 µL contains 2.5 U of *Taq* DNA polymerase, 10 mM Tris-HCl (pH 9.0), 50 mM KCl, 1.5 mM $MgCl_2$, and 200 µM of each dNTP. Each primer/loading dye mix includes the appropriate primer pair (0.26 pmol/µL of each primer), 13.8% sucrose, and 0.0081% cresol red.

The lyophilized *Taq* DNA polymerase in the Ready-To-Go PCR Bead becomes active immediately after adding primer/loading dye mix. In the absence of thermal cycling, nonspecific priming at room temperature allows *Taq* DNA polymerase to begin generating erroneous products that can show up as extra bands in gel analysis. Therefore, work quickly! Make sure that the thermal cycler is set and have all experimenters set up their PCRs as a coordinated effort. Add primer/loading dye mix to all reaction tubes, add each template, and begin thermal cycling as quickly as possible. Hold the reactions on ice until all students are ready to load into the thermal cycler.

For convenience, these laboratories are designed to be used with Ready-To-Go PCR Beads and primer/loading dye mix. If desired, all PCR reaction components can be prepared from scratch by following the recipes included in Recipes for Reagents and Stock Solutions. Each 25-µL reaction should contain the following:

PCR buffer (10x)	2.5 µL
dNTPs (10 mM)	1.0 µL
Forward primer (20 µM)	1.25 µL
Reverse primer (20 µM)	1.25 µL
Taq polymerase (1–5 U/µL)	1.0 µL
Template DNA	2.5 µL
Distilled or deionized water	15.5 µL

Surprisingly, a large number of failures in this experiment are due to the simple fact that students did not get template DNA into their PCRs! Observe each student adding DNA to the primer/loading dye mix. Students must see the small volume of template DNA in their pipette tip, then eject the DNA directly into the primer/loading dye mix, and finally confirm that the tip is empty before removing it from the PCR tube.

If your thermal cycler does not have a heated lid, add a drop of mineral oil on top of each reaction before thermal cycling. Be careful to avoid touching the dropper tip to the tube or sample; this will contaminate the mineral oil.

PCR amplification from crude cell extracts is biochemically demanding and requires the precision of automated thermal cycling. However, amplification of the loci in these experiments is not complicated by the presence of repeated units. Therefore, the recommended amplification times and temperatures will work adequately for most common thermal cyclers, which ramp between temperatures within a single heating/cooling block. A different cycling profile may be requried for extremely fast cyclers (such as the ABI Veriti) or brands of thermal cyclers that physically move PCR tubes between multiple temperature blocks (such as the Stratagene RoboCycler). Consult the manufacturers' recommendations when using these products.

DIGESTING PCR PRODUCTS WITH RESTRICTION ENZYMES

In Laboratory 2.3, restriction digests of PCR products generate restriction-fragment-length polymorphisms (RFLPs) that differentiate the alleles of the *TAS2R38* gene. Restriction digests are not performed in any of the other laboratory exercises.

Before the lab, divide the HaeIII enzyme (1–10 units/µL) into 10-µL aliquots, each of which can be shared by seven to eight students. Keep the aliquots on ice. The PCR buffer provides adequate salt and pH conditions for the HaeIII enzyme, so no additional restriction buffer is required for the reaction. Thirty minutes is the minimum time needed for complete digestion. If time permits, incubate the reactions for 1 or more hours. After several hours, the enzyme will denature and lose activity. Stop the reactions whenever it is convenient and store them in a freezer (–20°C).

ANALYZING DNA BY GEL ELECTROPHORESIS

Preparing and Loading Gels

The cresol red and sucrose in the primer mix function as loading dye, so that amplified samples can be loaded directly into an agarose gel. This is a nice time-saver. However, because the concentrations of sugar and cresol red are low, this mix is more difficult to use than typical loading dyes. Students should be encouraged to load carefully.

If the Ready-To-Go PCR Beads and primer/loading dye mix are not used and, instead, a traditional PCR is run, then 2 µL of 10x loading dye should be added to each reaction before it is run out on the gel.

Plasmid pBR322 digested with the restriction endonuclease BstNI is an inexpensive marker and produces fragments that are useful as size markers in this experiment. Use 20 µL of a 0.1 µg/µL stock solution of this DNA ladder per gel. Other markers or a 100-bp ladder may be substituted.

If mineral oil was used during PCR, pierce the pipette tip through the layer of mineral oil to withdraw the PCR sample. Do not pipette any mineral oil; leave the mineral oil behind in the original tube.

If you are planning to sequence student samples in Laboratory 2.1, make sure that students electrophorese only 15 µL of the amplified product. The remaining 10 µL must be retained for use in cycle sequencing.

Staining Gels

In Part III of Laboratories 2.1 and 2.2 and Part IV of Laboratory 2.3, SYBR Green, a fluorescent staining dye, is added to each sample before electrophoresis. According to the Ames test (a method to estimate the mutagenic properties of a chemical), SYBR Green is much less mutagenic than the classical stain ethidium bromide. Furthermore, bands are immediately visible after electrophoresis and no poststaining steps are necessary. SYBR Green can be imaged with the same filter set and UV transillumination used with ethidium bromide, but SYBR Green is more sensitive than ethidium bromide and produces much less background on a stained agarose gel.

As an alternative to SYBR Green, the gel can be stained with ethidium bromide or *Carolina*BLU after electrophoresis but before viewing and photographing. To stain the gel using ethidium bromide or *Carolina*BLU, omit Step 7 from the procedure (Part III of Laboratories 2.1 and 2.2 and Part IV of Laboratory 2.3) and then, after Step 9, follow one of the methods described under "Staining a Gel with Ethidium Bromide or *Carolina*BLU" below.

Always view and photograph gels as soon as possible after electrophoresis or staining/destaining. Over time, the small-sized PCR products will diffuse through the gel, lose sharpness, and disappear. Refrigeration will slow diffusion somewhat, so if absolutely necessary, gels can be wrapped in plastic wrap and stored for up to 24 h at 4°C. For best results, view and photograph gels immediately after staining/destaining is complete.

Staining a Gel with Ethidium Bromide or Carolina*BLU*

REAGENTS, SUPPLIES, & EQUIPMENT

*Carolina*BLU Final Stain (for *Carolina*BLU staining only)
Digital camera or photodocumentary system
Ethidium bromide <!> (1 µg/mL) (for ethidium bromide staining only)
Gel containing DNA fragments that have been separated by electrophoresis
Staining trays
UV transilluminator <!> and eye protection (for ethidium bromide staining only)
White light transilluminator (for *Carolina*BLU staining only)

See Cautions Appendix for appropriate handling of materials marked with <!>.

A. Staining with Ethidium Bromide

1. Place your gel in a staining tray and cover it with ethidium bromide solution. Allow it to stain for 10–15 min.
2. Decant the stain back into the storage container for reuse and rinse the gel in tap water.
3. View the gel using UV transillumination. Photograph the gel using a digital camera or photodocumentary system.

B. Staining with Carolina*BLU*

1. Place your gel in a staining tray and cover it with *Carolina*BLU Final Stain. Allow it to sit for 20–30 min with (optional) gentle agitation.

2. After staining, pour the stain back into the bottle for future use. (The stain can be used six to eight times.)
3. To destain the gel, cover it with deionized or distilled water.
4. Change the water three to four times over the course of 30–40 min. Agitate the gel occasionally.
5. View the gel using white light transillumination. Photograph the gel using a digital camera or photodocumentary system.

If desired, *Carolina*BLU Gel and Buffer Stain can be used to stain the DNA while it is being electrophoresed. This prestaining procedure will allow you to visualize your results before the end of the gel run. However, the standard staining procedure outlined in "Staining with *Carolina*BLU" above is still required for optimum viewing.

To prestain the gel during electrophoresis, add *Carolina*BLU Gel and Buffer Stain in the amounts indicated in the tables below. Note that the amount of stain added is dependent on the voltage used for electrophoresis. Do not use more stain than recommended. This may precipitate the DNA in the wells and create artifact bands.

Volume of *Carolina*BLU stain to add to the agarose gel:

Voltage	Agarose gel volume	Stain volume
<50 Volts	30 mL	40 μL (1 drop)
	200 mL	240 μL (6 drops)
	400 mL	520 μL (13 drops)
>50 Volts	50 mL	80 μL (2 drops)
	300 mL	480 μL (12 drops)
	400 mL	640 μL (16 drops)

Volume of *Carolina*BLU stain to add to the 1x TBE buffer:

Voltage	1x TBE buffer volume	Stain volume
<50 Volts	500 mL	480 μL (12 drops)
	3000 mL	3 mL (72 drops)
>50 Volts	500 mL	960 μL (24 drops)
	2600 mL	5 mL (125 drops)

Gels containing *Carolina*BLU may be prepared 1 d ahead of the lab day, if necessary. However, gels stored longer tend to fade and lose their ability to stain DNA bands during electrophoresis.

SEQUENCING PCR PRODUCTS

In Laboratory 2.1, single-nucleotide polymorphisms in the mitochondrial control region can only be detected by sequencing the amplicon (PCR product).

Using GENEWIZ DNA Sequencing Services

We recommend using GENEWIZ, Inc. for DNA barcode sequencing. GENEWIZ has optimized reaction conditions for producing the barcode sequences in this laboratory and produces excellent quality sequence with rapid turnaround—usually within 48 hours of receipt of samples. GENEWIZ sequences are automatically uploaded to the DNALC's *DNA Subway* website.

Before submitting samples for sequencing, consult the GENEWIZ guide at http://www.GENEWIZ.com/public/Sample-Submission-Guideline.aspx.

Prepare PCR products

1. Verify that you can see a PCR product of the correct predicted length on an agarose gel. DNA sequence can be obtained from virtually every PCR product that is visible on the gel. (You can take a chance on samples that do not produce bands, because a fair proportion of these will also produce sequences.)
2. Prepare eight tube strips of 0.2-mL PCR tubes appropriate for the number of samples that you wish to submit. If you are submitting a large number of samples (≥48), submit the samples on a 96-well plate, arranging them vertically (A1–H1). For more details, see the "Tubes and Plates" tab at www.GENEWIZ.com.
3. Submit 10 µL of PCR product for each sequencing reaction. Forward and reverse sequencing reactions must be submitted in separate tubes.

Register and Submit Samples for Sequencing

1. Go to www.GENEWIZ.com and click "Register" to create a user account.
2. When creating your account, it is very important to enter your institution name followed by "–DNALC." The suffix "–DNALC" must be added exactly to your name; otherwise, your sequence will not be processed properly or may be delayed.
3. Obtain a valid purchase order number from your purchasing department or use a valid credit card.
4. Log in to your user account to place your sequencing order. Under "Place an Order," select "Create Sequencing Order."
5. Under "Service Priority:" select "Standard."
6. Under "Create Order by:" select "Online Form." (Alternatively, you can select "Upload Excel Form," download the "Custom" GENEWIZ Excel template, fill in the information in Steps 10–18, and upload the file.)
7. Under "Sample Type," select "Custom."
8. When prompted to "Create an online form for," enter the number of samples that you will be sending for sequencing. (If you elect to do bidirectional sequencing, you must count separate forward and reverse reactions for each sample).
9. Click on "Create New Form" and a sample submission form will be displayed.
10. For "DNA Name," enter a name for each sample. This may be a number or initials.
11. For "DNA Type," select "Un-Purified PCR" from the drop-down menu.
12. For "DNA Length (vector + insert in bp)," enter 450 bp.
13. Leave "DNA Conc. (ng/µL)" blank. It is best to send in a gel image of representative samples. This will be used by GENEWIZ to calculate the correct amount of cleanup reagents and product to use in the sequencing reaction. If a gel image is

not supplied, GENEWIZ will use default amounts to set up the sequencing reactions.

14. Leave "My Primer Name" and "My Primer Conc." blank.
15. For "GENEWIZ Primer," select from the drop-down menu.

    If you are doing mitochondrial DNA, enter LCMtF; this is the standard primer. If you experience problems sequencing the forward strand or wish to get the reverse read, you may enter LCMtR. Remember you will have to reverse complement the read to align it to mtDNA sequenced with the forward primer.
16. Under "Special Request," be sure that "PCR-Clean Up" has been automatically selected. (This is the default when unpurified PCR is selected as the DNA type).
17. In the "Comments" box at the bottom of the form, type "Primer stored at GENEWIZ under DNALC."
18. Click on "Save & Next."
19. Carefully review your form and then click on "Next Step."
20. Enter your payment information and click on "Next Step."
21. Review your order and click on "Submit."

Ship Samples to GENEWIZ

1. Print a copy of the order form and mail it along with your samples.
2. Be sure that the tubes are labeled exactly the same in the gel photo and on the order form. Failure to do so may delay sequencing or make it impossible to complete. Email DNALCSeq@cshl.edu if you need help.
3. Ship your samples via standard overnight delivery service (Federal Express, if possible).
4. Pack your samples for a letter or small shipping box, padding samples to prevent too much shifting. Room temperature shipping—with no ice or ice pack—is expected. PCR products are stable at ambient temperature, even if shipped on a Friday for Monday delivery.
5. Address the shipment to GENEWIZ at the following location:

    GENEWIZ, Inc.
    115 Corporate Blvd.
    South Plainfield, NJ 07080
6. You may be able to reduce shipping costs by using a GENEWIZ drop box. Call 1-877-436-3949 to determine whether one is available in your area.

Sequencing On Your Own

To have student amplicons sequenced at your own institution or at a local university or company, you may perform the cycle-sequencing reaction and DNA purification in the instructions that follow, but check with a sequencing technician that these methods are compatible with the protocols in use at their facility. For speed, reproducibility, and ease of tracking samples, the following procedure is optimized for a 96-well microtiter plate, which is the most common format used for DNA sequencing.

Perform the Cycle-Sequencing Reaction

REAGENTS, SUPPLIES, & EQUIPMENT

Container with cracked or crushed ice
Cycle-sequencing master mix
Micropipettes and tips (1–20 μL)
Microtiter plate (96 well)
Mitochondrial control region amplicons* (PCR products) from Part III of Laboratory 2.1
Sealing foil
Thermal cycler

*Store on ice.

1. Label the microtiter plate with an identifying name and date.
2. Add 8 μL of cycle-sequencing master mix to each well of the microtiter plate that will receive a reaction.
3. To each well containing master mix, add 2 μL of an amplicon from Part III of Laboratory 2.1. (Each well should receive an amplicon from a different student.)
4. Cover the microtiter plate with sealing foil. Carefully seal the plate by pressing on each well of the plate with your thumb and by running your finger around the edge of the foil.
5. Store the reactions on ice or in the freezer until ready to begin thermal cycling.

Make sure to create a tight seal, otherwise some samples will evaporate during cycling.

6. Place the microtiter plate in a thermal cycler that has been programmed for one 2-min step at 96°C, followed by 34 cycles of the following profile:

Denaturing step:	10 sec	96°C
Annealing step:	5 sec	50°C
Extending step:	4 min	60°C

The profile may be linked to a 4°C hold program after the 34 cycles are completed.

Precipitate the Dye-Labeled DNA

REAGENTS, SUPPLIES, & EQUIPMENT

Clinical centrifuge with spin trays
Cycle sequencing reactions in 96-well microtiter plate
Ethanol (70%) <!>
Ethanol:sodium acetate (30:1)
Freezer
Micropipette and tips (20–100 μL)
Paper towels
Sealing foil
Thermal cycler (optional)
Vortexer

See Cautions Appendix for appropriate handling of materials marked with <!>.

Samples should turn a cloudy, yellowish white.

Ethanol precipitates the extended PCR products but leaves the primer and unincorporated nucleotides in solution. Centrifugation pellets the precipitated sequencing products on the bottom of the microtiter well.

1. Add 30 μL of 30:1 ethanol:sodium acetate to each well of the microtiter plate. Mix by pipetting in and out several times. Do not vortex.
2. Cover the microtiter plate with sealing foil and carefully seal. Incubate the plate in the freezer for 30 min at –20°C.
3. Balance the microtiter plate in a clinical centrifuge equipped with spin trays. Centrifuge the plate for 30 min at room temperature.
4. Remove the sealing foil from the microtiter plate. Invert the plate over the sink to drain off most of the ethanol solution. Then, gently tap the plate on the surface of a clean paper towel to wick off as much liquid as possible.

5. Add 50 µL of 70% ethanol into each well to wash the DNA pellet.
6. Carefully seal the microtiter plate with clean foil.
7. Balance the microtiter plate in a clinical centrifuge and centrifuge the plate for 15 min at room temperature.
8. Remove the sealing foil from the microtiter plate. Invert the plate over the sink to drain off most of the ethanol solution. Then, gently tap the plate on the surface of a clean paper towel to wick off as much liquid as possible.

To prevent cross-contamination between wells, move the plate to a clean area of toweling between each tap.

9. Repeat Steps 5–8 to wash, centrifuge, and drain the DNA pellets a second time.
10. If possible, air-dry the DNA pellets overnight.

DNA pellets are not soluble in ethanol and will not resuspend during washing.

11. After the microtiter plate is thoroughly dry, add 20 µL of autoclaved, distilled water to each well on the microtiter plate.

As an alternative to Step 11, incubate the plate without foil in a thermal cycler for 10 min at 37°C. Do not close thermal cycler lid!

12. Carefully seal the microtiter plate with clean foil.
13. Dissolve the DNA in the wells by vortexing the microtiter plate at half speed. Tap the plate on the benchtop to bring the DNA droplets to the bottom of the wells.
14. Freeze the microtiter plate at –20°C until ready to read the DNA samples on an automated sequencer such as an ABI 3700 DNA Analyzer.

Every well on the microtiter plate—including empty wells without DNA samples—must be filled with 20 µL of water before being loaded onto the DNA sequencer.

Recipes for Reagents and Stock Solutions

Laboratories 2.1–2.3

THE SUCCESS OF THE LABORATORY EXERCISES DEPENDS on the use of high-quality reagents. Follow the recipes with care and pay attention to cleanliness. Use a clean spatula for each ingredient or carefully pour each ingredient from its bottle.

The recipes are organized alphabetically within five sections, according to experimental procedure. Stock solutions that are used for more than one procedure are listed once, according to their first use in the laboratories.

CAUTION: See Cautions Appendix for appropriate handling of materials marked with <!>.

- Isolating Human DNA from Cheek Cells
  - Chelex (10%)
  - Methylene Blue Staining Solution (1%)
  - Saline Solution (0.9% Sodium Chloride [NaCl])
  - Sodium Hydroxide (4 N)
  - Tris-HCl (1 M, pH 8.0 and 8.3)
- Isolating Human DNA from Hair Sheaths
  - Proteinase K (100 mg/mL)
- Amplifying DNA by PCR
  - Cresol Red Dye (1%)
  - Cresol Red Loading Dye
  - dNTPs (10 mM)
  - Magnesium Chloride ($MgCl_2$) (1 M)
  - PCR Buffer (10x)
  - Potassium Chloride (KCl) (5 M)
  - Primer/Loading Dye Mix
- Analyzing DNA by Gel Electrophoresis
  - 100-bp DNA Ladder (0.125 µg/µL)
  - Agarose (2%)
  - Ethidium Bromide Staining Solution (1 µg/mL)
  - Loading Dye (10x)
  - pBR322/BstNI Size Marker (0.1 µg/µL)
  - SYBR Green DNA Stain
  - Tris/Borate/EDTA (TBE) Electrophoresis Buffer (1x)
  - Tris/Borate/EDTA (TBE) Electrophoresis Buffer (20x)
- Sequencing PCR Products
  - Cycle-Sequencing Master Mix
  - Ethanol:Sodium Acetate (30:1)
  - Sodium Acetate (3 M, pH 4.8)

General Notes

- Typically, solid reagents are dissolved in a volume of deionized or distilled water equivalent to 70%–80% of the finished volume of buffer. This leaves room for the addition of acids or base to adjust the pH. Finally, water is added to bring the solution up to the final volume.

- Buffers are typically prepared as 1x or 10x solutions. Solutions at a concentration of 10x are diluted when mixed with other reagents to produce a working concentration of 1x.
- Storage temperatures of 4°C and –20°C refer to normal refrigerator and freezer temperatures, respectively.

ISOLATING HUMAN DNA FROM CHEEK CELLS

Chelex (10%)

Makes 10 mL.
Store for up to 3 mo at room temperature.

1. Weigh out 1 g of Chelex 100 Resin (100–200 dry mesh, sodium form from Bio-Rad).
2. Add 7 mL of 50 mM Tris-HCl to the dry Chelex.
3. Adjust the pH to 11 using 4 N NaOH <!>.
4. Add deionized or distilled water to bring the total volume to 10 mL.
5. For long-term storage, sterilize the solution by autoclaving it for 15 min at 121°C or by passing it through a 0.45-µm or 0.22-µm sterile filter. However, if the solution is to be used immediately, it need not be sterile.
6. Before the lab, aliquot 100 µL of the solution into either a 0.2-mL or 0.5-mL tube (use the size accommodated by your thermal cycler) for each student. If you are planning to use a heat block or water bath instead of a thermal cycler, aliquot the solution into 1.5-mL microcentrifuge tubes. Because the Chelex resin settles quickly, make sure to shake the stock tube to resuspend the Chelex before pipetting each student aliquot.

Methylene Blue Staining Solution (1%)

Makes 100 mL.
Store indefinitely at room temperature.

1. Weigh out 1 g of methylene blue trihydrate <!> (MW 373.90).
2. Dissolve the methylene blue in 100 mL of deionized or distilled water.
3. Transfer the solution to dropper bottles.

Saline Solution (0.9% Sodium Chloride [NaCl])

Makes 1000 mL.
Store indefinitely at room temperature.

1. In a clean container, dissolve 9 g of NaCl (MW 58.44) in 700 mL of deionized or distilled water.
2. Add water to bring total solution volume to 1000 mL.
3. Make 10-mL aliquots in sterile 15-mL culture tubes.

Note: Because this solution will be used for a mouthwash, make sure that all glassware and plasticware are clean and free of chemical residue.

Sodium Hydroxide (4 N)

Makes 100 mL.
Store indefinitely at room temperature.

1. Slowly add 16 g of sodium hydroxide (NaOH) <!> pellets (MW 40.00) to 80 mL of deionized or distilled water, with stirring. The solution will get very hot.
2. After the NaOH pellets are completely dissolved, add water to a final volume of 100 mL.

Tris-HCl (1 M, pH 8.0 and 8.3)

Makes 100 mL.
Store indefinitely at room temperature.

1. Dissolve 12.1 g of Tris base <!> (MW 121.10) in 70 mL of deionized or distilled water.
2. Adjust the pH by slowly adding concentrated hydrochloric acid (HCl) <!> for the desired pH listed below.

pH 8.0	5.0 mL
pH 8.3	4.5 mL

   Monitor with a pH meter or strips of pH paper. (If neither is available, adding the volumes of concentrated HCl listed here will yield a solution with approximately the desired pH.)
3. Add deionized or distilled water to make a total volume of 100 mL of solution.
4. Make sure that the bottle cap is loose and autoclave for 15 min at 121°C.
5. After autoclaving, cool the solution to room temperature and tighten the lid for storage.

Notes: A yellow-colored solution indicates poor-quality Tris. If your solution is yellow, discard it and obtain a Tris solution from a different source. The pH of Tris solutions is temperature dependent, so make sure to measure the pH at room temperature. Many types of electrodes do not accurately measure the pH of Tris solutions; check with the manufacturer to obtain a suitable one.

ISOLATING HUMAN DNA FROM HAIR SHEATHS

Proteinase K (100 mg/mL)

Makes 50 mL.
Store indefinitely at –20°C.

1. Dissolve 5 g of proteinase K <!> in deionized or distilled water to a final volume of 50 mL.
2. Make 10-mL aliquots in sterile 15-mL culture tubes. These aliquots can be stored indefinitely in the freezer.
3. Before beginning the lab, aliquot 100 μL of the solution into either a 0.2-mL or 0.5-mL PCR tube (use the size accommodated by your thermal cycler) for each

student. If you are planning to use a heat block or water bath instead of a thermal cycler, aliquot the solution into 1.5-mL microcentrifuge tubes.

AMPLIFYING DNA BY PCR

Cresol Red Dye (1%)

Makes 50 mL.
Store indefinitely at room temperature.

1. Weigh out 500 mg of cresol red dye <!>.
2. In a 50-mL tube, mix the cresol red dye with 50 mL of distilled water.

Cresol Red Loading Dye

Makes 50 mL.
Store indefinitely at –20°C.

1. In a 50-mL tube, dissolve 17 g of sucrose in 49 mL of distilled water.
2. Add 1 mL of 1% cresol red dye <!> and mix well.

dNTPs (10 mM)

Makes 100 µL.
Store for up to 1 yr at −20°C

1. Add 10 µL each of 100 mM dTTP, dATP, dGTP, and dCTP.
2. Add 60 µL of deionized or distilled water and mix.

Magnesium Chloride ($MgCl_2$) (1 M)

Makes 100 mL.
Store indefinitely at room temperature.

1. Dissolve 20.3 g of $MgCl_2 \cdot 6H_2O$ <!> (MW 203.30) in 80 mL of deionized or distilled water.
2. Add deionized or distilled water to make a total volume of 100 mL of solution.
3. Make sure that the bottle cap is loose and autoclave for 15 min at 121°C.
4. After autoclaving, cool the solution to room temperature and tighten the lid for storage.

PCR Buffer (10x)

Makes 10 mL.
Store indefinitely at –20°C.

1. Combine the following in a 15-mL tube:

Deionized or distilled water	7.85 mL
Tris-HCl (1 M, pH 8.3)	1 mL
KCl (5 M) <!>	1 mL
$MgCl_2$ (1 M)	0.15 mL

2. Mix well.

Potassium Chloride (KCl) (5 M)

Makes 100 mL.
Store indefinitely at room temperature.

1. Dissolve 37.3 g of KCl <!> (MW 74.55) in 70 mL of deionized or distilled water.
2. Add deionized or distilled water to make a total volume of 100 mL of solution.
3. Make sure that the bottle cap is loose and autoclave for 15 min at 121°C.
4. After autoclaving, cool the solution to room temperature and tighten the lid for storage.

Primer/Loading Dye Mix

Makes enough for 50 PCRs.
Store for up to 1 yr at –20°C.

1. Obtain the primers for the laboratory exercise and dissolve each at a concentration of 15 pmol/μL. The primer sequences are as follows:
   - **Laboratory 2.1:** Mitochondrial (mt) Control Region

     5′-TTAACTCCACCATTAGCACC-3′ (forward primer)
     5′-GAGGATGGTGGTCAAGGGAC-3′ (reverse primer)

   - **Laboratory 2.2:** *PV92* Locus

     5′-GGATCTCAGGGTGGGTGGCAATGCT-3′ (forward primer)
     5′-GAAAGGCAAGCTACCAGAAGCCCCAA-3′ (reverse primer)

   - **Laboratory 2.3:** *PTC* Locus (*TAS2R38*)

     5′-CCTTCGTTTTCTTGGTGAATTTTTGGGATGTAGTGAAGAGGCGG-3′ (forward primer)
     5′-AGGTTGGCTTGGTTTGCAATCATC-3′ (reverse primer)

2. In a 1.5-mL tube, mix the following:

Distilled water	640 μL
Cresol red loading dye <!>	460 μL
Forward primer (15 pmol/μL)	20 μL
Reverse primer (15 pmol/μL)	20 μL

Note: This primer/loading dye mix is for use with Ready-to-Go PCR Beads. puRe *Taq* Ready-to-Go PCR Beads (GE Healthcare) are available in individual thin-wall tubes (0.2 or 0.5 mL) or in 96-well plates.

3. Vortex to mix.

ANALYZING DNA BY GEL ELECTROPHORESIS

100-bp DNA Ladder (0.125 μg/μL)

Makes 100 μL.
Store for up to 1 yr at –20°C.

1. Obtain a stock solution of 100-bp DNA Ladder (0.5 μg/μL) from New England BioLabs (N3231). Store the solution at –20°C.

2. Dilute a small amount of the stock at a time by combining the following in a 1.5-mL tube:

Deionized or distilled water	50 µL
Cresol red loading dye <!>	25 µL
DNA Ladder	25 µL

Agarose (2%)

Makes 200 mL.
Use fresh or store solidified agarose for several weeks at room temperature.

1. To a 600-mL beaker or Erlenmeyer flask, add 200 mL of 1× TBE electrophoresis buffer and 4 g of agarose (electrophoresis grade).
2. Stir to suspend the agarose.
3. Dissolve the agarose using one of the following methods:
   - Cover the flask with aluminum foil and heat the solution in a boiling water bath (double boiler) or on a hot plate until all of the agarose is dissolved (~10 min).
   - Heat the flask uncovered in a microwave oven at high setting until all of the agarose is dissolved (3–5 min per beaker).
4. Swirl the solution and check the bottom of the beaker to make sure that all of the agarose has dissolved. (Just before complete dissolution, particles of agarose appear as translucent grains.) Reheat for several minutes if necessary.
5. Cover the agarose solution with aluminum foil and hold in a hot water bath (at ~60°C) until ready for use. Remove any "skin" of solidified agarose from the surface before pouring the gel.

Notes: Samples of agarose powder can be preweighed and stored in capped test tubes until ready for use. Solidified agarose can be stored at room temperature and then remelted over a boiling water bath (15–20 min) or in a microwave oven (3–5 min per beaker) before use. When remelting, evaporation will cause the agarose concentration to increase; if necessary, compensate by adding a small volume of water. Always loosen the cap when remelting agarose in a bottle.

Ethidium Bromide Staining Solution (1 µg/mL)

Makes 500 mL.
Store in the dark indefinitely at room temperature.

1. Add 100 µL of 5 mg/mL ethidium bromide <!> to 500 mL of deionized or distilled water.
2. Store the ethidium bromide in a dark (preferably opaque) unbreakable container or wrap the container in aluminum foil.
3. Label the container **"CAUTION: Ethidium Bromide. Mutagen and cancer-suspect agent. Wear rubber gloves when handling."**

Note: Ethidium bromide is light sensitive.

Loading Dye (10x)

Makes 100 mL.
Store indefinitely at room temperature.

1. Dissolve the following ingredients in 60 mL of deionized or distilled water:

Bromophenol blue (MW 669.96) <!>	0.25 g
Xylene cyanol (MW 538.60)<!>	0.25 g
Sucrose (MW 342.3)	50.0 g
Tris-HCl (1 M, pH 8.0)	1 mL

2. Add deionized or distilled water to make a total volume of 100 mL.

pBR322/BstNI Size Marker (0.1 μg/μL)

Makes 100 μL.
Store for up to 1 yr at –20°C.

1. Add 1 μL of 10 μg/μL pBR322 to 84 μL of deionized or distilled water.
2. Add 10 μL of 10x buffer (provided by the supplier of the BstNI enzyme).
3. Add 5 μL of 10 U/μL BstNI restriction enzyme and incubate for 60 min at 60°C.
4. Electrophorese 5 μL (plus 1 μL cresol red loading dye <!>) in a 1%–2% agarose gel to check for complete digestion. Exactly five bands should be visible, corresponding to 1857, 1058, 929, 383, and 121 bp. Any additional bands indicate incomplete digestion; if this is the case, add additional enzyme and incubate again at 60°C.

Note: pBR322 precut with restriction enzyme BstNI is also available from New England BioLabs (N3031).

SYBR Green DNA Stain

Makes 100 μL.
Store in the dark for up to 3 mo at 4°C.

1. Obtain a stock solution of 10,000x SYBR Green I <!>, which comes dissolved in DMSO <!> and is stored at –20°C. Allow the dye to thaw for ~10 min at room temperature.
2. Add 1 μL of 10,000x SYBR Green I in DMSO to 100 μL of sucrose solution. Mix thoroughly.

Note: SYBR Green is light sensitive.

Tris/Borate/EDTA (TBE) Electrophoresis Buffer (1x)

Makes 10 L.
Store indefinitely at room temperature.

1. In a spigoted carboy, add 9.5 L of deionized or distilled water to 0.5 L of 20x TBE Electrophoresis Buffer.
2. Stir to mix.

Tris/Borate/EDTA (TBE) Electrophoresis Buffer (20x)

Makes 1 L.
Store indefinitely at room temperature.

1. Add the following dry ingredients to 500 mL of deionized or distilled water in a 2-L flask:

NaOH <!> (MW 40.00)	2 g
Tris base <!> (MW 121.10)	216 g
Boric acid <!> (MW 61.83)	110 g
EDTA <!> (disodium salt, MW 372.24)	14.8 g

2. Stir to dissolve, preferably using a magnetic stir bar.
3. Add deionized or distilled water to bring the total volume to 1 L.

Note: If the stored 20x TBE comes out of solution, place the flask in a water bath (37°C–42°C) and stir occasionally until all solid matter goes back into solution.

SEQUENCING PCR PRODUCTS

Cycle-Sequencing Master Mix

Makes 8 µL for each sequencing reaction.
Use fresh.

1. Purchase a BigDye Terminator Cycle Sequencing Kit. Combine the following in a 1.5-mL tube:

	Per sample	Per 96-well plate
BigDye Terminator	1 µL	105 µL
Forward mt primer (6 pmol/µL)	1 µL	105 µL
Sequencing buffer (5x)	1.5 µL	157.5 µL
Distilled water	4.5 µL	472.5 µL

2. Mix by vortexing or pipetting in and out.

Ethanol:Sodium Acetate (30:1)

Makes 3.41 mL.
Use fresh.

1. Add the following to a 1.5-mL tube:

Ethanol <!>	3.3 mL
Sodium acetate (3 M, pH 4.8)	110.0 µL

2. Mix by vortexing or pipetting in and out.

Sodium Acetate (3 M, pH 4.8)

Makes 100 mL.
Store indefinitely at room temperature.

1. Dissolve 40.8 g of sodium acetate trihydrate (MW 136.08) in 70 mL of deionized or distilled water.
2. Adjust the pH to 4.8 by adding glacial acetic acid <!>; monitor it with a pH meter.
3. Add water to bring the total volume of solution to 100 mL.

Answers to Questions

Laboratories 2.1–2.3

LABORATORY 2.1: USING MITOCHONDRIAL DNA POLYMORPHISMS IN EVOLUTIONARY BIOLOGY

Answers to Bioinformatics Questions

I. Use BLAST to Find DNA Sequences in Databases (Electronic PCR)

2.i. **What do you notice about the lengths (and colors) of the matches (bars) as you look from the top to the bottom?**

Typically, all or most of the significant alignments will have complete matches to the forward and reverse primers. Partial alignments will correspond to only the forward or reverse primer.

2.ii. **What is the E value of the most significant hit and what does it mean?**

The lowest E value obtained for a match to both primers is typically in the range of 0.003–0.04, with a larger number of matches in the 0.2–0.7 range. This might seem high for a probability, but in fact each of these values means that a match of this quality would be expected to occur by chance less than once in this database! More precisely, a score of 0.33 would mean that a single match would be expected to occur by chance once in every three searches.

E values are based on the length of the search sequence, and thus the relatively short primers used in this experiment produce relatively high E values. Searches with longer primers or long DNA sequences return E values with negative exponents (such as E-4), which look more like the probabilities that one might expect. This search turns up many matches with the same E value because human mitochondrial sequences are the most common sequences in DNA databases, with tens of thousands of accessions.

Note the names of any significant alignments that have E values of less than 0.1. Do they make sense? What do they have in common?

Yes. All are matches to human (*Homo sapiens*) mitochondrial sequences; their descriptions may include the terms "complete genome," "D loop," "control region," or "hypervariable region."

3.i. **What is the source and size of the sequence in which your BLAST hit is located?**

These BLAST hits, all *Homo sapiens* mitochondrial genomes, are ~16,570 bp in length.

4.ii. To which positions do the primers match in the subject sequence?

The exact position will depend on the hit. For the hit with accession number AY275537.2, the forward primer matches positions 15,970–15,989 and the reverse primer matches positions 16,409–16,390.

4.iii. What is the difference between the coordinates?

16,409 – 15,970 = 439 nucleotides. These are the absolute nucleotide coordinates for these primer sequences. The coordinates may vary slightly among accessions, although the difference between the coordinates should not.

4.iv. Note that the actual length of the amplified fragment includes both ends, so add 1 nucleotide to the result that you obtained in Step 4.iii to obtain the exact length of the PCR product amplified by the two primers.

439 + 1 = 440 nucleotides.

II. Examine the Structure of the Mitochondrial Chromosome and Its Genes

3.i. How many features does the amplicon span? What are they called? What is the size of each feature?

The amplicon spans portions of two features: the tRNA-Pro sequence and the D-loop sequence. The complete tRNA-Pro sequence is ~70 bp and the complete D-loop sequence is ~1100 bp.

3.ii. The largest feature that you identified in Step 3.i has several names. Do some research to learn and understand these names.

This noncoding region is most frequently called the "D loop," but it is also called the "control region" or the "hypervariable region." A D loop (displacement loop) occurs at the origin of replication. A newly synthesized strand displaces one of the parental strands, forming a "bubble" or a loop. "Control region" refers to the fact that this region contains the signals that control RNA and DNA syntheses. A single promoter on each DNA strand initiates transcription in each direction, and a single origin initiates the replication of each strand. "Hypervariable region" refers to the fact that this part of the mitochondrial chromosome accumulates point mutations at ~10 times the rate of nuclear DNA. This region is relatively tolerant of the high mutation rate because binding sites for DNA and RNA polymerase are defined by only short nucleotide sequences.

4.i. List the genes that are transcribed from the forward strand and those that are transcribed from the reverse (complement) strand.

Forward strand
tRNA, phenylalanine
rRNA, 12S
tRNA, alanine
tRNA, valine
rRNA, 16S
tRNA, leucine (copy 1)
gene, *ND1* (NADH dehydrogenase subunit 1)
tRNA, isoleucine
tRNA, methionine
gene, *ND2* (NADH dehydrogenase subunit 2)
tRNA, tryptophan
gene, *COX1* (cytochrome *c* oxidase subunit I)
tRNA, aspartic acid
gene, *COX2* (cytochrome *c* oxidase subunit II)
tRNA, lysine
gene, *ATP8* (ATP synthase F0, subunit 8)
gene, *ATP6* (ATP synthase F0, subunit 6)
gene, *COX3* (cytochrome *c* oxidase subunit III)
tRNA, glycine
gene, *ND3* (NADH dehydrogenase subunit 3)
tRNA, arginine

(Continued on facing page.)

(continued)	*Reverse strand (complement)*
gene, *ND4L* (NADH dehydrogenase subunit 4L)	tRNA, glutamine
gene, *ND4* (NADH dehydrogenase subunit 4)	tRNA, asparagine
tRNA, histidine	tRNA, cysteine
tRNA, serine (copy 2)	tRNA, tyrosine
tRNA, leucine (copy 2)	tRNA, serine (copy 1)
gene, *ND5* (NADH dehydrogenase subunit 5)	gene, *ND6* (NADH dehydrogenase subunit 6)
gene, *CYTB* (cytochrome *b*)	tRNA, glutamic acid
tRNA, threonine	tRNA, proline

On which strand are most genes found?

There are 29 genes on the forward strand and eight genes on the reverse strand.

4.ii. What kind of product do most of the genes produce? For what biological function are these products used?

Most genes in the mitochondrial genome encode transfer RNAs, which are used by the mitochondria for translation. The tRNAs needed for all 20 amino acids are produced in the mitchondria; however, the proteins involved, including those in the ribosome, are encoded by nuclear genes, synthesized in the cytoplasm, and then imported into the mitochondria.

4.iii. What is the spacing like between genes? Is there much intergenic DNA between two adjacent genes?

The mitochondrial chromosome has very little intergenic DNA. Many genes are immediately adjacent to one another. About one-third of the genes are separated by gaps of 1–70 nucleotides, and a few overlap one another by one to seven nucleotides. Gaps and overlaps occur between adjacent genes on the same strand and on different strands.

4.iv. Focus on one of the protein-coding genes. Compare the nucleotide coordinates for the gene and the cd. What do you notice? What does this tell you about the structure and origin of mitochondrial genes?

In eukaryotes, genes encoded by nuclear DNA have several noncoding sequences: the 5′- and 3′-untranslated regions (UTRs) and the introns. The cd is the set of codons that is translated into protein—beginning with a start codon, ending with a stop codon, and omitting all introns. For each mitochondrial gene, the gene and the cds are identical, meaning that mitochondrial genes have no UTRs and no introns. These are characteristics of prokaryotic genes. The structure of the genes encoded by the mitochondria, as well as the circular form of the mitochondrial chromosome, provides DNA evidence that mitochondria evolved from free-living bacteria.

III. Use MITOMAP to Explore the Role of Mitochondrial Mutations in Human Health and Aging

4. What types of mitochondrial mutations generally cause diseases? What do many of these diseases have in common?

Most mitochondrial diseases are caused by point mutations. Many mitochondrial diseases affect tissues with high energy demands: nerves, skeletal muscles, and the heart. Many are related to advancing age.

5. Research the mitochondrial theory of aging.

The mitochondrial theory of aging generally states that as mitochondrial genes accumulate mutations over time, the efficiency of energy production declines.

Mutations in respiratory chain proteins make them less efficient at transporting electrons. The increased "leak" of electrons leads to the production of more reactive oxygen species; this, in turn, generates more mutations in mitochondrial genes. These mutations lead to age-related declines, especially in nerve, muscle, and heart function.

Answers to Results and Discussion Questions

I. Interpret Your Gel and Think About the Mitochondrial Genome

4. **How would you interpret a lane in which you observe primer dimer but no bands as described in Step 3?**

   The presence of primer dimer confirms that the reaction contained all components necessary for amplification but indicates that there was insufficient template to amplify the target sequence.

5. **Propose a biological reason for the high mutation rate of mtDNA.**

   The mitochondrial genome is housed within the cell's energy-producing factory, where it is exposed to reactive by-products of oxidative phosphorylation. These reactive oxygen species (oxygen free radicals) are potent mutagens, and they cause enzymes involved in energy production to accumulate mutations that make them function less efficiently. It is hypothesized that this decline in mitochondrial efficiency is a major contributor to aging.

6. **The high mutability of the mitochondrial genome means that it evolves more quickly than the nuclear genome. This makes the mitochondrial control region a laboratory for the study of DNA evolution. However, can you think of any drawbacks to this high mutation rate when studying evolution?**

   The mutation rate is so high that some nucleotides have mutated several times over evolutionary history. This makes it difficult to determine the actual mutation rate and to ascertain the ancestral (original) state of a particular nucleotide. These "back mutations" also make it difficult to accurately calibrate the "mutation clock."

7. **Would you expect any difference in the mutation rates of the control region sequence in the mitochondrial genome versus the chromosome 11 insertion? What implication does this have for the study of human evolution?**

   Once removed from the context of the mitochondrion, the insertion sequence is subject to the lower mutation rate of nuclear DNA. The mutation clock effectively slows dramatically, preserving the insertion as a "molecular fossil" from that era in DNA history. This means that one can study human evolution within oneself by comparing the mitochondrial control region sequence with a nuclear insertion of the same sequence.

II. Visualize and Assess Your DNA Sequence

10.i. **What does the amplitude (height) of each peak represent? What do you notice about the amplitude of the peaks at the beginning of the sequence?**

The height of each peak (fluorescence) is proportional to the number of DNA molecules that are terminated with dye at that position. It is important that the height of the labeled peaks is significantly greater than the background of erroneous fluorescent products. The initial peaks in the sequence tend to be very high, probably due to increased efficiency of labeling the nucleotides adjacent to the forward primer.

10.ii. What do you notice about the trace at N positions?

At N positions, two peaks of similar amplitude overlap or intersect; no prominent peak rises among the background "noise."

11. Do the nucleotide numbers (positions) correspond to the same nucleotide sequence in both reads? What trick did you use to align them?

Different chromatograms are typically out of phase by up to 30 nucleotides because the software begins calling nucleotides at different points in different files. Chromatograms can be manually aligned by matching a unique sequence of five or more nucleotides.

12. What sequence feature coincides with the abrupt transition from a clean to a poor-quality sequence?

A long "C track" coincides with the change in sequence quality. In clean reads, this sequences reads CCCCCTCCCC. In low-quality sequences, an internal T > C polymorphism, and often a flanking polymorphism, results in a string of 10–11 Cs. For unknown reasons, this "C track" confounds the *Taq* DNA polymerase, and it essentially falls off the template DNA at this point.

III. Assess the Extent of Mitochondrial Variation in Modern Humans

11. Use these data to determine the average number of mutations, as well as the average percent difference, between any two pairs.

This number has been experimentally determined to be about seven mutations per 375 nucleotides (or 1.87%) within this part of the mitochondrial control region.

What does this number represent?

This number represents the extent of mitochondrial sequence variation in modern human populations. This is equal to the average number of mutations that have accumulated since all modern humans diverged from a common maternal ancestor.

12. If this is so, what is the approximate mutation rate for the mitochondrial control region?

If this was the case, the mutation rate for the mitochondrial control region would be about one new mutation every 28,571 yr.

$$\frac{200{,}000 \text{ years}}{7 \text{ mutations}} = 28{,}571 \text{ yr/mutation.}$$

Why is your calculation only approximate?

This is only approximate because it is based on the average accumulation of mutations. Nothing has been done to correct for back mutations; that can only be determined statistically.

IV. Assess Mitochondrial Variation in Ancient Hominids

A. *Ötzi the Iceman*

7. **Pool the class data to determine the average number of differences and average percent difference between Ötzi and modern humans.**

   The average number of mutations should be close to that determined in Step 11 of Part III, or about seven mutations per 375 nucleotides (1.87%). About 10% of students of European ancestry are expected to have sequences identical to those of Ötzi over this portion of the mitochondrial control region.

   What does this tell you about Ötzi and the DNA mutational clock?

   The extent of mitochondrial sequence variation between Ötzi and modern humans is the same as between modern humans. Therefore, based on this sequence, Ötzi appears to be a modern human.

B. *Neandertal*

8. **Pool the class data to determine the average number of differences and average percent difference between Neandertals and modern humans—and between Neandertals.**

   In a Neandertal–human comparison, the average number of mutations should be about 28 mutations per 375 nucleotides (7.47%). In a Neandertal–Neandertal comparison, the average number of mutations should be about eight mutations per 331 nucleotides (2.42%).

   What does this tell you about the relationship between Neandertals and modern humans?

   Neandertals had about the same level of genetic diversity as modern humans (1.87% compared to 2.42%, respectively). However, the extent of mitochondrial sequence varation (or the locations and types of mutations) in Neandertals and modern humans is not the same, as indicated by the 7.47% value obtained when comparing the two. Therefore, Neandertals cannot be the direct ancestor of modern humans.

9. **If modern humans arose 200,000 BP, approximately how long ago did humans and Neandertals share a common mitochondrial ancestor?**

   If seven differences between modern humans equal 200,000 years to a common ancestor, then 28 differences between modern humans and Neandertal equal 800,000 years to a common ancestor.

10. **What does the level of genetic variation in Neandertals suggest about Neandertal population changes during the course of their existence on Earth?**

    Neandertals had about the same level of genetic diversity as modern humans. However, Neandertals achieved this level of diversity in approximately 300,000 years of evolution as a species, compared about 150,000 years for modern humans. This suggests that Neandertal populations, like human ones, encountered several bottlenecks that reduced genetic diversity.

V. Discover What DNA Says About Human Evolution

6. **What do your results show? Is this what you expected?**

   In most three-way alignments, the African sequences will differ at the majority of SNP (single-nucleotide polymorphism) locations. This is counterintuitive; one would expect the European or Asian sequences to be different—and for the African sequences to have greater similarity.

7. **Pool the class results to confirm or refute your own results.**

   Although students may feel that their individual observations might be the result of small sample sizes, the trend of Africans differing from one another at most SNP locations is upheld when data are pooled.

8. **What does your analysis say about African mtDNA diversity? What population factors could account for this?**

   There is greater mtDNA diversity among Africans. Because mutations accumulate over time, this suggests that African populations are older than European and Asian populations. Genetic diversity also increases with population size, suggesting that African populations have been maintained at higher levels during long periods of human evolution.

9. **What does your analysis say about European and Asian mtDNA diversity? What population factors could account for this?**

   Europeans and Asians share the majority of their mitochondrial mutations with Africans. Most European and Asian mtDNA diversity is a subset of African diversity. This is mtDNA evidence that humans arose in Africa; founders of European and Asian populations carried a portion of the accumulated African diversity with them when they migrated out of Africa. Relatively few unique mutations have accumulated in European and Asian populations during the 50,000–70,000 years since they left Africa. This mtDNA evidence is supported by analyses of Y chromosome and autosomal polymorphisms.

LABORATORY 2.2: USING AN *ALU* INSERTION POLYMORPHISM TO STUDY HUMAN POPULATIONS

Answers to Bioinformatics Questions

I. Use BLAST to Find DNA Sequences in Databases (Electronic PCR)

2.i. **What do you notice about the lengths (and colors) of the matches (bars) as you look from the top to the bottom?**

There is only one complete match to the forward and reverse primers, followed by a number of partial matches.

2.ii. **What is the E value of the most significant hit and what does it mean?**

The most significant hit has a very low E value (e.g., 6E-12). An E value of 6E-12 denotes 6×10^{-12} or a 6 preceded by 11 decimal places! This means that the query has found strong matches in the database.

Note the names of any significant alignments that have E values of less than 0.1. Do they make sense?

It makes sense because it is from a human (chromosome 16).

3.i. What is the source and size of the sequence in which your BLAST hit is located?

The source of the sequence, which is 16,627 bp long, is a clone of part of human chromosome 16.

4.ii. To which positions do the primers match in the subject sequence?

The forward and reverse primers match positions 56,722–56,746 and positions 57,137–57,112, respectively.

4.iii. What is the difference between the coordinates?

57,137 – 56,722 = 415 nucleotides. These are the coordinates in accession AC009028.3.

4.iv. Note that the actual length of the amplified fragment includes both ends, so add 1 nucleotide to the result that you obtained in Step 4.iii to obtain the exact length of the PCR product amplified by the two primers.

415 + 1 = 416 nucleotides.

4.v. At this stage, do you have enough information to determine whether this is the + or the – allele?

There is not yet enough information.

II. Use BLAST to Identify Additional Alleles at the PV92 Locus

4. Compare the E values obtained by this search to those values that you obtained in Step 2.ii of Part I. What do you notice? Why is this so?

Three hits have extremely low E values (they have many decimal places). This is because the query sequence is longer.

5. Why does the first hit have an E value of 0?

This hit completely matches the query. It is the same human chromosome 16 clone identified in Part I.

6.i. What is the length of the *Alu* repeat (insert) at PV92?

The PV92 *Alu* is 308 bp long.

6.ii. What do you notice about the 3′ end of the *Alu* repeat sequence?

There is a poly(A) tail composed of a string of 28 adenines (As).

6.iii. What appears to have happened to this sequence when the *Alu* repeat inserted itself into the PV92 locus?

The target sequence GAAAGAA was duplicated during the insertion of the *Alu* element.

6.iv. If you assume that the amplicon that you identified in Part I is the – allele, what is the length of the + allele?

The + allele would appear to be the sum of 416 bp + 308 bp = 724 bp. However, the + allele also includes the 7-bp duplication of the target sequence. So the actual length of the + allele is 731 bp.

7. **What is going on here? How are the three repeated sequences related to one another?**

   There are annotations for *Alu*s belonging to two different subfamilies: Ya5 is an older group that includes PV92, and Yb8 is a younger group. The younger *Alu* jumped inside the original *Alu* at the PV92 locus. One can easily see two poly(A) tails in the sequence—one belonging to each *Alu*. This *Alu*-within-an-*Alu* allele is rare and inserted so recently that it has only been found in a few people, notably from the Basque region of Spain and northern Morocco.

III. Use Map Viewer to Determine the Chromosomal Location of the PV92 Insertion

6. **On which chromosome have you landed?**

   Chromosome 16.

8. **Which gene contains the amplicon? What is the function of that gene?**

   The amplicon lies within the cadherin H 13 (*CDH13*) gene. This gene produces a cell-adhesion protein that mediates interactions between cells in the heart.

10.i. **Determine the size of the gene using the map coordinates to the left of the "Contig" map.**

CDH13 is ~1.2 million nucleotides in length.

10.ii. **How many introns and exons does the gene have?**

CDH13 has 13 exons and 12 introns.

10.iii. **Where in the gene has the PV92 *Alu* inserted: an exon or intron? How does this explain the fact that the PV92 insertion is believed to be neutral, i.e., to have no phenotypic effect?**

The PV92 *Alu* has inserted within the second intron. Mutations within introns generally have no phenotypic effect.

Answers to Results and Discussion Questions

I. Determine Your PV92 Genotype

5. **How would you interpret a lane in which you observe primer dimer but no bands as described in Step 3?**

   The presence of primer dimer confirms that the reaction contained all components necessary for amplification, but it also indicates that there was insufficient template to amplify the target sequence.

6. **How does this relate to the most basic unit of information stored in digital form by a computer? What equivalent in digital information is provided by an *Alu* genotype?**

   An *Alu* allele (+ or –) is equivalent to one bit of information (a digital 0 or 1). An *Alu* genotype (+/+, –/–, or +/–) contains two bits of information.

II. Determine the Observed Genotype and Allele Frequencies for Your Class

5. **Is the + allele confined to any particular racial or ethnic group in your class? What can you say about individuals in the class who have at least one + allele?**

   The + allele is not exclusive to any racial or ethnic group. All people who have at least one + allele inherited their allele(s) from a common ancestor.

III. Test for Hardy–Weinberg Equilibrium in Your Class

1.ii. **How do the genotype frequencies that you observed in your experiment compare with those expected by the Hardy–Weinberg equation? Would you say that they are very similar or very different?**

   Observed genotype frequencies are typically quite similar to the expected frequencies.

13. **Is your *p* value greater or less than the 0.05 cutoff? What does this mean?**

   Class results typically have *p* values greater than 0.05. This means that there is no significant difference between observed and expected frequencies and that the observed frequencies are consistent with Hardy–Weinberg equilibrium.

14. **What conditions are required for a population to come into genetic equilibrium? Does your class satisfy these requirements?**

   Genetic equilibrium requires a relatively large population, no migration in or out of the group, no new mutations at the locus under study, and random mating with respect to the locus. Although the class itself would be a very small population, its members are more or less representative of a larger population in your town or region. There is probably a relatively small amount of migration in and out of your town or region. There is no evidence of very recent, new mutations at the PV92 locus that would influence genotypes. Because there is no way to tell a person's PV92 genotype by looking at them, people mate randomly with respect to this polymorphism. Thus, perhaps surprisingly, the class may generally fulfill the requirements for Hardy–Weinberg equilibrium.

IV. Compare Genotype Frequencies in World Populations

7. **Do the pie charts look similar or different? Does the Chi-square statistic and associated *p* value support your visual impression?**

   The results vary depending on the class, the population to which they compare the class results, and the student's visual impression. Refer to the detailed results in *Allele Server* while discussing results with students. For small samples, it is common for students to observe differences in the pie charts that are not supported by the Chi-square test; these differences are probably due to chance.

9. **Which groups have significantly different genotype frequencies? What is the most frequent genotype in each group?**

   European, African, Australian, and American populations typically have similar genotype frequencies, with the –/– genotype being the most common. The +/+ genotype is most common in Asian populations.

V. Compare Allele Frequencies in World Populations

7. Do you notice any pattern in the allele frequencies?

The + allele frequency is high in all Asian groups (up to 90%) and generally decreases moving westward through the Middle East, with European and African populations having frequencies of 10%–35%. High + allele frequencies are also found in American Indian populations: Yanamamo (96%) and Maya (70%). The + allele frequencies (%) are listed below.

African American, 20%
Alaskan Natives, 29%
Australian Aborigine, 15%
Breton, 27%
Cajun, 21%
Chinese, 86%
Euro-American, 18%
Filipino, 80%
French, 23%
German, 10%
Greek, Cyprus, 18%
Hispanic American, 51%
Hungarian, 12%
Indian Christian, 48%
Indian Hindu, 52%
Indian Muslim, 30%
Italian, Northern, 24%
Javanese, 84%
!Kung (Bushmen), 20%
Malay, 72%
Mayan, 70%
Moluccan, 69%
Mvskoke (Creek American Indian), 53%
Nguni, 24%
Nigerian, 9%
Pakistani, 30%
Papua New Guinean, 14%
Papua New Guinean, Coastal, 19%
Pushtoon (Afgani), 33%
Pygmy (Central African Republic), 26%
Pygmy (Dem. Rep. Congo), 35%
Sardinian (Aritzo), 17%
Sardinian (Marrubiu), 0%
Sardinian (Ollolai), 0%
Sardinian (San Teodoro), 27%
Sotho, 29%
South Indian, 56%
Swiss, 20%
Syrian, 18%
Taiwanese, 90%
Turkish, Cyprus, 58%
United Arab Emirati, 30%
Yanomamo, 96%

8. Suggest a hypothesis about the origin and dispersal of the + allele that accounts for your observation.

Most students conclude that this pattern is consistent with the PV92 *Alu* insertion arising in Asia and then being diluted by gene flow to the west. Well-studied students, especially after doing Parts VI and VII (below), may understand that the pattern could also be the product of migration and genetic drift.

9. If this is so, in what sort of hominid did the jump occur, and what implications does this have for your hypothesis of Step 8 above?

The PV92 insertion would have occurred in a population of *H. erectus*, which then survived to give rise to modern humans. If this jump occurred in Asia, then H. erectus must have survived in Asia to give rise to modern populations there. This would be consistent with the regional development hypothesis. The accepted replacement hypothesis—also called "out of Africa"—supports the PV92 insertion occurring in a *H. erectus* population in Africa. The worldwide frequencies of ~20% suggest that the + allele drifted to approximately this frequency in Africa before the migrations that gave rise to European, Asian, and Australian populations. The frequency then drifted much higher among the migrants that founded Asian populations, several of which may have carried a high + allele frequency when they migrated across the Bering Strait to found American Indian populations.

VI. Simulate a New *Alu* Jump in an Ancient Hominid Population

3.i. How did hominids live 200,000 BP and what size population group would be supported?

Our hominid ancestors existed only by hunting and gathering, so that would limit the size of each group to ~50 individuals.

3.ii. What would be the allele frequency if a single new *Alu* jump occurred in one child born into a group of this size?

If there were 50 people in the hunter-gatherer group, they would have 100 alleles. Only one of those alleles would have the new *Alu* insertion—for an allele frequency of 1%.

7. What happens to the new *Alu* insertion in the 100 populations?

The + allele frequency decreases from 1% to 0% in most of the populations within about 10 generations. The new *Alu* mutation is lost from these populations. Typically, the + allele is maintained in several populations at the end of 100 generations. Occasionally, the + allele will be fixed in a population, when the frequency rises to 100%.

8. What happens to the allele frequency and what causes this?

The + allele frequency changes dramatically within one population. This random fluctuation in allele frequency is termed genetic drift.

VII. Simulate Population Expansion

8. What do you notice about the allele frequency in those populations that maintain the + allele over 200 generations?

The + allele frequency drifts during the first 100 generations, but it stabilizes in the expanded population. The larger population is nearing Hardy–Weinberg equilibrium.

LABORATORY 2.3: USING A SINGLE-NUCLEOTIDE POLYMORPHISM TO PREDICT BITTER-TASTE ABILITY

Answers to Bioinformatics Questions

I. Use BLAST to Find DNA Sequences in Databases (Electronic PCR)

2.i. What do you notice about the lengths (and colors) of the matches (bars) as you look from the top to the bottom?

Typically, most of the significant alignments will have nearly complete matches to the forward and reverse primers. Partial alignments at the bottom of the list may correspond to only the forward or reverse primer.

2.ii. What is the E value of the most significant hit and what does it mean?

The most significant hit has a very low E value (e.g., 6E-12). An E value of 6E-12 denotes 6×10^{-12} or a 6 preceded by 11 decimal places! This means that the query has found strong matches in the database.

Note the names of any significant alignments that have E values of less than 0.1. Do they make sense? What do they have in common?

Yes. They are all examples of the *TAS2R38* bitter-taste receptor in humans and other primates.

3.i. What is the source and size of the sequence in which your BLAST hit is located?

The first hit is to *Homo sapiens* bitter taste receptor TAS2R38. The sequence is 1002 bp long.

4.ii. Note that position 43 of the forward primer is missing. What does this mean? To which positions do the primers match in the subject sequence?

This nucleotide did not match the genomic sequence. The position results vary, depending on the hit examined. For accession AC073647.9, the forward and reverse primers match positions 55,239–55,280 and 55,459–55,435, respectively.

4.iii. What is the difference between the coordinates?

55,459 – 55,239 = 220 nucleotides. These are the coordinates in accession AC073647.9.

4.iv. Note that the actual length of the amplified fragment includes both ends, so add 1 nucleotide to the result that you obtained in Step 4.iii to obtain the exact length of the PCR product amplified by the two primers.

220 + 1 = 221 nucleotides.

II. Use Map Viewer to Determine the Chromosomal Location of the *TAS2R38* Gene

6. On which chromosome have you landed?

Chromosome 7.

8. What is the function of the *TAS2R38* gene?

The *TAS2R38* gene encodes a bitter-taste receptor.

9. What genes are found on either side of *TAS2R38*?

Its nearest neighbors are *CLEC5A* and *MGAM*.

10.i. Determine the size of the *TAS2R38* gene using the map coordinates to the left of the "Contig" map.

TAS2R38 is ~1140 bp in length.

10.ii. How many introns and exons does the *TAS2R38* gene have?

TAS2R38 has a single coding exon.

10.iii. How do the structures of the neighboring genes differ from the structure of the *TAS2R38* gene?

These genes have multiple coding exons with intervening introns.

11. What are the functions of the neighboring genes?

CLEC5A is a lectin domain gene and *MGAM* is a gene that encodes a starch-digesting enzyme.

12. What do some of these genes have in common with *TAS2R38*? What can you conclude about the chromosomal locations of genes with similar functions?

Many of the genes encode taste and olfactory receptors. *TAS2R38* is part of a "cluster" of sensory receptors, each having a single exon. As you scroll up, there is a gene for an olfactory receptor (OR), *OR9A4*, followed by three nonfunctional OR pseudogenes (all ending in "P"). Further up are three other members of the *TAS2R* family of taste receptors. Clustering of genes according to function is seen in many areas of the human and other genomes.

III. Use Multiple Sequence Alignment to Explore the Evolution of the *TAS2R38* Gene

7.i. Why is the initial stretch of the alignment (as well as the last stretch of the alignment) highlighted in yellow?

The beginning of the gene is not amplified by the primers in this experiment.

7.ii. At what positions does the amplicon track along with the two human alleles?

The amplicon tracks along with the taster and nontaster alleles from positions 101 to 321.

7.iii. At what position in the sequence is the SNP that you examined in this experiment? What is the nucleotide difference between the taster and nontaster alleles?

The SNP is at position 145; there is a C in the taster allele and a G in the nontaster allele.

8.i. In the left column of the table below, list the nucleotide position(s) of all SNP(s) that you see in the sequence, including the SNP that you identified in Step 7.iii. Then, in the columns labeled "Nucleotide," record the nucleotides that are present for each allele. What is one combination of three nucleotides called?

The complete table is below Step 8.iii. The three nucleotides are inherited as a unit, or a haplotype. The haplotype C-C-G correlates most strongly with bitter-tasting ability. In coding sequence, triplets of nucleotides are called codons. Each codon codes for an amino acid.

8.ii. In the table below (under "Codon"), list the codons that are affected by the SNP(s).

The complete table is below Step 8.iii.

8.iii. Use a standard genetic code chart to determine which amino acids use the codons that you identified in Step 8.ii. Record your findings in the table below (under "Amino Acid").

SNP position	PTC-taster allele			PTC-nontaster allele		
	Nucleotide	Codon	Amino acid	Nucleotide	Codon	Amino acid
145	C	CCA	proline	G	GCA	alanine
785	C	GCT	alanine	T	GTT	valine
886	G	GTC	valine	A	ATC	isoleucine

9.i. What nucleotide is present at each of these positions in the chimpanzee, bonobo, and gorilla sequences?

The nucleotides in the chimpanzee, bonobo, and gorilla sequences are C, C, and G at positions 145, 785, and 886, respectively.

9.ii. Based on your findings in Step 9.i., are chimpanzees, bonobos, and gorillas tasters or nontasters? What does this indicate about the functional significance of bitter-taste receptors over evolutionary time?

The ancestral state of the *TAS2R38* gene corresponds to the taster allele, so the nontasting alleles arose after the human lineage split from other primates. Other primates are PTC tasters, suggesting that the ability to detect bitter tastes has a selective advantage in avoiding poisonous plants, many of which are bitter.

9.iii. What patterns do you notice in SNPs at other locations in the gene?

At some positions, one of the apes shares the SNP with humans. At other positions, apes share one SNP and humans share another. The bonobo differs from humans and other apes at a number of positions.

10.i. At what nucleotide positions does the primer bind?

The forward primer binds within the *TAS2R38* gene from nucleotides 101–144.

10.ii. What discrepancy do you notice between the primer sequence and the human *TAS2R38* sequences? At what position is this discrepancy located?

There is a single mismatch at position 143, where the primer has a G and the gene has an A.

Answers to Results and Discussion Questions

I. Determine Your PTC Genotype

6. **How would you interpret a lane in which you observe primer dimer but no bands as described in Steps 3 and 4?**

   The presence of primer dimer confirms that the reaction contained all components necessary for amplification, but it also indicates that there was insufficient template to amplify the target sequence.

II. Correlate Your PTC Genotype with Your PTC Phenotype

3. **According to your class results, how well does the *TAS2R38* genotype predict the PTC-tasting phenotype? If the genotypes were not predictive of the phenotypes, explain why this might be the case.**

   The presence of a T allele generally predicts PTC tasting, although heterozygotes are more likely to be weak tasters. Even in a relatively simple genetic system, such as PTC tasting, one allele rarely has complete dominance over another. This experiment examined only one of several mutations in the *TAS2R38* gene that influence bitter tasting. Furthermore, variability in taste perception is likely affected by processing in the brain, which involves numerous other genes.

III. Discover How the HaeIII Enzyme Discriminates the PTC-taster and -nontaster Alleles

1. **How is the forward primer sequence different from the human *TAS2R38* sequence?**

   The primer has a G where the gene has an A.

2. **What characteristic of the PCR allows the primer sequence to "override" the natural gene sequence? Draw a diagram to support your contention.**

   The dynamics of replication demand that every PCR product incorporates each of the two primers and that the PCR products are complementary to the primer sequences. The G in the forward primer is carried forward into all products of PCR amplification, but the A (in the template) is not. Thus, the A in the gene sequence is replaced by a G in each of the amplified products. This G "creates" an HaeIII recognition sequence that is not naturally present in the *TAS2R38* gene.

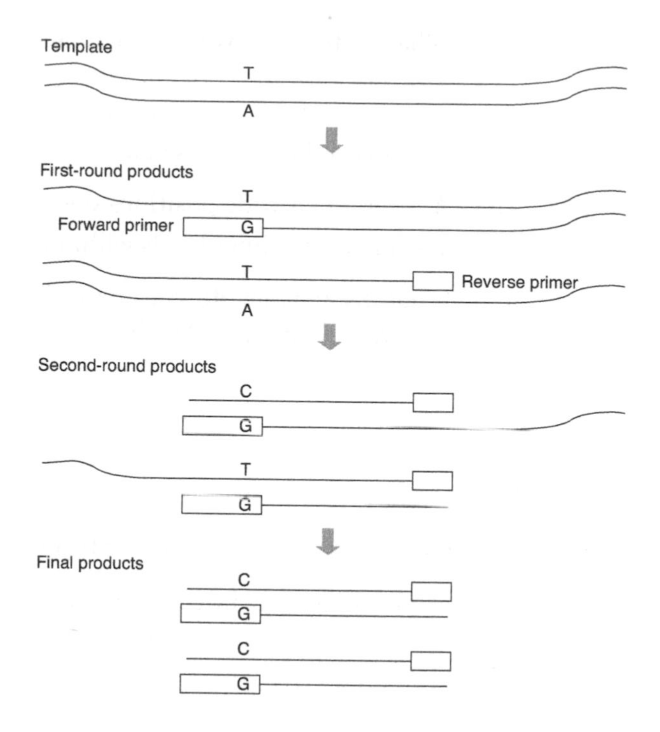

3. **The forward primer used in this experiment did not incorporate the fourth nucleotide of the HaeIII recognition sequence. Why was this important for the experiment? How did this enable the HaeIII enzyme to discriminate between the taster and nontaster alleles?**

   HaeIII cuts at the sequence GGCC. After PCR, this sequence is found at nucleotide positions 143–146 of the *TAS2R38* gene segment. A G nucleotide at position 145 of the nontaster allele changes the sequence into GGGC; this sequence is not recognized by the restriction enzyme. Therefore, HaeIII cuts the taster allele (GGCC) but not the nontaster allele (GGGC).

IV. Apply and Expand Your Knowledge

1. **Define the terms synonymous and nonsynonymous mutation. Is the C > G polymorphism in the *TAS2R38* gene a synonymous or nonsynonymous mutation? By what mechanism does this influence bitter-taste perception?**

   A synonymous mutation specifies the same amino acid as the wild-type allele; this is due to the redundancy of the genetic code. A nonsynonymous mutation creates a new codon that specifies a different amino acid. The G-to-C change at position 145 changes the codon CCA (proline) to GCA (alanine). This amino acid change alters the ability of the *TAS2R38* receptor to bind PTC in a lock-and-key fashion.

2. **Research other polymorphisms in the *TAS2R38* gene and how they may influence bitter-taste perception.**

   The *TAS2R38* gene contains five SNPs (single-nucleotide polymorphisms), three of which particularly influence bitter-taste perception. These three SNPs are inherited as a unit, with one combination, or haplotype—proline/alanine/valine (PAV)—correlating most strongly with bitter-tasting ability.

3. **Under some circumstances, balancing selection can produce heterozygote advantage, where heterozygotes are fitter than homozygous dominant or recessive individuals. What advantage might this be in this case?**

   Scientists are uncertain, but it may be that the nontasting alleles produce receptors that bind different kinds of bitter molecules. In this case, heterozygotes would have the advantage of detecting a greater range of potentially toxic molecules.

4. **Research how the methods of DNA typing used in this experiment differ from those used in forensic crime labs. Focus on (a) types of polymorphism used, (b) methods for separating alleles, and (c) methods for ensuring that samples are not mixed up.**

   (a) The Federal Bureau of Investigation (FBI) Combined DNA Index System (CODIS) uses a panel of 13 short tandem repeat (STR) polymorphisms for forensic DNA typing. (b) Each STR locus is labeled with one of four fluorescent dyes, and the alleles are differentiated by similar methods used in DNA sequencing. (c) Forensic DNA laboratories use a strict "chain of custody" to ensure that samples remain with their correct identifying label. Validated lab methods ensure that labels are checked during each step of the procedure.

5. **What ethical issues are raised by human DNA typing experiments?**

   Has the DNA sample been obtained with the willing consent of the donor, who understands the purpose for which it is being collected? Is the DNA used only for the express purpose for which it is collected, or is it also used for reasons other than those described to the donor? Is the DNA sample destroyed after its intended use, or is it stored for future use? Are the experimental results stored anonymously? Who has access to the results? Does the result of the experiment provide any unintended information—for example, about disease susceptibility or paternity?

CHAPTER 3

Plant Genomes

Vannevar Bush
(Courtesy of the Library of Congress, LC-USZ62-36967.)

Mary Lasker
(Courtesy of the Library of Congress, LC-USZ62-114993.)

As primary producers that perform photosynthesis to convert sunlight energy into chemical energy, plants are the ultimate source of all food, fiber, and carbon-based energy on Earth. Even so, the molecular genetic and genomic analysis of plant cells has lagged behind investigation of prokaryotic and animal cells. This is in large part due to research priorities and allocation of federal funds. In the United States, a national mandate to improve human health led to the ascendancy of the National Institutes of Health (NIH) during the second half of the 20th century. This focus on health included bacterial and viral pathogens, as well as mammalian models for cellular and physiological processes.

Humans are naturally self-absorbed with improving the quality and length of their own lives. After World War II, this desire was galvanized into political will by Vannevar Bush. As director for the Office of Scientific Research and Development (OSRD), he had marshaled the academic research effort in support of the war. In his 1945 report, *Science, The Endless Frontier*, Bush eloquently argued that the federal government should maintain the strong sponsorship of academic research it had established during the war—but bent it toward fostering improved public health. This led to the rapid expansion of the NIH and its rise as the preeminent medical research organization in the world. In the 1960s, Mary Lasker joined with Senators J. Lister Hill and John E. Fogarty to raise public and political awareness of cancer, which stimulat-

Senator J. Lister Hill (*left*) and Representative John E. Fogarty (*right*) with President John F. Kennedy at the 1960 signing of the International Health Research Act. Hill and Fogarty joined with private citizen Mary Lasker to form a legislative triumvirate that increased the power of the National Cancer Institute during the 1960s.
(Courtesy of Mary McAndrew.)

ed the ascension of the National Cancer Institute as the largest of all of the national institutes.

Plant research received early federal support through the system of land grant universities established by Congress in 1862. Eventually sponsored by the Department of Agriculture, this decentralized research program produced the "green revolution" that dramatically increased crop yields through a combination of improved plant cultivars, petrochemical fertilizers, and mechanization. However, despite the lobbying of congressmen from agricultural states, funding for the Department of Agriculture fell far behind the NIH in the post–World War II era. In 2009, a strategic move was made to put plant research on the same plane with disease-related research when a National Institute of Food and Agriculture was created within the Department of Agriculture. Even so, at $32.5 million, the Department of Agriculture's 2012 research budget for external research on crop and food development was only about 1% of the NIH's $25.5 billion budget for external research on health and disease.

Practically, plants present several barriers to genetic and genomic research. Most significantly, plant cells are surrounded by a "wooden box." The cellulose outer wall makes plant cells more difficult to grow in culture and can be a real barrier to transformation, a key step in the genetic manipulation of a species. Although reliable methods to transform bacterial and mammalian cells were developed in the 1970s, dicot plants were not readily transformed until the 1980s. A reliable transformation system for monocot plants, including all of the world's cereal crops, was not available until the 1990s.

Furthermore, most agriculturally important plants have large and complicated genomes. Although we think of living things as having diploid genomes with two copies of each chromosome, many common plants are polyploid, having four (tetraploid) or more copies of each chromosome. This complicates genetic analyses and makes recessive mutations difficult to observe. Polyploidy and high proportions of repetitive DNA make whole-genome sequencing difficult.

Ploidy and chromosome number of some crop plants

Plant	Chromosome number
Diploid (2*n*)	
Barley	14
Cabbage	18
Common bean	22
Maize	20
Pea	14
Pepper	24
Potato	24
Rye	14
Rice	24
Sorghum	20
Tomato	24
Diploid, triploid (2*n*, 3*n*)	
Apple	34, 51
Banana	22, 33
Pear	34, 51
Tetraploid (4*n*)	
Alfalfa	32
Cotton	52
Peanut	40
Potato	48
Soybean	40
Tobacco	48
Hexaploid (6*n*)	
Oats	42
Sweet potato	90
Heptaploid (8*n*)	
Sugarcane	80
Strawberry	56
Other	
Wheat (2*n*, 4*n*, 6*n*)	14, 28, 42
Coffee (2*n*, 4*n*, 6*n*, 8*n*)	22, 44, 66, 88

ORIGINS OF WHEAT AND RICE CULTIVATION

All organisms exploit their environments and temporarily alter natural landscapes when they eat or nest. Many develop natural collaborations, or *symbioses*, with other species. However, humans are unique in their capacity to restructure environmental resources on a massive scale and to transport them over long distances. Perhaps the most advanced human attribute is our ability to restructure the genetic and physical attributes of other living things purposefully. This human-directed evolution of other species gave rise to domestic plants and animals.

The domestication of plants and animals is perhaps the single greatest civilizing factor in human history. Increased production and performance of domesticated organisms made possible urbanization and task specialization in human society. Thus, the labor of fewer and fewer farmers produced enough food and clothing materials to satisfy the needs of growing numbers of nonfarmers—artisans, engineers, teachers, administrators, and merchants—freeing them to develop other elements of culture.

For these reasons, biologists and historians have long been interested in the origins of agriculture. Different modern food plants arose 5000–10,000 BP (years before present time) from three major centers of ancient cultivation: east Asia (producing rice, taro, soybean, millet, citrus, melon, cucumber, mango); the Fertile Crescent (wheat, barley, grapes, peas, chickpeas, lentils); and the Americas (corn, potato, dry beans, tomato, squash).

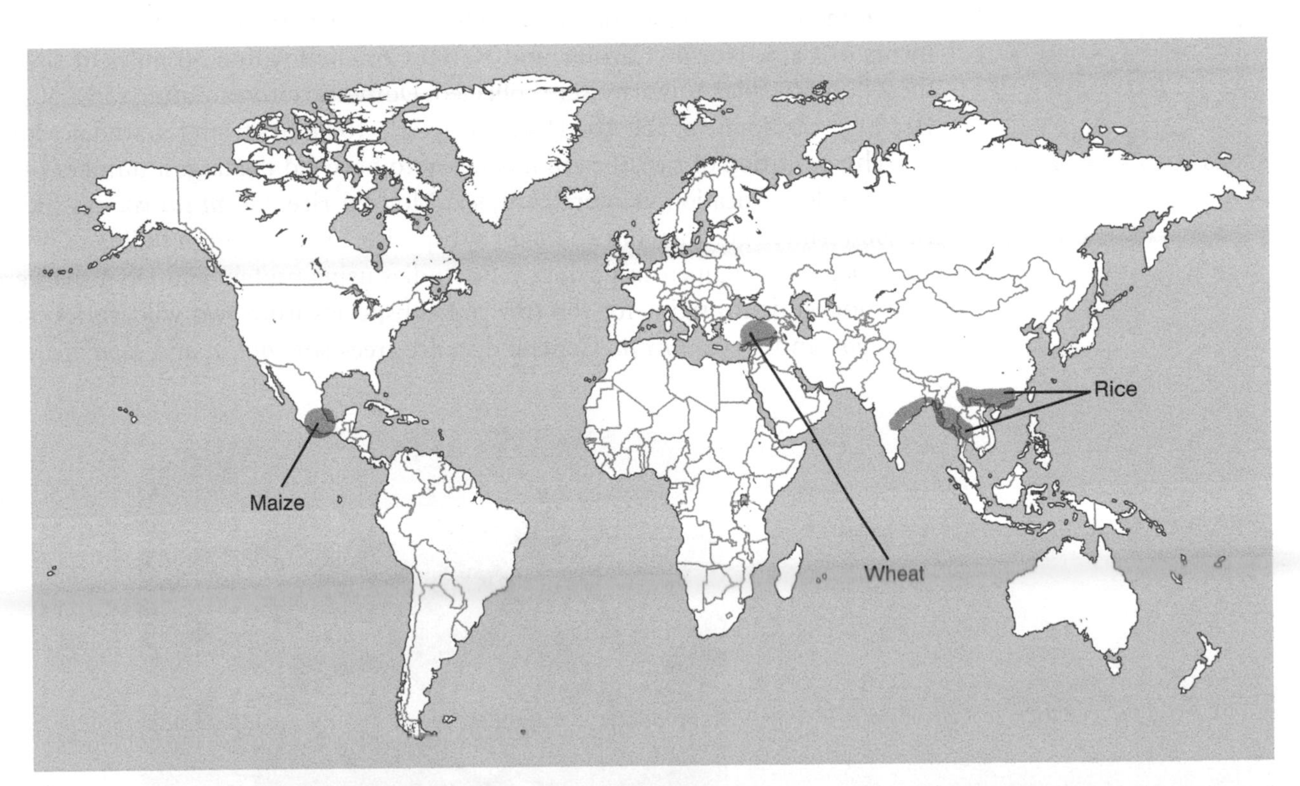

Cradles of maize, wheat, and rice domestication. Genetic studies reveal that the closest wild relatives of maize are in the Balsas River Valley of Central Mexico, and the closest wild relatives of wheat are near Karacadağ Mountain in southeastern Turkey. Wild relatives of *O. japonica* rice are found in southern China, whereas wild relatives of *O. indica* rice are found in eastern India, Thailand, and Myanmar.

(Adapted from www.Wikipedia.com.)

Grains are staple crops around which agrarian human societies have revolved. Wheat, rice, and corn account for almost 90% of grain production and nearly half of all calories consumed worldwide. Modern grains evolved from wild grasses. Sometime after the last Ice Age, people first had the idea to save some of the wild seeds they had gathered and plant them in some convenient place the following year. In watching the grain grow year after year, they would have selected plants with bigger seeds and, crucially, ones that were easier to harvest.

Studies on early agriculture have focused on the domestication of wheat (*Triticum* sp.) in the Fertile Crescent between the Tigris and Euphrates Rivers—including parts of modern-day Iran, Iraq, Syria, and Turkey—where many of the oldest human settlements have been found. Diploid einkorn and tetraploid emmer were the first domesticated wheats. Modern durum wheat—used in pasta, couscous, and breads—is a tetraploid derived from emmer. Common bread wheat is a hexaploid hybrid between emmer or durum and a wild diploid grass (*Aegilops cylindrica*).

Hakan Özkan
(Courtesy of Hakan Özkan.)

In 1997, Manfred Heun, of the Agricultural University of Norway, used 288 polymorphic markers to make DNA fingerprints of 194 wild and 68 domesticated lines of einkorn wheat from throughout the Fertile Crescent. A phylogenetic tree of genetic distances showed that domesticated einkorn is most closely related to wild varieties found near Karacadağ Mountain in southeastern Turkey. Hakan Özkan, of the University of Çukurova, Turkey, subsequently used similar methods to show that the closest wild ancestors of tetraploid emmer wheat are also found in this same small area of southeastern Turkey.

Domesticated einkorn dating from 9200–8200 BP was found at the ancient settlements of Cafer Höyük, Çayönü, and Nevalı Çori—all within 50 miles of sites where the wild Karacadağ einkorns were collected. Domestic emmer dating to 10,600 BP and 10,400 BP was found at Tell Abu Hureyra, Syria, ~100 miles from Karacadağ Mountain.

The evolution of rice, the grain believed to have fed the largest number of people since its domestication, is more complex. Modern rice, *Oryza sativa*, has more than 120,000 named cultivars of two major subspecies: lowland *Oryza indica* and highland *Oryza japonica*. Numerous phylogenetic studies using protein and DNA markers suggest independent origins for the two rice subspecies from two wild varieties, *Oryza rufipogon* and *Oryza nivara*. Genetic distance trees sort *indica* and *japonica* cultivars

Manfred Heun, 2008, collecting near Karacadağ Mountain in southeastern Turkey.
(Courtesy of Nadja Pöllath.)

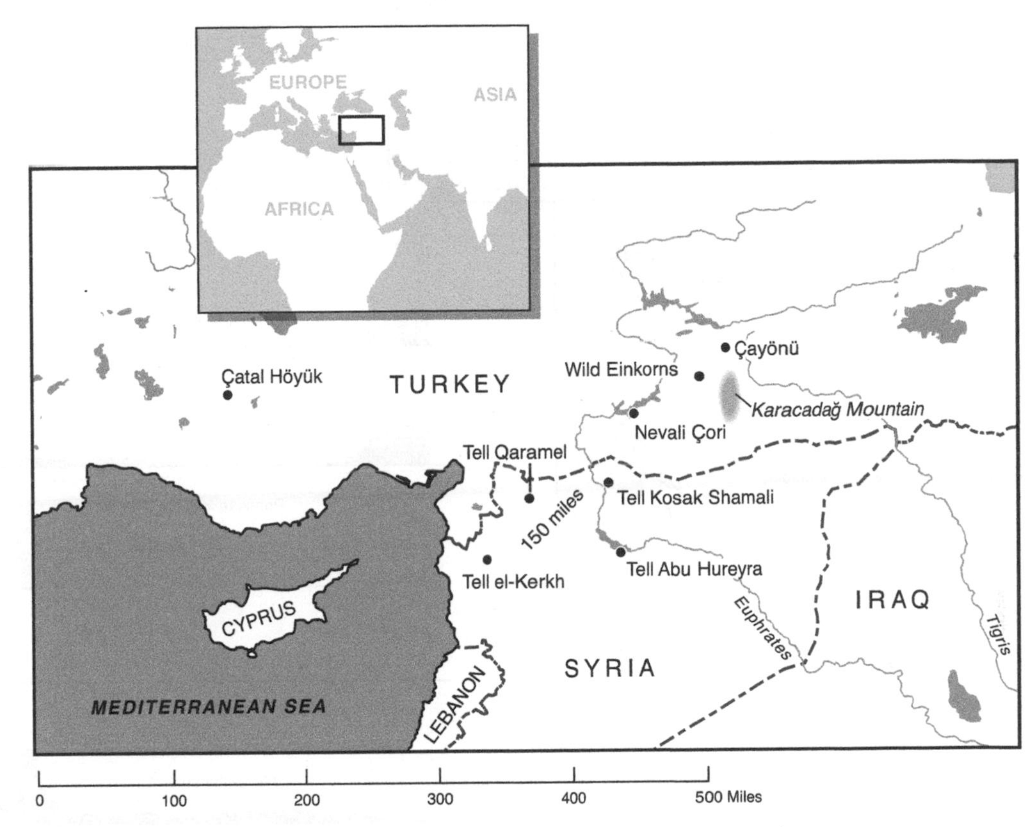

Cradle of wheat domestication. Genetic studies trace einkorn and emmer wheat to a core area near Karacadağ Mountain in southeastern Turkey, part of the Fertile Crescent between the Tigris and Euphrates Rivers. Domestic wheats dating as early as 10,600 BP have been found at several archeological sites in the region.

(Adapted from Lev-Yadun S, et al. 2000. *Science* 288: 1602–1603.)

into two distinct groups that are more closely related to wild varieties than they are to each other. Furthermore, molecular studies set the divergence of *indica* and *japonica* to 200,000–400,000 BP, also indicating their derivation from ancient wild stocks.

A 2006 study by Jason Londo and Barbara Schaal, of Washington University, St. Louis, examined sequence variation in 203 *japonica* and *indica* cultivars and 129 *rufipogon* varieties from India, China, and southeast Asia. They surveyed three gene regions representing different evolutionary histories: maternal (chloroplast *rbcL*), neutral (nuclear pseudogene *p-VATPase*), and functional (nuclear *S*-adnenosyl methionine, *SAM*). A correlation of common haplotypes indicated two major domestication events: *japonica* likely arose from wild rice populations in southern China, whereas *indica* originated in eastern India, Myanmar, or Thailand.

Jason Londo
(Courtesy of U.S. Environmental Agency, with permission from Jason Londo.)

The oldest evidence of rice cultivation was discovered in 2007 by Cheng Zong of Durham University. Paddy fields dating to 7700 BP were found at the early Neolithic site of Kuahuqiao, which is on the Yangtze River delta 100 miles south of Shanghai. The fields were next to earlier excavations of a Stone Age settlement of wooden houses built on stilts over a marshy area.

Assessing the pollen, microfossils, and charcoal in a soil profile extending over a 1500-year period showed that a brackish swamp had been cleared of scrub by burning, allowing grasses, including rice, to flourish. The disappearance of marine diatoms and their replacement by mainly freshwater forms suggested that the field had been man-

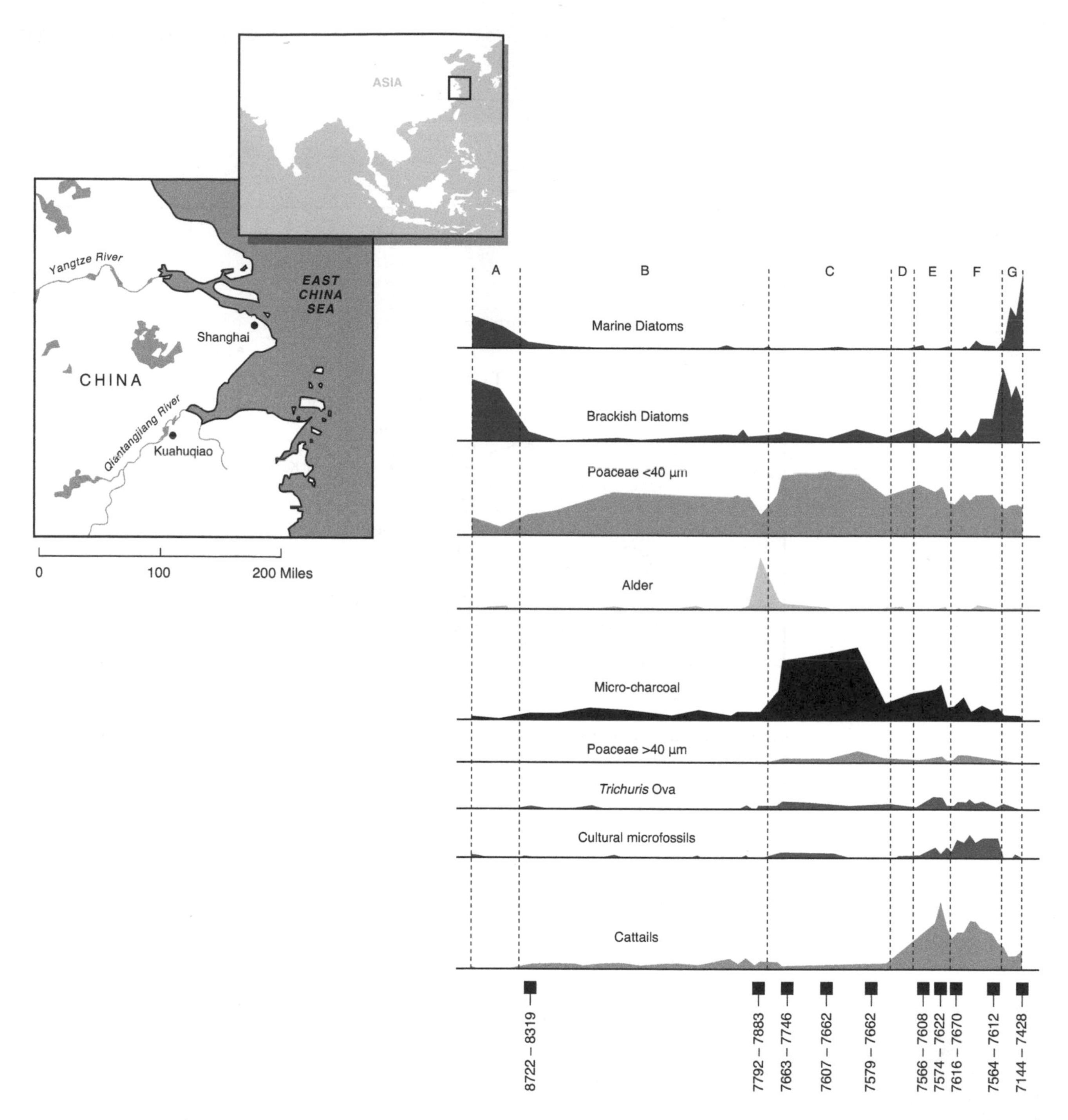

Early rice cultivation at Kuahuqiao, Southern China. The Kuahuqiao archeological site in the Qiantangjiang River delta (*left*) provides the earliest evidence of systematic rice cultivation, potentially including animal fertilizer. Phases A and B (*right*) show succession from a marine estuary to a freshwater swamp with scrubby trees. The sharp decline in marine and brackish diatoms indicates a change to mainly freshwater conditions ~9000 BP, with a gradual increase in grasses (pollen <40 µm) followed by a rapid rise in alder pollen ~7800 BP. Evidence of human management of the area begins ~7700 BP, with burning of the scrub marked by a sharp increase of charcoal. Large pollen (>40 µm) indicative of rice as well as ova of the human and pig parasite *Trichuris* appear. Cultural microfossils, including fungal spores associated with soil disturbance and decomposing animal dung, also appear. Continued presence of brackish diatoms suggests seasonal flooding of the fields at high tide. Controlled burning of brush not only made way for rice cultivation but also encouraged the growth of cattails, whose starchy rhizomes were also an important wild food. Human activities ended ~7550 BP, when the area was again inundated with saltwater.

(*Right* panel adapted from Zong Y, et al. 2007. *Nature* 449: 459–462.)

aged to exclude salt water and allow seasonal flooding—likely using a system of earthen bunds. The presence of ova of the parasitic *Trichuris* (trachina) worm, found in pig and human dung, indicates the use of fertilizer. Although only wild rice was found at the site, the high proportion of large grains suggested human selection for food. Thus, the Kuahuqiao site illustrates the early transition to agriculture, with hunter–gatherers settling down to manage the growth of a wild food resource.

DOMESTICATION SYNDROME

Karl Hammer, 1980, collecting barley in Libya.
(Courtesy of Karl Hammer.)

Nucleotide surveys show that domesticated grains have about two-thirds the genetic diversity of their wild ancestors. Thus, domestication involves decreasing natural diversity in favor of fixing a few important traits. In 1984, Karl Hammer, of the Central Institute for Genetics and Crop Plant Research in Gatersleben, Germany, coined the term *domestication syndrome* to describe the set of traits, selected either consciously or unconsciously by humans, that distinguishes a crop plant from its wild ancestor. These include larger fruits or grains, loss of natural seed dormancy and dispersion, apical dominance (growth from the central as opposed to side stems), synchronized flowering, and increased vigor and disease resistance.

The transition from wild to domesticated grain is best tracked by monitoring the loss of seed dispersion, or shattering. In wild grains, spikelets containing seeds detach, or shatter, from a supporting spike (rachis)—allowing dispersal by the wind. Nonshattering was an important trait in domestic grains, allowing seeds to stay attached to the rachis during harvest. Early agriculturalists would have unconsciously favored nonshattered grain, which stayed on the plant rather than falling to the ground.

Wild grains develop an abscission zone, a layer of thin-walled cells that break easily. This allows the seed to detach cleanly from the rachis, leaving a symmetrical abscission scar on the rachis and the base of the spikelet. Domestic grains were selected for mutations that produce an incomplete abscission zone—strong enough to allow the seeds to stay attached during harvest, but still weakened enough to be released by moderate threshing. This mechanical process leaves jagged scars on the rachis and the base of the spikelet.

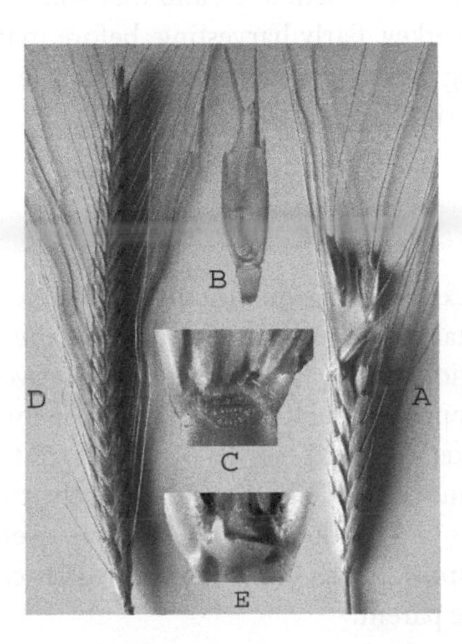

Evidence of nonshattering wheat. In wild einkorn wheat (*A*), the spikelet containing the seed (*B*) detaches (shatters) cleanly at the abscission zone, leaving a scar showing a circular pattern of vascular bundles (*C*). Domesticated wheat spikelets remain attached to the plant for easier harvesting (*D*). Threshing breaks off the spikelets above the abscission zone, leaving a jagged scar (*E*).
(Reprinted, with permission, from Tanno K, Willcox G. 2006. *Science* 311: 1886; ©AAAS.)

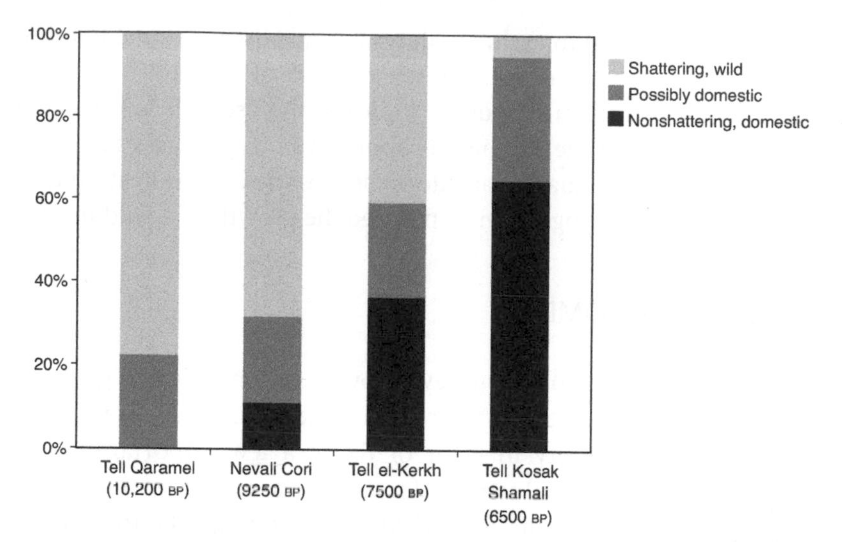

Domestication of wheat. Analysis of grain spikelets found at archeological sites in the Middle East dating from 10,200 to 6500 BP shows an increasing proportion of domestic wheat. This indicates that wheat was slowly domesticated in mixed fields with wild plants.
(Adapted, with permission, from Tanno K, Willcox G. 2006. *Science* 311: 1886; ©AAAS.)

A 2006 study by Ken-ichi Tanno, of the Research Institute for Humanity and Nature, Kyoto, and George Willcox, of the French Centre National de la Recherche Scientifique (CNRS), showed an increasing proportion of domestic-to-wild wheat spikelets collected from four ancient sites in northern Syria and southeastern Turkey dating from 10,200 to 6500 BP. Dorian Fuller, of University College London, found a similar situation in rice analyzed from the Neolithic site of Tianluoshan in the Yangtze Delta in 2009. Nonshattering spikelets increased from 27% to 39% over a 300-year period (6900–6600 BP), and of all plant remains, rice increased from 8% to 24%.

These findings confirmed an earlier analysis in barley, indicating that the cultivation of grains may have occurred gradually, with domesticated and wild plants growing in mixed fields over a period of several thousand years. This view is supported by the observations that there is no appreciable difference in the size of wild and domesticated grains found at ancient sites, and that wild wheat still occurs as a weed in cultivated fields in Turkey. Early harvesting, before many grains shattered, and a return to wild harvesting during crop failures would have made it difficult to select quickly for the nonshattering trait.

SHATTERING QTLS

Shattering is an example of a quantitative trait, one that varies continuously because of the incremental contributions of multiple genes. Quantitative trait loci (QTLs), chromosome regions associated with a quantitative trait, are typically identified by developing a mapping population from a cross between two homozygous plant varieties that show alternative expression of a given trait (e.g., + and −). F_1 progeny inherit one chromosome of each pair from each of the two parents. Chromosome regions from the + and − parents are exchanged (cross over at synapsis) during gamete formation in F_1 offspring. Each chromosome region carries markers that allow it to be traced back to the parent.

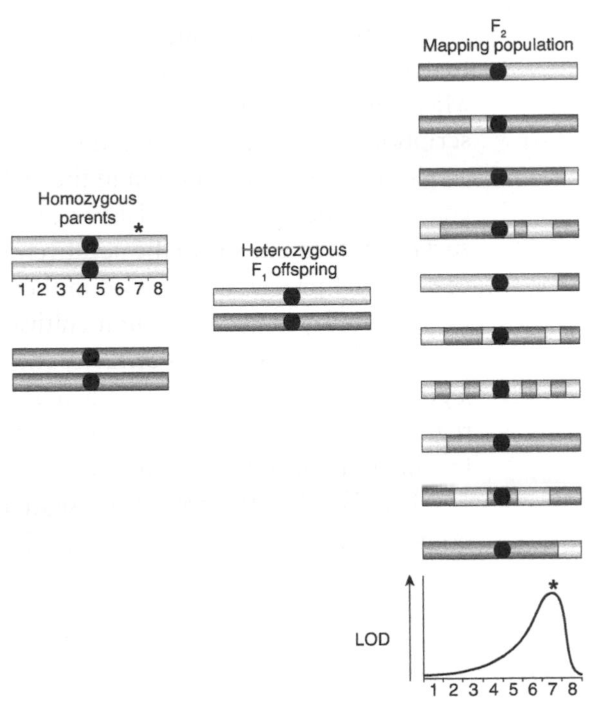

Quantitative trait locus (QTL) mapping. Homozygous parents, one of which expresses a trait of interest (*), are crossed. Then the heterozygous F_1 offspring are crossed to produce an F_2 mapping population with recombinant chromosomes. Inheritance of the trait correlates with markers from a specific region of an original parental chromosome (white). A logarithm of odds (LOD) score indicates the strength of linkage between the trait and markers on the parental chromosome.

Crossing F_1 plants produces F_2 offspring whose chromosomes are mosaics of the + and − grandparents. Alternately, a backcross between an F_1 plant and, for example, the − parent produces offspring in which chromosome fragments from the + parent chromosome are isolated against the background of a − genome. Either method produces a mapping population in which expression of the quantitative trait can be linked to chromosome segments derived from the + or − grandparent.

The relative contribution of each QTL, the *variance*, is based on the number of F_2 offspring having that QTL relative to the total number expressing the quantitative trait. The chromosome region that corresponds to a QTL that explains a significant amount of variation is scanned for candidate genes that bear a logical physiological or biochemical relationship to the trait. Then, plants are screened for SNPs or insertions/deletions (indels) that produce amino acid changes or may alter the regulation of a candidate gene.

Two QTLs for rice shattering were cloned in 2006; each of these explained 69% of the variance in separate mapping populations. The degree of seed shattering varies among rice varieties, with *O. indica* cultivars generally having greater shattering. Therefore, Saeko Konishi, of the Japanee Institute of the Society for Techno-Innovation of Agriculture, Forestry, and Fisheries, searched for QTLs in a cross between a shattering *indica* variety (Kasalath) and a nonshattering *japonica* variety (Nipponbare). He identified a strong QTL, *qSH1*, which is an ortholog of the *Arabidopsis* homeobox gene *REPLUMLESS* (*RPL*). A SNP (G>T) in the 5′ regulatory region was responsible for nonshattering in Nipponbare and other *japonica* cultivars. Crossing, or transforming the Kasalath *qSH1* allele into Nipponbare, rescued seed shattering, with formation of a complete abscission layer.

Multiple studies using crosses between *indica* and *japonica* strains and wild ancestors identified another locus, *sh4*, as a strong QTL. Changbao Li and Tao Sang, at Michigan State University, cloned the *sh4* gene, which encodes an Myb3 family transcription factor. They discovered that domestic cultivars differ from wild varieties by a single-nucleotide substitution in the DNA-binding domain, which replaces a positively charged lysine with a neutral asparagine. This loss of positive charge may be critical for surfaces that interact with the negatively charged DNA molecule. In vivo studies showed that the sh4 protein is expressed in the abscission layer. Domestic cultivars produce an incomplete layer, and a *japonica* cultivar engineered to express the wild-type *sh4* produced an improved abscission layer. The nonshattering allele *sh4* is fixed in all *indica* and *japonica* cultivars, and a 10-fold reduction in polymorphisms in the chromosomal regions that flank the nonshattering allele indicates a human-induced selective sweep for the nonshattering haplotype.

The fact that the crucial nonshattering allele of *sh4* arose only once during rice domestication conflicts with genetic distance studies suggesting that rice was domesti-

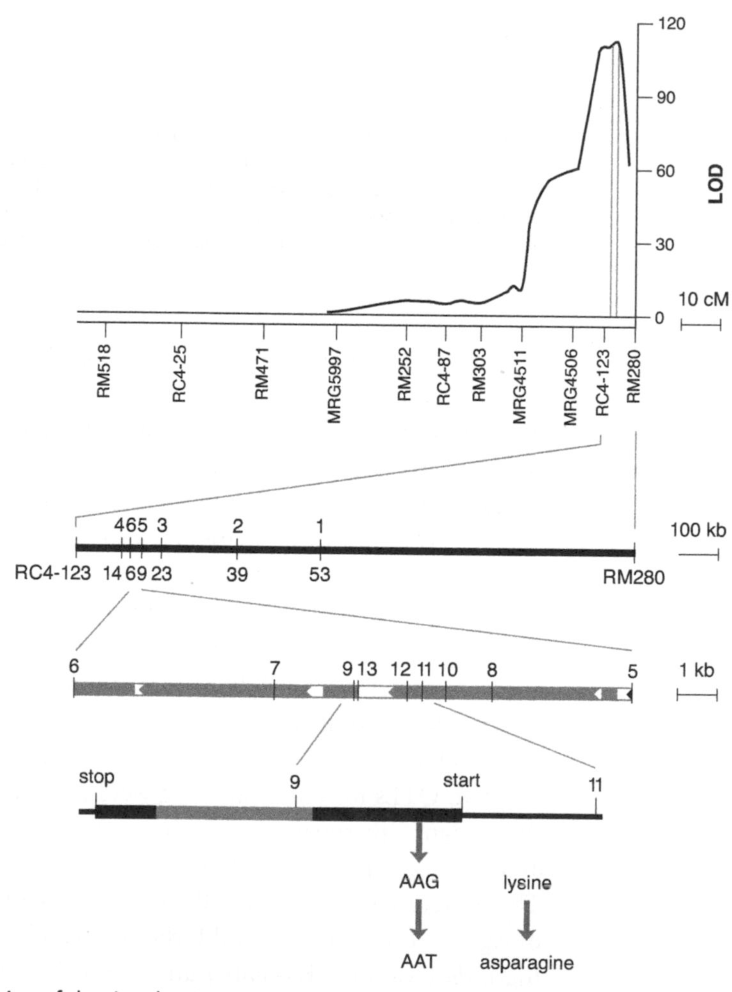

Molecular cloning of the rice shattering gene *sh4*. A strong QTL for shattering was identified on chromosome 4 from crosses between domestic *O. indica* and *O. japonica* strains with wild rice ancestors. The region of the highest LOD score was located between the simple sequence repeat (SSR) markers RC4-123 and RM280. Additional markers (1–12) were used to map the shattering-associated mutation to successively shorter regions. The *sh4* gene, centered on marker 9, has a G>T substitution found exclusively in nonshattering domestic strains.

(Adapted, with permission, from Li C, et al. 2006. *Science* 311: 1936–1939.)

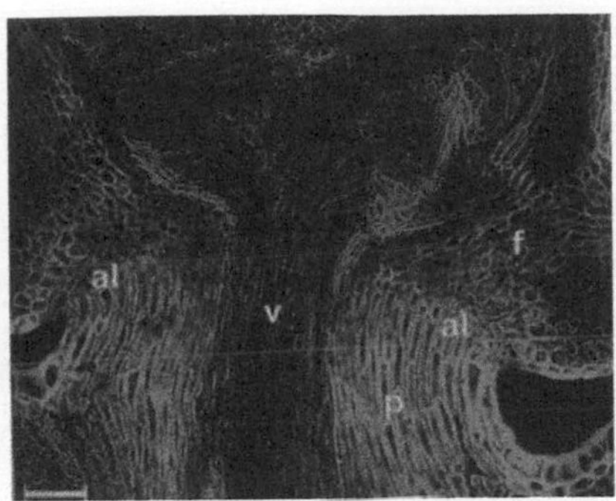

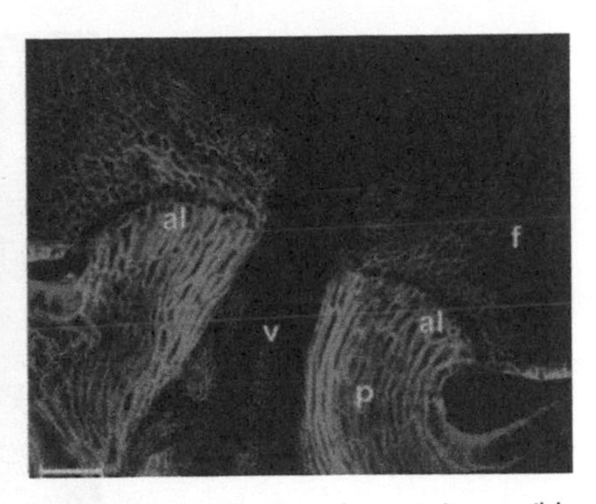

Effect of *sh4* allele on rice abscission layer. (*Left*) Micrograph of wild rice ancestor *O. nivara* (expressing a wild-type allele of *sh4*) shows a complete abscission layer (al) separating the pedicel from flower. (*Middle*) An *Oryza sativa japonica* cultivar (expressing a mutant *sh4* allele) shows an incomplete abscission layer. (*Right*) The same *japonica* cultivar transformed with a wild-type *sh4* allele shows an improved abscission layer.
(Reprinted, with permission, from Li C, et al. 2006. *Science* 311: 1936–1939; ©AAAS.)

cated twice from separate wild populations in southern China and eastern India/Indochina. These views can be reconciled by either of two models. A combination model proposes that early domestic cultivars developed separately from diverged wild ancestors in two regions. Crosses made by farmers allowed important alleles of domestication genes—including *sh4*—to flow between regions and become rapidly fixed across populations. An alternative "snowball" model proposes that a core set of domestication alleles—including *sh4*—was fixed in an early cultivar in one of the two major rice homelands. This cultivar then rolled along into the other regions, picking up traits through introgression with diverged local populations of wild rice. Hybridization with local plants then adapted the core domestication package to local climates and agricultural practices. The *qSH1* allele appears to have arisen independently in *japonica* cultivars, improving the initial nonshattering phenotype provided by *sh4*.

DOMESTICATION OF MAIZE

It could be argued that no single plant has been so inextricably tied to human culture as corn, known as *Zea mays* to scientists and as maize to the world outside of the United States. Maize gods and goddesses are central icons of every pre-Columbian culture of Mexico, including Mayan, Toltec, Zapotec, Mixtec, and Aztec. The Mayans believed that the gods literally created people from corn, and Aztecs named corn's wild ancestor, teosinte, "grain of the gods." There was good reason for this veneration. Maize was the basic source of sustenance for all pre-Columbian Indians. The large-scale cultivation of maize made possible the numerous and vast pre-Columbian cities—some with 100,000 or more residents—and the elaborate cultures that flourished in them. Social and religious life was organized around key events in maize cultivation—such as preparing the fields, sowing the seeds, blessing the new shoots, protecting the harvest, counting lost kernels, and preparing the first food from the harvest.

By any measure, maize is the most important crop in the Americas, accounting for $47 billion in 2010 revenue in the United States alone. Because of its economic importance and long history of cultivation, maize has been intensively studied as a model genetic organism since the early 1900s. Therefore, much of this chapter focuses on the genetics and genomics of this uniquely American plant.

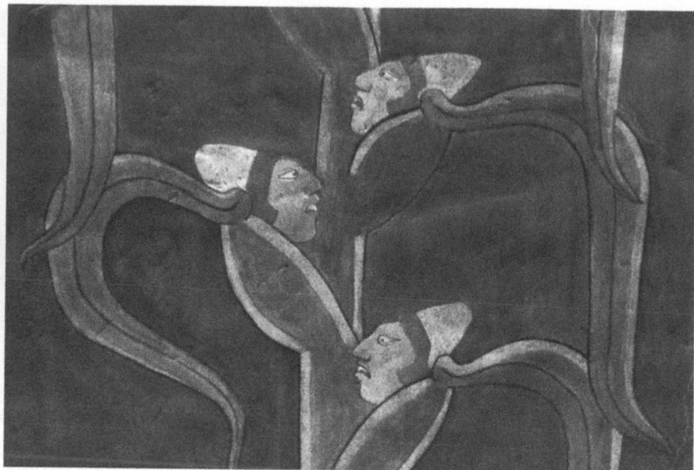

Pre-Columbian veneration of maize. (*Left*) Maize gods and goddesses are central icons of every pre-Columbian culture of Mexico, including Mayan, Toltec, Zapotec, Mixtec, and Aztec. (*Right*) A polychrome mural from the pyramid at Cacaxtla, with human heads emerging from maize husks, illustrates the Mayan belief that humans were literally "children of the corn."

Perhaps no other plant has undergone such a complete physical makeover at the hands of humans. Teosinte ears have only five to 12 small kernels, in a distichous arrangement of two opposing rows on a central rachis (stalk). The ripe kernels readily release (shatter) and fall to the ground. A flinty outer cover protects kernels from digestion by grazing animals and birds, allowing dispersal in their dung. In contrast, an ear of corn has 500–1200 polystichous kernels, with many ranks arranged radially around the rachis. The large, soft kernels are easily digested by animals. Attached to the ear when they fall to the ground, the hundreds of germinating kernels crowd one another for light and soil nutrients. Thus, modern corn is so adapted to cultivation that it cannot reproduce successfully without human intervention.

Teosinte and modern domestic maize. Teosinte has distichous ("two-rowed") ears with five to 12 kernels. Maize is polystichous ("many-rowed") with 500–1200 kernels. The teosinte stalk has many branches, whereas maize has a single dominant stalk.

(Photo from Nicolle Rager Fuller, National Science Foundation.)

Considering the striking differences in ear architecture, it is difficult to understand how teosinte could have evolved into modern corn. Moreover, how was the food potential of maize recognized, when the kernels of teosinte are covered with an inedible outer covering? Despite mounting evidence that maize evolved from teosinte, the close relationship between the two was vigorously debated throughout the first decades of the 20th century.

Teosinte and corn plants look very similar, with wide, strap-like leaves that are atypical of other grasses. Teosinte grows as a weed in cornfields across its range in Mexico and Central America. By the 1920s, it was known that crosses between teosinte and maize produce fertile hybrids. In the early 1930s, Rollins Emerson, of Cornell University, and George Beadle, of the California Institute of Technology, showed that the chromosomes of teosinte–maize hybrids pair completely and cross over at the same rate as maize–maize hybrids. This indicated that maize and teosinte are the same species, consistent with one evolving from the other.

In 1939, Beadle, then at Stanford University, provided an articulated hypothesis that teosinte is the sole progenitor of maize and that changes in only four to five major genes would be sufficient to produce the cultivated phenotype. Famously, he provided his key critics with a demonstration of how Mesoamericans could have first exploited teosinte as food: He heated teosinte kernels until they exploded, casting off their flinty coats and leaving behind edible popcorn.

Thus, one can envision a scenario in which Neolithic hunter–gathers burned sheaves of teosinte as fuel for their campfire, and popcorn conspicuously announced itself as the world's first fast food. Rather quickly, the ancient people would have identified stands of teosinte that made the biggest and best-tasting popcorn and brought them into camp for popping. Hunter–gatherers established seasonal movements and visited the same sites year after year. Thus, teosinte kernels that fell or were discarded

Cornell maize group, 1929. (Standing, *left to right*) Charles Burnham, Marcus Rhoades, Rollins Emerson, and Barbara McClintock. (Kneeling) George Beadle.
(Courtesy of the Archives, California Institute of Technology.)

Paul Mangelsdorf
(Courtesy of Economic Botany Archives of Oakes Ames, Harvard University.)

John Doebley, ca. 1993
(Courtesy of John Doebley.)

in trash near the campsite would be growing when the group returned the following year. In this way, stands of teosinte would have developed near permanent campsites.

The ancient Mesoamericans would have cultivated the teosinte stands by removing weed species and adding seeds from favored plants they came across in the wild. Eventually, they would have selected less bushy plants that made harvest easier, and, of course, ones with larger cobs and less-flinty kernels. For some time they would have continued to eat the cultivated maize as popcorn, but they would have also learned to grind the popped corn into a fine meal that could be mixed with water and eaten as porridge. Eventually they would have directly ground the maize kernels and learned the many uses of maize meal, such as making a sheet of flat, unleavened bread, or tortilla.

The early use of maize as popcorn is supported by the fact that popcorns are among the most ancient of cultivated maize varieties. A 1948 excavation at Bat Cave in New Mexico by Paul Mangelsdorf and C. Earle Smith of Harvard University identified ears of a popcorn variety in the oldest stratum, dated to 4500 BP. In more recent strata, they found strong evidence of introgression (exchange) of genes from teosinte, including ears with distichous kernels.

Frequent crossing with local populations of teosinte helped adapt the early maize varieties to local growing conditions, yet maintained their high genetic diversity. Sixty indigenous races of Mexican maize have been identified. With an overall nucleotide diversity of 2%–3% at neutral sites, any two races of indigenous maize are as genetically distinct from one another as a human is from a chimpanzee. Human genetic diversity is equivalent to self-fertilizing (selfing) maize for four generations.

Following up on Beadle's hypothesis for only a few major domestication genes, in 1993, John Doebley, of the University of Wisconsin, looked for QTLs in a mapping population from a cross between teosinte and a primitive maize. He reasoned that using a primitive maize would highlight genes involved in domestication, rather than genes responsible for later yield improvements. The analysis identified six major QTLs on

C. Earle Smith (standing, *second from right*) with members of the Bat Cave expedition, New Mexico, July 1948.
(Reprinted, with permission, from National Anthropological Archives, Smithsonian Institution.)

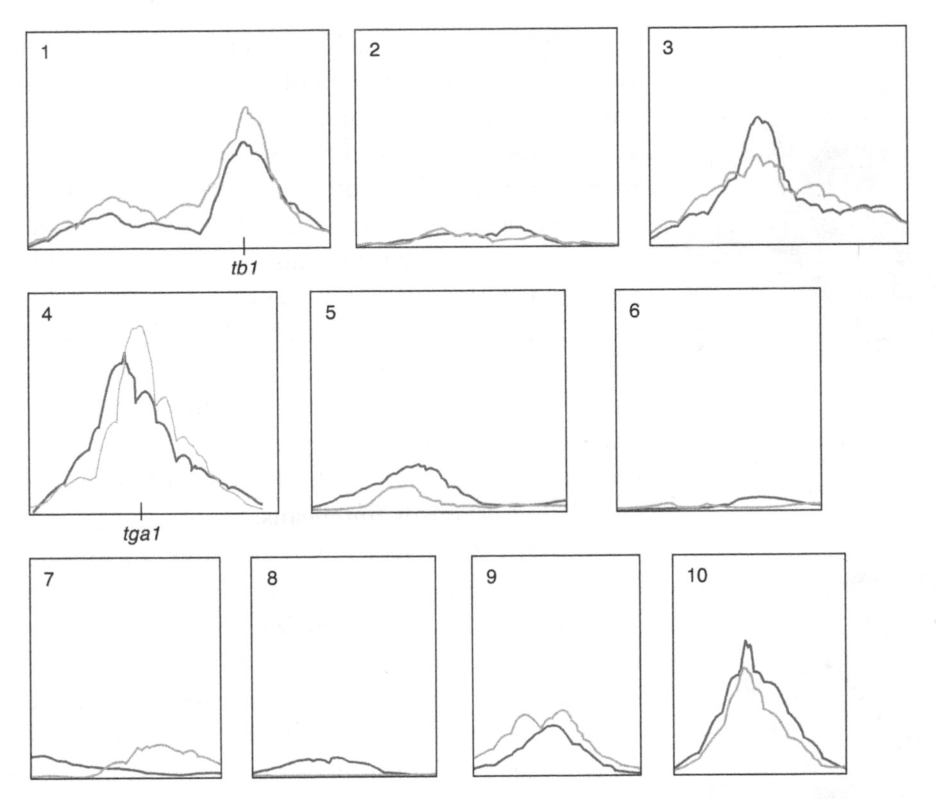

Domestication QTLs in maize. Plot of 10 teosinte–maize chromosomes showing the significance level of QTL effects on seven domestication traits influencing flower architecture. Plants were grown in two environments, Wisconsin (dark gray line) and Florida (gray line). Mutations in *tb1* (*teosinte branched 1*) reduce the large number of tillers, or auxiliary stalks, in teosinte to a single dominant stalk in maize. Mutations in *tga1* (*teosinte glume architecture 1*) reduce the stony glume that surrounds the teosinte kernel.
(Redrawn, with permission, from Briggs WH, et al. 2007. *Genetics* 177: 1915–1928; ©Genetics Society of America.)

chromosomes 1–5. Subsequent studies of teosinte–maize crosses identified similar patterns of QTL effects. Candidate gene studies and QTL mapping led to the cloning of genes at two loci that explain key elements of maize domestication syndrome:

- Teosinte branched 1 (*tb1*, chromosome 1), a member of the TCP family of transcriptional regulators of cell division, controls the production of tillers, or auxiliary stalks. In a classic case of apical dominance, the many stalks (with many ears) of teosinte were reduced to a single stalk in modern corn. This concentrated the plant's resources into producing a single, larger ear and allowed planting in rows for easier harvesting.
- Teosinte glume architecture 1 (*tga1*, chromosome 4) regulates construction of the glume, the hard outer coating that surrounds the teosinte kernel. This coating is softened and greatly reduced to only the base in the maize kernel. *tga1* also controls silica deposition in phytoliths, stony particles found in the glume and the cob. *tga1* encodes a transcription factor that appears to up-regulate a set of genes in teosinte and down-regulate the same genes in maize.

Transcription factors are overrepresented among the domestication genes identified thus far in grains, as well as in other agricultural plants. Transcription factors affect suites of genes in a coordinated fashion, with wide-ranging impacts on plant development. Thus, it makes sense that human selection would favor genes that affect broad developmental programs.

Richard MacNeish, ca. 1975
(Reprinted, with permission, ©Robert S. Peabody Museum of Archeology, Phillips Academy, Andover, Massachusetts.)

Dolores Piperno
(Courtesy of Don Hurlbert, Smithsonian Institution.)

A 2002 study by Doebley compared 99 microsatellite polymorphisms across 193 varieties of maize, representing its entire pre-Columbian distribution, and 67 teosintes from Mexico and Guatemala. Like the earlier studies on wheat, a tree of genetic distances pointed to a single maize domestication event. The lowland Balsas River subspecies of teosinte (*Zea mays* ssp. *parviglumis*) was at the base of the tree that led to all modern corn varieties; of these, the central highland varieties of maize were the most ancient. The study also showed that the highland teosinte subspecies *mexicana* contributed to up to 12% of the germline of highland maize varieties.

The antiquity of the central highland landraces had long been suggested by archeological finds. In the 1960s, Richard MacNeish, of the National Museum of Canada, surveyed 500 sites in the state of Puebla in Mexico. His excavation at Coxcatlán Cave uncovered 42 episodes of human occupation over a 10,000-year period, with ears of domesticated corn dating to 4700 BP. The cave also provided ancient examples of squash, bottle gourds, and beans.

In 2001, Bruce Benz, of Texas Wesleyan University, and Dolores Piperno, of the Smithsonian Institution, independently used accelerator mass spectrometry to carbon-date the oldest ears of maize, from Guilá Naquitz Cave in Oaxaca, to 6200 BP. The tiny ears—two inches or less in length—had nonshattering kernels with softened, reduced glumes. However, they possessed the distichous arrangement of teosinte, with two or four rows of opposing kernels. Although these tiny cobs retained their kernels for easy harvest, changes in cob architecture that greatly increased the number of kernels per cob had not yet been selected.

In 2009, Piperno found evidence for the first domesticated maize in the Balsas River Valley, 70 miles south of Mexico City. Microscopic analysis of the surfaces of simple grinding stones found at the Xihuatoxtla Rockshelter, dated to 8700 BP, identified large starch grains and rounded phytoliths with reduced silica that are characteristic of modern corn. These remains are more than 2000 years older than those of

Bruce Benz with Balsas River teosinte. Genetic analysis showed that the Balsas River subspecies of teosinte is the progenitor of all modern maize varieties.
(Courtesy of Bruce Benz.)

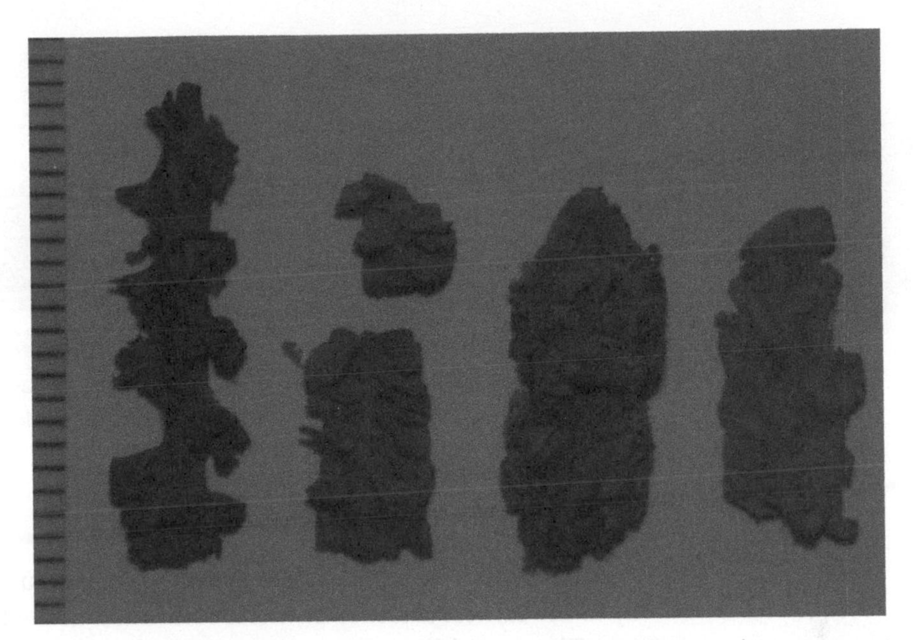

Earliest evidence of domesticated maize. Domesticated teosinte cobs from Guilá Naquitz, Oaxaca, dated to 6200 BP (*left* and adjacent two fragments), and two maize cobs from San Marcos Cave, Puebla, dated to 5300 BP (*right*).
(Courtesy of Bruce Benz.)

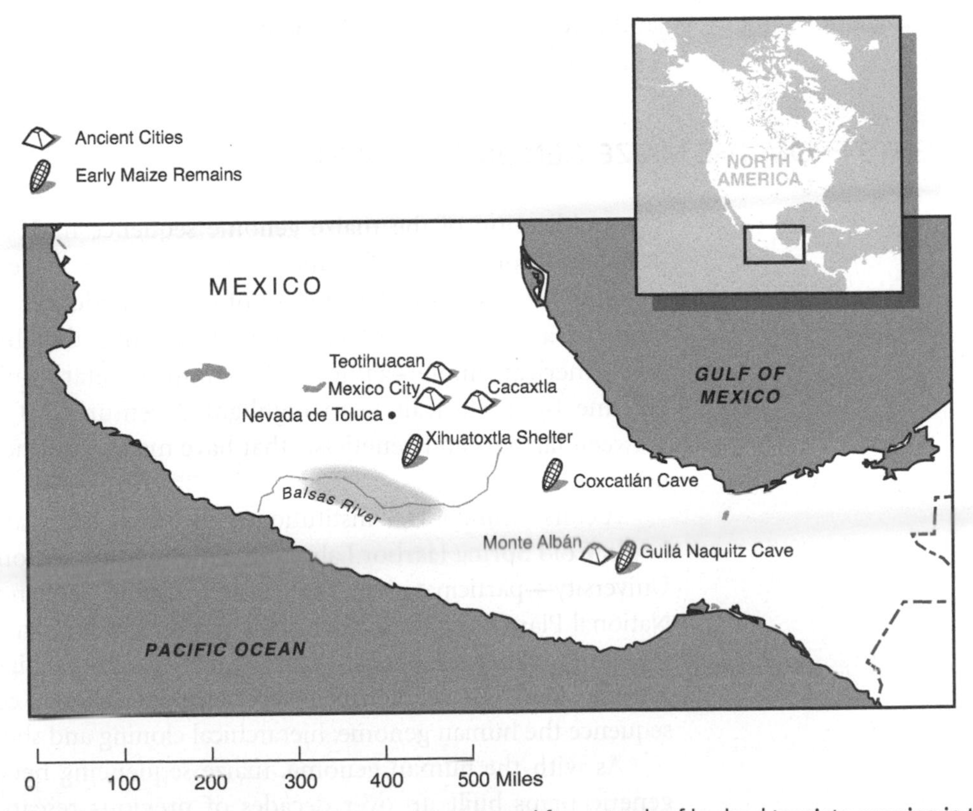

Cradle of maize domestication. Maize was domesticated from stands of lowland teosinte growing in the Balsas River Valley. However, many early examples of domesticated maize are found in the great Mesoamerican cultural centers that developed in the central highlands of Mexico.

Temple of the Sun at Teotihuacan and Pyramid Complex at Monte Albán. Teotihuacan (*left*) and Monte Albán (*right*) were early Mesoamerican metropolises made possible by the domestication of maize. (Photo of Monte Albán by Nick Saum, www.nicksaumphotography.com.)

Guilá Naquitz and push back the record of maize cultivation in the New World nearly to the earliest evidence of wheat cultivation in the Fertile Crescent.

These combined observations suggest a scenario in which hunter–gatherers living in the lowlands south of Mexico City selected key traits that adapted *parviglumis* teosinte for human consumption. Maize cultivation spread quickly to the highland regions to the north and east, where introgression of genes from *mexicana* teosinte adapted maize for upland growing conditions. Maize supported the growth of early highland cultures, represented by the ancient cities of Teotihuacan (near Mexico City) and Monte Albán (near Oaxaca). This established the widespread planting of highland races and their use as foundation stocks for other landraces in Mexico and Latin America.

INSIGHTS FROM THE MAIZE GENOME SEQUENCE

The publication of the maize genome sequence in 2009 was the culmination of a decade of coordinated effort under the National Science Foundation's National Plant Genome Initiative—and a century of corn genetics. In a larger sense, it was the end point of a continuum of agricultural breeding that began 10,000 years ago with Mesoamerican hunter–gathers. The complementary publication of a second maize genome by a Mexican team highlighted centuries of cross-country collaboration between farmers and geneticists that have made corn the most important crop in the new world.

A consortium of U.S. institutions—including teams at Washington University in St. Louis, Cold Spring Harbor Laboratory, the Arizona Genomics Institute, and Iowa State University—participated in the Maize Genome Sequencing Project. As part of the National Plant Genome Initiative, the project produced a finished sequence of B73, the major inbred line used in U.S. hybrids. To deal with the high repeat content of the maize genome, the 4-year sequencing project combined the two competing approaches used to sequence the human genome: hierarchical cloning and shotgun sequencing.

As with the human genome, maize sequencing benefited from the existence of genetic maps built up over decades of previous research. Three bacterial artificial chromosome (BAC) libraries were constructed that contained 300,000 cloned fragments of the maize genome, each averaging 150–180 kb in length. Partial sequences, genetic markers, and gene probes were used to sort the BAC clones according to shared sequence information. The genetic markers aligned clones to physical locations on chromosomes, while the gene probes provided the relative positions of genes.

A "tiling path" identified the most economical set of BACs needed to determine the genome sequence. Then each BAC clone along the tiling path was shotgun-sequenced. DNA fragments from each BAC were subcloned into plasmids, and 600–700 nucleotides of sequence were determined from the ends of each plasmid clone. A computer aligned the plasmid sequences to provide the entire sequence for each BAC clone. This provided 15x coverage—meaning that, on average, each nucleotide in the genome was sequenced 15 different times. The sequences of the finished BAC clones were then merged to provide a contiguous sequence along each maize chromosome.

Palomero Toluqueño

(Reprinted, with permission, from Vielle-Calzada JP, et al. 2009. *Science* *326:* 1078; ©AAAS.)

The Mexican team, headed by Jean-Philippe Vielle-Calzada, of the National Laboratory of Genomics for Biodiversity in Irapuato, sequenced an ancient popcorn variety, Palomero Toluqueño. They used Sanger and next-generation sequencing to generate short reads, providing 3x coverage of the genome. Then they scaffolded their genomic data—plus next-generation RNA sequence (RNA-seq) data from transcribed genes—onto the assembled B73 sequence.

The Palomero genome offered important insights into the domestication of maize. The Palomero genome has 20% less repetitive DNA than B73—mainly long terminal repeat (LTR) retrotransposons—suggesting that transposons may have acted as mutagens to rapidly alter modern cultivars. Indeed, nucleotide surveys show that chromosomes from any two maize cultivars differ from one another at ~1 of 100 nucleotide positions. Contrary to this extensive sequence diversity, Palomero and B73 genomes share more than 600 identical sequence regions (IDSRs), totaling 545 kb. Several IDSRs mapped adjacent to the major domestication genes *tga1* and *tb1*, providing evidence of a selective sweep. IDSRs showed significantly lower nucleotide diversity among Mexican landraces than among Balsas teosintes, suggesting recent, strong selection in maize. Coalescent simulations—which use sequence data to model population bottlenecks that occur during domestication—identified eight potential domestication loci.

Close examination revealed that several of these regions contain genes involved in metabolizing heavy metals or in responding to environmental (abiotic) stress. Because heavy metals are toxic at very low concentrations, high-affinity transporters and oxidizers tightly control their levels within cells. The Mexican team found three genes encoding cadmium and copper transporters located close together on the short arm of maize chromosome 5, which Doebley identified as one of the six major QTLs involved in maize domestication. At least 12 other IDSRs contain genes involved in heavy metal or abiotic responses, and which may have been important in adapting maize to the local environment.

The Palomero Toluqueño landrace takes its name from the region surrounding Nevado de Toluca, a 14,000-foot volcanic cone that dominates the landscape south of Mexico City. One of its Aztec names, "Lord of the Cornstalks," is appropriate because it looks down on the fields of the Balsas River Valley where maize was first domesticated. The Toluca volcano erupted most recently ~10,500 years ago, carpeting the region with ash and debris. This coincides closely with the early cultivation of maize

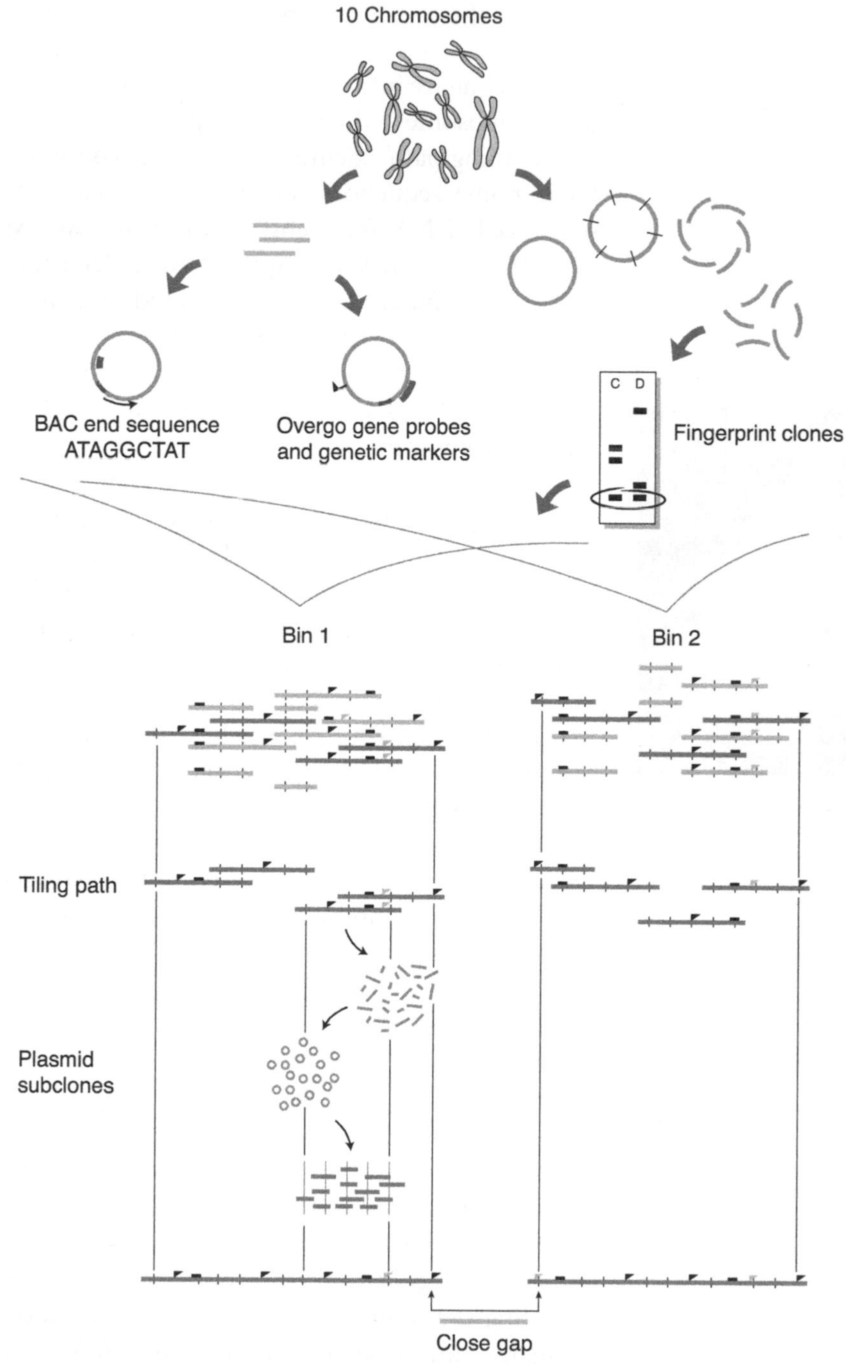

Hierarchical shotgun sequencing of the maize genome. Chromosome fragments were cloned into BACs. End sequences, gene probes, genetic markers, and fingerprint clones allowed related BAC sequences to be rapidly sorted into bins. A tiling path was made from the smallest number of overlapping BACs, which were then shotgun (randomly) cleaved and subcloned into plasmids. Contiguous DNA sequences were built up from the individual shotgun sequences derived from plasmids and were anchored to hierarchical chromosome positions provided by the BAC bins. Attention was then given to closing gaps to produce a contiguous chromosome sequence.

Nevado de Toluca. Known locally as "Lord of the Cornstalks," this volcano erupted most recently in 10,500 BP, adding heavy metals to the fields of the Balsas Valley where maize was first cultivated.

and provides a context for interpreting the conserved metal decontamination genes found by the Mexican team. Before industrialization, volcanic eruptions were the major source of heavy metals in the soil, and it appears that the cultivation of maize selected for plants that were tolerant to the volcanic soils in this region.

WHOLE-GENOME COMPARISONS

Comparing the maize genome with those of *Arabidopsis*, rice, sorghum, and soybean—completed in 2000, 2005, 2009, and 2010, respectively—provides an interesting study in genome evolution. The maize genome (2.3 billion nucleotides) is twice the size of soybean's genome (1.1 billion nucleotides), three times sorghum's (730 million nucleotides), six times rice's (390 million nucleotides), and 18 times that of *Arabidopsis* (125 million nucleotides). Despite the 18-fold difference in genome size, maize has only 1.22 times as many protein-coding genes (32,500) as *Arabidopsis* (26,700). This contradiction between genome size (C value) and gene number, also known as the C-value paradox, is primarily due to the proliferation of retrotransposons. This class of transposons, which moves via an mRNA intermediate, composes 75% of the maize genome, 55% of soybean, 42% of sorghum, 26% of rice, and only 2% of *Arabidopsis*. Similarly, transposons integrated within introns account for most of the differences in size between orthologous genes from different plant species.

Some of the differences in gene number and gene families arose from ancient paleopolyploidy events that occurred in the dicot lineage leading to *Arabidopsis* and soybean and in the monocot lineage leading to rice, sorghum, and maize. In each event, a whole-genome duplication resulted in a tetraploid set of chromosomes. Then, through a process called fractionation, chromosome segments were deleted to return to the diploid chromosome number. This process allowed genes on one chromosome pair to mutate, occasionally giving rise to new functionality, while retaining the genes from the other pair intact. Some of the duplicate genes—functioning ones and nonfunctioning pseudogenes—were retained on enlarged diploid chromosomes. Thus, whole-genome duplication and fractionation created a patchwork of chromosomes, with shared ancestral regions showing synteny (conserved gene order) between different plants.

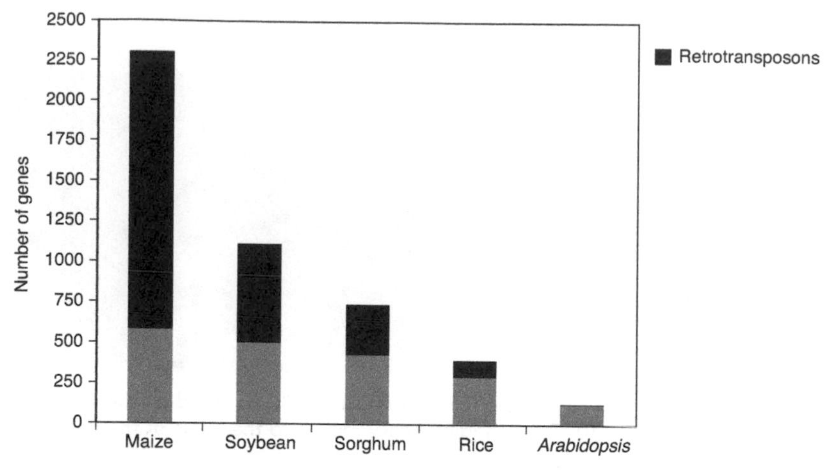

Retrotransposon content of plant genomes (million bp).

The maize and soybean genomes make an interesting comparative study of paleopolyploidy. Although only half the size, the soybean genome has 43% more genes (46,400) than maize (32,500). This is because ~75% of soybean genes are homeologs, that is, copies of genes retained following two genome duplication events that occurred ~59 and ~13 million BP. The maize genome underwent duplication within the last 12 million years, but only ~25% of maize genes are homeologs.

Sorghum and maize are closely related species, having shared a common ancestor ~5–12 million BP (the same order of magnitude as the human–chimp divergence). The sorghum genome appears to have changed little since diverging from the common ancestor. In contrast, shortly after diverging, the maize genome duplicated, creating two sets of chromosomes orthologous to the sorghum genome. Transcription factors are four times more frequent among retained homoeologous genes than among fractionated (lost) genes. This suggests that major phenotypic differences between maize and sorghum are due primarily to changes in the regulation of developmental programs

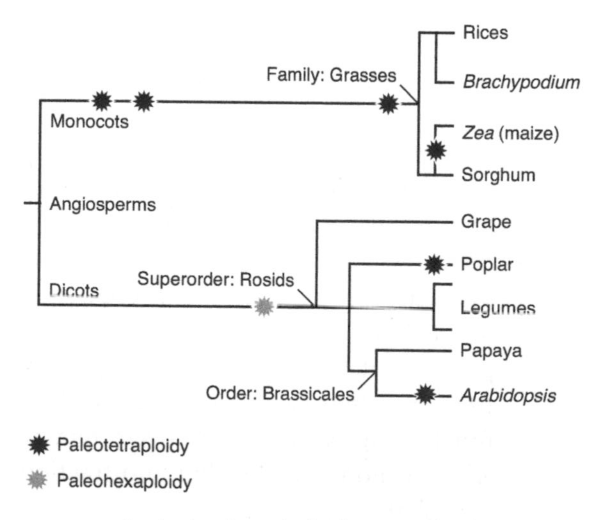

Ancient paleopolyploidy events in plants.

(Adapted, with permission, from Freeling M. 2009. *Annu Rev Plant Biol* 60: 433–453; Woodhouse MR, et al. 2010. *PLoS Biol* 8: e1000409.)

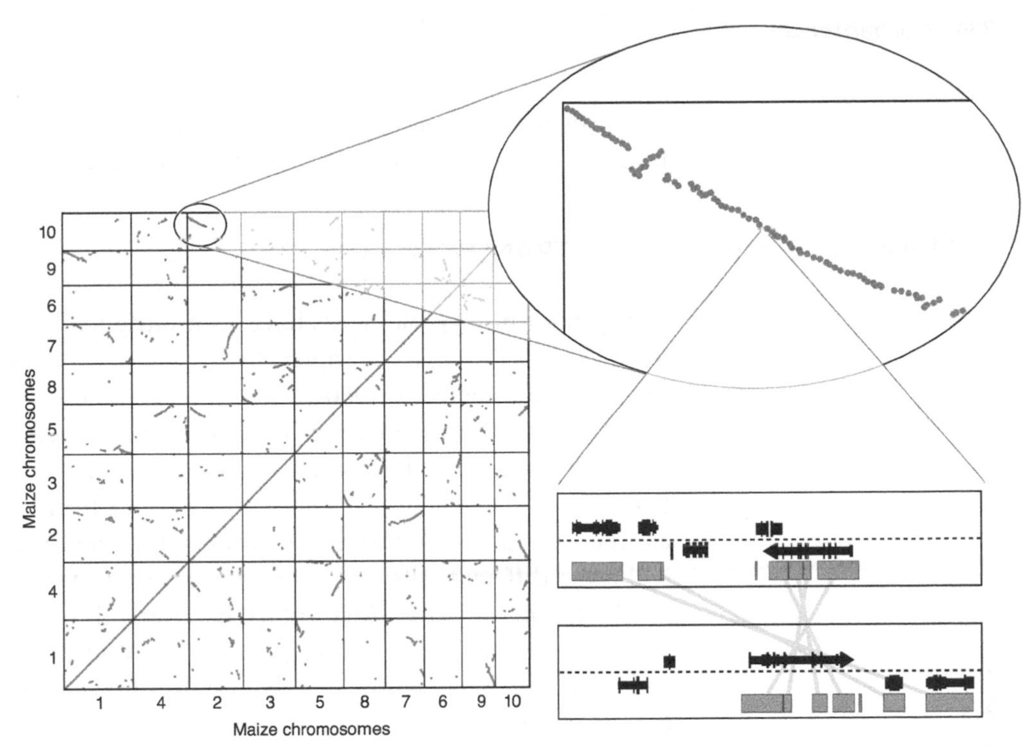

Bioinformatic evidence of whole-genome duplication in maize. A syntenic dotplot compares two whole-genome sequences to detect homologous gene regions that are derived from a common ancestor. Protein-coding regions are extracted from each genome and compared with each other, forming the "dots" of significant sequence similarity (hits). Syntenic regions appear as 45° diagonals running up the dotplot, while inversions run in the opposite direction. In this case, two haploid maize genomes are compared. As expected, the strong 45° diagonal shows a one–one correspondence between the diploid pair of genes from each corresponding chromosome. However, shorter diagonals show synteny—conserved gene order—between different chromosomes. These extra gene copies—retained homoeologs—are remnants of ancient whole-genome duplications. Circled is an inverted region of chromosome 2 that is homologous to a region of chromosome 10. A closeup of part of this region shows three homoeologs, in inverted orientation on opposite strands of chromosomes 2 and 10.

(Courtesy of Eric Lyons, CoGe, iPlant Collaborative, University of Arizona.)

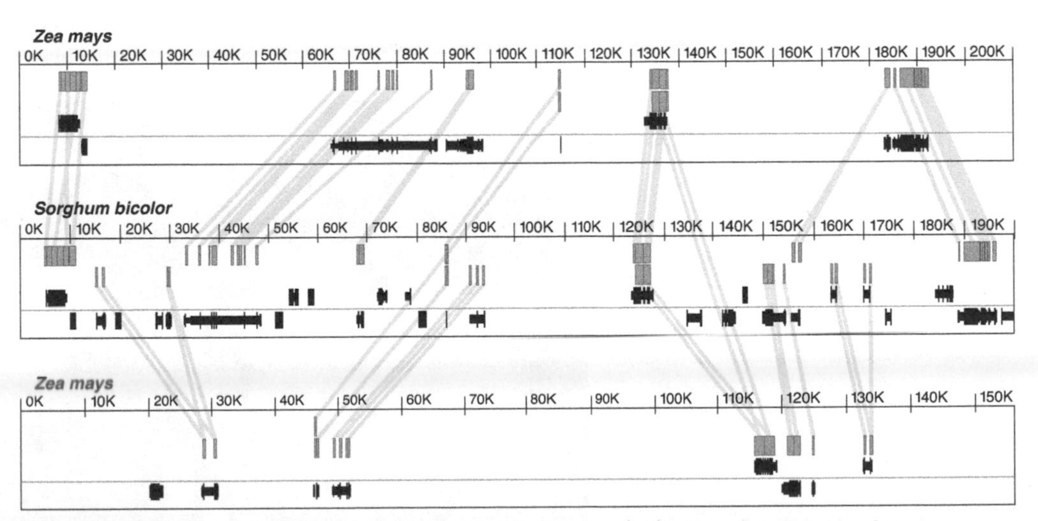

Genome fractionation in maize. Maize underwent whole-genome duplication after diverging from a common ancestor with sorghum. The tetraploid genome then returned to a diploid number by eliminating a fraction of duplicate genes from one or the other of the duplicated chromosomes. Diagram shows comparisons of genes from a region of a single sorghum chromosome (*middle*) and two homologous regions of maize (*top*, *bottom*). Thirteen sorghum genes have homologs on one of the two maize regions. Two sorghum genes are represented in both maize regions—these retained homeologs are remnants of the recent whole-genome duplication of maize.

(Courtesy of Eric Lyons, CoGe, iPlant Collaborative, University of Arizona.)

rather than to changes in the genes themselves. The same conclusion can be drawn from the extreme similarity between the human and chimp genomes.

BARBARA MCCLINTOCK AND TRANSPOSITION

Throughout history, biologists have been intrigued by variegated plants with unpigmented spots or stripes, in which chlorophyll or other pigments are not produced. In the late 1940s and early 1950s, Barbara McClintock, of the Carnegie Department of Genetics at Cold Spring Harbor Laboratory, developed a daring hypothesis regarding the genetic basis of color variations in the leaves and kernels of maize. She proposed that genetic "controlling elements" move about from one chromosomal location to another—resulting in unstable mutations that cause pigment production to cycle on and off in different cells.

Earlier in her career, McClintock had learned to distinguish maize chromosomes by size under the light microscope, as well as to identify the locations of centromeres and distinctive heterochromatin "knobs." In 1931, she teamed with Harriet Creighton, of Wellesley College, to explain the physical basis of genetic recombination. They correlated the movement of a knob on the short arm of chromosome 9 with the appearance of a set of linked recessive phenotypes; the chromosome fragments were exchanged when the homologous chromatids crossed over during meiosis.

McClintock continued to focus on this region, identifying a locus called *Dissociation* (*Ds*) as a frequent site of chromosome breakage. She then correlated chromosome breakage with the transposition of *Ds* to the end of the chromosome—and with changes in kernel pigmentation. She determined that the movement of *Ds* is activated by another locus, called *Activator* (*Ac*), on the long arm of chromosome 9.

Her model came together with observations of the *colored* (*C*) gene, which produces one of many pigments associated with red, bronze, and purple kernels in so-called Indian corn. Transposition of *Ds* into the *C* locus disables production of pur-

Barbara McClintock and Harriet Creighton at Cold Spring Harbor Laboratory, 1956. (Courtesy of Cold Spring Harbor Laboratory Archives.)

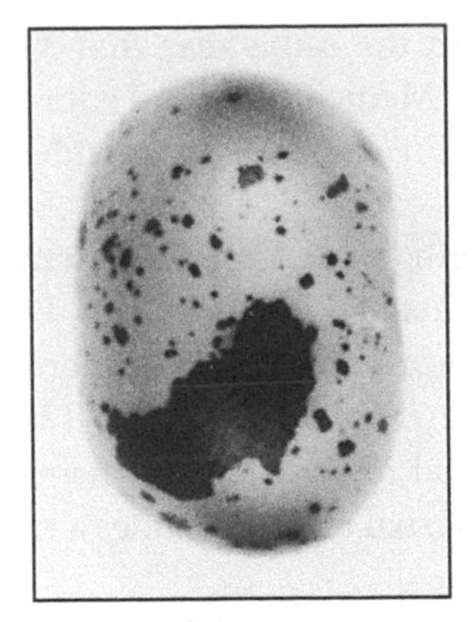

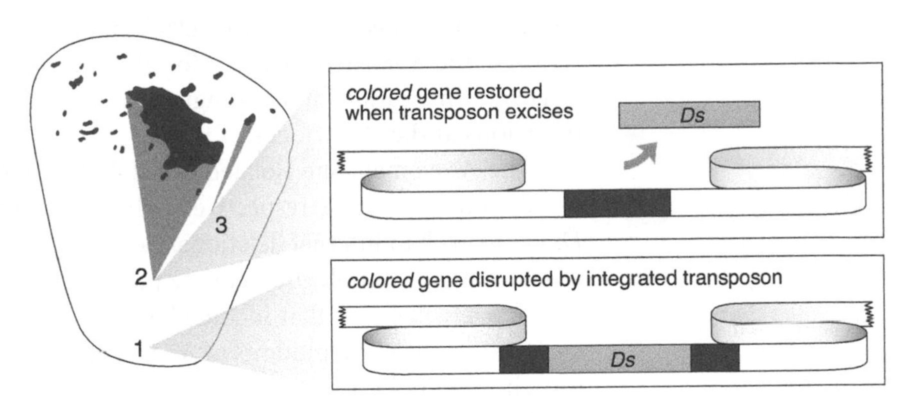

Ds transposition events in a single maize kernel (*left*). At the beginning of kernel development (*1*), the *Ds* transposon is inserted into the *colored* (*C*) gene, giving rise to a region of colorless (yellow or white) tissue. Later, *Ds* transposition restores the *C* gene, giving rise to colored sectors. A large sector results from a transposition event early in kernel development (*2*). A small sector results from a transposition event later in development (*3*).

ple pigment, producing a yellow or white kernel. Subsequent transposition of *Ds* restores function of the *C* gene and pigment production. McClintock realized that transposition during kernel development cycles pigment production in different areas, producing variegated sectors of purple and yellow tissues. The size of pigmented sectors within an unpigmented background indicates the timing of *Ds* transpositions that restored gene function. Transposition events early in development give rise to broad stripes or large patches of pigmented tissue, whereas events occurring late in development give rise to narrow stripes or small speckles.

Many high-caliber geneticists, especially those working in maize, accepted McClintock's work on the strength of her excellence in cytogenetics. However, some believed that transposition was peculiar to maize. Others thought that McClintock went too far in asserting that mutations caused by transposable elements provide a mechanism to reorganize the genome rapidly in response to stress. For the wider biological world, transposition contradicted the received dogma that genes are immutably fixed along the lengths of the chromosomes. McClintock did little to promote her ideas, and she stopped publishing in major scientific journals in 1950.

Thus, one can understand why McClintock's ideas did not gain widespread attention until the 1960s and 1970s, when transposons were found to be widespread in prokaryotic and eukaryotic organisms. By this time, insertion elements in bacteria and *P* elements in *Drosophila* had been discovered. McClintock's work is even more prescient when one considers that her key experiments predated knowledge of DNA structure and its function as the genetic molecule.

MOLECULAR MECHANISMS OF TRANSPOSITION

McClintock had found that *Ds* cannot transpose when *Ac* is bred out of a stock; however, *Ac* can transpose independently of *Ds*. Thus, *Ac* is an autonomous (complete) transposon, and *Ds* is a nonautonomous (incomplete) transposon. However, confirmation of the molecular details of transposition had to await the advent of restriction

Nina Fedoroff, 1980
(Courtesy of Cold Spring Harbor Laboratory Archives.)

enzymes as a means to dissect DNA molecules. In the same year that McClintock received the 1983 Nobel Prize for Physiology or Medicine, Nina Fedoroff, of the Carnegie Institution of Washington, used restriction enzymes to isolate *Ac* and *Ds* insertions at the *Waxy* locus of maize.

The *Ac* element she isolated was 4.3 kb in length, whereas two *Ds* elements were 4.1 kb and 2.0 kb long, respectively. All three elements had identical ends, but the two *Ds* elements had internal deletions. Genome hybridization revealed numerous *Ds* elements having homologous end sequences—terminal inverted repeats (TIRs)—but very few *Ds* elements that included an intact internal region. This confirmed that *Ds* elements are deletion mutants of *Ac* and that the internal region encodes a transposase protein required for transposition.

During transposition, the *Ac* transposase positions itself at the TIRs and makes double-stranded cuts on either side of an *Ac* or *Ds* element, releasing it from its current chromosome location. Then, another double-strand cut at a "target" site elsewhere in the genome enables the transposon to settle into a new locus. Like some restriction enzymes, the transposase makes staggered cuts with single-stranded extensions, or "sticky ends," that align the transposon at the target site. DNA polymerase then fills in single-stranded gaps at the sticky ends, and ligase covalently closes the double helix. This results in a target-site duplication, that is, a short sequence of direct repeats on either side of the transposon insertion.

Ac and other Class 2 transposons are also known as DNA transposons because the DNA sequence of the element actually excises from one chromosomal spot and inserts at another. This "cut-and-paste" mechanism explains why genomes have relatively few Class 2 transposons. Conversely, Class 1 transposons—including the plant *copia* elements and the vertebrate long interspersed elements (LINEs)—use a "copy-and-paste" mechanism. Class 1 transposons are also known as RNA transposons because they move via an mRNA intermediate. Like retroviruses, autonomous Class 1 transposons encode a reverse transcriptase (rt) enzyme. In a process called retrotransposition or retroposition, the transposon is first transcribed by RNA polymerase into mRNA. The rt enzyme then reverse-transcribes the mRNA to make a DNA copy, which is then inserted into a new genome location.

The transposition mechanism of L1 and its nonautonomous partner *Alu* has been closely examined. In addition to reverse-transcribing mRNA, the L1 rt enzyme also makes single-stranded nicks in DNA. The L1 rt associates with an *Alu* or L1 mRNA transcript, then catalyzes a nick in DNA within the consensus sequence AA^{v}TTTT. A poly(A) tail at the end of the *Alu* or L1 transcript hydrogen-bonds to the single-stranded TTTT sequence, creating a primer for the rt enzyme. Then, in a process called template-primed reverse transcription, the L1 rt synthesizes a complementary DNA and inserts the free end into a second nicked site on the chromosome. The insertion process creates direct repeats at either end of the transposon. These A-T-rich regions provide targets for the L1 nicking function, explaining why tandem arrays of retrotransposons are common in many genomes.

This copy-and-paste mechanism allows retrotransposons to accumulate over evolutionary time, accounting for much of the extreme variation in genome size across different plant and animal species. At any given time, only a small number of retrotransposons are transcriptionally active, and these give rise to families of transposons that can be identified by sequence similarity. As we have seen from the conversion of *Ac* into *Ds*, autonomous transposons are disabled when they accumulate mutations in

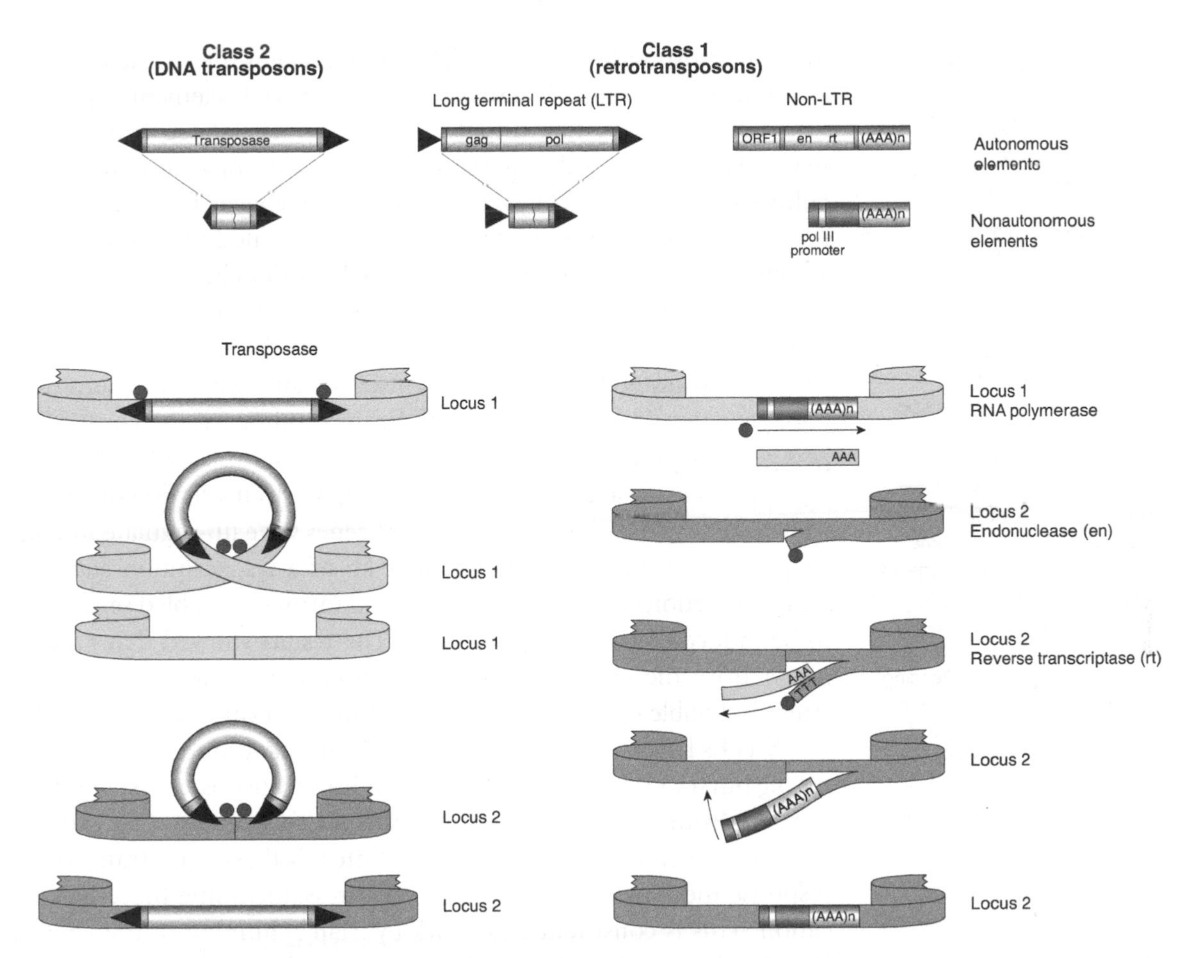

Major transposon classes and mechanisms. Autonomous elements encode one or more proteins required for transposition and, thus, can mobilize on their own. Nonautonomous elements lack mobilizing enzymes and, with the exception of SINEs, are deletion derivatives of autonomous elements. Class 2 transposons, also called DNA transposons, use a "cut-and-paste" mechanism. A transposon-encoded transposase makes double-stranded cuts on either side of the transposon to release it from its current position. The transposase then makes a cut at another chromosome locus, and the transposon inserts at the new location. Class 1 transposons, also called retrotransposons, use a "copy-and-paste" mechanism. A transposon-encoded reverse transcriptase (rt, or pol) converts an mRNA copy into a DNA copy, which then integrates into a new chromosome location. As illustrated, some non-LTR transposons encode an endonuclease that nicks a chromosome locus to provide an invasion site for the transposon mRNA. This mechanism is used by the ubiquitous human transposon system, L1 and *Alu*.

Susan Wessler, 2011
(Courtesy of Susan Wessler, University of California, Riverside.)

their transposase or rt genes. Thus, the vast majority of transposons in any genome are molecular fossils that have not jumped for millions of years.

Transposition events are rare and difficult to detect in real time. Just as McClintock deduced transposition from changes in kernel coloration, new transposition events can sometimes be detected as phenotypic differences between parents and offspring. For example, in 1991, Margaret Wallace, at the University of Michigan, found a de novo *Alu* transposition in the *NF1* gene of a neurofibromatosis patient that was not present in either parent.

In 2006–2009, Susan Wessler, of the University of Georgia, in collaboration with a research group at Kyoto University, Japan, caught the plant transposon *mPing* in a massive "burst" of activity and assessed the consequences of transposon-induced mutagenesis on the rice genome. Just as *Ds* is a deletion derivative of *Ac*, *miniature Ping* (*mPing*; 430 bp) is composed of the end portions of the *Ping* transposon (5430 bp), with the

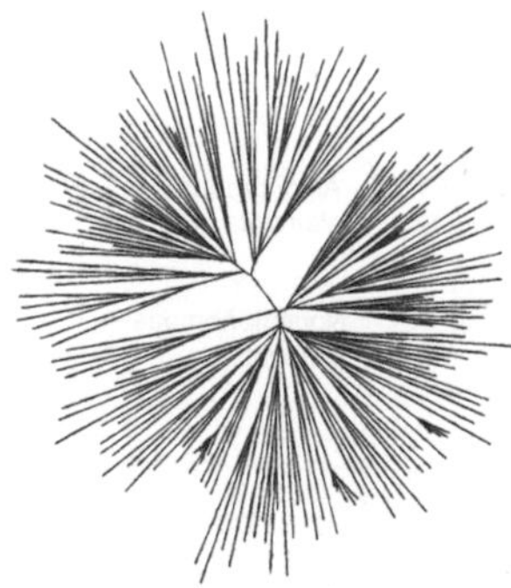
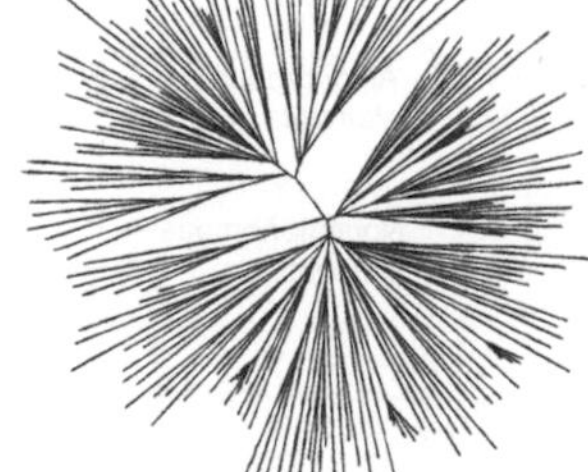

Cladogram of MITE transposons in rice. This unrooted cladogram shows the sequence relatedness of members of a family of MITE transposons in the rice genome. The star-like pattern suggests rapid expansion during a burst of activity from several active progenitors in the center of the tree. After silencing of the transposase, the inactive elements accumulated mutations and diverged from one another.
(Courtesy of Susan Wessler, University of California, Riverside.)

internal transposase gene deleted. Because it retains the characteristic TIRs, *mPing* is an example of a miniature inverted-repeat transposable element (MITE).

Wessler initially found that *mPing* elements isolated from various rice cultivars have virtually identical sequences. However, the copy number differs greatly among cultivars, with tropical *O. japonica* cultivars having as few as 25 and temperate *japonica* cultivars having up to 70 or more. This indicated that *mPing* was activated as rice adapted to the stress of cultivation in cooler, drier climates. Most strikingly, she identified several strains with more than 1000 copies of *mPing* and in which ~40 new transposition events occur in each plant. High-throughput sequencing of 24 individual plants showed that 91% of 1664 insertion sites were in regions of the genome with high gene density, but only 1% were within coding exons. Thus, although *mPing* targets genic regions, it minimizes deleterious mutations by avoiding exons.

A microarray analysis showed no change in transcription of 78% of 710 genes with nearby *mPing* insertions. About 16% of genes were up-regulated, roughly equaling the proportion of *mPing* inserts within 1 kb of a transcription start site. Furthermore, *mPing* insertions—both in promoters and introns—enabled seven of 10 resident genes to be induced by salt and cold stress. The results showed that *mPing* contains a promoter from the ORF1 of *Ping* (the transposase is encoded by ORF2) that acts as a stress-inducible enhancer when inserted into or near a resident plant gene.

MITEs have been identified in all of the eukaryotic genomes. The analysis of the *mPing* burst in rice provides a model for how transposon systems can protect the host genome from damage by avoiding transposition into coding exons. It also provides further confirmation of Barbara McClintock's thesis that transposons can provide a response mechanism to environmental stress, including those presented by domestication. This is consistent with work by Alan Schulman of the University of Helsinki showing how transposons respond to microclimates. He found that the number of copies of transcriptionally active *BARE-1* retrotransposons in a wild barley population increased with elevation and dryness in a single canyon in Israel.

In 2011, John Doebley found direct evidence for a transposon's role in the domestication of maize. He identified a *Hopscotch* retrotransposon and a *Tourist* MITE in the conserved 5′ regulatory region of the domestication gene *tb1*. In vitro assays of luciferase gene activity showed that *Tourist* decreases expression 50%, whereas *Hopscotch* increases expression twofold. The enhancer function of *Hopscotch* is consistent with increased *tg1* expression in maize. Measuring the accumulation of mutations suggested that the *Hopscotch* retrotransposon inserted into the teosinte genome 28,000–13,000 years BP, well before domestication by humans. This shows that the selection of apical dominance by early agriculturists acted on "standing variation"—the preexisting transposon mutation in teotsinte—as opposed to a new mutation.

The analysis of one transposon showed that the lines between science and culture blur in the Mexican story of maize. The Zapotec people of the highlands of Oaxaca south of Mexico City cultivate a race of maize called Zapalote Chico, from which they make dry, cracker-like totopos. Oral tradition held that if Zapalote Chico were stolen and planted by neighboring tribes, their maize fields would be poisoned.

Virginia Walbot, ca. 1998
(Courtesy of Virginia Walbot.)

Virginia Walbot, of Stanford University, confirmed this tradition on a molecular level when, in 1998, she found that Zapalote Chico contains multiple copies of *Mutator* (*Mu*) transposons with intact transposase genes. Transposition is held in check in the Zapalote Chico strain by an inhibitor of transposase transcription. However, outcrossing Zapalote Chico with other maize varieties releases inhibition (de-represses transcription) of the transposase, enabling active transposition. Transposon mutagenesis of genes regulating sex organs then induces sterility. This type of breeding-induced sterility (hybrid dysgenesis) was first discovered in the mid 1970s in the offspring of crosses between wild fruit flies containing *P* elements and laboratory fruit flies without *P* elements.

HETEROSIS AND MODERN CORN DEVELOPMENT

Each corn kernel is a seed that results from the fertilization of a female egg by a male pollen grain. (A long silk conducts each pollen grain to an egg.) Throughout most of the history of maize cultivation, plants were open-pollinated, with kernels arising from random fertilizations by pollen from throughout a corn field. Farmers saved kernels from the best plants as seed for the following year's crop.

Corn spread north and south from central Mexico to Indian cultures throughout the Americas. North American colonists learned how to plant maize from eastern Native American tribes, and pioneers and homesteaders brought maize with them during the western expansion of the 19th century. The movement of corn from place to place provided opportunities for selection and adaptation of new varieties. An estimated 800 drought-tolerant varieties were developed during the settlement of the key cornbelt states of Iowa, Illinois, and Missouri in the mid 1800s.

The best farmers experimented with ways to encourage cross-pollination between high-yielding plants. This could be accomplished by planting different varieties next to each other. Detasseling (cutting off the male flowering stalk before it matures) prevented inferior male plants from contributing their pollen to the field. Some crosses occurred serendipitously, when farmers planted seeds they carried with them when they moved to a new area.

Notably, Robert Reid brought a Southern Dent variety (*dent* referring to a depression, or "dimple," in the kernel) with him when he moved from Ohio to Illinois. After a poor

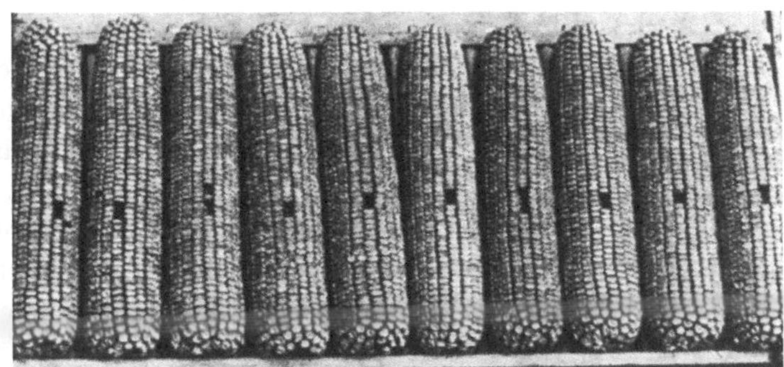

Troyer Reid hybrid corn. David and Chester Troyer developed and aggressively marketed the Troyer Reid corn variety, making it one of the most popular derivatives of Reid's Yellow Dent. David (*left*) detassels corn on his La Fontaine, Indiana farm, ca. 1905. This 10-ear sample of Troyer Reid (*above*) was grand champion at the 1932 International Grain and Livestock Show in Chicago.

(Reprinted, with permission, from Troyer AF, Palmer LS. 2006. *Crop Sci* 46: 2460–2467.)

William James Beal, 1910
(Reprinted, with permission, from Michigan State University Archives and Historical Collections.)

crop of the Southern Dent, he replanted patches of the same field with a local yellow variety in 1847, resulting in the Reid's Yellow Dent hybrid. It became known as "World's Fair Corn" after winning the international corn competition at the 1893 World Columbian Exposition. Reid's Yellow Dent was the most popular open-pollinated variety ever developed in the United States and is the largest single contributor to the genetic background of modern hybrids and inbred lines, including the most widely used inbred, B73.

Up to a point, this sort of selection improved traits such as kernel size, number of cob rows, and pest tolerance. However, from the advent of national statistics in 1866 until 1940, corn yield in the United States remained static at about 26 bushels per acre.

Charles Darwin did numerous experiments with maize hybrids, and, by 1876, concluded that cross-fertilized plants were taller and had "greater innate constitutional vigour" than self-pollinated ones. William James Beal, of Michigan State College, tested Darwin's thesis in 1878. He controlled breeding by crossing detasseled female corn plants with a different male variety. The resulting hybrids had a 53% higher yield than either of the open-pollinated parents. He then reproduced his experiment in field trials conducted with researchers in several states, showing the general usefulness of hybrids.

However, George Shull, at the Carnegie Station for Experimental Evolution on Long Island, was among the very first scientists to understand corn hybrids and yield in the context of Mendel's laws. In his 1908 paper, "The Composition of a Field of Maize," he realized that each plant is a complex hybrid of many genes found within the open-pollinated field. Seed that farmers planted from vigorous plants essentially came from crosses between various heterozygous parents—producing a mix of highly productive heterozygous offspring and less-productive homozygous offspring.

As had others, Shull found that inbred plants deteriorate. Maize plants that were selfed over five or more generations produced fewer rows and kernels and had less

George Shull (*above*) and Cold Spring Harbor (*right*), ca. 1908. Shull's cornfield is in the center of this photo, behind the chicken run and to the left of the main building of the Carnegie Station for Experimental Evolution.

(Courtesy of Cold Spring Harbor Laboratory Archives.)

resistance to the serious fungal disease corn smut. Some inbred strains were infertile or so weak that they failed to grow. However, he found that crosses between two inbred lines produced hybrid offspring with 30% more kernels than either inbred parent. Individually, the hybrid cobs were not much better than the best open-pollinated cobs. However, crossing produced a uniform population of high-yielding hybrids that would out-produce the mixed population in an open-pollinated field.

Shull reasoned that inbreeding had reduced traits to a homozygous, or "pure" condition, with detrimental traits masking the effects of beneficial ones. Crossing two

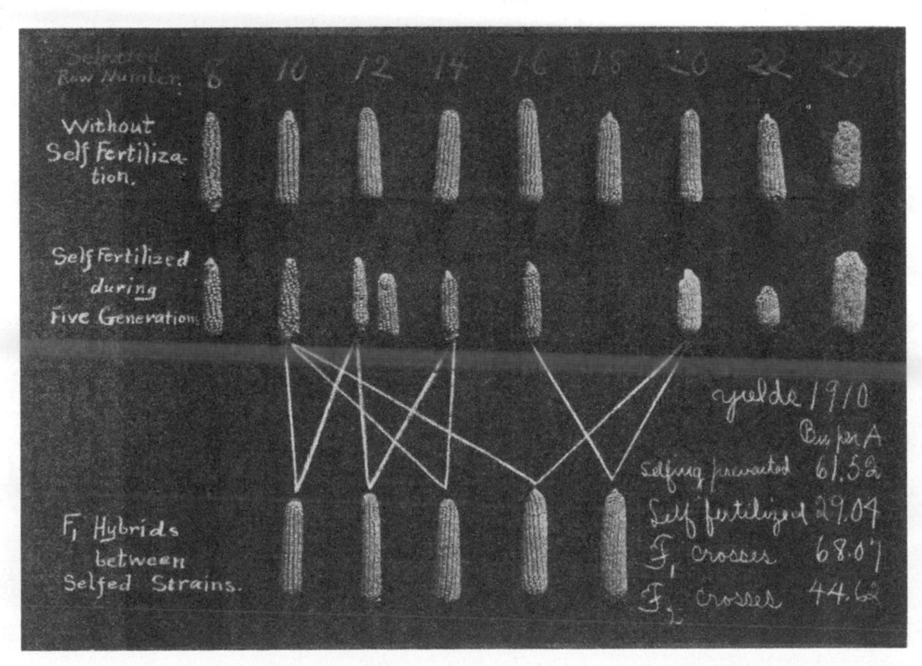

Shull's crosses with yields, 1910. Open-pollinated originators of inbred lines are in the top row, inbred lines are in the middle, and F_1 hybrids are at the bottom.

(Courtesy of Cold Spring Harbor Laboratory Archives.)

Edward East, 1904
(Courtesy of Harvard University Archives [HUP-1].)

Donald Jones, 1948
(Courtesy of The Connecticut Agricultural Experiment Station.)

"pure lines" then maximized beneficial heterozygous traits. In coining the term *heterosis*, Shull provided a genetic explanation for hybrid vigor. The leading maize geneticist of the era, Edward East, of Connecticut State College, corroborated Shull's conclusion with data from his own experiments. Shull initially thought that corn yield could be substantially improved if plant breeders adopted his pure-line method to produce hybrid seed. However, it proved too expensive to produce enough seed from the low-yielding inbred lines to make hybrid corn a commercial success.

By 1922, East's former student Donald Jones, of the Connecticut Agricultural Experiment Station, solved this problem when he essentially doubled Shull's pure-line method. His double cross, or four-way cross, mated two hybrids, each produced by inbred lines. These second-generation offspring (F_2's in genetic parlance) were nearly as vigorous as Shull's first-generation (F_1) hybrids. However, they were much less expensive to make because they used high-yielding hybrids as breeding stock.

More than any one person, Henry Wallace and his seed company, Pioneer Hi-Bred, popularized hybrid corn among American farmers. Using Jones's double cross, Wallace produced his own hybrid seed, Copper Cross, and began selling it off the back of his pickup truck in 1924. Two years later, he founded the Hi-Bred Corn Company in Des Moines, and by 1940 hybrid corn accounted for more than 90% of corn planted in major agricultural states. Between 1941 and 1961, corn yield doubled, from 31 to 62 bushels per acre.

Besides producing more grain, corn hybrids had other advantages that speeded their adoption. Hybrids incorporated valuable traits, and with much-improved drought resistance, proved their worth during the Dust Bowl years of the mid 1930s. The uniform height of hybrids was well suited to mechanical harvesting. Some also believe that farmers found the uniform field of hybrids to be an aesthetic reward for their labor.

Once hybrid corn caught on with farmers, yield increased at a rate of about two bushels per year, from 31 bushels per acre in 1941 to 163 bushels per acre in 2009. This fivefold explosion in corn yield coincided with improved cultivation methods—notably mechanization and increased use of fertilizers, herbicides, and insecticides. Wilbert Russell of Iowa State University grew hybrids developed from 1930 to 1970 under controlled conditions and concluded that hybridization accounted for ~60% of

Henry Wallace (*left*) founded Pioneer Hi-Bred in 1926. Salesmen sold Wallace's hybrid corn off the back of pickup trucks.
(© 2011 Pioneer Hi-Bred International, Inc.)

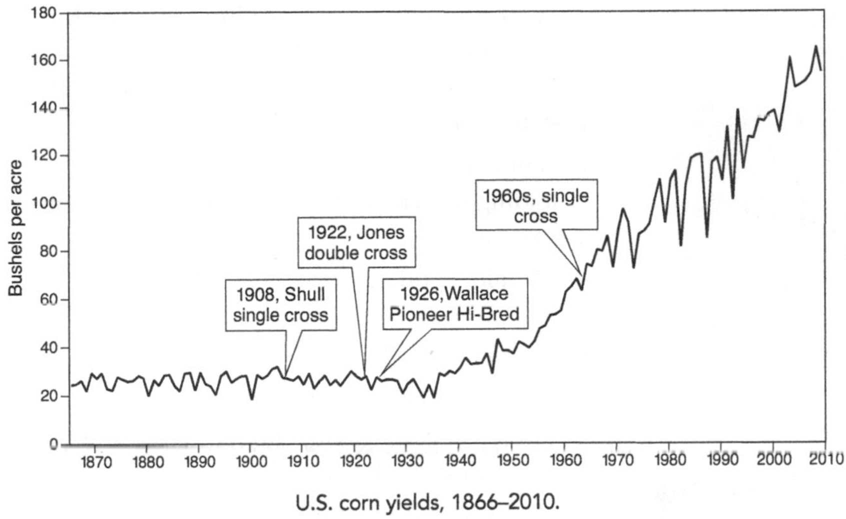

U.S. corn yields, 1866–2010.
(Data compiled from the National Agricultural Statistics Service.)

the increased yield in the 20th century. Based on this estimate, George Shull's method alone tripled corn yield.

With continuous improvement, even inbred lines became highly productive, and double crosses became unnecessary. By the mid 1960s, most breeders returned to making single crosses between two inbred lines, and the vast majority of modern hybrids are now produced using Shull's pure-line method.

THE MOLECULAR GENETIC BASIS OF HETEROSIS

Two major genetic explanations for heterosis have been debated for a century. Shull and East proposed that overdominance results from interactions between heterozygous alleles—with the heterozygous condition being phenotypically superior to either homozygous state of a given gene. A.B. Bruce of Cambridge University proposed that dominance results when sets of deleterious (recessive) mutations at different loci in each inbred parent are complemented in the offspring by dominant alleles from the other parent. Different studies have found evidence for dominance and overdominance mechanisms, as well as epistasis (gene-to-gene interactions).

Although many agricultural plants have been improved through hybridization between inbred lines, the molecular mechanism of heterosis remains unknown. QTL mapping and expression studies have identified hundreds of loci and genes involved in plant yield heterosis. Moreover, changes in gene expression are dispersed across several biochemical pathways. Thus, the situation is analogous to complex diseases in human: Heterosis appears to involve combinations of many, perhaps hundreds, of genes.

Studies have shown that the expression of genes involved in protein metabolism is frequently altered in hybrids. This makes sense considering that ~25% of eukaryotic cell metabolism is devoted to synthesizing, folding, and degrading proteins. A protein's biological activity depends on its proper processing following translation, including folding and assembly into complexes. Folding also stabilizes proteins, because linear regions are liable to attack by proteases. However, a significant percentage of newly synthesized proteins fail to fold properly and are unstable. Thus cells have quality-control systems—such as the ubiquitin pathway—to detect and degrade misfolded proteins.

Stephen Goff, 2011
(Courtesy of Stephen Goff, iPlant Collaborative, University of Arizona.)

Dani Zamir
(Courtesy of Dani Zamir.)

A 2005 transcriptome profiling study by Jun Yu, of Zhejiang University, compared gene expression in a superhybrid crop of rice and its two parents. A large proportion of the relatively few down-regulated genes in the hybrid were involved in key aspects of protein processing:

- Peptidyl-prolyl *cis–trans* isomerase (PPIase) performs a rate-limiting step of protein folding to achieve the long-range interactions that stabilize secondary and tertiary structure.
- Metallopeptidase M48 is a membrane-associated protease that participates in protein processing.
- UDP-glucose:glycoprotein glucosyltransferase (UGGT) detects misfolded proteins in the endoplasmic reticulum.
- Ubiquitin-conjugating enzyme (UBC2) helps to detect unfolded proteins and mark them for degradation.

Up to 5% of plant genes have alleles encoding unstable proteins. According to a heterosis model proposed by Stephen Goff, of the University of Arizona, a quality-control mechanism in hybrids detects and down-regulates these alleles. This allows hybrids to expend less energy on protein degradation and resynthesis—and apply more energy to biomass, seed, and fruit production.

Heterosis is usually achieved by full-genome hybridizations. Despite this seeming complexity, in 2010, Dani Zamir, of Hebrew University of Jerusalem, showed that heterozygosity for a single gene, *SINGLE FLOWER TRUSS* (*SFT*), increases tomato yield by up to 60%. Tomato plants homozygous for a missense mutation in *SFT* were crossed with a commercial inbred line to produce hybrids with a single normal copy of the gene. *SFT* produces the flowering hormone florigen, and hybrids produce significantly more flowers. However, *SFT* also suppresses the activity (is an antagonist) of

Single gene heterosis in tomato. A single copy of the *SINGLE FLOWER TRUSS* gene in the M82 x sft-4537 hybrid (*center*) dramatically increases yield over either parent.
(Reprinted from Krieger U, et al. 2010. *Nat Genet* 42: 459–463.)

SELF PRUNING (*SP*), which terminates flower growth. Thus, overdominance appears to result from tuning the opposing signals of *SFT* and *SP*. This indicates that additional cases of single-gene heterosis may be found in other agriculturally important plants.

THE GREEN REVOLUTION OF WHEAT AND RICE

Over the centuries, traditional varieties of wheat and rice, whose long stalks were useful for animal feed and bedding (straw), had been selected for quick growth to outcompete weeds. Adapted to relatively poor soils, these varieties grow too tall when treated with modern fertilizers, making them prone to lodging (falling over). Thus, dramatic increases in wheat yields in the 20th century were due primarily to selective breeding to produce semi-dwarf plants that devote more energy to grain production. Semi-dwarf varieties have thicker, stiffer stems that resist lodging. Critically, they respond efficiently to fertilizer, maturing faster and allowing multiple harvests per year. "Miracle" wheat and rice produced threefold to sixfold greater yields, making Mexico, India, Pakistan, and other poor countries self-sufficient in grain.

Henry Wallace, who had popularized hybrid corn through his company Pioneer Hi-Bred, also played a role in the Green Revolution of wheat and rice. Wallace became influential in agricultural policy, first as U.S. Secretary of Agriculture (1933–1940) and then as Vice President (1941–1945). In 1943, Wallace convinced the Rockefeller Foundation to establish the Mexican Agricultural Program outside of Mexico City, with the express purpose of improving cereal agriculture in the developing world. At the time, Mexico imported half of its wheat per year, and infection by stem rust kept local yields low.

Norman Borlaug, ca. 1954
(Courtesy of The World Food Prize Foundation.)

The American breeder Norman Borlaug joined the program in 1944, initially developing rust-resistant varieties. Combined with fertilizers and improved farming practices, this doubled wheat yield in a decade. In 1954, Borlaug began crossing local wheat varieties with Norin 10, a semi-dwarf strain developed in Japan. This yielded hardy, semi-dwarf strains with resistance to rust. In 1956, Mexico no longer imported wheat, and it was exporting 500,000 tons per year by 1964.

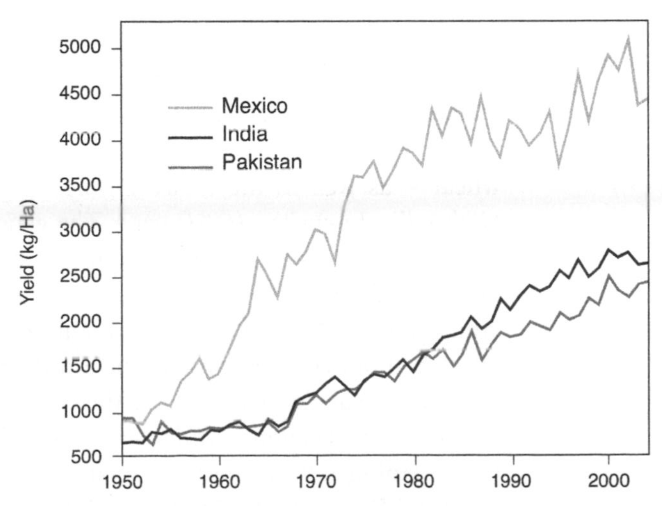

Wheat yields in Mexico, India, and Pakistan, 1950–2004.
(Courtesy of Food and Agriculture Organization, United Nations.)

President Lyndon Johnson with IR8 "Miracle Rice." Lyndon Johnson (crouching, *right*) and Philippine President Ferdinand Marcos (standing, *right*) visit the International Rice Research Institute (IRRI) on October 26, 1966. (*Left*) IR8 breeders Henry Beachell (crouching) and Walter Jennings (standing); (*center*) IRRI Director General Robert Chandler.
(Courtesy of International Rice Research Institute.)

With the specter of a rapidly increasing human population fanning the flames of chronic famine on the Indian subcontinent, in 1963, the Rockefeller Foundation gave the Mexican effort a worldwide mission as the International Maize and Wheat Improvement Center (CIMMYT). Borlaug introduced his semi-dwarf varieties into India and Pakistan in 1965. Wheat yields had nearly doubled by 1970 when Borlaug received the Nobel Peace Prize, gaining him recognition as the father of the Green Revolution.

The Rockefeller Foundation teamed with the Ford Foundation in 1960 to establish the International Rice Research Institute (IRRI) in the Philippines to do for rice what the Mexican Agricultural Program had done for wheat. Henry Beachell and Walter Jennings crossed a dwarf Chinese strain with a tall Indonesian strain to produce a hybrid, IR8, that was the foundation stock of modern semi-dwarf varieties. IR8 showed a strong fertilizer response, producing 50%–200% more grain than its dwarf and tall parents. Filipino farmers began planting IR8 in 1966, doubling yields by the mid 1970s.

Molecular genetic analysis subsequently showed that the semi-dwarf trait is caused by certain events in the biosynthesis and intracellular signaling of the growth hormone gibberellin. In 1999, Nicholas Harberd, of the John Innes Centre in Norwich UK, showed that two *Reduced height-1* genes (*Rht-B1* and *Rht-D1*) from semi-dwarf wheat— as well as the maize *Dwarf8* (*D8*) gene—are orthologs of the *Arabidopsis Gibberellin-Insensitive* (*GAI*) gene. The genes encode transcription factors with a conserved SH2 domain, which is common in many intracellular signaling proteins. The dwarfing al-leles of these genes all have alterations (deletions or stop codons) in the amino-terminal region, where a highly conserved amino acid sequence, abbreviated DELLA, modulates response to the growth hormone gibberellin.

The wheat *Rht-1* genes and the maize *D8* gene map to the same location of the ancestral cereal genome, and the amino acid sequences are 88% identical. Furthermore, transforming tall Basmati rice plants with the mutant *gai* allele produced dwarfed plants with decreased response to gibberellin. These results confirmed the equivalence of gibberellin response genes across the plant kingdom.

Packing IR8 Seed at the International Rice Research Institute. To launch IR8 in 1966, farmers came from all over the Philippines to pick up 2 kg of free seed.
(Courtesy of International Rice Research Institute.)

In 2002, teams led by Makoto Matsuoka, at Nagoya University, Japan, and Peter Chandler, at the Commonwealth Scientific and Industrial Research Organization, Australia, identified a defect in gibberellin biosynthesis as the cause of the semi-dwarf trait in IR8 varieties of rice. The Japanese group showed that IR8 plants respond to external gibberellin, indicating that the mutation must be in a gene for a gibberellin biosynthesis enzyme. Using biochemical analysis, they deduced that a nonfunctional gibberellin 20-oxidase gene (*GA20ox-2*) was the cause.

The Australian group identified the same gene by combining sequence data from the rice genome with evidence from previous linkage maps of the semi-dwarf locus on chromosome 1 (*sd-1*). Using these different approaches, each group identified the major causative *sd-1* mutation in IR8 as a several hundred-nucleotide deletion spanning the first and second exons of the *GA20ox-2* gene. Remarkably, each group also identified a C>T point mutation at position 798—which alters a conserved leucine (Leu 266) to phenylalanine—as the *sd-1* allele in another semi-dwarf strain, Calrose76.

Nicholas Harberd with semi-dwarf wheat.
(Reprinted, with permission, from Nicholas Harberd/JIC Photography.)

DIRECT GENETIC MODIFICATION OF PLANTS

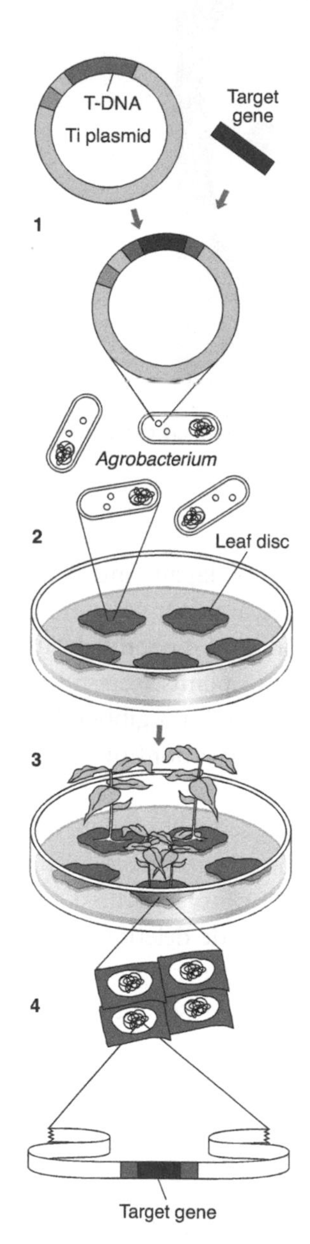

Agrobacterium transformation. First, the target gene encoding a desired trait is inserted into the T-DNA region of a Ti plasmid (*1*). Next, leaf discs are infected with *Agrobacterium* carrying the recombinant Ti plasmid (*2*). The plantlets are then regenerated in culture (*3*); the target gene has been integrated into the plant cell chromosomes (*4*).

Classical breeding over centuries slowly bent plants to human needs, turning humble weeds, such as teosinte, into modern agricultural wonders. Indirect methods of genetic alteration—selective breeding and hybridization—accounted for the majority of the dramatic increases during the Green Revolution after World War II, including hybrid maize and semi-dwarf varieties of wheat and rice. In the 1970s, direct genetic modification (GM)—using the tools of molecular genetics—offered the potential to fuel what some have called the second Green Revolution. Gene transfer technology made it theoretically possible to directly transfer the gene for any desirable trait into crop plants.

Critically, plant scientists today look for genes that will allow crops to adapt to shifting weather patterns induced by global climate change—notably, resistance to drought and waterlogging. Ultimately, multiple genes can be "stacked" in a single plant, endowing it at once with several discrete traits. In this way, the art of plant breeding may increasingly be converted into a precise science. Early success with direct gene transfer came for single-gene traits for pesticide and herbicide resistance. However, as in the case of heterosis, most of the important qualities we would like to "engineer" into crops are likely to be quantitative traits to which many genes contribute incrementally.

As had animal cell biologists before them, plant biologists turned to tumor-causing microbes to carry (transform) new DNA into plant cells. *Agrobacterium tumefaciens* causes crown gall disease, a tumor growth at the growing tip, or crown, of legumes (dicot plants such as beans, peas, and clover). *Agrobacterium* enters plant cells through wounds or abrasions in the epidermis. Then a bacterial DNA sequence integrates into the host-cell DNA, where it induces proliferation of cells and diverts the plant's metabolic machinery to produce opines, specialized molecules required for bacterial energy and growth.

The tumor-causing functions are performed by a giant tumor-inducing (Ti) plasmid that is separate from the main *Agrobacterium* chromosome. A 30–40-kb region of the Ti plasmid, the T-DNA, carries specific genes needed for tumor formation and opine synthesis. During infection, the T-DNA excises from the Ti plasmid and moves into the host-cell nucleus, where it integrates into the host chromosome.

To engineer a Ti-plasmid vector, the tumor and opine synthetase coding sequences are deleted from the T-DNA and replaced with a foreign gene of interest, or transgene. The transgene is linked to a strong promoter, such as that from cauliflower mosaic virus (CMV), which ensures high-level expression. Other sequences may also be incorporated into the T-DNA, such as a transit peptide signal, to ensure transport to and accumulation in the proper tissue. *Agrobacterium* cells containing the engineered Ti plasmid are then used to infect the target species.

Agrobacterium transformation is typically performed on plant protoplasts, cultured cells that have been stripped of their resistant cell walls. Antibiotic selection identifies clones of cells that have taken up the transgene. The cultured cells are then stimulated with plant hormones to regenerate plantlets. The transgene is inherited in a Mendelian fashion through successive generations of crosses, indicating that it has been stably integrated into the plant genome.

The utility of this system was first shown in 1983, when three groups nearly simultaneously reported that they had used *Agrobacterium* to transfer antibiotic-resistance genes into tobacco and petunias: Mary-Dell Chilton, at Washington University in St.

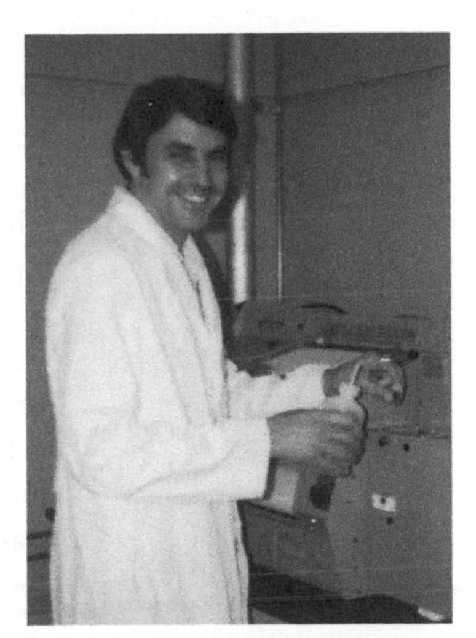

Pioneers of plant transformation. (*Bottom left*) Marc van Montagu and Jeff Schell. (From www. Wikipedia. com.) (*Upper left*) Mary-Dell Chilton. (Reprinted, with permission, from University Archives, Department of Special Collections, Washington University Libraries.) (*Center*) Robert Horsch and Robert Fraley. Horsch (*left*) holds cultured petunia cells, and Fraley (*right*) holds a culture of *Agrobacterium*. (Courtesy of Monsanto.) (*Upper right*) John Kemp, 1973. (Courtesy of University of Wisconsin-Madison Archives [Image ID #S04876].) (*Bottom right*) Theodore Klein. (Courtesy of Theodore Klein.)

Louis; Jeff Schell and Marc Van Montagu, at Ghent University; and Robert Fraley, Stephen Rogers, and Robert Horsch, at Monsanto. Later that year, John Kemp and Timothy Hall, at Agrigenetics Advanced Research Division, transferred the gene for the bean storage protein phaseolin into sunflower.

The Monsanto group simplified *Agrobacterium*-mediated gene transfer in 1985, with the discovery that transformed plantlets can be regenerated from simple leaf disks, whose cut edges provide entry sites for *Agrobacterium*. Transformation of *Arabidopsis* was revolutionized in 1998 with the floral dip method of Steven Clough and Andrew Bent, of the University of Illinois at Urbana–Champaign. In this method, flowers are dipped in an *Agrobacterium* solution for several seconds; ~1% of seeds then produce transgenic plants. In this way, transgenic plants can be developed in as little as 3 months.

Because *Agrobacterium* transformation was initially limited to dicot plants, additional methods were developed to introduce naked DNA directly into dicot and monocot protoplasts. In 1985, Michael Fromm and Virginia Walbot at Stanford University transformed plant protoplasts using electroporation, which applies pulses of high-voltage electricity. Particle bombardment was perfected in the late 1980s by

Theodore Klein and Michael Fromm, at the University of California at Berkeley. This method, in which DNA-coated tungsten or gold microprojectiles are "shot" into cells, can also transform cultured cells with intact cell walls.

Agrobacterium-mediated gene transfer was extended to key cereals a decade after the first transformation of dicots. In 1994, Yukoh Hiei, of Japan Tobacco, Inc., achieved stable expression and inheritance of a transgene in several cultivars of *japonica* rice.

Most people would be surprised to learn that the majority of fresh and processed foods have at least one genetically modified (GM) component. First consider the major GM food crops: soybeans, corn, beets, canola, tomatoes, potatoes, and papaya. Then consider how many processed foods contain components made from corn, soy, or beets: cornstarch, corn meal, high-fructose corn syrup, beet sugar, soy protein, soy sauce, tofu, and corn or soy oil. Finally, consider that 91% of soybeans, 90% of sugar beets, and 64% of corn planted in the United States in 2009 were genetically modified for herbicide resistance.

The most common GM crops are tolerant to the herbicide glyphosate (Roundup), which inhibits 5-enolpyruvylshikimate-3-phosphate synthase (EPSPS), an enzyme required for the biosynthesis of aromatic amino acids in the chloroplast. GM plants are transformed with a copy of a bacterial version of the *EPSPS* gene that is resistant to glyphosate and is expressed at high levels under the the direction of the 35S CMV promoter. Once the crop is established, treatment with Roundup kills off all of the weeds but spares the resistant plants. Herbicide-tolerant (HT) or Roundup Ready soybeans were introduced in 1996, followed in 1997 by Roundup Ready maize.

Proponents see many advantages to glyphosate-resistant crops. They increase yield and save time and fuel. Herbicide treatment enables reduced or no-till farming, which conserves soil fertility and lessens erosion. Glyphosate is safe, especially in comparison to the previous generation of herbicides, and quickly degrades in the environment. Opponents worry that herbicide resistance could "escape" if GM crops pollinate organic crops or wild relatives. Although not directly caused by genetic modification, more than 10 glyphosate-resistant weeds have emerged since 2000 and are now found in 22 states.

Insect resistance is another important GM trait. In the most common system, plants are transformed with a gene from the soil bacterium *Bacillus thuringiensis* (*Bt*). The Bt toxin produces crystalline proteins that paralyze the digestive system of susceptible insects. Believed to be an environmentally safe insecticide, *Bt* is active against

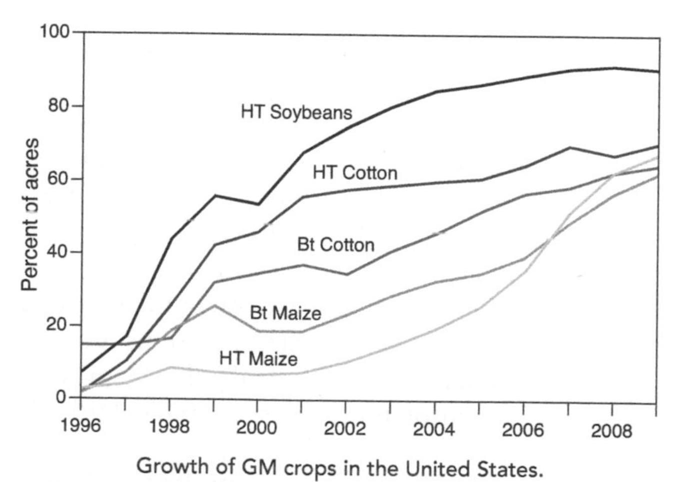

Growth of GM crops in the United States.
(Courtesy of Clive James, International Service for the Acquisition of Agri-Biotech Applications.)

several caterpillars, including the tobacco hornworm, corn rootworm, and cotton bollworm. *Bt* maize and cotton were introduced into the United States in 1996. By 2009, plantings of *Bt* transgenics comprised 63% of the U.S. corn crop, and 65% of cotton, the most important nonfood GM crop.

Despite their huge success in the United States and some major agricultural countries, GM crops are still unpopular in Europe. The European Union (EU) has approved only two GM crops for planting in Europe, neither of which is for human consumption. Monsanto's MON810, a variety of *Bt* maize, was approved in 1998, and BASF's Amflora, a potato that yields pure amylopectin used in textiles, papers, and adhesives, was approved in 2010.

QUALITY PROTEIN MAIZE

The maize kernel is composed of an embryo (or germ), endosperm, and surrounding layers of aleurone and pericarp (bran). The endosperm stores carbohydrates and proteins. The aleurone is one or several layers of thick-walled cells that secrete hydrolases to digest endosperm starches and proteins, making them available to the germinating embryo. The pericarp is composed of nonliving cells that protect the kernel.

Detailed knowledge of the pathways used to synthesize key amino acids and vitamins in the kernel is now being used to bolster the protein and vitamin content of corn. The process of using genetics and biochemistry to improve the nutritional properties of plants is termed "biofortification."

Amino acids are not stored by the human body and must be provided on a daily basis. The body can synthesize 11 of the 20 naturally occurring amino acids. The other nine are termed "essential" because they must be obtained in the foods we eat. Of plants, only soybeans produce "complete" protein containing all of the essential amino acids in the correct proportions needed in the human diet. Although protein comprises ~10% of corn's dry weight, corn is low in the essential amino acids lysine and tryptophan. Furthermore, ~98% of maize niacin (vitamin B3) is bound up with complex plant polysaccharides and peptides (as niacytin or niacinogen) that are not absorbed in the human intestine.

Mesoamericans countered these deficiencies with their methods of cultivating and cooking maize. Growing beans alongside maize not only provided nitrogen in the

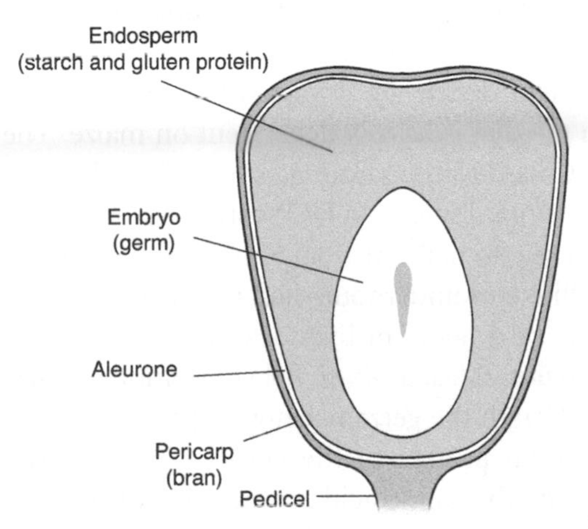

Structure of a maize kernel.

Hopi woman grinding corn, ca. 1897.
(Gelatin silver print by Adam Clark Vroman. Courtesy of SFMOMA, on behalf of the Sack Photographic Trust.)

soil to maintain fertility, but when beans and maize were consumed together, the beans supplemented the maize protein with lysine and tryptophan. Mesoamericans soaked maize in alkaline lime, softening the kernel to produce hominy, which could be eaten as porridge or dried and ground into meal (masa). While the maize is soaking, alkaline hydrolysis breaks down complex polysaccharides and peptides in the outer pericarp and aleurone layers, releasing the niacin concentrated in the aleurone. The traditional process, called *nixtamalization* (derived from the Aztec words for "ashes" and "corn dough"), produced lime by mixing wood ashes in water.

Some historians argue that the discovery of nixtamalized corn was key to the growth of large urban populations in Mesoamerica. Europeans had a different experience with maize, which was brought back to Spain by Christopher Columbus in 1492. Maize rapidly became the subsistence food of the masses, but Europeans failed to adopt nixtamalization. Chronic niacin deficiency was therefore first described in 1735 in peasants of southern Spain by Caspar Casal, physician to Philip V. The disease gained the name *pellagra* (from Italian for "rough skin") when corn cultivation spread to the fertile Po Valley of Italy, replacing peasants' barley with polenta (corn meal mush). An 1830 census found that 5% of the population of northern Italy was affected.

Protein deficiency (kwashiorkor) is a serious problem in developing countries in which the diet is heavily dependent on maize. The Food and Agriculture Organization of the United Nations estimates that 900 million people are undernourished. The World Health Organization (WHO) estimates that 200 million children suffer from stunted growth due to malnutrition; 5 million die annually. In addition to improving nutrition for millions of undernourished people worldwide, quality protein maize (QPM) could prove to be a boon for livestock producers. Pigs and chickens fed QPM put on weight about twice as fast as those fed conventional maize.

Edwin Mertz
(Courtesy of Purdue University Libraries, Karnes Archive & Special Collections.)

Although the germ has higher protein content, the endosperm contributes ~80% of the total protein of the kernel. Thus, efforts to produce QPM have focused on improving the amino acid composition of zein, the most abundant endosperm protein. In 1963, Lynn Bates, a graduate student working with Edwin Mertz at Purdue Univer-

Benjamin Burr, 2000
(Courtesy of Brookhaven National Laboratory.)

sity, discovered that corn plants with the *opaque2* mutation produce substantially more lysine and tryptophan. Benjamin Burr, of Brookhaven National Laboratory, cloned the *opaque2* locus in 1987 and found that it encodes a transcription factor that regulates expression of zein genes. Feeding trials in the developing world showed that the zein protein of *opaque2* mutants has an equivalent nutritive value as milk and can successfully treat children with kwashiorkor.

In 2000, CIMMYT scientists received the World Food Prize for their success in producing improved QPM for developing countries. By crossing *opaque2* mutants with local cultivars, they produced QPM varieties with yield improvements of up to 10%—and with drought and disease tolerance matching local maize varieties. Despite these efforts, nearly 50 years after its discovery, QPM accounts for less than 1% of maize harvested worldwide.

GOLDEN RICE AND ORANGE MAIZE

According to WHO, vitamin A deficiency is the leading cause of preventable blindness worldwide. An estimated 200 million preschool children are vitamin A deficient, and 250,000–500,000 become blind each year. Although vitamin supplements offer effective treatment, they must be manufactured and distributed on a regular basis, which is impractical on a global scale. Malnourished children are too poor to afford vitamin-rich foods, but in many areas they do eat large amounts of rice and corn.

Thus, the development of β-carotene-enriched golden rice and orange corn illustrates how biofortified foods can potentially provide a self-sustaining means to deliver the equivalent of a daily vitamin pill to the underdeveloped world. β-Carotene is also

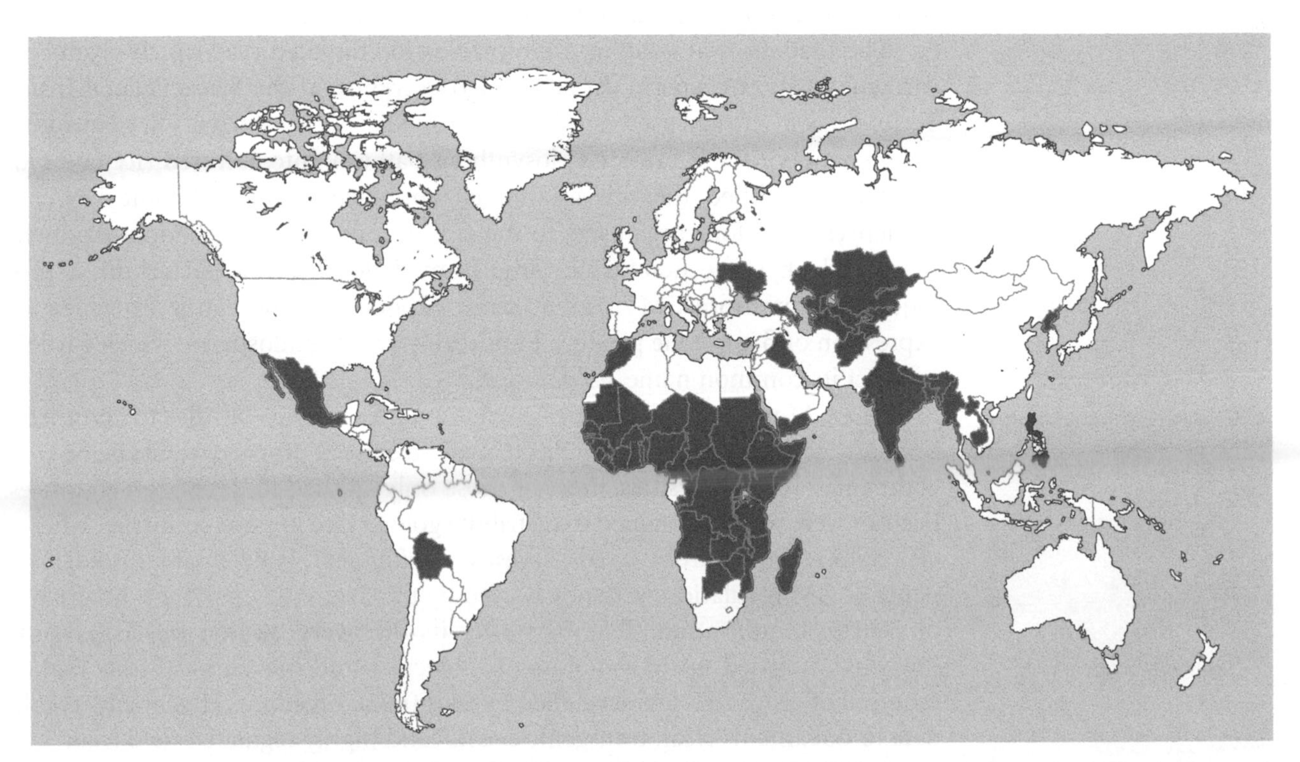

Vitamin A deficiency worldwide, 2009. Countries in which 20% or more of preschool-age children and women of childbearing age have severe Vitamin A deficiency (serum retinal level <0.70 μmol/L).
(Source: World Health Organization.)

called pro-vitamin A because it is cleaved in the intestines to produce two molecules of vitamin A. In contrast, closely related α-carotene produces only a single molecule of vitamin A.

β-Carotene is a member of the carotenoid pigment family, which includes several related red, yellow, and orange pigments found throughout the plant world. The brightly colored flowers and fruits attract animals needed to disperse pollen and seeds. However, the ubiquitous expression of carotenoids in plants only becomes apparent during leaf senescence, when decreasing production of chlorophyll unmasks these pigments and turns leaves shades of red, yellow, and orange.

The leaf surface is optimized to collect light needed for photosynthesis. This intense radiation generates free radicals, unstable molecules that contain an unpaired electron. These highly reactive molecules, which include oxygen free radicals, damage all types of biomolecules, including DNA. Carotenoids function as protective antioxidants by scavenging free radicals, especially those created when molecules are excited by light energy.

Interestingly, carotenoids perform an identical function in the human eye, which is also a light-harvesting apparatus. Carotenoids, especially zeaxanthin and lutein, are concentrated in the macula, the central portion of the retina that receives the most focused light. Diets in many developed nations are deficient in these carotenoids, and free radical damage contributes to age-related macular degeneration (AMD), a major cause of blindness in older people.

The Age-Related Eye Disease Study (AREDS), an ongoing clinical trial conducted by the National Eye Institute, showed that high-dose supplements of zeaxanthin and lutein significantly reduce the risk of AMD. Zeaxanthin and lutein are isomers; they have identical chemical formulas and differ only in the location of one double bond. Because zeaxanthin is most prevalent in the central macula, some believe that it may be the single most important protectant against AMD.

The first effort at vitamin A biofortification targeted rice, which produces no β-carotene in its endosperm. In 2000, Ingo Potrykus, at the Swiss Federal Institute of Technology in Zurich, used *Agrobacterium*-mediated transformation to introduce two key enzymes of the β-carotene biosynthesis pathway into cultured rice embryos. The project exemplified the mix-and-match approach of recombinant DNA, adding sequences from four organisms to the rice genome. Genes encoding phytoene synthase and phytoene desaturase came from daffodil and a bacterium, a promoter sequence came from CMV, and a transit peptide sequence came from pea. Correct expression of β-carotene produced noticeably yellow endosperm, which gave the GM product its common name, golden rice.

As a GM product, golden rice encountered numerous hurdles to commercialization. Although it entailed 70 intellectual and technical property rights belonging to 32 companies and universities, most of those only applied to developed countries. Free licenses were readily obtained to distribute golden rice in poor countries, where farmers could replant it from saved harvest year after year. Unlike patent holders, proponents of GM regulation were not swayed by this seemingly perfect humanitarian use of genetic modification. Ten years after its discovery, golden rice is, according to Potrykus, still tied up in a morass of molecular and biochemical tests that are not required of crops that are developed by traditional breeding. This included collecting data to document "clean transgenic events" and highly regulated field trials.

Torbert Rocheford, 2011
(Courtesy of Purdue University.)

In 2006, Torbert Rocheford, of Purdue University, began a systematic effort to optimize the maize carotenoid pathway to synthesize increased amounts of β-carotene.

Natural variation in maize creates flux in the carotenoid pathway that results in kernels ranging from white (colorless) to yellow to orange. Yellow kernels have more α-carotene and lutein, whereas orange kernels have more β-carotene and zeaxanthin. Rocheford exploited this natural variation to breed and visually select plants that produce deep orange kernels, an indicator of increased β-carotene content.

QTL Mapping

Edward Buckler
(Courtesy of Cornell University.)

QTL mapping had previously identified several chromosome loci involved in carotenoid biosynthesis. Edward Buckler and Carlos Harjes, of Cornell University, mapped the deep orange trait to a region of chromosome VIII containing the gene encoding lycopene ε cyclase. This enzyme sits close to a key bifurcation point in the carotenoid pathway at which lycopene is either converted to α-carotene and lutein or to β-carotene and zeaxanthin. They then correlated the orange trait with DNA poly-

OPP
phytoene synthase
phytoene desaturase
lycopene
lycopene ε cyclase
More yellow pigments
More orange pigments
α-carotene
β-carotene
β-carotene hydroxylase
OH
HO
lutein
OH
HO
zeaxanthin
OH
Vitamin A

Carotenoid biosynthesis pathway and sources of vitamin A. Mutations in genes encoding key biosynthetic enzymes control flux through the carotenoid pathway. Weak alleles of lycopene ε cyclase decrease production of α-carotene and shift flux toward the β-carotene (pro-vitamin A) side of the pathway. Strong alleles of β-carotene hydroxylase increase production of zeaxanthin, whereas weak alleles favor production of β-carotene. Cleavage of β-carotene yields two units of vitamin A, whereas α-carotene yields a single unit (shaded).

morphisms across several inbred lines. A transposon insertion was identified in the promoter of the ε cyclase gene; this "weak" allele decreased the conversion of lycopene to α-carotene, shunting lycopene to the other side of the pathway and producing more β-carotene. This polymorphism can be used in breeding experiments to rapidly identify varieties carrying the trait, a technique called "marker-assisted selection."

QTL mapping identified a second gene, β-carotene hydroxylase (*crtRB1*), which is needed to convert β-carotene into zeaxanthin. Several insertion polymorphisms in *crtRB1* decrease expression of the hydroxylase. Weak alleles correlate with high levels of β-carotene, whereas strong alleles correlate with increased production of zeaxanthin. Thus, it is possible to manipulate, or tune, carotenoid expression in maize to deal with vitamin deficiencies in two distinctly different populations: increasing β-carotene content to prevent pellagra in developing countries and increasing zeaxanthin content to prevent macular degeneration in developed countries.

Initial feeding studies in Zambia suggest that mothers will use orange corn, especially when they understand the health benefits for their children. However, the slow acceptance of biofortified grains thus far points to the difficulty of delivering superior agricultural products to the people who need them most. Agriculture is market-driven, and farmers are motivated primarily by increased yield. Invisible nutritional qualities offer little incentive to farmers or market premium to commercial seed companies.

BIOFUELS AND PLANT BIOMASS

The United States currently has about one-third of the world's automobiles and consumes 25% of the world's oil. According to the Department of Energy, the United States consumes ~3.5 TW (terawatts) of power on a continual basis—the output of 3500 coal-burning power plants. Demand for energy will continue to increase, upward of 50% for electricity alone by the year 2030. Dwindling supplies, high costs, and the environmental consequences of burning fossil fuels have inspired scientists to explore biofuels produced directly by plants or derived from plant materials—termed *biomass* or *feedstocks.*

Gasohol, a mixture containing anhydrous (dry) ethanol, is currently the only commercially available biofuel. A blend of 10% ethanol and 90% gas (labeled E10) is most common among the estimated 75% of U.S. gas stations that pump gasohol all or part of the year. Currently, ethanol for gasohol is almost exclusively produced by fermenting and distilling sugar obtained from corn, soybeans, or sugarcane.

Thus, ethanol production from corn currently competes with human consumption of corn, potentially pricing a key staple out of reach of developing countries. The challenge therefore is to find ways to convert nonfood biomass into biofuels. Most plant biomass is locked up in wood (plant cell walls), which consists of ~80% cellulose and related hemicellulose and 20% lignin. Cellulose, the most abundant biomaterial on Earth, is a complex polymer of glucose and is a large potential source of ethanol, termed *cellulosic ethanol.*

However, technology to produce cellulosic ethanol is currently too expensive to compete in the energy marketplace. Cellulose evolved as a protective barrier for plant cells and is highly resistant to chemical degradation. Thus, it is difficult to convert cellulose into sugars that can be fermented to produce alcohol. Because of its high aromatic (benzene-based) content, lignin is a poor source of sugar for fermentation. The cost-effective production of cellulosic biofuel will require a deeper understanding of how cell wall materials are synthesized and how they are degraded. Thus, bioinformatics is being used to "mine" plant and microbial genomes for naturally occurring enzymes—including cellulases, hemicellulases, and glycosyl hydrolases—that break down woody plant biomass.

A large portion of our future energy needs could be met by converting the billions of tons of nonfood biomass contained in agricultural, forestry, and municipal waste generated each year into cellulosic ethanol. Biomass can also be provided by "energy crops," including perennial grasses and fast-growing trees. Thus, the U.S. Department of Energy has funded several projects to sequence the genomes of feedstocks that can provide renewable sources of cellulose biomass. Poplar was the first tree species to have its genome sequenced, in 2006, and strides have been made in improving its large-scale culture and harvesting. Classical breeding has increased its growth rate, adaptability to different climates and soils, and resistance to disease and stress, while reducing its lignin content.

The model grass *Brachypodium distachyon* was sequenced in 2010, and the common pond plant duckweed (*Spirodela polyrhiza*) is currently being sequenced. A physical map of DNA markers for switchgrass was developed in 2010, allowing comparison of the switchgrass genome with the sequenced genomes of other grasses, including *Brachypodium* and rice. Switchgrass, a native perennial of temperate North America, can reach heights of greater than 10 feet in a single growing season.

Clearly, the continued development of human society is critically dependent on plants as sources of food and fuel. Although critics may argue that genetic modification of plants is new, genome science is only the current chapter in 10,000 years of human-controlled evolution of plants.

LABORATORY 3.1

Detecting a Transposon in Corn

▼ OBJECTIVES

This laboratory demonstrates several important concepts of modern biology. During the course of this laboratory, you will

- Learn about the relationship between genotype and phenotype.
- Learn how transposable elements can be used to mutagenize and tag genes.
- Conduct a literature search to discover how plant metabolism is under genetic control.
- Move between in vitro experimentation and in silico computation.

In addition, this laboratory utilizes several experimental and bioinformatics methods in modern biological research. You will

- Extract and purify DNA from plant tissue.
- Amplify a specific region of the genome by polymerase chain reaction (PCR) and analyze PCR products by gel electrophoresis.
- Use the Basic Local Alignment Search Tool (BLAST) to identify sequences in databases.
- Use the Map Viewer tool to picture genes on chromosomes.

INTRODUCTION

Barbara McClintock at Cold Spring Harbor Laboratory, ca. 1950.
(Courtesy Cold Spring Harbor Laboratory Archives.)

Throughout the first half of the 20th century, geneticists assumed that a stable genome was a prerequisite for faithfully transmitting genes from one generation to the next. Working at Cold Spring Harbor Laboratory in the post-WWII era, Barbara McClintock found quite a different story in *Zea mays* (maize, or corn). She observed numerous "dissociations"—broken and ring-shaped chromosomes—and traced the source of these mutations to two related loci, "dissociator" (*Ds*) and "activator" (*Ac*), on the short arm of chromosome 9.

Equipped with only her maize crosses and a light microscope, she showed that *Ac* and *Ds* are mobile genetic elements that transpose, or jump, from one chromosome location to another. McClintock also offered genetic proof that *Ac* moves independently, but *Ds* depends on *Ac* for transposition. She showed that transposons may inactivate gene expression by inserting into a gene or may reactivate expression by jumping out. Thus, McClintock explained color variegations, such as speckled kernels, that had intrigued botanists for centuries.

Today, the *Ac*/*Ds* system is an important tool in gene discovery, allowing scientists to characterize genes for which no biological role is known. In a process known as transposon mutagenesis, *Ac* and *Ds* elements are crossed into a corn strain to produce *Ds* insertions in genes. The *Ac*/*Ds* mutagenesis system also works well in a number of other plants including tobacco, tomato, and the model plant *Arabidopsis thaliana*.

This laboratory investigates the *Bronze* (*Bz*) gene of maize to show the molecular relationship between genotype and phenotype. The *Bz* gene encodes an enzyme that catalyzes an important step in the biosynthesis of anthocyanins—red, purple, and blue pigments in plants. The *bz* mutation, created by the insertion of a *Ds* transposon in the *Bz* gene, abolishes anthocyanin production. Wild-type *Bz* kernels and heterozygotes are dark purple, whereas the *Ds* transposon insertion leaves kernels dark yellow or "bronze." The color change affects all plant organs, including the stem and leaves.

In this lab, DNA is isolated from corn plants, and the *Bz* locus is amplified using PCR. Two primer sets are used: One is specific for the wild-type allele and the other for the *bz* mutant allele (bearing a *Ds* insertion). The PCR products can be readily differentiated by agarose gel electrophoresis.

The laboratory also includes bioinformatics exercises that complement the experimental methods. BLAST is used to identify nucleic acid and protein sequences in biological databases, predict the size of the PCR products for the wild-type allele, and learn about the function and evolutionary history of the BZ protein. Another bioinformatics tool, Map Viewer, is used to discover the chromosome location of the *Ds* insertion, and additional calculations are performed to determine where in the *Bz* gene the *Ds* transposon is inserted to produce the bronze phenotype.

(*A*) Wild-type (Wt) and *bronze* (*bz*) mutant maize plants. (*B*) Stems of Wt plants (*left*) are dark purple; stems of *bz* mutants (*right*) are dark yellow, or bronze. (*C*) Kernels from wild-type (*left*), *bz* mutant (*center*), and heterozygous (*right*) plants. Cob 637 (*Bz*/*Bz*) and cob 638 (*bz*/*bz*) were from homozygous self crosses. Cob 637/638 is a heterozygote (*Bz*/*bz*) from a cross between 637 and 638.

FURTHER READING

Dooner HK, Weck E, Adams S, Ralston E, Favreau M, English J. 1985. A molecular genetic analysis of insertions in the bronze locus in maize. *Mol Gen Genet* **200:** 240–246.

Edwards K, Johnstone C, Thompson C. 1991. A simple and rapid method for the preparation of plant genomic DNA for PCR analysis. *Nucleic Acids Res* **19:** 1349.

Fedoroff N, Wessler S, McClure M. 1983. Isolation of the transposable maize controlling elements *Ac* and *Ds*. *Cell* **35:** 235–242.

McClintock B. 1952. Chromosome organization and genic expression. *Cold Spring Harbor Symp Quant Biol* **16:** 13–47.

http://www.dnaftb.org/dnaftb/32/concept/index.html An animation explaining McClintock's elucidation of the *Ac*/*Ds* system.

PLANNING AND PREPARATION

The following table will help you to plan and integrate the different experimental methods.

Experiment	Day	Time	Activity	
I. Plant maize seeds	2–3 wk	15 min	Prelab:	Set up student stations.
	before Part II	15 min	Lab:	Plant maize seeds.
			Postlab:	Care for maize plants.
II. Isolate DNA from maize	1	30 min	Prelab:	Set up student stations.
		30–60 min	Lab:	Isolate DNA.
III. Amplify DNA by PCR	2	15 min	Prelab:	Aliquot primer/loading dye mixes.
				Set up student stations.
		15 min	Lab:	Set up PCRs.
		60–150 min	Postlab:	Amplify DNA in thermal cycler.
IV. Analyze PCR products by gel electrophoresis	3	30 min	Prelab:	Dilute TBE electrophoresis buffer.
				Prepare agarose gel solution.
				Set up student stations.
		30 min	Lab:	Cast gels.
	4	45+ min	Lab:	Load DNA samples into gel.
				Electrophorese samples.
				Photograph gels.

OVERVIEW OF EXPERIMENTAL METHODS

I. PLANT MAIZE SEEDS

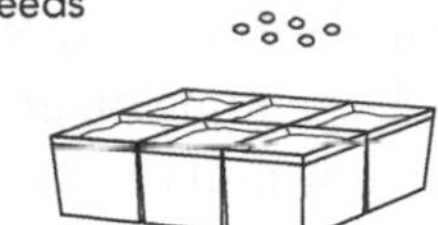

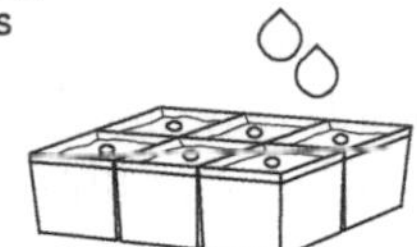

II. ISOLATE DNA FROM MAIZE

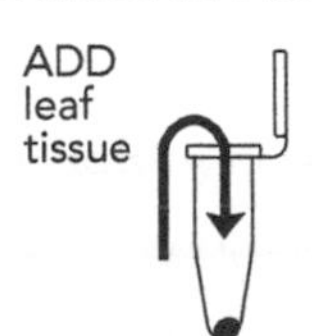

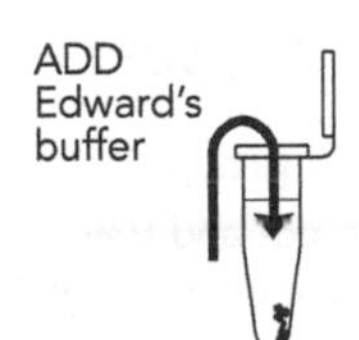

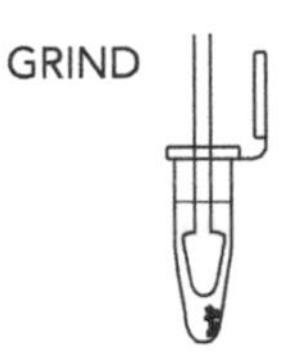

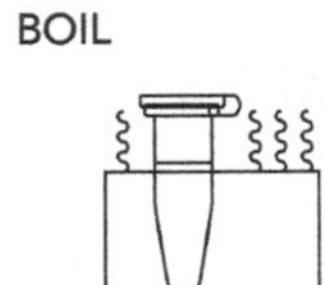

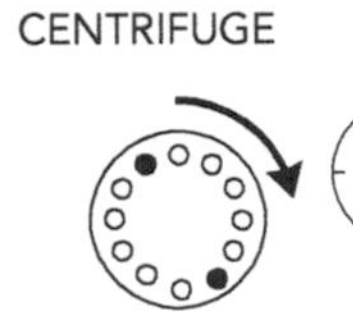

ADD and MIX
isopropanol

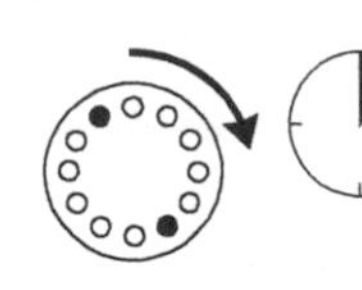

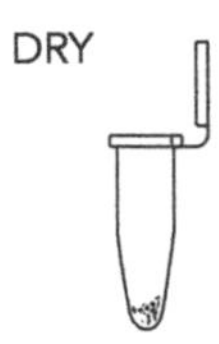

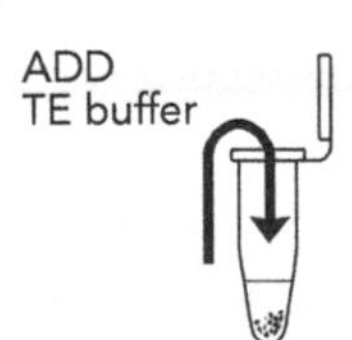

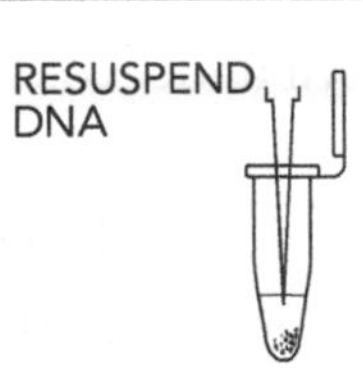

III. AMPLIFY DNA BY PCR

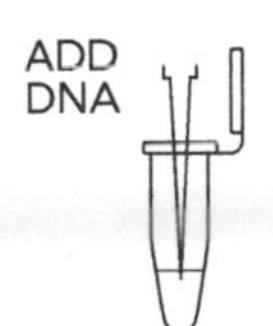

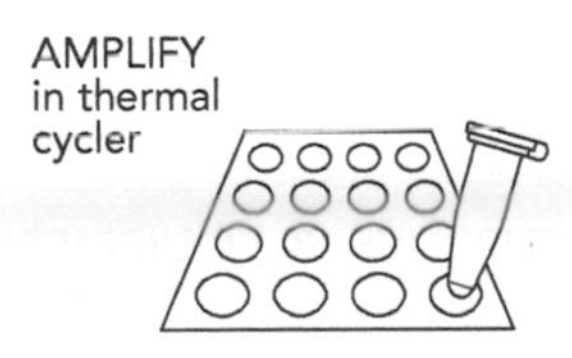

IV. ANALYZE PCR PRODUCTS BY GEL ELECTROPHORESIS

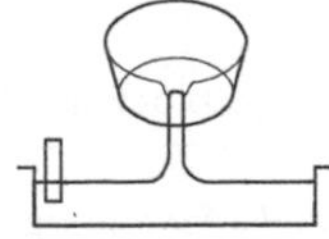

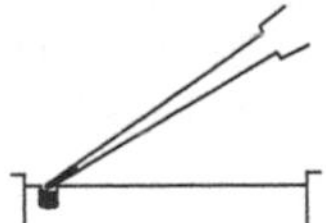

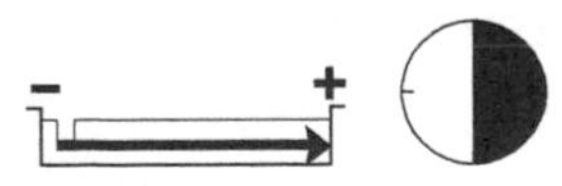

EXPERIMENTAL METHODS

I. Plant Maize Seeds

To extract DNA from plant tissue, you must plant the corn seeds 2–3 wk before DNA isolation and PCR. Two 1/4-in-diameter leaf disks are required for each experiment, but multiple small leaves and even whole plantlets can be used. Depending on growing conditions, you may observe the phenotypic differences between mutant and wild-type plants in as little as 1 wk.

REAGENTS, SUPPLIES, & EQUIPMENT

For each group

Planting pot and tray
Plastic dome lid or plastic wrap
Potting soil
3 Seeds from a self-cross of a *Bz/bz* heterozygous maize plant
Water

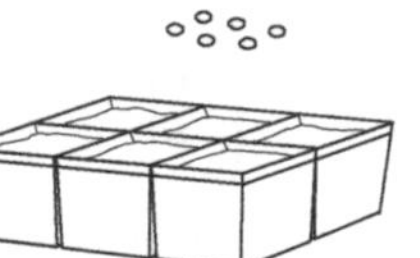

Planting one seed per pot allows optimal growth and easy observation of plant phenotypes.

1. Label the pot with your group number. Moisten the potting soil. Fill planting pot evenly with soil, but do not pack tightly.
2. Fit the pot into the tray, but leave one corner space empty to facilitate watering.
3. Use your finger to make a 0.5-in depression in the soil of the pot. Add three seeds to the hole and cover with soil.
4. Cover the pot with a plastic dome lid or plastic wrap to maintain humidity during germination. (Remove the cover 3–4 d after planting.)
5. Add 1/2 in of water to the tray, using the empty corner space. Water regularly to keep the soil damp, but do not allow the soil to remain soggy.
6. Grow the plants close to a sunny window at room temperature (20°C–22°C) or slightly warmer. If the plants are grown under cold conditions, it may take longer to discern the different phenotypes.
7. Plants homozygous for the *bz* mutation have bronze pigmented stalks, whereas wild-type and heterozygous plants have dark purple stalks. When the difference in phenotype becomes evident, continue to Part II.

For optimum growth, provide constant (24-h) fluorescent lighting about 1 ft directly above the plants. With 24-h fluorescent lighting, phenotypes can be discerned in 3 wk.

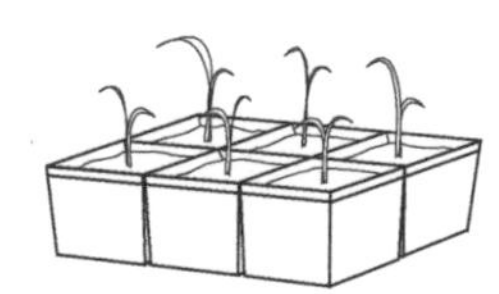

You may wish to continue to grow the plants after you have harvested tissue for DNA isolation and amplification. The phenotypic differences between mutant and wild-type plants become more obvious over time.

II. Isolate DNA from Maize

REAGENTS, SUPPLIES, & EQUIPMENT

For each group

Container with cracked or crushed ice
Edward's buffer (600 μL)
Isopropanol <!> (500 μL)
Microcentrifuge tube racks
2 Microcentrifuge tubes (1.5 mL)
Micropipettes and tips (100–1000 μL)
Permanent marker
Plastic pestle
Tris/EDTA (TE) buffer (200 μL)
Vortexer (optional)
Wild-type (*Bz*) or mutant (*bz*) maize plants from Part I

To share

Microcentrifuge
Water bath or heating block at 96°C–100°C

See Cautions Appendix for appropriate handling of materials marked with <!>.

Your instructor will assign a maize plant to your group.

Grinding the plant tissue breaks up the cell walls. When fully ground, the sample should be a green liquid.

Detergent in the Edward's buffer, called sodium dodecyl sulfate (SDS), dissolves lipids of the cell membrane.

Boiling denatures proteins, including enzymes that digest DNA.

The pellet may appear as a tiny teardrop-shaped smear or particles on the bottom side of the tube underneath the hinge. Do not be concerned if you cannot see a pellet. A large or greenish pellet is cellular debris carried over from the first centrifugation.

Dry the pellet quickly with a hair dryer. To prevent blowing the pellet away, direct the air across the tube mouth, not into the tube, for ~3 min.

The TE (Tris-EDTA) buffer provides conditions for stable storage of DNA. Tris provides a constant pH of 8.0, whereas EDTA binds cations (positive ions) that are required for DNase activity.

In Part III, you will use 2.5 µL of DNA for each PCR. This is a crude DNA extract and contains nucleases that will eventually fragment the DNA at room temperature. Keep the sample cold to limit this activity.

1. Obtain a wild-type or *bz* mutant maize plant and record its phenotype.
2. Harvest two pieces of leaf tissue that are ~1/4 in in diameter. (The large end of a 1000-µL pipette tip will punch disks of this size.) Place the tissue into a clean 1.5-mL tube and label the tube with your group number.
3. Add 100 µL of Edward's buffer to the tube. Grind the plant tissue by *forcefully* twisting it with a clean plastic pestle against the inner surface of the 1.5-mL tube for 1 min.
4. Add 400 µL of Edward's buffer to the tube. Grind briefly to remove tissue from the pestle and to liquefy any remaining pieces of tissue.
5. Vortex the tube for 5 sec, by hand or machine (if available).
6. Boil the sample for 5 min in a water bath or heating block.
7. Place the tube, along with those from other groups, in a balanced configuration in a microcentrifuge and centrifuge for 2 min to pellet any remaining cell debris. Centrifuge longer if unpelleted debris remains.
8. Label a fresh tube with your group number. Transfer 350 µL of supernatant to the fresh tube. Be careful not to disturb the pelleted debris when transferring the supernatant. Discard the old tube containing the debris.
9. Add 400 µL of isopropanol to the tube of supernatant.
10. Close the tube, mix by inverting several times, and then leave for 3 min at room temperature to precipitate nucleic acids, including DNA.
11. Place your tube and those of other groups in a balanced configuration in a microcentrifuge, with the hinges of the caps pointing outward. (During centrifugation, nucleic acids will gather on the side of the tube underneath the hinge.) Centrifuge for 5 min at maximum speed to pellet the DNA.
12. Carefully pour off and discard the supernatant from the tube. Then, *completely* remove the remaining liquid with a micropipette set at 100 µL.
13. Air-dry the pellet by allowing the tube to sit open for 10 min. Any remaining isopropanol will evaporate.
14. Add 100 µL of TE buffer to the tube. Dissolve the pelleted DNA by pipetting in and out, using a fresh tip for each tube. Make sure to wash the TE down the side of the tube underneath the hinge, where the pellet formed during centrifugation.
15. The DNA may be used immediately or stored at –20°C until you are ready to proceed to Part III. Keep the DNA on ice during use.

III. Amplify DNA by PCR

REAGENTS, SUPPLIES, & EQUIPMENT

For each group

BZ1/BZ2 primer/loading dye mix (25 µL)*
BZ1/Ds primer/loading dye mix (25 µL)*
Container with cracked or crushed ice
DNA from wild-type or *bz* mutant maize* from Part II
Microcentrifuge tube rack
Micropipettes and tips (1–100 µL)
Permanent marker
2 Ready-To-Go PCR Beads in 0.2- or 0.5-mL PCR tubes

To share

Thermal cycler

*Store on ice.

1. Obtain two PCR tubes containing Ready-To-Go PCR Beads. Label the tubes with your group number. Then, label each tube with the name of the primer set as follows:
   - For the BZ1/BZ2 PCR, write "Bz."
   - For the BZ1/Ds PCR, write "Ds."

The primer/loading dye mix will turn purple as the PCR bead dissolves.

2. Use a micropipette with a fresh tip to add 22.5 µL of the BZ1/BZ2 primer/loading dye mix to the "Bz" tube. Then, use a fresh tip to add 22.5 µL of the BZ1/Ds primer/loading dye mix to the "Ds" tube. Allow the beads to dissolve for ~1 min.
3. Use a micropipette to add 2.5 µL of maize DNA (from Part II) directly into each primer/loading dye mix. Use a fresh tip each time and ensure that no DNA remains in the tip after pipetting.
4. Store your samples on ice until your class is ready to begin thermal cycling.

If the reagents become splattered on the wall of the tube, pool them by pulsing the sample in a microcentrifuge or by sharply tapping the tube bottom on the lab bench.

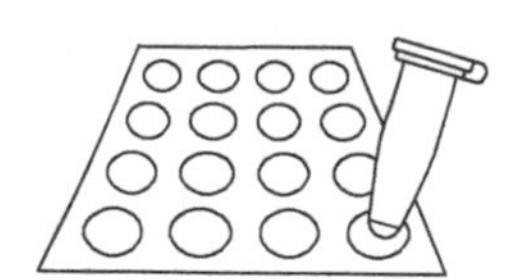

5. Place your PCR tubes, along with those of other groups, in a thermal cycler that has been programmed for 30 cycles of the following profile:

Denaturing step:	30 sec	94°C
Annealing step:	30 sec	55°C
Extending step:	30 sec	72°C

   The profile may be linked to a 4°C hold program after the 30 cycles are completed.

6. After thermal cycling, store the amplified DNA on ice or at –20°C until you are ready to proceed to Part IV.

IV. Analyze PCR Products by Gel Electrophoresis

REAGENTS, SUPPLIES, & EQUIPMENT

For each group

2% Agarose in 1x TBE (hold at 60°C) (50 mL per gel)
Container with cracked or crushed ice
Gel-casting tray and comb
Gel electrophoresis chamber and power supply
Latex gloves
Masking tape
Microcentrifuge tube rack
Micropipette and tips (1–100 µL)
pBR322/BstNI marker (20 µL per gel)*
PCR products* from Part III
SYBR Green DNA stain <!> (10 µL)
1x TBE buffer (300 mL per gel)

To share

Digital camera or photodocumentary system
UV transilluminator <!> and eye protection
Water bath for agarose solution (60°C)

*Store on ice.
See Cautions Appendix for appropriate handling of materials marked with <!>.

1. Seal the ends of the gel-casting tray with masking tape and insert a well-forming comb.
2. Pour the 2% agarose solution into the tray to a depth that covers about one-third the height of the open teeth of the comb.

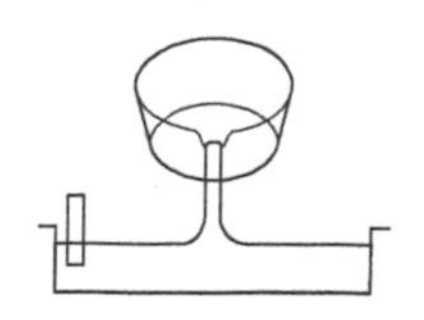

Avoid pouring an overly thick gel, which will be more difficult to visualize.

3. Allow the gel to completely solidify; this takes ~20 min.
4. Remove the masking tape, place the gel into the electrophoresis chamber, and add enough 1x TBE buffer to cover the surface of the gel.

The gel will become cloudy as it solidifies.

5. Carefully remove the comb and add additional 1x TBE buffer to fill in the wells and just cover the gel, creating a smooth buffer surface.

Do not add more buffer than necessary. Too much buffer above the gel channels electrical current over the gel, increasing the running time.

6. Orient the gel according to the diagram in Step 8, so that the wells are along the top of the gel.

7. Add 2 μL of SYBR Green DNA stain to the "Bz" tube and add 2 μL of SYBR Green DNA stain to the "Ds" tube. In addition, add 2 μL of SYBR Green DNA stain to 20 μL of pBR322/BstNI marker.

A 100-bp ladder may also be used as a marker.

8. Use a micropipette with a fresh tip to load each sample from Step 7 into your assigned wells according to the following diagram:

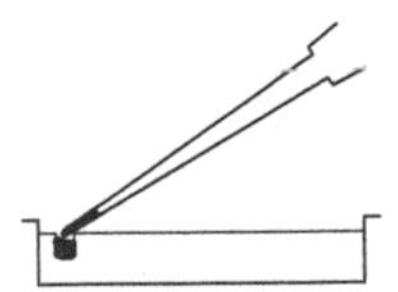

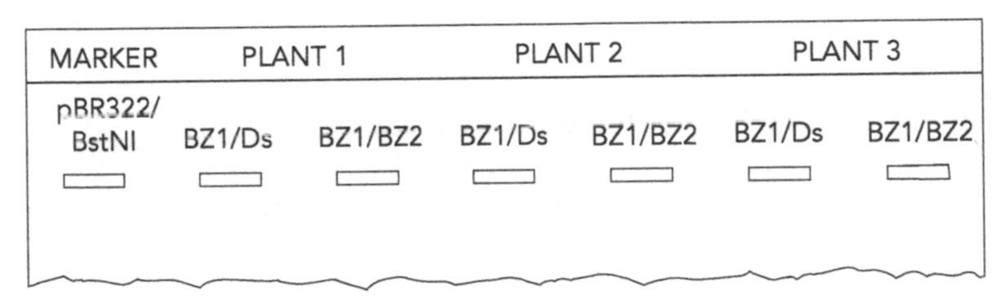

Expel any air from the tip before loading. Be careful not to push the tip of the pipette through the bottom of the sample well.

9. Run the gel for ~30 min at 130 V. Adequate separation will have occurred when the cresol red dye front has moved at least 50 mm from the wells.

Transillumination, where the light source is below the gel, increases brightness and contrast.

10. View the gel using UV transillumination. Photograph the gel using a digital camera or photodocumentary system.

BIOINFORMATICS METHODS

For a better understanding of the experiment, perform the following bioinformatics exercises before you analyze your results.

I. Use BLAST to Find DNA Sequences in Databases (Electronic PCR)

1. Perform a BLAST search as follows:

   i. Do an Internet search for "ncbi blast."

   ii. Click on the link for the result "BLAST: Basic Local Alignment Search Tool." This will take you to the Internet site of the National Center for Biotechnology Information (NCBI).

   iii. Click on the link "nucleotide blast" (blastn) under the heading "Basic BLAST."

   iv. Enter the BZ1/BZ2 primer sequences into the search window. These are the query sequences.

   > The following primer sets were used in the experiment:
   >
   > BZ1 5′-CGAATGGCTGTTGCATTTCCAT-3′ (forward primer)
   > BZ2 5′-ACGGGACGCAGTTGGCCAGGA-3′ (reverse primer)
   >
   > BZ1 5′-CGAATGGCTGTTGCATTTCCAT-3′ (forward primer)
   > Ds 5′-TCTACCGTTTCCGTTTCCGTTT-3′ (reverse primer)

   v. Omit any nonnucleotide characters from the window because they will not be recognized by the BLAST algorithm.

vi. Under "Choose Search Set," select the "Nucleotide collection (nr/nt)" database from the drop-down menu.

vii. Under "Program Selection," optimize for "Somewhat similar sequences (blastn)."

viii. Click on "BLAST." This sends your query sequences to a server at NCBI in Bethesda, Maryland. There, the BLAST algorithm will attempt to match the primer sequences to the millions of DNA sequences stored in its database. While searching, a page showing the status of your search will be displayed until your results are available. This may take only a few seconds or more than 1 min if many other searches are queued at the server.

2. Analyze the results of the BLAST search, which are displayed in three ways as you scroll down the page:

   i. First, a graphical overview illustrates how significant matches (hits) align with the query sequence. Matches of differing lengths are indicated by color-coded bars. What do you notice about the lengths (and colors) of the matches (bars) as you look from the top to the bottom?

   ii. This is followed by a list of significant alignments (hits) with links to the corresponding accession numbers. (An accession number is a unique identifier given to a sequence when it is submitted to a database such as GenBank.) Note the scores in the "E value" column on the right. The Expectation or E value is the number of alignments with the query sequence that would be expected to occur by chance in the database. The lower the E value, the higher the probability that the hit is related to the query. For example, an E value of 1 means that a search with your sequence would be expected to turn up one match by chance. Longer query sequences generally yield lower E values. An alignment is considered significant if it has an E value of less than 0.1. What is the E value of the most significant hit and what does it mean? Note the names of any significant alignments that have E values of less than 0.1. Do they make sense? What do they have in common?

   iii. Third is a detailed view of each primer (query) sequence aligned to the nucleotide sequence of the search hit (subject, abbreviated "Sbjct"). Note that the first match to the forward primer (nucleotides 1–22) and to the reverse primer (nucleotides 23–43) are within the same subject (accession number). Do the rest of the alignments produce good matches to both primers?

3. Click on the accession number link to open the data sheet for the first hit (the first subject sequence).

   i. At the top of the report, note basic information about the sequence, including its length (in base pairs, or bp), database accession number, source, and references to papers in which the sequence is published. What is the source and size of the sequence in which your BLAST hit is located?

   ii. In the middle section of the report, the sequence features are annotated, with their beginning and ending nucleotide positions ("xx..xx"). These features may include genes, coding sequences (cds), regulatory regions, ribosomal RNA (rRNA), and transfer RNA (tRNA). You examine these features more closely in Part II.

iii. Scroll to the bottom of the data sheet. This is the nucleotide sequence to which the term "Sbjct" refers.

iv. Use your browser's "Back" button to return to the BLAST results page.

4. Predict the length of the product that the primer set would amplify in a PCR (in vitro) as follows:

   i. Scroll down to the alignments section (third section) of the BLAST results page. Locate the alignment for DQ493652.1.

   ii. To which positions do the primers match in the subject sequence?

   iii. The lowest and highest nucleotide positions in the subject sequence indicate the borders of the amplified sequence. Subtract the lowest nucleotide position in the subject sequence from the highest nucleotide position in the subject sequence. What is the difference between the coordinates?

   iv. Note that the actual length of the amplified fragment includes both ends, so add 1 nucleotide to the result that you obtained in Step 4.iii to obtain the exact length of the PCR product amplified by the two primers.

5. Obtain the nucleotide sequence of the amplicon that the primer set would amplify in a PCR (in vitro) as follows:

   i. Open the sequence data sheet for the hit that you identified in Step 4.i (DQ493652.1) by clicking on the accession number link. Scroll to the bottom of the data sheet.

   ii. The bottom section of the data sheet lists the entire nucleotide sequence that contains the PCR product. Highlight all of the nucleotides between the coordinates that you identified in Step 4.iii, from the beginning of the forward primer to the end of the reverse primer.

   iii. Copy and paste the highlighted sequence into a text document. Then, delete all nonnucleotide characters and spaces. This is the amplicon, or amplified PCR product. Save this text document.

In Step 5.iii, you can retain the nucleotide coordinates and spaces to ease readability. Reduce the point size of the font so that each row of 60 nucleotides sits on one line. Then set the font to Courier or another nonproportional font to align the blocks of sequence.

II. Identify the Bz Amino Acid Sequence

1. Open the data sheet for GenBank accession number DQ493652.1. The middle of the report, under "FEATURES," contains annotations of sequence features such as promoters, genes, mRNA sequences, and coding sequences (cds), with their beginning and ending nucleotide positions ("xx..xx"). For example, the cds feature in this GenBank record is from positions 76 to 598 and from positions 699 to 1591. For each mRNA and cds feature, "join" shows the coordinates of coding exons that are spliced together for translation into a polypeptide chain. The use of angle brackets ("<" or ">") indicates that the feature extends beyond the nucleotide position indicated.
2. Look at the word(s) after "/gene=." What is the name of this gene?
3. Look at the "cds" feature for this gene. What do you think the comma in "(76..598,699..1591)" stands for?
4. Each range of numbers, such as "76..598," gives the first and last nucleotides of one exon. How many exons and introns are in this gene?

5. Identify the feature(s) that the two primers span, using the nucleotide coordinates that you identified in Step 4.iii of Part I. Where does the forward primer begin? Where does the reverse primer end? Which part of the gene does the amplicon span?
6. The name of the protein is shown after "/product=." What is the name of the protein product that this gene encodes?
7. The final part of the "cds" entry is labeled "/translation=." It lists the one-letter abbreviations for the amino acids specified by the coding sequence. Copy the entire amino acid sequence and paste it into a text document for use in Step 2 of Part V.
8. Open the data sheet for GenBank accession number AF391808.3 (a sequence of the *Bz* locus region) and scroll down to the "FEATURES" section.
   i. Directly under "FEATURES," find the word "source." On which chromosome ("/chromosome=") is this sequence located?
   ii. Note that this sequence contains many genes (and, therefore, many features). Identify the gene ("/gene=") features on either side of the *Bz* gene. What are the names of these neighboring genes?

III. Use Map Viewer to Determine the Chromosomal Location of the *Bz* Gene

1. Open "Map Viewer" (http://www.ncbi.nlm.nih.gov/mapview).
2. Find "Plants" in the table to the right and click on the triangle-shaped icon to the left of "Plants." Click on "Flowering Plants" and then on "Monocots." Next, find "*Zea mays*" (maize) in the list and click on the magnifying glass icon under the "Tools" column. If more than one build is displayed, select the one with the highest number; this will be the most recent version.
3. Enter "bz1" into the "Search for" window and click "Find" to see the chromosomal location of the *bz1* gene. Small horizontal bars on chromosomes indicate the positions of hits. On which chromosome is the *bz1* gene located?
4. Under search results, click on the link to the "bz1" map element for an STS (sequence-tagged site) hit to move to the *bz1* locus.
5. Barbara McClintock worked on a number of genes involved in corn anthocyanin biosynthesis, some of which she mapped to this region. Can you identify any genes in the vicinity of the *bz1* locus? You may refine your position on the chromosome by using the zoom toggle on the left or by clicking on the thin vertical gray line along which the genes are annotated. Click the names of those genes or, if provided, accession number links to find more information. What can you determine about some of them? (For an idea of what genes to look out for, view the animation at http://www.dnaftb.org/dnaftb/32/concept/index.html.)

IV. Determine the Insertion Site of the *Ds* Transposon

During PCR, you used the primer pair BZ1/Ds for one of the reactions. This amplicon, from *bz* mutants, is composed partly of sequence from the *Bz* gene and partly of sequence from the *Ds* transposon. In this exercise, you will determine how many nucleotides of the amplicon are from the *Ds* transposon and the position at which *Ds* inserted into the *Bz* gene.

1. Conduct a nucleotide BLAST search with the BZ1/Ds primers by following Steps 1.i–viii in Part I. Examine the alignments section (the third section) of the BLAST results page.
   i. From what species are the hits that match the BZ1 primer (nucleotides 1–22) only? Do their descriptions mention anything about the gene or locus?
   ii. From what species are the hits that match the Ds primer (nucleotides 23–44) only? Do their descriptions mention anything about the gene or locus?
   iii. Do any hits align with both the BZ1 and Ds primers?
   iv. Based on the information gathered in Steps 1.i–iii, is the sequence of your amplicon in the database? Explain how you came to your conclusion.
2. Locate the alignment for "*Zea mays* bz-m2(Ac) (gb | AF355378.1 | AF355378)." Determine the length of the BZ1/Ds amplicon by following Steps 4.ii–iv in Part I.
3. To determine the *Ds* insertion site in the *Bz* gene, answer the questions below and label the diagram with your answers.

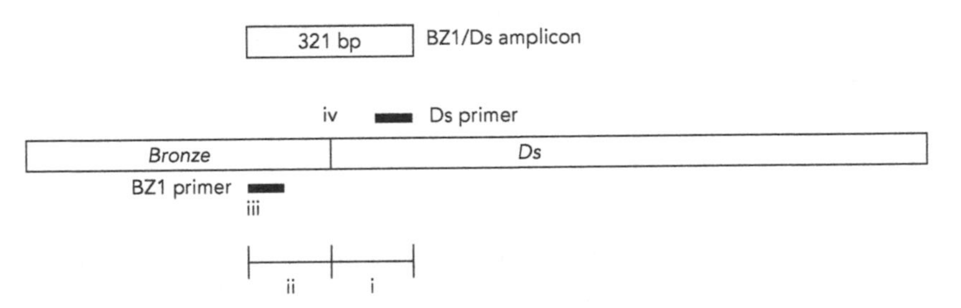

   i. Identify one complete sequence of the *Ds* transposon (e.g., AF332952.1), which is one of the hits that matches the Ds primer only. How far into this *Ds* sequence does the Ds primer reach?
   ii. Using values from Steps 2 and 3.i, calculate the number of *Bz* nucleotides in the BZ1/Ds amplicon.
   iii. Look at the hit DQ493652.1, which is the *Bz* gene GenBank entry that you examined in Parts I and II. What is the first nucleotide position of the BZ1 primer in the *Bz* gene?
   iv. What is the insertion position of *Ds* in the *Bz* gene? (Remember that the first nucleotide position of the BZ1 primer is the same as the first position of the amplicon.)
4. Refer to Step 3 in Part II above. In which part of the *Bz* gene has the *Ds* transposon inserted?
5. Examine the data sheet for AF391808.3 (the same as in Step 8 of Part II above). Scan the "gene" features and determine whether you can find additional transposons inserted into this sequence. What are some of the names of your finds?

V. Use BLAST to Determine the Function of the BZ Protein

1. Return to the BLAST page at NCBI (www.ncbi.nlm.nih.gov/BLAST). This time, click on "protein blast."
2. Copy the BZ amino acid sequence from your text file (from Step 7 in Part II) and paste it into the search window.

3. Click on "BLAST" to send the amino acid sequence to the NCBI server.
4. An algorithm quickly scans the query sequence to identify functional domains (regions of the protein) that are conserved in different organisms. If a conserved domain is displayed, click on it to get some quick information about it. What do you find?
5. Close the "Conserved Domain Search" window and return to the BLAST results page.
6. Examine the graphical overview. Do the BLAST hits show strong or weak homology with your query sequence?
7. Scroll down to "Sequences producing significant alignments" and "Alignments." Based on the E values, the titles of the hits, and the links to the accession numbers, what can you conclude about the conservation of the BZ protein during plant evolution?

Further Research: Secondary Plant Metabolites

Plants synthesize a number of substances that are not required to maintain basic metabolism. These "secondary plant metabolites" include phenolic acids, lignins, tannins, terpenoids (such as carotenoids), and flavonoids (such as anthocyanins). Secondary metabolites have important roles in the interaction between plants and the environment. Tannins, for example, deter predators. Anthocyanins and carotenoids produce a variety of colors in stalks, seeds, and fruits that may attract animals for pollination and seed dispersal.

Most of these compounds require complex biosynthetic pathways involving numerous genes. Some genes encode transcription factors that regulate the activity of enzyme-producing genes. Other genes encode enzymes that work at various steps in the pathway, converting one product into the substrate for another enzyme. The *bz* mutation disables the production of an enzyme involved in the final step of anthocyanin biosynthesis. Thus, *bz* mutants accumulate the unfinished precursor of the purple pigment, which gives them their distinctively colored stalks and kernels. Do an Internet search for anthocyanin biosynthesis to get an idea of the complexity of secondary metabolism.

As with many mutations in corn, *bz* is caused by a transposon insertion. In McClintock's *Ac/Ds* system, the *Ac* element encodes a functional transposase gene, whose product is required for transposition. *Ds* elements are mutated copies of *Ac*, typically with internal truncations that destroy the transposase gene. Thus, the *Ds* element is incapable of transposing on its own and produces stable mutants. However, *Ds* can still be *trans*-activated by a transposase encoded by an *Ac* element elsewhere in the genome. Thus, *Ds* mutations are unstable in the presence of an active *Ac* element, cycling between mutant and wild-type states. This cycling typically produces spotted or striped coloration.

RESULTS AND DISCUSSION

The following diagram shows how PCR amplification identifies the *Ds* insertion in the *Bz* gene. Because the *Ds* insertion is too large to amplify across, two sets of primers are used. One set of primers (BZ1/BZ2) straddles the *Ds* insertion site and only amplifies the wild-type (Wt) allele. The second set (BZ1/Ds) only amplifies the *Ds* insertion allele, with one primer located in the *Bz* gene and the other in the *Ds* transposon.

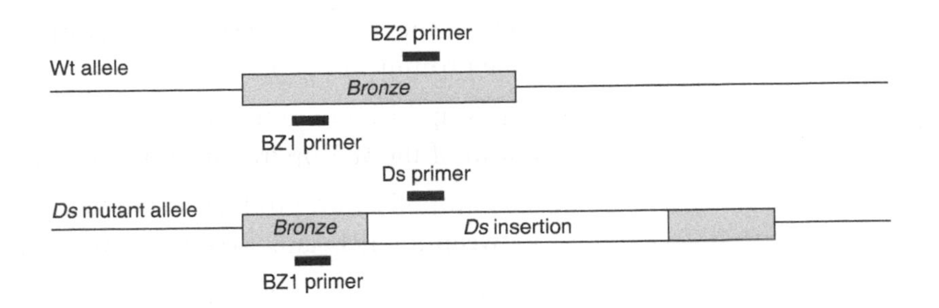

I. Think About the Experimental Methods

1. Describe the purpose of each of the following steps or reagents used in DNA isolation (Part II of Experimental Methods):
   i. Grinding with pestle.
   ii. Edward's buffer.
   iii. Boiling.
   iv. Tris-EDTA (TE) buffer.
2. What is the purpose of performing each of the following PCRs?
   i. BZ1/BZ2.
   ii. BZ1/Ds.

II. Interpret Your Gel and Think About the Experiment

1. Observe the photograph of the stained gel containing your PCR samples and those from other students. Orient the photograph with the sample wells at the top. Use the sample gel shown below to help interpret the band(s) in each lane of the gel.

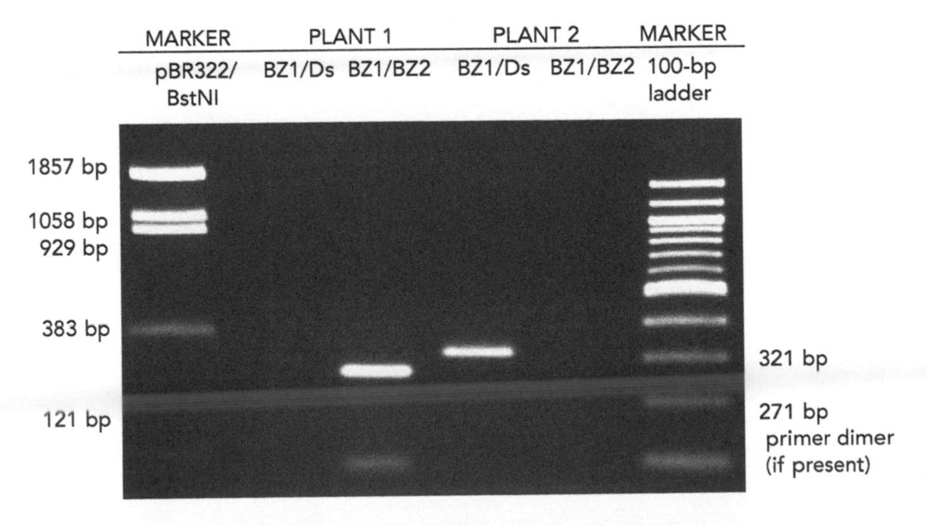

If a 100-bp ladder was used, as shown on the right-hand side of the sample gel, the bands increase in size in 100-bp increments starting with the fastest-migrating band of 100 bp.

2. Locate the lane containing the pBR322/BstNI markers on the left side of the gel. Working from the well, locate the bands corresponding to each restriction fragment: 1857, 1058, 929, 383, and 121 bp. The 1058- and 929-bp fragments will be very close together or may appear as a single large band. The 121-bp band may be very faint or not visible.

3. Scan across the row of student results. Notice that virtually all student lanes contain one prominent band.
   i. The amplification product of the mutant *bz* allele (321 bp) should align just ahead of the 383-bp fragment of the pBR322/BstNI marker.
   ii. The amplification product of the wild-type *Bz* allele (271 bp) should align between the 121- and 383-bp fragments of the pBR322/BstNI marker.
4. Carefully examine the banding patterns of the two samples in the gel above and complete the entries for genotype and phenotype in the chart below. Under "Sample," indicate which result is derived from Plant 1, which is derived from Plant 2, and which is not represented in the gel. (Your own samples may yield any of the banding patterns listed in the table.)

Additional faint bands at other positions occur when the primers bind to chromosome loci other than the exact locus and give rise to "nonspecific" amplification products.

321 bp (BZ1/Ds)	271 bp (BZ1/BZ2)	Plant genotype	Plant phenotype	Sample
present	present			
present	absent			
absent	present			

5. It is common to see a diffuse (fuzzy) band that runs ahead of the 121-bp marker. This is "primer dimer," an artifact of the PCR that results from the primers overlapping one another and amplifying themselves. How would you interpret a lane in which you observe primer dimer but no bands as described in Step 3?
6. Would you classify the *bz* mutation as recessive or dominant? Explain your reasoning.
7. Count the number of plants with the *Bz*/*Bz*, *Bz*/*bz*, and *bz*/*bz* genotypes in your class. Based on the genotype distribution of these plants, which genotype was the parental plant? Why may the genotype distribution observed by your class deviate from what is expected under Mendel's laws of inheritance?

LABORATORY
3.2

Detecting a Transposon in *Arabidopsis*

▼ OBJECTIVES

This laboratory demonstrates several important concepts of modern biology. During the course of this laboratory, you will

- Learn about the relationship between genotype and phenotype.
- Learn how transposable elements can be used to mutagenize and tag genes.
- Conduct a literature search to discover how homeotic genes have a role in plant and animal development.
- Move between in vitro experimentation and in silico computation.

In addition, this laboratory utilizes several experimental and bioinformatics methods in modern biological research. You will

- Extract and purify DNA from plant tissue.
- Amplify a specific region of the genome by polymerase chain reaction (PCR) and analyze PCR products by gel electrophoresis.
- Use the Basic Local Alignment Search Tool (BLAST) to identify sequences in databases.
- Use the Map Viewer tool to picture genes on chromosomes.

INTRODUCTION

As Barbara McClintock showed in maize, "dissociator" (*Ds*) elements are able to jump, or transpose, from one chromosomal location to another in the presence of "activator" (*Ac*) elements. *Ac* elements encode a transposase that recognizes repeat sequences at either end of *Ds* and *Ac* elements, promoting transposition. When *Ds* elements jump, they can insert into a gene and disrupt gene function, acting as a natural mutagen.

Today, the *Ac/Ds* system is an important tool in gene discovery, allowing scientists to identify the genes that are mutated after transposition. Although they occur naturally in maize, *Ds* elements and the *Ac* transposase have been transformed into other plant species so that they can be used for transposon-induced mutagenesis.

When using chemicals or radiation to mutate organisms, the sequence at the site of any mutations is unknown. Identifying the gene often requires careful genetic mapping to locate the mutation to a region of the genome and further characterization to locate the specific site that is mutated.

When transposition is induced, the sequence at the site of insertion is known: It is the sequence of the transposon. This allows rapid identification of the location by a process called inverse PCR. First, DNA from mutant plants is isolated. Next, the DNA

Wild-type (*left*) and *clf-2* mutant (*right*) *Arabidopsis* plants.

is digested with an enzyme that cuts DNA frequently but that does not cut the transposon, resulting in many fragments, one of which will include the transposon. These fragments are ligated under conditions that favor self-ligation, with each fragment forming a circle. PCR is then performed using primers that point outward from the ends of the transposon, rather than pointing toward each other. This results in amplification of the DNA on either side of the transposon insertion site around the opposite side of the circularized fragment. Once this DNA is sequenced, it can be used to determine the location of the insertion in the genome.

This laboratory investigates the *CURLY LEAF* (*CLF*) gene of *Arabidopsis thaliana* to analyze the molecular relationship between genotype and phenotype. The *CLF* gene encodes a protein that is involved in homeotic gene regulation; it helps to control the correct spatial and tissue-specific expression of genes during development. The recessive *clf-2* mutation in this lab was created through *Ds* transposon mutagenesis and produces a dwarf phenotype with curly leaves, early flowers, and fused flower parts.

In this lab, DNA is isolated from *Arabidopsis* plants and the *CLF* locus is amplified using PCR. Two primer sets are used: One is specific for the wild-type allele and the other for the *clf-2* mutant allele (bearing a *Ds* insertion). The PCR products can be readily differentiated by agarose gel electrophoresis.

The laboratory also includes bioinformatics exercises that complement the experimental methods. BLAST is used to identify nucleic acid and protein sequences in biological databases, predict the size of the PCR products for the wild-type allele, and learn about the function and evolutionary history of the CLF protein. Another bioinformatics tool, Map Viewer, is used to discover the chromosome location of the *Ds* insertion, and additional calculations are performed to determine where in the *CLF* gene the *Ds* transposon is inserted to produce the *clf-2* phenotype.

FURTHER READING

Edwards K, Johnstone C, Thompson C. 1991. A simple and rapid method for the preparation of plant genomic DNA for PCR analysis. *Nucleic Acids Res* **19:** 1349.

Fedoroff N, Wessler S, McClure M. 1983. Isolation of the transposable maize controlling elements *Ac* and *Ds*. *Cell* **35:** 235–242.

Feldmann KA, Marks MD, Christianson ML, Quantrano RS. 1989. A dwarf mutant of *Arabidopsis* generated by T-DNA insertion mutagenesis. *Science* **243:** 1351–1354.

Goodrich J, Puangsomlee P, Martin M, Long D, Meyerowitz EM, Coupland G. 1997. A Polycomb-group gene regulates homeotic gene expression in *Arabidopsis*. *Nature* **386:** 44–51.
Hsieh T-F, Hakim O, Ohad N, Fischer RL. 2003. From flour to flower: How Polycomb-group proteins influence multiple aspects of plant development. *Trends Plant Sci* **8:** 439–445.
Kim G-T, Tsukaya H, Uchimiya H. 1998. The *CURLY LEAF* gene controls both division and elongation of cells during the expansion of the leaf blade in *Arabidopsis thaliana*. *Planta* **206:** 175–183.
McClintock B. 1951. Chromosome organization and genic expression. *Cold Spring Harbor Symp Quant Biol* **16:** 13–47.

PLANNING AND PREPARATION

The following table will help you to plan and integrate the different experimental methods.

Experiment	Day	Time	Activity	
I. Plant *Arabidopsis* seeds	3–4 wk before Part II	15 min	Prelab:	Set up student stations.
		15 min	Lab:	Plant *Arabidopsis* seeds.
			Postlab:	Care for *Arabidopsis* plants.
II. Isolate DNA from *Arabidopsis*	1	30 min	Prelab:	Set up student stations.
		30–60 min	Lab:	Isolate DNA.
III. Amplify DNA by PCR	2	15 min	Prelab:	Aliquot primer/loading dye mixes. Set up student stations.
		15 min	Lab:	Set up PCRs.
		60–150 min	Postlab:	Amplify DNA in thermal cycler.
IV. Analyze PCR products by gel electrophoresis	3	30 min	Prelab:	Dilute TBE electrophoresis buffer. Prepare agarose gel solution. Set up student stations.
		30 min	Lab:	Cast gels.
	4	45+ min	Lab:	Load DNA samples into gel. Electrophorese samples. Photograph gels.

OVERVIEW OF EXPERIMENTAL METHODS

I. PLANT *ARABIDOPSIS* SEEDS

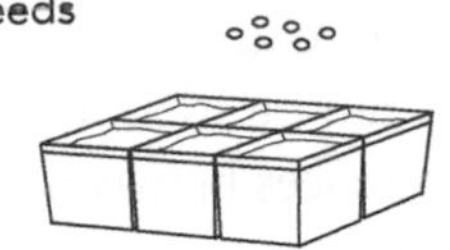

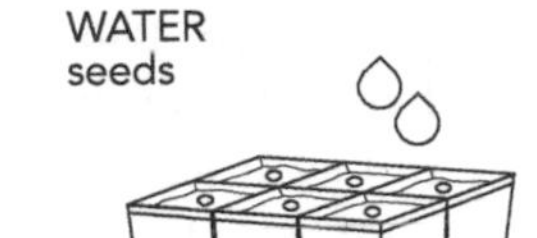

II. ISOLATE DNA FROM *ARABIDOPSIS*

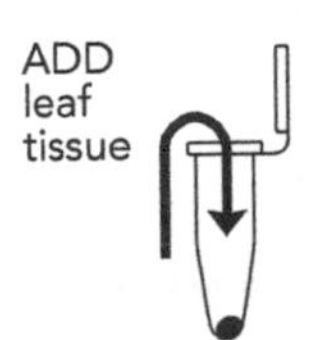

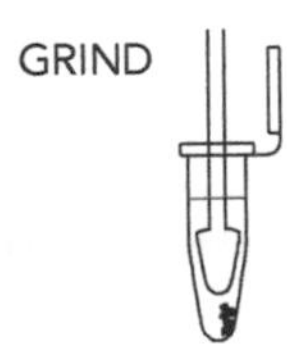

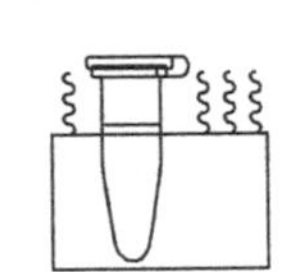

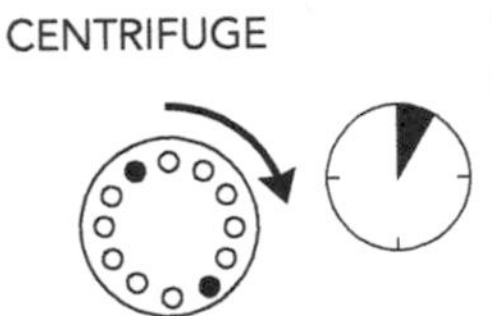

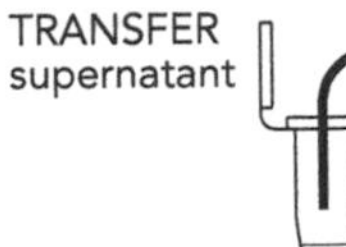
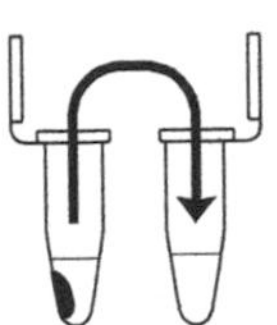

ADD and MIX
isopropanol

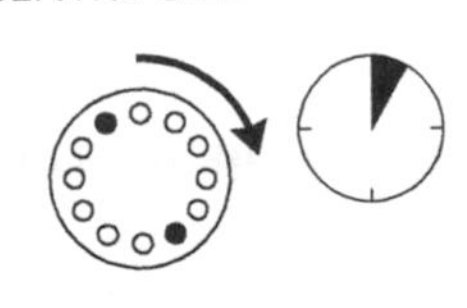

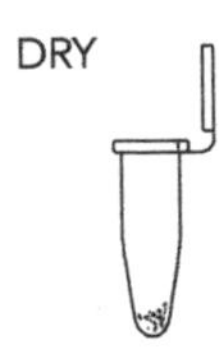

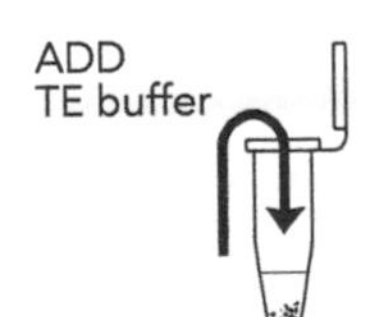

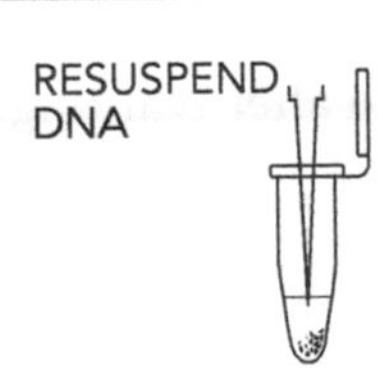

III. AMPLIFY DNA BY PCR

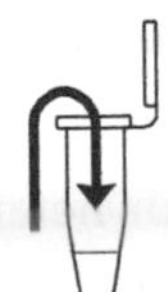

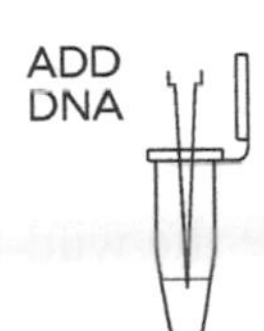

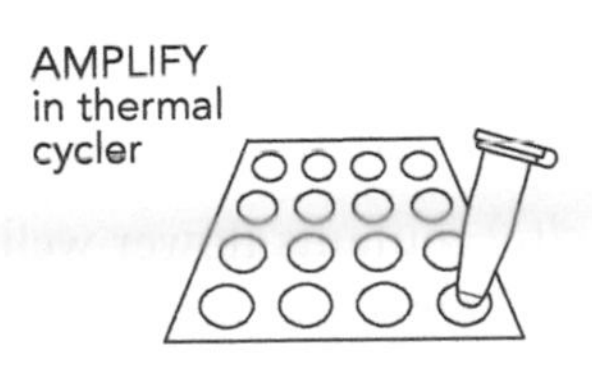

IV. ANALYZE PCR PRODUCTS BY GEL ELECTROPHORESIS

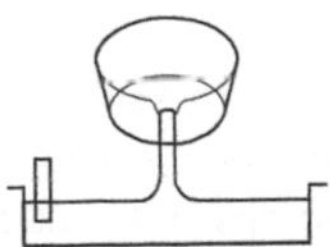

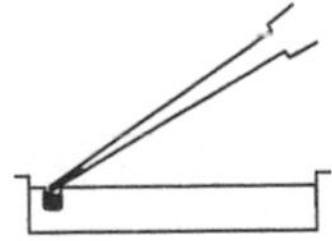

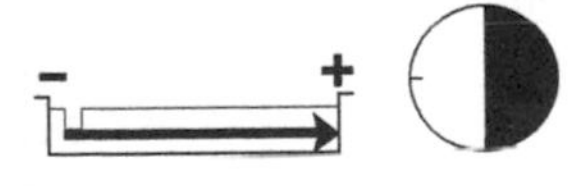

EXPERIMENTAL METHODS

I. Plant *Arabidopsis* Seeds

To extract DNA from plant tissue, you must plant the *Arabidopsis* seeds 3–4 wk before DNA isolation and PCR. Two 1/4-in-diameter leaf disks are required for each experiment, but multiple small leaves and even whole plantlets can be used. Depending on growing conditions, you may observe the phenotypic differences between mutant and wild-type plants in as little as 2 wk. For further information on cultivation, refer to The Arabidopsis Information Resource (TAIR) at http://www.arabidopsis.org.

REAGENTS, SUPPLIES, & EQUIPMENT

For each group
Planting pot and tray
Plastic dome lid or plastic wrap
Potting soil
10–20 Seeds from a self-cross of a *CLF/clf-2* heterozygous *Arabidopsis* plant
1 Sheet of paper (4 x 4 in)
Water

To share
Refrigerator

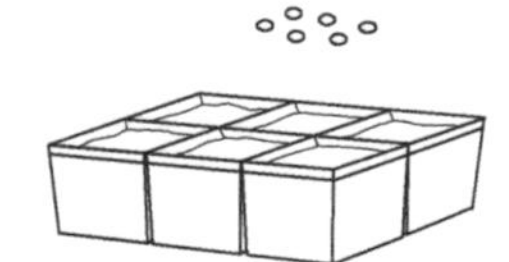

Arabidopsis seeds are very tiny and difficult to handle, so planting is not as simple as it may seem.

For optimum growth, provide constant (24-h) fluorescent lighting ~1 ft directly above the plants. With 24-h fluorescent lighting, phenotypes can be discerned in 2–3 wk.

1. Label the pot with your group number. Moisten the potting soil. Fill the planting pot evenly with soil, but do not pack tightly.
2. Fit the pot into the tray, but leave one corner space empty to facilitate watering.
3. Carefully scatter seeds evenly on top of the soil as follows:
   i. Fold a 4 x 4-in sheet of paper in half.
   ii. Place 10–20 seeds into the fold of the paper and gently tap them onto the soil. Provide space between seeds, so that they will grow better and plant phenotypes can be readily observed.
4. Cover the pot with a plastic dome lid or plastic wrap to maintain humidity during germination. (Remove the cover 3–7 d after planting.)
5. Add 1/2 in of water to the tray, using the empty corner space. Water regularly to keep the soil damp, but do not allow the soil to remain soggy.
6. Place the planted seeds in a refrigerator for 2–4 d. This "vernalization" step is needed for optimum germination.
7. Grow the plants close to a sunny window at room temperature (20°C–22°C) or slightly warmer. If the plants are grown under cold conditions, it may take longer to discern the different phenotypes.
8. The small *clf-2* mutants flower well before the wild-type or heterozygous plants. When the difference in phenotype becomes evident, continue to Part II.

You may wish to continue to grow the plants after you have harvested tissue for DNA isolation and amplification. The phenotypic differences between mutant and wild-type plants become more obvious over time.

II. Isolate DNA From *Arabidopsis*

REAGENTS, SUPPLIES, & EQUIPMENT

For each group
Container with cracked or crushed ice
Edward's buffer (600 µL)
Isopropanol <!> (500 µL)
Microcentrifuge tube racks
2 Microcentrifuge tubes (1.5 mL)
Micropipettes and tips (100–1000 µL)
Permanent marker

REAGENTS, SUPPLIES, & EQUIPMENT *(continued)*

For each group (continued)
Plastic pestle
Tris/EDTA (TE) buffer (200 μL)
Vortexer (optional)
Wild-type or *clf-2* mutant *Arabidopsis* plants from Part I

To share
Microcentrifuge
Water bath or heating block at 96°C–100°C

See Cautions Appendix for appropriate handling of materials marked with <!>.

Your instructor will assign an Arabidopsis *plant to your group.*

1. Obtain a wild-type or *clf-2* mutant *Arabidopsis* plant and record its phenotype.
2. Harvest two pieces of leaf tissue that are ~1/4 in in diameter. (The large end of a 1000-μL pipette tip will punch disks of this size.) If the plant is small, take multiple leaves to make an equivalent amount of tissue. The *clf-2* mutant may be so small that you need to use the entire plant. If so, carefully remove all soil from the roots. Place the tissue into a clean 1.5-mL tube and label the tube with your group number.

Grinding the plant tissue breaks up the cell walls. When fully ground, the sample should be a green liquid.

Detergent in the Edward's buffer, called sodium dodecyl sulfate (SDS), dissolves lipids of the cell membrane.

3. Add 100 μL of Edward's buffer to the tube. Grind the plant tissue by *forcefully* twisting it with a clean plastic pestle against the inner surface of the 1.5-mL tube for 1 min.
4. Add 400 μL of Edward's buffer to the tube. Grind briefly to remove tissue from the pestle and to liquefy any remaining pieces of tissue.
5. Vortex the tube for 5 sec, by hand or machine (if available).

Boiling denatures proteins, including enzymes that digest DNA.

6. Boil the sample for 5 min in a water bath or heating block.

The pellet may appear as a tiny teardrop-shaped smear or particles on the bottom side of the tube underneath the hinge. Do not be concerned if you cannot see a pellet. A large or greenish pellet is cellular debris carried over from the first centrifugation.

7. Place the tube, along with those from other groups, in a balanced configuration in a microcentrifuge and centrifuge for 2 min to pellet any remaining cell debris. Centrifuge longer if unpelleted debris remains.
8. Label a fresh tube with your group number. Transfer 350 μL of supernatant to the fresh tube. Be careful not to disturb the pelleted debris when transferring the supernatant. Discard the old tube containing the debris.
9. Add 400 μL of isopropanol to the tube of supernatant.

Dry the pellet quickly with a hair dryer. To prevent blowing the pellet away, direct the air across the tube mouth, not into the tube, for ~3 min.

10. Close the tube, mix by inverting several times, and then leave for 3 min at room temperature to precipitate nucleic acids, including DNA.
11. Place your tube and those of other groups in a balanced configuration in a microcentrifuge, with the hinges of the caps pointing outward. (During centrifugation, nucleic acids will gather on the side of the tube underneath the hinge.) Centrifuge for 5 min at maximum speed to pellet the DNA.

The TE (Tris-EDTA) buffer provides conditions for stable storage of DNA. Tris provides a constant pH of 8.0, whereas EDTA binds cations (positive ions) that are required for DNase activity.

12. Carefully pour off and discard the supernatant from the tube. Then, *completely* remove the remaining liquid with a micropipette set at 100 μL.

In Part III, you will use 2.5 μL of DNA for each PCR. This is a crude DNA extract and contains nucleases that will eventually fragment the DNA at room temperature. Keep the sample cold to limit this activity.

13. Air-dry the pellet by allowing the tube to sit open for 10 min. Any remaining isopropanol will evaporate.
14. Add 100 μL of TE buffer to the tube. Dissolve the pelleted DNA by pipetting in and out, using a fresh tip for each tube. Make sure to wash the TE down the side of the tube underneath the hinge, where the pellet formed during centrifugation.
15. The DNA may be used immediately or stored at –20°C until you are ready to proceed to Part III. Keep the DNA on ice during use.

III. Amplify DNA by PCR

REAGENTS, SUPPLIES, & EQUIPMENT

For each group

CLF1/CLF2 primer/loading dye mix (25 μL)*
CLF1/Ds primer/loading dye mix (25 μL)*
Container with cracked or crushed ice
DNA from wild-type or *clf-2* mutant *Arabidopsis** from Part II
Microcentrifuge tube rack
Micropipettes and tips (1–100 μL)
Permanent marker
2 Ready-To-Go PCR Beads in 0.2- or 0.5-mL PCR tubes

To share

Thermal cycler

*Store on ice.

1. Obtain two PCR tubes containing Ready-To-Go PCR Beads. Label the tubes with your group number. Then, label each tube with the name of the primer set as follows:
   - For the CLF1/CLF2 PCR, write "CLF."
   - For the CLF1/Ds PCR, write "Ds."

The primer/loading dye mix will turn purple as the PCR bead dissolves.

2. Use a micropipette with a fresh tip to add 22.5 μL of the CLF1/CLF2 primer/loading dye mix to the "CLF" tube. Then, use a fresh tip to add 22.5 μL of the CLF1/Ds primer/loading dye mix to the "Ds" tube. Allow the beads to dissolve for ~1 min.
3. Use a micropipette to add 2.5 μL of *Arabidopsis* DNA (from Part II) directly into each primer/loading dye mix. Use a fresh tip each time and ensure that no DNA remains in the tip after pipetting.
4. Store your samples on ice until your class is ready to begin thermal cycling.

If the reagents become splattered on the wall of the tube, pool them by pulsing the sample in a microcentrifuge or by sharply tapping the tube bottom on the lab bench.

5. Place your PCR tubes, along with those of other groups, in a thermal cycler that has been programmed for 30 cycles of the following profile:

Denaturing step:	30 sec	94°C
Annealing step:	30 sec	65°C
Extending step:	30 sec	72°C

   The profile may be linked to a 4°C hold program after the 30 cycles are completed.

6. After thermal cycling, store the amplified DNA on ice or at –20°C until you are ready to proceed to Part IV.

IV. Analyze PCR Products by Gel Electrophoresis

REAGENTS, SUPPLIES, & EQUIPMENT

For each group

2% Agarose in 1x TBE (hold at 60°C) (50 mL per gel)
Container with cracked or crushed ice
Gel-casting tray and comb
Gel electrophoresis chamber and power supply
Latex gloves
Masking tape
Microcentrifuge tube (1.5 mL)
Microcentrifuge tube rack
Micropipette and tips (1–100 μL)
pBR322/BstNI marker (20 μL per gel)*
PCR products* from Part III
SYBR Green <!> DNA stain (10 μL)
1x TBE buffer (300 mL per gel)

REAGENTS, SUPPLIES, & EQUIPMENT *(continued)*

To share

Digital camera or photodocumentary system
UV transilluminator <!> and eye protection
Water bath for agarose solution (60°C)

*Store on ice.
See Cautions Appendix for appropriate handling of materials marked with <!>.

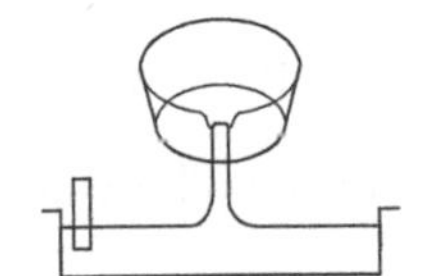

Avoid pouring an overly thick gel, which will be more difficult to visualize.

The gel will become cloudy as it solidifies.

Do not add more buffer than necessary. Too much buffer above the gel channels electrical current over the gel, increasing the running time.

A 100-bp ladder may also be used as a marker.

1. Seal the ends of the gel-casting tray with masking tape and insert a well-forming comb.
2. Pour the 2% agarose solution into the tray to a depth that covers about one-third the height of the open teeth of the comb.
3. Allow the gel to completely solidify; this takes ~20 min.
4. Remove the masking tape, place the gel into the electrophoresis chamber, and add enough 1× TBE buffer to cover the surface of the gel.
5. Carefully remove the comb and add additional 1× TBE buffer to fill in the wells and just cover the gel, creating a smooth buffer surface.
6. Orient the gel according to the diagram in Step 8, so that the wells are along the top of the gel.
7. Combine 20 μL of each PCR product from Part III ("CLF" and "Ds") in a fresh 1.5-mL tube. Add 2 μL of SYBR Green DNA stain to the combined PCR product. In addition, add 2 μL of SYBR Green DNA stain to 20 μL of pBR322/BstNI marker.

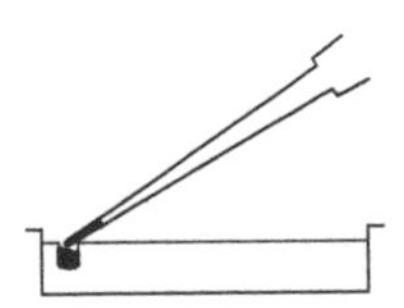

Expel any air from the tip before loading. Be careful not to push the tip of the pipette through the bottom of the sample well.

8. Use a micropipette with a fresh tip to load each sample from Step 7 into your assigned well according to the following diagram.

MARKER	PLANTS					
pBR322/ BstNI	1	2	3	4	5	6

Transillumination, where the light source is below the gel, increases brightness and contrast.

9. Run the gel for ~30 min at 130 V. Adequate separation will have occurred when the cresol red dye front has moved at least 50 mm from the wells.
10. View the gel using UV transillumination. Photograph the gel using a digital camera or photodocumentary system.

BIOINFORMATICS METHODS

For a better understanding of the experiment, perform the following bioinformatics exercises before you analyze your results.

I. Use BLAST to Find DNA Sequences in Databases (Electronic PCR)

1. Perform a BLAST search as follows:
   i. Do an Internet search for "ncbi blast."

ii. Click on the link for the result "BLAST: Basic Local Alignment Search Tool." This will take you to the Internet site of the National Center for Biotechnology Information (NCBI).

iii. Click on the link "nucleotide blast" (blastn) under the heading "Basic BLAST."

iv. Enter the CLF1/CLF2 primer sequences into the search window. These are the query sequences.

> The following primer sets were used in the experiment:
>
> CLF1 5′-TTAACCCGGACCCGCATTTGTTTCGG-3′ (forward primer)
> CLF2 5′-AGAGAAGCTCAAACAAGCCATCGA-3′ (reverse primer)
>
> CLF1 5′-TTAACCCGGACCCGCATTTGTTTCGG-3′ (forward primer)
> Ds 5′-GTCGGCGTGCGGCTGGCGGCG-3′ (reverse primer)

v. Omit any nonnucleotide characters from the window because they will not be recognized by the BLAST algorithm.

vi. Under "Choose Search Set," select the "Nucleotide collection (nr/nt)" database from the drop-down menu.

vii. Under "Program Selection," optimize for "Somewhat similar sequences (blastn)."

viii. Click on "BLAST." This sends your query sequences to a server at NCBI in Bethesda, Maryland. There, the BLAST algorithm will attempt to match the primer sequences to the millions of DNA sequences stored in its database. While searching, a page showing the status of your search will be displayed until your results are available. This may take only a few seconds or more than 1 min if many other searches are queued at the server.

2. Analyze the results of the BLAST search, which are displayed in three ways as you scroll down the page:

i. First, a graphical overview illustrates how significant matches (hits) align with the query sequence. Matches of differing lengths are indicated by color-coded bars. What do you notice about the lengths (and colors) of the matches (bars) as you look from the top to the bottom?

ii. This is followed by a list of significant alignments (hits) with links to the corresponding accession numbers. (An accession number is a unique identifier given to a sequence when it is submitted to a database such as GenBank.) Note the scores in the "E value" column on the right. The Expectation or E value is the number of alignments with the query sequence that would be expected to occur by chance in the database. The lower the E value, the higher the probability that the hit is related to the query. For example, an E value of 1 means that a search with your sequence would be expected to turn up one match by chance. Longer query sequences generally yield lower E values. An alignment is considered significant if it has an E value of less than 0.1. What is the E value of the most significant hit and what does it mean?

iii. Third is a detailed view of each primer (query) sequence aligned to the nucleotide sequence of the search hit (subject, abbreviated "Sbjct"). Note that the first match to the forward primer (nucleotides 1–26) and to the reverse primer (nucleotides 27–50) are within the same subject (accession number). Do the rest of the alignments produce good matches to both primers?

3. Click on the accession number link to open the data sheet for the first hit (the first subject sequence).
   i. At the top of the report, note basic information about the sequence, including its length (in base pairs, or bp), database accession number, source, and references to papers in which the sequence is published. What is the source and size of the sequence in which your BLAST hit is located?
   ii. In the middle section of the report, the sequence features are annotated, with their beginning and ending nucleotide positions ("xx..xx"). These features may include genes, coding sequences (cds), regulatory regions, ribosomal RNA (rRNA), and transfer RNA (tRNA). You examine these features more closely in Part II.
   iii. Scroll to the bottom of the data sheet. This is the nucleotide sequence to which the term "Sbjct" refers.
   iv. Use your browser's "Back" button to return to the BLAST results page.
4. Predict the length of the product that the primer set would amplify in a PCR (in vitro) as follows:
   i. Scroll down to the alignments section (third section) of the BLAST results page. Locate the alignment for AC003040.3.
   ii. To which positions do the primers match in the subject sequence?
   iii. The lowest and highest nucleotide positions in the subject sequence indicate the borders of the amplified sequence. Subtract the lowest nucleotide position in the subject sequence from the highest nucleotide position in the subject sequence. What is the difference between the coordinates?
   iv. Note that the actual length of the amplified fragment includes both ends, so add 1 nucleotide to the result that you obtained in Step 4.iii to obtain the exact length of the PCR product amplified by the two primers.
5. Obtain the nucleotide sequence of the amplicon that the primer set would amplify in a PCR (in vitro) as follows:
   i. Open the sequence data sheet for the hit that you identified in Step 4.i (AC003040.3) by clicking on the accession number link. Scroll to the bottom of the data sheet.
   ii. The bottom section of the data sheet lists the entire nucleotide sequence that contains the PCR product. Highlight all of the nucleotides between the coordinates that you identified in Step 4.iii, from the beginning of the forward primer to the end of the reverse primer.
   iii. Copy and paste the highlighted sequence into a text document. Then, delete all nonnucleotide characters and spaces. This is the amplicon, or amplified PCR product. Save this text document for use in Step 3 of Part III.

In Step 5.iii, you can retain the nucleotide coordinates and spaces to ease readability. Reduce the point size of the font so that each row of 60 nucleotides sits on one line. Then set the font to Courier or another nonproportional font to align the blocks of sequence.

II. Identify the CLF Amino Acid Sequence

1. Open the data sheet for GenBank accession number AC003040.3. The middle of the report, under "FEATURES," contains annotations of sequence features such as

promoters, genes, mRNA sequences, and coding sequences (cds), with their beginning and ending nucleotide positions (“xx..xx”). Note that this sequence contains many genes (and, therefore, many features). For each mRNA and cds feature, “join” shows the coordinates of coding exons that are spliced together for translation into a polypeptide chain. The use of angle brackets (“<” or “>”) indicates that the feature extends beyond the nucleotide position indicated.

2. Scan the nucleotide positions (“xx..xx”) of all gene features (“/gene=”) in the sequence. Use the nucleotide coordinates of the primers (as determined in Step 4.iii of Part I above) to identify the gene feature that contains at least one of the primers. What is the name of this gene?
3. Look at the “cds” feature for this gene. What do you think the comma in “6935..6997,7275..7367” stands for?
4. Each range of numbers, such as “6935..6997,” gives the first and last nucleotides of one exon. How many exons and introns are in this gene?
5. Identify the feature(s) that the two primers span, using the nucleotide coordinates that you identified in Step 4.iii of Part I. Where does the forward primer begin? Where does the reverse primer end? Which part of the gene does the amplicon span?
6. The name of the protein is shown after “/product=.” What is the name of the protein product that this gene encodes?
7. The final part of the “cds” entry is labeled “/translation=.” It lists the one-letter abbreviations for the amino acids specified by the coding sequence. Copy the entire amino acid sequence and paste it into a text document for use in Step 2 of Part V.

III. Use Map Viewer to Determine the Chromosomal Location of the *CLF* Gene

1. Open “Map Viewer” (http://www.ncbi.nlm.nih.gov/mapview).
2. Find “Plants” in the table to the right and click on the triangle-shaped icon to the left of “Plants.” Click on “Flowering Plants” and then on “Eudicots.” Next, find “*Arabidopsis thaliana*” (thale cress) in the list and click on the “B” icon under the “Tools” column. If more than one build is displayed, select the one with the highest number; this will be the most recent version.
3. Paste the amplicon from Step 5.iii of Part I into the search window. (Usually, primers are not long enough to produce a result using map BLAST.) Omit any nonnucleotide characters from the window because they will not be recognized by the BLAST algorithm.
4. Use the default settings and click on “Begin Search.”
5. Click on “View report” to retrieve the results.
6. In “Other reports,” click on the “[Thale cress genome view]” link to see the chromosomal location of the BLAST hit. Small horizontal bars on chromosomes indicate the positions of hits. On which chromosome have you landed?
7. Click on the marked chromosome number to move to the *CLF* locus.

8. The chromosome is represented by one or more vertical lines. Click on the small blue arrow after the vertical line labeled "Genes_seq" to display the genes. The amplicon (vertical red bar) occupies most of the default view. What is the function of the *CLF* gene? Click on the name of the gene under the "Symbol" track and follow the links to find out.
9. Use the zoom toggle on the left to zoom out for a better perspective of the *CLF* gene. You may refine your position on the chromosome by clicking on the thin vertical line under "Genes_seq" and choosing a specific zoom level.
10. Introns and noncoding sequences of the *CLF* gene are denoted by a thin line, whereas exons are denoted by thick bars along the line.
    i. Determine the size of the *CLF* gene using the map coordinates to the left of the "Contig" map.
    ii. How many exons does the *CLF* gene have?
    iii. Where does the amplicon (vertical red bar) line up with the *CLF* gene?
11. Click the names of several neighboring genes and follow the links for more information about them. Can you figure out what the prefix "AT2G" means?

IV. Determine the Insertion Site of the *Ds* Transposon

During PCR, you used the primer pair CLF1/Ds for one of the reactions. This amplicon, from *clf-2* mutants, is composed partly of sequence from the *CLF* gene and partly of sequence from the *Ds* transposon. In this exercise, you will determine how many nucleotides of the amplicon are from the *Ds* transposon and the position at which *Ds* inserted into the *CLF* gene.

1. Conduct a nucleotide BLAST search with the CLF1/Ds primers by following Steps 1.i–viii in Part I. Examine the alignments section (the third section) of the BLAST results page.
    i. From what species are the hits that match the CLF1 primer (nucleotides 1–26) only? Do their descriptions mention anything about the gene or locus?
    ii. From what species are the hits that match the Ds primer (nucleotides 27–47) only? Do their descriptions mention anything about the gene or locus?
    iii. Do any hits align with both the CLF1 and Ds primers?
    iv. Based on the information gathered in Steps 1.i–iii, is the sequence of your amplicon in the database? Explain how you came to your conclusion.
2. To determine the *Ds* insertion site in the *CLF* gene, answer the questions below and label the diagram with your answers.

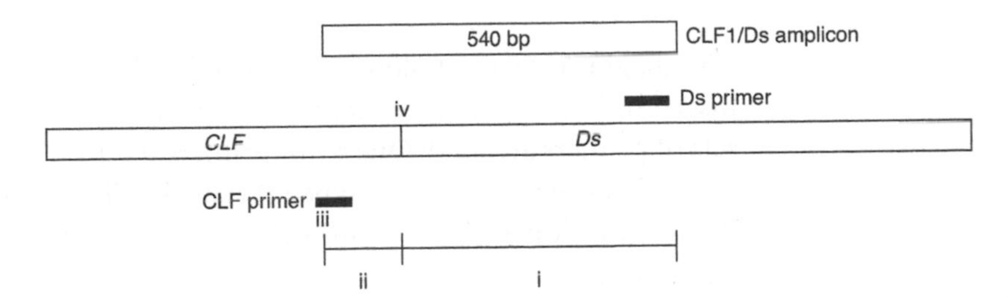

i. Identify one complete sequence of the *Ds* transposon (e.g., AF332952.1), which is one of the hits that matches the Ds primer only. How far into this *Ds* sequence does the Ds primer reach?

ii. A photograph of a gel of the PCR results shows that the CLF1/Ds amplicon is ~540 nucleotides long. Use this number and the answer from Step 2.i to calculate the number of *CLF* nucleotides in the CLF1/Ds amplicon.

iii. Look at the hit AC003040.3, which is the *CLF* gene GenBank entry that you examined in Parts I and II. What is the first nucleotide position of the CLF1 primer in the *CLF* gene sequence?

iv. What is the insertion position of *Ds* in the *CLF* gene? (Remember that the first nucleotide position of the CLF1 primer is the same as the first position of the amplicon.)

3. Refer to Step 3 in Part II above. In which part of the *CLF* gene has the *Ds* transposon inserted?

V. Use BLAST to Determine the Function of the CLF Protein

1. Return to the BLAST page at NCBI (www.ncbi.nlm.nih.gov/BLAST). This time, click on "protein blast."
2. Copy the CLF amino acid sequence from your text file (from Step 7 in Part II) and paste it into the search window.
3. Click on "BLAST" to send the amino acid sequence to the NCBI server.
4. An algorithm quickly scans the query sequence to identify functional domains (regions of the protein) that are conserved in different organisms. If a conserved domain is displayed, click on it to get some quick information about it. What do you find?
5. Close the "Conserved Domain Search" window and return to the BLAST results page.
6. Examine the graphical overview. Do the BLAST hits show strong or weak homology with your query sequence?
7. Scroll down to "Sequences producing significant alignments" and "Alignments." Based on the E values, the titles of the hits, and the links to the accession numbers, what can you conclude about the conservation of the CLF protein during plant evolution?

RESULTS AND DISCUSSION

The diagram on the following page shows how PCR amplification identifies the *Ds* insertion in the *CLF* gene. Because the *Ds* insertion is too large to amplify across, two sets of primers are used. One set of primers (CLF1/CLF2) straddles the *Ds* insertion site and only amplifies the wild-type (Wt) allele. The second set (CLF1/Ds) only amplifies the *Ds* insertion allele, with one primer located in the *CLF* gene and the other in the *Ds* transposon.

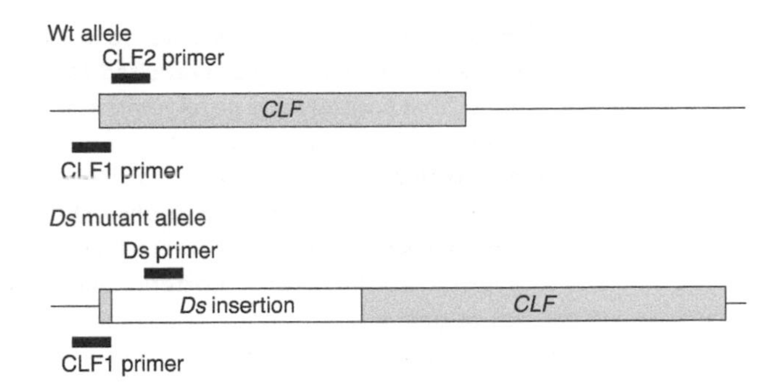

I. Think About the Experimental Methods

1. Describe the purpose of each of the following steps or reagents used in DNA isolation (Part II of Experimental Methods):
   i. Grinding with pestle.
   ii. Edward's buffer.
   iii. Boiling.
   iv. Tris-EDTA (TE) buffer.
2. What is the purpose of performing each of the following PCRs?
   i. CLF1/CLF2.
   ii. CLF1/Ds.

II. Interpret Your Gel and Think About the Experiment

1. Observe the photograph of the stained gel containing your PCR samples and those from other students. Orient the photograph with the sample wells at the top. Use the sample gel shown below to help interpret the band(s) in each lane of the gel.

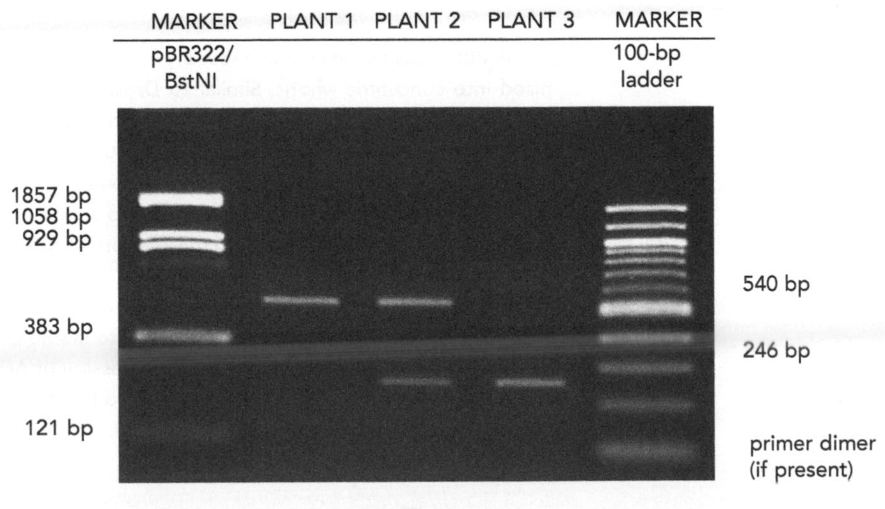

If a 100-bp ladder was used, as shown in the right-hand slide of the sample gel, the bands increase in size in 100-bp increments starting with the fastest-migrating band of 100 bp.

2. Locate the lane containing the pBR322/BstNI markers on the left side of the gel. Working from the well, locate the bands corresponding to each restriction fragment: 1857, 1058, 929, 383, and 121 bp. The 1058- and 929-bp fragments will be very close together or may appear as a single large band. The 121-bp band may be very faint or not visible.

Further Research: Homeotic Genes in Development

Homeotic genes were first discovered in *Drosophila*, but they have since been found in vertebrates and plants. They are required for the correct spatial and tissue-specific expression of genes that control early development of an organism. Do an Internet search to gain insight into how CLF and other homeotic proteins control development throughout the plant and animal worlds.

During development in *Drosophila*, the embryo is divided into smaller and more specialized domains through the action of a hierarchy of homeotic genes. The first genes in this hierarchy establish the anterior/posterior and dorsal/ventral axes of the embryo. The next three series of genes—the gap, pair rule, and segment polarity genes—divide the anterior/posterior axis into segments. The last genes are expressed throughout the remainder of development and adult life.

In *Drosophila*, the combined activities of the homeotic proteins specify the identities of the three body segments: head, thorax, and abdomen. For these body segments to develop normally, correct patterns of homeotic gene expression must be maintained through the multiple mitotic divisions that occur during their formation. These expression patterns can be thought of as a molecular memory system. In homeotic gene mutants, one body part is exchanged for a different one. For example, the *Antennapedia* mutant has legs in the place of antennae. Genes related to *Antennapedia* establish initial patterns of homeotic gene expression. These patterns are maintained later in development by genes of the Polycomb group (PcG) and trithorax group (trxG).

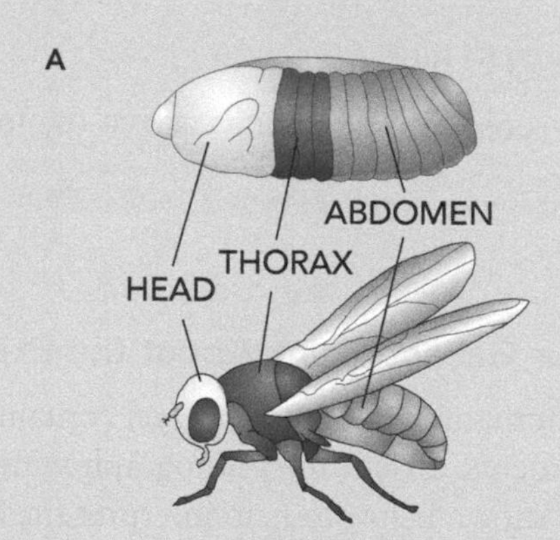

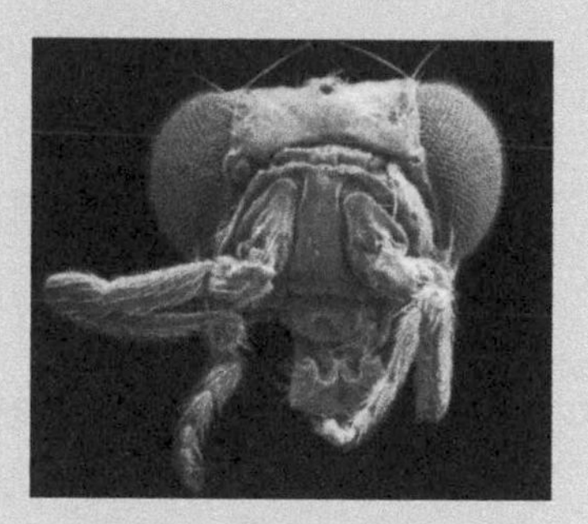

Drosophila body plan (*A*) and *Antennapedia* mutant (*B*).
(Courtesy of F.R. Turner, Indiana University.)

Arabidopsis flowers contain four organs—sepals, petals, stamens, and carpels—that are organized into concentric whorls. Similar to *Drosophila* segments, the identities of these whorls are established by the combined action of homeotic genes.

The *Arabidopsis* homeotic gene *AGAMOUS* is normally expressed in the third or fourth flower whorl and is required for the correct development of stamens and carpels. It is not expressed in the first or second whorls, which develop into sepals and petals. This is because the CLF protein represses *AGAMOUS* expression in these whorls. Thus, like other PcG proteins, CLF helps to maintain patterns of homeotic gene expression after whorl identity has already been established. However, plants that are homozygous for the *clf-2* mutation fail to repress *AGAMOUS* and exhibit partial transformation of sepals and petals into stamens and carpels. Although the exact mechanism is not known, the ectopic (unusual) expression of *AGAMOUS* is also responsible for the more visible elements of the *clf-2* phenotype: curled leaves, dwarf size, and early flowering.

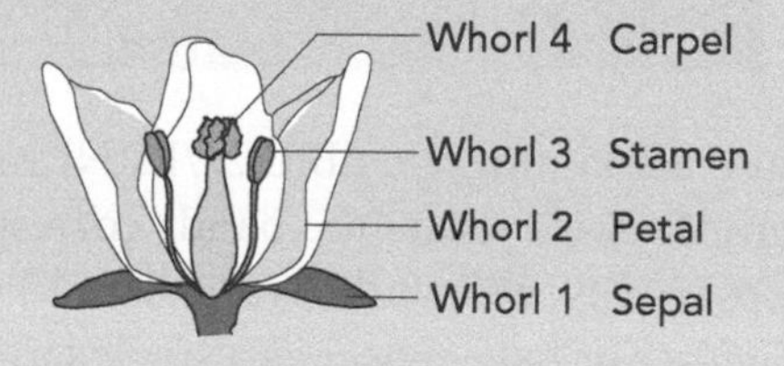

Arrangement of flower organs in *Arabidopsis*.

3. Scan across the row of student results. Notice that virtually all student lanes contain one or two prominent bands.
   i. The amplification product of the mutant *clf-2* allele (540 bp) should align between the 383- and 929-bp fragments of the pBR322/BstNI marker.
   ii. The amplification product of the wild-type *CLF* allele (246 bp) should align between the 121- and 383-bp fragments of the pBR322/BstNI marker.
4. Carefully examine the banding patterns of the samples in the gel above and complete the entries for genotype and phenotype in the chart below. Under "Sample," indicate which result was derived from Plant 1, Plant 2, and Plant 3. (Your own samples may yield any of the banding patterns listed in the table.)

540 bp (CLF1/Ds)	246 bp (CLF1/CLF2)	Plant genotype	Plant phenotype	Sample
present	present			
present	absent			
absent	present			

Additional faint bands at other positions occur when the primers bind to chromosome loci other than the exact locus and give rise to "nonspecific" amplification products.

5. It is common to see a diffuse (fuzzy) band that runs ahead of the 121-bp marker. This is "primer dimer," an artifact of the PCR that results from the primers overlapping one another and amplifying themselves. How would you interpret a lane in which you observe primer dimer but no bands as described in Step 3?
6. Would you classify the *clf-2* mutation as recessive or dominant? Explain your reasoning.
7. Count the number of plants with the +/+, –/+, and –/– genotypes in your class. Based on the genotype distribution of these plants, which genotype was the parental plant? Why may the genotype distribution observed by your class deviate from what is expected under Mendel's laws of inheritance?

LABORATORY 3.3

Linkage Mapping a Mutation

▼ OBJECTIVES

This laboratory demonstrates several important concepts of modern biology. During the course of this laboratory, you will

- Use linkage and recombination to map a genetic mutation in *Arabidopsis*.
- Learn how restriction-fragment-length polymorphisms (RFLPs) are used as molecular markers.
- Learn about the relationship between genotype and phenotype.
- Move between in vitro experimentation and in silico computation.

In addition, this laboratory utilizes several experimental and bioinformatics methods in modern biological research. You will

- Extract and purify DNA from plant tissue.
- Amplify a specific region of the genome by polymerase chain reaction (PCR).
- Digest PCR products with restriction enzymes.
- Analyze PCR products by gel electrophoresis.
- Use the Basic Local Alignment Search Tool (BLAST) to identify sequences in databases.
- Use the Map Viewer tool to electronically map genes on chromosomes.

INTRODUCTION

Since Alfred Sturtevant constructed the first genetic map of a *Drosophila* chromosome in 1913, new mutations have been mapped using his method of linkage analysis. Determining the map position of a new mutation—and its corresponding gene—consists of testing for linkage with a number of previously mapped genes or DNA markers. Linkage is the principle that the closer that two genes or markers are located to each other on a chromosome, the greater the chance that they will be inherited together as a unit (linked). Conversely, loci farther apart on the chromosome are more likely to be separated by chromosome recombination during meiosis. Thus, the frequency of recombination with previously mapped genes or markers allows one to determine the map position of a gene of interest.

This laboratory is adapted from Cold Spring Harbor *Arabidopsis* Molecular Genetics Course Manual at http://www.arabidopsis.org/cshl-course/3-genetic_mapping.html and EMBO Practical Course on Genetic and Molecular Analysis of *Arabidopsis*, Module 2: Mapping mutations using molecular markers at http://www.isv.cnrs-gif.fr/embo99/manuals/.

The increasing availability of whole-genome sequences and sophisticated computer software has made it possible to map genes using bioinformatics approaches. However, traditional mapping techniques are still used to map genes for which no sequence information is available (e.g., mutant phenotypes produced by chemical mutagenesis). Although early gene maps relied on genes and mutations with observable phenotypes, modern gene maps are populated with DNA polymorphisms that are detected by molecular methods. In *Arabidopsis*, molecular markers exploit the natural differences between distinct ecotypes, such as the widely used Landsberg (Ler) and Columbia (Col) ecotypes, which differ by about 1% at the DNA level.

This laboratory uses linkage to map *ago1* (argonaute), a mutation that produces a dwarf phenotype and serrated leaves. In preparation for this experiment, a Ler plant homozygous for the *ago1* mutation (–/–) was crossed with a wild-type Col plant (+/+). A heterozygous F_1 plant was then allowed to self-pollinate. The resulting F_2 progeny conform to the expected 1:2:1 genotypic ratio of homozygous wild-type (+/+):heterozygous wild type (+/–):homozygous mutant (–/–) individuals. However, due to recombination events during gamete formation, each of the F_2 chromosomes is a mixture of the Ler and Col ecotypes.

F_2 plants that are homozygous for the *ago1* mutation are used for mapping. Because *ago1* is always in a Ler background, the number of recombination events is equivalent to the number of times that the Col ecotype is found on the chromosome. Recombination events are scored by evaluating a panel of four DNA polymorphisms called CAPS (cleaved amplified polymorphic sequences) markers that identify homologous regions of the Ler and Col chromosomes. At each CAPS location (locus), a point mutation creates a restriction site in either the Ler or Col ecotype.

Each marker region is amplified by PCR and then cut with a specific restriction enzyme to identify a restriction-fragment-length polymorphism (RFLP) that distinguishes the two ecotypes. After agarose gel electrophoresis, the banding pattern indi-

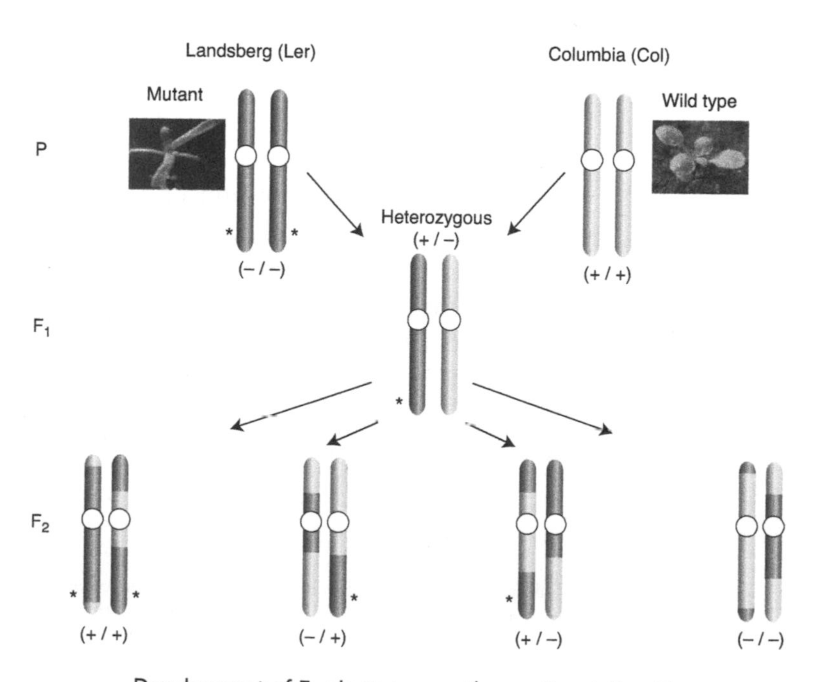

Development of F_2 plants to map the *ago1* mutation (*).

cates the ecotype source of each of the paired chromosomes: homozygous Ler/Ler, heterozygous Ler/Col, or homozygous Col/Col. By pooling class data, recombination frequencies can be calculated between *ago1* and each of the CAPS markers, allowing one to map *ago1*.

This laboratory includes bioinformatics exercises that complement the experimental methods. BLAST is used to identify sequences in biological databases and to predict the sizes of the PCR products amplified by the four CAPS primer sets. The sizes of the fragments generated by restriction enzyme digestion are then predicted using an online tool hosted by the DNA Learning Center (http://www.dnalc.org/bioinformatics). Another bioinformatics tool, Map Viewer, is used to discover the chromosome location of the *AGO1* gene, learn about the function of the argonaute protein, and electronically map *ago1* and the four CAPS markers in *Arabidopsis*.

FURTHER READING

Bohmert K, Camus I, Bellini C, Bouchez D, Caboche M, Benning C. 1998. *AGO1* defines novel locus of *Arabidopsis* controlling leaf development. *EMBO J* **17:** 170–180.

Edwards K, Johnstone C, Thompson C. 1991. A simple and rapid method for the preparation of plant genomic DNA for PCR analysis. *Nucleic Acids Res* **19:** 1349.

Konieczny A, Ausbel FM. 1993. A procedure for mapping *Arabidopsis* mutations using co-dominant ecotype-specific PCR-based markers. *Plant J* **4:** 403–410.

Korneef M, Alonso-Blanco C, Stam P. 1998. Genetic analysis. *Methods Mol Biol* **82:** 105–117.

http://www.dnaftb.org/dnaftb/11/concept/index.html Sturtevant's linkage experiments explained as an animation in DNA from the Beginning, concept 11: Genes get shuffled when chromosomes exchange pieces. DNA Learning Center, Cold Spring Harbor Laboratory, Cold Spring Harbor, N.Y.

PLANNING AND PREPARATION

The following table will help you to plan and integrate the different experimental methods.

Experiment	Day	Time		Activity
I. Plant *Arabidopsis* seeds	3–4 wk before Part II	15 min	Prelab:	Set up student stations.
		15 min	Lab:	Plant *Arabidopsis* seeds.
			Postlab:	Care for *Arabidopsis* plants.
II. Isolate DNA from *Arabidopsis*	1	30 min	Prelab:	Set up student stations.
		30–60 min	Lab:	Isolate DNA.
III. Amplify DNA by PCR	2	15 min	Prelab:	Aliquot primer/loading dye mixes. Set up student stations.
		15 min	Lab:	Set up PCRs.
		60–150 min	Postlab:	Amplify DNA in thermal cycler.
IV. Digest PCR products to produce CAPS markers	3	30–60 min	Prelab:	Aliquot restriction enzymes. Set up student stations.
		60 min	Lab:	Set up restriction digests.
		60+ min	Postlab:	Incubate restriction digests.
V. Analyze PCR products by gel electrophoresis	4	30 min	Prelab:	Dilute TBE electrophoresis buffer. Prepare agarose gel solution. Set up student stations.
		30 min	Lab:	Cast gels.
	5	45+ min	Lab:	Load DNA samples into gel. Electrophorese samples. Photograph gels.

OVERVIEW OF EXPERIMENTAL METHODS

I. PLANT *ARABIDOPSIS* SEEDS

PLANT
seeds

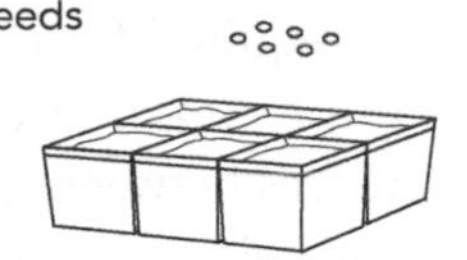

WATER
seeds

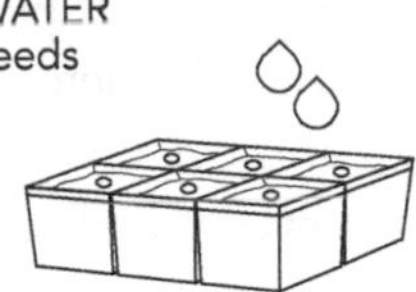

GROW
plants for
3–4 wk

II. ISOLATE DNA FROM *ARABIDOPSIS*

ADD
leaf
tissue

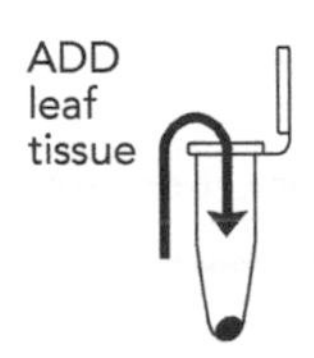

GRIND

ADD
Edward's
buffer

GRIND

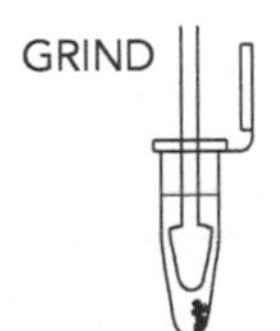

VORTEX

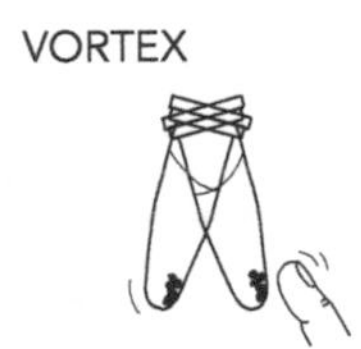

BOIL

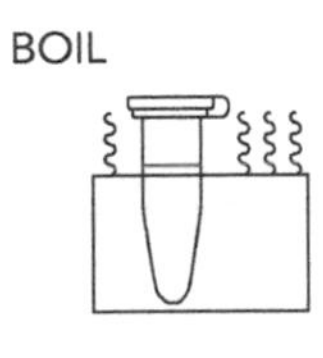

CENTRIFUGE

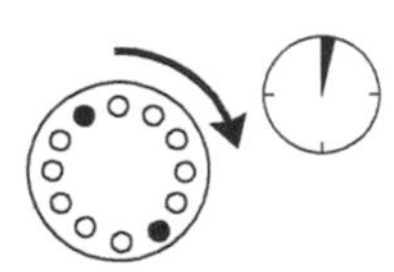

TRANSFER
supernatant

ADD and MIX
isopropanol

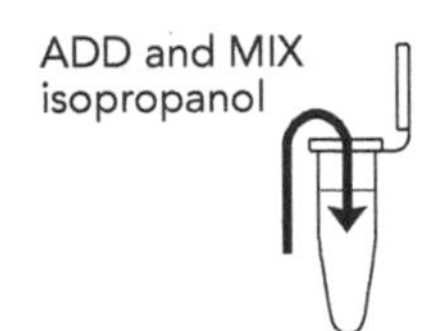

CENTRIFUGE

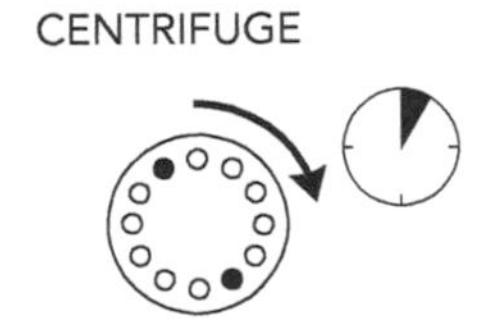

POUR OFF
supernatant

REMOVE
supernatant

DRY

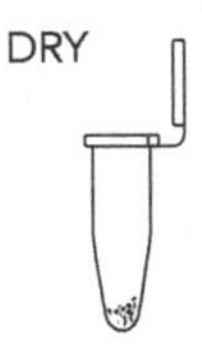

ADD
TE buffer

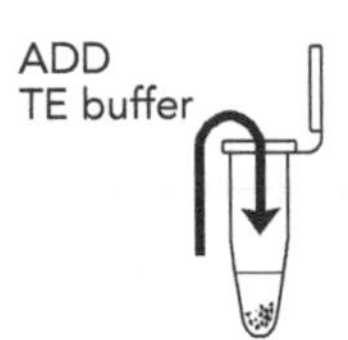

RESUSPEND
DNA

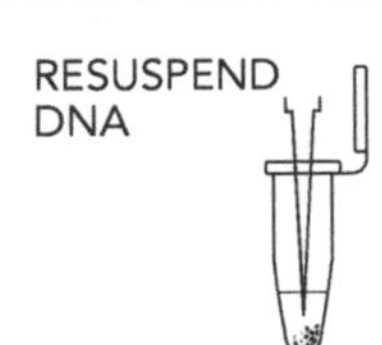

leaf
DNA
extract

III. AMPLIFY DNA BY PCR

ADD
primer/
loading
dye mix

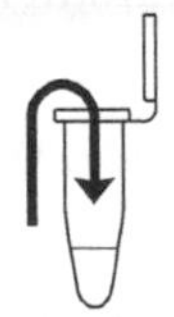

ADD
DNA

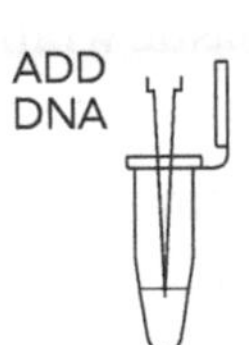

AMPLIFY
in thermal
cycler

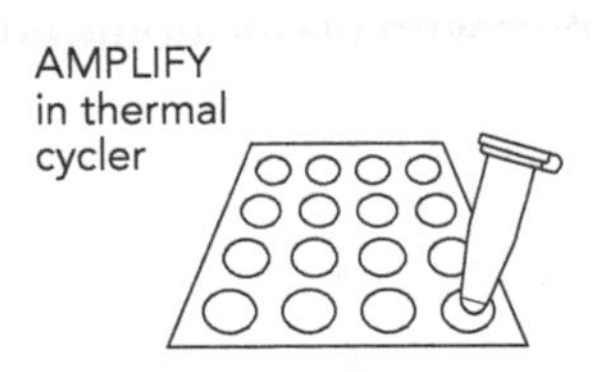

IV. DIGEST PCR PRODUCTS TO PRODUCE CAPS MARKERS

TRANSFER
restriction
enzyme/
buffer

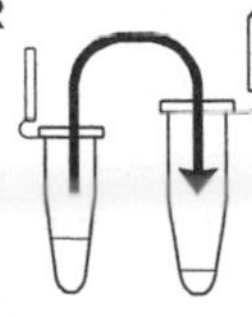

ADD
PCR
products

MIX

INCUBATE

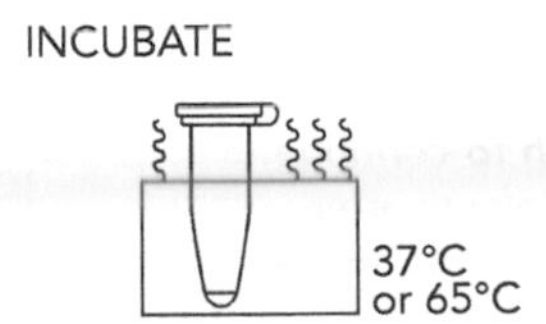

V. ANALYZE PCR PRODUCTS BY GEL ELECTROPHORESIS

POUR
gel

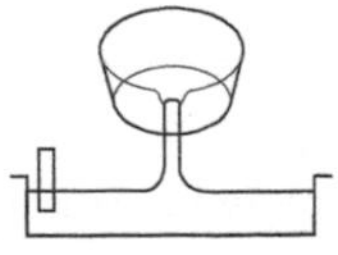

SET
20 min

LOAD
gel

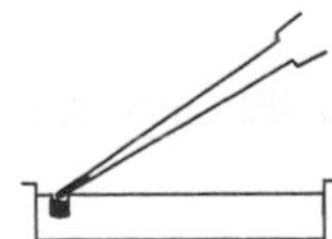

ELECTROPHORESE
130 V

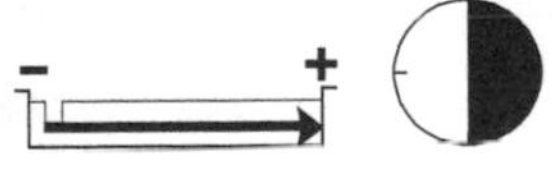

EXPERIMENTAL METHODS

I. Plant *Arabidopsis* Seeds

To extract DNA from plant tissue, you must plant the *Arabidopsis* seeds 3–4 wk before DNA isolation and PCR. Two 1/4-in-diameter leaf disks are required for each experiment, but multiple small leaves and even whole plantlets can be used. Depending on growing conditions, you may observe the phenotypic differences between mutant and wild-type plants in as little as 2 wk. For further information on cultivation, refer to The Arabidopsis Information Resource (TAIR) at http://www.arabidopsis.org.

REAGENTS, SUPPLIES, & EQUIPMENT

For each group

2 Planting pots and 1 tray
Plastic dome lid or plastic wrap
Potting soil
10–20 Seeds from a self-cross of an *ago1* heterozygous *Arabidopsis* plant
10–20 Seeds from Col or Ler controls
1 Sheet of paper (4 x 4 in)
Water

To share

Refrigerator

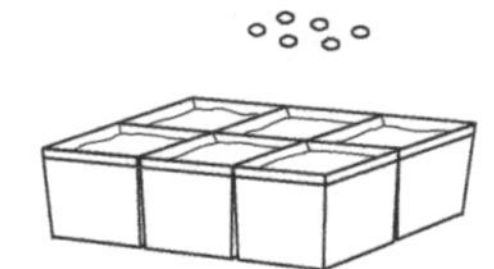

Arabidopsis seeds are very tiny and difficult to handle, so planting is not as simple as it may seem.

1. Label each pot with your group number and plant type (use "A" for "*ago1*," "C" for "Col," and "L" for "Ler"). Moisten the potting soil. Fill planting pots evenly with soil, but do not pack tightly.
2. Fit pots into the tray, but leave one corner space empty to facilitate watering.
3. Carefully scatter seeds evenly on top of the soil as follows:
   i. Fold a 4 x 4-in sheet of paper in half.
   ii. Place 10–20 seeds into the fold of the paper and gently tap them onto the soil. Provide space between seeds, so that they will grow better and plant phenotypes can be readily observed.
4. Cover the pots with plastic dome lids or plastic wrap to maintain humidity during germination. (Remove covers 3–7 d after planting.)
5. Add 1/2 in of water to the tray, using the empty corner space. Water regularly to keep the soil damp, but do not allow the soil to remain soggy.
6. Place the planted seeds in a refrigerator for 2–4 d. This "vernalization" step is needed for optimum germination.
7. Grow the plants close to a sunny window at room temperature (20°C–22°C) or slightly warmer. If the plants are grown under cold conditions, it may take longer to discern the different phenotypes.
8. The dwarf *ago1* mutants are very small compared to the wild-type and heterozygous plants, and they have serrated leaves. When the difference in phenotype becomes evident, continue to Part II.

For optimum growth, provide constant (24-h) fluorescent lighting ~1 ft directly above the plants. With 24-h fluorescent lighting, phenotypes can be discerned in 2–3 wk.

You may wish to continue to grow the plants after you have harvested tissue for DNA isolation and amplification. The phenotypic differences between mutant and wild-type plants become more obvious over time.

II. Isolate DNA From *Arabidopsis*

REAGENTS, SUPPLIES, & EQUIPMENT

For each group

ago1 and Col or Ler control *Arabidopsis* plants from Part I
Container with cracked or crushed ice
Edward's buffer (1.2 mL)
Isopropanol <!> (1 mL)

REAGENTS, SUPPLIES, & EQUIPMENT *(continued)*

For each group (continued)

Microcentrifuge tube racks
4 Microcentrifuge tubes (1.5 mL)
Micropipettes and tips (100–1000 μL)
Permanent marker
2 Plastic pestles
Tris/EDTA (TE) buffer (400 μL)
Vortexer (optional)

To share

Microcentrifuge
Water bath or heating block at 96°C–100°C

See Cautions Appendix for appropriate handling of materials marked with <!>.

Your instructor will assign either a Col or Ler Arabidopsis plant to your group. These plants are the controls. You will share your results with a group that tested the other control.

1. Obtain one *ago1* mutant and one Col or Ler *Arabidopsis* plant and record their phenotypes.
2. From each plant, harvest two pieces of leaf tissue that are ~1/4 in in diameter. (The large end of a 1000-μL pipette tip will punch disks of this size.) If the plant is small, take multiple leaves to make an equivalent amount of tissue. The *ago1* mutant may be so small that you need to use the entire plant. If so, carefully remove all soil from the roots. Place the tissue into clean 1.5-mL tubes (use one tube for each plant). Label the tubes with your group number and plant phenotype code (use "A" for "*ago1*," "C" for "Col," and "L" for "Ler").

Grinding the plant tissue breaks up the cell walls. When fully ground, the sample should be a green liquid.

3. Add 100 μL of Edward's buffer to each tube. Grind the plant tissue by *forcefully* twisting it with a clean plastic pestle against the inner surface of each 1.5-mL tube for 1 min. Use a clean pestle for each tube.

Detergent in the Edward's buffer, called sodium dodecyl sulfate (SDS), dissolves lipids of the cell membrane.

4. Add 400 μL of Edward's buffer to each tube. Grind briefly to remove tissue from the pestle and to liquefy any remaining pieces of tissue.
5. Vortex the tubes for 5 sec, by hand or machine (if available).

Boiling denatures proteins, including enzymes that digest DNA.

6. Boil the samples for 5 min in a water bath or heating block.
7. Place the tubes, along with those from other groups, in a balanced configuration in a microcentrifuge and centrifuge for 2 min to pellet any remaining cell debris. Centrifuge longer if unpelleted debris remains.
8. Label two fresh tubes with your group number and plant phenotype code. Transfer 350 μL of each supernatant to the appropriate fresh tube. Be careful not to disturb the pelleted debris when transferring the supernatant. Discard the old tubes containing the debris.
9. Add 400 μL of isopropanol to each tube of supernatant.
10. Close the tubes, mix by inverting several times, and then leave for 3 min at room temperature to precipitate nucleic acids, including DNA.

The pellet may appear as a tiny teardrop-shaped smear or particles on the bottom side of the tube underneath the hinge. Do not be concerned if you cannot see a pellet. A large or greenish pellet is cellular debris carried over from the first centrifugation.

11. Place your tubes and those of other groups in a balanced configuration in a microcentrifuge, with the hinges of the caps pointing outward. (During centrifugation, nucleic acids will gather on the side of the tube underneath the hinge.) Centrifuge for 5 min at maximum speed to pellet the DNA.
12. Carefully pour off and discard the supernatant from each tube. Then, *completely* remove the remaining liquid with a micropipette set at 100 μL.

Dry the pellets quickly with a hair dryer. To prevent blowing the pellets away, direct the air across the tube mouths, not into the tubes, for ~3 min.

13. Air-dry the pellets by allowing the tubes to sit open for 10 min. Any remaining isopropanol will evaporate.

The TE (Tris-EDTA) buffer provides conditions for stable storage of DNA. Tris provides a constant pH of 8.0, whereas EDTA binds cations (positive ions) that are required for DNase activity.

14. Add 100 µL of TE buffer to each tube. Dissolve the pelleted DNA by pipetting in and out, using a fresh tip for each tube. Make sure to wash the TE down the side of the tube underneath the hinge, where the pellet formed during centrifugation.
15. The DNA may be used immediately or stored at –20°C until you are ready to proceed to Part III. Keep the DNA on ice during use.

In Part III, you will use 2.5 µL of DNA for each PCR. These are crude DNA extracts and contain nucleases that will eventually fragment the DNA at room temperature. Keep the samples cold to limit this activity.

III. Amplify DNA by PCR

REAGENTS, SUPPLIES, & EQUIPMENT

For each group

Container with cracked or crushed ice
DNA from *ago1* mutant *Arabidopsis** from Part II
DNA from Col or Ler control *Arabidopsis** from Part II
Microcentrifuge tube rack
Micropipettes and tips (1–100 µL)
Permanent marker
Primer/loading dye mix* (50 µL) for g4026
Primer/loading dye mix* (50 µL) for H77224
Primer/loading dye mix* (50 µL) for m235
Primer/loading dye mix* (50 µL) for UFO
8 Ready-To-Go PCR Beads in 0.2- or 0.5-mL PCR tubes

To share

Thermal cycler

*Store on ice.

You will set up ago1 and control (Col or Ler) reactions for each of the four CAPS markers. Carry on with either the Col or Ler control as assigned by your instructor in Part II. You will share your results with a team that tested the other control.

1. Obtain eight PCR tubes containing Ready-To-Go PCR Beads. Label the tubes with your group number. Then, label each tube with the code in the first column of the following table:

Code on tube	Primer/loading dye mix				DNA sample	
	g4026	H77224	m235	UFO	*ago1*	Col or Ler
GA	22.5 µL				2.5 µL	
GC or GL	22.5 µL					2.5 µL
HA		22.5 µL			2.5 µL	
HC or HL		22.5 µL				2.5 µL
MA			22.5 µL		2.5 µL	
MC or ML			22.5 µL			2.5 µL
UA				22.5 µL	2.5 µL	
UC or UL				22.5 µL		2.5 µL

The primer/loading dye mix will turn purple as the PCR bead dissolves.

2. Use a micropipette with a fresh tip to add 22.5 µL of the appropriate primer/ loading dye mix (listed in the table) to each tube. Allow the beads to dissolve for ~1 min.
3. Use a micropipette to add 2.5 µL of *ago1*, Col, or Ler *Arabidopsis* DNA (from Part II) directly into the appropriate reactions (listed in the table). Use a fresh tip each time and make sure that no DNA remains in the tip after pipetting.

If the reagents become splattered on the wall of the tube, pool them by pulsing the sample in a microcentrifuge or by sharply tapping the tube bottom on the lab bench.

4. Store your samples on ice until your class is ready to begin thermal cycling.
5. Place your PCR tubes, along with those of other groups, in a thermal cycler that has been programmed for 30 cycles of the following profile:

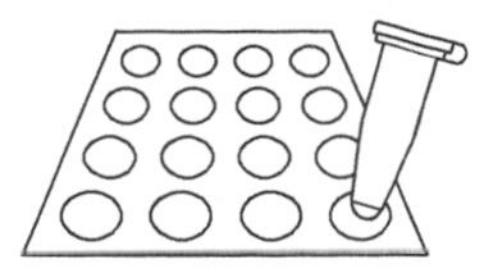

Denaturing step:	30 sec	94°C
Annealing step:	30 sec	55°C
Extending step:	30 sec	72°C

The profile may be linked to a 4°C hold program after the 30 cycles are completed.

6. After thermal cycling, store the amplified DNA on ice or at –20°C until you are ready to proceed to Part IV.

IV. Digest PCR Products to Produce CAPS Markers

REAGENTS, SUPPLIES, & EQUIPMENT

For each group

Container with cracked or crushed ice
Microcentrifuge tube rack
8 Microcentrifuge tubes (1.5 mL)
Micropipette and tips (1–20 μL)
PCR products* from Part III
Permanent marker
Restriction enzyme/buffer mix* (25 μL) for HindIII
Restriction enzyme/buffer mix* (25 μL) for RsaI
Restriction enzyme/buffer mix* (50 μL) for TaqI

To share

Microcentrifuge (optional)
Water baths at 37°C and 65°C

*Store on ice.

Note that the RsaI and HindIII reactions will be incubated at 37°C, whereas the TaqI reaction will be incubated at 65°C. This is because TaqI is isolated from the thermophilic (heat-loving) bacterium Thermus aquaticus, which inhabits hot springs.

1. Label eight 1.5-mL microcentrifuge tubes with your group number and the code in the first column of the table below:

Code on tube	PCR product	Restriction enzyme/buffer mix			Incubation temp.
		HindIII	TaqI	RsaI	
GA	10 μL			10 μL	37°C
GC or GL	10 μL			10 μL	37°C
HA	10 μL		10 μL		65°C
HC or HL	10 μL		10 μL		65°C
MA	10 μL	10 μL			37°C
MC or ML	10 μL	10 μL			37°C
UA	10 μL		10 μL		65°C
UC or UL	10 μL		10 μL		65°C

2. Use a micropipette with a fresh tip to transfer 10 μL of the appropriate restriction enzyme/buffer mix to the appropriate 1.5-mL tube as listed in the table. Use a fresh tip for each enzyme/buffer mix.

A lengthy incubation helps to ensure that the restriction enzymes digest the CAPS markers completely. A partial digest can make a homozygous plant appear heterozygous, confounding the analysis.

3. Use a micropipette with a fresh tip to add 10 μL of each PCR product into the appropriate tube from Step 2. Use a fresh tip for each tube.
4. Mix and pool the reagents in each tube by pulsing in a microcentrifuge or by sharply tapping the tube bottom on the lab bench.
5. Place your tubes, along with those from other students, in the appropriate water bath (37°C or 65°C, as listed in the table) and incubate for at least 1 h.
6. Store your samples on ice or at –20°C until you are ready to begin Part V.

V. Analyze CAPS Markers by Gel Electrophoresis

REAGENTS, SUPPLIES, & EQUIPMENT

For each group

2% Agarose in 1x TBE (hold at 60°C) (50 mL per gel)
CAPS markers from Part IV
Container with cracked or crushed ice
Gel-casting tray and comb
Gel electrophoresis chamber and power supply
Latex gloves
Masking tape
Microcentrifuge tube rack
Micropipette and tips (1–100 μL)
pBR322/BstNI marker (20 μL per gel)*
SYBR Green <!> DNA stain (20 μL)
1x TBE buffer (300 mL per gel)

To share

Digital camera or photodocumentary system
UV transilluminator <!> and eye protection
Water bath for agarose solution (60°C)

*Store on ice.
See Cautions Appendix for appropriate handling of materials marked <!>.

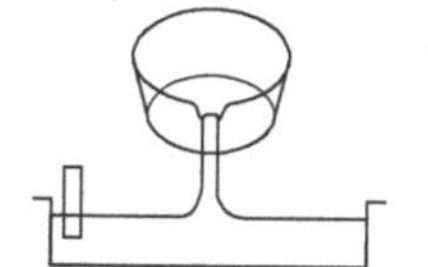

Avoid pouring an overly thick gel, which will be more difficult to visualize.

1. Seal the ends of the gel-casting tray with masking tape and insert a well-forming comb.
2. Pour the 2% agarose solution into the tray to a depth that covers about one-third the height of the open teeth of the comb.
3. Allow the gel to completely solidify; this takes ~20 min.

The gel will become cloudy as it solidifies.

4. Remove the masking tape, place the gel into the electrophoresis chamber, and add enough 1x TBE buffer to cover the surface of the gel.

Do not add more buffer than necessary. Too much buffer above the gel channels electrical current over the gel, increasing the running time.

5. Carefully remove the comb and add additional 1x TBE buffer to fill in the wells and just cover the gel, creating a smooth buffer surface.
6. Orient the gel according to the diagram in Step 8, so that the wells are along the top of the gel.

A 100-bp ladder may also be used as a marker.

7. Use a micropipette with a fresh tip to load 2 μL of SYBR Green DNA stain to each CAPS marker (from Part IV). In addition, add 2 μL of SYBR Green DNA stain to 20 μL of pBR322/BstNI marker for each gel.

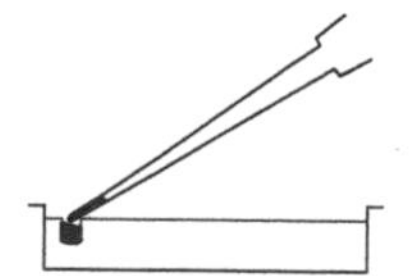

Expel any air from the tip before loading. Be careful not to push the tip of the pipette through the bottom of the sample well.

8. Use a micropipette with a fresh tip to load each sample from Step 7 into your assigned wells, according to the following diagram (the example is for g4026). Use a fresh tip for each sample.

MARKER	GROUP 1	GROUP 2	GROUP 1	GROUP 2
pBR322/ BstNI	Ler control (GL)	Col control (GC)	*ago1* mutant (GA)	*ago1* mutant (GA)

Transillumination, where the light source is below the gel, increases brightness and contrast.

9. Run the gel for ~30 min at 130 V. Adequate separation will have occurred when the cresol red dye front has moved at least 50 mm from the wells.
10. View the gel using UV transillumination. Photograph the gel using a digital camera or photodocumentary system.

BIOINFORMATICS METHODS

For a better understanding of the experiment, perform the following bioinformatics exercises before you analyze your results.

I. Use BLAST to Find DNA Sequences in Databases (Electronic PCR)

1. Perform a BLAST search as follows:
   - i. Do an Internet search for "ncbi blast."
   - ii. Click on the link for the result "BLAST: Basic Local Alignment Search Tool." This will take you to the Internet site of the National Center for Biotechnology Information (NCBI).
   - iii. Click on the link "nucleotide blast" (blastn) under the heading "Basic BLAST."
   - iv. Enter the g4026 primer sequences into the search window. These are the query sequences.

The following primer sets were used in the experiment:

g4026	5′-GGGGTCAGTTACATTACTAGC-3′ (forward primer) 5′-GTACGGTTCTTCTTCCCTTA-3′ (reverse primer)
H77224	5′-GGATTTGGGGAAGAGGAAGTAA-3′ (forward primer) 5′-TCCTTAGCCTTGCTTTGATAGT-3′ (reverse primer)
m235	5′-GAATCTGTTTCGCCTAACGC-3′ (forward primer) 5′-AGTCCACAACAATTGCAGCC-3′ (reverse primer)
UFO	5′-GTGGCGGTTCAGACGGAGAGG-3′ (forward primer) 5′-AAGGCATCATGACTGTGGTTTTTC-3′ (reverse primer)

   - v. Omit any nonnucleotide characters from the window because they will not be recognized by the BLAST algorithm.
   - vi. Under "Choose Search Set," select the "Nucleotide collection (nr/nt)" database from the drop-down menu.
   - vii. Under "Program Selection," optimize for "Somewhat similar sequences (blastn)."
   - viii. Click on "BLAST." This sends your query sequences to a server at NCBI in Bethesda, Maryland. There, the BLAST algorithm will attempt to match the primer sequences to the millions of DNA sequences stored in its database. While searching, a page showing the status of your search will be displayed until your results are available. This may take only a few seconds or more than 1 min if many other searches are queued at the server.

2. Analyze the results of the BLAST search, which are displayed in three ways as you scroll down the page:
   - i. First, a graphical overview illustrates how significant matches (hits) align with the query sequence. Matches of differing lengths are indicated by color-coded bars. What do you notice about the lengths (and colors) of the matches (bars) as you look from the top to the bottom?
   - ii. This is followed by a list of significant alignments (hits) with links to the corresponding accession numbers. (An accession number is a unique identifier given to a sequence when it is submitted to a database such as GenBank.)

Note the scores in the "E value" column on the right. The Expectation or E value is the number of alignments with the query sequence that would be expected to occur by chance in the database. The lower the E value, the higher the probability that the hit is related to the query. For example, an E value of 1 means that a search with your sequence would be expected to turn up one match by chance. Longer query sequences generally yield lower E values. An alignment is considered significant if it has an E value of less than 0.1. What is the E value of the most significant hit and what does it mean? Note the names of the most significant hits. Do they make sense? What do they have in common?

iii. Third is a detailed view of each primer (query) sequence aligned to the nucleotide sequence of the search hit (subject, abbreviated "Sbjct"). Note that the first match to the forward primer (nucleotides 1–21) and to the reverse primer (nucleotides 22–41) are within the same subject (accession number). Do the rest of the alignments produce good matches to both primers?

3. Click on the accession number link to open the data sheet for the first hit (the first subject sequence).

   i. At the top of the report, note basic information about the sequence, including its length (in base pairs, or bp), database accession number, source, and references to papers in which the sequence is published. What is the source and size of the sequence in which your BLAST hit is located?

   ii. In the middle section of the report, the sequence features are annotated, with their beginning and ending nucleotide positions ("xx..xx"). These features may include genes, coding sequences (cds), regulatory regions, ribosomal RNA (rRNA), and transfer RNA (tRNA). You examine these features more closely in Part II.

   iii. Scroll to the bottom of the data sheet. This is the nucleotide sequence to which the term "Sbjct" refers.

   iv. Use your browser's "Back" button to return to the BLAST results page.

4. Predict the length of the product that the primer set would amplify in a PCR (in vitro) as follows:

   i. Scroll down to the alignments section (third section) of the BLAST results page. Locate the alignment for the g4026 accession numbers listed in the table below.

   ii. To which positions do the primers match in the subject sequence?

   iii. The lowest and highest nucleotide positions in the subject sequence indicate the borders of the amplified sequence. Record these numbers in the table below.

   iv. Subtract the lowest nucleotide position in the subject sequence from the highest nucleotide position in the subject sequence. Note that the actual length of the amplified fragment includes both ends, so add 1 nucleotide to the result to obtain the exact length of the PCR product amplified by the two primers. Record the length of the amplicon in the table below.

   v. The two hits for g4026 give slightly different amplicon sizes. Why?

Marker	Accession no.	Lowest position	Highest position	Amplicon length (bp)
g4026	AC002292			
	AM748036			
H77224	NM_105947			
m235	AF000657			
UFO	AC000107			

5. Obtain the nucleotide sequence of the amplicon that the primer set would amplify in a PCR (in vitro) as follows:
   i. Open the sequence data sheet for the hit that you identified in Step 4.i by clicking on the accession number link. Scroll to the bottom of the data sheet.
   ii. The bottom section of the data sheet lists the entire nucleotide sequence that contains the PCR product. Highlight all of the nucleotides between the coordinates that you identified in Step 4.iii, from the beginning of the forward primer to the end of the reverse primer.
   iii. Copy and paste the highlighted sequence into a text document. Then, delete all nonnucleotide characters and spaces. This is the amplicon, or amplified PCR product. Save this text document for use in Step 2 of Part II and/or in Step 3 of Part IV.
6. Before proceeding to Part II, repeat Steps 1–5 for the second, third, and fourth primer sets.

II. Digest CAPS Markers Electronically (In Silico)

1. Open http://www.dnalc.org/bioinformatics/nucleotide_analyzer.htm#clipper.
2. Copy the g4026 marker from the sequence repository that you generated in Part I and paste it into the sequence window.
3. Enter the name of the restriction enzyme used to produce the CAPS marker in the experiment.
4. Click "run" and analyze the results.
   i. The total length of the amplicon in nucleotides is at the top left-hand corner of the screen. Record the length in the table below.
   ii. The predicted restriction fragment sizes are listed on the right-hand side of the results screen, under "Fragment sequence" (they are also listed to the right of the "Agarose gel" diagram). Record the lengths of the restriction fragments in the table below.

Marker	Restriction enzyme	Amplicon length (bp)	Restriction fragment lengths (bp)
g4N026	RsaI		
H77224	TaqI		
m235	HindIII		
UFO	TaqI		

5. Repeat Steps 1–4 with the sequences of the remaining three markers.

III. Use Map Viewer to Characterize the *AGO1* Gene

1. Return to the NCBI home page and then click on "Map Viewer," which is located in the "Hot Spots" column on the right.
2. Find "Plants" in the table to the right and click on the triangle-shaped icon to the left of "Plants." Click on "Flowering Plants" and then on "Eudicots." Next, find "*Arabidopsis thaliana*" (thale cress) in the list and click on the magnifying glass icon under the "Tools" column. If more than one build is displayed, select the one with the highest number because this will be the most recent version.
3. Enter "ago1" into the "Search for" window and click "Find" to see the chromosomal location of the *AGO1* gene. Small horizontal bars on chromosomes indicate the positions of hits. To the right of the hits, select "Gene" from the "Quick Filter" list and click "Filter." On which chromosome is the *AGO1* gene located?
4. Click on the marked chromosome number to move to the *AGO1* locus.
5. Examine the structure of the *AGO1* gene. Introns and noncoding sequences of the genes are denoted by thin lines and exons are denoted by thick bars along the line. If needed, you may refine your position on the chromosome by using the zoom toggle on the left or by clicking on the thin vertical gray line along which the genes are annotated.
   i. Determine the size of the *AGO1* gene using the map coordinates to the left of the map, under "Genes_seq."
   ii. How many exons and introns does the *AGO1* gene have?
6. What is the function of the *AGO1* gene? Click on "AGO1" in the pink speed bar (under the "Symbol" track) to find out.
7. Scroll down the page to "Related Sequences" and click on mRNA "U91995.1."
8. At the bottom of the data sheet for U91995.1, extract (highlight and copy) the coding sequence at the bottom of the page. Paste this *AGO1* sequence into the sequence repository that you started in Step 5.iii of Part I. Do not forget to delete all nonnucleotide characters and spaces. You will use this sequence in Step 3 of Part IV.

IV. Use Map Viewer to Determine the Chromosomal Locations of the *AGO1* Gene and the CAPS Markers

1. Open "Map Viewer" (http://www.ncbi.nlm.nih.gov/mapview).
2. Find "Plants" in the table to the right and click on the triangle-shaped icon to the left of "Plants." Click on "Flowering Plants" and then on "Eudicots." Next, find "*Arabidopsis thaliana*" (thale cress) in the list and click on the "B" icon under the "Tools" column. If more than one build is displayed, select the one with the highest number; this will be the most recent version.
3. For each CAPS marker (from Step 5.iii of Part I) and the *AGO1* sequence (from Step 8 of Part III), copy the sequence from your repository and paste it into the search window.
4. Use the default settings and click on "Begin Search."
5. Click on "View report" to retrieve the results.
6. In "Other reports," click on the "[Thale cress genome view]" link to see the chro-

mosomal locations of the BLAST hits. Small horizontal bars on chromosomes indicate the positions of hits. Click on the link to chromosome 1.

7. Click on the pink "Blast hit" links to identify the locations of the *AGO1* sequence and CAPS markers. Record your answers in the table below to the questions that follow.

Locus	Locus order	Map position (Mb)
AGO		
g4026		
H77224		
m235		
UFO		

   i. What is the order of the five search sequences on the chromosome, from top to bottom?

   ii. Use the ruler to the left of the Contig map to determine the map positions of the CAPS markers and the *AGO1* gene. (Keep in mind that M = Mb = mega base pairs = 1 million base pairs.)

   iii. Which CAPS marker is closest to *AGO1*? How many megabases separate them?

RESULTS AND DISCUSSION

I. Interpret Your Gel and Think About the Experiment

1. Observe the photograph of the stained gel containing your PCR samples and those from other students. Orient the photograph with the sample wells at the top. Use the sample gels shown below to help interpret the band(s) in each lane of the gel.

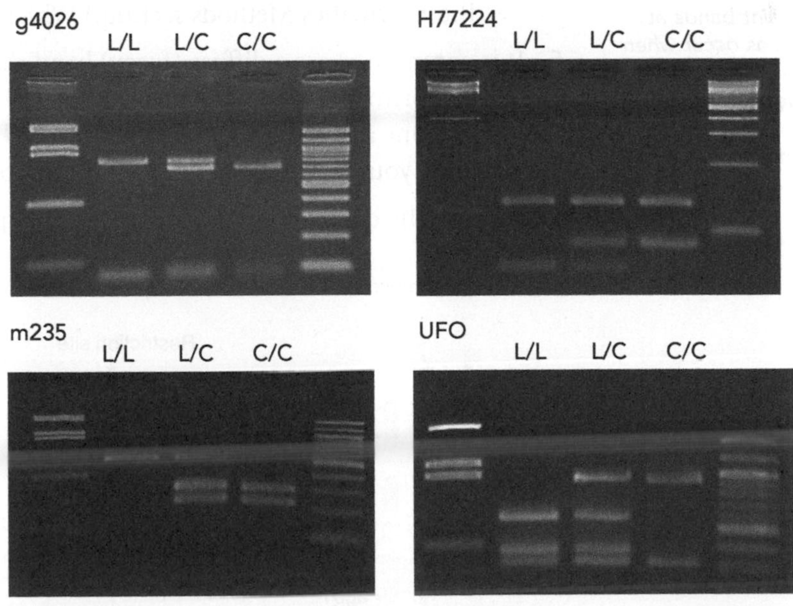

If a 100-bp ladder was used, as shown on the right-hand side of the sample gels, the bands increase in size in 100-bp increments starting with the fastest-migrating band of 100 bp.

2. Locate the lane containing the pBR322/BstNI markers on the left side of the gel. Working from the well, locate the bands corresponding to each restriction fragment: 1857, 1058, 929, 383, and 121 bp. The 1058- and 929-bp fragments will be very close together or may appear as a single large band. The 121-bp band may be very faint or not visible.

3. Carefully examine the banding patterns of the samples in the gels above and identify each of the following fragments in the Ler (L/L) and Col (C/C) controls:

CAPS marker	Col fragments (bp)	Ler fragments (bp)
g4026	668, 94,[a] 78,[a] 4[a]	762, 78,[a] 4[a]
H77224	137, 91	137, 71, 20[a]
m235	301, 230	531
UFO	980, 315	600, 390, 315

[a]May not be visible.

4. Scan across the row of student results. Notice that virtually all lanes contain bands.

   i. Compare the bands present in your *ago1* plant with the Ler and Col controls above and then score the genotype for each of the four CAPS markers as Ler/Ler, Ler/Col, or Col/Col.

CAPS marker	Fragments (bp)	Genotype (Ler/Ler, Ler/Col, or Col/Col)
g4026		
H77224		
m235		
UFO		

   (The incomplete digest of a homozygous marker may leave an additional faint band in the position of the larger parent fragment [higher on the gel], which might be mistaken for a heterozygous result.)

   ii. Compare the results of your electronic (in silico) and electrophoretic (in vitro) analyses. (You generated your in silico results in Part II of the Bioinformatics Methods section.) Which fragments are missing and why?

Additional faint bands at other positions occur when the primers bind to chromosome loci other than the exact locus and give rise to "nonspecific" amplification products.

5. It is common to see a diffuse (fuzzy) band that runs ahead of the 121-bp marker. This is "primer dimer," an artifact of the PCR that results from the primers overlapping one another and amplifying themselves. How would you interpret a lane in which you observe primer dimer but no bands as described in Step 3?

6. Refer to the diagram below and answer the questions that follow.

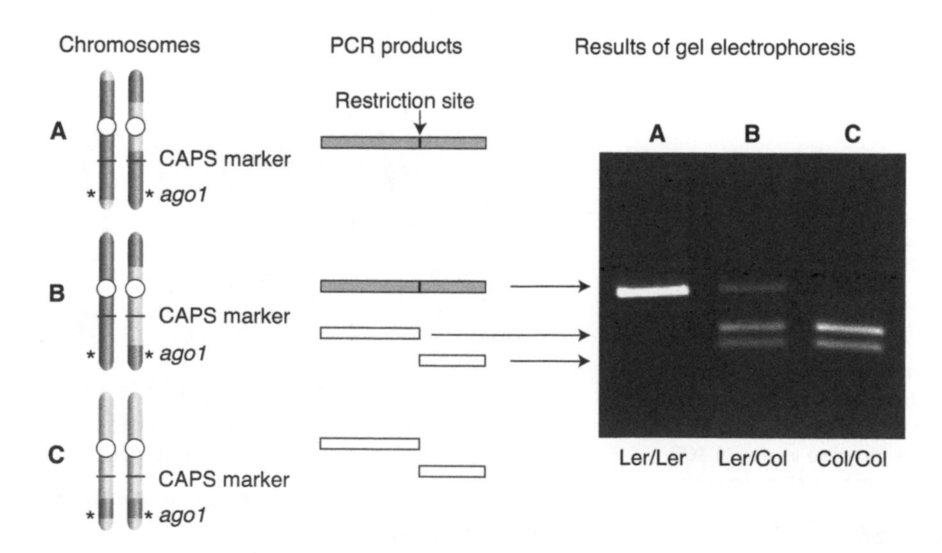

i. From which *Arabidopsis* parent (P) ecotype did all of the mutants analyzed in this experiment inherit the *ago1* mutation?

ii. How many copies of chromosome 1 are assayed in each sample tube?

iii. What is the significance of each of the following genotypes for a given CAPS marker?

- Ler/Ler.
- Col/Col.
- Ler/Col.

II. Calculate Recombination Frequencies and Map the *ago1* Locus

Use the chart below to record your answers to the questions that follow.

	g4026		H77224		m235		UFO	
Genotype	No. of plants	No. of Col alleles	No of plants	No. of Col alleles	No. of plants	No. of Col alleles	No. of plants	No. of Col alleles
Ler/Ler								
Ler/Col								
Col/Col								
Totals								

1. For each of the four CAPS markers, count the number of plants of each genotype: Ler/Ler, Ler/Col, and Col/Col. Exclude from the analysis any samples whose genotypes could not be determined.
2. For each marker, determine the total number of Col alleles as follows:

   i. Multiply the total number of plants of each genotype by the number of Col alleles in that genotype. Remember that each Col/Col plant contributes two copies of the Col allele, whereas each Ler/Col plant contributes one copy of the Col allele.

   ii. Add up the total number of Col alleles.
3. For each marker, determine the total number of alleles by multiplying the total number of plants in the table above by 2. Organize your answers to this and the remaining questions in the table below.

	g4026	H77224	m235	UFO
Total no. of alleles				
Recombination frequency (%)				
Map distance (cM)				

4. The recombination frequency (r) between each CAPS marker and *ago1* is proportional to the number of chromosomes with the Col allele or the Col allele frequency. Calculate the recombination frequency (%) using the following formula:

$$r = \frac{\text{number of Col alleles}}{\text{total number of alleles}} \times 100.$$

5. Next, convert the recombination frequency (%) for each CAPS marker into a map distance (D, in cM) from the *ago1* locus. In *Arabidopsis*, a reasonable estimate of map distance is given by the Kosambi function (ln = natural log):

$$D = 25 \times \ln \left[\frac{(100 + 2r)}{(100 - 2r)}\right].$$

6. The following map shows the locations of the four CAPS makers on *Arabidopsis* chromosome 1. Use the map distances (D) to locate *ago1* on chromosome 1.

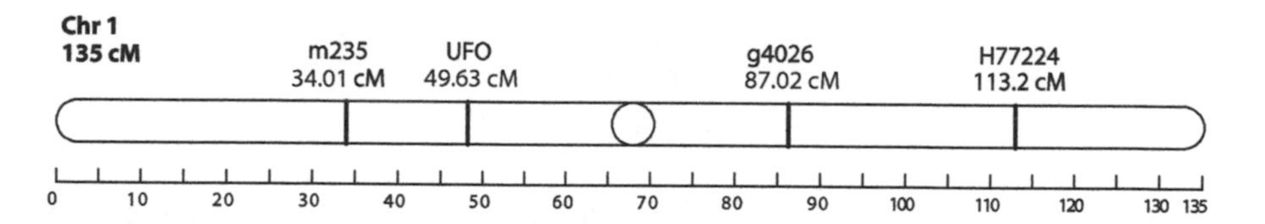

7. Do the results of your mapping agree with the location of *ago1* from the bioinformatics exercises? Explain your reasoning.

LABORATORY 3.4

Detecting Genetically Modified Foods by Polymerase Chain Reaction

▼ OBJECTIVES

This laboratory demonstrates several important concepts of modern biology. During the course of this laboratory, you will

- Learn about the relationship between genotype and phenotype.
- Learn methods for identifying transgenic crops.
- Move between in vitro experimentation and in silico computation.

In addition, this laboratory utilizes several experimental and bioinformatics methods in modern biological research. You will

- Extract and purify DNA from plant tissue.
- Amplify a specific region of the genome by polymerase chain reaction (PCR) and analyze PCR products by gel electrophoresis.
- Use the Basic Local Alignment Search Tool (BLAST) to identify sequences in databases.

INTRODUCTION

During the "green revolution" of the 1950s through the 1970s, high-yielding strains of wheat, corn, and rice, coupled with extensive use of chemical fertilizers, irrigation, mechanized harvesters, pesticides, and herbicides, greatly increased the world food supply. The results were especially dramatic in underdeveloped countries. Now, genetic engineering is fueling a second "green revolution." Genes that encode herbicide resistance, insect resistance, drought tolerance, frost tolerance, and other traits have been added to many plants of commercial importance. In 2003, 167 million acres of farmland worldwide were planted in genetically modified (GM) crops—equal to one-fourth of the total land under cultivation. The most widely planted GM crops are soybeans, corn, cotton, canola, and papaya.

Two important transgenes ("transferred genes") have been widely introduced into crop plants. The *Bt* gene, from *Bacillus thuringiensis*, produces a toxin that protects against caterpillars, reducing applications of insecticides and increasing yields. The glyphosate resistance gene protects food plants against the broad-spectrum herbicide Roundup, which efficiently kills invasive weeds in the field. The major advantages of the "Roundup Ready" system include better weed control, less crop injury, higher yield, and lower environmental impact than traditional weed-control systems. Notably, fields treated with Roundup require less tilling; this preserves soil fertility by lessening soil runoff and oxidation.

Most Americans would probably be surprised to learn that more than 60% of fresh vegetables and processed foods sold in supermarkets today are genetically modified by gene transfer. In 2004, ~85% of soy and 45% of corn grown in the United States were grown from Roundup Ready seed.

This laboratory uses a rapid method to isolate DNA from plant tissue and food products. PCR is then used to assay for evidence of the 35S promoter, which drives expression of the glyphosate resistance gene and many other plant transgenes. Herbicide resistance correlates with the presence of the insertion allele, which is composed of the transgene and the 35S promoter. The presence or absence of this insertion allele is readily detected by electrophoresis on an agarose gel. Amplification of the gene encoding tubulin, a protein found in all plants, provides evidence of amplifiable DNA in the preparation. Tissue samples from wild-type and Roundup Ready soy plants are negative and positive controls, respectively, for the 35S promoter. Because soy and corn are ingredients in many processed foods, it is not difficult to detect the 35S promoter in a variety of food products.

The laboratory includes bioinformatics exercises that complement the experimental methods. BLAST is used to identify sequences in biological databases, make predictions about the outcome of the experiments, and discover some of the genes and functions that are transferred into GM plants.

FURTHER READING

Castle LA, Siehl DL, Gorton R, Patten PA, Chen YH, Bertain S, Cho HJ, Wong ND, Liu D, Lassner MW. 2004. Discovery and directed evolution of a glyphosate tolerance gene. *Science* **304:** 1151–1154.

Edwards K, Johnstone C, Thompson C. 1991. A simple and rapid method for the preparation of plant genomic DNA for PCR analysis. *Nucleic Acids Res* **19:** 1349.

Stalker DM, McBride KE, Malyj LD. 1988. Herbicide resistance in transgenic plants expressing a bacterial detoxification gene. *Science* **242:** 419–423.

Vollenhofer S, Burg K, Schmidt J, Kroath H. 1999. Genetically modified organisms in food screening and specific detection by polymerase chain reaction. *J Agric Food Chem* **47:** 5038–5043.

PLANNING AND PREPARATION

The following table will help you to plan and integrate the different experimental methods.

Experiment	Day	Time	Activity	
I. Plant soybean seeds	2–3 wk before Part II	15 min	Prelab:	Set up student stations.
		15 min	Lab:	Plant soybean seeds.
			Postlab:	Care for soybean plants.
II. Isolate DNA from soybean and food products	1	30 min	Prelab:	Set up student stations.
		30–60 min	Lab:	Isolate DNA.
III. Amplify DNA by PCR	2	15 min	Prelab:	Aliquot primer/loading dye mix. Set up student stations.
		15 min	Lab:	Set up PCRs.
		60–150 min	Postlab:	Amplify DNA in thermal cycler.
IV. Analyze PCR products by gel electrophoresis	3	30 min	Prelab:	Dilute TBE electrophoresis buffer. Prepare agarose gel solution. Set up student stations.
		30 min	Lab:	Cast gels.
	4	45+ min	Lab:	Load DNA samples into gel. Electrophorese samples. Photograph gels.

OVERVIEW OF EXPERIMENTAL METHODS

I. PLANT SOYBEAN SEEDS

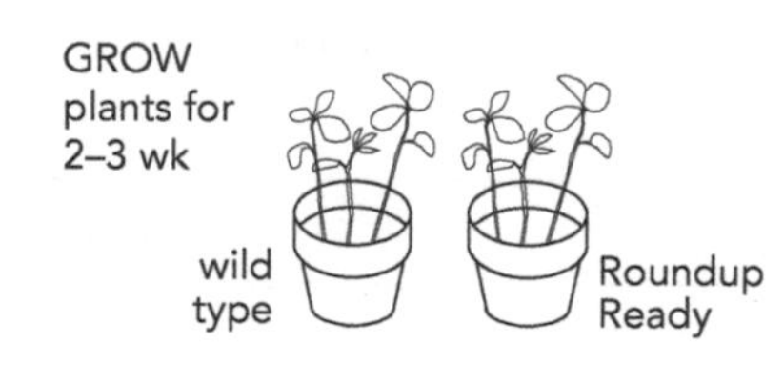

II. ISOLATE DNA FROM SOYBEAN AND FOOD PRODUCTS

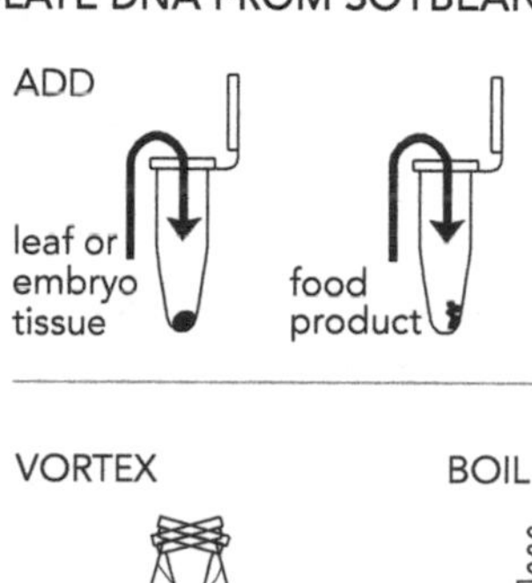

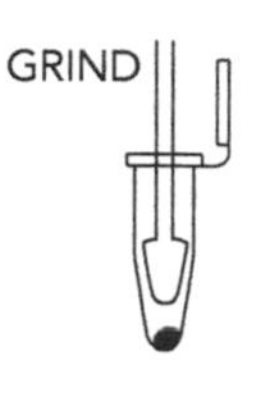

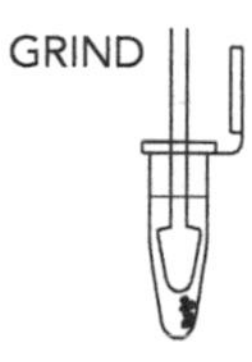

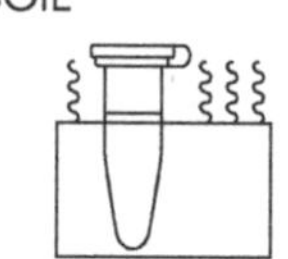

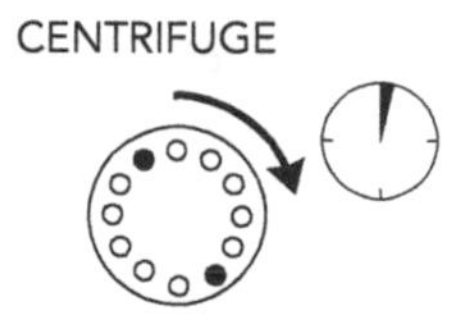

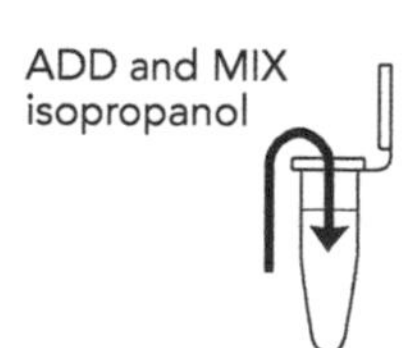

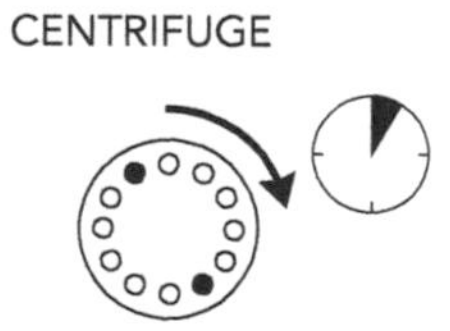

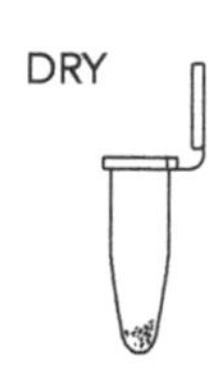

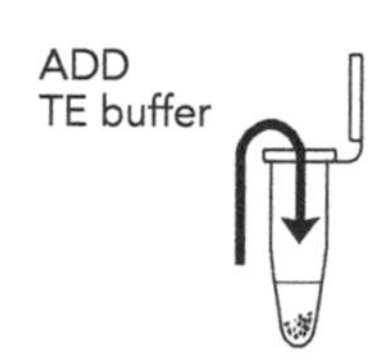

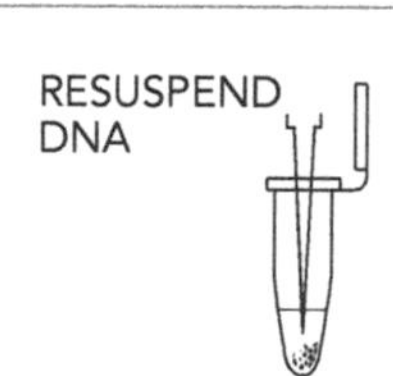

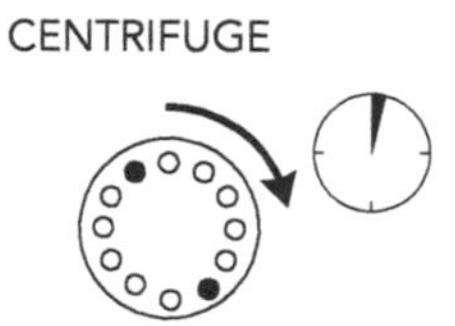

III. AMPLIFY DNA BY PCR

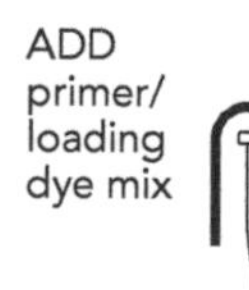

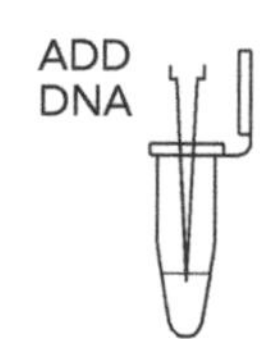

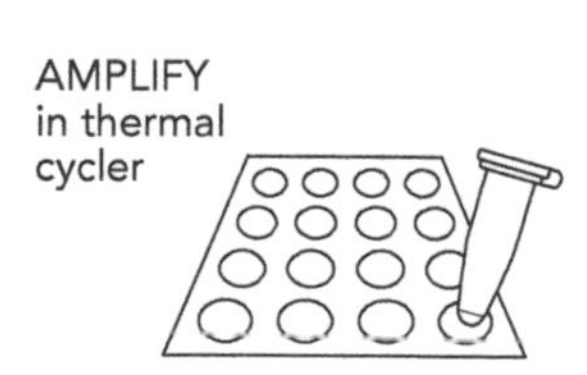

IV. ANALYZE PCR PRODUCTS BY GEL ELECTROPHORESIS

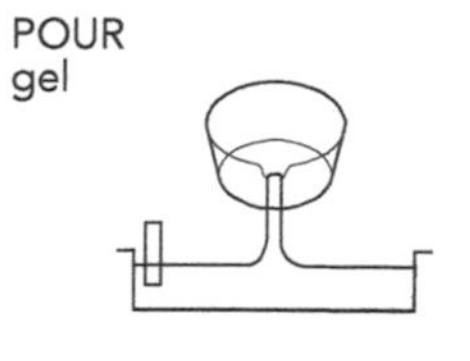

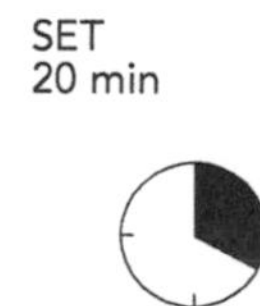

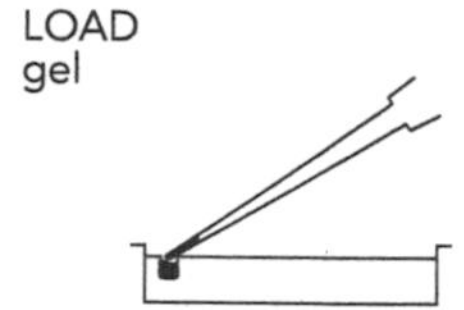

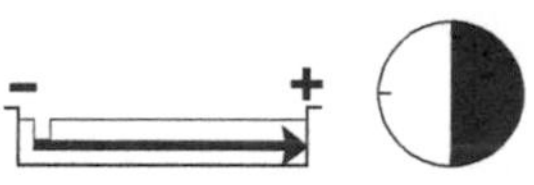

EXPERIMENTAL METHODS

I. Plant Soybean Seeds

To extract DNA from leaf tissue, you must plant the soybean seeds 2–3 wk before DNA isolation and PCR. Two 1/4-in-diameter leaf disks are required for each experiment. Alternatively, DNA can be extracted from the seed embryo without planting; however, this method is less reproducible than obtaining DNA from leaf tissue. To prepare seed embryos for DNA extraction, soak the wild-type and Roundup Ready seeds separately in water for at least 30 min before proceeding to Part II.

REAGENTS, SUPPLIES, & EQUIPMENT

For each group

Planting pot
Plastic dome lid or plastic wrap
Potting soil
3 Seeds from wild-type or Roundup Ready soybean plants
Water

Planting three seeds per pot allows optimal growth and easy observation.

For optimum growth, provide constant (24-h) fluorescent lighting ~1 ft directly above the plants.

1. Label your pot "RR" (for "Roundup Ready") or "WT" (for "wild type"). Moisten the potting soil. Fill the planting pot evenly with soil, but do not pack tightly.
2. Use your finger to make a 0.5-in depression in the soil of your pot. Add three seeds to the hole and cover with soil.
3. Cover the pot with a plastic dome lid or plastic wrap to maintain humidity during germination. (Remove the cover 3–4 d after planting.)
4. Water the plants from above to prevent the soil from drying out. Drain off excess water and do not allow the pot to sit in water.
5. Grow the plants close to a sunny window at room temperature (20°C–22°C) or slightly warmer. If the plants are grown under cold conditions, it may take longer to grow the plants.
6. The first true leaves may be visible 2 wk after planting, depending on light and temperature conditions. These will follow the cotyledons, or seed leaves. When the first true leaves become visible, continue to Part II.

If you plan to test for Roundup sensitivity/resistance, you may wish to continue to grow the plants after you have harvested tissue for DNA isolation and amplification.

II. Isolate DNA from Soybean and Food Products

REAGENTS, SUPPLIES, & EQUIPMENT

For each group

Container with cracked or crushed ice
Edward's buffer (2.2 mL)
Food product containing soy or corn
Isopropanol <!> (1 mL)
Microcentrifuge tube racks
4 Microcentrifuge tubes (1.5 mL)
Micropipettes and tips (100–1000 µL)
Permanent marker
Plastic bag or sheet of paper
2 Plastic pestles
Scalpel or razor blade (for seed embryos only)
Tris/EDTA (TE) buffer (400 µL)
Vortexer (optional)
Wild-type or Roundup Ready soybean plant from Part I or seed embryos

To share

Microcentrifuge
Water bath or heating block at 96°C–100°C

See Cautions Appendix for appropriate handling of materials marked with <!>.

Your instructor will assign either a wild-type or a Roundup Ready soybean plant (or seed embryo) to you. These are the controls. You will share your results with a group that tested the other control.

1. Obtain one soybean plant (or seed embryo) and one food product and record their phenotypes.
2. Prepare tissue from your plant (or seed embryo) and food product as follows:
   i. For soy leaves: From each plant, harvest two pieces of leaf tissue that are ~1/4 in in diameter. (The large end of a 1000-μL pipette tip will punch disks of this size.) Place the tissue into clean 1.5-mL tubes (use one tube for each plant). Label the tubes with your group number and soybean type (use "WT" for "wild type" and "RR" for "Roundup Ready").

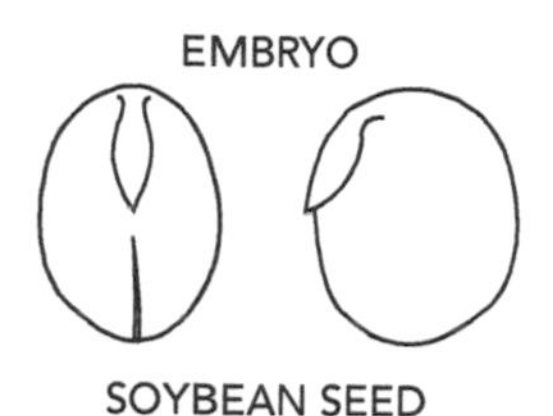

SOYBEAN SEED
with seed coat removed

   ii. For seed embryos: The embryo is a small (3-mm) flap of tissue located beneath the hilum, the light- or dark-colored scar marking where the seed was attached to the pod. Gently remove the seed coat by rubbing the seed between your fingers. Remove the embryo flap with a scalpel or razor blade and place it in a 1.5 mL tube. Label the tube with your group number and soybean type (use "WT" for "wild type" and "RR" for "Roundup Ready").
   iii. For food products: Crush a small amount of dry product on a clean sheet of paper or in a clean plastic bag to produce a coarse powder. Add the crushed food product to a clean 1.5-mL tube to a level that is about halfway to the 0.1-mL mark. Label the tube with your group number and "FP" (for "food product").

Grinding the plant tissue breaks up the cell walls. When fully ground, the sample should be a green liquid.

Detergent in the Edward's buffer, called sodium dodecyl sulfate (SDS), dissolves lipids of the cell membrane.

3. Add 100 μL of Edward's buffer to each tube. Grind the plant tissue by *forcefully* twisting it with a clean plastic pestle against the inner surface of each 1.5-mL tube for 1 min. Use a clean pestle for each tube.
4. Add 900 μL of Edward's buffer to each tube. Grind briefly to remove tissue from the pestle and to liquefy any remaining pieces of tissue (not all of the dry food will liquefy).

Boiling denatures proteins, including enzymes that digest DNA.

5. Vortex the tubes for 5 sec, by hand or machine (if available).
6. Boil the samples for 5 min in a water bath or heating block.

The pellet may appear as a tiny teardrop-shaped smear or particles on the bottom side of the tube underneath the hinge. Do not be concerned if you cannot see a pellet. A large or greenish pellet is cellular debris carried over from the first centrifugation.

7. Place the tubes, along with those from other groups, in a balanced configuration in a microcentrifuge and centrifuge for 2 min to pellet any remaining cell debris. Centrifuge longer if unpelleted debris remains.
8. Label two fresh tubes with your group number and sample code ("WT" or "RR" and "FP"). Transfer 350 μL of each supernatant to the appropriate fresh tube. Be careful not to disturb the pelleted debris when transferring the supernatant. Discard the old tubes containing the debris.
9. Add 400 μL of isopropanol to each tube of supernatant.
10. Close the tubes, mix by inverting several times, and then leave for 3 min at room temperature to precipitate nucleic acids, including DNA.
11. Place your tubes and those of other groups in a balanced configuration in a microcentrifuge, with the hinges of the caps pointing outward. (During centrifugation, nucleic acids will gather on the side of the tube underneath the hinge.) Centrifuge for 5 min at maximum speed to pellet the DNA.

Dry the pellets quickly with a hair dryer. To prevent blowing the pellets away, direct the air across the tube mouths, not into the tubes, for ~3 min.

12. Carefully pour off and discard the supernatant from each tube. Then, *completely* remove the remaining liquid with a micropipette set at 100 μL.
13. Air-dry the pellets by letting the tubes sit open for 10 min. Any remaining isopropanol will evaporate.

The TE (Tris-EDTA) buffer provides conditions for stable storage of DNA. Tris provides a constant pH of 8.0, whereas EDTA binds cations (positive ions) that are required for DNase activity.

In Part III, you will use 2.5 µL of DNA for each PCR. These are crude DNA extracts and contain nucleases that will eventually fragment the DNA at room temperature. Keep the samples cold to limit this activity.

14. Add 100 µL of TE buffer to each tube. Dissolve the pelleted DNA by pipetting in and out, using a fresh tip for each tube. Make sure to wash the TE down the side of the tube underneath the hinge, where the pellet formed during centrifugation. Check that all DNA is dissolved and that no particles remain in the tip or on the side of the tube.
15. Incubate the sample for 5 min at room temperature.
16. Microcentrifuge the tubes for 1 min at maximum speed to pellet any material that did not go into solution. (When using this DNA in Part III, make sure to pipette only the supernatant. This will prevent you from transferring contaminating debris from the pellet to your PCR.)
17. The DNA may be used immediately or stored at –20°C until you are ready to proceed to Part III. Keep the DNA on ice during use.

III. Amplify DNA by PCR

REAGENTS, SUPPLIES, & EQUIPMENT

For each group

Container with cracked or crushed ice
DNA from food product* from Part II
DNA from wild-type or Roundup Ready soybeans* from Part II
Microcentrifuge tube rack
Micropipettes and tips (1–100 µL)
Permanent marker
Primer/loading dye mix* (50 µL) for 35S promoter
Primer/loading dye mix* (50 µL) for tubulin
4 Ready-To-Go PCR Beads in 0.2- or 0.5-mL PCR tubes

To share

Thermal cycler

*Store on ice.

You will set up 35S and tubulin reactions for the food product and the wild-type ("WT") or Roundup Ready ("RR") soybean samples. Carry on with either the WT or RR soybean sample as assigned by your instructor in Part II. These are the controls. You will share your results with a team that tested the other control.

The primer/loading dye mix will turn purple as the PCR bead dissolves.

1. Obtain four PCR tubes containing Ready-To-Go PCR Beads. Label the tubes with your group number. Then, label each tube with the name of the primer set and DNA sample as follows:
    - For the 35S PCRs, write "35S FP" (for "food product") on one tube and either "35S WT" (for "wild-type soy plant") or "35S RR" (for "Roundup Ready soy plant") on the other tube.
    - For the tubulin PCRs, write "T FP" on one tube and either "T WT" or "T RR" on the other tube.
2. Use a micropipette with a fresh tip to add 22.5 µL of the 35S primer/loading dye mix to each of the tubes marked "35S." Then, use a fresh tip to add 22.5 µL of the tubulin primer/loading dye mix to each of the tubes marked "T." Allow the beads to dissolve for ~1 min.
3. Use a micropipette to add 2.5 µL of food product DNA (from Part II) directly into each of the tubes marked "FP." Then, add 2.5 µL of either wild-type or Roundup Ready soybean DNA (from Part II) directly into each of the tubes marked "WT" or "RR," respectively. Use a fresh tip each time and make sure that no DNA remains in the tip after pipetting.
4. Store your samples on ice until your class is ready to begin thermal cycling.
5. Place your PCR tubes, along with those of other groups, in a thermal cycler that has been programmed for 32 cycles of the following profile:

If the reagents become splattered on the wall of the tube, pool them by pulsing the sample in a microcentrifuge or by sharply tapping the tube bottom on the lab bench.

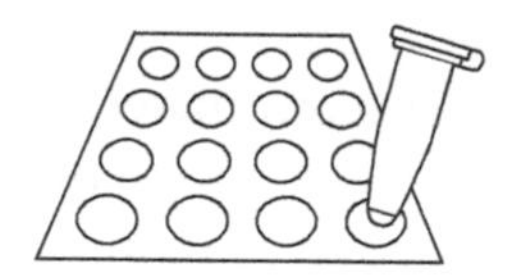

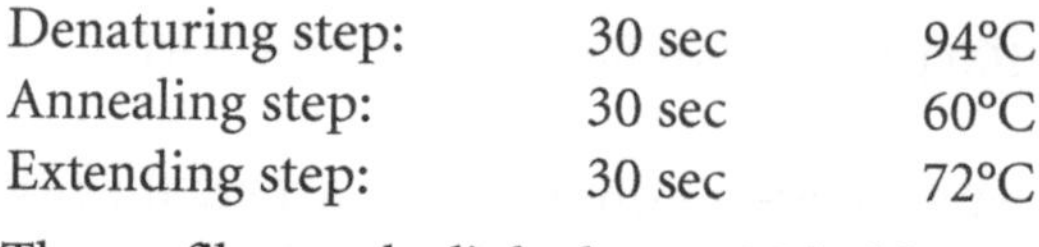

Denaturing step:	30 sec	94°C
Annealing step:	30 sec	60°C
Extending step:	30 sec	72°C

The profile may be linked to a 4°C hold program after the 32 cycles are completed.

6. After thermal cycling, store the amplified DNA on ice or at –20°C until you are ready to proceed to Part IV.

IV. Analyze PCR Products by Gel Electrophoresis

REAGENTS, SUPPLIES, & EQUIPMENT

For each group

2% Agarose in 1x TBE (hold at 60°C) (50 mL per gel)
Container with cracked or crushed ice
Gel-casting tray and comb
Gel electrophoresis chamber and power supply
Latex gloves
Masking tape
Microcentrifuge tube rack
Micropipette and tips (1–100 µL)
pBR322/BstNI marker* (20 µL per gel)
PCR products* from Part III
SYBR Green <!> DNA stain (12 µL)
1x TBE buffer (300 mL per gel)

To share

Digital camera or photodocumentary system
UV transilluminator <!> and eye protection
Water bath for agarose solution (60°C)

*Store on ice.
See Cautions Appendix for appropriate handling of materials marked with <!>.

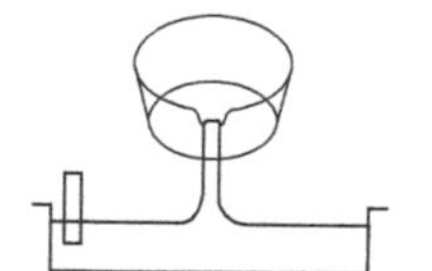

Avoid pouring an overly thick gel, which will be more difficult to visualize.

The gel will become cloudy as it solidifies.

Do not add more buffer than necessary. Too much buffer above the gel channels electrical current over the gel, increasing the running time.

A 100-bp ladder may also be used as a marker.

1. Seal the ends of the gel-casting tray with masking tape and insert a well-forming comb.
2. Pour the 2% agarose solution into the tray to a depth that covers about one-third the height of the open teeth of the comb.
3. Allow the gel to completely solidify; this takes ~20 min.
4. Remove the masking tape, place the gel into the electrophoresis chamber, and add enough 1x TBE buffer to cover the surface of the gel.
5. Carefully remove the comb and add additional 1x TBE buffer to fill in the wells and just cover the gel, creating a smooth buffer surface.
6. Orient the gel according to the diagram in Step 8, so that the wells are along the top of the gel.
7. Use a micropipette with a fresh tip to load 2 µL of SYBR Green DNA stain to each PCR product (from Part III). In addition, add 2 µL of SYBR Green DNA stain to 20 µL of pBR322/BstNI marker for each gel.
8. Use a micropipette with a fresh tip to load each sample from Step 7 into your assigned wells, according to the following diagram. Use a fresh tip for each PCR product.

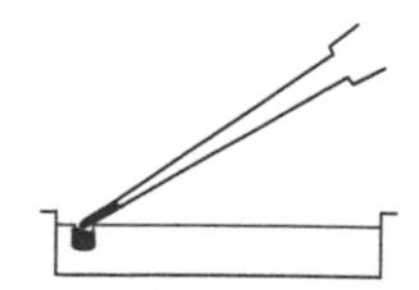

Expel any air from the tip before loading. Be careful not to push the tip of the pipette through the bottom of the sample well.

MARKER	WILD-TYPE SOY		ROUNDUP READY SOY	
pBR322/ BstNI	tubulin	35S	tubulin	35S

MARKER	FOOD 1		FOOD 2	
pBR322/ BstNI	tubulin	35S	tubulin	35S

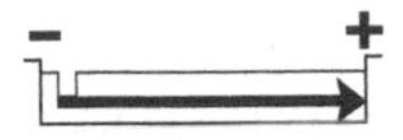

Transillumination, where the light source is below the gel, increases brightness and contrast.

9. Run the gel for ~30 min at 130 V. Adequate separation will have occurred when the cresol red dye front has moved at least 50 mm from the wells.
10. View the gel using UV transillumination. Photograph the gel using a digital camera or photodocumentary system.

BIOINFORMATICS METHODS

For a better understanding of the experiment, perform the following bioinformatics exercises before you analyze your results.

I. Use BLAST to Find DNA Sequences in Databases (Electronic PCR)

1. Perform a BLAST search as follows:
   i. Do an Internet search for "ncbi blast."
   ii. Click on the link for the result "BLAST: Basic Local Alignment Search Tool." This will take you to the Internet site of the National Center for Biotechnology Information (NCBI).
   iii. Click on the link "nucleotide blast" (blastn) under the heading "Basic BLAST."
   iv. Enter the first primer set of sequences into the search window. These are the query sequences.

   > The following primer sets were used in the experiment:
   >
   > 5′-CCGACAGTGGTCCCAAAGATGGAC-3′ (forward primer)
   > 5′-ATATAGAGGAAGGGTCTTGCGAAGG-3′ (reverse primer)
   >
   > 5′-GGGATCCACTTCATGCTTTCGTCC-3′ (forward primer)
   > 5′-GGGAACCACATCACCACGGTACAT-3′ (reverse primer)

   v. Omit any nonnucleotide characters from the window because they will not be recognized by the BLAST algorithm.
   vi. Under "Choose Search Set," select the "Nucleotide collection (nr/nt)" database from the drop-down menu.
   vii. Under "Program Selection," optimize for "Somewhat similar sequences (blastn)."
   viii. Click on "BLAST." This sends your query sequences to a server at NCBI in Bethesda, Maryland. There, the BLAST algorithm will attempt to match the primer sequences to the millions of DNA sequences stored in its database. While searching, a page showing the status of your search will be displayed until your results are available. This may take only a few seconds or more than 1 min if many other searches are queued at the server.

2. Analyze the results of the BLAST search, which are displayed in three ways as you scroll down the page:
   i. First, a graphical overview illustrates how significant matches (hits) align with the query sequence. Matches of differing lengths are indicated by color-coded bars. What do you notice about the lengths (and colors) of the matches (bars) as you look from the top to the bottom?
   ii. This is followed by a list of significant alignments (hits) with links to the corresponding accession numbers. (An accession number is a unique identifier given to a sequence when it is submitted to a database such as GenBank.) Note the scores in the "E value" column on the right. The Expectation or E value is the number of alignments with the query sequence that would be expected to occur by chance in the database. The lower the E value, the higher the probability that the hit is related to the query. For example, an E value of 1 means that a search with your sequence would be expected to turn up one match by chance. Longer query sequences generally yield lower E values. An alignment is considered significant if it has an E value of less than 0.1. What is the E value of the most significant hit and what does it mean? Note the names of any significant alignments that have E values of less than 0.1. Do they make sense? What do they have in common?
   iii. Third is a detailed view of each primer (query) sequence aligned to the nucleotide sequence of the search hit (subject, abbreviated "Sbjct"). Note that some of the hits match to the forward primer (nucleotides 1–24) and to the reverse primer (nucleotides 25–49) within the same subject (accession number). Do the rest of the alignments produce good matches to both primers?
3. Click on the accession number link to open the data sheet for the first hit (the first subject sequence).
   i. At the top of the report, note basic information about the sequence, including its length (in base pairs, or bp), database accession number, source, and references to papers in which the sequence is published. Use the information given here to help you to determine which DNA sequence this primer set amplifies. Does this primer set detect a GM product or is it used to amplify the control sequence?
   ii. In the middle section of the report, the sequence features are annotated, with their beginning and ending nucleotide positions ("xx..xx"). These features may include genes, coding sequences (cds), regulatory regions, ribosomal RNA (rRNA), and transfer RNA (tRNA). You examine these features more closely in Part II.
   iii. Scroll to the bottom of the data sheet. This is the nucleotide sequence to which the term "Sbjct" refers.
   iv. Use your browser's "Back" button to return to the BLAST results page.
4. Predict the length of the product that the primer set would amplify in a PCR (in vitro) as follows:
   i. Scroll down to the alignments section (third section) of the BLAST results page. Examine the alignments for a sequence to which both primers align once each.

ii. To which positions do the primers match in the subject sequence?

iii. The lowest and highest nucleotide positions in the subject sequence indicate the borders of the amplified sequence. Subtract the lowest nucleotide position in the subject sequence from the highest nucleotide position in the subject sequence. What is the difference between the coordinates?

iv. Note that the actual length of the amplified fragment includes both ends, so add 1 nucleotide to the result that you obtained in Step 4.iii to obtain the exact length of the PCR product amplified by the two primers.

5. Obtain the nucleotide sequence of the amplicon that the primer set would amplify in a PCR (in vitro) as follows:

    i. Open the sequence data sheet for the hit that you identified in Step 4.i by clicking on the accession number link. Scroll to the bottom of the data sheet.

    ii. The bottom section of the data sheet lists the entire nucleotide sequence that contains the PCR product. Highlight all of the nucleotides between the coordinates that you identified in Step 4.iii, from the beginning of the forward primer to the end of the reverse primer.

    iii. Copy and paste the highlighted sequence into a text document. Then, delete all nonnucleotide characters and spaces. This is the amplicon, or amplified PCR product. Save this text document for use in Step 1.ii of Part II.

6. Before proceeding to Part II, repeat Steps 1–5 for the second primer set.

II. Use BLAST to Identify Transgenes Driven by the 35S Promoter

In this exercise, you will perform another BLAST search with the sequence of the PCR product produced with the GM primer set. The BLAST search with the GM primer set in Part I above identified numerous cloning vectors that incorporate a 35S promoter. By using the sequence of the PCR product (and not just the primers), you will limit hits from cloning vectors and find out more about transgenes driven by the promoter.

1. Perform a BLAST search as follows:

    i. Follow Steps 1.i–iii in Part I above.

    ii. Copy the sequence of the amplicon for the GM product from the text document (from Step 5.iii in Part I) and paste it into the BLAST search window.

    iii. Under "Choose Search Set," select the "Nucleotide collection (nr/nt)" database from the drop-down menu.

    iv. To limit this search to the green plants, type "viridiplantae" in the "Organism" search box.

    v. Under "Program Selection," optimize for "Somewhat similar sequences (blastn)."

    vi. Click on "BLAST."

2. What do you notice about the E values obtained by this search compared to the E values that you obtained in Step 2.ii of Part I? Why is this so?

3. Scroll down to the alignments section (third section) of the BLAST results page, select one of the hits, and carefully examine the alignment for that hit. The lowest

and highest nucleotide positions in the subject sequence indicate the borders of the PCR product produced with the GM primer set. What are these positions?

4. Click on the accession number link to open the data sheet for the hit. The middle of the report, under "FEATURES," contains annotations of sequence features such as promoters, genes, mRNA sequences, and coding sequences (cds), with their beginning and ending nucleotide positions ("xx..xx"). The use of angle brackets ("<" or ">") indicates that the feature extends beyond the nucleotide position indicated. If this data sheet does not list any features, click on your browser's "Back" button to return to the BLAST results page and repeat Step 3 for a different hit until you find a data sheet with features.
    i. Scan the nucleotide positions ("xx..xx") of all features in the sequence. Use the nucleotide positions that you determined in Step 3 to identify the feature that contains the PCR product. What is the name of this feature?
    ii. Look at the features that are located immediately after the feature that you identified in Step 4.i. Identify the name of the transgene by looking at the annotations. For example, the name of the gene is after "/gene=," and the name of its protein product is after "/product=." What is the name of the transgene in this sequence?
5. Learn more about the transgene that you identified in Step 4.ii by performing a Google search with the name of the gene (or protein). Refine your search by adding the word "gene" to any abbreviation. What did you learn?

RESULTS AND DISCUSSION

I. Think About the Experimental Methods

1. Describe the purpose of each of the following steps or reagents used in DNA isolation (Part II of Experimental Methods):
    i. Grinding with pestle.
    ii. Edward's buffer.
    iii. Boiling.
    iv. Tris-EDTA (TE) buffer.
2. What is the purpose of performing each of the following PCRs?
    i. Tubulin.
    ii. Wild-type soybean.
    iii. Roundup Ready soybean.

II. Interpret Your Gel and Think About the Experiment

If a 100-bp ladder was used, as shown on the right-hand side of the sample gel, the bands increase in size in 100-bp increments starting with the fastest-migrating band of 100 bp.

1. Observe the photograph of the stained gel containing your PCR samples and those from other students. Orient the photograph with the sample wells at the top. Use the sample gels shown below to help interpret the band(s) in each lane of the gel.
2. Locate the lane containing the pBR322/BstNI markers on the left side of the gel. Working from the well, locate the bands corresponding to each restriction frag-

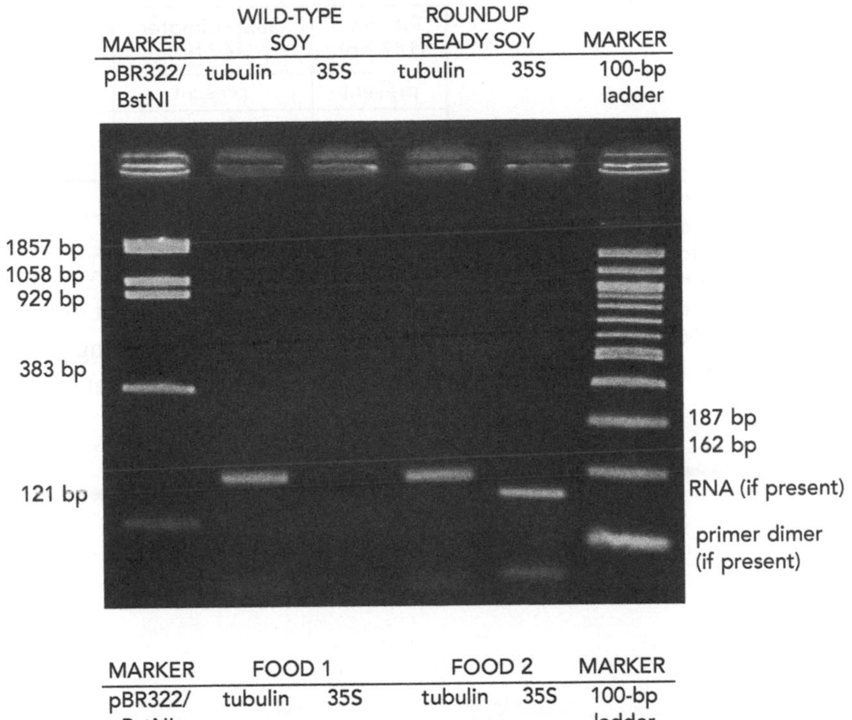

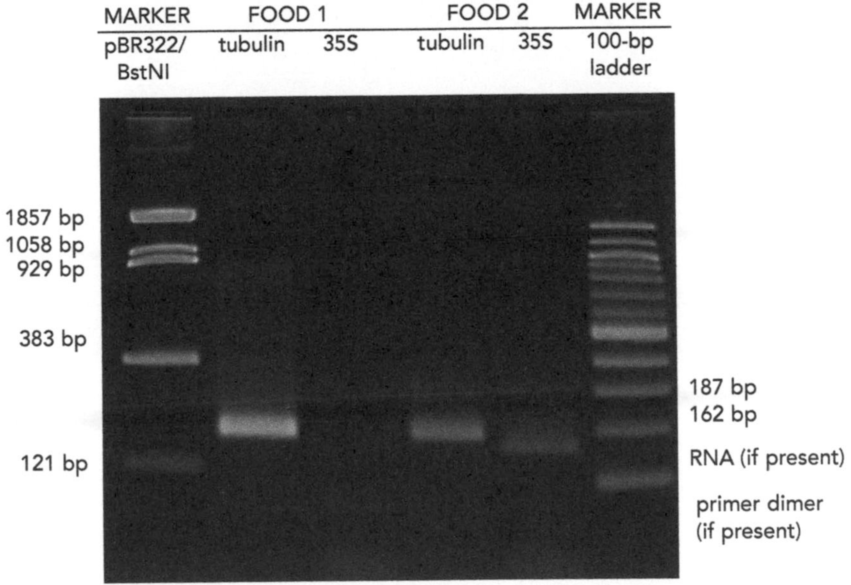

ment: 1857, 1058, 929, 383, and 121 bp. The 1058- and 929-bp fragments will be very close together or may appear as a single large band. The 121-bp band may be very faint or not visible.

3. Scan across the row of student results. Notice that virtually all student lanes contain one prominent band. The amplification products of the 35S promoter (162 bp) and tubulin gene (187 bp) should align between the 121- and 383-bp fragments of the pBR322/BstNI marker.

4. Carefully examine the banding patterns of the four samples in the gels above; foods 1 and 2 are examples of expected results. In the table below, indicate which sample(s) in the gels shows the banding patterns. Explain what each pattern means. (Your own samples may yield any of the banding patterns listed in the table.)

Tubulin (187 bp)	35S Promoter (162 bp)	Sample(s)	Explanation
present	present		
present	absent		
absent	absent		
absent	present		

Additional faint bands at other positions occur when the primers bind to chromosome loci other than the exact locus and give rise to "nonspecific" amplification products.

5. It is common to see a diffuse (fuzzy) band that runs ahead of the 121-bp marker. This is a "primer dimer," an artifact of the PCR that results from the primers overlapping one another and amplifying themselves. How would you interpret a lane in which you observe primer dimer but no bands as described in Step 3?

LABORATORY 3.5

Using DNA Barcodes to Identify and Classify Living Things

▼ OBJECTIVES

This laboratory demonstrates several important concepts of modern biology. During the course of this laboratory, you will

- Collect and analyze sequence data from plants or animals—or products from them.
- Use DNA sequence to identify species.
- Explore relationships among species.

In addition, this laboratory utilizes several experimental and bioinformatics methods in modern biological research. You will

- Collect plants, animals, or products in your local environment or neighborhood.
- Extract and purify DNA from tissue or processed material.
- Amplify a specific region of the chloroplast or mitochondrial genome by polymerase chain reaction (PCR) and analyze PCR products by gel electrophoresis.
- Use the Basic Local Alignment Search Tool (BLAST) to identify sequences in databases.
- Use multiple sequence alignment and tree-building tools (MUSCLE and PHYLIP) to analyze phylogenetic relationships.

INTRODUCTION

Taxonomy, the science of classifying living things according to shared features, has always been a part of human society. Carl Linnaeus formalized biological classification with his system of binomial nomenclature that assigns each organism a genus and species name.

Identifying organisms has grown in importance as we monitor the biological effects of global climate change and attempt to preserve species diversity in the face of accelerating habitat destruction. We know very little regarding the diversity of plants and animals—let alone microbes—living in many unique ecosystems on Earth. Fewer than 2 million of the estimated 5 to 50 million plant and animal species have been identified. Scientists agree that the yearly rate of extinction has increased from about 1 species per million to 100–1000 species per million. This means that thousands of plants and animals are lost each year. Most of these have not yet been identified.

Classical taxonomy falls short in this race to catalog biological diversity before it disappears. Specimens must be carefully collected and handled to preserve their distinguishing features. Differentiating subtle anatomical differences between closely related species requires the subjective judgment of highly trained specialists, and few of these specialists are being educated in colleges today.

Now, DNA barcodes allow nonexperts to identify species objectively, even from small, damaged, or industrially processed material. Just as the unique pattern of bars in a universal product code (UPC) identifies each consumer product, a "DNA barcode" is a unique pattern of DNA sequence that identifies each living thing. Short DNA barcodes, ~700 nucleotides in length, can be quickly processed from thousands of specimens and unambiguously analyzed by computer programs.

The International Barcode of Life (iBOL) organizes collaborators from more than 150 countries to participate in a variety of "campaigns" to census diversity among plant and animal groups—including ants, bees, butterflies, fish, birds, mammals, fungi, and flowering plants—and within ecosystems—including the seas, poles, rain forests, kelp forests, and coral reefs. The 10-year Census of Marine Life, completed in 2010, provided the first comprehensive list of more than 190,000 marine species and identified 6000 potentially new species.

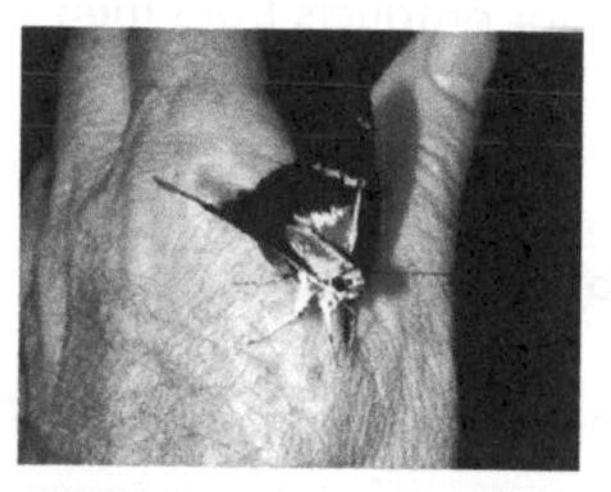

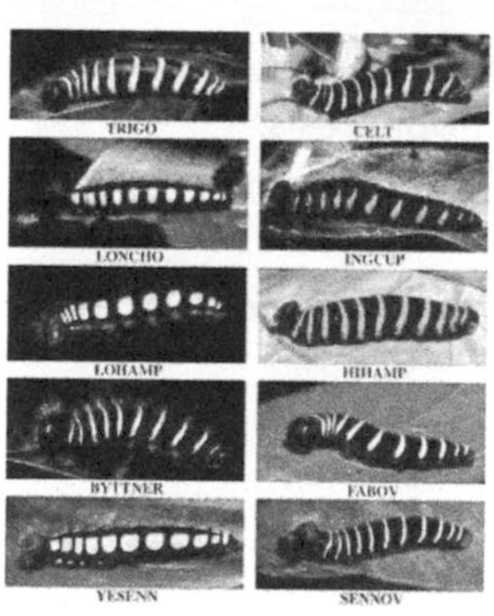

DNA barcoding revealed that what was once thought to be one species of butterfly is really 10 species, with caterpillars that eat different plants. (Hebert et al. 2004.)

There is a surprising level of biological diversity literally in front of our eyes. DNA barcodes have shown that a well-known skipper butterfly (*Astraptes fulgerator*), identified in 1775, is actually 10 distinct species. DNA barcodes have revolutionized the classification of orchids, a complex and widespread plant family with an estimated 20,000 members. DNA barcodes have also been used to catalog 54 species of bees and 24 species of butterflies in community gardens in New York City, showing that the urban environment is unexpectedly diverse.

DNA barcodes have also been used to detect food fraud and products taken from conserved species. Working with researchers from The Rockefeller University and the American Museum of Natural History, students from Trinity High School in New York City found that 25% of 60 seafood items purchased in grocery stores and restaurants in New York City were mislabeled as more expensive species. One mislabeled fish was the Acadian redfish, an endangered species. Another group identified three protected whale species as the source of sushi sold in California and Korea. However, using DNA barcodes to identify potential biological contraband among products seized by customs is still in its infancy.

Barcoding relies on short, highly variable regions of the mitochondrial and chloroplast genomes. With thousands of copies per cell, mitochondrial and chloroplast sequences are readily amplified by PCR, even from very small or degraded specimens. A region of the chloroplast gene *rbcL*, which encodes the RuBisCo large subunit, is used for barcoding plants. The most abundant protein on Earth, RuBisCo (ribulose-1,5-bisphosphate carboxylase oxygenase) catalyzes the first step of photosynthesis. A region of the mitochondrial gene *COI*, which encodes cytochrome *c* oxidase subunit I, is used for barcoding animals. Cytochrome *c* oxidase is involved in the electron transport phase of respiration. Thus, barcode genes are involved in the key reactions of life: storing energy in glucose and releasing it to form ATP.

This laboratory uses DNA barcoding to identify plants or animals (or products made from them) and includes bioinformatics exercises that complement the experimental methods. First, a sample of tissue (e.g., a small leaf disc, a whole insect, or a piece of muscle) is collected; the specimen is preserved whenever possible and its geographical location and local environment are noted. DNA is extracted from the tissue sample, and the barcode portion of the *rbcL* or *COI* gene is amplified by PCR. The amplified sequence (amplicon) is submitted for sequencing in one or both directions.

The sequencing results are then used to search DNA databases. A close match quickly identifies a species that is already represented in a database. However, some barcodes will be entirely new, and identification may rely on placing the unknown species in a phylogenetic tree with close relatives. Novel DNA barcodes can be submitted to the database at the Barcode of Life Data System (BOLD) (http://www.boldsystems.org) at the University of Guelph.

FURTHER READING

Hebert PD, Cywinska A, Ball SL, DeWaard JR. 2003. Biological identifications through DNA barcodes. *Proc R Soc B Biol Sci* **270:** 313–321.

Hebert PD, Penton EH, Burns JM, Janzen DH, Hallwachs W. 2004. Ten species in one: DNA barcoding reveals cryptic species in the neotropical skipper butterfly *Astraptes fulgerator. Proc Natl Acad Sci* **101:** 14812–14817.

Hollingsworth PM, Forrest LL, Spouge JL, Hajibabaei M, Ratnasingham S, van der Bank M, Chase MW, Cowan RS, Erickson DL, Fazekas AJ, et al. (CBOL Plant Working Group). 2009. A DNA barcode for land plants. *Proc Natl Acad Sci* **106:** 12794–12797.

Ratnasingham S, Hebert PDN. 2007. BOLD: The Barcode of Life Data System (http://www.barcodinglife.org). *Mol Ecol Notes* **7:** 355–364.

Stoeckle M. 2003. Taxonomy, DNA, and the bar code of life. *BioScience* **53:** 2–3.

Van Den Berg C, Higgins WE, Dressler RL, Whitten WM, Soto-Arenas MA, Chase MW. 2009. A phylogenetic study of Laeliinae (Orchidaceae) based on combined nuclear and plastid DNA sequences. *Ann Bot* **104:** 417–430.

PLANNING AND PREPARATION

The following table will help you to plan and integrate the different experimental methods.

Experiment	Day	Time		Activity
I. Collect, document, and identify specimens	1	Varies	Lab:	Collect tissue or processed material.
II. Isolate DNA from plant or animal tissue	2	30–60 min	Prelab:	Aliquot reagents and solutions.
				Set up student stations.
		60 min	Lab:	Isolate DNA.
II. (Alternate) Isolate DNA from plant tissue	2	30–60 min	Prelab:	Prepare and aliquot buffers, reagents, and solutions.
				Set up student stations.
		60 min	Lab:	Isolate DNA.
III. Amplify DNA by PCR	3	15 min	Prelab:	Aliquot primer/loading dye mixes.
				Set up student stations.
		15 min	Lab:	Set up PCRs.
		60–150 min	Postlab:	Amplify DNA in thermal cycler.
IV. Analyze PCR products by gel electrophoresis	4	30 min	Prelab:	Dilute TBE electrophoresis buffer.
				Prepare agarose gel solution.
				Set up student stations.
		30 min	Lab:	Cast gels.
	5	45+ min	Lab:	Load DNA samples into gel.
				Electrophorese samples.
				Photograph gels.

OVERVIEW OF EXPERIMENTAL METHODS

I. COLLECT, DOCUMENT, AND IDENTIFY SPECIMENS

COLLECT specimen

DOCUMENT specimen

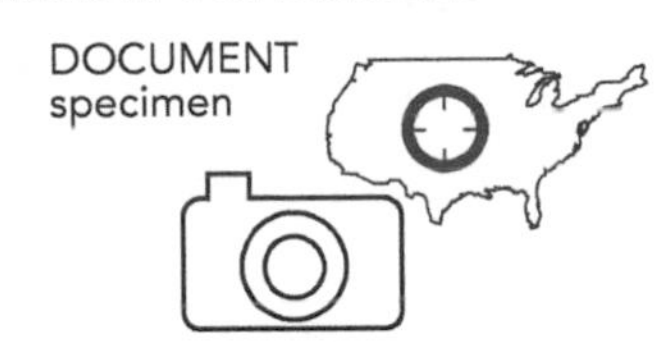

IDENTIFY specimen

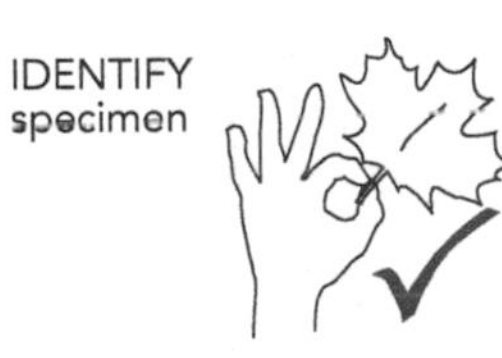

COLLECT tissue sample

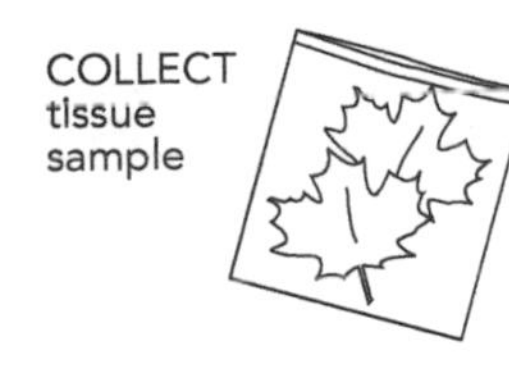

II. ISOLATE DNA FROM PLANT OR ANIMAL TISSUE

ADD tissue

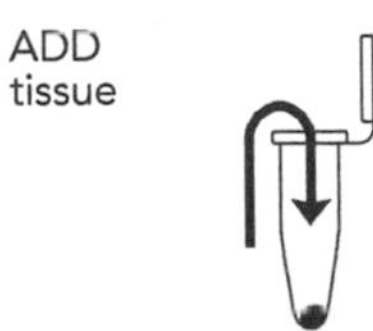

ADD nuclei lysis solution

GRIND sample

INCUBATE at 65°C

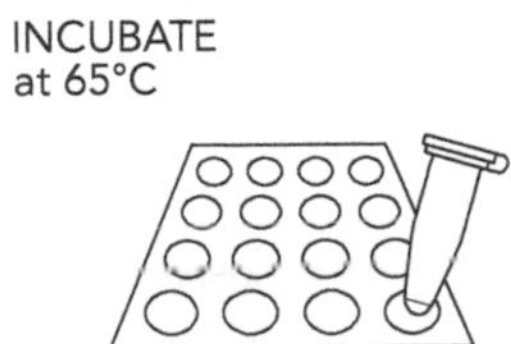

ADD RNase

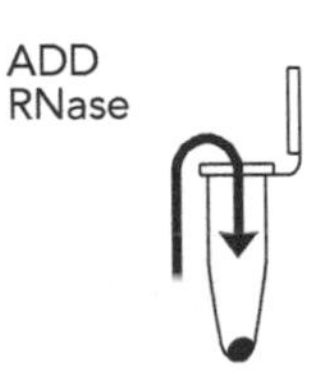

INCUBATE at room temperature

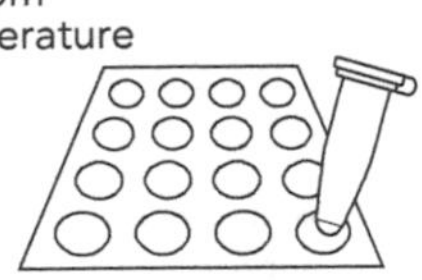

ADD protein precipitation solution

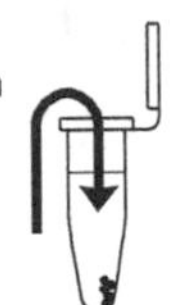

VORTEX

CHILL on ice

CENTRIFUGE

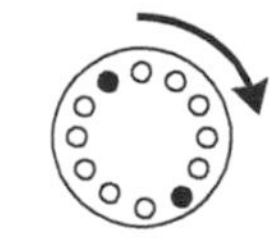

TRANSFER supernatant

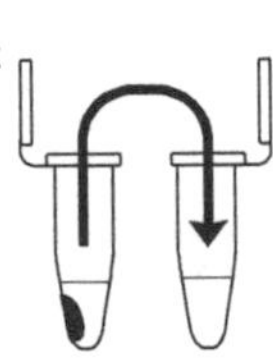

ADD and MIX isopropanol

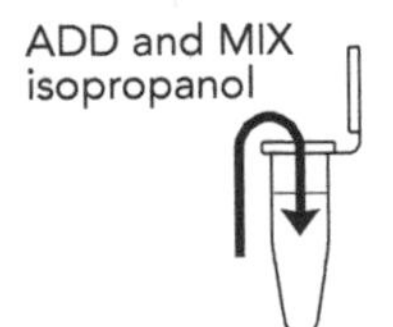

CENTRIFUGE

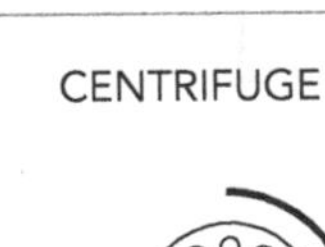
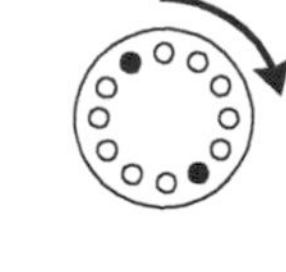

REMOVE supernatant

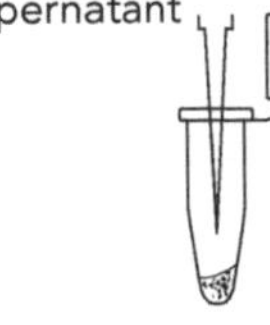

ADD ethanol

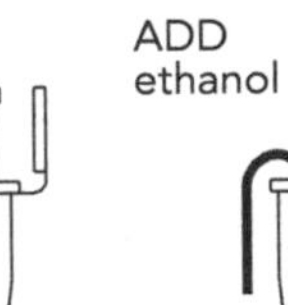
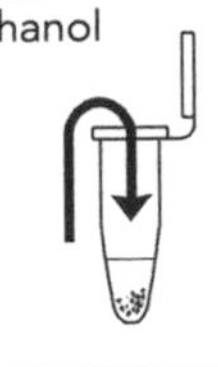

CENTRIFUGE

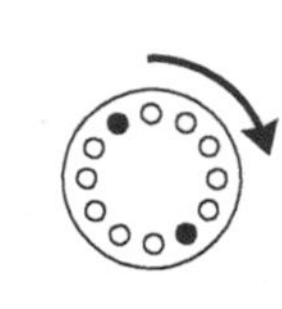

REMOVE ethanol

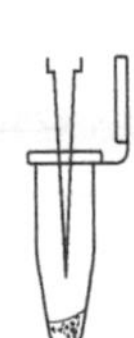
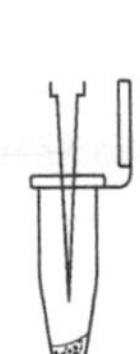

DRY pellet

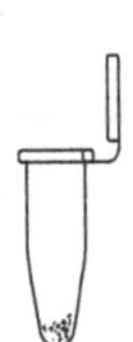
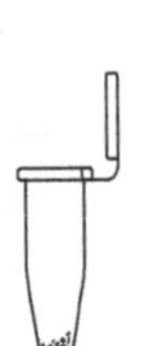

ADD rehydration solution

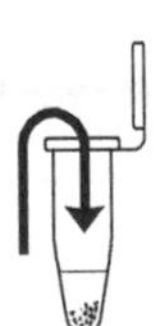

INCUBATE at 65°C or 4°C

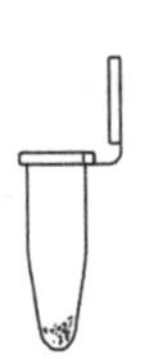

STORE at −20°C

II. (ALTERNATE) ISOLATE DNA FROM PLANT TISSUE

ADD plant tissue

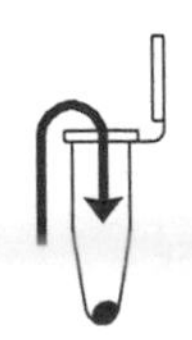

GRIND

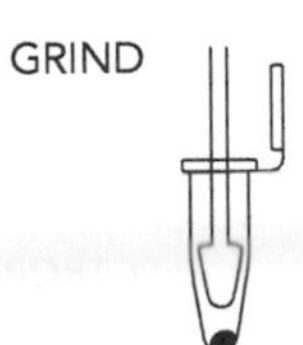

ADD Edward's buffer

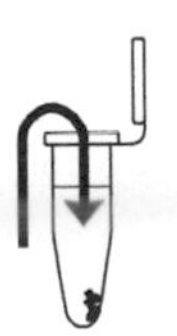

VORTEX

BOIL

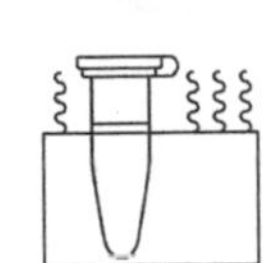

CENTRIFUGE

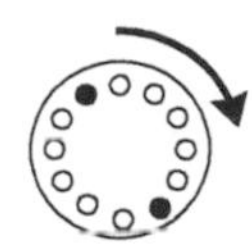

TRANSFER supernatant

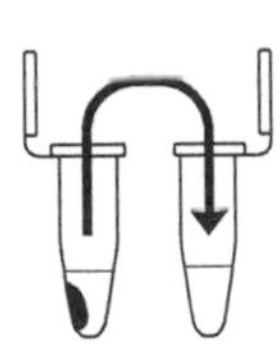

ADD and MIX isopropanol

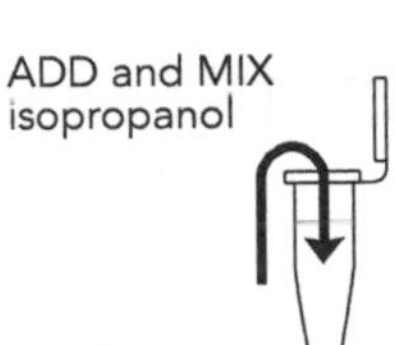

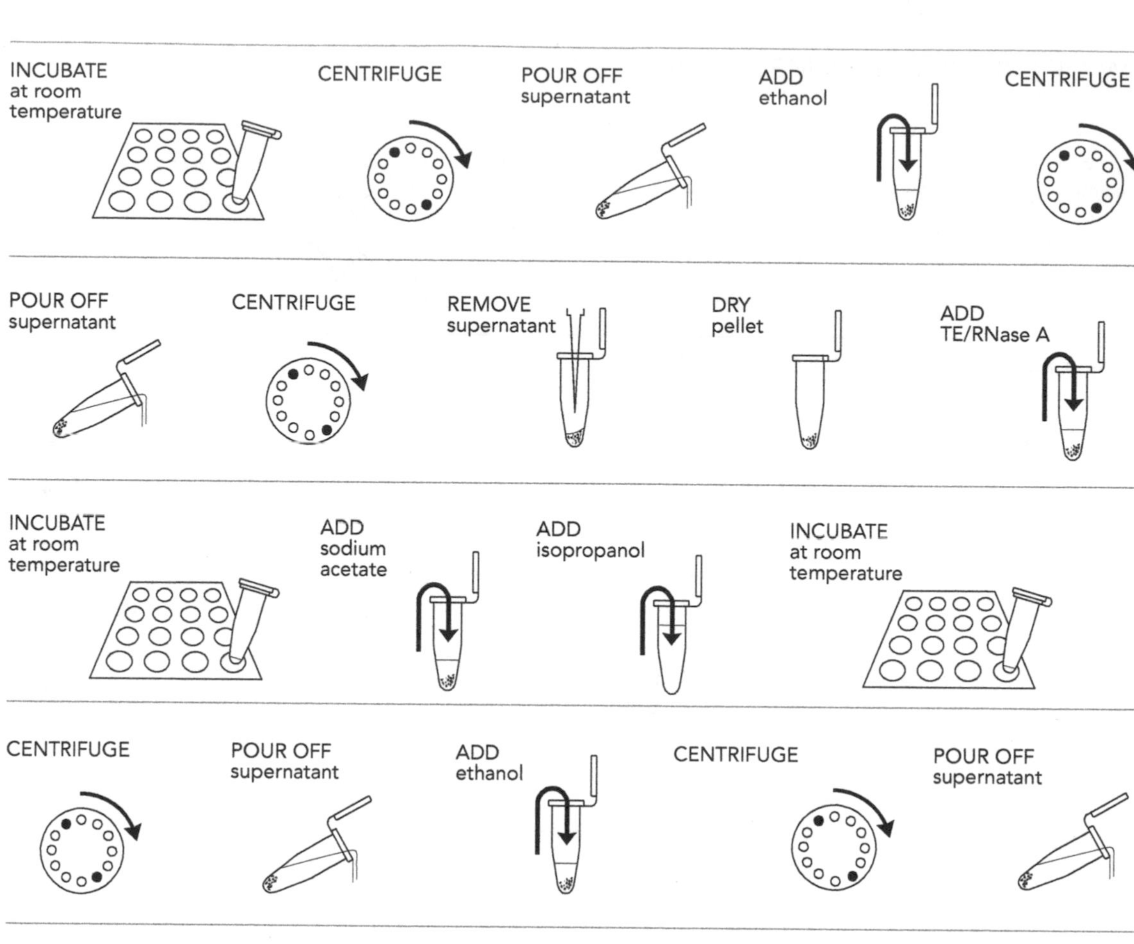

CENTRIFUGE

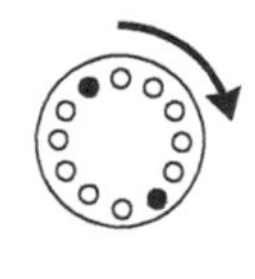

REMOVE
supernatant

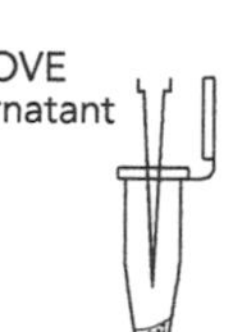

DRY
pellet

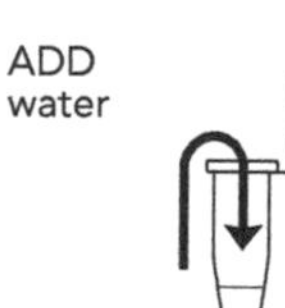
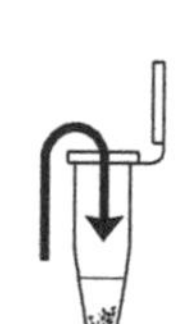

ADD
water

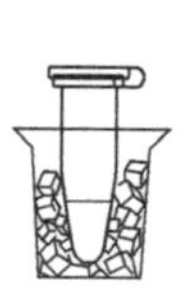

STORE

III. AMPLIFY DNA BY PCR

ADD
primer/
loading
dye mix

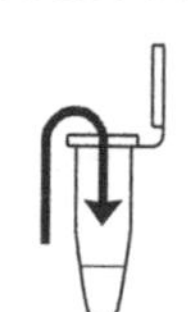

ADD
DNA

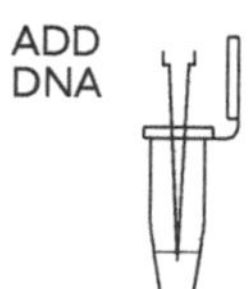

AMPLIFY
in thermal
cycler

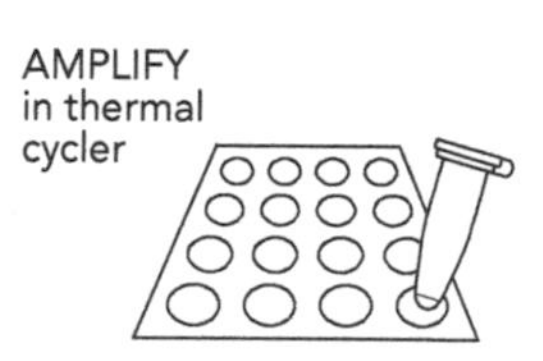

IV. ANALYZE PCR PRODUCTS BY GEL ELECTROPHORESIS

POUR
gel

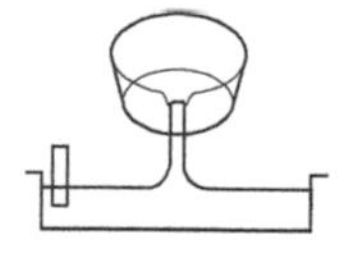

SET
20 min

LOAD
gel

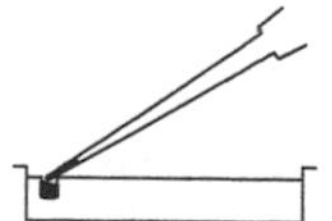

ELECTROPHORESE
130 V

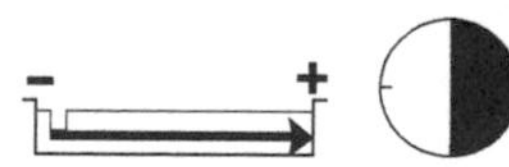

EXPERIMENTAL METHODS

I. Collect, Document, and Identify Specimens

REAGENTS, SUPPLIES, & EQUIPMENT

To share

Collection tubes, jars, or bags
Computer with Internet access
Field guide or taxonomic key
GPS-enabled digital camera or smartphone with camera
Ruler or coin
Tweezers, scalpel, or scissors

If you are participating in a collaborative project, you may be asked to follow a specific procedure to document and identify your specimens.

1. Use a digital camera or smartphone with camera to photograph your specimen in its natural environment or where it was obtained or purchased. Take wide, medium, and close-up views. For scale, include a person in wide and medium shots, and include a ruler or coin in close-ups.
2. Collect the specimen according to a strategy or campaign outlined by your teacher. Use a tweezers, scalpel, or scissors to collect a small sample of tissue. Place it in a tube, collection jar, or bag. Freeze your sample at –20°C until you are ready to begin Part II.
3. A global positioning system (GPS)-enabled phone or camera stores latitude, longitude, and altitude coordinates along with other metadata for each photo taken in Step 1. Using your computer or smartphone, extract and visualize this geotag information.
   - In Apple iPhoto, click on "i" (image properties) to plot the photo on a map. Click on "Photo" and then "Show extended photo info" to find the GPS coordinates.
   - GeoSetter (photo metadata freeware for PCs) will plot your photo on a map.
   - In Google Picasa photo editor, click on "i" to find GPS coordinates.
   - Your smartphone's manual should explain how to use the GPS feature to obtain coordinates. Many smartphones also have applications (apps) that make it easy to harvest GPS coordinates. A smartphone app can record your location continuously, making it easy to document a collection trip or a sampling transect.
4. Share your collection location by dropping a pin on a Google map.
   i. Sign in to your Google Maps account.
   ii. Create and name a new map.
   iii. Zoom in as much as possible on the collection location.
   iv. Click on the blue pin icon to create a pin and then drag it to the location.
   v. Give a title to the pin and add any collection notes in the description field.
   vi. To add a link to a photo or other URL, click on the picture icon under the "Rich text" option.
   vii. Click on "Done" to save your pin drop.
   viii. Click on "Collaborate" to share your map with others.
5. Use a field guide or taxonomic key to identify your specimen as precisely as possible: kingdom > phylum > class > order > family > genus > species.

As an alternative to Step 6.i, click on "Taxonomy" to explore barcode records by group.

6. Check to see whether your specimen is represented in the Barcode of Life Database, BOLD (http://www.boldsystems.org).
   i. From the Bold Systems home page, click on "Public Data Portal."
   ii. Search by entering taxonomic names in the search bar. If the species or genus is represented in the database, the search will list relevant records. Follow links to records and to access published and released data.
   iii. Click on the box beside records to select the data. To download sequences, click on "FASTA" to the right of the search bar and save the FASTA file. You will use these sequences in Part III of the bioinformatics exercises.

II. Isolate DNA from Plant or Animal Tissue

REAGENTS, SUPPLIES, & EQUIPMENT

For each group

Container with cracked or crushed ice
DNA rehydration solution (250 µL)
Ethanol (70%) <!> (1.3 mL)
Isopropanol <!> (1.3 mL)
Microcentrifuge tube racks
4 Microcentrifuge tubes (1.5 mL)
Micropipettes and tips (1–1000 µL)
Nuclei lysis solution (1.2 mL)*
Permanent marker
2 Plastic pestles
Protein precipitation solution (0.5 mL)
RNase solution (7 µL)
2 Specimens from Part I
Vortexer (optional)

To share

Microcentrifuge
Water baths or heating blocks at 37°C and 65°C

*Store on ice.
See Cautions Appendix for appropriate handling of materials marked with <!>.

If you only have one specimen, make a duplicate prep to provide a balance for centrifuge steps.

Nuclei lysis solution dissolves membrane-bound organelles, including the nucleus, mitochondria, and chloroplast.

1. Harvest plant tissue (two pieces, ~1/4 in in diameter) or animal tissue (~10–20 mg). (The large end of a 1000-µL pipette tip will punch plant tissue discs of ~1/4 in in diameter. Animal tissue should be about the size of a pencil eraser.) Place the tissue into clean 1.5-mL tubes (use one tube for each specimen). Label the tubes with your group number and a specimen identification number.
2. Add 100 µL of nuclei lysis solution to each tube.
3. Grind the tissue by forcefully twisting it with a clean plastic pestle against the inner surface of each 1.5-mL tube for 1 min. Use a clean pestle for each tube.
4. Add 400 µL of nuclei lysis solution to each tube.
5. Incubate the tubes in a water bath or heat block for 15 min at 65°C.
6. Add 3 µL of RNase solution to each tube. Close the caps and mix by rapidly inverting the tubes several times.
7. Incubate the tubes in a water bath or a heating block for 15 min at 37°C. Then, leave the tubes for 5 min at room temperature.
8. Add 200 µL of protein precipitation solution to each tube. Vortex the tubes for 5 sec by hand or machine (if available).
9. Incubate the tubes for 5 min on ice.
10. Place your tubes and those of other groups in a balanced configuration in a microcentrifuge, with the hinges of the caps pointing outward. Centrifuge for 4 min at maximum speed to pellet the protein. Discard the old tubes containing the debris.

11. Label two fresh tubes with your group number and specimen identification number. Transfer 600 µL of each supernatant to the appropriate fresh tube. Be careful not to disturb the pelleted debris when transferring the supernatant. Discard old tubes containing the debris.

12. Add 600 µL of isopropanol to each tube of supernatant, close the tubes, and mix by rapidly inverting several times.

The pellet may appear as a tiny teardrop-shaped smear or particles on the bottom side of the tube underneath the hinge. Do not be concerned if you cannot see a pellet.

13. Place your tubes and those of other groups in a balanced configuration in a microcentrifuge, with the hinges of the caps pointing outward. Centrifuge for 1 min at maximum speed to pellet the DNA.

14. Carefully pour off and discard the supernatant from each tube. Then, add 600 µL of 70% ethanol to each tube, close the caps, and flick the bottom of each tube several times to "wash" the pellet.

Nucleic acids are not soluble in ethanol and will not dissolve during washing.

15. Place your tubes and those of other groups in a balanced configuration in a microcentrifuge, with the hinges of the caps pointing outward. Centrifuge for 1 min at maximum speed.

16. Carefully pour off and discard the supernatant from each tube. Then, completely remove the remaining liquid with a micropipette set at 100 µL.

Dry the pellets quickly with a hair dryer. To prevent blowing the pellets away, direct the air across the tube mouths, not into the tubes, for ~3 min.

17. Air-dry the pellets by allowing the tubes to sit open for 10 min. Any remaining ethanol will evaporate.

18. Add 100 µL of DNA rehydration solution to each tube. Dissolve the pelleted DNA by pipetting in and out, using a fresh tip for each tube. Make sure to wash the solution down the side of the tube underneath the hinge, where the pellet formed during centrifugation.

19. Incubate the DNA for 45–60 min at 65°C, or overnight at 4°C.

In Part III, you will use 2 µL of DNA for each PCR. This is a crude DNA extract and contains nucleases that will eventually fragment the DNA at room temperature. Keep the sample cold to limit this activity.

20. The DNA may be used immediately or stored at –20°C until you are ready to proceed to Part III. Keep the DNA on ice during use.

II. (Alternate) Isolate DNA from Plant Tissue

REAGENTS, SUPPLIES, & EQUIPMENT

For each group

Container with cracked or crushed ice
Deionized or distilled H_2O (1.2 mL)
Edward's buffer (2.2 mL)
Ethanol (70%) <!> (2.2 mL)
Isopropanol <!> (2.2 mL)
Microcentrifuge tube racks
4 Microcentrifuge tubes (1.5 mL)
Micropipettes and tips (100–1000 µL)
Permanent marker
2 Plant specimens from Part I
2 Plastic pestles
Sodium acetate (3 M; 150 µL)
Tris/EDTA (TE) buffer with RNase A (250 µL)
Vortexer (optional)

To share

Microcentrifuge
Water bath or heating block at 96°C–100°C

See Cautions Appendix for appropriate handling of materials marked with <!>.

If you only have one specimen, make a duplicate prep to provide a balance for centrifuge steps.

1. Harvest two pieces of tissue ~1/4 in in diameter from each plant. (The large end of a 1000-µL pipette tip will punch discs of this size.) Place the tissue into clean 1.5-mL tubes (use one tube for each plant). Label the tubes with your group number and a specimen identification number.

Grinding the plant tissue breaks up the cell walls. When fully ground, the sample should be a green liquid.

Detergent in the Edward's buffer, called sodium dodecyl sulfate (SDS), dissolves lipids of the cell membrane.

Boiling denatures proteins, including enzymes that digest DNA.

2. Add 100 μL of Edward's buffer to each tube. Grind the plant tissue by *forcefully* twisting it with a clean plastic pestle against the inner surface of each 1.5-mL tube for 1 min. Use a clean pestle for each tube.
3. Add 900 μL of Edward's buffer to each tube. Grind briefly to remove tissue from the pestle and to liquefy any remaining pieces of tissue.
4. Vortex the tubes for 5 sec, by hand or machine (if available).
5. Boil the samples for 5 min in a water bath or heating block.
6. Place the tubes, along with those from other groups, in a balanced configuration in a microcentrifuge, and centrifuge for 2 min to pellet any remaining cell debris. Centrifuge longer if unpelleted debris remains.
7. Label two fresh tubes with your group number and specimen identification number. Transfer 350 μL of each supernatant to the appropriate fresh tube. Be careful not to disturb the pelleted debris when transferring the supernatant. Discard old tubes containing the debris.
8. Add 400 μL of isopropanol to each tube of supernatant.
9. Close the tubes, mix by inverting several times, and then leave for 3 min at room temperature to precipitate nucleic acids, including DNA.

The pellet may appear as a tiny teardrop-shaped smear or particles on the bottom side of the tube underneath the hinge. Do not be concerned if you cannot see a pellet. A large or greenish pellet is cellular debris carried over from the first centrifugation.

10. Place your tubes and those of other groups in a balanced configuration in a microcentrifuge, with the hinges of the caps pointing outward. (During centrifugation, nucleic acids will gather on the side of the tube underneath the hinge.) Centrifuge for 5 min at maximum speed to pellet the DNA.
11. Carefully pour off and discard the supernatant from each tube. Then, add 500 μL of 70% ethanol to each tube, close the caps, and flick the bottom of each tube several times to "wash" the pellet.
12. Place your tubes and those of other groups in a balanced configuration in a microcentrifuge, with the hinges of the caps pointing outward. Centrifuge for 1 min at maximum speed.
13. Carefully pour off and discard the supernatant from each tube. Centrifuge the tubes again for 30 sec at maximum speed to force any remaining ethanol to the bottom.
14. Completely remove the remaining liquid with a micropipette set at 100 μL. Air-dry the pellets by allowing the tubes to sit open for 10 min. Any remaining ethanol will evaporate.
15. Add 100 μL of TE/RNase A buffer to each tube. Dissolve the pelleted DNA by pipetting in and out, using a fresh tip for each tube. Make sure to wash the buffer down the side of the tube underneath the hinge, where the pellet formed during centrifugation.
16. Incubate the tubes for 5 min at room temperature.
17. Add 400 μL of H_2O, 50 μL of 3 M sodium acetate, and 550 μL of isopropanol to each tube to precipitate the DNA.
18. Close the caps, mix by inverting several times, and then leave the tubes for 3 min at room temperature.

19. Place your tubes and those of other groups in a balanced configuration in a microcentrifuge, with the hinges of the caps pointing outward. Centrifuge for 5 min at maximum speed.

Nucleic acids are not soluble in ethanol and will not dissolve during washing.

20. Carefully pour off and discard the supernatant from each tube. Then, add 500 µL of 70% ethanol to each tube, close the caps, and flick the bottom of each tube several times to "wash" the pellet.

21. Place your tubes and those of other groups in a balanced configuration in a microcentrifuge, with the hinges of the caps pointing outward. Centrifuge for 1 min at maximum speed.

22. Carefully pour off and discard the supernatant from each tube. Centrifuge the tubes again for 30 sec at maximum speed to force any remaining ethanol to the bottom.

Dry the pellets quickly with a hair dryer. To prevent blowing the pellets away, direct the air across the tube mouths, not into the tubes, for ~3 min.

23. Completely remove the remaining liquid with a micropipette set at 100 µL. Air-dry the pellets by allowing the tubes to sit open for 10 min. Any remaining ethanol will evaporate.

24. Add 100 µL of H_2O to each tube and dissolve the DNA pellet by pipetting in and out several times.

In Part III, you will use 2 µL of DNA for each PCR. This is a crude DNA extract and contains nucleases that will eventually fragment the DNA at room temperature. Keep the sample cold to limit this activity.

25. The DNA may be used immediately or stored at –20°C until you are ready to proceed to Part III. Keep the DNA on ice during use.

III. Amplify DNA by PCR

REAGENTS, SUPPLIES, & EQUIPMENT

For each group

Container with cracked or crushed ice
DNA from specimen of interest* from Part II
Microcentrifuge tube rack
Micropipettes and tips (1–100 µL)
Permanent marker
Primer/loading dye mix (25 µL)*
2 Ready-To-Go PCR Beads in 0.2- or 0.5-mL PCR tubes

To share

Thermal cycler

*Store on ice.

1. Obtain two PCR tubes containing Ready-To-Go PCR Beads. Label the tubes with your group number and specimen identification number.

The primer/loading dye mix will turn purple as the PCR bead dissolves.

2. Use a micropipette with a fresh tip to add 23 µL of the primer/loading dye mix to each tube. Allow the beads to dissolve for ~1 min.

If the reagents become splattered on the wall of the tube, pool them by pulsing the sample in a microcentrifuge or by sharply tapping the tube bottom on the laboratory bench.

3. Use a micropipette with a fresh tip to add 2 µL of your DNA (from Part II) directly into the appropriate primer/loading dye mix. Ensure that no DNA remains in the tip after pipetting.

4. Store your samples on ice until your class is ready to begin thermal cycling.

5. Place your PCR tubes, along with those of other groups, in a thermal cycler that has been programmed for 35 cycles of the following profile:

Denaturing step:	15 sec	94°C
Annealing step:	15 sec	54°C
Extending step:	30 sec	72°C

The profile may be linked to a 4°C hold program after the 35 cycles are completed.

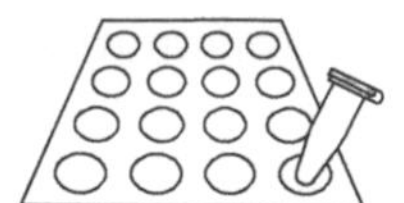

6. After thermal cycling, store the amplified DNA on ice or at –20°C until you are ready to proceed to Part IV.

IV. Analyze PCR Products by Gel Electrophoresis

REAGENTS, SUPPLIES, & EQUIPMENT

For each group

2% Agarose in 1x TBE (hold at 60°C) (50 mL per gel)
Container with cracked or crushed ice
Gel-casting tray and comb
Gel electrophoresis chamber and power supply
Latex gloves
Masking tape
Microcentrifuge tube rack
2 Microcentrifuge tubes (1.5 mL)
Micropipette and tips (1–100 μL)
pBR322/BstNI marker (20 μL per gel)*
PCR products* from Part II
SYBR Green <!> DNA stain (10 μL)
1x TBE buffer (300 mL per gel)

To share

Digital camera or photodocumentary system
UV transilluminator <!> and eye protection
Water bath for agarose solution (60°C)

*Store on ice.
See Cautions Appendix for appropriate handling of materials marked with <!>.

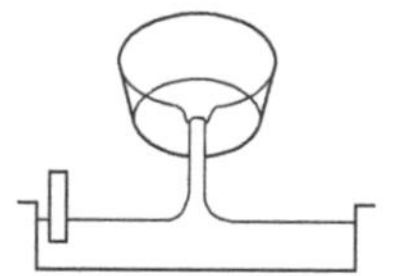

Avoid pouring an overly thick gel, which will be more difficult to visualize.

The gel will become cloudy as it solidifies.

Do not add more buffer than necessary. Too much buffer above the gel channels electrical current over the gel, increasing the running time.

A 100-bp ladder may also be used as a marker.

1. Seal the ends of the gel-casting tray with masking tape and insert a well-forming comb.
2. Pour the 2% agarose solution into the tray to a depth that covers about one-third the height of the open teeth of the comb.
3. Allow the gel to completely solidify; this takes ~20 min.
4. Remove the masking tape, place the gel into the electrophoresis chamber, and add enough 1X TBE buffer to cover the surface of the gel.
5. Carefully remove the comb and add additional 1X TBE buffer to fill in the wells and just cover the gel, creating a smooth buffer surface.
6. Orient the gel according to the diagram in Step 8, so that the wells are along the top of the gel.
7. Use a micropipette with a fresh tip to transfer 5 μL of each of your PCR products (from Part III) to fresh 1.5-mL microcentrifuge tubes. Add 2 μL of SYBR Green DNA stain to each tube. In addition, add 2 μL of SYBR Green DNA stain to 20 μL of pBR322/BstNI marker.
8. Use a micropipette with a fresh tip to load each sample from Step 7 into your assigned wells, according to the following diagram:

MARKER	SAMPLES					
pBR322/ BstNI	1	2	3	4	5	6

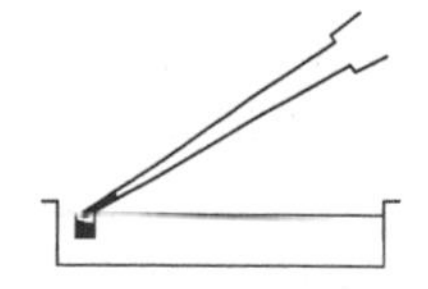

Expel any air from the tip before loading, and be careful not to push the tip of the pipette through the bottom of the sample well.

Store the remaining 20 μL of your PCR product on ice for subsequent use in DNA sequencing.

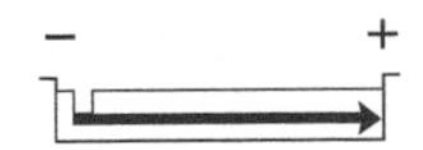

Transillumination, where the light source is below the gel, increases brightness and contrast.

9. Run the gel for ~30 min at 130 V. Adequate separation will have occurred when the cresol red dye front has moved at least 50 mm from the wells.
10. View the gel using UV transillumination. Photograph the gel using a digital camera or photodocumentary system.

BIOINFORMATICS METHODS

I. Use BLAST to Find DNA Sequences in Databases (Electronic PCR)

1. Perform a BLAST search as follows:
   i. Perform an Internet search for "ncbi blast."
   ii. Click on the link for the result "BLAST: Basic Local Alignment Search Tool." This will take you to the Internet site of the National Center for Biotechnology Information (NCBI).
   iii. Click on the link "nucleotide blast" (blastn) under the heading "Basic BLAST."
   iv. Enter the primer set you used into the search window. These are the query sequences.

> The following primer sets may have been used in the experiment:
>
> Plant *rbcL*
> rbcLAf, 5′-ATGTCACCACAAACAGAGACTAAAGC-3′ (forward primer)
> rbcLa rev, 5′-GTAAAATCAAGTCCACCRCG-3′ (reverse primer)
>
> Animal *COI*
> LepF1, 5′-ATTCAACCAATCATAAAGATATTGG-3′ (forward primer)
> LepR1, 5′-TAAACTTCTGGATGTCCAAAAAATCA-3′ (reverse primer)
>
> VF1F, 5′-TCTCAACCAACCACAAAGACATTGG-3′ (forward primer)
> VF1R, 5′-TAGACTTCTGGGTGGCCAAAGAATCA-3′ (reverse primer)

   v. Omit any nonnucleotide characters from the window because they will not be recognized by the BLAST algorithm.
   vi. Under "Choose Search Set," select "NCBI Genomes (chromosome)" database from the drop-down menu.
   vii. Under "Program Selection," optimize for "Somewhat similar sequences (blastn)."
   viii. Click on "BLAST." This sends your query sequences to a server at NCBI in Bethesda, Maryland. There, the BLAST algorithm will attempt to match the primer sequences to the millions of DNA sequences stored in its database. While searching, a page showing the status of your search will be displayed until your results are available. This may take only a few seconds or more than 1 min if many other searches are queued at the server.

If a 100-bp ladder was used, as shown on the right-hand side of the sample gel, the bands increase in size in 100-bp increments starting with the fastest-migrating band of 100 bp.

Additional faint bands at other positions occur when the primers bind to chromosome loci other than the exact locus and give rise to "nonspecific" amplification products.

2. Analyze the results of the BLAST search, which are displayed in three ways as you scroll down the page:
   i. First, a graphical overview illustrates how significant matches (hits) align with the query sequence. Matches of differing lengths are indicated by color-coded

bars. What do you notice about the lengths (and colors) of the matches (bars) as you look from the top to the bottom?

ii. This is followed by a list of significant alignments (hits) with links to the corresponding accession numbers. (An accession number is a unique identifier given to a sequence when it is submitted to a database such as GenBank.) Note the scores in the "E value" column on the right. The Expectation or E value is the number of alignments with the query sequence that would be expected to occur by chance in the database. The lower the E value, the higher the probability that the hit is related to the query. For example, an E value of 1 means that a search with your sequence would be expected to turn up one match by chance. Longer query sequences generally yield lower E values. An alignment is considered significant if it has an E value of less than 0.1. What is the E value of the most significant hit and what does it mean? Note the names of any significant alignments that have E values of less than 0.1. Do they make sense? What do they have in common?

iii. Third is a detailed view of each primer (query) sequence aligned to the nucleotide sequence of the search hit (subject, abbreviated "Sbjct"). Note that some matches to the forward primer (nucleotides 1–26 of rbcL; nucleotides 1–25 of Lep or VF) or to the reverse primer (nucleotides 27–46 of rbcL; nucleotides 26–53 of Lep or VF) are within the same subject (accession number). Do the alignments produce good matches to both primers?

3. Predict the length of the product that the primer set would amplify in a PCR (in vitro) as follows:

   i. Scroll down to the alignments section (third section) of the BLAST results page. Select a hit that matches both primer sequences.

   ii. To which positions do the primers match in the subject sequence?

   iii. The lowest and highest nucleotide positions in the subject sequence indicate the borders of the amplified sequence. Subtract the lowest nucleotide position in the subject sequence from the highest nucleotide position in the subject sequence. What is the difference between the coordinates?

   iv. Note that the actual length of the amplified fragment includes both ends. Thus, add 1 nucleotide to the result that you obtained in Step 3.iii to obtain the exact length of the PCR product amplified by the two primers.

4. Click on the accession number link to open the data sheet for the hit used in Step 3.

   i. At the top of the report, note basic information regarding the sequence, including its length (in base pairs, or bp), database accession number, source, and references to papers in which the sequence is published.

   ii. In the middle section of the report, the sequence features are annotated, with their beginning and ending nucleotide positions ("xx..xx"). These features may include genes, coding sequences (cds), regulatory regions, ribosomal RNA (rRNA), and transfer RNA (tRNA). Identify the feature(s) located between the nucleotide positions of the primers identified in Step 3.ii.

   iii. Scroll to the bottom of the data sheet. This is the nucleotide sequence to which the term "Sbjct" refers.

II. Upload and Assemble Your DNA Sequences in *DNA Subway*

In Parts II–IV, you will use the Blue Line of *DNA Subway* to analyze sequences generated by DNA sequencing. Generally, you will progress in a stepwise fashion through the "stops" on each "branch line." If your DNA sample has not been sequenced, you can follow these directions to assess another sequence.

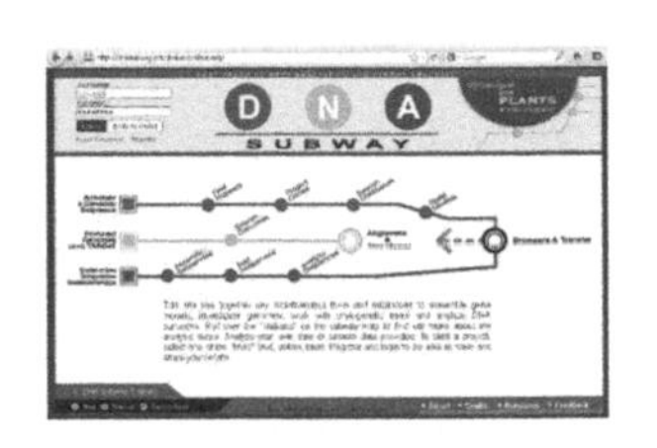

Like many other bioinformatics tools, DNA Subway works best on Mozilla/Firefox and a recent version of Java. For details, follow the "Manual" link at the bottom left of the home page.

1. Open *DNA Subway* (http://www.dnasubway.org). If you are a registered user, log in with your username and password (only registered users can save and share work). Alternatively, enter as a guest to gain temporary access.
2. Click on the blue square to generate a new phylogenetic project.
3. Under "Select Project Type," check either "rbcL" or "COI." (Plant *rbcL* sequences must be analyzed separately from animal *COI* sequences.)
4. Under "Select Sequence Source," there are several ways to obtain sequences for barcode analysis:
   - To upload .ab1 trace files or sequences in FASTA format, click "Browse" to navigate to a folder on your desktop or drive containing your sequence. Select the sequence by clicking on its file name and then click "Open."
   - To enter a sequence in FASTA format, paste the sequence into the window in the following format:

     >sequence name

     ATCGCCCCTTAATATTGCCTT...
   - To import a sequence/trace from the DNALC, click on "Import trace files from DNALC" and then find your tracking number. Select one or more files from the list and then click "Add Selected Files."
   - Select a sample sequence by clicking on the name of the desired sequence in the provided window.
5. Provide a project title (required) and click on "Continue."

The electropherogram view is only available for .ab1 files that include a trace file. FASTA files do not include trace information needed to output an electropherogram.

6. On the "Assemble Sequences" branch line, click on "Sequence Viewer." Click on a sequence name to view an electropherogram with quality scores for each nucleotide.
   i. The DNA sequencing software measures the fluorescence emitted in each of four channels—A, T, C, and G—and records it as a trace, or electropherogram. In a good sequencing reaction, the nucleotide at a given position will be fluorescently labeled far in excess of background (random) labeling of the other three nucleotides, producing a "peak" at that position in the trace. Thus, peaks in the electropherogram correlate to nucleotide positions in the DNA sequence. You can increase and decrease peak height by clicking the + and – buttons for the *y* axis.

   ii. A software program called Phred analyzes the sequence file and "calls" a nucleotide (A, T, C, or G) for each peak. If two or more nucleotides have relatively strong signals at the same position, the software calls an "N" for an undetermined nucleotide.

iii. Phred also examines the peaks around each call and assigns a quality score for each nucleotide. The electropherogram viewer represents each Phred score as a blue bar. The quality score corresponds to a logarithmic error probability that the nucleotide call is wrong or, conversely, to the accuracy of the call. A Phred score of 20, represented by the thin horizontal blue line, is considered the cutoff for high-quality sequence. What is the error rate and accuracy associated with a Phred score of 20?

Phred score	Error	Accuracy
10	one in 10	90%
20	one in 100	99%
30	one in 1000	99.9%
40	one in 10,000	99.99%
50	one in 100,000	99.999%

iv. Every sequence read begins with nucleotides (As, Ts, Cs, and Gs) interspersed with Ns. In clean sequences, where experimental conditions were near optimal, the initial Ns will end within the first 25 nucleotides. The remaining sequence will have very few, if any, internal Ns. Then, at the end of the read, the sequence will abruptly change back over to Ns. Large numbers of Ns scattered throughout the sequence indicate a poor-quality sequence. Sequences with average Phred scores below 20 will be flagged with a "Low Quality Score Alert." You will need to be careful when drawing conclusions from analyses made with poor-quality sequences. What do you notice regarding the electropherogram peaks and quality scores at nucleotide positions that are labeled "N"?

7. Click on "Sequence Trimmer" to automatically remove Ns from the 5′ and 3′ ends of selected sequences. The blinking "R" bullet indicates that the program is running. The bullet changes to a "V" when results are ready to view.

8. Once the program has finished, click on "Sequence Trimmer" again to view the trimmed sequences.

9. If you have good-quality forward and reverse reads for a sample, click on "Pair Builder" to associate a forward read with its corresponding reverse read. This will generate a consensus sequence, which is the best agreement between multiple sequences—in this case, between a forward and reverse read. If you have only a single read (sequence from only one primer), skip to Part III.

   i. Check the boxes for two sequences that you wish to pair and confirm your selection in the pop-up.

   ii. Click on the "F" to the right of the reverse sequence. The entry will change to "R," indicating that the sequence has been transformed into its reverse complement.

   iii. Click on "Save" to save your pair assignments.

   iv. Click on "Consensus Builder" to align the paired forward and reverse reads. Once the program has finished, click on "Consensus Builder" again to view the consensus sequence. How does the consensus sequence optimize the amount of sequence information available for analysis? Why does this occur?

DNA is composed of two antiparallel strands that "read" in opposite orientations. The reverse complement makes the reverse strand sequence equivalent to that of the forward strand by reversing the sequence order and by complementing each nucleotide (changing A→T, T→A, G→C, and C→G).

If you have two reads with one primer, you can also build a consensus for those reads. Ensure that both sequences are oriented cor-

rectly: For a forward primer, "F" should be displayed for both primers, whereas for a reverse primer, both sequences should display "R."

10. In nature, the forward strand and its reverse complement are a perfect match. However, because the sequencing process is not perfect, there are often differences between forward and reverse reads.
    i. Positions highlighted in yellow mark differences in nucleotide calls between the forward and reverse reads. When there is a discrepancy at a nucleotide position in two or more reads, the consensus software selects the nucleotide with the highest-quality Phred score. Do differences tend to occur in certain areas of your sequence? Explain.
    ii. Check the consensus sequence at yellow mismatches and, if necessary, override the judgment made by the software. To do this, click on a highlighted mismatch to see the electropherograms and Phred scores for each read, click on the desired nucleotide in the gray rectangle to change the consensus sequence at that position, and then click the button to save your change(s).

Change the consensus only if you have a strong reason to believe that the consensus is wrong. Changing the consensus sequence arbitrarily is likely to create a change in the sequence that does not represent the sequence in the organism.

11. Identify a sequence to be used in the BLAST analysis in Part III.
    i. Few or no internal mismatches indicate good-quality sequence from both the forward and reverse reads, and the consensus sequence may be used. If this is the case, continue to Part III.
    ii. If your consensus sequence contains a large number of mismatches, one or both sequences may be of poor quality. Often, one of the sequencing reactions produces a high-quality read that can be used on its own. Return to "Sequence Viewer" and examine the distribution of Ns in the electropherograms to determine whether they are mainly confined to one of the two sequences. If one of the sequences seems to be of good quality, you can use it in Part III. But first, return to "Pair Builder," click on the red "X" to undo the pairing, and click "Save."

To pair two different sequences, return to "Pair Builder," click on the red "X" to undo a pairing, and then pair two new sequenc-es and generate a consensus sequence as described in Steps 9.i–iv.

III. Use BLAST to Identify Your Species

A BLAST search can quickly identify any close matches to your sequence in sequence databases. In this way, you can often quickly identify an unknown sample to the genus or species level. It also provides a means to add samples for the phylogenetic analysis that you will perform in Part IV.

1. On the "Add Sequences" branch line, click on "BLASTN." Then, click on the "BLAST" button next to the sequence that you want to query against DNA databases. Once the search has finished, click on "View" to see a list that has information regarding the 20 most significant alignments (hits).
    i. The "Name" column includes the *accession number*, a unique identifier given to each sequence submitted to a database. Prefixes indicate the database name; these include "gb" for GenBank, "emb" for European Molecular Biology Laboratory, and "dbj" for DNA Databank of Japan.
    ii. The "Details" column includes the organism and sequence description or gene name of the hit. Click on the genus and species name for a link to an image of the organism as well as additional links to detailed descriptions at Wikipedia and Encyclopedia of Life (EOL).

iii. The last few columns show statistics that allow comparison of hits across different searches. The number of mismatches over the length of the alignment gives a rough idea of how closely two sequences match. The bit score formula takes into account gaps in the sequence; the higher the score, the better the alignment. The Expectation or E value is the number of alignments with the query sequence that would be expected to occur by chance in the database. The lower the E value, the higher the probability that the hit is related to the query. For example, an E value of 0 means that a search with your sequence would be expected to turn up no matches by chance. Why do most of these significant hits typically have E values of 0? (This is generally not the case with BLAST searches with primers, as illustrated in Step 2.ii of Part I.) What does it mean when there are multiple BLAST hits with similar E values?

2. Add BLAST sequence data to your phylogenetic analysis by checking the box(es) above the accession number(s), and then clicking on "Add BLAST hits to project."
3. Add additional sequences to your analysis. Click on "Upload Data" to include additional data. Add your own .ab1 files or sequences in FASTA format, or import data from other sources. If you identified sequences in BOLD, you may upload them now.
4. Click on "Reference Data" to select data that will let you compare your barcode sequence in an appropriate phylogenetic context. If you are fairly certain about the taxonomy of your sample based on your BLAST results, you may want to select reference data from a narrow taxonomic group of plants or animals that includes the source of your sample. If you have little idea about the taxonomy of your sample, include a very broad selection of reference data.

IV. Perform a Phylogenetic Analysis

Many unknown species can be rapidly identified by a BLAST search. A phylogenetic analysis adds depth to your understanding by showing how your sequence fits into a broader taxonomy of living things. In this exercise, a *phylogenetic tree* (a graphical representation of relationships among taxonomic groups) is built by analyzing the similarities and differences among DNA sequences. If your BLAST search fails to identify your sequence, a phylogenetic analysis can usually identify it to at least the family level.

1. Click on "Select Data." Then, check boxes to select any or all of the sequences that you have uploaded from your own sequencing projects, from BLAST searches, and from reference data sets. Click "Save."
2. Click on "MUSCLE" to align the sequences.
3. Once the program has finished, click on "MUSCLE" again to view the alignment in Jalview.

   i. Scroll through your alignments to view similarities among sequences. Nucleotides are color-coded, and each row of nucleotides is the sequence of a single organism or sequencing reaction. Columns are matches (or mismatches) at a single nucleotide position across all sequences. Dashes (–) are gaps in sequence, where nucleotides in one sequence are not represented in other sequences.

MUSCLE is a multiple sequence alignment program that aligns two or more sequences in a manner that produces the fewest gaps. Jalview is a Java utility for viewing and editing the alignments produced by MUSCLE; it also calculates and displays phylogenetic trees.

   ii. The 5′ (leftmost) and 3′ (rightmost) ends of the sequences are usually misaligned. Why does this occur?

   iii. Note any sequence that introduces large, internal gaps (– – –) in the alignment. This is either poor-quality or unrelated sequence that should be excluded from the analysis. To remove it, return to "Select Data," uncheck that sequence, and save your change. Then click on "MUSCLE" to recalculate.

It is important to remove sequence gaps and unaligned ends before constructing a phylogenetic tree. Gaps and unaligned ends are scored as mismatches by tree-building algorithms, making sequences appear less related than they actually are and forcing related sequences into different clades.

4. Trim unaligned ends off the sequences.

   i. Identify the leftmost point at which all or most sequences show corresponding nucleotide color bars. (There should be few or no gaps in the vertical column of nucleotides at this point.)

   ii. Click on the nucleotide coordinate bar directly above this nucleotide in the top sequence. This will activate a red cursor and a pop-up menu.

   iii. Click on "Remove Left" to trim the leftmost sequences to this nucleotide position.

   iv. Trim the rightmost sequences by repeating Steps 4.i–iii and clicking "Remove Right."

   v. Click "Submit trimmed alignment."

Tree-building algorithms attempt to reconstruct the order in which sequence mutations accumulated as different lineages diverged from a common ancestor. A number of plausible trees can be constructed from any set of sequences; thus, an algorithm presents what it determines to be the optimal one. The maximum-likelihood algorithm evaluates possible trees and determines which is most likely to have been produced by the observed data. Because it fits mutations to a tree, the maximum-likelihood method produces the most parsimonious tree—one that accounts for the data with the shortest branch lengths.

5. Click on "PHYLIP ML" to generate a phylogenetic tree using the maximum-likelihood method. Once the program has finished, click on "PHYLIP ML" again to open the tree in a new window. The MUSCLE alignment used to produce the tree will also open. Look at your tree.

   i. The branch tips are the DNA sequences of individual species or samples that you analyzed. Two branches are connected to each other by a node (black square), which represents the common ancestor of the two sequences.

   ii. The length of each branch is a measure of the evolutionary distance from the ancestral sequence at the node. (The tree visualization software may assign a numerical value to each branch, which is proportional to its length.) Species or sequences with short branches from a node are closely related; those with longer branches are more distantly related.

6. A group formed by a common ancestor and its descendants is called a "clade." Related clades, in turn, are connected by nodes to make larger, less-closely related clades. Click on a node (black square) to highlight sequences in that clade. Click the node again to deselect the clade.

   i. What assumptions are made when one infers evolutionary relationships from sequence differences?

   ii. Generally, the clades will follow established taxonomic relationships: genus < family < order < class < phylum. However, traditional taxonomic relationships and newer phylogenetic (gene) trees disagree on some placements, and much research is focused on reconciling these differences. Why do taxonomic relationships and phylogenetic trees sometimes disagree?

7. Find and evaluate your sequence's position in the tree. If your sequence is closely related to any of the reference or uploaded sequences, it will share a single node

The neighbor-joining algorithm builds a tree from the bottom up by comparing the evolutionary distance between pairs of DNA sequences. Sequences with best-matching sequences are linked as "neighbors" that share common nodes in the tree. Because the branch distances are produced in a pairwise manner, neighbor-joining does not optimize branch length and tree parsimony. The chief advantage of neighbor-joining—that it is less computationally intensive than maximum likelihood—has become less important as the processing power of computers has increased.

with those species. If your sequence is identical to another sequence, the two will diverge directly from the node without branches. If your sequence is distantly related to all of the species in your tree, your sequence will sit on a branch by itself—with the other sequences grouping together as a clade.

i. To identify the smallest clade that includes your sequence, click on the node that is directly connected to your sequence. The sequences that are highlighted are the closest relatives of your sequence in the tree.

ii. Look at the scientific names of sequences within the most closely associated clade. If all members share the same genus name, you have identified your sequence as belonging to that genus. If different genus names are represented, check whether they belong to the same family or order.

8. Return to the menu and click on "PHYLIP NJ" to generate a phylogenetic tree using the neighbor-joining method. How does it compare with the maximum-likelihood tree? What does this tell you?

9. If neither tree places your sequence within an identifiable clade—or if that clade is only at order level—you can add more sequences to increase the resolution of your analysis. Return to Steps 3 and 4 of Part III and add more reference sequences or obtain sequences within the order or family clade that contained your sequence. Then, repeat Steps 1–7 to generate and analyze trees from your refined data set.

V. Submit DNA Barcodes to GenBank

Once you have analyzed rbcL or COI barcoding sequences, you can submit your results to GenBank.

1. Under "Browsers and Transfer," click on "Export to GenBank." For submission, quality sequence for both the forward and reverse strand is required, and this sequence must be trimmed. If you have not met these requirements, *DNA Subway* will alert you.

2. Once you have met these initial requirements, only sequences that are paired and have a consensus sequence will be available for submission. To submit, record the name of any collectors, the institution storing your specimen, the individual that identified the specimen using traditional taxonomy, the date and location of collection, the sex and life stage (if known), and the names of the PCR primers used to amplify the region.

3. To create a submission, click "New submission." A list of sequence pairs in the project will be displayed. Pairs unavailable for submission will be displayed in gray with a description of why they are unavailable. To submit these sequences, the indicated problems must be addressed.

4. Select any available pair that you wish to submit and click "Continue." Follow the online instructions to complete the submission.

RESULTS AND DISCUSSION

I. Interpret Your Gel

1. Observe the photograph of the stained gel containing your PCR samples and those from other students. Orient the photograph with the sample wells at the top. Use the sample gel shown below to help interpret the band(s) in each lane of the gel.

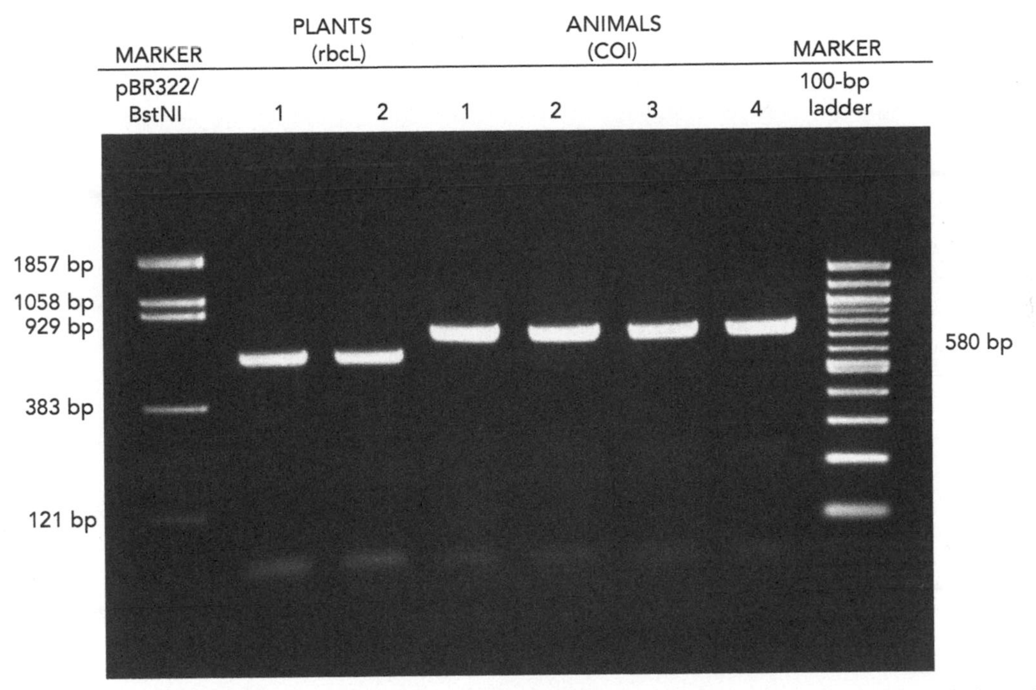

2. Locate the lane containing the pBR322/BstNI markers on the left side of the gel. Working from the well, locate the bands corresponding to each restriction fragment: 1857, 1058, 929, 383, and 121 bp. The 1058- and 929-bp fragments will be very close together or may appear as a single large band. The 121-bp band may be very faint or not visible.

3. Scan across the row of student results. Notice that virtually all student lanes contain one prominent band. Which samples amplified well, and which ones did not? If your samples did not amplify, describe several reasons why this might be the case.

4. It is common to see a diffuse (fuzzy) band that runs ahead of the 121-bp marker. This is a "primer dimer," an artifact of the PCR that results from the primers overlapping one another and amplifying themselves. How would you interpret a lane in which you observe primer dimer but no bands as described in Step 3?

5. Based on your results, your teacher will help you to decide whether your sample should be sent for sequencing. Generally, DNA sequence can be obtained from any sample that gives an obvious band on the gel.

LABORATORY 3.6

Detecting Epigenetic DNA Methylation in *Arabidopsis*

▼ OBJECTIVES

This laboratory demonstrates several important concepts of modern biology. During the course of this laboratory, you will

- Relate genotype and phenotype.
- Explore the epigenetic effect of DNA methylation of a gene.
- Move between in vitro experimentation and in silico computation.

In addition, this laboratory utilizes several experimental and bioinformatics methods in modern biological research. You will

- Extract and purify DNA from tissue.
- Amplify the regulatory region of the *FWA* gene by polymerase chain reaction (PCR) and analyze the PCR products by gel electrophoresis.
- Use the Basic Local Alignment Search Tool (BLAST) to search databases for related sequences.

INTRODUCTION

Although each cell of a multicellular organism carries the same genome, different sets of genes are expressed in different cell types and at different times. The spatial and temporal expression of genes must be tightly controlled for proper growth and development. Furthermore, expression patterns must also be stably inherited from generation to generation. Many aspects of gene regulation are controlled by specific DNA sequences within or outside genes. These regulatory DNA sequences include promoters, which are typically located immediately "upstream" of the genes they control, and enhancers, which may be distant.

In the 1940s, Conrad Waddington, a geneticist and developmental biologist at the University of Edinburgh, coined the term *epigenetics* to describe changes in gene expression that are not dependent on the DNA sequence itself. The prefix "epi" means "on" or "above," and thus *epigenetics* describes genetic modifications that occur "on top of" the DNA molecule rather than within it.

Today, epigenetics is a growing field encompassing several different mechanisms that regulate gene expression—including DNA methylation, histone modification, and regulatory RNAs. DNA methylation, in which a methylase enzyme adds a methyl group to a cytosine nucleotide, is an important epigenetic mechanism in bacteria, animals, and plants. Typically, methylation of a promoter prevents binding by transcrip-

tion factors that are needed to initiate transcription, thus "silencing" a gene. Patterns of methylation can be inherited, maintaining correct gene silencing from generation to generation. The inheritance of an epigenetic effect through one parental line is termed "imprinting."

An epigenetic locus (epiallele) is usually maintained in a fully methylated state, meaning that the corresponding nucleotide is methylated on both strands of the DNA molecule. During semiconservative replication, the methyl group is retained on each of the original strands. Then, a methylase adds a new methyl group to the corresponding nucleotide on the newly synthesized strands. This explains why methyl modifications ("marks") are heritable, even though they do not alter the DNA sequence.

Time to flowering is an important phenotype in plants. This experiment relates an epigenetic genotype of the *FWA* gene to a late-flowering phenotype in the model plant *Arabidopsis thaliana*. Although *FWA* is expressed in the seed endosperm, normal flowering requires that both copies of the gene are silenced in the tissues of wild-type plants. *FWA* is silenced by DNA methylation of repeated sequences in the promoter and 5′ region of the gene. Although *fwa-1* mutants have an identical gene sequence, demethylation of this region allows transcription of the *FWA* gene and causes late flowering.

In this laboratory, wild-type and *fwa-1* mutant plants are grown, and genomic DNA is extracted from normal and late-flowering plants. The purified DNA is then incubated with the restriction enzyme McrBC. Unlike classical restriction enzymes, McrBC is modification dependent, recognizing and cutting regions with methylated GC and/or AC dinucleotides. Thus, the methylated wild-type allele is cut by McrBC, but the demethylated mutant allele is not.

The McrBC-digested DNA and uncut controls are then amplified by PCR, and the products are analyzed by agarose gel electrophoresis. Primers spanning the *FWA* promoter and 5′ repeat region amplify a PCR product from the intact DNA from *fwa-1* mutant plants but fail to amplify DNA from the digested wild-type DNA. Amplification of the tubulin gene provides a positive control for DNA isolation.

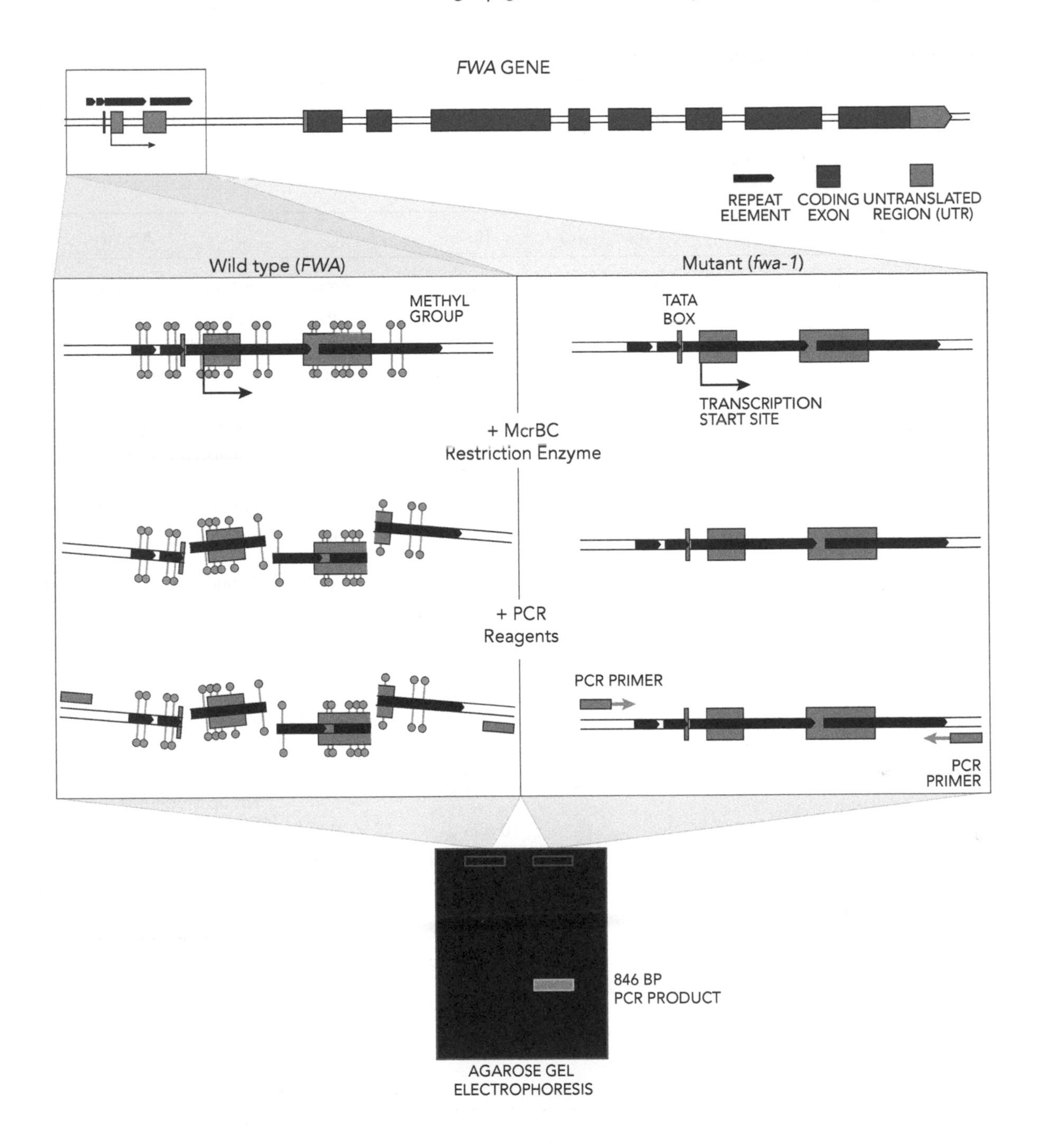

FURTHER READING

Kinoshita Y, Saze H, Kinoshita T, Miura A, Soppe WJ, Koornneef M, Kakutani T. 2007. Control of *FWA* gene silencing in *Arabidopsis thaliana* by SINE-related direct repeats. *Plant J* **49:** 38–45.

Martienssen RA, Colot V. 2001. DNA methylation and epigenetic inheritance in plants and filamentous fungi. *Science* **293:** 1070–1074.

Soppe WJJ, Jacobsen SE, Alonso-Blanco C, Jackson JP, Kakutani T, Koornneef M. 2000. The late flowering phenotype of *fwa* mutants is caused by gain-of-function epigenetic alleles of a homeodomain gene. *Mol Cell* **6:** 791–802.

Waddington CH. 1942. The epigenotype. *Endeavour* **1:** 18–20.

PLANNING AND PREPARATION

The following table will help you to plan and integrate the different experimental methods.

Experiment	Day	Time		Activity
I. Plant *Arabidopsis* seeds	4–6 wk before Part II	15–30 min	Prelab:	Set up student stations.
			Lab:	Plant *Arabidopsis* seeds and water regularly.
			Postlab:	Care for *Arabidopsis* plants.
II. Isolate DNA from *Arabidopsis*	1	30–60 min	Prelab:	Aliquot nuclei lysis solution, RNase solution, protein precipitation solution, isopropanol, ethanol, and DNA rehydration solution.
				Set up student stations.
		110 min	Lab:	Isolate DNA.
III. Digest DNA with McrBC	2	30 min	Prelab:	Aliquot McrBC enzyme, restriction buffer master mix, and dH_2O.
				Set up student stations.
		60 min to overnight	Lab:	Set up restriction digests.
			Postlab:	Incubate restriction digests at 37°C.
IV. Amplify DNA by PCR	3	15 min	Prelab:	Prepare and aliquot primer mix.
				Set up student stations.
		10 min	Lab:	Set up PCRs.
		60–150 min	Postlab:	Amplify DNA in thermal cycler.
V. Analyze PCR products by gel electrophoresis	4	30 min	Prelab:	Dilute TBE electrophoresis buffer.
				Prepare agarose gel solution.
				Set up student stations.
		30 min	Lab:	Cast gels.
		45+ min		Load DNA samples into gel.
				Electrophorese samples.
				Photograph gels.

OVERVIEW OF EXPERIMENTAL METHODS

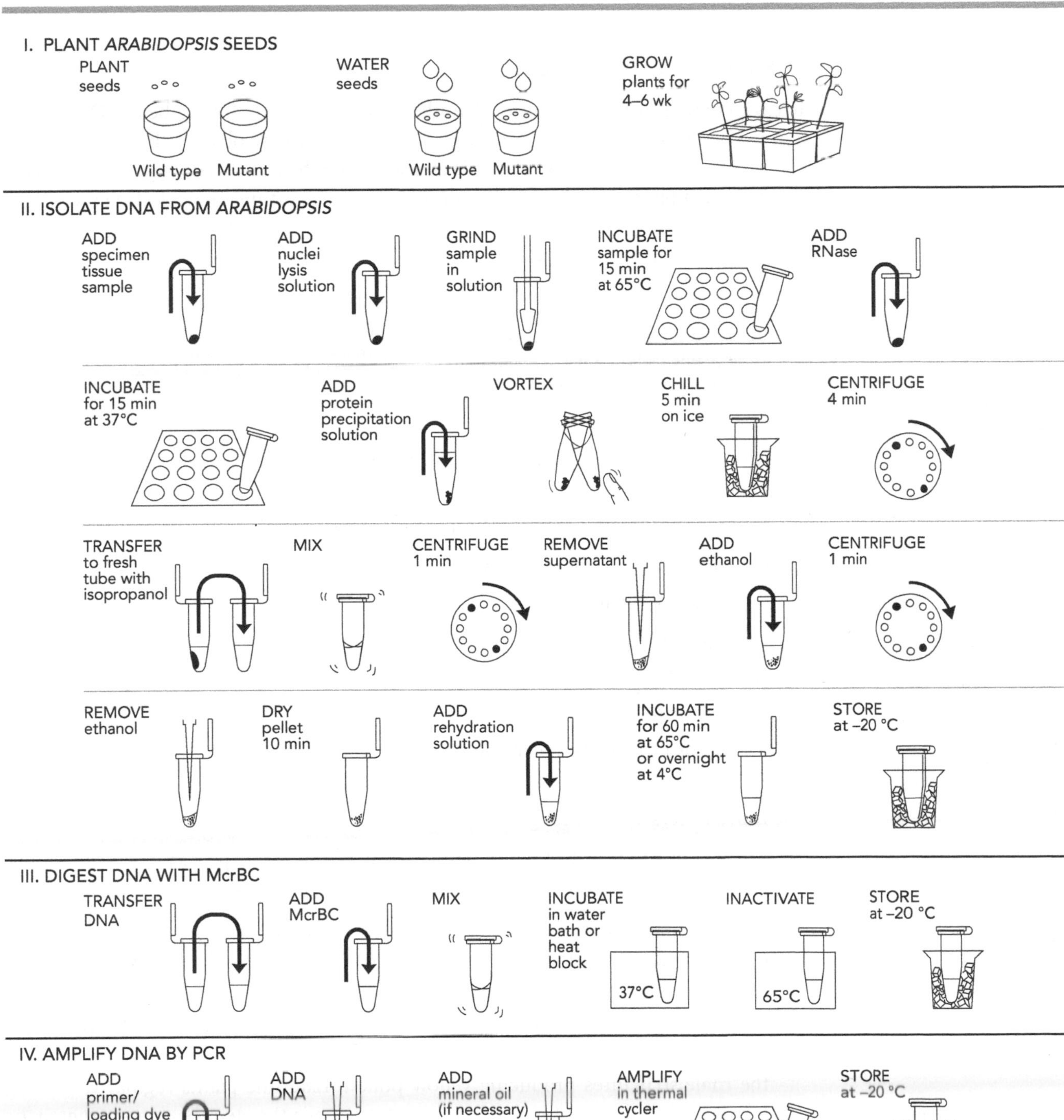

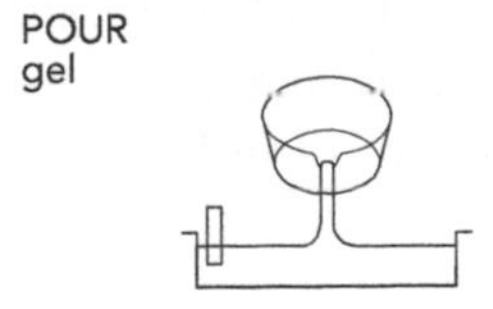

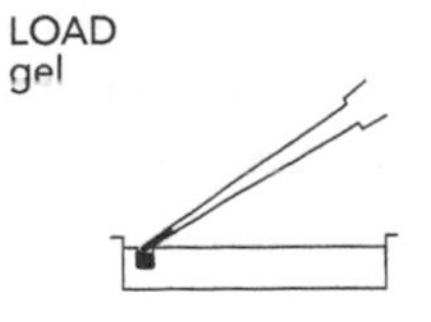

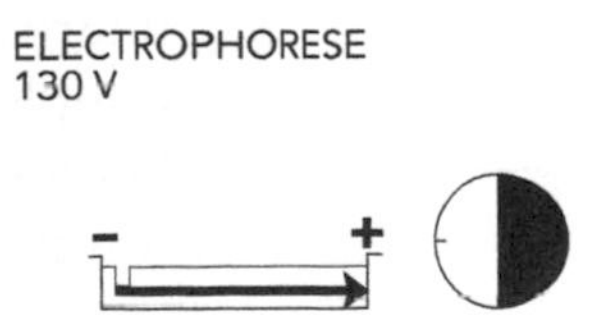

EXPERIMENTAL METHODS

I. Plant *Arabidopsis* Seeds

To extract DNA from plant tissue, plant the *Arabidopsis* seeds 4–6 wk before DNA isolation and PCR. Depending on growing conditions, you may observe the late-flowering phenotype in *fwa-1* mutants in as little as 4 wk. Go to http://arabidopsis.info/, Growing *Arabidopsis* (upper right) for further information on planting and collecting seeds.

REAGENTS, SUPPLIES, & EQUIPMENT

To share

Arabidopsis seeds: wild-type *Ler* (Landsberg *erecta*) and mutant *fwa-1*
Planting pots and tray
Plastic dome lid or plastic wrap
Potting soil
1 Sheet of paper (4 x 4 in)

For best results, use a potting soil formulated specifically for Arabidopsis.

1. Moisten the potting soil. Fill planting pots evenly with soil, but do not pack tightly.
2. Fit pots into the tray, but leave one corner space empty to facilitate watering.
3. Carefully scatter seeds evenly on top of the soil.
   i. Fold a 4 x 4-in sheet of paper in half.
   ii. Place the seeds into the fold of the paper and gently tap them onto the soil.
   iii. Provide space between seeds (0.75-in radius around the seeds), so that they will grow better and plant phenotypes can be readily observed.
   iv. Record the date of planting.

Arabidopsis *seeds are very tiny and difficult to handle; thus, planting is not as simple as it may seem.*

Germination requires a humid environment.

4. Cover the pots with plastic dome lids or plastic wrap to assist germination. (Remove covers 1–2 d after germination.)
5. Add 1/2 in of water to the tray, using the empty corner space. Water regularly to keep the soil damp, but do not allow the soil to remain soggy.
6. Grow the plants close to a sunny window at room temperature (20°C–22°C). For optimum growth, provide a constant (24 h) fluorescent light source ~1 ft directly above the plants.
7. Measure flowering time for wild-type *Ler* and *fwa-1* mutant plants. Record flowering time as the number of days from planting to the emergence of anthers, with the male structures producing yellow pollen. Examine plants regularly, looking for signs that the plant is developing a stem. Once the stem begins to emerge, or bolt, check daily for flowers.
8. Continue to Part II when the late-flowering phenotype is detected in *fwa-1* mutant plants. You may wish to continue to grow the plants to harvest seeds for future use. To prepare for harvest, stop watering the plants and allow seeds to dry completely. Directions for harvesting and storing seeds can be found at the European Arabidopsis Stock Centre: http://arabidopsis.info/InfoPages?template=newgrow;web_section=arabidopsis.

With 24-h fluorescent lighting, the phenotype can be detected in 2–3 wk.

II. Isolate DNA from *Arabidopsis*

REAGENTS, SUPPLIES, & EQUIPMENT

For each group

Container with cracked or crushed ice
DNA rehydration solution (250 µL)
Ethanol (70%) <!> (1.5 mL)
Isopropanol <!> (1.5 mL)
Microcentrifuge tube rack
4 Microcentrifuge tubes (1.5 mL)
Micropipettes and tips (1–1000 µL)
Nuclei lysis solution (1.5 mL)*
Permanent marker
2 Plastic pestles
Protein precipitation solution (0.5 mL)
RNase solution (10 µL)
Vortexer (optional)
Wild-type *Ler* and mutant *fwa-1 Arabidopsis* plants from Part I

To share

Microcentrifuge
Water baths or heating blocks at 65°C and 37°C

*Store on ice.
See Cautions Appendix for appropriate handling of materials marked with <!>.

1. Obtain wild-type *Ler* and mutant *fwa-1 Arabidopsis* plants.

Using more than the recommended amount can inhibit the DNA extraction or amplification.

2. Cut a piece of tissue from each plant that is ~1/4 in in diameter. Be careful not to cross-contaminate different specimens.

   The large end of a 1000-µL pipette tip will punch leaf disks of this size.

Step 4 dissolves membrane-bound organelles, including the nucleus, mitochondria, and chloroplast.

3. Place each sample into a clean 1.5-mL tube labeled with genotype and group number.
4. Add 100 µL of nuclei lysis solution to each sample tube.

Grinding the plant tissue breaks up the cell walls. When fully ground, the sample should be a green liquid. There may be some particulate matter remaining.

5. Twist a clean plastic pestle against the inner surface of one of the 1.5-mL tubes to grind the tissue forcefully for 1 min. Repeat for the other tube, using a clean pestle.
6. Add 500 µL of nuclei lysis solution to each tube.
7. Incubate the tubes in a water bath or heat block for 15 min at 65°C.

Step 8 degrades RNA that could interfere with PCR.

8. Add 3 µL of RNase solution to each tube. Close the caps and mix by rapidly inverting the tubes several times.
9. Incubate the tubes in a water bath or heat block for 15 min at 37°C. Then, stand the tubes for 5 min at room temperature.
10. Add 200 µL of protein precipitation solution to each tube. Vortex the tubes for 5 sec. Use a vortexer if available; if not, securely grasp the upper part of each tube and vigorously hit the bottom end with the index finger of the opposite hand. (If necessary, make a balance tube with the appropriate volume of water for the centrifuge steps.)
11. Stand the tubes for 5 min on ice.

The pellet will collect on the bottom side of the tube under the cap hinge.

12. Place your tubes and those of other groups in a balanced configuration in a microcentrifuge with cap hinges pointing outward. Centrifuge for 4 min at maximum speed to pellet protein and cell debris.
13. Label clean 1.5-mL tubes with your group number and genotype for each sample. Use a fresh tip to transfer 600 µL of supernatant to the clean tubes. Be careful not to disturb the pelleted debris when transferring the supernatants. Discard old tubes containing precipitate.

The pellet may appear as a tiny teardrop-shaped smear or particles on the bottom side of the tube under the cap hinge. Do not be concerned if you cannot see a pellet. A large or greenish pellet is cell debris carried over from the first centrifugation.

Dry the pellets quickly with a hair dryer. To prevent blowing the pellet away, direct the air across the tube mouth, not into the tube, for ~3 min.

This is a crude DNA extract and contains nucleases that will eventually fragment the DNA at room temperature. Keep the sample cold to limit this activity.

14. Add 600 µL of isopropanol to the supernatant in each tube. Close caps and mix by rapidly inverting the tubes several times.
15. Place your tubes and those of other groups in a balanced configuration in a microcentrifuge with cap hinges pointing outward. Centrifuge for 1 min at maximum speed to pellet the DNA.
16. Carefully pour off the supernatant from the tubes. Add 600 µL of 70% ethanol. Close the caps and flick the bottom of the tube several times to "wash" the pellets.
17. Centrifuge the tubes for 1 min at maximum speed.
18. Carefully pour off the supernatant from each tube and discard. Use a micropipette with fresh tips to remove any remaining ethanol, being careful not to disturb the pellets.
19. Air-dry the pellets for 10–15 min to evaporate any remaining ethanol.
20. Add 100 µL of DNA rehydration solution to each tube and dissolve the DNA pellets by pipetting in and out several times.
21. Incubate the DNA for 45–60 min at 65°C or overnight at 4°C.
22. Store your samples on ice or at –20°C until you are ready to begin Part III.

III. Digest DNA with McrBC

REAGENTS, SUPPLIES, & EQUIPMENT

For each group

Container with cracked or crushed ice
Distilled or deionized water (dH_2O)
McrBC enzyme (10,000 U/mL) (16 µL)*
Microcentrifuge tube rack
4 Microcentrifuge tubes (1.5 mL)
Micropipette and tips (1–20 µL)
Permanent marker
Restriction buffer master mix (65 µL)*
Wild-type *Ler* and mutant *fwa-1* DNA* from Part II

To share

Water bath or heat block at 65°C and 37°C

*Store on ice.

1. Label four 1.5-mL tubes with your group number and DNA type as follows:

   "WU" (wild-type undigested DNA)
   "WD" (wild-type digested DNA)
   "MU" (mutant undigested DNA)
   "MD" (mutant digested DNA)

2. Use the matrix below as a checklist while adding reagents to each reaction. Read down each column, adding the same reagent to all appropriate tubes. Use a fresh tip for each reagent. Store tubes on ice as you set up the reactions. Refer to the detailed directions that follow.

Tube	dH_2O	Restriction buffer master mix	McrBC	Wild-type DNA	Mutant DNA
WU	6 µL	13 µL	—	6 µL	—
WD	4 µL	13 µL	2 µL	6 µL	—
MU	6 µL	13 µL	—	—	6 µL
MD	4 µL	13 µL	2 µL	—	6 µL

3. Add 6 µL of dH_2O to the WU and MU tubes. Add 4 µL of dH_2O to the WD and MD tubes.
4. Use a fresh tip to add 13 µL of restriction buffer master mix to each tube.
5. Use a fresh tip to add 2 µL of McrBC to the WD and MD tubes.
6. Use a fresh tip to add 6 µL of wild-type *Ler* DNA to the WU and WD tubes.
7. Use a fresh tip to add 6 µL of mutant *fwa-1* DNA to the MU and MD tubes.
8. Mix and pool reagents in all test tubes by pulsing in a microcentrifuge or by sharply tapping the tube bottoms on the laboratory bench.
9. Incubate the reactions for 60 min or overnight in a water bath or heat block set to 37°C .
10. Transfer your samples for 20 min to a 65°C water bath or heat block to inactivate the McrBC.
11. Store your samples on ice or in the freezer until you are ready to begin Part IV.

Sixty min is the minimum time needed for complete digestion. If time permits, incubate reactions overnight.

Step 10 is critical. The McrBC must be denatured before setting up PCR reactions.

IV. Amplify DNA by PCR

REAGENTS, SUPPLIES, & EQUIPMENT

For each group

Container with cracked or crushed ice
Digested and undigested wild-type and mutant DNA* from Part III
FWA primer/loading dye mix (110 µL)*
Micropipettes and tips (1–100 µL)
Permanent marker
4 Ready-To-Go PCR Beads in 0.2- or 0.5-mL PCR tubes

To share

Thermal cycler

*Store on ice.

1. Obtain four PCR tubes containing Ready-To-Go PCR Beads. Label the tubes with your group number and DNA type as follows:
   - "WU" (wild-type undigested).
   - "WD" (wild-type digested).
   - "MU" (mutant undigested).
   - "MD" (mutant digested).
2. Use the matrix below as a checklist while adding reagents to each reaction. Read down each column, adding the same reagent to all appropriate tubes. Use a fresh tip for each reagent. Store tubes on ice as you set up the reactions. Refer to the detailed directions that follow.

Tube	*FWA* primer mix	Wild-type undigested DNA	Wild-type digested DNA	Mutant undigested DNA	Mutant digested DNA
WU	22.5 µL	2.5 µL	—	—	—
WD	22.5 µL	—	2.5 µL	—	—
MU	22.5 µL	—	—	2.5 µL	—
MD	22.5 µL	—	—	—	2.5 µL

The primer/loading dye mixes will turn purple as the PCR bead dissolves.

The small volume of DNA will be drawn out of the tip by capillary action of the tip in contact with the large volume of primer/loading dye mix.

3. Use a fresh tip to add 22.5 µL of the *FWA* primer/loading dye mix to each tube. Allow the beads to dissolve for 1 min.
4. Use a fresh tip to add 2.5 µL of wild-type undigested DNA to the WU tube. Pipette directly into the primer/loading dye mix and ensure that no DNA remains in the tip after pipetting.
5. Use a fresh tip to add 2.5 µL of wild-type digested DNA to the WD tube. Pipette directly into the primer/loading dye mix and ensure that no DNA remains in the tip after pipetting.
6. Use a fresh tip to add 2.5 µL of mutant undigested DNA to the MU tube. Pipette directly into the primer/loading dye mix and ensure that no DNA remains in the tip after pipetting.

If reagents become splattered on the wall of the tube, pool them by pulsing in a microcentrifuge or by sharply tapping the tube bottom on the laboratory bench.

Perform the following if your thermal cycler does not have a heated lid: Before thermal cycling, add a drop of mineral oil on top of your PCR reaction. Be careful not to touch the dropper tip to the tube or reaction or the oil will be contaminated with your sample.

7. Use a fresh tip to add 2.5 µL of mutant digested DNA to the MD tube. Pipette directly into the primer/loading dye mix and ensure that no DNA remains in the tip after pipetting.
8. Store reactions on ice until your class is ready to begin thermal cycling.
9. Place your PCR tubes, along with those of the other students, in a thermal cycler that has been programmed for 35 cycles of the following profile:

Denaturing step:	30 sec	94°C
Annealing step:	30 sec	60°C
Extending step:	30 sec	72°C

   The profile may be linked to a 4°C hold program after the 35 cycles are completed.
10. After thermal cycling, store the amplified DNA on ice or at –20°C until you are ready to proceed to Part V.

V. Analyze PCR Products by Gel Electrophoresis

REAGENTS, SUPPLIES, & EQUIPMENT

For each group

2% Agarose in 1x TBE (hold at 60°C) (50 mL per gel)
Container with cracked or crushed ice
DNA ladder, 100 bp (20 µL per gel)*
Gel-casting tray and comb
Gel electrophoresis chamber and power supply
Latex gloves
Masking tape
Microcentrifuge tube rack
Micropipette and tips (1–100 µL)
PCR products* from Part IV
SYBR Green <!> DNA stain (10 µL per group)
1x TBE buffer (300 mL per gel)
1x TBE buffer (500 mL per gel)

To share

Digital camera or photodocumentary system
Microwave, hot plate, or boiling water bath
UV transilluminator <!> and eye protection
Water bath for agarose solution (60°C)

*Store on ice.
See Cautions Appendix for appropriate handling of materials marked with <!>.

1. Seal the ends of the gel-casting tray with masking tape—or another method appropriate for the gel electrophoresis chamber used—and insert a well-forming comb.

2. Pour the 2% agarose solution into the tray to a depth that covers about one-third the height of the comb teeth.

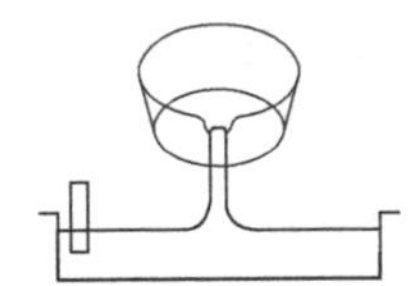

Avoid pouring an overly thick gel, which makes visualization of the DNA more difficult.

3. Allow the gel to completely solidify; this takes ~20 min.

The gel will become cloudy as it solidifies.

4. Remove thc masking tape, place the gel into the electrophoresis chamber, and add enough 1x TBE buffer to cover the surface of the gel.

Do not add more buffer than necessary. Too much buffer above the gel channels electrical current over the gel, increasing running time.

5. Carefully remove the comb, add additional 1x TBE buffer to fill in the wells, and cover the gel, creating a smooth buffer surface.
6. Add 1 µL of SYBR Green DNA stain to 20 µL of the 100-bp DNA ladder.
7. Orient the gel according to the diagram in Step 9 below, so that the wells are along the top. Use a fresh tip to load 20 µL of the stained 100-bp DNA ladder into the far left well.
8. Use fresh tips to add 1 µL of SYBR Green DNA stain to each PCR reaction.
9. Use fresh tips to load 15 µL of each sample in your assigned wells according to the following diagram.

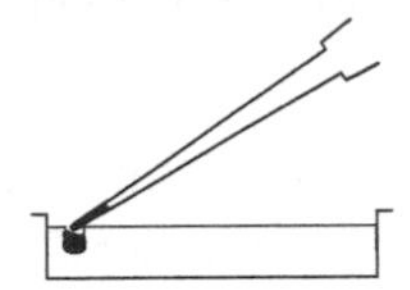

Expel any air from the tip before loading and be careful not to push the pipette through the bottom of the sample well.

MARKER	WILD TYPE (*Ler*)		MUTANT (*fwa-1*)	
100 bp	WU	WD	MU	MD

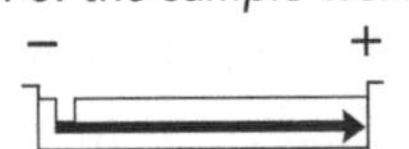

10. Run the gel for ~30 min at 130 V. Adequate separation will have occurred when the cresol red dye front has moved at least 50 mm from the wells.
11. View the gel using UV transillumination. Photograph the gel using a digital camera or photo-documentary system.

Transillumination, where the light source is below the gel, increases brightness and contrast.

BIOINFORMATICS METHODS

For a better understanding of the experiment, do the following bioinformatics exercises before you analyze your results.

I. Use BLAST to Find DNA Sequences in Databases (Electronic PCR)

1. Perform a BLAST search as follows:
   i. Do an Internet search for "ncbi blast."
   ii. Click on the link for the result "BLAST: Basic Local Alignment Search Tool." This will take you to the Internet site of the National Center for Biotechnology Information (NCBI).
   iii. Under the heading "Basic BLAST," click on the link "nucleotide blast" (blastn).
   iv. Enter the *FWA* primer sequences into the search window. These are the query sequences.

The following primer sets were used in this experiment:

FWA 5′-TCAGCGTCTACCAAATCTACACTTTT-3′ (forward primer)

FWA 5′-CAGACAAATCGGGAACCAAA-3′ (reverse primer)

v. Omit any nonnucleotide characters from the window because they will not be recognized by the BLAST algorithm.

vi. Under "Choose Search Set," select the "Nucleotide collection (nr/nt)" database from the drop-down menu.

vii. Under "Program Selection," optimize for "Somewhat similar sequences (blastn)."

viii. Click on "BLAST." This sends your query sequences to a server at NCBI in Bethesda, Maryland. There, the BLAST algorithm will attempt to match the primer sequences to the millions of DNA sequences stored in its database. While searching, a page showing the status of your search will be displayed until your results are available. This may take only a few seconds or more than 1 min if many other searches are queued at the server.

2. The results of the BLAST search are displayed in three ways as you scroll down the page:

   i. First, a *Graphic Summary* illustrates how significant matches, or "hits," align with the query sequence. Why are some alignments longer than others?

   ii. This is followed by *Descriptions of sequences producing significant alignments*, a table with links to database reports. The accession number is a unique identifier given to a sequence when it is submitted to a database, such as GenBank. The accession link leads to a detailed report on the sequence. Note the scores in the "E value" column on the right. The Expectation or E value is the number of alignments with the query sequence that would be expected to occur by chance in the database. The lower the E value, the higher the probability that the hit is related to the query. For example, an E value of 1 means that a search with your sequence would be expected to turn up one match by chance. What is the E value of your most significant hit and what does it mean? What does it mean if there are multiple hits with similar E values? What do the descriptions of significant hits have in common?

   iii. Next is an *Alignments* section, which provides a detailed view of each primer sequence (*Query*) aligned to the nucleotide sequence of the search hit (*Subject*). Notice that hits have matches to one or both of the primers: *FWA* forward = nucleotides 1–26 and *FWA* reverse = nucleotides 27–46.

3. Predict the length of the product that the primer set would amplify in a PCR (in vitro).

   i. Scroll down to the alignments section (third section) of the BLAST results page. Select a hit that matches both primer sequences.

   ii. To which positions do the primers match in the subject sequence?

   iii. The lowest and highest nucleotide positions in the subject sequence indicate the borders of the amplified sequence. Subtract the lowest nucleotide position in the subject sequence from the highest nucleotide position in the subject sequence. What is the difference between the coordinates?

   iv. Note that the actual length of the amplified fragment includes both ends. Thus, add 1 nucleotide to the result that you obtained in Step 3.iii to obtain the exact length of the PCR product amplified by the two primers.

4. Click on the accession link to open the data sheet for the hit used in Step 3.
   i. The top section of the report contains basic information regarding the sequence, including its base pair (bp) length, database accession number, source, and references to papers in which the sequence is published.
   ii. The middle section contains annotations of gene and regulatory features with their beginning and ending nucleotide positions ("xx..xx"). These features may include chromosome location, genes, coding sequences (cds), regulatory regions, ribosomal RNA (rRNA), and transfer RNA (tRNA).
   iii. Identify the features located between the nucleotide positions identified by the primers as determined in Step 3.ii. What feature is specifically related to gene transcription, what is its function, and how many nucleotides is it from the transcription start site?
5. Obtain the nucleotide sequence of the amplicon outlined by the *FWA* primer set.
   i. The bottom section of the data sheet lists the entire nucleotide sequence that contains the PCR product. Highlight all of the nucleotides between the coordinates that you identified in Step 3.ii, from the beginning of the forward primer to the end of the reverse primer.
   ii. Copy and paste the highlighted sequence into a text document. Then, delete all nonnucleotide characters and spaces. This is the amplicon, or amplified PCR product. Save this text document.
   iii. Experimentation has shown that 90% of DNA methylation in *FWA* occurs at CpG dinucleotides. ("CpG" designates two nucleotides joined by a phosphodiester bond on the same strand, not a C = G base pair.) Examine the nucleotide sequence of the amplicon and highlight each CpG dinucleotide. How many CpG dinucleotides can you identify? (Don't forget to look for the dinucleotide between line breaks!) What does this suggest regarding the wild-type gene in this region?
   iv. The McrBC enzyme recognizes methylated cytosines at the sequence G^{m}C or A^{m}C. Combining this information with the methylation sites from Step 5.iii above means that McrBC recognizes the half-sites GCG and ACG. Go back to the highlighted sequence from Step 5.iii and circle each place where a G or an A precedes the highlighted CG dinucleotide. These are the recognition sequences for McrBC. How many do you identify? How would you go about determining how many McrBC recognition sequences are on the reverse (complementary) strand? How many are there?
   v. What does all of this suggest regarding the relationship among repeats, the TATA box, and methylation?

II. Use Map Viewer to Determine the Chromosome Location of the *FWA* Gene

1. Open "Map Viewer" (http://www.ncbi.nlm.nih.gov/mapview).
2. Find "Plants" in the table to the right and click the drop-down menu; select "flowering plants." Under "eudicots," choose "*Arabidopsis thaliana* (common name of

thale cress)" and click on the "B" icon under the "Tools" header. If more than one build is displayed, select the one with the highest number because this will be the most recent version.

3. Select "BLASTN" from the top menu and paste the 846-bp amplicon defined by the primers into the search window. (Primers usually are not long enough to produce a result in the map BLAST.) Omit any nonnucleotide characters from the window because they will not be recognized by the BLAST algorithm.
4. Use the default settings and click on "BLAST."
5. In "Other reports," click on the "[Thale cress genome view]" link to see the chromosome location of the BLAST hit. A small horizontal bar on one or more chromosomes indicates the position of each hit. On which chromosome have you landed?
6. Click on the marked chromosome number to move to the *FWA* gene.
7. The chromosome is represented by one or more vertical lines. Click on the small blue arrow after the vertical line labeled "Genes_seq" to display the genes. The amplicon (vertical red bar) occupies the whole field of the default view.
8. Use the zoom toggle on the left to zoom out for a better perspective of the gene. You can refine your position on the chromosome by clicking on the thin vertical line under "Genes_seq" and choosing a specific zoom level. Which gene is closest to the amplicon?
9. Introns in the transcript are denoted by thin lines in between exons, which are represented by rectangular boxes along the line. Noncoding sequences in exons are shown with unfilled boxes and coding sequences with filled boxes.
10. Discover the properties of the *FWA* gene and amplicon.
    i. Determine the size of the *FWA* gene using the map coordinates to the left of the "Contig" map.
    ii. How many exons and introns does the gene have?
    iii. The direction of the transcript is marked with a small black arrow to the right of the gene symbol. Where does the amplicon appear in relation to the gene: in an exon, an intron, or upstream of or downstream from the transcript?

III. Identify the FWA Amino Acid Sequence and Function

1. Open the NCBI Internet site http://www.ncbi.nlm.nih.gov/.
2. Obtain the GenBank record for the *FWA* gene by pasting the *FWA* gene ID (AF178688.1) into the Search window at the top of the page. In the drop-down menu, change "All Databases" to "Nucleotide" and click search.
3. Follow the "*FWA*" link to the GenBank record.
4. Scroll down to translation, a sequence of letter abbreviations for the amino acids. Extract (highlight and copy) the protein sequence for the *FWA* gene.
5. Return to the NCBI home page: http://www.ncbi.nlm.nih.gov/.
6. Click on "BLAST" in the top speed bar.
7. Click on "protein-protein BLAST (blastp)."

8. Paste the FWA protein sequence into the Search window. Under "Choose Search Set," click on the database drop-down menu and select "non-redundant protein sequences." Next, under Program Selection, choose "blastp (protein-protein BLAST)."
9. Click on "BLAST" to send the amino acid sequence to the NCBI server.
10. An algorithm quickly scans the query sequence to identify functional domains (regions of the protein) that are conserved in different organisms. Follow links to get some quick information regarding the domains. What do you find?
11. Once the BLAST results appear, examine the results.
12. What can you tell from the graphical overview?
13. Scroll down to the list of significant alignments. What can you conclude from the E values, the titles of the hits, and the accession links?
14. Scroll down to the Alignments section to see just how well the FWA amino acid sequence matches that from other organisms. Follow the Accession links to sequence data sheets and then proceed to journal abstracts and articles to gain insight into how FWA and other homeobox proteins control development.

RESULTS AND DISCUSSION

I. Observe Plant Growth

1. Do you notice any growth or size differences between the wild-type *Ler* and mutant *fwa-1* plants?
2. How many days after planting did the *Ler* and *fwa-1* plants flower? What does this suggest is the normal function of *FWA*?

II. Think About the Experimental Methods

1. Describe the purpose of each of the following steps or reagents used in DNA isolation:
   - i. Using only a small amount of tissue.
   - ii. Grinding tissue with pestle.
   - iii. Nuclei lysis buffer.
   - iv. Protein precipitation solution.
   - v. Heating DNA to 65°C.
   - vi. McrBC incubation at 37°C.
   - vii. McrBC incubation at 65°C.
2. What is the purpose of performing each of the following PCR reactions?
   - i. Wild-type *Ler* undigested (WU).
   - ii. Wild-type *Ler* digested (WD).
   - iii. Mutant *fwa-1* undigested (MU).
   - iv. Mutant *fwa-1* digested (MD).

3. The recognition site for McrBC is G/A^{m}C...55–103 bp... G/A^{m}C, where ^{m}C represents a methylated cytosine. Why is McrBC an unusual restriction endonuclease? Why is it so effective in identifying regions that are imprinted by methylation?

III. Interpret Your Gel and Think About the Experiment

1. Observe the photograph of the stained gel containing your PCR samples and those from other students. Orient the photograph with the sample wells at the top. Use the sample gel shown below to help interpret the band(s) in each lane of the gel.

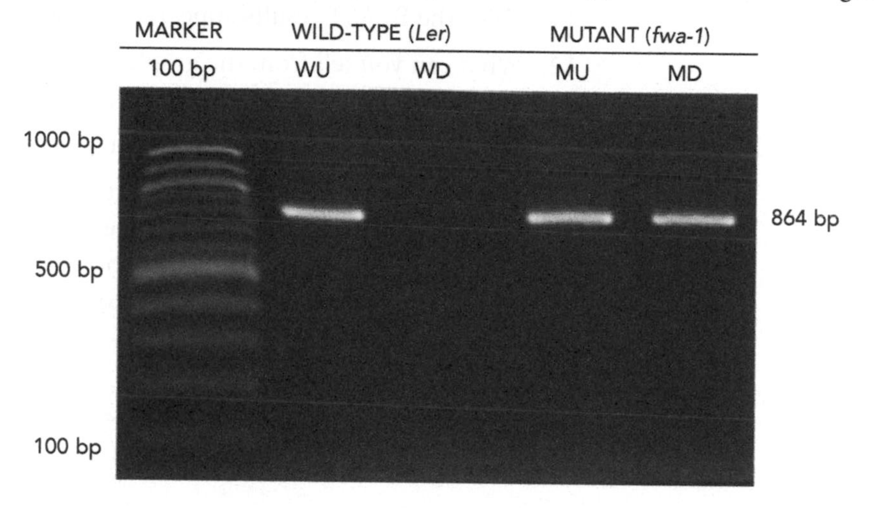

2. Locate the lane containing the 100-bp DNA ladder on the left side of the gel. Working from this well, locate the lowest band. This is a 100-bp fragment. Fragments increasing in increments of 100 bp—from 100 bp to 1000 bp—form the "rungs" of the ladder. The two largest bands are 1200 and 1517 bp. Fragments in the sample lanes may be close together, depending on how long the gel was run. The PCR product from the *FWA* gene is 846 bp; thus, it should align between the 800- and 900-bp markers.
3. It is common to see a diffuse (fuzzy) band that runs ahead of the 100-bp marker. This is a "primer dimer," an artifact of the PCR that results from the primers overlapping one another and amplifying themselves.
4. Which samples amplified and which ones did not? Give reasons why some samples may not have amplified; some of these may be errors in procedure.

Laboratory Planning and Preparation

Laboratories 3.1–3.6

AT THE OUTSET OF EACH LABORATORY EXERCISE, divide the class into teams and assign each team a number. This will make it easier to mark and identify the tubes used in each experiment.

OBTAINING SEEDS AND SPECIMENS

Seeds for Laboratories 3.1–3.3 may be obtained from the DNA Learning Center at www.dnalc@cshl.edu. Seeds for Laboratories 3.4 and 3.6 may be obtained from Carolina Biological Supply Company at www.carolina.com.

Wild-type *Ler* (Landsberg *erecta*) and mutant *fwa-1* can be ordered from The Arabidopsis Information Resource (TAIR; http://arabidopsis.org/). Click on Search> Seed/Germplasm and then search for CS28445 and CS106 under "germplasm/stock name."

Fluorescent light fixtures for growing *Arabidopsis* should be fitted with at least two 40-W "daylight" bulbs (not cool white). The following products suggested for *Arabidopsis* cultivation are available from Carolina Biological Supply Company (www.carolina.com). Products from garden stores may be substituted.

24-Cell Tray, item #66-5669

36-Cell Tray, item #66-5670

Poly-Flats (6-cm-deep cells that can be separated into individual pots)

Redi-Earth Soil (8-lb. bag), item #15-9701

Standard Poly-Tray without Holes (54 x 27 x 6 cm tray), item #66-5666

For detailed instructions on growing *Arabidopsis*, visit the Arabidopsis Biological Resource Center (ABRC; www.biosci.ohio-state.edu/~plantbio/Facilities/abrc/handling.htm). Should you wish to collect seeds for future experiments, the site also describes how to harvest and store seeds.

For Laboratory 3.4, have students bring in foods that they want to test for transgenes. Fresh or dry food products work well. Products that have been tested successfully using this procedure include corn and tortilla chips, artificial bacon bits, corn muffin mix, granola and energy bars, protein powder, and pet food. Food products should contain either soy or corn as an ingredient. Tissue samples from wild-type and Roundup Ready soybean plants are used as negative and positive controls in the experiment.

An extension of the laboratory would be to test plants for glyphosate sensitivity or resistance. To do this, obtain a bottle of commercially available glyphosate, such as

Roundup. Treat wild-type and Roundup Ready plants with glyphosate following the manufacturer's instructions, being careful to spray each plant with the same amount of herbicide. Avoid skin contact by wearing disposable gloves. As a control, leave some wild-type and genetically modified plants untreated. Be careful not to spray control plants accidentally. Observe any effects for 10 d. A further extension would be to do a dose-response curve by testing the effect of different concentrations of herbicide on plants.

For Laboratory 3.5, the collection of specimens may support a census of life in a specific area or habitat, an evaluation of products purchased in restaurants or supermarkets, or contribute to a larger "campaign" to assess biodiversity across large areas. It may make sense for you to use sampling techniques from ecology. For example, a quadrat samples the plant and/or animal life in one square meter (or 1/4 square meter) of habitat, whereas a transect collects samples along a fixed path through a habitat.

Use common sense when collecting specimens. Respect private property; obtain permission to collect in nonpublic places. Respect the environment; protect sensitive habitats and collect only enough of a sample for barcoding. Do not collect specimens that may be threatened or endangered. Be wary of poisonous or venomous plants and animals. Do not take more sample than you need. Only a small amount of tissue is needed for DNA extraction—a piece of plant leaf ~1/4 in in diameter or a piece of animal tissue the size of a pencil eraser. Keep in mind that you will also need a small sample for classical taxonomic analysis and to act as a reference sample if you plan to submit your data to GenBank. Taxonomic keys for local plants or animals are often available online, at libraries, or from universities, natural history museums, and botanical gardens.

Minimize damage to living plants by collecting a single leaf or bud or several needles. When possible, use young, fresh leaves or buds. Flexible, nonwaxy leaves work best. New leaves and buds have about the same number of cells as mature leaves, so they contain about the same amount of DNA in a smaller volume of tissue. The cell walls are thinner than in mature plant materials, making them easier to break during the mechanical grinding used for DNA extraction. Tougher materials, such as pine needles or holly leaves, can work if the sample is kept small and is well ground. Dormant leaf buds can often be obtained from bushes and trees that have dropped their leaves. Fresh frozen leaves work well. Dried leaves and herbarium samples are variable. Avoid twigs or bark. If woody material must be used, select flexible twigs with soft pith inside. As a last resort, scrape a small sample of the softer, growing cambium just beneath the bark. Roots and tubers are a poor choice, because high concentrations of storage starches and other sugars can interfere with DNA extraction. "Field Techniques Used by Missouri Botanical Garden" contains many good methods for collecting and preparing plant specimens (www.mobot.org/mobot/molib/fieldtechbook/handbook.pdf).

Small invertebrate animals, such as insects, can be collected whole and euthanized by placing them in a freezer for several hours. Samples of muscle tissue can be taken from animal foods such as fish, poultry, or red meat. Blood, internal organs, and bone marrow are all good sources of DNA. Bone and skin are difficult. Fresh and frozen samples work equally well.

Other than fish, do not collect vertebrate animals. Use care if collecting from road-kill animals and avoid animal droppings that are possible vectors for disease.

ISOLATING DNA

To extract DNA, tissues are broken up by grinding them with a mortar and pestle. This can be accomplished directly in a 1.5-mL tube using a plastic pestle. A no-cost pestle can be made by heating a 1-mL pipette tip in a gas flame until it just melts, forcing the melted tip into a 1.5-mL tube, and twisting to obtain a smooth surface.

These experiments also require a heating block or a boiling water bath at 96°C–100°C. A boiling water bath can be made by placing the tubes in a floating rack within a beaker of water on top of a hot plate or by filling a beaker with water, covering it tightly with a double layer of aluminum foil, and using a pencil to punch holes in the aluminum foil to hold the tubes. If using aluminum foil, make sure that the tubes are immersed and add hot water as necessary to maintain the water level. Regulate the temperature to maintain a low boil and carefully watch for lids opening as the tubes heat.

For Laboratories 3.1 and 3.2, make sure that each student team extracts DNA from either a wild-type or mutant (*bz* or *clf-2*) plant and that both types of plants are being examined by the class as a whole. For Laboratory 3.3, each team should extract DNA from both a control (either Col or Ler) and mutant (*ago1*) plant; teams ultimately share results for the Col and Ler controls.

For Laboratory 3.4, each group should extract DNA from the food product of choice, ideally one that contains either soy or corn as an ingredient. Tissue samples from wild-type and Roundup Ready soybean plants are used as negative and positive controls in the experiment.

In Laboratory 3.5, two methods for DNA extraction are provided. The first is a universal DNA extraction method that uses a commercial kit. Although it is more expensive than the alternate method for plants using Edward's buffer, it has the advantage of working reproducibly with almost any kind of plant or animal specimen. The Wizard Genomic DNA Purification kit is available from the Promega Company, catalog #A1120 (100 preps) or #A1125 (500 preps). This extraction method is also used in Laboratory 3.6. The second method is optimized for plants. Although it takes about 20 min longer than the previous method, it uses readily available reagents.

Do not use more tissue than recommended for these DNA extraction procedures. Using a small amount of tissue reduces carry-forward of PCR inhibitors present in the sample. These include metal ions (from plants and animals) and polysaccharides and secondary metabolites (from plants).

AMPLIFYING DNA BY PCR

Each Ready-To-Go PCR Bead contains reagents so that a final reaction volume of 25 μL contains 2.5 U *Taq* DNA polymerase, 10 mM Tris-HCl (pH 9.0), 50 mM KCl, 1.5 mM $MgCl_2$, and 200 μM of each dNTP. Each primer/loading dye mix includes the appropriate primer pair (0.26 pmol/μL of each primer), 13.8% sucrose, and 0.0081% cresol red.

The lyophilized *Taq* DNA polymerase in the Ready-To-Go PCR Bead becomes active immediately after adding primer/loading dye mix. In the absence of thermal cycling, nonspecific priming at room temperature allows *Taq* DNA polymerase to begin generating erroneous products that can show up as extra bands in gel analysis.

Therefore, work quickly! Make sure that the thermal cycler is set and have all experimenters set up their PCRs as a coordinated effort. Add primer/loading dye mix to all reaction tubes, add each template, and begin thermal cycling as quickly as possible. Hold the reactions on ice until all groups are ready to load into the thermal cycler.

For convenience, these laboratories are designed to be used with Ready-To-Go PCR Beads and primer/loading dye mix. If desired, all PCR reaction components can be prepared from scratch by following the recipes included in Recipes for Reagents and Stock Solutions. Each 25-μL reaction should contain the following:

PCR buffer (10X)	2.5 μL
dNTPs (10 mM)	1.0 μL
Forward primer (20 μM)	1.25 μL
Reverse primer (20 μM)	1.25 μL
Taq polymerase (1–5 U/μL)	1 μL
Template DNA	2.5 μL
Distilled or deionized water	15.5 μL

Surprisingly, a large number of failures in this experiment are due to the simple fact that students did not get template DNA into their PCRs! Observe each student adding DNA to the reaction. Students must see the small volume of template DNA in their pipette tip, then eject the DNA directly into the reaction, and finally confirm that the tip is empty before removing it from the PCR tube.

If your thermal cycler does not have a heated lid, add a drop of mineral oil on top of each reaction before thermal cycling. Be careful to avoid touching the dropper tip to the tube or sample; this will contaminate the mineral oil.

PCR amplification from crude cell extracts is biochemically demanding and requires the precision of automated thermal cycling. However, amplification of the loci in these experiments is not complicated by the presence of repeated units. Therefore, the recommended amplification times and temperatures will work adequately for all types of thermal cyclers.

In Laboratories 3.1 and 3.2, the *Ds* insertion is too large to amplify across, so a single primer set cannot amplify both the wild-type and insertion alleles. Thus, wild-type and insertion alleles are amplified in separate PCRs using two different sets of primers. One primer set spans the *Ds* insertion site and amplifies only the wild-type allele. One of the primers adjacent to the insertion site is paired with a primer located within the *Ds* transposon to amplify the insertion allele.

Laboratory 3.3 uses four PCRs to analyze each plant. Each PCR requires a specific primer set to amplify one of four CAPS markers on chromosome 1: g4026, H77224, m235, and UFO.

For Laboratory 3.4, each team tests a food product of their choice and either a positive or a negative control. Two PCRs are performed for each food sample or control plant; therefore, each lab team will set up a total of four reactions. One primer set amplifies the 35S promoter from cauliflower mosaic virus. The presence of a 35S product is diagnostic for the presence of a transgene, because the 35S promoter is used to drive expression of the glyphosate (Roundup) resistance gene or the *Bt* gene in edible crops. A second primer set amplifies a fragment of a tubulin gene and controls for the presence of plant template DNA. Because the tubulin gene is found in all plant genomes, the presence of a tubulin product indicates amplifiable DNA in the sample isolated.

For Laboratory 3.5, each team performs two PCRs. Each gel should contain one lane with a size marker and the rest of the lanes can be used for amplified fragments. Be sure that students save 20 μL of amplified fragment; this is required for sequencing. DNA barcoding relies on finding a universal chromosome location (locus) that has retained enough sequence conservation through evolutionary history that it can be identified in many organisms, but it also has enough sequence diversity to differentiate organisms to at least the family level. Regions of the chloroplast *rbcL* gene and the mitochondrial *COI* gene fulfill these requirements.

Primers are designed to target conserved sequences that flank the variable *rbcL* and *COI* and barcode regions. However, even the conserved flanking regions have accumulated enough sequence differences over evolutionary time that it is impossible to identify a universal primer set for *rbcL* or *COI* or that will work across all taxonomic groups of plants and animals. Thus, barcode primers often need to accommodate sequence variation, or degeneracy, at one or several nucleotide positions.

The degeneracy problem is solved when the oligonucleotides primers are synthesized or primer mixes are made. Traditionally, a mixture of primers is synthesized—each having a different nucleotide at variable positions. However, synthetic nucleotides are now available that pair with multiple nucleotides, which are incorporated at variable positions in a single primer.

The table below shows the letter abbreviation given for degenerate nucleotides. For example, a primer with the sequence "ATCCR" contains both ATCCA and ATCCG.

W = A or T	K = G or T
B = C or G or T	V = A or C or G
S = G or C	R = A or G
D = A or G or T	N = A or C or G or T
M = A or C	Y = C or T
H = A or C or T	

Even degenerate primers cannot ensure amplification in taxonomic groups in which all or part of a particular primer sequence is deleted. Thus, broad surveys of unknown plants or animals typically use multiple primer sets.

For Laboratory 3.6, each team amplifies DNA from four samples: wild-type and *FWA* DNA that is uncut, as controls, or cut with the methylation-sensitive restriction enzyme McrBC to test for DNA methylation. The primer set amplifies a region that spans the *FWA* promoter and a 5′ repeat region that is methylated in wild type, but demethylated in *FWA* plants. The presence of amplicon for undigested samples confirms that the DNA isolation was successful. After digestion, the wild-type allele is cut by McrBC and should not amplify, and the DNA from *FWA* plants should not be cut and should amplify.

DIGESTING PCR PRODUCTS WITH RESTRICTION ENZYMES

In Laboratory 3.3, restriction digests of PCR products generate restriction-fragment-length polymorphisms (RFLPs) that differentiate chromosomes of the Col and Ler ecotypes. Restriction digests are not performed in any of the other laboratory exercises.

Before restriction enzyme digestion, it is always prudent to analyze PCR products by agarose gel electrophoresis to confirm that the amplification has produced the expected amplicons. However, gel analysis adds one or two additional lab periods to the experiment. Thus, as a time-saver, this step has been omitted from the laboratory exercise. If you like, you may have students electrophorese 5 µL of each PCR product on a 2% agarose gel and save the remaining 20 µL of each PCR product for the restriction enzyme digestion. The expected PCR product sizes are 844 bp (for g4026), 222 bp (for H77224), 531 bp (for m235), and 1294 bp (for UFO).

Students set up separate restriction digests for each CAPS marker, using DNA isolated from an *ago1* mutant and a control (either Col or Ler). Three restriction enzymes are used, and it is imperative to use the correct reaction conditions (buffer and incubation temperature) for each. To simplify the method and minimize error, we recommend that a combination restriction enzyme/buffer mix be made up for each enzyme according to the recipe provided on page 481. Note that HindIII and RsaI are incubated at 37°C, whereas TaqI is incubated at 65°C. Incubate the reactions for 60 min or more to ensure complete cutting. If you prefer, reactions can be incubated for several hours or even overnight. After several hours, the enzyme will denature and lose activity. Stop the reactions whenever it is convenient and store them in a freezer (–20°C).

Laboratory 3.6 relies on efficient cutting of methylated DNA by the methylation-dependent enzyme McrBC. If digestion is not complete, even a small amount of template DNA may be amplified during PCR, which could make it difficult to determine the difference in methylation in wild-type and *fwa-1* plants. For this reason, it is important to incubate the restriction digest for at least 1 h, but overnight digestion is recommended. McrBC requires guanosine 5′-triphosphate and BSA for efficient cutting. To minimize the effect of pipetting errors, make up a restriction buffer mix according the recipe on page 481.

ANALYZING DNA BY GEL ELECTROPHORESIS

Preparing and Loading Gels

The cresol red and sucrose in the primer mix function as loading dye, so that amplified samples can be loaded directly into an agarose gel. This is a nice time-saver. However, because the concentrations of sugar and cresol red are low, this mix is more difficult to use than typical loading dyes. Students should be encouraged to load carefully.

If the Ready-To-Go PCR Beads and primer/loading dye mix are not used and, instead, a traditional PCR is run, 10x loading dye should be added to each reaction before it is run out on the gel.

Plasmid pBR322 digested with the restriction endonuclease BstNI is an inexpensive marker and produces fragments that are useful as size markers in this experiment. Use 20 µL of a 0.1 µg/µL stock solution of this DNA ladder per gel. Other markers or a 100-bp ladder may be substituted.

In Laboratory 3.6, if mineral oil was used during PCR, pierce the pipette tip through the layer of mineral oil to withdraw the PCR sample. Do not pipette any mineral oil; leave the mineral oil behind in the original tube.

In Laboratories 3.1, 3.2, and 3.4, two PCRs are performed for each DNA sample. Because the amplified fragments expected in the two PCRs in Laboratory 3.2 differ sufficiently in size, they are mixed together before electrophoresis to simplify the analysis of the three genotypes.

For Laboratory 3.3, teams ultimately share Col and Ler controls for analysis on agarose gels. For each CAPS marker, the electrophoretic panel should include one Col control, one Ler control, and one size marker. The rest of the available lanes in the gel can be filled with the *ago1* mutants of different lab teams. For example, up to five teams could share one Col control and one Ler control on a gel with eight lanes. The experiment is set up so that two teams ultimately share one Col control and one Ler control for analysis on agarose gels, but the number of PCRs can be reduced, depending on the number of lanes available in the agarose gels cast by your apparatus.

Staining Gels

In Part IV of Laboratories 3.1, 3.2, 3.4, and 3.5 and Part V of Laboratories 3.3 and 3.6, SYBR Green, a fluorescent staining dye, is added to each sample before electrophoresis. According to the Ames test (a method to estimate the mutagenic properties of a chemical), SYBR Green is much less mutagenic than the classical stain ethidium bromide. Furthermore, bands are immediately visible after electrophoresis and no post-staining steps are necessary. SYBR Green can be imaged with the same filter set and UV transillumination used with ethidium bromide, but SYBR Green is more sensitive than ethidium bromide and produces much less background on a stained agarose gel.

As an alternative to SYBR Green, the gel can be stained with ethidium bromide or *Carolina*BLU after electrophoresis but before viewing and photographing. To stain the gel using ethidium bromide or *Carolina*BLU, omit Step 7 from the procedure (Part IV of Laboratories 3.1, 3.2, 3.4, and 3.5 and Part V of Laboratories 3.3 and 3.6) and then, after Step 9, follow one of the methods described under "Staining a Gel with Ethidium Bromide or *Carolina*BLU" below.

Always view and photograph gels as soon as possible after electrophoresis or staining/destaining. Over time, the small-sized PCR products will diffuse through the gel, lose sharpness, and disappear. Refrigeration will slow diffusion somewhat, so if absolutely necessary, gels can be wrapped in plastic wrap and stored for up to 24 h at 4°C. For best results, view and photograph gels immediately after staining/destaining is complete.

Staining a Gel with Ethidium Bromide or Carolina*BLU*

REAGENTS, SUPPLIES, & EQUIPMENT

*Carolina*BLU Final Stain (for *Carolina*BLU staining only)
Digital camera or photodocumentary system
Ethidium bromide <!> (1 µg/mL) (for ethidium bromide staining only)
Gel containing DNA fragments that have been separated by electrophoresis
Staining trays
UV transilluminator <!> and eye protection (for ethidium bromide staining only)
White light transilluminator (for *Carolina*BLU staining only)

See Cautions Appendix for appropriate handling of materials marked with <!>.

A. Staining with Ethidium Bromide

1. Place your gel in a staining tray and cover it with ethidium bromide solution. Allow it to stain for 10–15 min.

2. Decant the stain back into the storage container for reuse and rinse the gel in tap water.
3. View the gel using UV transillumination. Photograph the gel using a digital camera or photodocumentary system.

B. Staining with Carolina*BLU*

1. Place your gel in a staining tray and cover it with *Carolina*BLU Final Stain. Allow it to sit for 20–30 min with (optional) gentle agitation.
2. After staining, pour the stain back into the bottle for future use. (The stain can be used six to eight times.)
3. To destain the gel, cover it with deionized or distilled water.
4. Change the water three to four times over the course of 30–40 min. Agitate the gel occasionally.
5. View the gel using white light transillumination. Photograph the gel using a digital camera or photodocumentary system.

SEQUENCING PCR PRODUCTS

DNA sequencing of the *rbcL* or *COI* amplicon is required to determine the nucleotide sequence that constitutes the DNA barcode. The forward, reverse, or both DNA strands of the amplified barcode region may be sequenced. A single, good-quality barcode from the forward strand is sufficient to identify an organism. The majority of database sequences are from the forward strand, so sequencing only the forward strand reduces sequencing cost and simplifies analysis. If you only do a forward read, save the remaining 10 μL of amplicon. If the forward read fails and time permits, send the remainder out to sequence the reverse strand.

Bidirectional sequencing, however, is important for several reasons. (1) A reverse sequence may provide a readable barcode when the forward sequence fails. (2) Good forward and reverse reads can be combined to produce a consensus sequence that extends the read by up to 40 or more nucleotides. This is because the primer itself is not sequenced for either strand, and additional nucleotides downstream from the primer are typically unreadable. Thus, good forward and reverse primers complement these missing sequences, adding most of the primer sequences on either end. (3) One direction may provide a read through a region that is refractory to sequencing in the other direction, such as a homopolymeric region containing a long string of C residues. Thus, the insurance provided by bidirectional sequencing may be worth the added cost, especially if you must complete an analysis in a limited amount of time.

Sequencing different barcode regions—*rbcL* and *COI*—and using degenerate and multiplex primers complicate DNA sequencing. Strictly speaking, each different primer would have to be provided for forward and reverse sequencing reactions. As a work-around for this problem, the primers used in this experiment incorporate a universal M13 primer sequence. In addition to a sequence specific to the *rbcL* or *COI* barcode locus, the 5′ end of each primer has an identical 17- or 18-nucleotide sequence from the bacteriophage vector M13.

In the traditional approach to genome sequencing, genomic DNA is cloned into an M13 vector. A universal M13 primer is then used to sequence the genomic insert just downstream from the primer. This same strategy is used in sequencing *rbcL* and *COI* barcodes in this experiment. During the first cycle of PCR, the M13 portion of the primer does not bind to the template DNA. However, the entire primer sequence is covalently linked to the newly synthesized DNA and is amplified in subsequent rounds of PCR. Thus, the M13 sequence is included in every full-length PCR product. This allows a sequencing center to use universal forward and reverse M13 primers for the PCR-based reactions that prepare any *rbcL* or *COI* amplicon for sequencing.

The sequences of the M13 forward and reverse primers are

M13F(–21): TGTAAAACGACGGCCAGT
M13R(–27): CAGGAAACAGCTATGAC

Using GENEWIZ DNA Sequencing Services

We recommend using GENEWIZ, Inc. for DNA barcode sequencing. GENEWIZ has optimized reaction conditions for producing the barcode sequences in this laboratory and produces excellent-quality sequence with rapid turnaround—usually within 48 h of receipt of samples. GENEWIZ sequences are automatically uploaded to the DNALC's *DNA Subway* website.

Before submitting samples for sequencing, consult the GENEWIZ guide at http://www.GENEWIZ.com/public/Sample-Submission-Guideline.aspx.

Prepare PCR products

1. Verify that you can see a PCR product of the correct predicted length on an agarose gel. DNA sequence can be obtained from virtually every PCR product that is visible on the gel. (You can take a chance on samples that do not produce bands, because a fair proportion of these will also produce sequences.)
2. Prepare eight strips of 0.2-mL PCR tubes appropriate for the number of samples that you wish to submit. If you are submitting a large number of samples (≥48), submit the samples on a 96-well plate, arranging them vertically (A1–H1). For more details, see the "Tubes and Plates" tab at http://www.GENEWIZ.com.
3. Submit 10 µL of PCR product for *each* sequencing reaction. Forward and reverse sequencing reactions must be submitted in separate tubes.

Register and Submit Samples for Sequencing

1. Go to http://www.GENEWIZ.com and click "Register" to create a user account.
2. When creating your account, it is very important to enter your institution name followed by "–DNALC." The suffix "–DNALC" must be added exactly to your name; otherwise, your sequence will not be processed properly or may be delayed.
3. Obtain a valid purchase order number from your purchasing department or use a valid credit card.
4. Log in to your user account to place your sequencing order. Under "Place an Order," select "Create Sequencing Order."

5. Under "Service Priority:," select "Standard."
6. Under "Create Order by:," select "Online Form." (Alternatively, you can select "Upload Excel Form," download the "Custom" GENEWIZ Excel template, fill in the information in Steps 10–18, and upload the file.)
7. Under "Sample Type," select "Custom."
8. When prompted to "Create an online form for," enter the number of samples that you will be sending for sequencing. (If you elect to do bidirectional sequencing, you must count separate forward and reverse reactions for each sample).
9. Click on "Create New Form" and a sample submission form will be displayed.
10. For "DNA Name," enter a name for each sample. This may be a number or initials.
11. For "DNA Type," select "Un-Purified PCR" from the drop-down menu.
12. For "DNA Length (vector + insert in bp)," enter 650 for *rbcL* or 800 for *COI*.
13. Leave "DNA Conc. (ng/µl)" blank. It is best to send in a *gel image* of representative samples. This will be used by GENEWIZ to calculate the correct amount of cleanup reagents and product to use in the sequencing reaction. If a gel image is not supplied, GENEWIZ will use default amounts to set up the sequencing reactions.
14. Leave "My Primer Name" and "My Primer Conc." blank.
15. For "GENEWIZ Primer," select one from the following drop-down menu:
    - "M13F(–21)" to sequence the forward strand.
    - "M13R" to sequence the reverse strand.
16. Under "Special Request," be sure that "PCR-Clean Up" has been automatically selected. (This is the default when unpurified PCR is selected as the DNA type).
17. In the "Comments" box at the bottom of the form, type "Primer stored at GENEWIZ under DNALC."
18. Click on "Save & Next."
19. Carefully review your form and click on "Next Step."
20. Enter your payment information and click on "Next Step."
21. Review your order and click on "Submit."

Ship Samples to GENEWIZ

1. Print a copy of the order form to mail along with your samples.
2. Be sure that the tubes are labeled exactly the same in the gel photo and on the order form. Failure to do so may delay sequencing or make it impossible to complete. E-mail DNALCSeq@cshl.edu if you need help.
3. Ship your samples via standard overnight delivery service (use Federal Express, if possible).
4. Pack your samples in a letter pack or small shipping box, making sure to pad samples to prevent too much shifting. Room temperature shipping—with no ice or

ice pack—is expected. PCR products are stable at ambient temperature, even if shipped on a Friday for Monday delivery.

5. Address the shipment to GENEWIZ at the following location:

   GENEWIZ, Inc.
   115 Corporate Blvd.
   South Plainfield, NJ 07080

6. You may be able to reduce shipping costs by using a GENEWIZ drop box. Call 1-877-436-3949 to determine whether one is available in your area.

Sequencing On Your Own

To have student amplicons sequenced at your own institution or at a local university or company, you may perform the cycle-sequencing reaction and DNA purification in the instructions that follow, but check with a sequencing technician that these methods are compatible with the protocols in use at their facility. For speed, reproducibility, and ease of tracking samples, the following procedure is optimized for a 96-well microtiter plate, which is the most common format used for DNA sequencing.

Perform the Cycle-Sequencing Reaction

REAGENTS, SUPPLIES, & EQUIPMENT

Amplicons (PCR products)* from Part III of Laboratory 3.5
Container with cracked or crushed ice
Cycle-sequencing master mix
Micropipettes and tips (1–20 µL)
Microtiter plate (96 well)
Sealing foil
Thermal cycler

*Store on ice.

1. Label the microtiter plate with an identifying name and date.
2. Add 8 µL of cycle-sequencing master mix to each well of the microtiter plate that will receive a reaction.
3. To each well containing master mix, add 2 µL of an amplicon from Part III of Laboratory 3.5. (Each well should receive an amplicon from a different group.)
4. Cover the microtiter plate with sealing foil. Carefully seal the plate by pressing on each well of the plate with your thumb and by running your finger around the edge of the foil.

Be sure to create a tight seal, otherwise some samples will evaporate during cycling.

5. Store the reactions on ice or in the freezer until ready to begin thermal cycling.
6. Place the microtiter plate in a thermal cycler that has been programmed for one 2-min step at 96°C, followed by 34 cycles of the following profile:

Denaturing step:	10 sec	96°C
Annealing step:	5 sec	50°C
Extending step:	4 min	60°C

   The profile may be linked to a 4°C hold program after the 34 cycles are completed.

Precipitate the Dye-Labeled DNA

REAGENTS, SUPPLIES, & EQUIPMENT

Clinical centrifuge with spin trays
Cycle-sequencing reactions in 96-well microtiter plate
Ethanol (70%) <!>
Ethanol:sodium acetate (30:1)
Freezer
Micropipette and tips (20–100 μL)
Paper towels
Sealing foil
Thermal cycler (optional)
Vortexer

See Cautions Appendix for appropriate handling of materials marked with <!>.

Samples should turn a cloudy, yellowish white.

1. Add 30 μL of 30:1 ethanol:sodium acetate to each well of the microtiter plate. Mix by pipetting in and out several times. Do not vortex.

Ethanol precipitates the extended PCR products but leaves the primer and unincorporated nucleotides in solution. Centrifugation pellets the precipitated sequencing product on the bottom of the microtiter well.

2. Cover the microtiter plate with sealing foil and carefully seal. Incubate the plate in the freezer for 30 min at –20°C.
3. Balance the microtiter plate in a clinical centrifuge equipped with spin trays. Centrifuge the plate for 30 min at room temperature.
4. Remove the sealing foil from the microtiter plate. Invert the plate over the sink to drain off most of the ethanol solution. Then, gently tap the plate on the surface of a clean paper towel to wick off as much liquid as possible.

To prevent cross-contamination between wells, move the plate to a clean area of toweling between each tap.

5. Add 50 μL of 70% ethanol into each well to wash the DNA pellet.
6. Carefully seal the microtiter plate with clean foil.
7. Balance the microtiter plate in a clinical centrifuge and centrifuge the plate for 15 min at room temperature.

DNA pellets are not soluble in ethanol and will not resuspend during washing.

8. Remove the sealing foil from the microtiter plate. Invert the plate over the sink to drain off most of the ethanol solution. Then, gently tap the plate on the surface of a clean paper towel to wick off as much liquid as possible.
9. Repeat Steps 5–8 to wash, centrifuge, and drain the DNA pellets a second time.

As an alternative to Step 11, incubate the plate without foil in a thermal cycler for 10 min at 36°C. Do not close thermal cycler lid!

10. If possible, air-dry the DNA pellets overnight.
11. After the microtiter plate is thoroughly dry, add 20 μL of autoclaved, distilled water to each well on the microtiter plate.
12. Carefully seal the microtiter plate with clean foil.

Every well on the microtiter plate—including empty wells without DNA samples—must be filled with 20 μL of water before being loaded onto the DNA sequencer.

13. Dissolve the DNA in the wells by vortexing the microtiter plate at half speed. Tap the plate on the benchtop to bring the DNA droplets to the bottom of the wells.
14. Freeze the microtiter plate at –20°C until ready to read the DNA samples on an automated sequencer such as an ABI 3700 DNA Analyzer. If desired, *Carolina*BLU Gel and Buffer Stain can be used to stain the DNA while it is being electrophoresed. This prestaining procedure will allow you to visualize your results before the end of the gel run. However, the standard staining procedure outlined in "Staining with *Carolina*BLU" above is still required for optimum viewing.

Recipes for Reagents and Stock Solutions

Laboratories 3.1–3.6

THE SUCCESS OF THE LABORATORY EXERCISES DEPENDS on the use of high-quality reagents. Follow the recipes with care and pay attention to cleanliness. Use a clean spatula for each ingredient or carefully pour each ingredient from its bottle.

The recipes are organized alphabetically within four sections, according to experimental procedure. Stock solutions that are used for more than one procedure are listed once, according to their first use in the laboratories.

CAUTION: See Cautions Appendix for appropriate handling of materials marked with <!>.

- Isolating DNA
    - Edward's Buffer
    - Ethylenediaminetetraacetic Acid (EDTA) (0.5 M, pH 8.0)
    - Sodium Chloride (NaCl) (5 M)
    - Sodium Dodecyl Sulfate (SDS) (10%)
    - Tris/EDTA (TE) Buffer
    - Tris-HCl (1 M, pH 8.0 and 8.3)
- Amplifying DNA by PCR
    - Cresol Red Dye (1%)
    - Cresol Red Loading Dye
    - dNTPs (10 mM)
    - Magnesium Chloride ($MgCl_2$) (1 M)
    - PCR Buffer (10x)
    - Potassium Chloride (KCl) (5 M)
    - Primer/Loading Dye Mix
- Digesting PCR Products with Restriction Enzymes
    - Restriction Enzyme/Buffer Mix
    - Restriction Buffer Mix
- Analyzing DNA by Gel Electrophoresis
    - 100-bp DNA Ladder (0.125 µg/µL)
    - Agarose (2%)
    - Ethidium Bromide Staining Solution (1 µg/mL)
    - Loading Dye (10x)
    - pBR322/BstNI Size Marker (0.1 µg/µL)
    - SYBR Green DNA Stain
    - Tris/Borate/EDTA (TBE) Electrophoresis Buffer (1x)
    - Tris/Borate/EDTA (TBE) Electrophoresis Buffer (20x)

General Notes

- Typically, solid reagents are dissolved in a volume of deionized or distilled water equivalent to 70%–80% of the finished volume of buffer. This leaves room for the addition of acids or base to adjust the pH. Finally, water is added to bring the solution up to the final volume.

- Buffers are typically prepared as 1x or 10x solutions. Solutions at a concentration of 10x are diluted when mixed with other reagents to produce a working concentration of 1x.
- Storage temperatures of 4°C and –20°C refer to normal refrigerator and freezer temperatures, respectively.

ISOLATING DNA

Edward's Buffer

Makes 50 mL.
Store indefinitely at room temperature.

1. Combine the following in a 50-mL tube:

Deionized or distilled water	32.5 mL
Tris-HCl (1 M, pH 8.0)	10 mL
NaCl (5 M)	2.5 mL
EDTA (0.5 M, pH 8.0) <!>	2.5 mL
SDS (10%) <!>	2.5 mL

2. Mix thoroughly.

Ethylenediaminetetraacetic Acid (EDTA) (0.5 M, pH 8.0)

Makes 100 mL.
Store indefinitely at room temperature.

1. Add 18.6 g of EDTA <!> (disodium salt dihydrate, MW 372.24) to 80 mL of deionized or distilled water.
2. Adjust the pH by slowly adding ~2.2 g of sodium hydroxide (NaOH) <!> pellets (MW 40.00); monitor with a pH meter or strips of pH paper. (If neither is available, adding 2.2 g of NaOH pellets will make a solution of ~pH 8.0.)
3. Mix vigorously with a magnetic stirrer or by hand.
4. Add deionized or distilled water to make a total volume of 100 mL of solution.
5. Make sure that the bottle cap is loose and autoclave for 15 min at 121°C.
6. After autoclaving, cool the solution to room temperature and tighten the lid for storage.

Notes: Use only the disodium salt of EDTA. EDTA will only dissolve after the pH has reached 8.0 or higher.

Sodium Chloride (NaCl) (5 M)

Makes 500 mL.
Store indefinitely at room temperature.

1. Dissolve 146.1 g of NaCl (MW 58.44) in 250 mL of deionized or distilled water.
2. Add deionized or distilled water to make a total volume of 500 mL of solution.

Sodium Dodecyl Sulfate (SDS) (10%)

Makes 100 mL.
Store indefinitely at room temperature.

1. Dissolve 10 g of electrophoresis-grade SDS <!> (MW 288.37) in 80 mL of deionized or distilled water.
2. Add deionized or distilled water to make a total volume of 100 mL of solution.

Note: SDS is the same as sodium lauryl sulfate.

Tris/EDTA (TE) Buffer

Makes 100 mL.
Store indefinitely at room temperature.

1. Combine the following:

Deionized or distilled water	99 mL
Tris-HCl (1 M, pH 8.0)	1 mL
EDTA (0.5 M, pH 8.0) <!>	200 µL

2. Mix thoroughly.

Tris-HCl (1 M, pH 8.0 and 8.3)

Makes 100 mL.
Store indefinitely at room temperature.

1. Dissolve 12.1 g of Tris base <!> (MW 121.10) in 70 mL of deionized or distilled water.
2. Adjust the pH by slowly adding concentrated hydrochloric acid (HCl) <!> for the desired pH listed below.

pH 8.0	5.0 mL
pH 8.3	4.5 mL

   Monitor with a pH meter or strips of pH paper. (If neither is available, adding the volumes of concentrated HCl listed here will yield a solution with approximately the desired pH.)
3. Add deionized or distilled water to make a total volume of 100 mL of solution.
4. Make sure that the bottle cap is loose and autoclave for 15 min at 121°C.
5. After autoclaving, cool the solution to room temperature and tighten the lid for storage.

Notes: A yellow-colored solution indicates poor-quality Tris. If your solution is yellow, discard it and obtain a Tris solution from a different source. The pH of Tris solutions is temperature dependent, so make sure to measure the pH at room temperature. Many types of electrodes do not accurately measure the pH of Tris solutions; check with the manufacturer to obtain a suitable one.

AMPLIFYING DNA BY PCR

Cresol Red Dye (1%)

Makes 50 mL.
Store indefinitely at room temperature.

1. Weigh out 500 mg of cresol red dye <!>.
2. In a 50-mL tube, mix the cresol red dye with 50 mL of distilled water.

Cresol Red Loading Dye

Makes 50 mL.
Store indefinitely at –20°C.

1. In a 50-mL tube, dissolve 17 g of sucrose in 49 mL of distilled water.
2. Add 1 mL of 1% cresol red dye <!> and mix well.

dNTPs (10 mM)

Makes 100 µL.
Store for up to 1 yr at –20°C.

1. Add 10 µL each of 100 mM dTTP, dATP, dGTP, and dCTP.
2. Add 60 µL of deionized or distilled water and mix.

Magnesium Chloride ($MgCl_2$) (1 M)

Makes 100 mL.
Store indefinitely at room temperature.

1. Dissolve 20.3 g of $MgCl_2 \cdot 6H_2O$ <!> (MW 203.30) in 80 mL of deionized or distilled water.
2. Add deionized or distilled water to make a total volume of 100 mL of solution.
3. Make sure that the bottle cap is loose and autoclave for 15 min at 121°C.
4. After autoclaving, cool the solution to room temperature and tighten the lid for storage.

PCR Buffer (10x)

Makes 10 mL.
Store indefinitely at –20°C.

1. Combine the following in a 15-mL tube:

Deionized or distilled water	7.85 mL
Tris-HCl (1 M, pH 8.3)	1 mL
KCl (5 M) <!>	1 mL
$MgCl_2$ (1 M)	0.15 mL

2. Mix well.

Potassium Chloride (KCl) (5 M)

Makes 100 mL.
Store indefinitely at room temperature.

1. Dissolve 37.3 g of KCl <!> (MW 74.55) in 70 mL of deionized or distilled water.
2. Add deionized or distilled water to make a total volume of 100 mL of solution.
3. Make sure that the bottle cap is loose and autoclave for 15 min at 121°C.
4. After autoclaving, cool the solution to room temperature and tighten the lid for storage.

Primer/Loading Dye Mix

Makes enough for 50 PCRs.
Store for up to 1 yr at –20°C.

1. Obtain the primers for the laboratory exercise and dissolve each at a concentration of 15 pmol/μL. The primer sequences are as follows:

   - **Laboratory 3.1:** *Bz* Locus (Maize)

     primers for the *Bz* (wild-type) allele
     BZ1 5′-CGAATGGCTGTTGCATTTCCAT-3′ (forward primer)
     BZ2 5′-ACGGGACGCAGTTGGGCAGGA-3′ (reverse primer)

     primers for the *bz* (mutant) allele
     BZ1 5′-CGAATGGCTGTTGCATTTCCAT-3′ (forward primer)
     Ds 5′-TCTACCGTTTCCGTTTCCGTTT-3′ (reverse primer)

   - **Laboratory 3.2:** *CLF* Locus (*Arabidopsis*)

     primers for the *CLF* (wild-type) allele
     CLF1 5′-TTAACCCGGACCCGCATTTGTTTCGG-3′ (forward primer)
     CLF2 5′-AGAGAAGCTCAAACAAGCCATCGA-3′ (reverse primer)

     primers for the *clf-2* (mutant) allele
     CLF1 5′-TTAACCCGGACCCGCATTTGTTTCGG-3′ (forward primer)
     Ds 5′-GTCGGCGTGCGGCTGGCGGCG-3′ (reverse primer)

   - **Laboratory 3.3:** CAPS Markers (*Arabidopsis*)

     primers for the g4026 marker
     5′-GGGGTCAGTTACATTACTAGC-3′ (forward primer)
     5′-GTACGGTTCTTCTTCCCTTA-3′ (reverse primer)

     primers for the H77224 marker
     5′-GGATTTGGGGAAGAGGAAGTAA-3′ (forward primer)
     5′-TCCTTAGCCTTGCTTTGATAGT-3′ (reverse primer)

     primers for the m235 marker
     5′-GAATCTGTTTCGCCTAACGC-3′ (forward primer)
     5′-AGTCCACAACAATTGCAGCC-3′ (reverse primer)

     primers for the UFO marker
     5′-GTGGCGGTTCAGACGGAGAGG-3′ (forward primer)
     5′-AAGGCATCATGACTGTGGTTTTTC-3′ (reverse primer)

- **Laboratory 3.4:** GMO Testing

primers for the 35S promoter (transgene) 5′-CCGACAGTGGTCCCAAAGATGGAC-3′ (forward primer) 5′-ATATAGAGGAAGGGTCTTGCGAAGG-3′ (reverse primer) primers for tubulin 5′- GGGATCCACTTCATGCTTTCGTCC-3′ (forward primer) 5′- GGGAACCACATCACCACGGTACAT-3′ (reverse primer)

- **Laboratory 3.5**

primers for rbcL barcoding (plants) 5′-TGTAAAACGACGGCCAGTATGTCACCACAAACAGAGACTAAAGC-3′ (forward primer, rbcLaf-M13) 5′-CAGGAAACAGCTATGACGTAAAATCAAGTCCACCRCG-3′ (reverse primer, rbcLa-revM13) primers for COI barcoding (fish) 5′-TGTAAAACGACGGCCAGTCAACCAACCACAAAGACATTGGCAC-3′ (forward primer, VF2_t1) 5′-TGTAAAACGACGGCCAGTCGACTAATCATAAAGATATCGGCAC-3′ (forward primer, FishF2_t1) 5′-CAGGAAACAGCTATGACACTTCAGGGTGACCGAAGAATCAGAA-3′ (reverse primer, FishR2_t1) 5′-CAGGAAACAGCTATGACACCTCAGGGTGTCCGAARAAYCARAA-3′ (reverse primer, FR1d_t1) primers for COI barcoding (mammals and insects) 5′-GTAAAACGACGGCCAGTATTCAACCAATCATAAAGATATTGG-3′ (forward primer, LepF1_t1) 5′-TGTAAAACGACGGCCAGTTCTCAACCAACCACAAAGACATTGG-3′ (forward primer, VF1_t1) 5′-TGTAAAACGACGGCCAGTTCTCAACCAACCACAARGAYATYGG-3′ (forward primer, VF1d_t1) 5′-TGTAAAACGACGGCCAGTTCTCAACCAACCAIAAIGAIATIGG-3′ (forward primer, VF1i_t1) 5′-CAGGAAACAGCTATGACTAAACTTCTGGATGTCCAAAAAATCA-3′ (reverse primer, LepR1_t1) 5′-CAGGAAACAGCTATGACTAGACTTCTGGGTGGCCRAARAAYCA-3′ (reverse primer, VR1d_t1) 5′-CAGGAAACAGCTATGACTAGACTTCTGGGTGGCCAAAGAATCA-3′ (reverse primer, VR1_t1) 5′-CAGGAAACAGCTATGACTAGACTTCTGGGTGICCIAAIAAICA-3′ (reverse primer, VR1i_t1) primers for COI barcoding (metazoans) 5′-TGTAAAACGACGGCCAGGTCAACAAATCATAAAGATATTGG-3′ LCO1490 (forward) 5′-CAGGAAACAGCTATGACTAAACTTCAGGGTGACCAAAAAATCA-3′ HCO2198 (reverse)

- **Laboratory 3.6**

primers for *FWA*
5′-TCAGCGTCTACCAAATCTACACTTTT-3′ (forward primer)
5′- CAGACAAATCGGGAACCAAA-3′ (reverse primer)

2. In a 1.5-mL tube, mix the following for each primer mix:

Cresol red loading dye <!>	460 µL
Forward primer (15 pmol/µL)	20 µL
Reverse primer (15 pmol/µL)	20 µL

   Bring volume up to 1140 µL with distilled water.

   Note: This primer/loading dye mix is for use with Ready-to-Go PCR Beads. puRe *Taq* Ready-to-Go PCR Beads (GE Healthcare) are available in individual thin-wall tubes (0.2 or 0.5 mL) or in 96-well plates. Barcoding primer mixes have multiple forward and reverse primers. Ensure that you include them all.

3. Vortex to mix.

DIGESTING PCR PRODUCTS WITH RESTRICTION ENZYMES

Laboratory 3.4: Restriction Enzyme/Buffer Mix

Makes 10 µL for each reaction.
Store for up to several hours on ice until use.

1. For each restriction enzyme to be used in the experiment (HindIII, RsaI, and TaqI), combine the following in a 1.5-mL tube:

	Per Sample
Bovine serum albumin (1 mg/mL)	1 µL
Restriction buffer (10x) (from supplier)	1 µL
Distilled water	7 µL
Restriction enzyme (10 U/µL)	1 µL

2. Mix by pipetting in and out and store on ice.

Note: Do not freeze the restriction enzyme/buffer mix, because the water in the mix will form ice crystals that will destroy enzyme activity.

Laboratory 3.6: Restriction Buffer Mix

Makes enough for 120 reactions.
Store at –20°C.

1. Combine the following in a 1.5-mL tube:

Restriction buffer (2x)	1500 µL
Bovine serum albumin (BSA) (100x) (provided with enzyme)	30 µL
Guanosine 5′-triphosphate (GTP) (provided with enzyme)	30 µL

ANALYZING DNA BY GEL ELECTROPHORESIS

100-bp DNA Ladder (0.125 µg/µL)

Makes 100 µL.
Store for up to 1 yr at –20°C.

1. Obtain a stock solution of 100-bp DNA Ladder (0.5 µg/µL) from New England BioLabs (N3231). Store the solution at –20°C.
2. Dilute a small amount of the stock at a time by combining the following in a 1.5-mL tube:

Deionized or distilled water	50 µL
Cresol red loading dye <!>	25 µL
DNA Ladder	25 µL

Agarose (2%)

Makes 200 mL.
Use fresh or store solidified agarose for several weeks at room temperature.

1. To a 600-mL beaker or Erlenmeyer flask, add 200 mL of 1x TBE electrophoresis buffer and 4 g of agarose (electrophoresis grade).
2. Stir to suspend the agarose.
3. Dissolve the agarose using one of the following methods:
   - Cover the flask with aluminum foil and heat the solution in a boiling water bath (double boiler) or on a hot plate until all of the agarose is dissolved (~10 min).
   - Heat the flask uncovered in a microwave oven at high setting until all of the agarose is dissolved (3–5 min per beaker).
4. Swirl the solution and check the bottom of the beaker to ensure that all of the agarose has dissolved. (Just before complete dissolution, particles of agarose appear as translucent grains.) Reheat for several minutes if necessary.
5. Cover the agarose solution with aluminum foil and hold in a hot water bath (at ~60°C) until ready for use. Remove any "skin" of solidified agarose from the surface before pouring the gel.

Notes: Samples of agarose powder can be preweighed and stored in capped test tubes until ready for use. Solidified agarose can be stored at room temperature and then remelted over a boiling water bath (15–20 min) or in a microwave oven (3–5 min per beaker) before use. When remelting, evaporation will cause the agarose concentration to increase; if necessary, compensate by adding a small volume of water. Always loosen the cap when remelting agarose in a bottle.

Ethidium Bromide Staining Solution (1 µg/mL)

Makes 500 mL.
Store in the dark indefinitely at room temperature.

1. Add 100 µL of 5 mg/mL ethidium bromide <!> to 500 mL of deionized or distilled water.
2. Store the ethidium bromide in a dark (preferably opaque) unbreakable container or wrap the container in aluminum foil.
3. Label the container **"CAUTION: Ethidium Bromide. Mutagen and cancer-suspect agent. Wear rubber gloves when handling."**

Note: Ethidium bromide is light sensitive.

Loading Dye (10x)

Makes 100 mL.
Store indefinitely at room temperature.

1. Dissolve the following ingredients in 60 mL of deionized or distilled water:

Bromophenol blue (MW 669.96)	0.25 g
Xylene cyanol (MW 538.60)	0.25 g
Sucrose (MW 342.3)	50.0 g
Tris-HCl (1 M, pH 8.0)	1 mL

2. Add deionized or distilled water to make a total volume of 100 mL.

pBR322/BstNI Size Marker (0.1 µg/µL)

Makes 100 µL.
Store for up to 1 yr at –20°C.

1. Add 1 µL of 10 µg/µL pBR322 to 84 µL of deionized or distilled water.
2. Add 10 µL of 10x buffer (provided by the supplier of the BstNI enzyme).
3. Add 5 µL of 10 U/µL BstNI restriction enzyme and incubate for 60 min at 60°C.
4. Electrophorese 5 µL (plus 1 µL cresol red loading dye <!>) in a 1%–2% agarose gel to check for complete digestion. Exactly five bands should be visible, corresponding to 1857, 1058, 929, 383, and 121 bp. Any additional bands indicate incomplete digestion; if this is the case, add additional enzyme and incubate again at 60°C.

Note: pBR322 precut with restriction enzyme BstNI is also available from New England BioLabs (N3031).

SYBR Green DNA Stain

Makes 100 µL.
Store in the dark for up to 3 mo at 4°C.

1. Obtain a stock solution of 10,000x SYBR Green I <!>, which comes dissolved in DMSO <!> and is stored at –20°C. Allow the dye to thaw for ~10 min at room temperature.

2. Add 1 µL of 10,000x SYBR Green I in DMSO to 100 µL of sucrose solution. Mix thoroughly.

Note: SYBR Green is light sensitive.

Tris/Borate/EDTA (TBE) Electrophoresis Buffer (1x)

Makes 10 L.
Store indefinitely at room temperature.

1. In a spigoted carboy, add 9.5 L of deionized or distilled water to 0.5 L of 20x TBE Electrophoresis Buffer.
2. Stir to mix.

Tris/Borate/EDTA (TBE) Electrophoresis Buffer (20x)

Makes 1 L.
Store indefinitely at room temperature.

1. Add the following dry ingredients to 500 mL of deionized or distilled water in a 2-L flask:

NaOH <!> (MW 40.00)	2 g
Tris base <!> (MW 121.10)	216 g
Boric acid <!> (MW 61.83)	110 g
EDTA <!> (disodium salt, MW 372.24)	14.8 g

2. Stir to dissolve, preferably using a magnetic stir bar.
3. Add deionized or distilled water to bring the total volume to 1 L.

Note: If the stored 20x TBE comes out of solution, place the flask in a water bath (37°C–42°C) and stir occasionally until all solid matter goes back into solution.

Answers to Questions

Laboratories 3.1–3.6

LABORATORY 3.1: DETECTING A TRANSPOSON IN CORN

Answers to Bioinformatics Questions

I. Use BLAST to Find DNA Sequences in Databases (Electronic PCR)

2.i. What do you notice about the lengths (and colors) of the matches (bars) as you look from the top to the bottom?

Typically, most of the significant alignments will have complete matches to the forward and reverse primers. Partial alignments at the bottom of the list may correspond to only one primer or parts of both primers.

2.ii. What is the E value of the most significant hit and what does it mean?

The lowest E value obtained for a match to both primers is typically about 0.030. A score of 0.030 denotes a probability of 30×10^{-3} (or 30 times out of 1000) to match the hit sequence just by chance.

Note the names of any significant alignments that have E values of less than 0.1. Do they make sense? What do they have in common?

Many hits have E values of less than 0.1. Most are in corn and contain the Bz locus. Thus, they make sense.

2.iii. Do the rest of the alignments produce good matches to both primers?

There are many BLAST hits that contain good matches to both primers.

3.i. What is the source and size of the sequence in which your BLAST hit is located?

The first hit may vary as sequences are added to the database. When this volume went to press, the first hit was to the *bz* locus in a teosinte species, *Zea diploperennis*, which is 139,360 bp long.

4.ii. To which positions do the primers match in the subject sequence?

The forward primer matches positions 668–689 in the subject sequence, and the reverse primer matches positions 918–938.

4.iii. Subtract the lowest nucleotide position in the subject sequence from the highest nucleotide position in the subject sequence. What is the difference between the coordinates?

938 – 668 = 270 nucleotides (these are the coordinates in accession DQ493652.1).

4.iv. Note that the actual length of the amplified fragment includes both ends, so add 1 nucleotide to the result that you obtained in Step 4.iii to obtain the exact length of the PCR product amplified by the two primers.

270 + 1 = 271 nucleotides.

II. Identify the Bz Amino Acid Sequence

2. What is the name of this gene?

It is the *Bz* gene.

3. What do you think the comma in "(76..598,699..1591)" stands for?

The comma in "(76..598,699..1591)" signifies an intron.

4. How many exons and introns are in this gene?

The *Bz* gene contains two exons and one intron.

5. Where does the forward primer begin? Where does the reverse primer end? Which part of the gene does the amplicon span?

The forward primer begins at position 668, in the intron; the reverse primer ends at position 938, in the second exon. The two primers therefore span a sequence from the intron into the second exon.

6. What is the name of the protein product that this gene encodes?

Its product is UDP-glucose flavonoid-3-*O*-glucosyltransferase.

8.i. On which chromosome ("/chromosome=") is this sequence located?

The *Bz* gene is located on chromosome 9.

8.ii. What are the names of these neighboring genes?

The *Bz* gene is flanked by *stk1*, a serine threonine kinase gene, and *stc1*, a sesquiterpene cyclase gene.

III. Use Map Viewer to Determine the Chromosomal Location of the *Bz* Gene

3. On which chromosome is the *bz1* gene located?

The *bz1* gene is on chromosome 9, as expected.

5. Can you identify any genes in the vicinity of the *bz1* locus?

Yes.

What can you determine about some of them?

The *c1* gene, for example, was studied by Barbara McClintock, who found this gene to be crucial for anthocyanin biosynthesis in corn.

IV. Determine the Insertion Site of the *Ds* Transposon

1.i. From what species are the hits that match the BZ1 primer (nucleotides 1–22) only? Do their descriptions mention anything about the gene or locus?

Most hits are from corn (*Zea mays*) and mention the *Bz* locus.

1.ii. From what species are the hits that match the Ds primer (nucleotides 23–44) only? Do their descriptions mention anything about the gene or locus?

Most hits are from corn (*Zea mays*) and mention *Ds* or the related transposon *Ac*. Several are cloning vectors used in *Ds* mutagenesis.

1.iii. Do any hits align with both the BZ1 and Ds primers?

Yes. There are hits that contain matches to both primers simultaneously.

1.iv. Based on the information gathered in Steps 1.i–iii, is the sequence of your amplicon in the database? Explain how you came to your conclusion.

Yes. The *bz* mutant sequence is in the database because there are hits that contain matches to both primers simultaneously.

2. Determine the length of the BZ1/Ds amplicon by following Steps 4.ii–iv in Part I.

The length of the BZ1/Ds amplicon is 321 bp.

3. To determine the *Ds* insertion site in the *Bz* gene, answer the questions below and label the diagram with your answers.

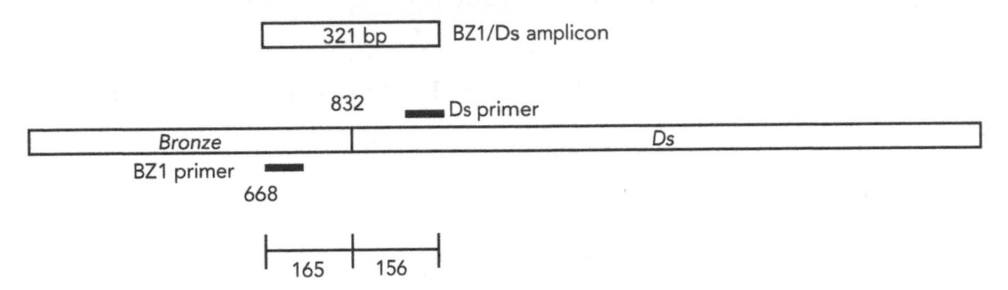

3.i. How far into this *Ds* sequence does the Ds primer reach?

156 bp.

3.ii. Using values from Steps 2 and 3.i, calculate the number of *Bz* nucleotides in the BZ1/Ds amplicon.

321 bp – 156 bp = 165 bp.

3.iii. What is the first nucleotide position of the BZ1 primer in the *Bz* gene?

668.

3.iv. What is the insertion position of *Ds* in the *Bz* gene?

668 + 164 = 832. Nucleotide 668 in the gene and the first nucleotide of the amplicon align with each other; therefore, the insertion point stretches only 164 nucleotides beyond position 668.

4. In which part of the *Bz* gene has the *Ds* transposon inserted?

The transposon inserted into the second exon of the gene.

5. What are some of the names of your finds?

Corn chromosomes seem to be riddled with transposons and retrotransposons. The sequence in AF391808.3 contains transposons such as *Misfit CACTA*, *Tourist-Zm1*, and *Stowaway-Zm3*. Examples of retrotransposons in this sequence are *Huck2*, *Opie2b*, *Zeon1*, and *RIRE1*.

V. Use BLAST to Determine the Function of the BZ Protein

4. What do you find?

The protein product of the *Bz* gene (BZ) contains a UDPGT (UDP-glucosyl transferase) domain.

6. **Do the BLAST hits show strong or weak homology with your query sequence?**
   The BLAST hits show strong homology across the entire protein.

7. **Based on the E values, the titles of the hits, and the links to the accession numbers, what can you conclude about the conservation of the BZ protein during plant evolution?**
   Extremely low E values across a range of organisms—*Arabidopsis, Petunia, Oryza* (rice), *Vitis* (grape), and *Fragaria* (strawberries)—indicate that the flavonoid-3-*O*-glucosyltransferase domain, and other parts of the BZ protein, have been strongly conserved among plants through evolutionary time.

Answers to Results and Discussion Questions

I. Think About the Experimental Methods

1. **Describe the purpose of each of the following steps or reagents used in DNA isolation (Part II of Experimental Methods):**

1.i. **Grinding with pestle.**
Grinding with the pestle breaks the cell walls of the maize tissue.

1.ii. **Edward's buffer.**
The detergent component of Edward's buffer, sodium dodecyl sulfate (SDS), dissolves lipids that comprise the cell membrane.

1.iii. **Boiling.**
Boiling denatures proteins, including enzymes that degrade DNA (DNases).

1.iv. **Tris-EDTA (TE) buffer.**
TE buffer provides conditions for stable storage of DNA. Tris provides a constant pH of 8.0, and EDTA binds cations (positive ions) that are required for DNase activity.

2. **What is the purpose of performing each of the following PCRs?**

2.i. **BZ1/BZ2.**
The BZ1/BZ2 primer pair detects the presence of the wild-type *Bz* allele.

2.ii. **BZ1/Ds.**
The BZ1/Ds primer pair detects the presence of the mutant *bz* allele bearing a *Ds* transposon insertion.

II. Interpret Your Gel and Think About the Experiment

4. **Carefully examine the banding patterns of the two samples in the gel above and complete the entries for genotype and phenotype in the chart below. Under "Sample," indicate which result is derived from Plant 1, which is derived from Plant 2, and which is not represented in the gel.**
   If you observe only the smaller band, the plant is homozygous for the wild-type allele (–/–). If you observe only the larger band, the plant is homozygous for the mutant *bz* allele (+/+). If you observe both bands, the plant is heterozygous (+/–).

321 bp (BZ1/Ds)	271 bp (BZ1/BZ2)	Plant genotype	Plant phenotype	Sample
present	present	*bz/Bz* (+/–)	wild type	not represented
present	absent	*bz/bz* (+/+)	mutant	plant 2
absent	present	*Bz/Bz* (–/–)	wild type	plant 1

5. **How would you interpret a lane in which you observe primer dimer but no bands as described in Step 3?**

   The presence of primer dimer confirms that the reaction contained all components necessary for amplification, but that there was insufficient template to amplify the target sequence.

6. **Would you classify the *bz* mutation as recessive or dominant? Explain your reasoning.**

   The *bz* mutation is recessive. Plants that are heterozygous do not demonstrate the mutant phenotype. Only plants that are homozygous for the *bz* mutation demonstrate the bronze phenotype.

7. **Count the number of plants with the *Bz/Bz*, *Bz/bz*, and *bz/bz* genotypes in your class. Based on the genotype distribution of these plants, which genotype was the parental plant? Why may the genotype distribution observed by your class deviate from what is expected under Mendel's laws of inheritance?**

   First, count the number of *Bz/Bz*, *Bz/bz*, and *bz/bz* plants and express it as a ratio. Then, draw Punnett squares for the six kinds of parental crosses, which produce three different genotype ratios. The observed ratio of *Bz/Bz*, *Bz/bz*, and *bz/bz* plants should be closest to the 1:2:1 ratio predicted by a cross between two heterozygous parents. (Maize has the ability to self-pollinate. The seeds for this experiment, in fact, came from a heterozygous parental plant that self-pollinated.) Small sample size or bias toward picking one phenotype will cause deviations from the expected ratio.

LABORATORY 3.2: DETECTING A TRANSPOSON IN *ARABIDOPSIS*

Answers to Bioinformatics Questions

I. Use BLAST to Find DNA Sequences in Databases (Electronic PCR)

2.i. **What do you notice about the lengths (and colors) of the matches (bars) as you look from the top to the bottom?**

There is only one complete match to the forward and reverse primers, followed by a number of partial matches to one primer or parts of both primers.

2.ii. **What is the E value of the most significant hit and what does it mean?**

The most significant hit has a low E value (e.g., 2E-04). An E value of 2E-04 denotes a probability of 2×10^{-04} or 0.0002. This means that the query has found a strong match in the database.

2.iii. **Do the rest of the alignments produce good matches to both primers?**

All of the rest of the alignments are shorter, partial matches to one primer or parts of both primers.

3.i. What is the source and size of the sequence in which your BLAST hit is located?

The first hit may vary as sequences are added to the database. When this volume went to press, the first hit was *Arabidopsis thaliana* chromosome 2, which is 19,698,289 bp long.

4.ii. To which positions do the primers match in the subject sequence?

The forward primer matches positions 6888–6913, and the reverse primer matches positions 7110–7133.

4.iii. Subtract the lowest nucleotide position in the subject sequence from the highest nucleotide position in the subject sequence. What is the difference between the coordinates?

7133 – 6888 = 245 nucleotides. These are the coordinates in accession AC003040.3.

4.iv. Note that the actual length of the amplified fragment includes both ends, so add 1 nucleotide to the result that you obtained in Step 4.iii to obtain the exact length of the PCR product amplified by the two primers.

245 + 1 = 246 nucleotides.

II. Identify the CLF Amino Acid Sequence

2. Use the nucleotide coordinates of the primers (as determined in Step 4.iii of Part I above) to identify the gene feature that contains at least one of the primers. What is the name of this gene?

The gene is identified as *At2g23380*.

3. What do you think the comma in "6935..6997,7275..7367" stands for?

The comma in "6935..6997,7275..7367" signifies an intron.

4. How many exons and introns are in this gene?

The gene contains 17 exons and 16 introns.

5. Where does the forward primer begin? Where does the reverse primer end? Which part of the gene does the amplicon span?

The forward primer begins at position 6888, before the first exon. The reverse primer ends at position 7133, between the first and second exons. Therefore, the amplicon spans the first exon and includes some noncoding sequence on either side.

6. What is the name of the protein product that this gene encodes?

Its product is the curly leaf (or CLF) protein (Polycomb group).

III. Use Map Viewer to Determine the Chromosomal Location of the *CLF* Gene

6. On which chromosome have you landed?

The *CLF* gene is on chromosome 2, as expected.

8. What is the function of the *CLF* gene?

The *CLF* gene encodes a protein that functions as a repressor of *AGAMOUS* gene expression. *AGAMOUS* is a homeotic gene that is involved in the development of the flower structure, including the stamens and carpels. *CLF*, therefore, has an indirect role in flower development.

10.i. Determine the size of the *CLF* gene using the map coordinates to the left of the "Contig" map.

CLF is approximately 4800 bp in length.

10.ii. How many exons does the *CLF* gene have?

The *CLF* gene has 17 exons.

10.iii. Where does the amplicon (vertical red bar) line up with the *CLF* gene?

It lines up with the first exon and part of the first intron.

11. Can you figure out what the prefix "AT2G" means?

AT2G stands for *Arabidopsis thaliana* chromosome 2.

IV. Determine the Insertion Site of the *Ds* Transposon

1.i. From what species are the hits that match the CLF1 primer (nucleotides 1–26) only? Do their descriptions mention anything about the gene or locus?

There is only one strong match to the CLF1 primer: the same *Arababidopsis thaliana* chromosome 2 clone identified in the BLAST search with the CLF1 and CLF2 primers in Part I.

1.ii. From what species are the hits that match the Ds primer (nucleotides 27–47) only? Do their descriptions mention anything about the gene or locus?

Most hits are from corn (*Zea mays*) and mention *Ds* or the related transposon *Ac*. Several are cloning vectors used in *Ds* mutagenesis.

1.iii. Do any hits align with both the CLF1 and Ds primers?

No hits contain matches to both primers simultaneously.

1.iv. Based on the information gathered in Steps 1.i–iii, is the sequence of your amplicon in the database? Explain how you came to your conclusion.

No. The *Arababidopsis thaliana* hit contains only the CLF primer and not the Ds primer. Thus, it appears that the *clf-2* mutant sequence is not in the database.

2. To determine the *Ds* insertion site in the *CLF* gene, answer the questions below and label the diagram with your answers.

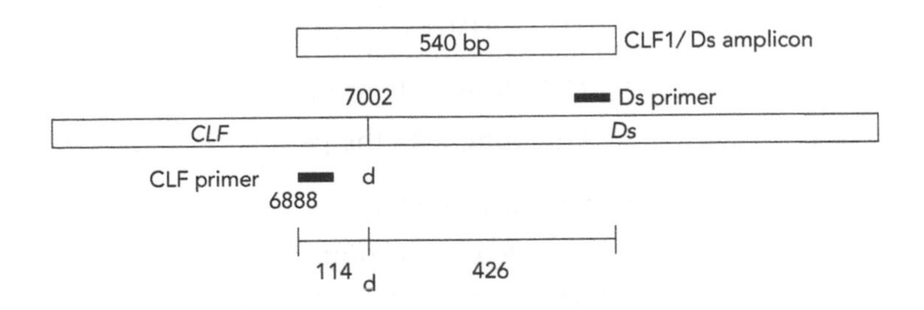

2.i. How far into this *Ds* sequence does the Ds primer reach?

426 bp.

2.ii. The CLF1/Ds amplicon is ~540 nucleotides long. Use this number and the answer from Step 2.i to calculate the number of *CLF* nucleotides in the CLF1/Ds amplicon.

540 bp – 426 bp = 114 bp.

2.iii. What is the first nucleotide position of the CLF1 primer in the *CLF* gene sequence?

6888.

2.iv. What is the insertion position of *Ds* in the *CLF* gene?

6888 + 113 = 7002. Nucleotide 6888 in the gene sequence and the first nucleotide of the amplicon align with each other; therefore, the insertion point stretches only 113 nucleotides beyond position 6888.

3. In which part of the *CLF* gene has the *Ds* transposon inserted?

The transposon inserted into the first intron of the *CLF* gene.

V. Use BLAST to Determine the Function of the CLF Protein

4. What do you find?

The protein product of the *CLF* gene contains a SET domain that has methyltransferase activity (adds methyl groups to other proteins).

6. Do the BLAST hits show strong or weak homology with your query sequence?

The BLAST hits show strong homology with the query, especially to more than 200 amino acid residues in the second half of the CLF protein (which includes the SET domain).

7. Based on the E values, the titles of the hits, and the links to the accession numbers, what can you conclude about the conservation of the CLF protein during plant evolution?

Extremely low E values across a range of organisms—*Arabidopsis, Petunia, Oryza* (rice), *Solanum lycopersicon* (tomato), *Triticum* (wheat), *Vitis* (grape), and *Zea mays* (corn)—indicate that the SET domain, and other parts of the CLF protein, have been strongly conserved among plants through evolutionary time.

Answers to Results and Discussion Questions

I. Think About the Experimental Methods

1. Describe the purpose of each of the following steps or reagents used in DNA isolation (Part II of Experimental Methods):

1.i. Grinding with pestle.

Grinding with the pestle breaks the cell walls of the *Arabidopsis* tissue.

1.ii. Edward's buffer.

The detergent component of Edward's buffer, sodium dodecyl sulfate (SDS), dissolves lipids that comprise the cell membrane.

1.iii. **Boiling.**

Boiling denatures proteins, including enzymes that degrade DNA (DNases).

1.iv. **Tris/EDTA (TE) buffer.**

TE buffer provides conditions for stable storage of DNA. Tris provides a constant pH of 8.0, and EDTA binds cations (positive ions) that are required for DNase activity.

2. **What is the purpose of performing each of the following PCRs?**

2.i. **CLF1/CLF2.**

The CLF1/CLF2 primer pair detects the presence of the wild-type *CLF* allele.

2.ii. **CLF1/Ds.**

The CLF1/Ds primer pair detects the presence of the mutant *clf-2* allele bearing a *Ds* transposon insertion.

II. Interpret Your Gel and Think About the Experiment

4. **Carefully examine the banding patterns of the samples in the gel above and complete the entries for genotype and phenotype in the chart below. Under "Sample," indicate which result was derived from Plant 1, Plant 2, and Plant 3. (Your own samples may yield any of the banding patterns listed in the table.)**

If you observe only the smaller band, the plant is homozygous for the wild-type allele (–/–). If you observe only the larger band, the plant is homozygous for the mutant *clf-2* allele (+/+). If you observe both bands, the plant is heterozygous (+/–).

540 bp (CLF1/Ds)	246 bp (CLF1/CLF2)	Plant genotype	Plant phenotype	Sample
present	present	*CLF/clf-2* (+/–)	wild type	plant 2
present	absent	*clf-2/clf-2* (+/+)	mutant	plant 1
absent	present	*CLF/CLF* (–/–)	wild type	plant 3

5. **How would you interpret a lane in which you observe primer dimer but no bands as described in Step 3?**

The presence of primer dimer confirms that the reaction contained all components necessary for amplification, but that there was insufficient template to amplify the target sequence.

6. **Would you classify the *clf-2* mutation as dominant or recessive? Explain your reasoning.**

The *clf-2* mutation is recessive. Plants that are heterozygous do not demonstrate the mutant phenotype. Only plants that are homozygous for the *clf-2* mutation demonstrate the curly leaf phenotype.

7. **Count the number of plants with the +/+, –/+, and –/– and genotypes in your class. Based on the genotype distribution of these plants, which genotype was the parental plant? Why may the genotype distribution observed by your class deviate from what is expected under Mendel's laws of inheritance?**

First, count the number of +/+, +/–, and –/– plants and express it as a ratio. Then, draw Punnett squares for the six kinds of parental crosses, which produce three different genotype ratios. The observed ratio of +/+, +/–, and –/– plants should be closest to the 1:2:1 ratio predicted by a cross between two heterozygous parents. (*Arabidopsis thaliana* has the ability to self-pollinate. The seeds for this experiment, in fact, came from a heterozygous parental plant that self-pollinated.) Small sample size or bias toward picking one phenotype will cause deviations from the expected ratio.

LABORATORY 3.3: LINKAGE MAPPING A MUTATION

Answers to Bioinformatics Questions

I. Use BLAST to Find DNA Sequences in Databases (Electronic PCR)

2.i. What do you notice about the lengths (and colors) of the matches (hits) as you look from the top to the bottom?

Typically, most of the significant alignments will have complete matches to the forward and reverse primers. Partial alignments at the bottom of the list may correspond to only one primer or parts of both primers.

2.ii. What is the E value of the most significant hit and what does it mean?

The most significant hit has an E value of about 0.1. This might seem high for a probability, but in fact this means that a match of this quality would be expected to occur by chance less than once in this database!

Note the names of the most significant hits. Do they make sense? What do they have in common?

The most significant alignments are in *Arabidopsis*. Thus, they make sense.

2.iii. Do the rest of the alignments produce good matches to both primers?

All of the rest of the alignments are shorter, partial matches to one primer or parts of both primers.

3.i. What is the source and size of the sequence in which your BLAST hit is located?

The first hit may vary as sequences are added to the database. At the time that this volume went to press, the first hit was to *Arabidopsis thaliana* chromosome 1, which is 30,427,671 bp long.

4.ii. To which positions do the primers match in the subject sequence?

For accession number AC002292.1, the forward primer matches from position 23 to 43 and the reverse primer matches from position 847 to 866. For AM748026.1, the forward primer matches from position 158,027 to 158,047 and the reverse primer matches from position 158,852 to 158,871.

4.iii. The lowest and highest nucleotide positions in the subject sequence indicate the borders of the amplified sequence. Record these numbers in the table below.

See the table below.

4.iv. Note that the actual length of the amplified fragment includes both ends, so add 1 nucleotide to the result to obtain the exact length of the PCR product amplified by the two primers. Record the length of the amplicon in the table below.

Marker	Accession no.	Lowest position	Highest position	Amplicon length (bp)
g4026	AC002292	23	866	844
	AM748036	158027	158871	845
H77224	NM_105947	716	943	228
m235	AF000657	3920	4450	531
UFO	AC000107	50367	51661	1295

4.v. The two hits for g4026 give slightly different amplicon sizes. Why?

The two hits are from the two ecotypes (Ler and Col). The DNA sequence in the two ecotypes differs in length by 1 nucleotide.

II. Digest CAPS Markers Electronically (In Silico)

4.i. The total length of the amplicon in nucleotides is at the top left-hand corner of the screen. Record the length in the table below.

See the table below.

4.ii. Record the lengths of the restriction fragments in the table below.

Marker	Restriction enzyme	Amplicon length (bp)	Restriction fragment lengths (bp)
g4026	RsaI	844 (Col)	668, 94, 78, 4
		845 (Ler)	763, 78, 4
H77224	TaqI	228 (Col)	137, 91
m235	HindIII	531 (Col)	301, 230
UFO	TaqI	1295 (Col)	980, 315

III. Use Map Viewer to Characterize the *AGO1* Gene

3. On which chromosome is the *AGO1* gene located?

The *AGO1* gene is on chromosome 1.

5.i. Determine the size of the *AGO1* gene using the map coordinates to the left of the map, under "Genes_seq."

The *AGO1* gene is 6489 bp in length.

5.ii. How many exons and introns does the *AGO1* gene have?

The *AGO1* gene has 22 exons and 21 introns.

6. What is the function of the *AGO1* gene?

AGO1 is a key component of the RNAi pathway and has a role in leaf and petal development.

IV. Use Map Viewer to Determine the Chromosomal Locations of the *AGO1* Gene and the CAPS Markers

7. Click on the pink "Blast hit" links to identify the locations of the *AGO1* sequence and CAPS markers. Record your answers in the table below and answer the questions that follow.

Locus	Locus order	Map position (Mb)
AGO	3	17.9
g4026	4	22.3
H77224	5	27.5
m235	1	8.1
UFO	2	11

7.i. What is the order of the five search sequences on the chromosome, from top to bottom?

See the table above.

7.ii. Use the ruler to the left of the Contig map to determine the map positions of the CAPS markers and the *AGO1* gene.

See the table above.

7.iii. Which CAPS marker is closest to *AGO1*? How many megabases separate them?

g4026 is the closest marker; the distance between *AGO1* and g4026 is ~4.4 Mb.

Answers to Results and Discussion Questions

I. Interpret Your Gel and Think About the Experiment

4.ii. Compare the results of your electronic (in silico) and electrophoretic (in vitro) analyses. (You generated your in silico results in Part II of the Bioinformatic Methods section.) Which fragments are missing and why?

The smallest fragments may be missing. These fragments travel through the gel quickly and can run off the gel. In addition, short fragments of DNA are difficult to stain and visualize, making them impossible to see.

6.i. From which *Arabidopsis* parent (P) ecotype did all of the mutants analyzed in this experiment inherit the *ago1* mutation?

Ler.

6.ii. How many copies of chromosome 1 are assayed in each sample tube?

Two.

6.iii. What is the significance of each of the following genotypes for a given CAPS marker?

- **Ler/Ler.**

  There was no recombination between *ago1* and the CAPS marker in either chromosome; the *ago1* mutation remained linked to the CAPS marker on both chromosomes.

- **Col/Col.**

  A recombination event occurred between *ago1* and the CAPS marker on each of the two chromosomes.

- **Ler/Col.**

  A recombination event occurred between *ago1* and the CAPS marker on one chromosome but not on the other.

The likelihood of recombination occurring between two loci (e.g., between a CAPS marker and *ago1*) increases in proportion to the distance between the markers on the chromosome. The CAPS marker that is closest to *ago1* should show the fewest recombination events and be most frequently found linked to the Ler allele for the marker.

II. Calculate Recombination Frequencies and Map the *ago1* Locus

Use the chart below to record your answers to the questions that follow.

	g4026		H77224		m235		UFO	
Genotype	No. of plants	No. of Col alleles	No of plants	No. of Col alleles	No. of plants	No. of Col alleles	No. of plants	No. of Col alleles
Ler/Ler								
Ler/Col								
Col/Col								
Totals								

1. **For each of the four CAPS markers, count the number of plants of each genotype: Ler/Ler, Ler/Col, and Col/Col. Exclude from the analysis any samples whose genotypes could not be determined.**

   For example, the following data can be used:

g4026:	8 Ler/Ler	2 Ler/Col	0 Col/Col
H77224:	5 Ler/Ler	3 Ler/Col	2 Col/Col
m235:	5 Ler/Ler	3 Ler/Col	2 Col/Col
UFO:	6 Ler/Ler	3 Ler/Col	1 Col/Col

2.i. **Multiply the total number of plants of each genotype by the number of Col alleles in that genotype. Remember that each Col/Col plant contributes two copies of the Col allele, whereas each Ler/Col plant contributes one copy of the Col allele.**

See answer to 2.ii below.

2.ii. **Add up the total number of Col alleles.**

g4026: $(2 \times 1) + (0 \times 2) = 2$
H77224: $(3 \times 1) + (2 \times 2) = 7$
m235: $(3 \times 1) + (2 \times 2) = 7$
UFO: $(3 \times 1) + (1 \times 2) = 5$

3. **For each marker, determine the total number of alleles by multiplying the total number of plants in the table above by 2.**

   g4026: $2 \times (8 + 2 + 0) = 20$
   H77224: $2 \times (5 + 3 + 2) = 20$
   m235: $2 \times (5 + 3 + 2) = 20$
   UFO: $2 \times (6 + 3 + 1) = 20$

   Organize your answers to this and the remaining questions in the table below.

	g4026	H77224	m235	UFO
Total no. of alleles	20	20	20	20
Recombination frequency (%)	10	35	35	25
Map distance (cM)	10	43	43	27

4. **Calculate the recombination frequency (%) using the following formula:**

$$r = \frac{\text{number of Col alleles}}{\text{total number of alleles}} \times 100.$$

g4026: $r = \frac{2}{20} \times 100 = 10.$

H77224: $r = \frac{7}{20} \times 100 = 35.$

m235: $r = \frac{7}{20} \times 100 = 35.$

UFO: $r = \frac{5}{20} \times 100 = 25.$

5. **Next, convert the recombination frequency (%) for each CAPS marker into a map distance (D, in cM) from the *ago1* locus. In *Arabidopsis*, a reasonable estimate of map distance is given by the Kosambi function (ln = natural log):**

$$D = 25 \times \ln \left[\frac{(100 + 2r)}{(100 - 2r)}\right].$$

g4026: $D = 25 \times \ln \left[\frac{(100 + 2 \times 10)}{(100 - 2 \times 10)}\right] = 10 \text{ cM}.$

H77224: $D = 25 \times \ln \left[\frac{(100 + 2 \times 35)}{(100 - 2 \times 35)}\right] = 43 \text{ cM}.$

m235: $D = 25 \times \ln \left[\frac{(100 + 2 \times 35)}{(100 - 2 \times 35)}\right] = 43 \text{ cM}.$

UFO: $D = 25 \times \ln \left[\frac{(100 + 2 \times 25)}{(100 - 2 \times 25)}\right] = 27 \text{ cM}.$

6. **The following map shows the locations of the four CAPS makers on *Arabidopsis* chromosome 1. Use the map distances (D) to locate *ago1* on chromosome 1.**

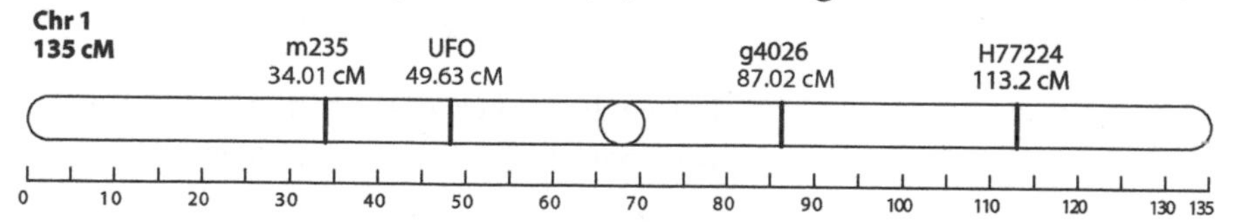

g4026: 87.02 +/– 10 cM = 77 cM or 97 cM.
H77224: 113.2 +/– 43 cM = 156 cM or 70 cM.
m235: 34.01 +/– 43 cM = –9 cM or 77 cM.
UFO: 49.63 +/– 27 cM = 23 cM or 77 cM.

Conclusion: The average location from all four markers places *ago1* between UFO and g4026 at about 75 cM, which is the genetic map location of *ago1*.

7. **Do the results of your mapping agree with the location of *ago1* from the bioinformatics exercises? Explain your reasoning.**

   The map position for *ago1* is at approximately 75 cM according to our example data, in between UFO and g4026. From the bioinformatics, UFO is at 11 Mb in the sequence, *ago1* is at 17.9 Mb, and g4026 is at 22.3 Mb.

   To compare the two, the genetic position needs to be converted to a physical position, assuming that the physical and genetic positions are proportional to each other.

   To convert the physical position, we can look at this interval. First, the total interval is 84.9 cM – 47.5 cM, or 37.4 cM. This represents 22.3 Mb – 11 Mb, or 11.3 Mb.

   This suggests that each Mb is approximately 37.4 cM/11.3 Mb, or 3.3 cM/Mb.

   From the mapping, *ago* is located 84.9 cM – 75 cM, or 9.9 cM, from g4026.

   This suggests *ago1* is located 9.9 cM/3.3 cM/Mb, or 3 Mb, from g4026.

   This would give a map position of 22.3 Mb – 3 Mb, or 19.3 Mb.

   Thus, the mapping results get the relative position of *ago1* on the chromosome correct, but the predicted location is not precise. Sampling error in the mapping and differences between the physical map and genetic map could explain this discrepancy.

LABORATORY 3.4: DETECTING GENETICALLY MODIFIED FOODS BY POLYMERASE CHAIN REACTION

Answers to Bioinformatics Questions

I. Use BLAST to Find DNA Sequences in Databases (Electronic PCR)

2.i. **What do you notice about the lengths (and colors) of the matches (bars) as you look from the top to the bottom?**

Typically, most or all of the significant alignments will have complete matches to the forward and reverse primers. Partial alignments may correspond to only the forward or reverse primer.

2.ii. **What is the E value of the most significant hit and what does it mean?**

The most significant hit for the first primer pair has an E value of about 6E-04. (An E value of 6E-04 denotes 6×10^{-4} or 0.0006.) The most significant hit for the second primer pair has an E value of ~0.002. This means that both queries have found strong matches in the database.

Note the names of any significant alignments that have E values of less than 0.1. Do they make sense? What do they have in common?

Many hits have E values of less than 0.1. For the first primer set, most are for cloning vectors that incorporate a 35S promoter; for the second primer set, most are for tubulin. Thus, they make sense.

2.iii. **Do the rest of the alignments produce good matches to both primers?**

Yes. However, some hits have multiple alignments for each primer, suggesting that the sequences are duplicated.

3.i. **Does this primer set detect a GM product or is it used to amplify the control sequence?**

The first primer set amplifies the 35S promoter of cauliflower mosaic virus (CaMV) and detects a GM product. The second primer set amplifies tubulin, the positive control for plant DNA.

4.ii. **To which positions do the primers match in the subject sequence?**

The positions will depend on the hit that each student selects.

4.iii. **Subtract the lowest nucleotide position in the subject sequence from the highest nucleotide position in the subject sequence. What is the difference between the coordinates?**

For most hits, the difference is 161 bp for the first primer set. For the second, it is 184 bp.

4.iv. **Note that the actual length of the amplified fragment includes both ends, so add 1 nucleotide to the result that you obtained in Step 4.iii to obtain the exact length of the PCR product amplified by the two primers.**

The first primer set amplifies a 162-bp product. The second primer set amplifies a 185–187-bp product, depending on the species.

II. Use BLAST to Identify Transgenes Driven by the 35S Promoter

2. **What do you notice about the E values obtained by this search compared to the E values that you obtained in Step 2.ii of Part I? Why is this so?**

The E values are much smaller (having many more decimal places) because the query sequence is longer.

3. **The lowest and highest nucleotide positions in the subject sequence indicate the borders of the PCR product produced with the GM primer set. What are these positions?**

The positions will depend on the hit each student takes.

4.i. **Use the nucleotide positions that you determined in Step 3 to identify the feature that contains the PCR product. What is the name of this feature?**

The feature is a promoter (specifically, the 35S promoter from CaMV).

4.ii, 5. **What is the name of the transgene in this sequence? What did you learn?**

The transgenes are different depending on the hit. Many include an antibiotic resistance gene such as *neomycin phosphotransferase* or *chloramphenicol acetyltransferase* downstream from the promoter. These selectable markers are required to identify transgenic plants after transformation. Other examples include Cry1Aa, which encodes the insect resistance protein Bacillus thuringiensis (BT) toxin, virus resistance conferred by expressing a viral protein such as pea early-browning virus coat protein, herbicide resistance genes such as *CP4-EPSPS* that confers resistance to glyphosate, and a sodium pump that confers salt tolerance.

Answers to Results and Discussion Questions

I. Think About the Experimental Methods

1. **Describe the purpose of each of the following steps or reagents used in DNA isolation (Part II of Experimental Methods):**

1.i. **Grinding with pestle.**

Grinding with the pestle pulverizes dry food and breaks the cell walls of the soybean or other fresh tissue.

1.ii. **Edward's buffer.**

The detergent component of Edward's buffer, sodium dodecyl sulfate (SDS), dissolves lipids that compose the cell membrane.

1.iii. **Boiling.**

Boiling denatures proteins, including enzymes that degrade DNA (DNases).

1.iv. **Tris/EDTA (TE) buffer.**

TE buffer provides conditions for stable storage of DNA. Tris provides a constant pH of 8.0 and EDTA binds cations (positive ions) that are required for DNase activity.

2. **What is the purpose of performing each of the following PCRs?**

2.i. **Tubulin.**

Tubulin serves as a positive control for the presence of amplifiable plant DNA in the sample.

2.ii. **Wild-type soybean.**

Wild-type soybean does not possess the 35S promoter; therefore, it serves as a negative control for the absence of the 35S promoter.

2.iii. **Roundup Ready soybean.**

Roundup Ready soybean possesses the 35S promoter; therefore, it serves as a positive control for the presence of the 35S promoter.

II. Interpret Your Gel and Think About the Experiment

4. **In the table below, indicate which sample(s) in the gels show the banding patterns. Explain what each pattern means.**

Tubulin (187 bp)	35S Promoter (162 bp)	Sample(s)	Explanation
present	present	Roundup Ready soybean and food 2	transgene present
present	absent	wild-type soybean and food 1	transgene absent
absent	absent	not shown	no template DNA
absent	present		transgene present, tubulin reaction failed

LABORATORY 3.5: USING DNA BARCODES TO IDENTIFY AND CLASSIFY LIVING THINGS

Answers to Bioinformatics Questions

I. Use BLAST to Find DNA Sequences in Databases (Electronic PCR)

2.i. **What do you notice about the lengths (and colors) of the matches (bars) as you look from the top to the bottom?**

The lengths and colors give you information about how much of your query matched sequences in the database. When both the forward and reverse primer match a sequence, you will see a black vertical line between the forward and reverse primer in the graphic summary. Typically, most of the significant alignments will have complete matches to the forward and reverse primers. There may be a color difference between the forward and reverse primer matches.

2.ii. What is the E value of the most significant hit and what does it mean?

The lowest E value obtained for a match to both primers should be in the range of 0.001–2E-04 (0.0002). This might seem high for a probability, but in fact these values mean that a match of this quality would be expected to occur by chance in only two of 10,000 searches in this database! More precisely, a score of 0.33 would mean that a single match would be expected to occur by chance once in every three searches. E values are based on the length of the search sequence; thus, the relatively short primers used in this experiment produce relatively high E values.

Note the names of any significant alignments that have E values of less than 0.1. Do they make sense? What do they have in common?

For the *rbcL* primers, the sequence sources should all be chloroplast genomes. For *COI* primers, the hits should all be mitochondrial genomes.

2.iii. Do the alignments produce good matches to both primers?

Some of the matching subjects (accession numbers) may only match the forward primer. When there is a match to the forward and reverse primer, one match may be poorer than the other (as indicated by the color coding). The reverse primer for *rbcL* is shorter in sequence, so any BLAST match is necessarily lower in significance. The reverse primer also contains an ambiguous nucleotide R, which means A or G.

3.ii. To which positions do the primers match in the subject sequence?

The answers will vary for each hit and primer set. For *Phoenix dactylifera* (NC_013991.2), the *rbcL* primers match 56,930–56,955 and 57,509–57,528, respectively. For *Spathius agrili* (NC_014278.1), the *Lep* primers match 2035–2060 and 2718–2743, respectively. For *Rattus tunneyi* (NC_014861.1), the *VF1* primers match 5339–5363 and 6022–6048, respectively.

3.iii. Subtract the lowest nucleotide position in the subject sequence from the highest nucleotide position in the subject sequence. What is the difference between the coordinates?

For the *rbcL* primers, using *Phoenix dactylifera* (NC_013991.2) as an example gives 56,955 – 57,509 = 554 nucleotides. The length will vary by accession number (hit). The range in possible lengths for this primer set should be between 550 and 600 nucleotides.

For the *Lep* primers, using *Spathius agrili* (NC_014278.1) as an example yields 2743 – 2035 = 708 nucleotides. The range of possible lengths for these primers is between 670 and 750 nucleotides.

For the *VF1* primers, using *Rattus tunneyi* (NC_014861.1) as an example gives 6048 – 5339 = 709 nucleotides. These primers usually have hits ranging from 700 to 720 nucleotides.

3.iv. Note that the actual length of the amplified fragment includes both ends. Thus, add 1 nucleotide to the result that you obtained in Step 3.iii to obtain the exact length of the PCR product amplified by the two primers.

4.ii. Identify the feature(s) located between the nucleotide positions of the primers identified in Step 3.ii.

For the *rbcL* primers, the hits are in the coding region (cds) for the ribulose-1,5-bisphosphate carboxylase/oxygenase large subunit. For the *COI* primers, the hits are in the CDS for cytochrome oxidase subunit 1.

II. Upload and Assemble Your DNA Sequences in *DNA Subway*

6.iii. What is the error rate and accuracy associated with a Phred score of 20?

A Phred score of 20 equals one error in 100, or 99% accuracy.

6.iv. What do you notice regarding the electropherogram peaks and quality scores at nucleotide positions that are labeled "N"?

At "N" positions, peaks for different channels of similar amplitude often overlap. There may be no prominent peaks above low-amplitude background ("noise"). Quality scores are less than 20.

9.iv. How does the consensus sequence optimize the amount of sequence information available for analysis? Why does this occur?

The consensus sequence extends the ends, producing a longer contiguous sequence. The 5′ sequence immediately following the forward primer has many sequencing errors and is trimmed, and the reverse read extends this low-quality region. Likewise, the forward read extends the low-quality region at the 3′ end. (The sequence quality also drops as the distance from the primer increases.) This usually results in regions of the consensus sequence where only the forward or reverse sequence will be of high quality, but the overall quality of the consensus sequence will be better than either the forward or reverse sequence alone.

10.i. Do differences tend to occur in certain areas of your sequence? Explain.

Differences clustering at the 5′ and 3′ ends are typically a combination of low-quality sequence near the primers and loss of signal as the distance from the primer increases. Few or no internal mismatches indicate good-quality sequence from both the forward and reverse reads. Large numbers of yellow mismatches—especially in long blocks—may indicate that you have incorrectly paired sequences from two different sources (organisms) or that you failed to reverse-complement the reverse strand. A large number of mismatches in properly paired and reverse-complemented sequences indicates that one or both sequences is of poor quality.

III. Use BLAST to Identify Your Species

1.iii. Why do the most of these significant hits typically have E values of 0?

The lower the E value, the lower the probability of a random match and the higher the probability that the BLAST hit is related to the query. Searching with a long (500 bp or more) barcode sequence increases the number of significant alignments with low E values (compared to searches with short primers, as observed in Step 2.ii of Part I).

What does it mean when there are multiple BLAST hits with similar E values?

It is common to have multiple hits with identical or very similar E values. Of course, identical matches to the same species would be expected to have an E value of 0. However, other hits with 0 or very low E values are often found for members of the same genus. In some families of plants or animals, the barcode regions used in this experiment are not variable enough to make a conclusive species determination. Similar E values would also be obtained when two sequences have the same number of sequence differences, but at different positions.

IV. Perform a Phylogenetic Analysis

3.ii. The 5′ (leftmost) and 3′ (rightmost) ends of the sequences are usualy misaligned. Why does this occur?

The quality of sequences is usually low at both ends, due to gaps (–) or undetermined nucleotides (Ns). The lengths of the sequences in the databases may also vary, leading to gaps in the alignment.

6.i. What assumptions are made when one infers evolutionary relationships from sequence differences?

The major assumption is that mutations occur at a constant rate, that the "molecular clock" provides the measure of evolutionary time. Because branch lengths of a phylogenetic tree represent mutations per unit of time, an increase in the mutation rate at some point in evolutionary time would artificially lengthen branch lengths. If the barcode region mutates more frequently in one clade, a larger number of differences would be incorrectly interpreted as increased phylogenetic distance between it and other clades. In addition, although there is a chance that any given nucleotide has undergone multiple substitutions (e.g., A > T > C or A > T > A), tree-building algorithms only evaluate nucleotide positions as they occur in the sequences being compared. If the sequences being evaluated do not include a variation that happened during evolution, it will not be taken into account. The algorithm will assume only the minimum number of substitutions. Because the chance of multiple substitutions increases over time, the phylogenetic tree will tend to overestimate the relatedness among distantly related species that diverged extremely long ago.

6.ii. Why do taxonomic relationships and phylogenetic trees sometimes disagree?

Traditional taxonomic relationships are primarily based on morphological (physical) features. Related clades share morphological features by descent from a common ancestor. However, unrelated groups may develop a similar morphological feature when they independently adapt to similar challenges or environments. (For example, bats and birds have wings, but this feature arose independent of a common ancestor.) Phylogenetic (gene) trees can call attention to situations—at many taxonomic levels—where morphological similarities have been misinterpreted as a close taxonomic relationship. Also, gene trees may identify new species that cannot be differentiated by morphology alone.

8. **How does it compare to the maximum-likelihood tree? What does this tell you?**

   The trees will likely have a different arrangement of nodes and place some sequences on different nodes. This tells you that there are multiple possible solutions for most phylogenetic relationships and that different algorithms will calculate different optimum trees.

Answers to Results and Discussion Questions

I. Interpret Your Gel

3. **Which samples amplified well and which ones did not?**

   Each barcode primer set is optimized to amplify the same region across a range of species. Although the size of products can vary, the majority of PCR products will be of similar size and, therefore, will migrate to the same position on the gel. Of course, barcodes amplified using different primer sets (e.g., *rbcL* vs. *COI*) will produce different-sized products that will migrate to different positions on the gel. The intensity of staining (thickness of bands) will vary among reactions. This is related to the mass of the product that is produced by the PCR and to the volume of the reaction that is successfully loaded into the well.

   If your samples did not amplify, describe several reasons why this might be the case.

   Major problems in PCR amplification typically occur due to problems with the DNA extraction procedure: The grinding step may not have sufficiently disrupted the tissue, the supernatant transferred after protein precipitation may have carried forward too many inhibitors, or the nucleic acid pellet may have been lost after the precipitation step. In addition, it may be difficult to extract DNA from tough leaves or dry materials. Another consideration is that the small volume of DNA template may not have been pipetted directly into the PCR (it may have been left in the pipette or on the wall of the PCR tube). Finally, some primer sets may not work with certain groups of organisms (e.g., *rbcL* primers work less efficiently with nonvascular plants such as mosses and liverworts).

4. **How would you interpret a lane in which you observe primer dimer but no bands as described in Step 3?**

   The presence of primer dimer confirms that the reaction contained all components necessary for amplification but that there was insufficient template to amplify the target sequence.

LABORATORY 3.6: DETECTING EPIGENETIC DNA METHYLATION IN *ARABIDOPSIS*

Answers to Bioinformatics Questions

I. Use BLAST to Find DNA Sequences in Databases (Electronic PCR)

2.i. **Why are some alignments longer than others?**

   The longest matches match both primers, whereas most of the shorter matches match to partial sequences of the *FWA* gene and only match one primer.

2.ii. What is the E value of your most significant hit and what does it mean? What does it mean if there are multiple hits with similar E values?

The lowest E value obtained for a match to both primers should be 2E-04, or 0.0002. This might seem high for a probability, but, in fact, this value means that a match of this quality would be expected to occur by chance in only two of 10,000 searches of this database! E values are based on the length of the search sequence, and the relatively short primers used in this experiment produce relatively high E values. Searches with longer DNA sequences return E values with smaller values. Hits with similar E values are often from closely related species.

What do the descriptions of significant hits have in common?

They all are from *Arabidopsis thaliana*, and many mention the *FWA* gene and promoter.

3.ii. To which positions do the primers match in the subject sequence?

The answers vary for each hit. For CP002687.1, the primers match positions 13,037,952–13,037,977 and 13,038,770–13,038,789, respectively. For AF178688.1, the primers match positions 757–782 and 1583–1602.

3.iii. Subtract the lowest nucleotide position in the subject sequence from the highest nucleotide position in the subject sequence. What is the difference between the coordinates?

For AF178688.1, the difference is 847; for CP002687.1, the difference is 837.

3.iv. Note that the actual length of the amplified fragment includes both ends. Thus, add 1 nucleotide to the result that you obtained in Step 3.iii to obtain the exact length of the PCR product amplified by the two primers.

4.iii. Identify the features located between the nucleotide positions identified by the primers, as determined in Step 3.ii.

The amplicon contains two repeat regions and TAT signal.

What feature is specifically related to gene transcription, what is its function, and how many nucleotides is it from the transcription start site?

The TATA signal (or TATA box) is a common binding site for transcription factors, which recruit RNA polymerase to the promoter to initiate transcription. The TATA box is located 25 nucleotides from the transcription start site, the textbook position for this transcriptional element.

5.iii. How many CpG dinucleotides can you identify?

The amplicon contains 25 CpG dinucleotides.

What does this suggest regarding the wild-type gene in this region?

The wild-type gene is heavily methylated in the promoter region.

5.iv. These are recognition sequences for McrBC. How many do you identify?

The amplicon contains three GCG and five ACG sites.

How would you go about determining how many McrBC recognition sequences are on the reverse (complementary) strand?

You can generate the reverse complement sequence of the published sequence and search for the GCG and ACG sequences. More cleverly, you can deduce that these sequences on the reverse strand sequence appear as CGC and CGT on the forward strand.

How many are there?

The amplicon contains six CGC and four CGT sites.

5.v. What does all of this suggest regarding the relationship among repeats, the TATA box, and methylation?

Repeats join with other elements, such as the TATA box, as binding sites for transcription factors in the promoter region of a gene. Methylation of these elements blocks the binding and silences the gene.

II. Use Map Viewer to Determine the Chromosome Location of the *FWA* Gene

5. On which chromosome have you landed?

The amplicon is on chromosome 4.

8. Which gene is closest to the amplicon?

As expected, it is the *FWA* gene.

10.i. Determine the size of the *FWA* gene using the map coordinates to the left of the "Contig" map.

13,042,000 – 13,038,000 = 4000 bp.

10.ii. How many exons and introns does the gene have?

There are 10 exons and nine introns.

10.iii. Where does the amplicon appear in relation to the gene: in an exon, an intron, or upstream of or downstream from the transcript?

The amplicon is adjacent to the first exon, just upstream of the transcript.

III. Identify the FWA Amino Acid Sequence and Function

10. Follow links to get some quick information regarding the domains. What do you find?

The FWA protein includes a START domain and a homeobox domain. The START domain is found in lipid-binding proteins and forms a deep lipid-binding domain that functions in lipid sensing or lipid exchange. Homeobox domains fold into structures that bind to DNA and are generally found in transcription factors. Given these domains, it is possible that *FWA* senses lipid levels and modulates the transcription level of target genes in response to these levels.

12. What can you tell from the graphical overview?

All of the hits match the query sequence over all, or nearly all, of the sequence. The red color indicates that they all have very high alignment scores, suggesting that they are very good matches.

13. **What can you conclude from the E values, the titles of the hits, and the accession links?**

    All of the E values are very small—either at or near 0. This suggests that the database has several proteins that are highly related to the query sequence. On closer inspection, many of the BLAST hits are to FWA or HDG6 (an alternative name for the same protein) in *Arabidopsis thaliana*, showing that the database contains multiple protein entries for FWA. Other hits are to FWA from closely related plants. Additional hits are to related proteins in *Arabidopsis* and more distantly related plants, suggesting that *FWA* is part of a plant-specific gene family.

Answers to Results and Discussion Questions

I. Observe Plant Growth

1. **Do you notice any growth or size differences between the wild-type *Ler* and mutant *fwa-1* plants?**

   Except for late flowering, the *Ler* and *fwa-1* plants should look very similar. The number of the rosette leaves, at the base of the plant, is correlated with flowering time. Because flowering is delayed, *fwa-1* plants grow more and larger leaves.

2. **How many days after planting did the *Ler* and *fwa-1* plants flower?**

   The *fwa-1* mutants should flower 2 wk or longer after the *Ler* plants.

   What does this suggest is the normal function of *FWA*?

   FWA is normally expressed only in endosperm (seed storage) tissue. The *fwa-1* epiallele causes ectopic, or inappropriate, expression of the *FWA* gene in the developing plant. Therefore, the normal function of *FWA* actually depends on it being turned off in the developing plant, which decreases flowering time.

II. Think About the Experimental Methods

1. **Describe the purpose of each of the following steps or reagents used in DNA isolation:**

1.i. **Using only a small amount of tissue.**

Using a small amount of tissue reduces carry-forward of PCR inhibitors present in the sample. These include metal ions, polysaccharides, and secondary metabolites.

1.ii. **Grinding tissue with pestle.**

Grinding disrupts plant cell walls. It also produces small clumps of cells that are more easily lysed to release DNA.

1.iii. **Nuclei lysis buffer.**

Nuclei lysis buffer contains a detergent that dissolves lipids in the cell membrane and membrane-bound organelles that contain DNA: the nucleus, mitochondria, and chloroplasts.

1.iv. **Protein precipitation solution.**

Cellular proteins are removed by a salt precipitation, which leaves the high-molecular-weight genomic DNA in solution.

1.v. **Heating to 65°C.**

Heating to 65°C helps to break down the cell and nuclear membranes and also denatures enzymes that can degrade the purified DNA.

1.vi. **McrBC incubation at 37°C.**

The McrBC endonuclease is active at 37°C.

1.vii. **McrBC incubation at 65°C.**

Incubation at 65°C denatures the McrBC protein, deactivating its enzymatic function. It is important to deactivate the enzyme before starting the PCR reaction; otherwise, it will continue to cut during the initial cycles of PCR.

2. **What is the purpose of performing each of the following PCR reactions?**

2.i. **Wild-type *Ler* undigested (WU).**

This is a positive control for amplification of the *FWA* amplicon in the wild-type plant before McrBC digestion.

2.ii. **Wild-type *Ler* digested (WD).**

This tests whether the wild-type *FWA* gene is cleaved by McrBC, indicating that it is methylated.

2.iii. **Mutant *fwa-1* undigested (MU).**

This is a positive control for amplification of the *FWA* gene in the mutant plant before McrBC digestion.

2.iv. **Mutant *fwa-1* digested (MD).**

This tests whether the mutant *FWA* gene is not cleaved by McrBC, indicating that it is not methylated.

3. **The recognition site for McrBC is G/A^{m}C...55–103 bp...G/A^{m}C, where ^{m}C represents a methylated cytosine. Why is McrBC an unusual restriction endonuclease?**

In traditional restriction-modification systems, methylation of the recognition sequence prevents the restriction enzyme from binding and cutting. Conversely, McrBC is modification dependent because it requires methylated sequences (although it does not actually cut at the methylated nucleotides).

Why is it so effective in identifying regions that are imprinted by methylation?

The half-recognition sites, G/A^{m}C, are very short and can be located at variable distances from one another. This "relaxed" recognition sequence is extremely common, ensuring that virtually all methylated DNA molecules are efficiently cleaved.

III. Interpret Your Gel and Think About the Experiment

4. **Which samples amplified and which ones did not? Give reasons why some samples may not have amplified; some of these may be errors in procedure.**

The WU and MU lanes each should show a single band that aligns between the 800- and 900-bp markers. These act as positive controls to confirm that the *FWA* locus can be amplified and are required to interpret the results for the digested DNA. No bands indicate that the amplification failed, most likely because the extracted DNA is contaminated or at too low a concentration. Multiple faint bands indicate nonspecific amplification of other sequences in the genome, due to suboptimal reaction conditions.

The WD lane should show no band. The *FWA* locus is heavily methylated in wild-type *Ler* plants, and it is therefore digested by McrBC. There is no intact template to amplify. A faint band suggests incomplete digestion, which left some template available to amplify.

The MU lane should show a single band that aligns with the 800-bp marker. Lack of methylation in *fwa-1* plants protects the *FWA* locus from cleavage by McrBC and provides a template for amplification.

CHAPTER 4

Genome Function

EVEN WITH COMPLETE SEQUENCES OF ENTIRE genomes in hand, deriving meaningful knowledge from DNA sequences remains a major challenge. Scientists have used computer algorithms to predict hundreds of thousands of genes in many different organisms. Many of these genes have little or no sequence similarity to well-understood genes, making it difficult to deduce what they might do. For example, more than half the genes discovered in the human genome sequencing projects initially had unknown functions.

Even genes that are related to other well-characterized genes require experimental confirmation of their predicted functions. As DNA sequencing becomes cheaper and faster, more and more predicted genes have unknown biological functions. Biologists have therefore used a number of approaches in a variety of organisms to assay the function of predicted genes. It is now becoming possible to quickly assay the function of virtually any gene.

These investigations are providing new insights into many different aspects of biology, including how organisms develop, how the genome is organized, how different organisms are related to one another, and even some glimpses into how life emerged.

STUDYING GENE FUNCTION IN MODEL ORGANISMS

A human is a complicated organism, and it would be unethical or impossible to do many experiments on human subjects. Therefore, biologists often use simple "model" organisms that are easy to keep and manipulate in the laboratory. Despite obvious differences, model animals share with humans many key biochemical and physiological functions that have been conserved (maintained) throughout evolution. For instance, invertebrates lack a skeleton and are much smaller than vertebrates, but they have nervous systems, muscles, and behaviors that can be studied easily in the lab.

With the rediscovery of Mendel's principles of inheritance, it became clear that traits are determined by discrete factors that we now call genes. Following Mendel's example, genes were identified by tracing traits through genetic crosses over several generations. In many cases, the presence of a gene could only be deduced by a mutation that produces a trait contrasting with the normal condition. In this way, different alleles, or forms of a gene, were related to different traits—genotypes were related to phenotypes.

Initially interested in how animals develop from zygote to adult, Thomas Hunt Morgan pursued the physical basis of heredity during the first decades of the 20th century. With limited space and funds for research, he needed to find an inexpensive, small organism to study. This led Morgan to develop the fruit fly *Drosophila* as the first laboratory animal model. Fruit flies are easy to maintain, have large numbers of offspring, and grow quickly—from embryo to adult in 12 days. These features make *Drosophila* a powerful tool for studying gene function and animal development.

Morgan and his students at Columbia University identified the first fruit fly mutations in the early 1900s and confirmed that Mendel's laws operate in this small animal. They identified a number of spontaneous mutants and used these to first establish that genes are located on chromosomes. They also discovered that alleles of different genes located on the same chromosome sometimes recombine through crossing-over and that the rate of recombination is proportional to the distance between the genes. This allowed them to map the relative positions of genes on chromosomes, producing the first genetic maps.

Hermann Müller, ca. 1927
(Courtesy of The Lilly Library, Indiana University, Bloomington.)

Spontaneous mutants are rare, and this initially limited progress in genetics. In 1926, while working at the University of Texas in Austin, Hermann Müller first used X rays to induce new mutations. Increasing the frequency of mutations using chemical mutagens or radiation made identifying mutants in model organisms much easier. The right dose of X rays produces unique mutations in virtually all of the offspring of irradiated flies. X-ray mutagenesis was joined by chemical and transposon mutagenesis to provide an arsenal of rapid means to produce new gene variations, accelerating the rate of gene discovery.

HOMEOTIC GENES IN *DROSOPHILA*

Because of the pioneering work by Morgan and Müller, *Drosophila* became a popular model organism. In the 1970s and 1980s, Christiane Nüsslein-Volhard and Eric Wieschaus were studying the process through which an egg develops into a complex organism. While working together at the European Molecular Biology Laboratory in Heidelberg in 1979, they used chemical mutagenesis screens to identify mutations affecting *Drosophila* development, including a number that affected the patterning of segments along the body.

Christiane Nüsslein-Volhard and Eric Wieschaus
(Courtesy of Christiane Nüsslein-Volhard.)

Fruit flies have distinct segments within three major body regions: head, thorax, and abdomen. Their work showed that early in development, proteins inherited from the mother (maternal factors) set down developmental patterns within these segments. The concentrations of these proteins are graded, with some concentrated at one end of the embryo and others at the opposite end. At any position in the embryo, there is a different concentration and mix of these proteins, which acts as a biochemical code for that location. These proteins, in turn, regulate specific genes needed to define functions for different regions and segments of the body.

First, the general regions of the head, thorax, and abdomen are mapped out by expressing a set of "gap" genes. Flies with mutations in this class of genes are missing segments, which produce gaps in the normal body plan. For instance, *Krüppel* mutants are missing the thoracic segments. The gap genes encode proteins that control the expression of "pair-rule" genes, which subdivide the body into seven bands. Mutants in pair-rule genes are missing every other segment. The segments are further defined by "segment polarity" genes, which define the anterior/posterior (head/tail) orientation of 14 segments. In mutants of segment polarity genes, the posterior end of each segment is replaced by a mirror image of the anterior end. This cascade of regulatory interactions determines the segmental pattern of the fly embryo, from head to tail.

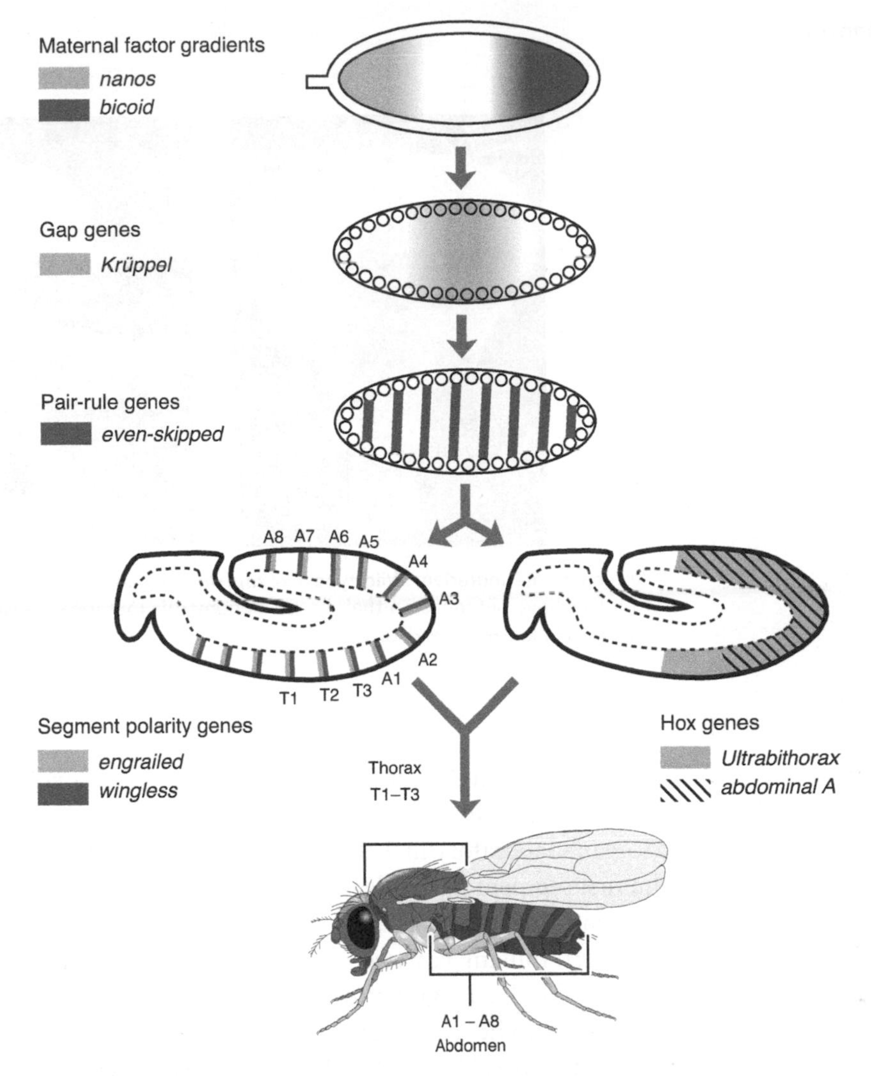

A cascade of patterning genes defines body pattern. Stages of patterning of the anterior/posterior axis in *Drosophila* embryo. First, gradients of maternal factors define regions. The relative levels of maternal factors determine where gap genes are expressed. Gap genes, in turn, regulate pair-rule genes. The pair-rule genes activate the segment polarity genes and cooperate with gap genes to pattern the expression of Hox genes. Segment polarity and Hox genes control the differentiation of segments.

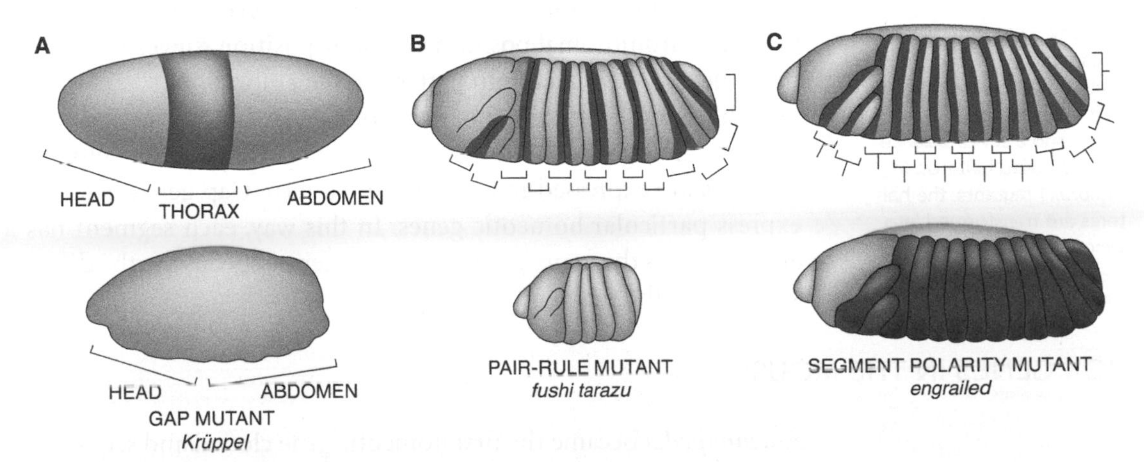

Three types of body patterning mutations in *Drosophila*. (*A*) Gap genes control differentiation along the head-to-tail axis. In this example, a *Krüppel* mutant is missing the thoracic segments. (*B*) Pair-rule genes define each segment: *fushi tarazu* mutants are missing every other segment. (*C*) Segment polarity genes define the anterior and posterior of each segment: *engrailed* mutants have segments that are entirely anterior.

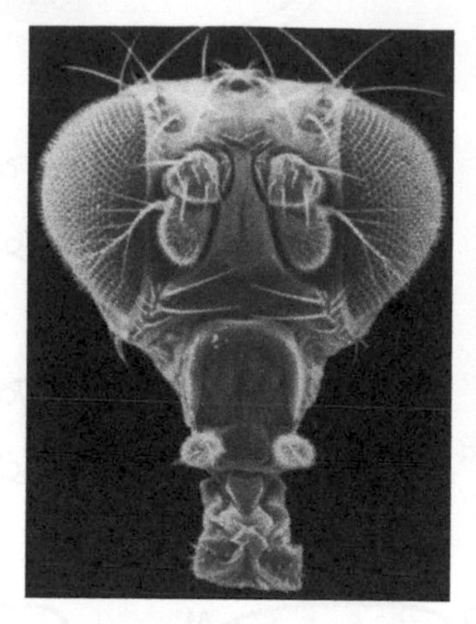

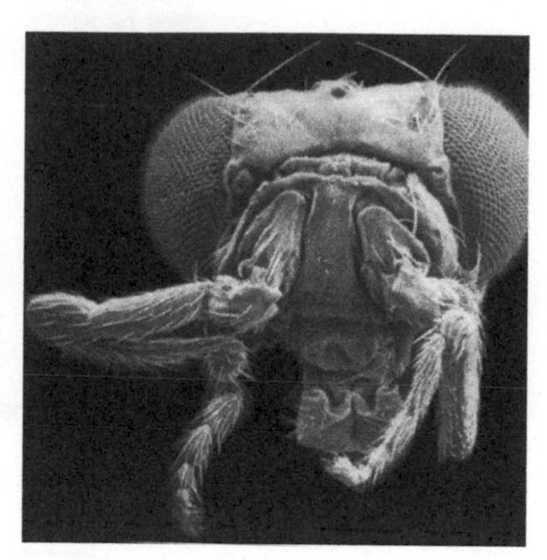

Antennapedia mutation. Wild-type *Drosophila* (*left*) and *Antennapedia* mutant fly (*right*). Wild-type fruit flies have antennae on their heads. A homeotic mutation transforms antennae into legs.
(Courtesy of F.R. Turner, Indiana University.)

Edward Lewis, 1951
(Courtesy of Cold Spring Harbor Laboratory Archives.)

Bithorax mutation. Wild-type *Drosophila* (*top*) and *bithorax1* mutant fly (*bottom*). Wild-type fruit flies have one set of wings and a set of balance organs, called halteres. In *bithorax1* mutants, the halteres are transformed into wings.
(Courtesy of the Archives, California Institute of Technology.)

Beginning in the late 1940s, while at the California Institute of Technology, Edward Lewis worked with a *Drosophila* mutant called *bithorax1*, which had been discovered in Morgan's laboratory in 1915. The fruit fly thorax is composed of three segments, with the first segment having legs, the second segment having legs and wings, and the third segment having legs and halteres (small wing-like structures used for balance). The *bithorax1* mutant has two sets of wings and no halteres. Lewis showed that this is because the third segment of the thorax is transformed into a duplicate of the second segment. Another striking mutant is *Antennapedia*, where antennae are replaced by legs. Mutants such as *bithorax1* and *Antennapedia*, where one part of the body is transformed into another, are called "homeotic" (from the Greek *homoīōsis*, meaning similar).

Using a genetic screen, Lewis identified mutants in eight homeotic genes. Mapping these genes by recombination showed that they line up along *Drosophila* chromosome 3 in the same order as the body regions that they affect. The genes at one end of the chromosome affect segments at the anterior of the fly, whereas those further along the chromosome act toward the posterior. The parallelism, or colinearity, between chromosomal position and body position suggested that pattern on the chromosome was translated into pattern in the embryo.

The expression pattern of homeotic genes is governed by gap, pair-rule, and segment polarity genes. Pair-rule and segment polarity genes help to ensure that homeotic genes are expressed in a segmental pattern. Gap genes determine which segments express particular homeotic genes. In this way, each segment has a defined identity and expresses the appropriate homeotic genes, specifying the differentiation of the tissues within the segment.

HOX GENES IN THE MOUSE

Antennapedia became the first homeotic gene cloned and sequenced in 1984 by Walter Gehring at the University of Basel. Using Southern blotting, he showed that part of the *Antennapedia* gene hybridizes to other regions of the fruit fly genome, suggesting that other genes contained sequences related to *Antennapedia*. A careful search revealed

Walter Gehring, 1985
(Courtesy of Cold Spring Harbor Laboratory Archives.)

that all homeotic genes contain this conserved region, which they called the homeobox. This region encodes the DNA-binding domain of a family of transcription factors that regulate the expression of genes involved in development. Each homeodomain binds to a different DNA sequence, determining which downstream genes are turned on or off and which segment develops a particular set of traits.

To identify downstream genes, researchers use microarrays to measure changes in gene expression in homeotic mutants. First, messenger RNAs (mRNAs) from wild-type and mutant organisms are labeled with different fluorescent dyes. The labeled RNAs are hybridized to chips, with spots of DNA representing every gene. After hybridization, the relative amounts of wild-type and mutant message bound to each spot are measured, revealing which genes are up- or down-regulated in the mutant. In this way, thousands of downstream genes have been identified. Direct sequencing of mRNAs, RNA seq, is now replacing microarray studies of gene expression. Bioinformatic searches of genome sequence have revealed additional homeodomain-binding sites in hundreds of genes, suggesting that they are direct targets of regulation. Although the details still remain to be revealed, this concerted regulation explains how expression of a single homeotic gene, such as *bithorax1* or *Antennapedia*, can orchestrate the development of a complex body part, such as a wing or antenna.

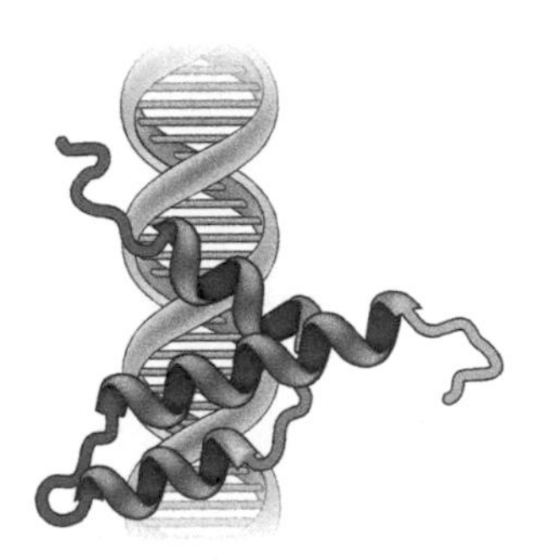

The *Antennapedia* homeodomain bound to DNA. The homeodomain consists of three α-helices and an amino-terminal arm. The third helix and the amino terminus contact the major and minor grooves of the DNA, respectively.

Walter Gehring and other investigators were curious about whether homeobox genes are found in other organisms. Using DNA from *Drosophila* homeotic genes as probes, vertebrate homologs, called Hox genes, were quickly identified in frogs, mice, and humans. Like their *Drosphila* counterparts, Hox genes in these organisms are in clusters, arranged along the chromosome in the same order in which they are expressed in the embryo, from head to tail. This is most clearly seen in the vertebrate neural tube, where the expression pattern of Hox genes corresponds to segmental structures called rhombomeres.

To study the function of Hox genes, researchers turned to mice, which were established as a vertebrate model organism in the early 1900s. Like fruit flies, mice are (relatively) small, breed quickly, and are amenable to genetic analysis. However, they are much more like human beings than fruit flies and so are an important tool for studying mammalian physiology, development, and disease. The discovery of Hox genes by sequence homology rather than by phenotype underscored a limitation of classical genetics. How could mutations be made in these genes to study their function in mammals? As a random process, mutagenesis is ineffective in targeting mutations in specific genes, so a new technique was needed.

Mario Capecchi
(Courtesy of Mario Capecchi, Eccles Institute of Human Genetics.)

Mario Capecchi, at the University of Utah, and Oliver Smithies, at the University of North Carolina, independently developed "gene targeting" as a way to precisely mutate, or "knock out," a gene of interest in mice. Their basic method takes advantage of homologous recombination, a natural exchange between sections of DNA with very similar sequences. Homologous recombination is the mechanism for crossing-over between chromatids during meiosis and for repairing double-strand breaks in DNA.

To create a knockout, a targeting vector is constructed by inserting a portion of the target gene (the transgene) into a plasmid. The flanking portions of the transgene have strict homology with the genomic sequence of the target gene. However, the central part of the construct includes a selectable marker, most commonly a neomycin resistance gene, *neo*. The targeting vector is then transformed into cultured mouse cells. Subsequent selection with neomycin allows only cells that have taken up the vec-

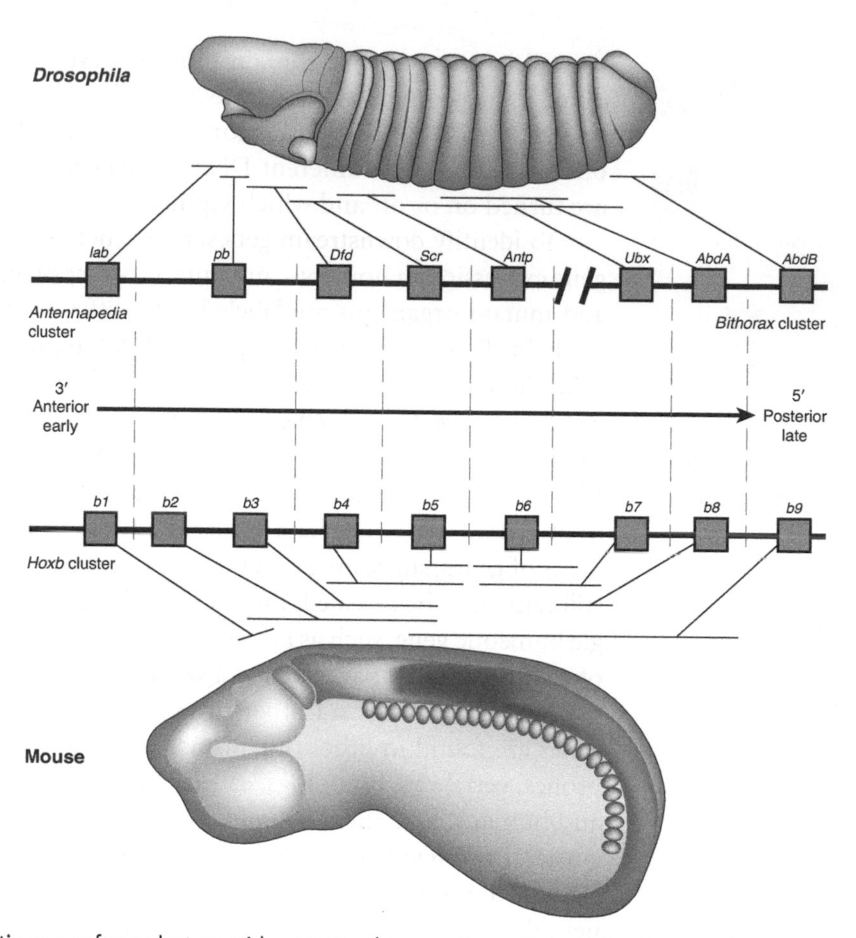

Homeotic genes form clusters with conserved expression patterns. The *Drosophila Antennapedia* and *bithorax* clusters align with the mouse *Hoxb* cluster. The homeotic genes are arranged along the chromosome in the same relative position as their position of expression and function along the body axis.

Oliver Smithies, 1984
(Courtesy of Cold Spring Harbor Laboratory Archives.)

Martin Evans
(Courtesy of Cardiff University.)

tor (with *neo*) to survive. Homologous recombination at the target site replaces the genomic sequence with the altered sequence, including the *neo* gene. In this context, the *neo* gene performs its second function. Because the *neo* gene is integrated within a coding exon, the function of the target gene is disrupted, hence the term "knockout."

Complementing their work, Martin Evans, at Cambridge University, isolated mouse embryonic stem (ES) cells. These cells are multipotent; they have the ability to develop and differentiate into a variety of cell types. Evans showed that these cells can be genetically altered and then reintroduced into embryos that give rise to genetically modified mice. In 1987, Capecchi and Smithies showed that their "gene targeting" works with ES cells, providing the capability to knock out essentially any gene in the mouse genome. For their work, Capecchi, Smithies, and Evans received the 2007 Nobel Prize in Physiology or Medicine.

Capecchi quickly used the new method to knock out Hox genes in mice. It initially proved to be difficult to assign functions to mouse Hox genes, because the mouse has four clusters of duplicated Hox genes (A, B, C, and D) rather than the single set found in *Drosophila*. Equivalent genes in different clusters often have overlapping or redundant functions, so, for example, single knockouts of *HoxA10*, *HoxC10*, or *HoxD10* have subtle phenotypes. However, triple knockouts transform the posterior lumbar vertebrae and sacrum into additional anterior vertebrae that form ribs.

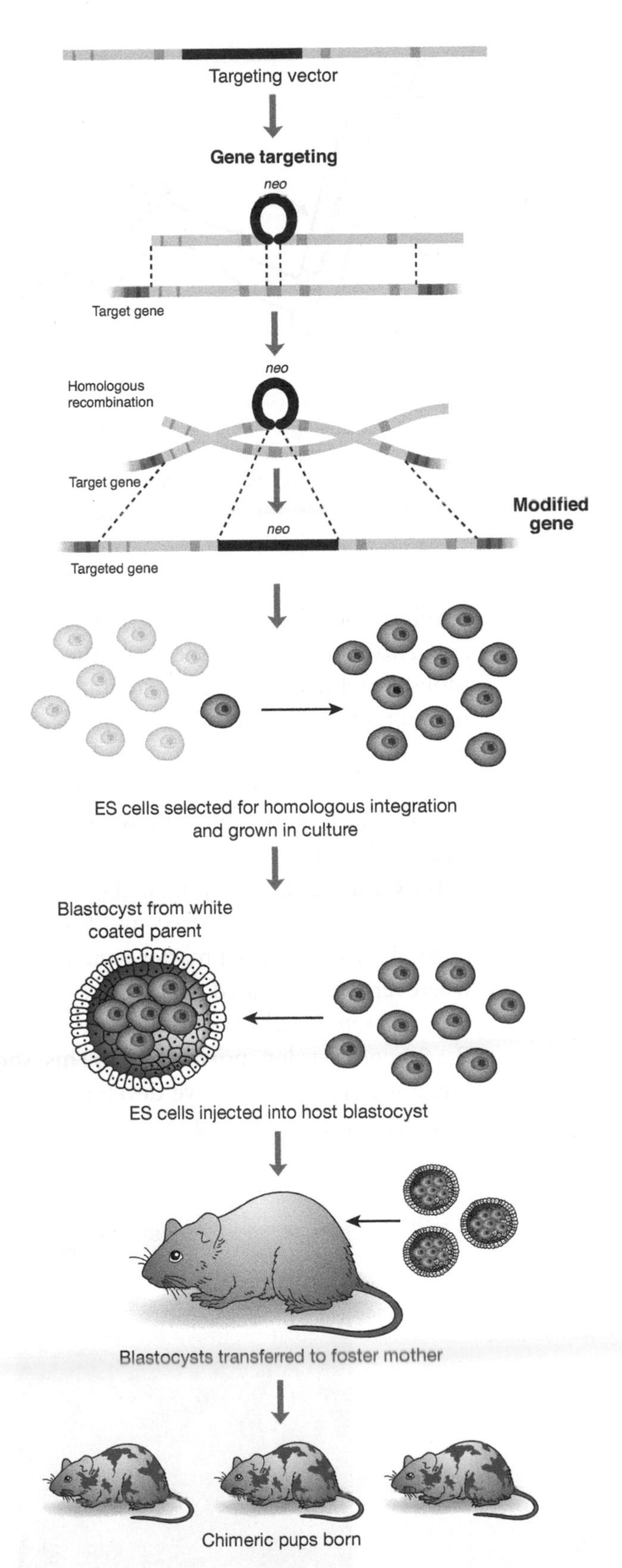

Homologous recombination. Targeting vectors for homologous recombination contain a marker gene, neomycin resistance (*neo*), flanked by sequence from the target. Correct targeting by homologous recombination results in neomycin-resistant cells. ES cells with the desired change are combined with a blastocyst and transferred to a foster mother. Chimeric pups have a mixture of cells from the blastocyst and ES cells.

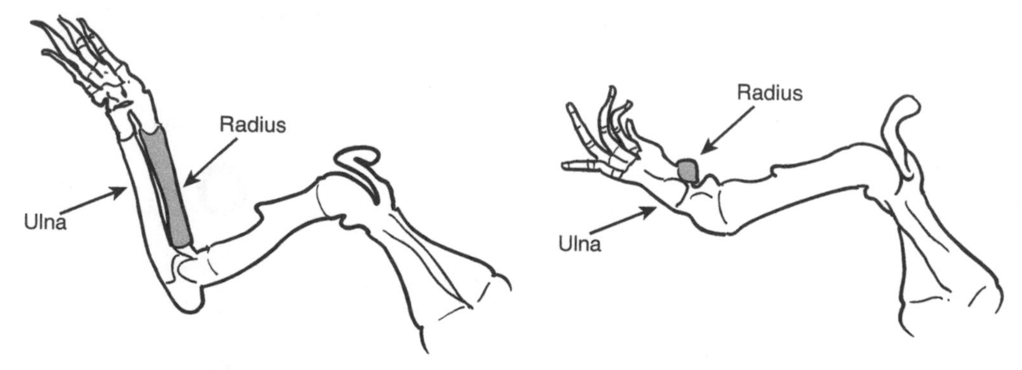

Hox genes in limb development. Mutations in paralogous Hox genes cause dramatic disruptions of limb skeletal morphology. Forelimbs of a control (*left*) newborn and *HoxA11/HoxD11* double mutant (*right*) mouse missing normal radius and ulna.

(Redrawn, with permission, from Boulet AM, Capecchi MR. 2004. *Development* 131: 299–309; ©Company of Biologists.)

Similarly, deleting individual *Hox11* genes has little effect, but deleting both *HoxA11* and *HoxD11* produces the dramatic phenotype of deleted radius and ulna in the mouse forelimb. Thus, the vertebrate Hox genes define the identity of body parts in a manner analogous to their *Drosophila* counterparts, illustrating evolutionary conservation of genetic control of animal body plan development.

Using radioactively labeled DNA probes for the *Drosophila* homeobox gene paired, Gehring, in 1991, discovered the vertebrate *Pax6* gene. *Small eye* mutants in mice and humans with aniridia (missing the iris) were shown to have mutations in *Pax6*. Gehring then identified the *Drosophila* homolog of *Pax6*. Called *eyeless*, this mutant had been discovered by Morgan's group in 1915. Gehring showed that the normal *eyeless* gene is a master regulator of eye development, inducing eye formation wherever it is expressed. *Pax6* and *eyeless* are so closely related as to be interchangeable. *Drosophila* whose normal *eyeless* gene has been replaced with a mouse or squid *Pax6* gene develop normal eyes. This showed a conservation of function in the homeotic regulation of eye development stretching back the estimated 500 million years to a common ancestor of insects, molluscs, and vertebrates.

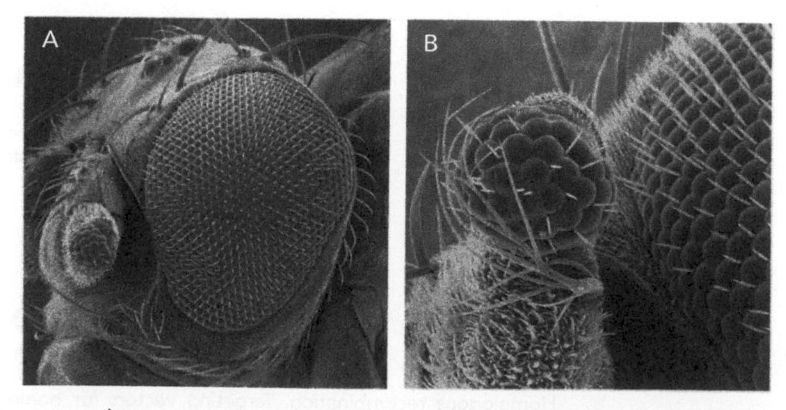

Eyeless is a conserved master control gene for eyes. (*A*) In *Drosophila*, expression of the mouse *Pax6* gene induces ectopic eyes, to the left of the normal eye. (*B*) Higher magnification of the ectopic eye.

(Reproduced, with permission, from Gehring WJ. 1996. *Genes Cells* 1: 11–15; ©Wiley.)

DEVELOPMENT IN *C. ELEGANS*

Sydney Brenner, 1980
(Courtesy of Cold Spring Harbor Laboratory Archives.)

Working at Oxford University in the early 1960s, Sydney Brenner made major discoveries in molecular genetics, including identifying mRNAs and determining that the genetic code is read in triplets. After solving the genetic code in 1966, Brenner joined Francis Crick in believing that it was time to move on to one of the great unanswered questions in biology: How does the brain develop and function? But the brains of mice—with millions of cells—or even those of fruit flies—with hundreds of thousands of cells—seemed to be too complex. When Brenner moved to the MRC Laboratory of Molecular Biology in Cambridge, he looked for a simpler animal, but one still having a nervous system and other distinct, differentiated tissues—including epidermis (skin), muscles, and intestine. After careful groundwork, he selected the microscopic roundworm *Caenorhabditis elegans* as his model for nervous system development.

Although some roundworms are parasitic, *C. elegans* is free-living. This worm grows quickly—from embryo to adult in three days—is easy to culture and can be stored in a freezer. *C. elegans* has only ~1000 cells, including 302 neurons, a number Brenner believed was small enough to allow detailed analysis. Furthermore, because the body is transparent, every cell in embryos and adult worms can be examined in detail. Mating animals, isolating genes, and introducing foreign DNA are much easier in worms than in more complicated animals. These features make *C. elegans* a great model for understanding how cells divide, develop, and take on specialized tasks in higher (eukaryotic) organisms.

John Sulston
(From *PLoS*.)

Brenner and his colleague John Sulston showed that *C. elegans* also has a small, compact genome, which proved to be important when large-scale sequencing became possible. At only 100 megabases (100 million bases), the *C. elegans* genome is ~1/30 the size of the human genome. For this reason, the worm genome was chosen as the "trial run" for the Human Genome Project, providing the first complete genome sequence of a multicellular organism in 1998. Despite the much smaller genome, *C. elegans* has a similar number of protein-coding genes (19,500) as humans (23,700), and 40% of worm genes have human homologs.

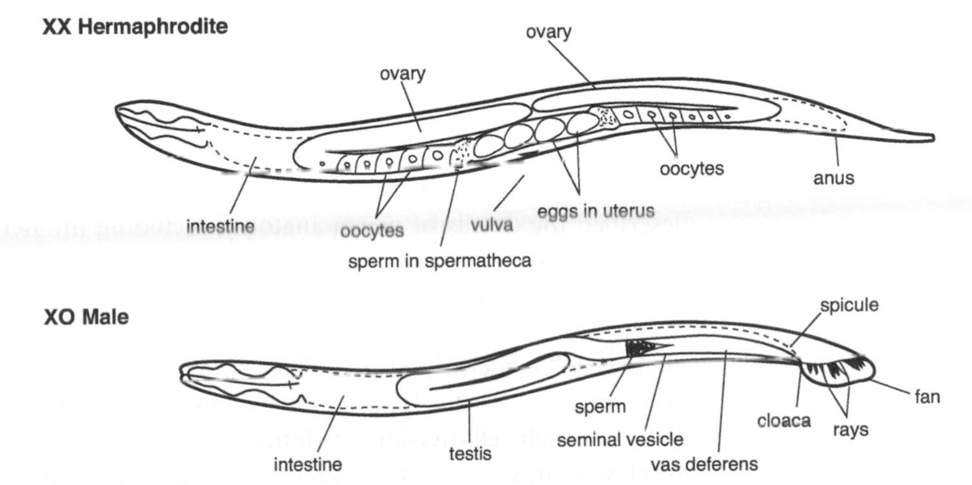

The nematode *C. elegans*. Diagram of adult hermaphrodite and male *C. elegans*.
(Redrawn from Zarkower D. 2006. http://www.wormbook.org.)

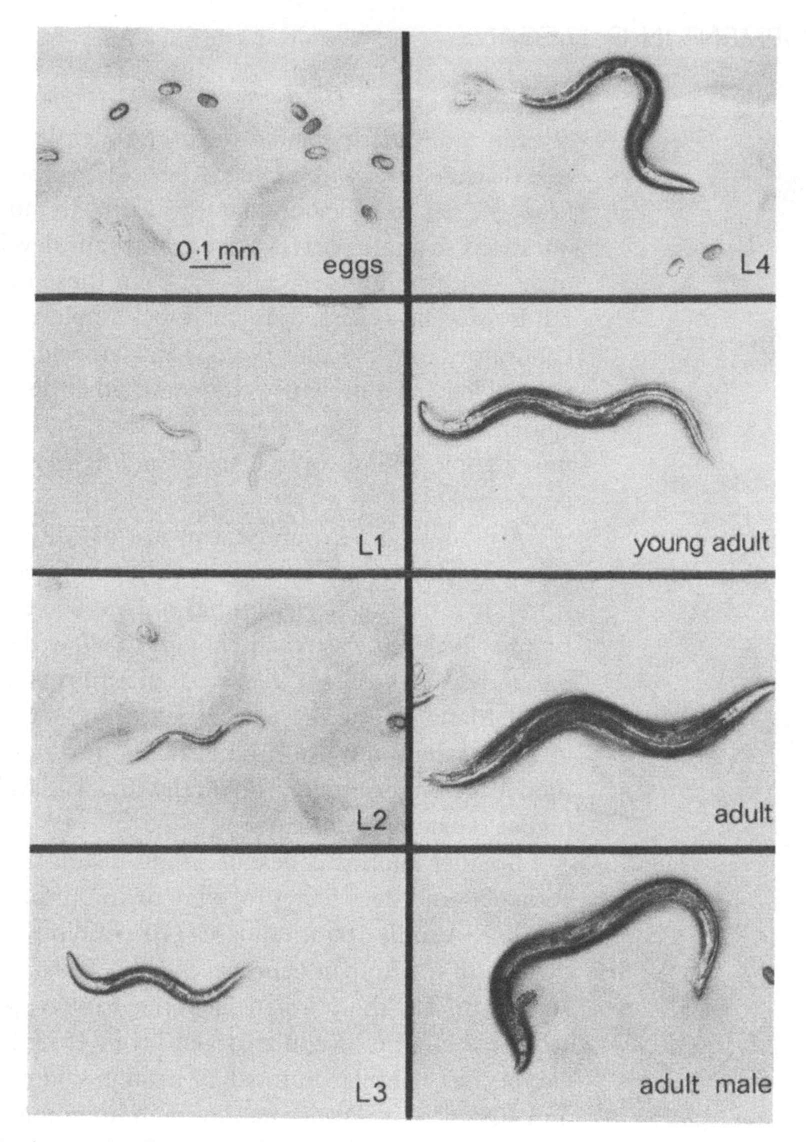

Stages in *C. elegans* development. After hatching, *C. elegans* passes through four progressively larger larval stages (L1–L4) marked by molts, followed by adulthood.

(Reprinted from Wood WB. 1988. *The nematode* Caenorhabditis elegans. Cold Spring Harbor Laboratory Press, Cold Spring Harbor, NY; photo courtesy of J. Sulston.)

Sydney Brenner's group identified hundreds of mutants and developed genetic techniques to map and characterize them. Using light and electron microscopes, they described the details of worm anatomy, including the nervous system. These efforts took more than a decade, but they laid the foundation for research by thousands of "worm geneticists" throughout the world.

As Brenner proposed in 1963, an important step toward understanding development was "to identify every cell in the worm and trace (their) lineages." Sulston observed cells in developing embryos under a light microscope and sketched the results of each cell division. He found that 1090 cells are produced during *C. elegans* development, with most cell divisions occurring identically in every animal. This is different from humans and many other organisms, where the cell division pattern is

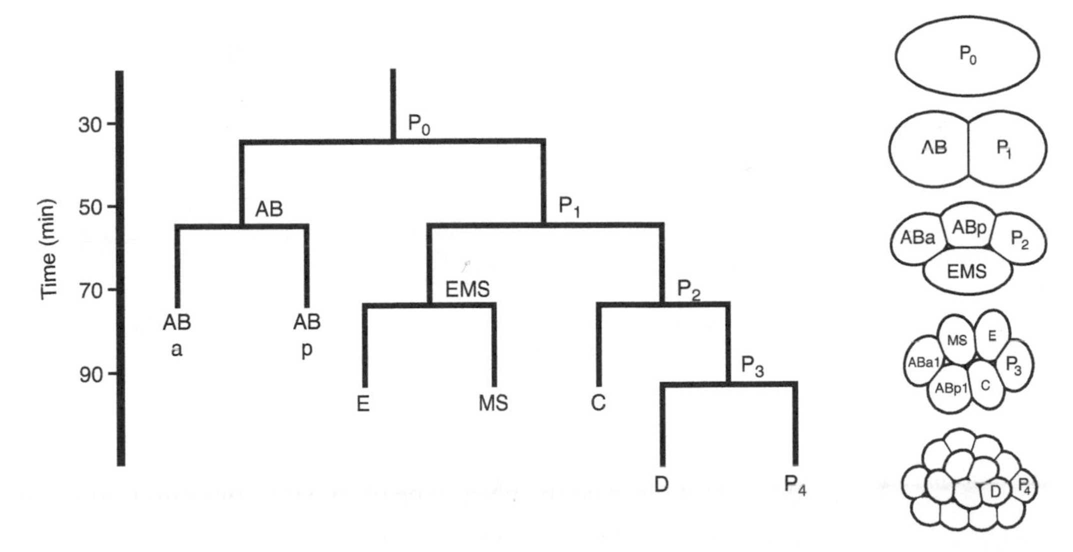

Cell lineage. (*Left*) Cell division pattern in early *C. elegans* embryogenesis. Each cell is represented by a vertical line. Horizontal lines connect daughter cells to their parent. Timing in minutes is shown on the *y* axis. (*Right*) Schematic of cell positions within embryo.

(*Left*, Based on Sulston et al. 1983. *Dev Biol* 100: 64–119. *Right*, Adapted from Gönczy P, Rose LS. 2005. http://www.wormbook.org.)

slightly different in each embryo. The invariant lineage allowed Sulston to figure out the origin and relationships among every cell in the developing *C. elegans* embryo. In the 1980s, he discovered that 131 of the 1090 cells die during worm development, leaving "corpses" that are easy to identify. These cells undergo programmed cell death, or apoptosis, committing "suicide" as a normal part of development.

CELL DEATH MUTANTS

Members of Brenner's group systematically screened for mutations in genes that control programmed cell death. They called the genes that they identified *ced*, for cell death abnormal. (In *C. elegans*, genes and proteins are given three-letter designations

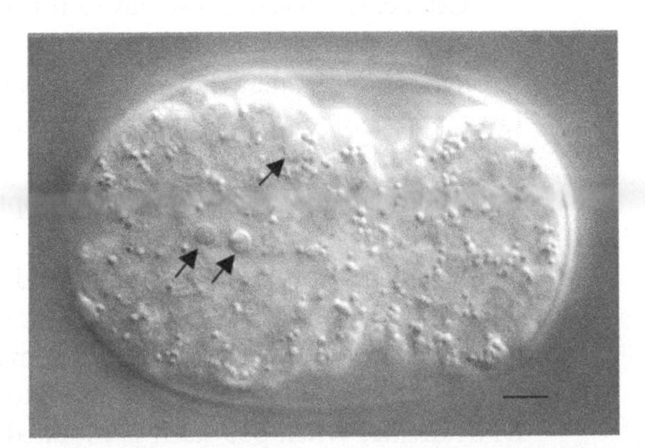

Programmed cell death. Nomarski image of a normal embryo with apoptotic cells. Three cells indicated by arrows underwent programmed cell death and exhibit a raised-button-like appearance. Bar, 5 μm.

(Reprinted from Conradt B, Xue D. 2005. http://www.wormbook.org.)

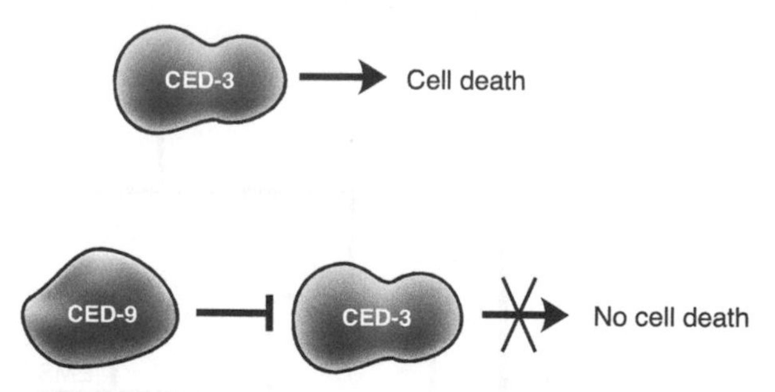

Model for the activation of programmed cell death in *C. elegans*. CED-3 activation leads to cell death. CED-9 prevents activation of CED-3, blocking cell death.

Robert Horvitz, 1993
(Photo by Donna Coveney, MIT.)

that describe the phenotype of mutants. By convention, gene names are in lowercase italics, whereas protein names are capitalized.) After training with Brenner, Robert Horvitz continued these screens at the Massachusetts Institute of Technology, where his lab identified numerous *ced* genes.

There is interplay between *ced* genes. For example, *ced-3* encodes a protein required for programmed cell death. This protein, CED-3, is a protease that actively degrades other proteins, and any cell in which it is produced dies. *ced-3* mutants produce no protease, and cells that normally undergo apoptosis survive. In contrast, the CED-9 protein prevents programmed cell death. In *ced-9* worm mutants, cells that normally live undergo apoptosis. In fact, CED-9 prevents CED-3 from killing cells. The more CED-9 protein present, the better a cell is protected against apoptosis. When CED-9 is present in high enough levels, apoptosis is prevented even in cells that are usually programmed to die.

Programmed cell death is also important for normal human development. For example, apoptosis eliminates cells between developing digits, sculpting fingers and toes. The apoptosis pathway has been conserved through evolution, so worm *ced* genes have homologs in the human genome. Human *CASP9* is the homolog of worm *ced-3*, and human *BCL2* is the homolog of worm *ced-9*. These genes encode proteins with equivalent cell death functions in humans. The genes are so similar that replacing *ced-9* with the human *BCL2* gene still protects worm cells against death.

Cell death processes also have a role in cancer. For example, in B-cell lymphomas, a translocation between chromosomes 14 and 18 brings *BCL2* under the control of a strong enhancer that normally activates expression of immunoglobulin genes. Overexpression of *BCL2* prevents normal cell death, contributing to the overproliferation of B lymphocytes.

Another important function of apoptosis is to maintain genome stability. In normal cells, the "checkpoint" protein p53 monitors the state of DNA. When p53 senses DNA damage (mutations) during cell division, it activates one of two parallel pathways. If damage is limited, p53 activates DNA-repair proteins and arrests cell division while the mutations are repaired. If damage is severe, such as UV-induced mutations in sunburned skin, p53 initiates apoptosis. This ensures that only cells with intact DNA divide. When p53 itself is mutated, this damage-sensing system fails, and cells survive to pass on severe mutations to daughter cells. This allows the accumulation of mutations that can lead to cancer.

Because of the important role that apoptosis has in normal and disease processes, Sydney Brenner, Robert Horvitz, and John Sulston were awarded the 2002 Nobel Prize in Physiology or Medicine.

ANTISENSE RNA

Attempts to use Capecchi's gene targeting by recombination failed in *C. elegans*. Less costly and time-consuming methods, based on RNA, were developed to knock out genes in a sequence-specific manner. In 1978, Paul Zamecnik at Harvard Medical School had shown that short RNA oligonucleotides bind to a complementary mRNA from Rous sarcoma virus and inhibit gene function in cultured mouse cells. The bound RNA was thought to interfere with translation, possibly at the ribosome. Because the olignonucleotides that achieve the knockout are complementary ("anti") to the protein-coding, or "sense," RNA message, this strategy became known as "antisense" RNA.

In the mid 1980s, antisense RNA was adapted to study gene function in tissue culture, plants, fungi, and animals. To disrupt a gene, researchers produced antisense RNA by in vitro transcription (using purified RNA polymerase) or within a genetically modified organism. This was accomplished by placing the DNA sequence for a target gene in a "backward" orientation, such that a promoter initiates RNA transcription from the opposite strand as the normal mRNA. Numerous studies showed that an antisense RNA could produce the same phenotype as a DNA mutation affecting the same target gene.

Antisense technology was adopted, with varying success, by industry. The first genetically modified (GM) plant designed for human consumption was the "Flavr-Savr" tomato, developed by Calgene in 1991. Calgene scientists created transgenic tomatoes that expressed an antisense RNA for polygalacturonase (PG), an enzyme that initiates fruit rotting. The resulting transgenic tomatoes had reduced PG levels and softened more slowly than traditional tomatoes. The plan was to harvest the transgenic tomatoes after they ripened on the vine, giving them better flavor. Flavr-

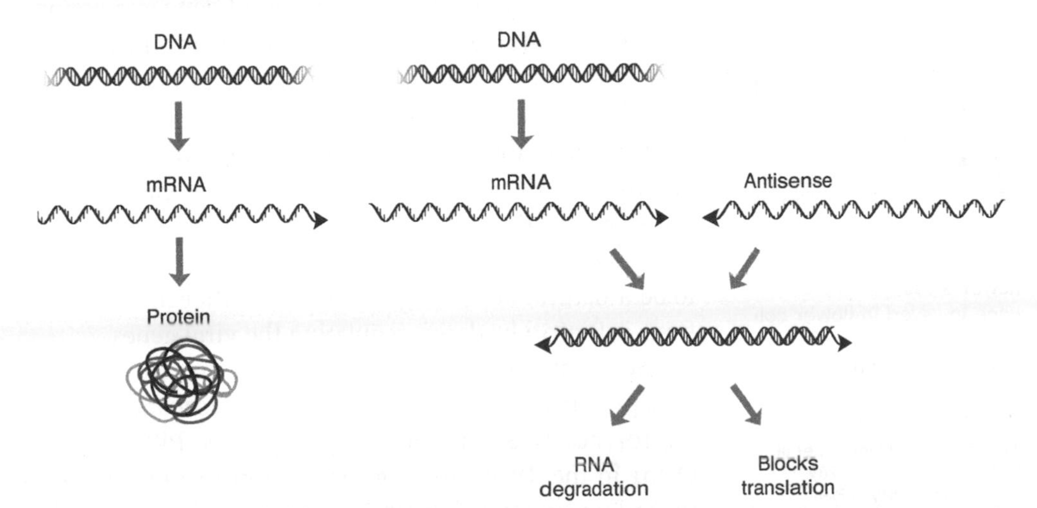

Antisense technology. (*Left*) Normally, DNA is transcribed into mRNA and translated into protein. (*Right*) Introducing antisense oligonucleotides that are complementary to an mRNA blocks gene function. Hybridization of an antisense oligonucleotide to a complementary mRNA results in RNase H cleavage of the message or interference with translation, both of which block protein production.

Savr tomatoes were approved for sale in the United States in 1994. As expected, the transgenic tomatoes had a longer shelf life. However, they were prone to bruising, were more expensive, and did not taste better than conventional tomatoes. Anti-GM sentiment kept Flavr-Savr out of Europe. After several years in limited United States markets, the GM tomato proved to be unprofitable and was withdrawn.

In 1998, an antisense RNA drug called fomivirsen, or Vitravene, was approved for treatment of cytomegalovirus (CMV) retinitis, which can lead to blindness in immune-deficient patients. Fomivirsin is a short oligonucleotide that is complementary to CMV mRNA. Injected directly into the eye, it blocks translation of key CMV genes.

NOT-SO-PURPLE PETUNIAS

A major breakthrough in understanding the mechanism of gene silencing came from unexpected experimental results in plants. In 1990, Rich Jorgensen's group at DNA Plant Technology, Inc., attempted to make petunias a deeper shade of purple by introducing an extra copy of the gene encoding chalcone synthase (CHS), the rate-limiting enzyme in the production of anthocyanin (purple) pigments. They reasoned that additional copies of the CHS gene would produce more mRNA, encoding more of the key enzyme, which, in turn would lead to an excess of purple pigment. Their hope was to produce an intensely purple flower approaching true black, which has yet to be found naturally in plants.

To their surprise, instead of making darker flowers, plants with the extra CHS gene had different types of flowers—some purple, some with purple and white patches, and some pure white. How could this be? The level of expression of the CHS gene correlated with color, with white flowers or patches producing 50-fold less CHS mRNA than purple ones. This suggested that, in white areas, the introduced gene (the transgene) somehow interfered with its own expression and with the expression of the endogenous gene, an effect they called cosuppression. The variability in flower color on one plant showed that genetically identical cells could exhibit different phenotypes that are due to reversible suppression of gene expression.

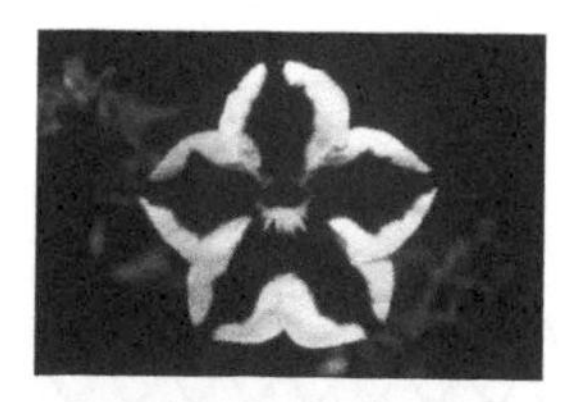

Unexpected gene silencing. White sections in petunia flower areas are where a gene involved in flower coloration, chalcone synthase, was silenced when extra copies of the gene were introduced to petunias. (Reprinted from Napoli C, et al. 1990. *Plant Cell* 2: 279–289. ©American Society of Plant Biology; photo courtesy of R. Jorgensen.)

Antoine Stuitje, at the Free University in Amsterdam, later showed that the transgene for another plant pigment causes similar color suppression, proving that the silencing effect is not specific to CHS. Cosuppression by transgenes was also observed in fungi, nematodes, and some protozoa. Work on plant viruses also uncovered examples of gene silencing. Recombinant viruses carrying sequences related to host genes can suppress the expression of homologous host genes. Similarly, plants containing sequences related to viral genes can suppress viral gene expression. This effect seemed to be a natural defense against viruses: After infection by viruses, some plants recover from infection by down-regulating the viral genes.

Scientists used this phenomenon to save the papaya industry from papaya ringspot virus (PRSV). Papaya is Hawaii's second most important fruit crop, accounting for nearly $50 million in yearly sales. PRSV destroyed the papaya industry on Oahu in the 1950s, forcing production to move to the Big Island of Hawaii. PRSV turned up on Hawaii in 1992, again threatening the industry. Fortunately, by 1995 the GM "Rainbow" strain had been developed, which expressed a transgene of the coat protein of the ringspot virus. Rainbow is highly resistant to PRSV, producing 125,000 pounds of fruit per acre compared to 5000 for nonresistant plants.

Transgenic papaya. Aerial view of transgenic "Rainbow" papaya (dark) surrounded by nontransgenic variety infected with ringspot virus on the island of Hawaii.
(Courtesy of Dennis Gonsalves.)

THE MECHANISM OF RNA SILENCING

The initial proposed mechanism—that an antisense RNA strand binds to and silences a sense strand—found little experimental support. Sir David Baulcombe at the Sainsbury Laboratory in Norwich, England, showed that silencing affects mRNA levels in plants. Other experiments suggested that some forms of silencing affect protein translation, rather than mRNA levels. No simple model seemed to explain all of the results.

Andrew Fire
(From Linda A. Cicero/ Stanford News Service.)

Significant progress came from work in worms. In 1987, while working at the Carnegie Institution of Washington in Baltimore, Andy Fire and Susan White-Harrison tested antisense technology in *C. elegans*. They created DNA constructs with a strong promoter driving the expression of "sense" or "antisense" versions of a gene called *unc-22*, mutations that cause worms to twitch rather than move in a coordinated manner. As expected, they observed the twitching phenotype when they introduced the "antisense" construct into wild-type worms. However, the "sense" version of the *unc-22* construct also induced twitching. They hypothesized that this puzzling result was due to the unintended production of antisense transcripts from the sense construct.

In 1995, Susan Guo and Ken Kemphues at Cornell University had the same result when they attempted to silence a gene called *par-1* by injecting RNA directly into *C. elegans*. As expected, injecting antisense RNA produced the *par-1* mutant phenotype. However, control worms injected with "sense" RNA also showed the mutant phenotype. These experiments with naked RNA preparations stripped away any confounding contribution of a DNA construct, provoking further interest in finding the trigger for gene silencing by RNA.

Craig Mello
(Used with permission of the University of Massachusetts Medical School.)

In 1998, Fire and Craig Mello, at Johns Hopkins University, explained a decade of confusing results when they proposed that double-stranded RNA (dsRNA) is the trigger for silencing. dsRNA is more stable than single-stranded RNA (ssRNA), which could explain why an RNA injection could have long-lasting effects. A small amount of contaminating dsRNA also could explain silencing by DNA transgenes and sense RNA preparations.

To test their hypothesis, Fire and Mello carefully isolated RNAs for the sense and antisense strands of selected genes, making sure to avoid any dsRNA contamination. They injected worms with one or both strands and checked for loss of gene function. Dramatically, when they injected complementary strands, the resulting dsRNA was more than 100 times more effective at gene silencing than the antisense RNA strand alone. They called this phenomenon of dsRNA-dependent gene silencing RNA interference, or RNAi. Fire and Mello were awarded the 2006 Nobel Prize in Physiology or Medicine for proving that dsRNA is the trigger for a fundamental mechanism for controlling gene expression in many eukaryotes.

Only a few dsRNA molecules need to be introduced into a cell to induce RNAi. It was soon realized that it is not even necessary to inject dsRNA to silence a target gene. Worms can take up dsRNA merely soaking in a solution of dsRNA or by feeding on bacteria that express a dsRNA identical to a target gene. RNAi can spread throughout the body of the worm, so that injecting dsRNA into intestinal cells also silences genes in muscle tissue. Moreover, gene silencing can often be inherited from one generation to the next, presumably by dsRNA that is passed on from the egg cell as it divides.

Sir David Baulcombe, 2006
(From www.wikimedia.org.)

Once the trigger was identified, attention turned to the mechanism of RNAi. How can a small number of dsRNA molecules lead to silencing that spreads and amplifies throughout the body? Although a likely mode of action was through complementary base pairing of one strand of the dsRNA to the target mRNA, how this worked remained a mystery. It was also unclear how RNAi affected transcription in some systems and translation in others.

In 1999, Sir David Baulcombe reasoned that any long RNA molecules that were active in RNAi already would have been observed. He searched for and found short dsRNA molecules that are homologous to plant genes that had been silenced by RNAi.

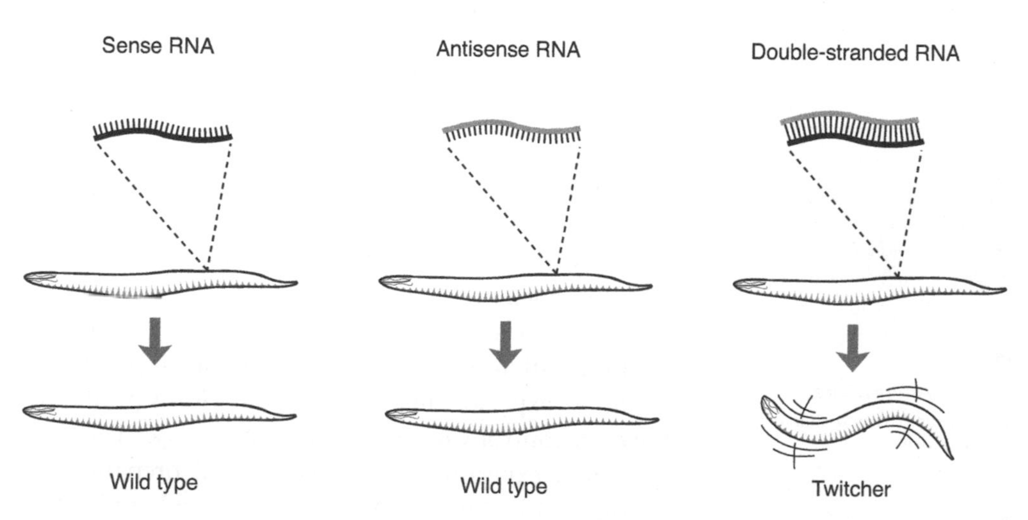

Double-stranded RNA (dsRNA) induces silencing. Purified sense and antisense RNAs targeting the *unc-22* gene have little effect. dsRNA induces twitching, the same phenotype as mutations that disrupt *unc-22*.

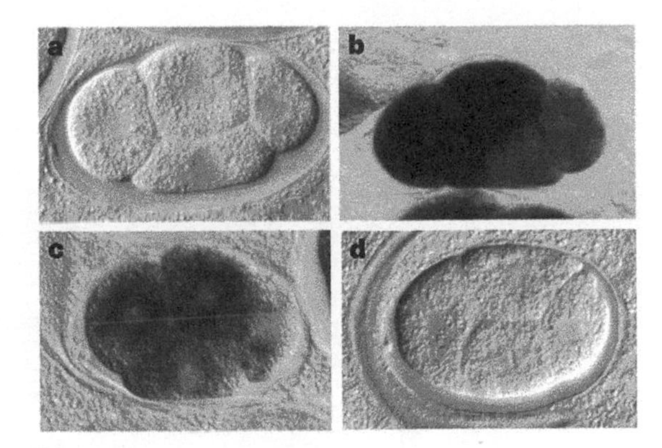

dsRNA treatment reduces mRNA levels. Interference contrast images of in situ hybridization of *mex-3* probe in *C. elegans* embryos. (*a*) Negative control showing lack of staining in the absence of probe. (*b*) Embryo from uninjected parent showing normal pattern of *mex-3* mRNA expression. (*c*) Embryo from a parent injected with purified *mex-3* antisense RNA showing slightly lower levels of *mex-3*. (*d*) Embryo from a parent injected with dsRNA showing no *mex-3* RNA expression.
(Reprinted, with permission, from Fire A, et al. 1998. *Nature* 391: 806–811; ©Macmillan.)

These fragments are short enough to move between plant cells, yet long enough to confer specificity by base pairing to a target transcript. Consistent with this, Fire and Mello found that short dsRNAs induce silencing in *C. elegans*.

In 2000, using fruit fly extracts in cell-free biochemical systems, Greg Hannon, at Cold Spring Harbor Laboratory, and David Bartel, at Massachusetts Institute of Technology, independently showed that long dsRNAs are processed into small dsRNA molecules, 21–23 nucleotides in length. Working in parallel, Tom Tuschl, at the Max Planck Institute for Biophysical Chemistry, introduced short dsRNAs into *Drosophila* cells and showed that they triggered cleavage of target mRNAs. Hannon found that these short interfering RNAs (siRNAs) copurify along with a nuclease required for mRNA degradation, forming an RNA-induced silencing complex (RISC). This work strongly suggested that long dsRNAs are processed into siRNAs, which then guide a nuclease to degrade target RNA transcripts.

IDENTIFYING DICER AND SLICER

Greg Hannon, 2004
(Courtesy of Cold Spring Harbor Laboratory Archives.)

Hannon used a candidate gene approach to search newly sequenced genomes for the enzyme that cleaves long RNAs into siRNAs. Previous research had identified a small number of dsRNA-specific nucleases, including members of the RNase III family. So Hannon used RNase III gene sequences to identify several homologs in the *Drosophila* and *C. elegans* genomes. Candidate genes from three different RNase III families were cloned into cultured *Drosophila* cells, and their encoded proteins were purified. One of the purified nucleases efficiently cut long dsRNA into siRNAs. Hannon called it Dicer.

Somewhat ironically, Hannon then used gene silencing to confirm that Dicer is required for RNAi in vivo. He designed dsRNAs homologous to the two Dicer family members and introduced them into *Drosophila*. This knocked out all RNAi activity.

Although RISC clearly has enzymatic activity that cleaves target mRNA, it proved to be difficult to determine which protein in RISC possesses this “slicing” activity. Genetic and biochemical experiments showed that Argonaute family members are required for RNAi and are core components of RISC in various species. However, X-

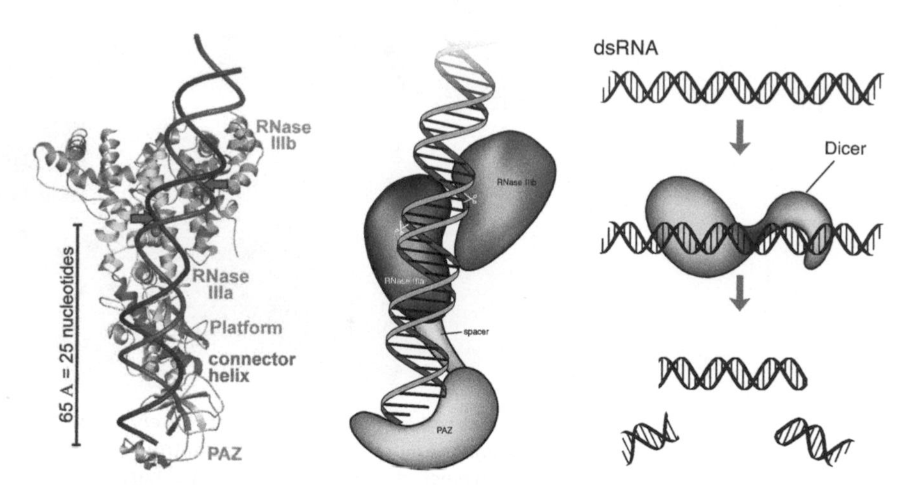

Dicer cuts long dsRNA. (*Left*) View of the crystal structure of *Giardia* Dicer, indicating the RNase III and PAZ domains, with a modeled dsRNA substrate. (*Center*) Illustration representing the domains of Dicer. A PAZ domain binds to the end of dsRNA. Two RNase III domains cut the strands of RNA, releasing a short dsRNA. The length of the spacer region between the PAZ and RNase III domains determines the length of the short RNA. (*Right*) Schematic showing Dicer cleaving a long dsRNA into a short dsRNA.
(*Left*, reprinted from Macrae IJ, et al. 2006. *Cold Spring Harbor Symp Quant Biol* 71: 73–80.)

Leemor Joshua-Tor, 2004
(Courtesy of Cold Spring Harbor Laboratory Archives.)

ray crystallography provided the key insight. In 2004, Leemor Joshua-Tor's group at Cold Spring Harbor Laboratory showed that the crystal structure of the conserved PIWI domain in Argonaute resembles the active site of RNase H, which cleaves ssRNA. This was consistent with Argonaute slicing single-stranded messages. The crystal structure of full-length Argonaute showed that a PAZ domain binds to a single strand of an siRNA, facing it toward a groove in RISC through which mRNAs can slide. When a complementary mRNA hybridizes to the siRNA, the interaction places the message in just the right orientation to allow cleavage by the PIWI domain.

The discovery of Dicer and Slicer provided a complete mechanism for how dsRNA triggers cleavage of target mRNAs. First, Dicer processes exogenous dsRNAs into siRNAs. Then, Argonaute loads a single-stranded siRNA, which, in turn, binds to a complementary mRNA. Once positioned, the mRNA is cleaved by Slicer (the PIWI domain of Argonaute). This pathway is distinctly different from the antisense hypothesis, where ssRNA complementary to a message was thought to directly inhibit translation.

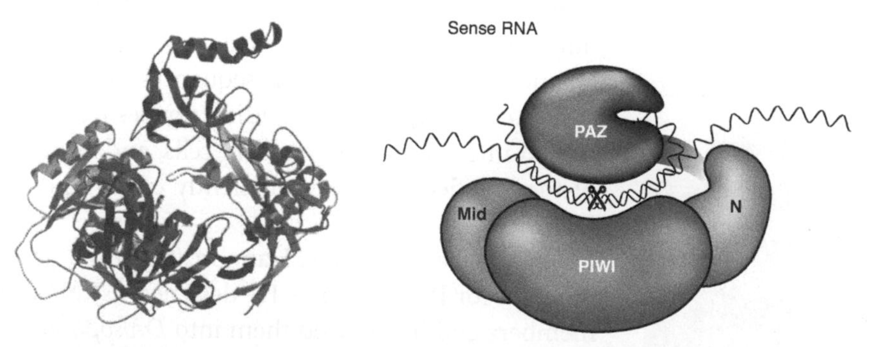

Argonaute structure and function. (*Left*) The crystal structure of Argonaute revealed a groove that binds to single-stranded siRNA and allows mRNA to slide past. (*Right*) Illustration representing the domains in Argonaute. The siRNA is held in place by the PAZ domain. mRNA complementary to the siRNA fits between the middle (Mid), PIWI, and amino-terminal (N) domains and is cleaved by the RNase activity of the PIWI domain.
(Reprinted, with permission, from Song JJ, et al. 2004. *Science* 305: 1434–1437 [AAAS]; structure courtesy of Leemor Joshua-Tor.)

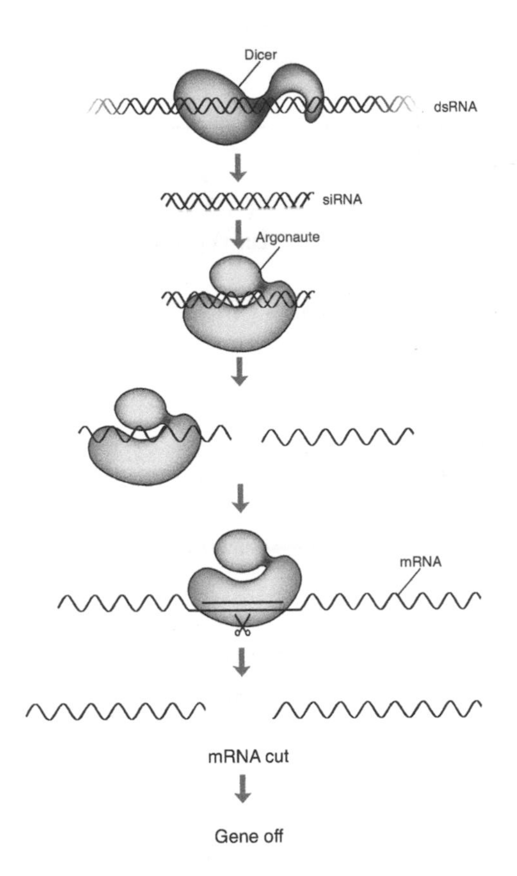

RNA pathway. First, dsRNA is recognized by Dicer and cleaved into an siRNA. The siRNA binds to Argonaute in RISC. After one strand of the siRNA is removed, the other strand is used to target cleavage of complementary mRNAs. Cleaved mRNAs are completely degraded.

MICRORNAS REGULATE GENES

Victor Ambros
(Courtesy of Victor Ambros.)

Gary Ruvkun
(Courtesy of Gary Ruvkun.)

In 1993, Victor Ambros, of Dartmouth University, and Gary Ruvkun, of Harvard Medical School, discovered that endogenous genes can trigger RNAi. Their work centered on the cell lineage genes *lin-4* and *lin-14*, mutants of which disrupt the timing of cell divisions during *C. elegans* development. Originally identified by Sydney Brenner, *lin-4* mutants repeat a pattern of early cell divisions and fail to progress to late cell divisions. In contrast, *lin-14* mutants skip early cell divisions and progress directly to late divisions.

Ruvkun discovered that the LIN-14 protein is abundant early in development but later diminishes. He showed that when LIN-14 protein is present, cells adopt early fates and when it is absent, they adopt late fates. Ambros found that *lin-4* is needed to lower LIN-14 protein concentrations, allowing the switch from early to late fates. Furthermore, the *lin-4* gene product is a tiny RNA, which he called a small temporal RNA (stRNA). Meanwhile, Ruvkun found that mutations in the 3′-untranslated region (UTR) of *lin-14* lead to overproduction of LIN-14 protein late in development.

Ambros and Ruvkun shared their sequence data to determine whether the *lin-4* stRNA and the *lin-14* mRNA might hybridize with each other. Although the two sequences are not completely complementary, they predicted that *lin-4* binds to several sequences in the 3′ UTR of *lin-14*. Ruvkun confirmed that *lin-4* binding inhibits *lin-14* translation, explaining how *lin-4* regulates LIN-14 protein levels during development.

lin-4 was the first example of an endogenous gene coding for a regulatory RNA to be identified in eukaryotes. However, it remained an oddity until 2000, when Ruvkun discovered a second small regulatory RNA, *let-7*, which also controls cell division timing in *C. elegans* by binding to the 3′ UTR of target genes. Ruvkun then used BLAST

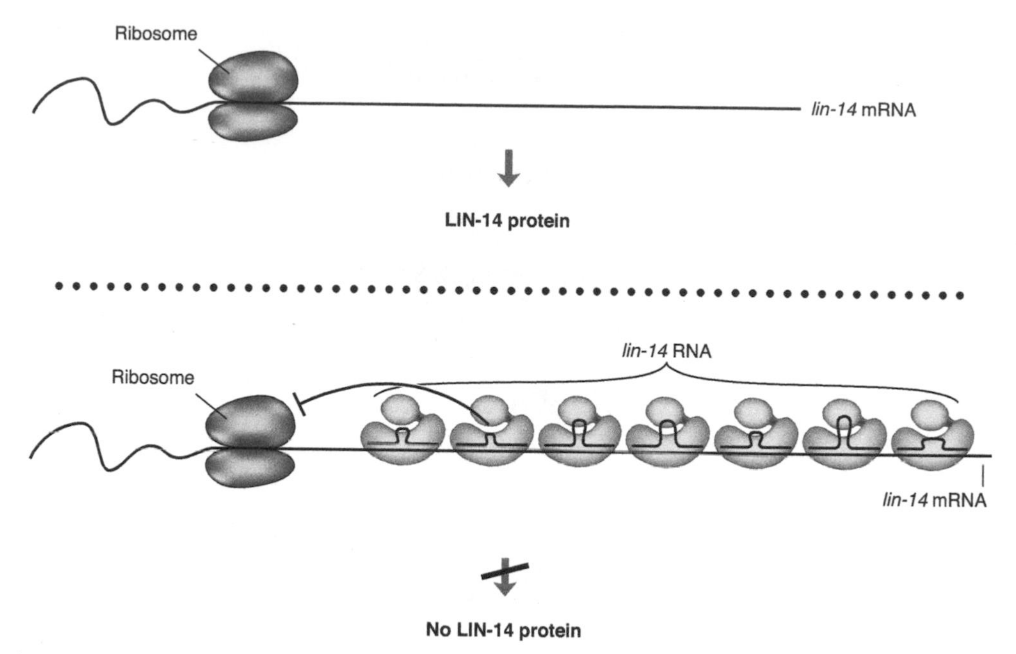

lin-4 mRNA binds to the *lin-14* mRNA UTR. (*Top*) In the absence of *lin-4, lin-14* mRNA is translated to produce LIN-14 protein. (*Bottom*) *lin-4* RNA binds to the UTR of *lin-14* mRNA, which inhibits translation.

to find identical *let-7* sequences in fruit flies and humans. This immediately suggested that tiny RNAs such as *lin-4* and *let-7*—microRNAs (miRNAs)—have important functions that have been conserved throughout the hundreds of millions of years since worms, flies, and humans diverged from a common ancestor.

During the next several years, bioinformatic, genetic, and biochemical methods were used to identify miRNAs in a variety of organisms, and it became clear that most genomes contain many genes that encode miRNAs. At present, more than 1400 miRNAs have been identified in humans, nearly 700 in mice, 200 in fruit flies, and 175 in *C. elegans*. With each potentially controlling the expression of multiple genes, miRNAs are an important, and nearly universal, mechanism to control gene expression.

Both *lin-4* and *let-7* miRNAs are derived from longer pre-miRNAs that fold into stem-loop secondary structures, in a manner similar to transfer RNAs. Several lines of evidence show that Dicer processes pre-miRNAs. First, miRNAs are the same size as siRNAs. Second, loss of Dicer causes developmental defects in worms, fruit flies, and plants. Remarkably, Dicer mutants in *C. elegans* include cell division defects similar to those caused by loss of *lin-4* or *let-7*. In 2001, several groups found that the absence of Dicer activity blocks miRNA processing, leading to the accumulation of pre-miRNAs.

Worm	*let-7*	UGAGGUAGUAGGUUGUAUAGU**U**
Fly	*let-7*	UGAGGUAGUAGGUUGUAUAGU
Mouse	*let-7*	UGAGGUAGUAGGUUGUAUAGU
Human	*let-7*	UGAGGUAGUAGGUUGUAUAGU**U**

microRNA conservation. Sequence alignment of *let-7* among diverse animals. The sequence of *let-7* is nearly identical in nematodes, insects, and mammals. The 5′ region of the miRNAs is marked by the box. This seed region binds to the target mRNA and therefore must be completely complementary to it. The 3′ U (bold) in human and worm *let-7* is removed in the mature miRNA.

The miRNA pathway actually extends back to the nucleus where longer, primary (pri)-miRNA transcripts are produced. In 2003, Yoontae Lee, of Seoul National University, discovered an enzyme called Drosha that cleaves pri-miRNAs into pre-miRNAs. Like Dicer, Drosha is related to RNase III and recognizes double-stranded portions of RNA transcripts.

TWO DISTINCT RNAi PATHWAYS

In less than a decade, Dicer and Argonaute proteins were shown to be at the core of two hitherto unknown pathways for RNA-based regulation. The RNAi pathway begins with a long dsRNA (of foreign origin) that is cleaved into siRNAs by Dicer. A single strand of each siRNA guides Argonaute to complementary mRNAs. In the animal miRNA pathway, transcripts from miRNA genes are processed by Drosha and Dicer into miRNAs that also guide Argonaute to complementary mRNAs.

An important distinction between the two pathways is the effect on target mRNAs. siRNAs are completely complementary to their targets. In contrast, in animals miRNAs are incompletely complementary to their targets, creating bulges in the complex of the miRNA and mRNA target. These bulges inhibit the ability of Argonaute proteins to cleave the target mRNA and promote translational repression. In many organisms, multiple Argonaute proteins have specialized functions, and this also affects target regula-

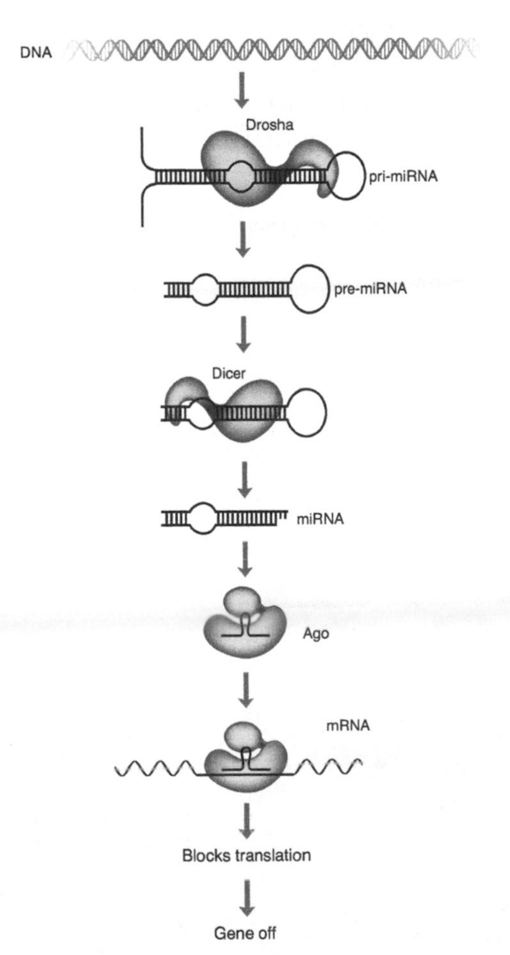

miRNA processing. Primary miRNA (pri-miRNA) transcripts fold into hairpin structures that are recognized and cleaved into pre-miRNAs by Drosha in the nucleus. pre-miRNAs are exported to the cytoplasm, where they are cleaved by Dicer into miRNAs, which then bind to Argonaute (Ago). Ago uses one strand of the miRNA to target partially complementary target mRNAs, turning off genes largely by blocking translation.

tion. For example, miRNAs in *Drosophila* are loaded onto AGO1, which cuts mRNA poorly, and siRNAs bind to AGO2, which cuts target transcripts efficiently.

Although the first siRNAs were exogenous and miRNAs are endogenous, the source of the RNA is immaterial. Plant miRNAs are usually fully complementary to their targets and promote mRNA cleavage. Likewise, siRNAs in animals are produced from endogenous dsRNA and are completely complementary to their targets, which are cleaved by Argonaute proteins. Conversely, miRNAs that are introduced into cells are processed by the miRNA pathway and their targets are not cleaved.

RNAi EVOLVED TO SILENCE VIRAL AND TRANSPOSON DNA

Components of the RNAi machinery have been identified throughout the eukaryotic lineage, suggesting that it is more than 1 billion years old. RNAi most likely evolved as a mechanism to protect cells from self-replicating elements—viruses and transposons—within their own genomes. According to the interpretation put forward by Richard Dawkins in his 1976 book, *The Selfish Gene*, these sorts of sequences replicate "selfishly," without regard for the host genome. Work done since the early 1990s has established that the genome can respond to viruses and transposons using both RNAi pathways—siRNAs and miRNAs.

The first evidence that RNAi is an antiviral mechanism came from work in plants. In 1993, William Dougherty, of Oregon State University, reported that tobacco plants transformed with a gene sequence from the tobacco etch potyvirus (TEV) are resistant to the disease. By 1996, Dougherty and Sir David Baulcombe had both shown that viral genes are still transcribed in resistant plants, but that the virally encoded mRNA is degraded post-transcriptionally (after transcription). Baulcombe showed that plants naturally recover from infection by down-regulating viral mRNA levels.

JOURNAL OF AGRICULTURAL RESEARCH

Vol. 37 Washington, D. C., August 1, 1928 No. 3

HOSTS AND SYMPTOMS OF RING SPOT, A VIRUS DISEASE OF PLANTS[1]

By S. A. Wingard[2]

Associate Plant Pathologist, Virginia Agricultural Experiment Station

INTRODUCTION

Early report of silencing. As reported in 1928, tobacco recovers from viral infection over time. The virus causing the initial symptoms activates viral RNA silencing that inhibits spread of the infection into the upper leaves and causes them to be specifically immune to tobacco ringspot virus secondary infection.

(Reprinted, with permission, from Baulcombe D. 2004. *Nature* 431: 356–363; ©Macmillan.)

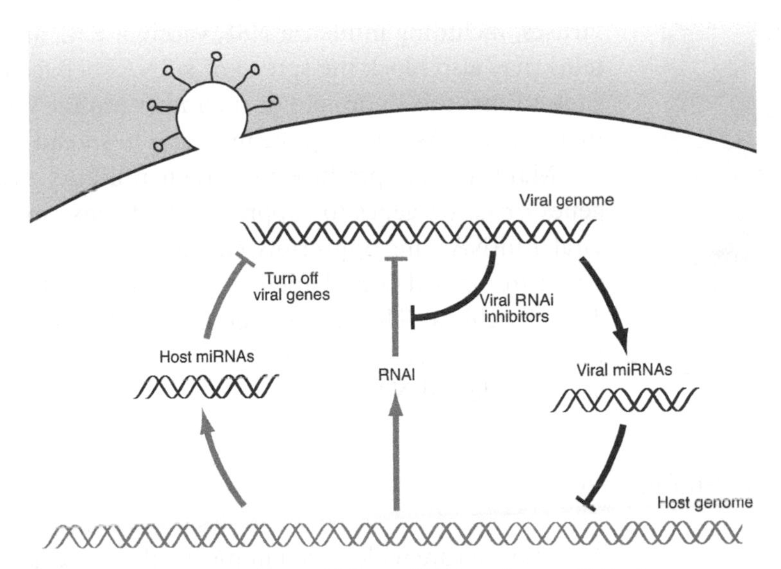

An RNAi arms race between virus and host. Cells use RNAi to destroy RNA virus genomes and produce miRNAs that turn off viral genes. However, viruses have evolved defenses to RNAi. Many viruses produce protein inhibitors of RNAi. Also, virus genomes encode miRNAs, which use the host RNAi machinery to down-regulate host genes important for immune response.

With the discovery of the RNAi machinery, the mechanism of viral immunity in plants became clear. Viral dsRNAs are cleaved by Dicer, and the resulting siRNAs target viral RNAs for degradation by RISC. Many siRNAs can move between cells, allowing immunity to build in uninfected cells and outpacing the spread of the viral infection through the plant.

Vicki Vance, 1990s
(Courtesy of Vicki Vance.)

In 2002, Shou Wei Ding, at the University of California, Riverside, reported that infection by the insect flock house virus (FHV) leads to the accumulation of virus-derived siRNAs in *Drosophila*. The accumulation of these siRNAs is dependent on Argonaute, confirming a role of the RNAi machinery in antiviral defense. In 2006, Louisa Wu, at the University of Maryland, showed that fruit flies lacking components of the RNAi machinery are highly susceptible to X virus. Similar responses also occur in mosquitoes, nematodes, and shrimp, showing that RNAi is also a conserved antiviral mechanism in invertebrates.

Antiviral responses are also mounted by miRNAs encoded in mammalian genomes. In 2005, Irene Pederson, at the University of California, San Diego, showed that five miRNAs predicted to target the hepatitis C virus genome can inhibit viral replication. These miRNAs are up-regulated in liver cells by interferon-β, part of the innate response to viral infection in mammals. Similarly, T cells express five miRNAs that target HIV genes. In addition to directly targeting viruses, miRNAs also regulate genes involved in the development and function of the immune system.

James Carrington
(Courtesy of James Carrington.)

Reminiscent of an escalating arms race, many plant and animal viruses produce proteins that suppress the host RNAi response. In 1998, Vicki Vance, at the University of South Carolina, and James Carrington, at Oregon State University, independently reported that the HC-Pro protein of TEV inhibits silencing. HC-Pro binds siRNAs and inhibits the formation of activated RISC. Similar siRNA-binding inhibitors have been identified in diverse viruses and appear to have evolved independently at four different times.

Other virally encoded proteins bind dsRNAs or bind directly to Dicer, inhibiting cleavage of dsRNAs into siRNAs. Silencing suppressor proteins are found in human

Tom Tuschl
(Courtesy of Tom Tuschl.)

viruses, including influenza NS1, vaccinia E3L, and Ebola VP35. In plants, viral proteins may also block the spread of siRNAs between cells. For example, in 2002, Ding showed that the 2b protein of cucumber mosaic virus inhibits the intercellular movement of siRNAs, allowing the infection to spread to uninfected tissues.

Many viruses produce their own miRNAs that control the expression of viral genes or target genes to suppress host defense mechanisms. As with other miRNAs, viral miRNAs are transcripts encoded in viral genomes that fold to form hairpin structures in self-complementary regions. The first viral miRNAs were identified by Tom Tuschl, at The Rockefeller University, in a lymphoma cell line infected with Epstein-Barr virus. Other herpesviruses, as well as adenoviruses and polyomaviruses, also encode miRNAs.

TRANSPOSON SILENCING

Ronald Plasterk
(©KNAW, Amsterdam, the Netherlands.)

Rob Martienssen, 1998

Transposons are widespread in most eukaryotic genomes, and RNAi silences transposition to maintain genome integrity. The first clear evidence of this came in 1999, when Craig Mello and Ronald Plasterk, at the University of Amsterdam, independently showed that *C. elegans* mutants affecting the RNAi pathway have high rates of transposition. Similar effects were quickly reported in fruit flies, fungi, and plants. Consistent with this, small RNAs homologous to transposons were identified in *Arabidopsis*, *Trypanosoma*, and *Drosophila* in 2004.

In plants, siRNAs are produced by Dicer cleavage of dsRNA derived from transposons. In mammals, another similar class of small RNAs, called Piwi-interacting RNAs (piRNAs), are produced without Dicer. Instead, piRNAs are produced by cleavage of single-stranded transcripts by Piwi proteins, which are members of the Argonaute family. In either case, siRNAs and piRNAs are loaded onto Argonaute family members and guide cleavage of RNAs transcribed from transposons. This is analogous to siRNA-directed cleavage of mRNAs.

Many transposons and repetitive DNA elements are sequestered in heterochromatin—compact and transcriptionally inactive regions of chromosomes (including the centromeres). DNA and histone methylation, along with other modifications, maintain heterochromatin and transposon silencing. In 2002, Rob Martienssen, at Cold Spring Harbor Laboratory, showed that RNAi is essential for transposon silencing in yeast. When the homologs of Dicer or Argonaute are deleted, histone methylation of heterochromatin is lost and transposition is massively activated. Contrary to the dogma that heterochromatin is always kept silent, Martienssen found that during DNA replication, heterochromatin regions—including transposons—are briefly transcribed. These ssRNA transcripts are then copied by an RNA-dependent RNA polymerase to form dsRNA. Dicer cleaves the dsRNA to form siRNAs. Paradoxically, this brief transcription is the key step for transcriptional silencing of transposons by RNAi.

In 2004, Danesh Moazed at Harvard identified a protein complex called RITS (RNA-induced initiation of transcriptional gene silencing) that mediates these chromatin modifications by RNAi. RITS includes Argonaute, which binds to siRNAs that are derived from transposons in heterochromatin. The siRNA targets RITS back to transcripts while they are still associated with DNA, and RITS mediates histone and DNA modifications that silence transcription. Although the mechanism giving rise to small RNAs is different in animals, Piwi-bound piRNAs are also thought to target DNA and histone methylation of transposons in animal genomes.

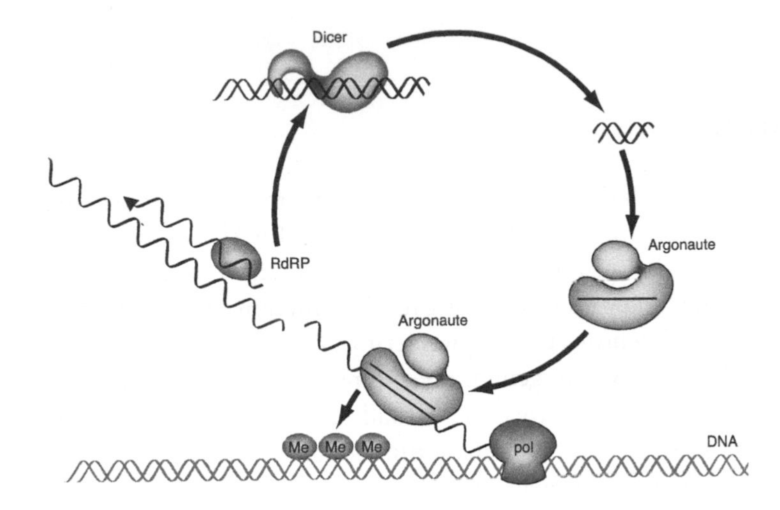

Dicer and Argonaute mediate DNA methylation. Single-stranded transcripts are copied by RNA-dependent RNA polymerase (RdRP), forming dsRNA. Dicer in the nucleus cleaves the dsRNA, producing siRNAs. The siRNAs are bound to Argonaute in a complex called RITS, which recruits DNA or histone methylases to the transcribed region, silencing transcription.

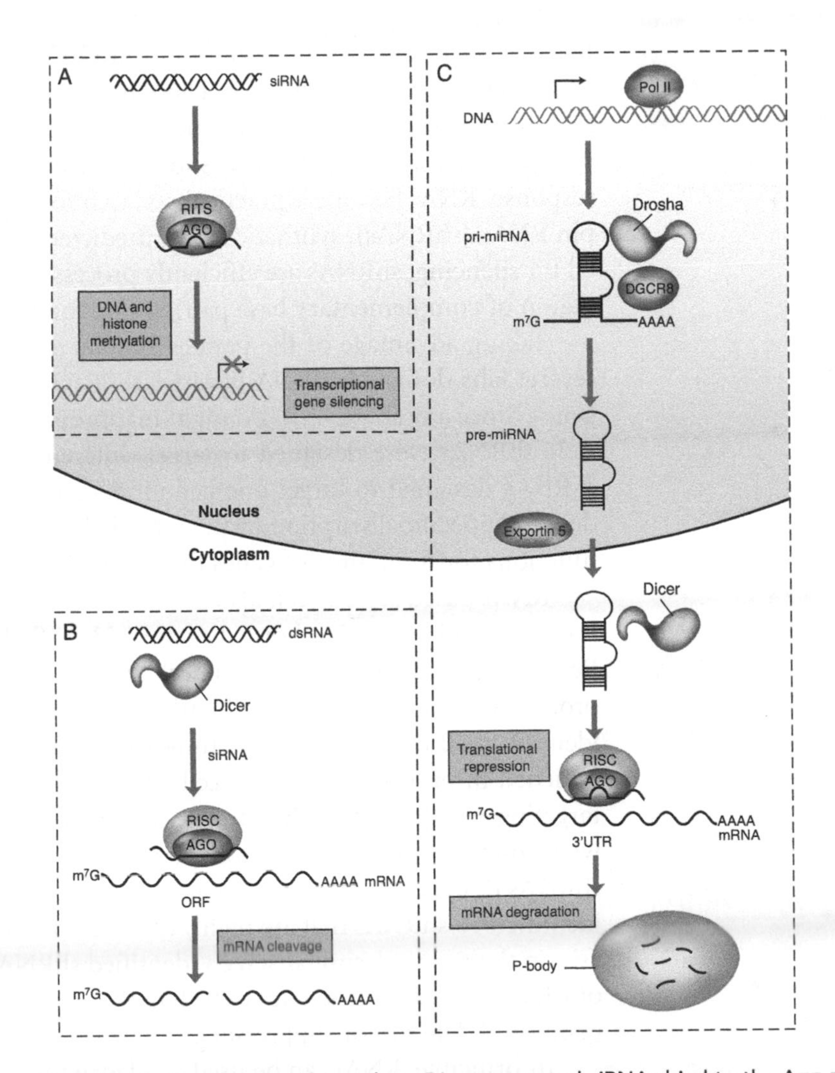

Comparison of RNAi pathways. (*A*) In the nucleus, Dicer-generated siRNAs bind to the Ago protein in the RITS complex and direct DNA or histone methylation of target genes, turning off transcription. (*B*) In the cytoplasm, Dicer cleaves long dsRNA into siRNAs. These bind to Ago in the RISC, directing cleavage of completely complementary target mRNAs. (*C*) miRNAs are produced from transcripts in the nucleus. pri-miRNAs form hairpins that are cleaved by Drosha. The resulting pre-miRNAs are exported to the cytoplasm by Exportin-5, where they are cleaved into miRNAs by Dicer. The miRNAs direct binding of RISC to incompletely complementary mRNAs that are sequestered in P-bodies, inhibiting translation and leading to mRNA degradation.

RNAi AS RESEARCH TOOL AND THERAPEUTIC

Clearly, RNAi is an important mechanism to defend the genome from viral attack, regulate transposon activity, and control gene expression. However, RNAi is also a powerful tool for biological research. Because RNAi disrupts gene function by a sequence-specific mechanism, it provides a key tool to investigate the function of genes that are identified in large-scale sequencing projects. In a pioneering 2003 study, Julie Ahringer, at the University of Cambridge, fed worms genetically engineered bacterial strains producing dsRNAs corresponding to the sequences of virtually every protein-coding gene identified in the *C. elegans* genome. This systematic functional analysis by RNAi successfully silenced 86% of 19,500 predicted genes. Genome-wide RNAi screens in worms subsequently identified functions for genes involved in aging, cell division, apoptosis, and many other processes.

Julie Ahringer
(Courtesy of Julie Ahringer.)

Initial attempts to use dsRNA to trigger gene-specific RNAi in mammalian cells failed because long dsRNA induces an antiviral response, including degrading mRNAs and arresting protein synthesis. These nonspecific cellular responses mask any gene-specific responses that might be induced by RNAi. With the discovery of siRNAs and miRNAs, biologists realized that short RNAs introduced into cells bypass the antiviral response. RNAi became a practical research tool with the development of short hairpin RNAs (shRNAs), synthetic RNAs predicted from the gene sequences to be targeted for silencing. shRNAs are efficiently processed by Dicer because they have a looped region of complementary base pairing that mimics naturally occurring pre-miRNAs.

Taking advantage of the parallel human and mouse genome sequencing efforts, several labs designed shRNA libraries to inactivate essentially all human and mouse genes. Since any single shRNA might inappropriately target an unintended gene, multiple shRNAs were designed to target different regions of each gene. When several shRNAs designed to target one gene induce the same phenotype, this provides evidence of specific disruption of the intended gene. RNAi screens have identified novel functions for thousands of genes that affect many different biological processes.

Scott Lowe

RNAi-based screens can also be very specific. For example, a 2009 screen by Scott Lowe, at Cold Spring Harbor Laboratory, identified new candidate tumor suppressors. Because tumor suppressors inhibit cell growth, loss of their function leads to increased proliferation, an important hallmark of tumorigenesis. Lowe reasoned that he could identify small RNAs that silence tumor suppressors, because they would be selectively enriched in proliferating tumor cells. To test this hypothesis, he showed that shRNAs targeting p53 (a well-studied tumor suppressor) promote lymphomas in mice, even when diluted 100-fold with nonspecific shRNAs. He then screened mixtures of 48 shRNAs designed to silence mouse orthologs of 1000 putative human cancer genes. He identified 10 shRNAs that are highly enriched in tumors and that promote tumorigenesis on their own. A similar screen identified shRNAs that promote liver tumors, some of which were also turned up in the lymphoma screen. The targets of these shRNAs are good candidate tumor suppressors.

In principle, RNAi can be used to silence the activity of a disease-causing gene or to "fine-tune" gene expression to improve cell metabolism or relieve disease symptoms. Thus, scientists are actively pursuing small RNAs as a new class of therapeutic molecules. They are easy to design, cheap to produce, and can be rapidly tested in animal models. RNAi therapies are being developed for cancers, viral infections, and age-related macular degeneration (a common cause of blindness).

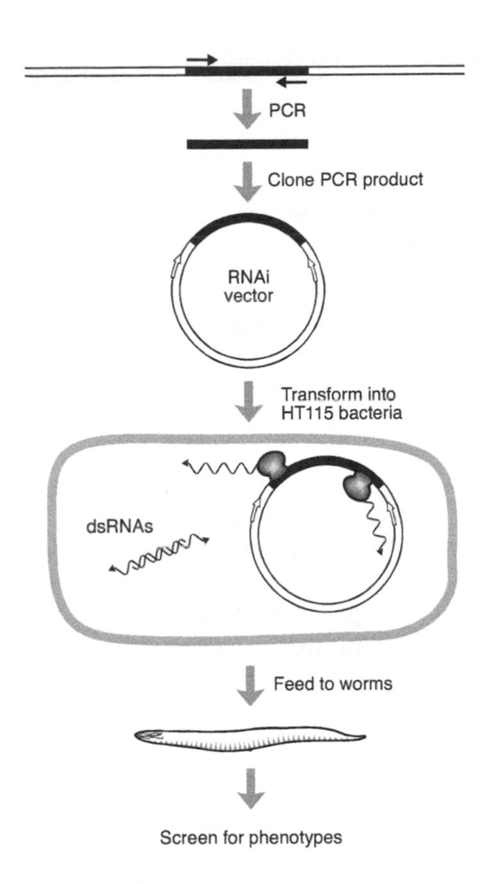

Functional genomics screens. Gene function can be assayed by targeting gene silencing with dsRNA. To produce dsRNA, a portion of each gene is amplified by PCR and cloned into a bacterial vector designed to transcribe both strands of the cloned amplicon. The resulting constructs are transformed into bacterial strain HT115, which expresses T7 RNA polymerase. The polymerase, in turn, initiates transcription at promoters (white arrows) that flank the insert in the vector. The complementary transcripts hybridize to form dsRNA. When worms are fed each bacterial strain, they absorb the dsRNA, inducing RNAi. The RNAi-treated worms are then screened for phenotypes.

Sampling of genes organized by functional groups identified by RNAi screen

Functional category	Human genes in the category	Human shRNA clones	Mouse genes in the category	Mouse shRNA clones
Apoptosis	587	2,525	584	1,834
Cancer relevant	871	3,793	937	3,005
Cell cycle	536	2,646	505	1,815
Checkpoint	124	650	121	434
DNA repair	118	620	139	413
DNA replication	240	1,168	258	841
Enzymes	3,010	12,928	2,934	9,703
GPCR	690	2,632	716	2,132
Kinases	625	3,163	579	2,440
Phosphatases	210	933	198	725
Proteases	470	1,761	467	1,384
Proteolysis	305	1,778	286	1,021
Signal transduction	2,724	11,250	2,690	8,660
Trafficking	488	1,993	483	1,560
Transcription	841	3,576	799	2,604

Adapted, with permission, from Chang K, et al. 2006. *Nat Methods* 3: 707–714 (©Macmillan).

RNAi-based therapeutics in clinical studies

Type	Target	Disease
siRNA	VEGF	AMD
siRNA	VEGFR1	AMD
siRNA	nucleocapsid	RSV infection
siRNA	RTP801	AMD
siRNA	keratin 6a	pachyonychia congenita
siRNA	ribonucleotide reductase	solid tumors
anti-miRNA	miR-122	hepatitis C
siRNA	p53	acute renal failure
siRNA	immunoproteasome β-subunits	metastatic melanoma
shRNA	Tat/rev	AIDS lymphoma

Although very promising, RNAi therapies face many practical challenges, and none has yet been approved for humans. RNA is difficult to deliver into cells or to target particular cell types, such as tumor and macula cells. Then, chemical modifications are required to increase RNA stability within the target cell type. It also hard to design miRNAs that target only a single gene, and "off-target" effects—where other genes are unintentionally silenced—are common.

Bevasiranib, developed by OPKO Health, provided an object lesson of the difficulties of bringing an RNAi-based therapeutic to market. Bevasiranib was designed to treat wet age-related macular degeneration (AMD), in which abnormal blood vessels grown behind the retina leak fluid and damage the macula (the central part of the retina). Bevasiranib is an siRNA that targets vascular endothelial growth factor (VEGF), and early trials showed that it inhibits blood vessel formation in the retina. In 2007, Thomson Scientific named it one of the top five drugs entering Phase III clinical trials. However, subsequent research revealed that siRNAs that were not designed to target VEGF work equally well to reduce vascular growth by inducing a nonspecific response through the cell surface receptor TRL3. Phase III clinical trials were halted in 2009, when it became clear that Bevasiranib was unlikely to achieve its intended effect of improving vision in wet AMD patients.

Interestingly, lower levels of Dicer activity occur in patients with AMD. Deliberately lowering Dicer activity in the retinal pigment epithelium (RPE) of mice causes the tissue to degenerate, suggesting that the low Dicer levels may contribute to AMD. When Dicer is absent, the levels of *Alu* transcripts in the RPE increase to toxic levels. This shows that Dicer acts to control retrotransposons by transcript degradation.

In 2010, Mark Davis, at the California Institute of Technology, provided the first evidence of successful RNAi-based therapy in humans. Melanoma patients were infused with nanoparticles containing siRNAs designed to inactivate RRM2 (M2 subunit of ribonucleotide reductase), which is needed to form new blood vessels in growing tumors. The nanoparticles are composed of cyclodextrin (a sugar-based polymer), polyethylene glycol (a stabilizer), and human transferrin protein, which targets transferrin receptors that are overexpressed on the surface of tumor cells. Biopsies of tumors after treatments showed that nanoparticles accumulate preferentially in melanoma cells, decreasing RRM2 mRNA and protein levels. Notably, Davis detected a specific RNA fragment produced by siRNA-directed cleavage of the RRM2 mRNA. At present, it

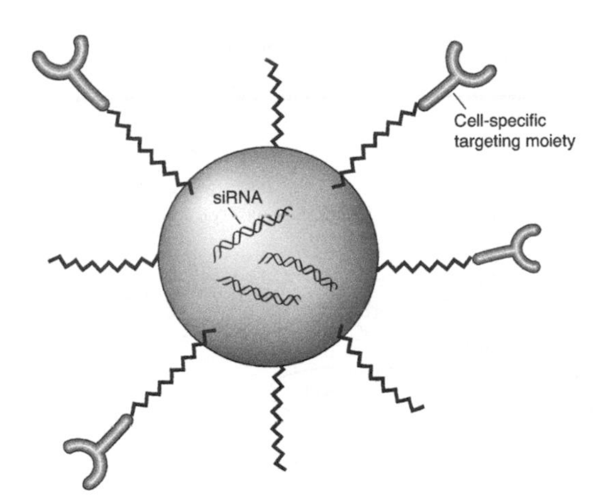

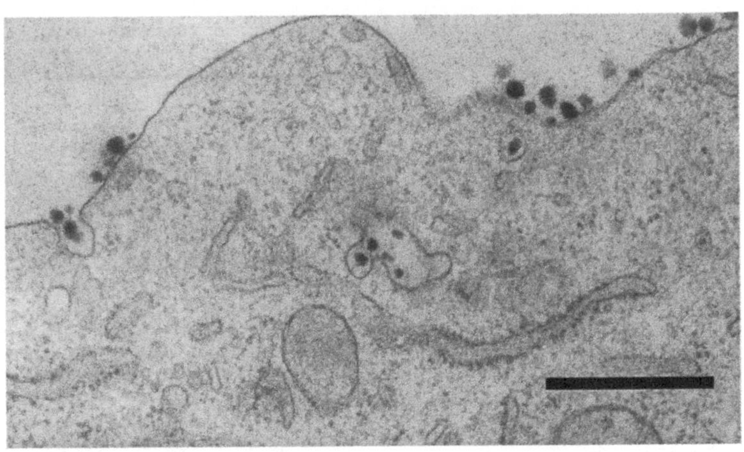

Targeted RNAi therapy. Nanoparticles targeted to human tumor cells can specifically deliver an siRNA drug. (*Left*) A schematic of a targeting nanoparticle. These particles bind specifically to target cells. (*Right*) An electron micrograph of siRNA-containing nanoparticles entering and within a tumor cell. Bar, 500 nm. (From Caltech/Swaroop Mishra.)

remains unclear if the initial excitement about RNAi-based technologies will be borne out. Research continues, and only time will tell.

In addition to their important role in controlling gene expression in normal tissue, changes in the expression or sequence of more than 400 miRNA genes have been associated with more than 250 diseases. These diseases include numerous cancers, cardiovascular disease, and neurobiological disorders.

For example, miRNAs are clearly involved in the biogenesis, or development, of cancer. In 2002, Carlo Croce and colleagues reported that two miRNAs, *miR-15* and *miR-16*, are deleted or down-regulated in more than half of chronic lymphocytic leukemias. All cancers examined to date have many miRNAs that are either up- or down-regulated. This, in turn, has profound effects on the expression of their target genes. Overexpressed miRNAs often act as oncogenes, promoting cancer, whereas underexpressed miRNAs often act to suppress cell growth. Different cancers, even those affecting the same tissue, can have very different miRNA changes. For instance, different forms of pancreatic tumors differ greatly regarding the miRNAs that are up- or down-regulated. Pancreatic endocrine tumors generally have higher levels of miRNAs 23a, 342, 26a, 30d, 26b, 103, and 107 and lower levels of miRNAs 155, 326, 339, and 326, whereas pancreatic insulinomas have high levels of miRNAs 203, 204, and 211. These differences can be used for diagnostic purposes and may have important implications for the treatment of these cancers.

In addition to characterizing the overproduction or underproduction of miRNAs in tumors, miRNAs are also being investigated as powerful therapies for cancers. One area of interest is in overcoming drug resistance. Some cells in a tumor may have mutations that cause resistance to a drug (primary resistance). Over time, these cells dominate within the tumor as the nonresistant cells are killed, and they can acquire new mutations that increase their resistance. Eventually, the treatment will lose its effectiveness and the cancer will progress.

One approach to overcome resistance has been to treat patients with two or more anticancer agents that target different molecular pathways important for cancer. Simultaneously targeting more than one defect within a tumor reduces the likelihood that cells will be resistant to both treatments or acquire this resistance before dying.

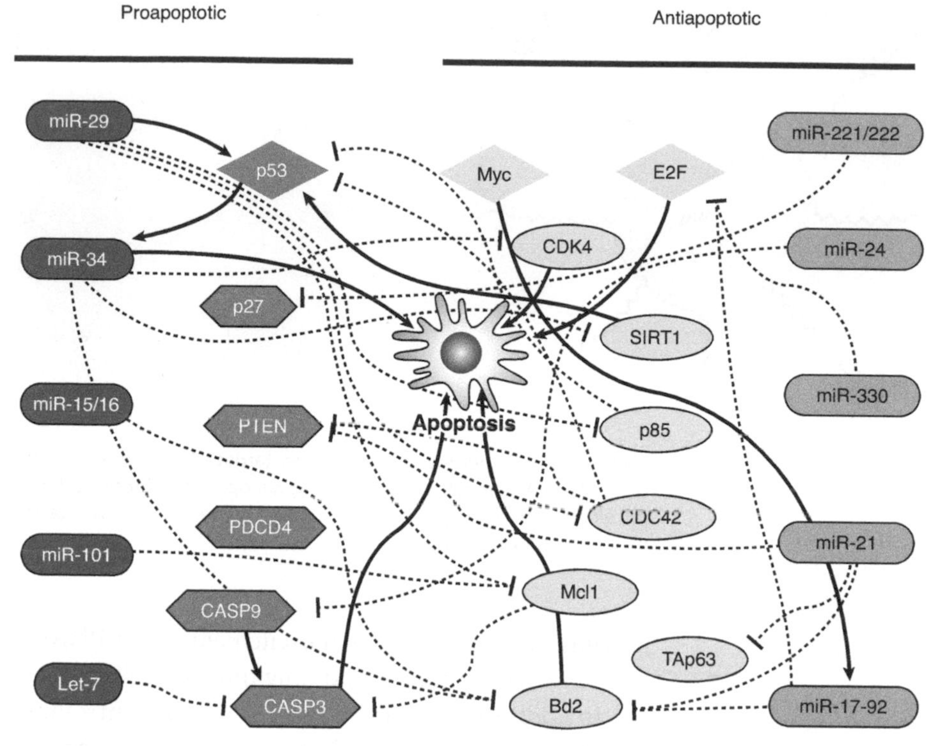

Multiple roles of miRNAs in cancer. miRNA gene regulatory networks involved in the control of apoptosis (programmed cell death) of cancer cells are shown. miRNAs are in general either proapoptotic (dark gray, *left*) or antiapoptotic (light gray, *right*). miRNA expression is often regulated as part of this network. The activation of genes/miRNAs is indicated by arrows and the targeting of mRNA by miRNA is shown by lines with blocked ends. Oncogenes such as *p53* are indicated in dark gray (angular shapes) and tumor-suppressor genes such as *TAp63* are shown in light gray.

(Adapted, with permission, from *World J Biol Chem.* 2010. 1: 41–54.)

Work by numerous groups shows that changing the expression levels of different miRNAs, including *hsa-Let-7g, hsa-miR-181b, Let-7* family members, *miR-30c, miR-130a, miR-335,* and *miR34,* increases the effectiveness of other treatments, in effect acting as an extra drug in the cocktail.

miRNAs can inhibit the expression of multiple genes and simultaneously downregulate more than one important biochemical pathway. In a sense, they are the equivalent of combined therapies—just one RNA drug can be used to target multiple cancer genes. For example, in 2008, Ji Qing and colleagues at the Peking Union Medical College showed that *miR-34* targets *Notch, HMGA2,* and *Bcl-2*—genes that all affect the self-renewal and survival of cancer stem cells. Increasing the levels of *miR-34* in human gastric cancer cells inhibits their growth and leads to programmed cell death, suggesting that *miR-34* could act as a potent drug for certain cancers.

RNA COMPLEXITY

One way to understand the function of the genome is to identify those parts that are transcribed into RNA and what these transcripts do. Until recently, most transcripts were thought to be mRNAs that code for protein, along with a small number of transfer RNAs (tRNAs) and ribosomal RNAs (rRNAs). The majority of the genome was thought to be transcriptionally silent.

Fine-resolution microarrays and high-throughput sequencing are now revealing remarkable complexity in the transcriptome—the entire collection of transcripts in a cell or organism. These studies show that protein-coding genes produce many more transcripts than previously imagined, with extensive alternative splicing and highly variable 5′ and 3′ ends. More surprising is the finding that ~90% of the human genome is transcribed, far more than the 25% occupied by genes. This means that transcription occurs extensively in intergenic regions (between genes), antisense to genes, and within introns of genes.

More than 30,000 long noncoding RNAs at least 200 nucleotides in length have been identified in mammals. The vast majority of these noncoding transcripts have yet to be characterized. Many short, unstable transcripts may be transcriptional "noise" of transcribing adjacent genes. However, a number of noncoding RNAs are known to guide chromatin modifications, interfere with transcription of nearby genes, affect splicing, induce RNAi of complementary messages, or bind to proteins to alter their activity or cellular localization. Many noncoding RNAs are differentially expressed, suggesting that they have important roles in development. Tens of thousands of piRNAs have been identified; these are often encoded in clusters and help to silence transposons in plant and animal genomes. The fact of pervasive transcription has radically changed our view of the genome and suggests that new aspects of gene regulation and genome organization await discovery.

RNA ENZYMES

Sidney Altman, 1989
(Courtesy of Michael Marsland, Office of Public Affairs, Yale University.)

Until the late 1970s, proteins were considered the only enzymes in cells. At the time, Sidney Altman, of Yale University, was studying an enzyme called RNase P that cleaves the 5′ leader sequence of pre-tRNA to produce a mature tRNA. After purifying the enzyme from *Escherichia coli*, he found that it was composed of a protein and an RNA component. In 1978, Benjamin Stark, a graduate student in Altman's lab, found that the RNA component is essential to the enzymatic activity of RNase P. Work ultimately showed that the RNA component catalyzes cleavage on its own—in the absence of any protein.

At about the same time, Thomas Cech, at the University of Colorado, was working on transcription of rRNA in the protozoan *Tetrahymena*. In 1981, he discovered an rRNA molecule that performs RNA splicing without any additional enzyme. This "self-splicing" RNA contains a catalytic site that cleaves itself to remove an intervening sequence (intron) and then splices together two exons to yield a mature rRNA. The splicing reaction requires only guanosine and magnesium ions; no external energy source is needed.

Thomas Cech
(Courtesy of the Albert and Mary Lasker Foundation.)

Altman and Cech received the 1989 Nobel Prize in Chemistry. Their discovery of "ribozymes"—enzymes composed of RNA—changed the prevailing dogma that RNA is merely a messenger between DNA and proteins. RNA was now seen as a multifaceted molecule that can also do things on its own.

Group I ribozymes (introns), related to those discovered by Cech, were later found in tRNA, rRNA, and mRNA—particularly in the organelles of fungi and plants. Another type of ribozyme emerged, called Group II ribozymes, in which self-splicing takes place without guanosine. During splicing, sequences within Group II ribozymes bind to form a characteristic loop or "lariat" structure, which is then excised. The vast majority of eukaryotic mRNAs are processed by the spliceosome, a huge complex composed of five small nuclear RNAs (snRNAs) and ~300 proteins. The mechanism for splicing these

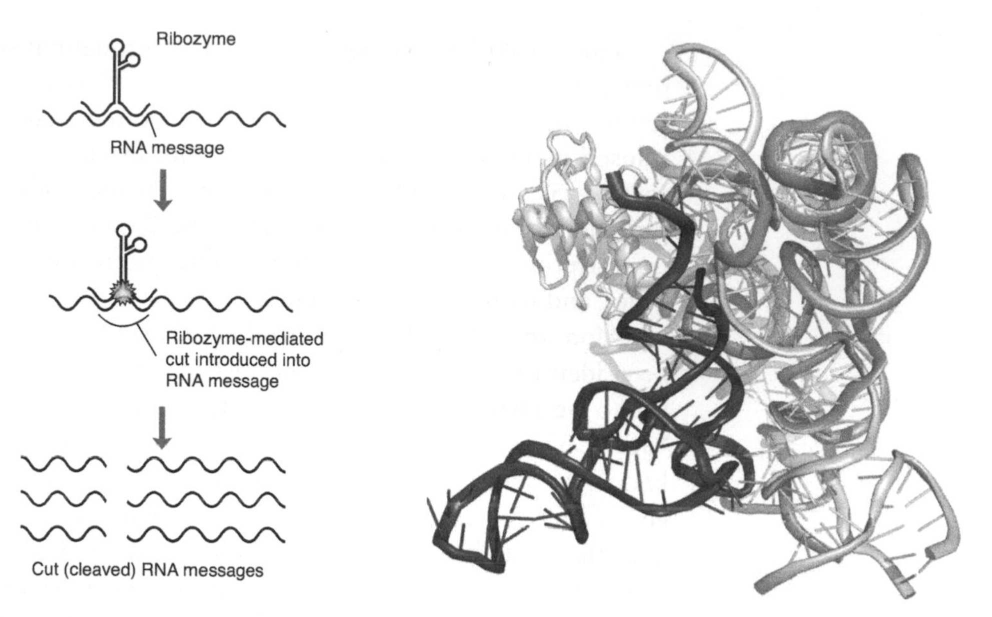

RNase P RNA binds and cleaves mRNA. (*Left*) Schematic of RNase P binding and cleaving an mRNA. (*Right*) Crystal structure of RNase P (light gray) bound to substrate mRNA (dark gray).
(Structure reproduced from Kazantsev AV, et al. 2005; doi: 10.1073/pnas.0506662102; www.wikipedia.org.)

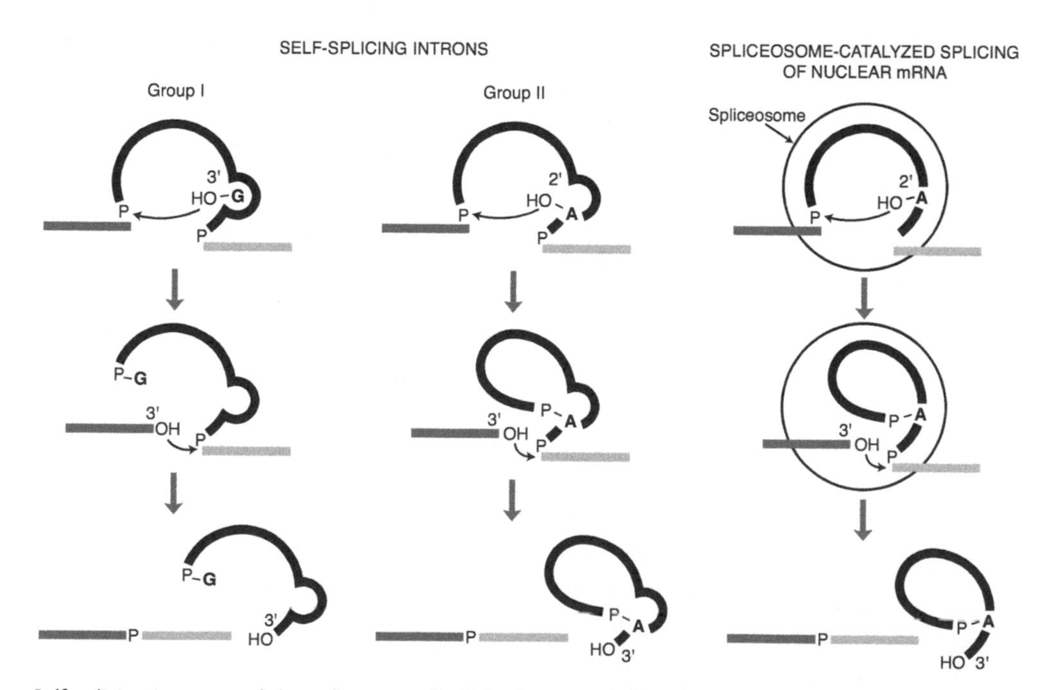

Self-splicing introns and the splicesome. (*Left*) In Group I introns, the 3′ OH group of a free guanosine attacks the 5′ phosphate at the splice site, displacing it from the exon. The 3′ OH at one end of the exon attacks the phosphodiester bond, removing the other end of the intron and reforming the phosphodiester bond with itself. (*Center*) In Group II introns, a similar mechanism is used, except that an internal 2′ OH of an adenoside attacks the 5′ splice site of the exon to produce a lariat structure. In the final step, the 3′ OH of the 5′ exon attacks the phosphodiester bond to remove the lariat. (*Right*) The chemical mechanism for splicing Group III introns is very similar to Group II introns, but the spliceosomal RNAs and proteins are required to stabilize the reaction.

introns also includes the formation of a lariat, suggesting that "spliceosomal" introns are related to Group II self-splicing introns. The U2 and U6 snRNAs contain conserved motifs found in Group II ribozymes and bind to complementary sequences in introns and at exon boundaries. They are required to catalyze splicing and can do so in the absence of proteins. These observations support a model in which Group II self-splicing introns evolved into spliceosomal introns, which cannot self-splice, by moving catalytic activity from within the introns to snRNAs at the core of the spliceosome.

The prokaryotic ribosome is another large complex composed of ~50 proteins and three RNAs. Although eukaryotic ribosomes are larger and have more protein components, it is thought that translation occurs via a highly conserved mechanism in all organisms. After the discovery of ribozymes, a long-standing question was whether the peptidyl transferase activity that forms peptide bonds between adjacent amino acids is found within the protein or RNA component of the ribosome.

In 2000, the X-ray structures of large and small subunits of the bacterial ribosome were determined by Ada Yonath, of the Weizmann Institute of Science and Max Planck Research Unit for Structural Biology, and Venkatraman Ramakrishnan, of the MRC Laboratory of Molecular Biology. Peptidyl transferase activity is found at the core of the large subunit, where 27 proteins surround a tightly folded mass of 5S and 23S rRNAs. The same year, Thomas Steitz, of Yale University, completed a high-resolution study of the large subunit bound to a substrate analog that mimics the chemical intermediate of peptide bond formation. All contacts to the analog are buried deep within the 23S RNA portion of the ribosome, far from the protein components, confirming that rRNA is the catalytic component of the ribosome. Ramakrishnan, Steitz, and Yonath shared the Nobel Prize for Chemistry in 2009.

Another class of noncoding RNAs, the small nucleolar RNAs (snoRNAs), guide chemical modifications of rRNAs, tRNAs, snRNAs, and other RNAs. snoRNAs localize in the nucleolus or in Cajal bodies, where RNA processing occurs. snoRNAs are conserved in archaea and eukaryotes, suggesting that they arose more than 2 billion years ago. Although not catalytic themselves, snoRNAs are important for the folding and function of catalytic RNAs. For example, more than 100 sites in 23S rRNA are methylated by snoRNPs (small nucleolar ribonucleoproteins), snoRNAs bound to proteins, and these modifications are important for rRNA folding and translation. There is evidence that some snoRNAs are recognized by the RNAi machinery and processed into miRNAs, and that some miRNAs may have evolved from snoRNAs.

Ada E. Yonath
(Photo by Micheline Pelletie.)

Venkatraman Ramakrishnan
(Courtesy of Venkatraman Ramakrishnan.)

Thomas A. Steitz, ca. 2001
(Courtesy of Thomas Steitz.)

THE RNA WORLD AND RNA PROTOCELLS

In its current state, life relies on a complex biochemical network composed of three main interactions between three types of molecules:

DNA is *replicated* (assisted by protein and RNA).
DNA is *transcribed* into RNA (assisted by proteins).
RNA is *translated* into proteins (assisted by RNA and proteins).

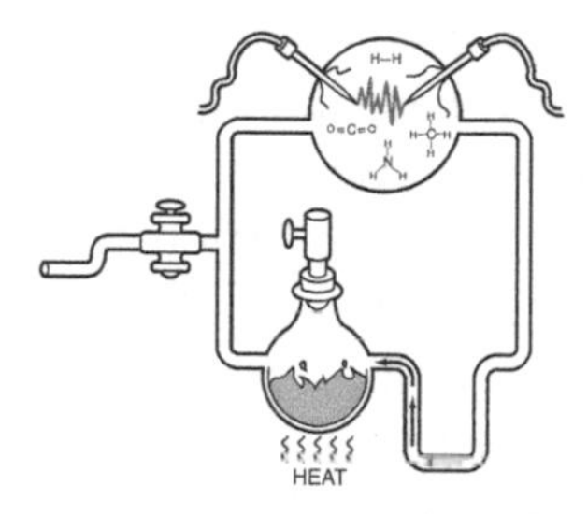

Creating the building blocks of life. Stanley Miller placed methane (CH_4), ammonia (NH_3), hydrogen (H_2), and water (H_2O) into a closed system to simulate prebiotic Earth. Gases accumulated in an upper chamber, simulating the atmosphere. Electric discharges in this chamber simulated lightning, which provides energy to stimulate reactions. The gases from the upper chamber condensed and dissolved in water in the lower chamber, mimicking an ocean. Nucleotides, sugars, amino acids, and other organic compounds accumulated in the "ocean" over time.

Each element is dependent on the others, and it is difficult to imagine that this complex network emerged as a whole. If one assumes that these molecules and functions evolved in a stepwise fashion, the first genetic molecule would have had to have two key abilities: the ability to self-replicate and the ability to catalyze reactions that could create the other key molecules of the network.

The discovery of ever-expanding functions for RNA, and especially the catalytic activity in ribozymes and ribosomes, resurrected the hypothesis that RNA was the first genetic molecule and the predominant molecule of life. In the early 1960s, Leslie Orgel, Francis Crick, and Carl Woese each imagined an ancient "RNA world" in which RNA predated DNA as the original genetic material and proteins were the first biological catalysts.

Darwin encouraged generations of biologists to consider that life evolved in a "warm little pond," which concentrated the precursor molecules of life. Experiments done in the 1950s by Stanley Miller and Harold Urey at the University of Chicago prompted an ongoing exploration of "prebiotic" evolution that originated RNA and other key organic molecules. They applied energy—in the form of electrical discharges, UV radiation, or heat—to a closed system containing gases representing the primitive Earth's atmosphere: ammonia, hydrogen, carbon dioxide, and methane. Molecules condensing out of the "atmosphere" circulated through a flask of water, representing the primitive ocean. Simple organic molecules such as hydrogen cyanide, formaldehyde, and acetaldehyde first formed in the "atmosphere." These further reacted in the "ocean" to form amino acids, nucleotides, sugars, and other building blocks of life. Other investigators later showed that hot, deep sea vents or meteor impact sites could have been sources of both early organic precursors and the energy needed for them to react together.

Stanley Lloyd Miller
(Courtesy of Donald Miller.)

Harold Clayton Urey (standing)
(Reprinted, with permission, from Special Research Center, University of Chicago Library.)

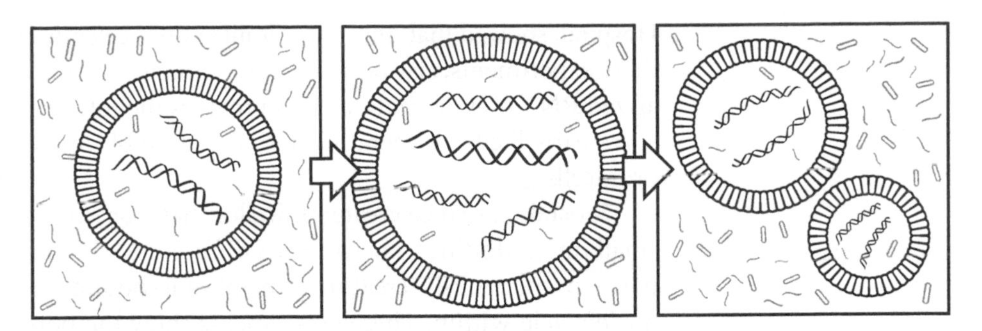

Simple protocells with fatty acid membranes may have enclosed building blocks for a replicating genome. Fatty acids and nucleotides from the environment could be added to the membrane and genome, allowing these primitive cells to grow. Physical forces could break large protocells into smaller ones in a primitive form of cell division.

A key evolutionary step would have been to surround a replicating RNA within a simple "protocell" composed of fatty acids. This encapsulated RNA would have had a competitive advantage over other replicating RNAs, first by sequestering nucleotides needed to quickly replicate itself and, later, by concentrating amino acids needed to build proteins. Fatty acids related to the phospholipids that form cellular membranes are likely to have existed on ancient Earth. Hydrophobic fatty acids naturally self-assemble into spherical micelles, grow larger by incorporating additional fatty acids, and split into smaller ones, mimicking the growth and replication of cells.

In 1947, the X-ray crystallographer John Desmond Bernal, of Birkbeck College, London, first proposed that clay minerals may have provided substrates for catalyzing reactions that produced RNA and other key organic molecules. Montmorillonite clay, formed from the weathering of volcanic ash and abundant on the early Earth, received particular scrutiny. Montmorillonite is composed of alternating layers of silicon and aluminum oxides. Its unique chemical properties can catalyze many types of reactions, including those driven by electron and proton exchanges. Each layer has negative regions that readily bind nucleotides under acidic conditions, and the confined environment between layers creates a natural "flow cell" in which reactants are brought into contact with the catalytic surfaces.

Working at Rensselaer Polytechnic Institute, in 1989 James Ferris demonstrated that montmorillonite can catalyze the formation of RNA polymers, providing a possible source for the first polynucleotides. In 2003, Jack Szostak, at Massachusetts General

John Desmond Bernal
(Photo by Ramsey and Muspratt. ©Peter Lofts.)

James Ferris
(Courtesy of James Ferris.)

Jack Szostak, 2004
(From www.wikipedia.com.)

Hospital, showed that montmorillonite also accelerates the rate of micelle formation. In these experiments, clay particles were sometimes found within micelles, where they catalyzed RNA polymerization. This provides the intriguing possibility that clay helped to spark early life by accelerating both the assembly and encapsulation of ancient RNA in protocells.

Szostak has since created micelles that trap large molecules but are permeable to small ones. Modified nucleic acids diffuse into these micelles and assemble into polymers spontaneously, creating a membrane-bound structure with growing strands of nucleic acid within. These experiments show the potential of simple micelles to perform many of the functions necessary for an early form of life.

SELF-REPLICATING RNA AND RNA STABILITY

Could RNA have been the first genetic molecule? Could a primitive RNA self-replicate? How could have the downstream protein world evolved? How could the "upstream" DNA world have evolved?

Evidence has mounted during the past two decades that shows that a primitive RNA could have had the ability to self-replicate. In 1988, Cech demonstrated that Group I ribozymes have a ligation function that can add nucleotides to an RNA chain. However, this ligation reaction is very limited, adding only guanosines to the 3′ end of the ribozyme.

In 1993, Szostak and David Bartel used in vitro selection to artificially "evolve" a Group I ribozyme that ligates two RNA molecules when they are bound to a complementary sequence. Bartel and others have since improved on this template-dependent ligation, creating RNA polymerase ribozymes that can extend an RNA primer using an RNA template as a guide. However, the best of these enzymes can copy just 20 nucleotides—far shorter than the smallest ribozyme (~200 nucleotides). Thus, a self-replicating ribozyme has yet to be identified.

In 2009, two advances brought the "holy grail" of a self-replicating RNA closer. Tracey Lincoln and Gerald Joyce, of The Scripps Research Institute, created an RNA system that can self-replicate and evolve. In this system, two ribozymes copy each other through template-directed ligation of short polynucleotide building blocks. Given a mixture of polynucleotides, the ribozymes occasionally incorporate muta-

David Bartel
(Courtesy of David Bartel.)

Tracey Lincoln
(Photo by BioMedical Graphics, The Scripps Research Institute.)

Gerald Joyce
(Photo by BioMedical Graphics, The Scripps Research Institute.)

tions, and variants that are better at replicating are selected over time. Cech and Quentin Vicens identified a Group I ribozyme that can form new phosphodiester bonds in RNA by attacking the 3′ hydroxyl group on a 5′ triphosphate group. This is distinct from ligation chemistry and is analogous to the reaction needed to add single nucleotides to a growing RNA molecule.

Exonucleases, which sequentially digest nucleotides from the free ends of polynucleotide chains, would likely have coevolved with replicating RNA genomes. The evolving RNA world would have generated numerous noncompetitive molecules—early evolutionary dead ends—and exonucleases would have provided an efficient means to recycle nucleotides as "spare" parts for new RNA molecules. However, exonucleases would present a challenge for RNA genomes, because their free ends would also be exposed to digestion. Indeed, one argument against an RNA world has been the notion that RNA, being single stranded, is inherently less stable than DNA. Although the half-lives of most prokaryotic mRNAs are only minutes, many eukaryotic mRNAs have half-lives of hours or days. However, mRNA only comprises ~5% of the total cellular RNA. Abundant rRNAs (equaling ~85% of cellular RNA) and tRNAs (~10%) are stabilized by stem-loop and hairpin structures formed by regions of complementary nucleotides, and they have half-lives of several days. When one considers that bacteria can replicate as frequently as once every 20 min, an individual RNA molecule would certainly be stable enough to perform catalytic activity for a reasonable period of time.

Hairpin and stem-loop structures would have protected regions of early RNA genomes from degradation by exonucleases. However, joining the free ends of a linear RNA genome into a covalently closed circle would have provided greater protection against exonucleases. Viroids may resemble early replicating RNA genomes. Composed of as few as 220 nucleotides, viroids are closed circles of RNA, essentially opposing hairpins with extensive internal regions of complementary base pairing or stem loops. Viroids do not encode proteins, and some are ribozymes that self-cleave and ligate additional units to produce multimeric genomes. This is a very simple example of the duplication needed to create diploid or polyploid genomes.

A

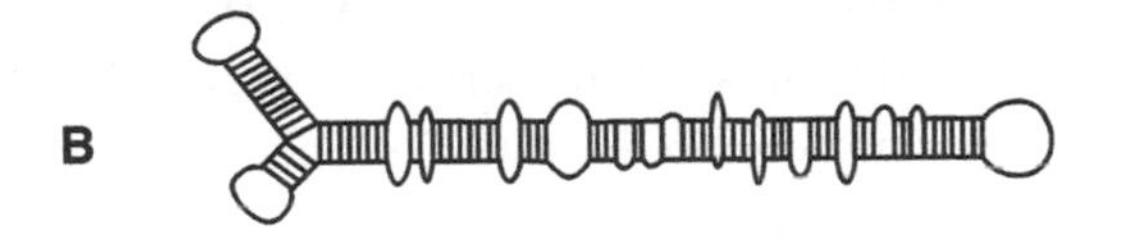

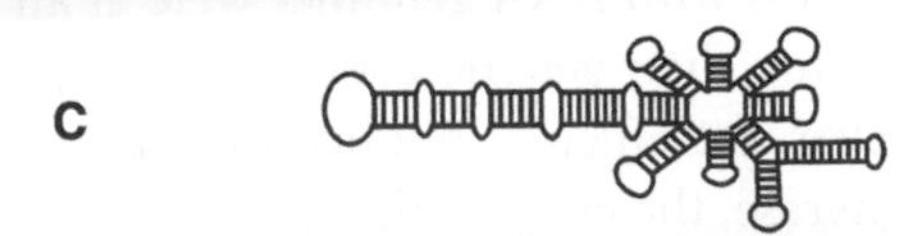

Viroids are closed circular single-stranded RNAs. Depending on the arrangement of complementary sequence, viroids adopt different secondary structures. (*A*) Rod-like secondary structure for a member of family *Pospiviroidae*. (*B*) Quasi-rod-like secondary structure proposed for a member of family *Avsunviroidae*. (*C*) Complex branched conformation proposed for a *Pelamoviroid*.

(Redrawn from Góra-Sochacka A. 2004. *Acta Biochim Pol* 51: 587–607.)

GETTING TO THE PROTEIN WORLD

Given RNA's ability to function independently as an enzyme, it is not too hard to imagine that the "downstream" process of protein translation could have evolved from an initial RNA ancestor. A first step may have been the evolution of RNAs that could be translated into short peptides. This may have initially involved linking free peptides using a peptidyl transferase (ribosome-like) activity contained within a self-replicating RNA genome itself. Thus, a simple RNA genome would have directly translated part of its own genetic code.

As peptides gained in complexity, this would have favored selection of RNA molecules specifically adapted for translation. The first simple ribosome would likely have been made entirely of RNA, with protein components added later. Experiments done by Cech in 1992 showed that some catalytic RNAs have weak tRNA synthetase activity, the reaction that couples an amino acid to a tRNA. This suggests that as the translation apparatus evolved, ribozymes could have initially had the role of charging primitive tRNAs with amino acids. However, an expanding repertoire of protein-containing ribosomes and protein tRNA synthetases would have conferred a selective advantage to early cells. Protein enzymes proved to be more stable than RNA and catalyzed a broader range of reactions, allowing new functions to emerge.

Jay Mittenthal
(Reprinted, with permission, from School of Molecular and Cellular Biology, University of Illinois; photo by Brian Stauffer.)

Phylogenetic and structural analyses of proteins from diverse organisms done by Jay Mittenthal at the University of Illinois at Urbana-Champaign suggest that the first proteins to evolve were enzymes involved in nucleotide metabolism. John Baross, at the University of Washington, found that proteins involved in translation are among the most ancient. Proteins used in DNA metabolism and replication are less ancient, consistent with the first emergence of an RNA world.

GETTING TO A DNA WORLD

A plausible mechanism to transition from an RNA world to a DNA world came in 1970 from work on Rous sarcoma virus (RSV) by Howard Temin, at the University of Wisconsin Medical School, and David Baltimore, at the Massachusetts Institute of Technology. Temin proposed that RSV converts its RNA genome into a DNA copy, a provirus, which then integrates into the host genome to complete its life cycle. Using cell-free systems of RSV components, Baltimore and Satoshi Mizutani, in Temin's lab, each isolated an enzyme that converts RNA into a DNA copy. Because this runs opposite to the traditional process of transcription, they called the enzyme reverse transcriptase. Temin and Baltimore shared the 1975 Nobel Prize in Physiology or Medicine.

In addition to being more stable than RNA, DNA is naturally well adapted for replication—each strand can act as a template for the replication of the other. Organisms with DNA genomes were at an evolutionary advantage for protecting and replicating their genetic codes, resulting in a near-total conversion to a DNA world. However, the efficient duplication of DNA genomes required the evolution of DNA polymerase, the enzyme that makes a DNA copy from a DNA template.

Early DNA genomes were likely circular, without free ends subject to digestion by exonucleases. This is strongly supported by the fact that the vast majority of bacterial genomes and plasmids are circular. The few examples of linear bacterial genomes have adopted different strategies to protect chromosome ends. Hairpin sequences protect

Howard Temin, 1981
(Courtesy of Cold Spring Harbor Laboratory Archives.)

David Baltimore, 1985
(Courtesy of Cold Spring Harbor Laboratory Archives.)

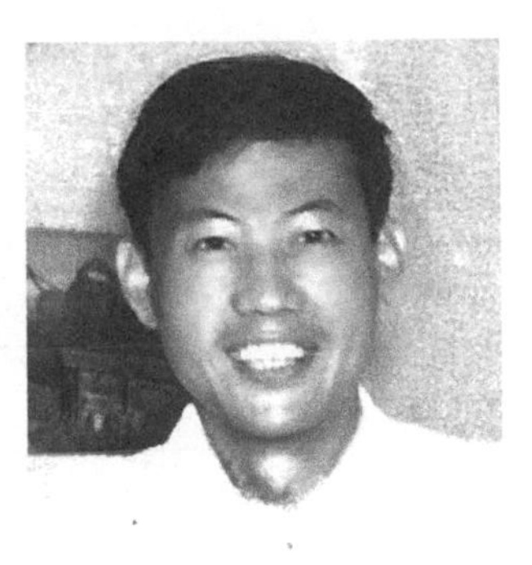
Satoshi Mizutani
(Courtesy of McArdle Laboratory for Cancer Research, University of Wisconsin, Madison.)

chromosome ends in *Borrelia burgdorferi*, the causative agent of Lyme disease, and covalently linked proteins protect the ends of soil microbes of the genus *Streptomyces*. However, there is probably some physical limit to the size of a circular genome that can be reliably replicated. At 13 million nucleotides, the "slime bacterium" *Sorangium cellulosum* is the largest bacterial genome sequenced to date.

Elizabeth H. Blackburn
(From UCSF Archives; photo by Elisabeth Fall/fallfoto.com.)

The advent of linear DNA chromosomes allowed the duplication of larger and greater numbers of chromosomes. The evolution of a second lipid bilayer encapsulating DNA in a nucleus—the hallmark of eukaryotes—would have provided the increased biochemical organization required to support more complex genomes. Linear genomes were also adapted to the development of sexuality and multicellularity, which placed increasing demands for correct chromosome sorting during meiosis and mitosis.

Eukaryotes required a new mechanism to protect the free ends of linear chromosomes from exonucleases. This is provided by the telomere, a complicated looped structure at each end of the chromosome. The telomere is composed of a 10,000–15,000-bp region of GGGTTA telomere repeats, followed by a single-stranded overhang of 150–200 nucleotides rich in Gs and the telomere repeat (the G-strand overhang). The G-strand overhang loops back and invades the double-stranded repeat region, displacing one strand and base pairing to complementary TAACCC repeats. This buries the single-stranded end of the chromosome within the loop. Numerous protein complexes bind to and stabilize the loop.

Carol Greider, 1985
(Courtesy of Carol Greider.)

Replicating the ends of linear chromosomes also presented a problem for DNA polymerases, which had initially evolved to replicate circular genomes. Perhaps representing another remnant of the RNA world, DNA polymerases initiate replication with a short RNA primer annealed to one end of the DNA to be copied. The RNA primer is degraded after replication, leaving a short sequence of DNA uncopied. If not corrected, a DNA genome would thus shorten with each cycle of replication.

Telomere and chromosome length are maintained by the enzyme telomerase, a ribonucleoprotein with reverse transcriptase activity. Szostak and Elizabeth Blackburn, at the University of California, Berkeley, cloned the telomere structure in 1982, leading Blackburn's student, Carol Greider, to purify telomerase in 1985. They showed that the RNA component of telomerase is a sequence template (UAACCC) that is reverse-transcribed by the enzyme component to add new GGGTTA repeats to the 3′

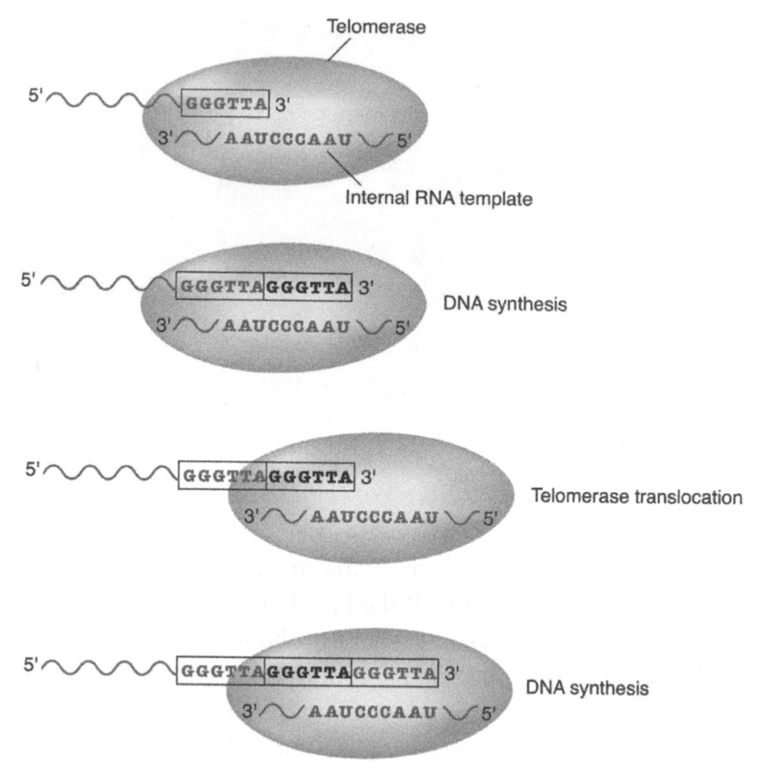

Telomerase contains an RNA template that hybridizes to the end of the chromosome. The template allows telomerase to add DNA to the end of the chromosome in a sequence-specific way. After DNA synthesis, telomerase translocates to the newly added repeat and begins a new cycle of synthesis.

end of the G-strand overhang. This elongated template compensates when DNA replication shortens the 5′ strand, maintaining the number of double-strand repeats in the telomere. Szostak later showed that progressive loss of telomere length in yeast mutants leads to chromosome instability and cell death, providing a link between telomeres and aging.

Blackburn, Greider, and Szostak shared the 2009 Nobel Prize in Physiology or Medicine for showing the critical role that RNA and a ribonucleoprotein enzyme have in maintaining the ends of chromosomes in the DNA world.

LABORATORY
4.1

Culturing and Observing *C. elegans*

▼ OBJECTIVES

This laboratory demonstrates several important concepts of modern biology. During the course of this laboratory, you will

- Learn about the use of model organisms in research.
- Observe development and identify specific developmental stages in *Caenorhabditis elegans*.
- Explore the relationship between genotype and phenotype.

In addition, this laboratory utilizes several experimental methods in modern biological research. You will

- Use sterile technique to isolate and grow pure cultures of bacteria and *C. elegans*.
- Use antibiotic selection to maintain a recombinant bacterial culture.
- Use dissecting microscopes to observe and analyze cultures of bacteria and *C. elegans*.

INTRODUCTION

Sydney Brenner won a Nobel prize for establishing *C. elegans* as a model organism.

(Photo courtesy of Matthew Meselson.)

A human is a complicated organism, and most molecular genetic experiments would be either technically difficult or unethical to perform on human subjects. For these reasons, biologists often use simpler "model" organisms that are easy to culture and manipulate in the laboratory. Despite obvious physical differences, model organisms and humans share many key biochemical and physiological functions that have been conserved (preserved) during evolution. The nematode worm *C. elegans* is one of several organisms commonly studied by biological researchers today.

C. elegans is a microscopic roundworm. Although some roundworms are parasitic, *C. elegans* is a free-living worm that feeds on soil bacteria. These worms grow quickly, developing from embryo to adult in 3 d. *C. elegans* is a simple animal with only ~1000 cells, and scientists know exactly how each of those cells develops from the fertilized egg. *C. elegans* was the first multicellular organism to have its entire genome sequenced, with the surprising finding that 40% of its genes have human matches. Mating animals, isolating genes, and introducing foreign DNA are much easier in *C. elegans* than in more complicated animals. All of these features make *C. elegans* a great model for understanding how cells divide, develop, and take on specialized tasks in higher (eukaryotic) organisms. Recently, the discovery that any of the organism's genes can be "silenced" using a technique called RNA interference (RNAi) has made *C. elegans* an ideal organism to quickly determine the functions of genes identified by sequencing the genome.

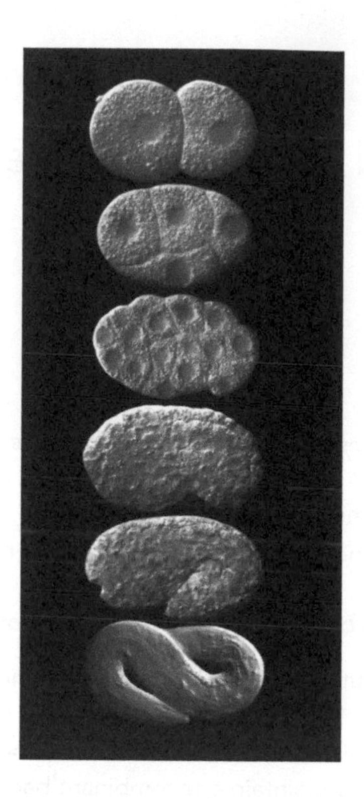

Embryonic development in *C. elegans*. This series of images shows different stages during embryonic development. Fertilization to hatching takes just 14 h.

(Reprinted, with permission, from O'Rourke M, Bowerman B. 2005. *Nature* 434: 444–445; ©Macmillan.)

This laboratory introduces *C. elegans* and describes methods required for its culture; these techniques and familiarity with *C. elegans* are prerequisites for RNAi experiments. Included are techniques for growing *Escherichia coli* cells and preparing plates to feed *C. elegans*. The most common strain of bacteria used to feed worms is the *E. coli* strain OP50, which is grown on standard LB plates or in LB broth. For RNAi experiments, specialized strains of bacteria, each containing a plasmid that expresses a gene-specific double-stranded RNA (dsRNA), are fed to worms to trigger gene silencing.

A technique for spreading bacteria onto standard plates to isolate single cells from one another is described. Each cell then reproduces to form a visible colony composed of genetically identical clones. Small-scale suspension cultures of *E. coli* are then grown by overnight incubation using cells derived from a single colony, which minimizes the chance of using a cell mass contaminated with a foreign microorganism. These overnight cultures are used to inoculate, or seed, specialized "NGM (nematode growth medium)-lite" agar plates on which the worms are grown. *E. coli* OP50 is seeded to NGM-lite plates. In addition, RNAi strains expressing dsRNA corresponding to three genes, *dpy-11, bli-1*, and *unc-22*, are seeded to NGM-lite plates with ampicillin and isopropyl-β-D-thiogalactopyranoside (IPTG). Ampicillin selects for bacteria carrying the RNAi plasmid, and IPTG triggers dsRNA expression from the plasmid in the bacteria. In Laboratory 4.2, the function of these three genes is assayed when RNAi is induced by culturing worms on these plates.

Two methods to propagate worms are described. Worms growing on an NGM-lite plate are transferred by cutting out a section of the medium and placing it on the surface of a new plate. This "chunking" method is a rapid way to move multiple worms from a plate where the *E. coli* food source has been consumed to a plate with a fresh lawn of bacteria. Individual worms can be transferred to fresh plates using the flattened tip of a platinum wire; this technique for "picking" individual worms is the starting point for genetic crosses and RNAi experiments.

Finally, wild-type *C. elegans* hermaphrodites are observed using microscopy, and their morphology, behavior, and life cycle are analyzed. Abnormal morphology and behavior of mutant worms are also observed.

FURTHER READING

Brenner S. 1974. The genetics of *Caenorhabditis elegans*. *Genetics* **77:** 71–94.

C. elegans Sequencing Consortium. 1998. Genome sequence of the nematode *C. elegans:* A platform for investigating biology. *Science* **282:** 2012–2018.

De Melo JV, De Souza W, Peixoto CA. 2002. Ultrastructural analyses of the *Caenorhabditis elegans* DR 847 *bli-1(n361)* mutant which produces abnormal cuticle blisters. *J Submicrosc Cytol Pathol* **34:** 291–297.

Fire A, Xu S, Montgomery MK, Kostas SA, Driver SE, Mello CC. 1998. Potent and specific genetic interference by double-stranded RNA in *Caenorhabditis elegans*. *Nature* **391:** 806–811.

Ko FCF, Chow KL. 2002. A novel thioredoxin-like protein encoded by the *C. elegans dpy-11* gene is required for body and sensory organ morphogenesis. *Development* **129:** 1185–1194.

Kramer JM, French RP, Park EC, Johnson JJ. 1990. The *Caenorhabditis elegans rol-6* gene, which interacts with the *sqt-1* collagen gene to determine organismal morphology, encodes a collagen. *Mol Cell Biol* **10:** 2081–2089.

Moerman DG, Benian GM, Waterston RH. 1986. Molecular cloning of the muscle gene *unc-22* in *Caenorhabditis elegans* by Tc1 transposon tagging. *Proc Natl Acad Sci* **83:** 2579–2583.

Moerman DG, Benian GM, Barstead RJ, Schriefer LA, Waterston RH. 1988. Identification and intracellular localization of the *unc-22* gene product of *Caenorhabditis elegans*. *Genes Dev* **2:** 93–105.

PLANNING AND PREPARATION

The following table will help you to plan and integrate the different experimental methods.

Experiment	Day	Time		Activity
Stage A: Culturing *E. coli*				
	2 or more d before Part I	90 min	Prelab:	Prepare LB, LB/amp, NGM-lite, and NGM-lite/amp + IPTG plates.
	1–2 d before Part I	30 min	Prelab:	Streak starter plates for OP50 and RNAi feeding strains (*dpy-11*, *bli-1*, *unc-22*).
I. Streak *E. coli* to obtain single colonies	1	20 min	Prelab:	Set up student stations.
		30 min	Lab:	Streak plates.
		15–20 h	Postlab:	Incubate plates.
II. Grow *E. coli* overnight cultures	2	15 min	Prelab:	Aliquot LB and LB/amp.
				Set up student stations.
		15 min	Lab:	Prepare overnight cultures.
		12–48 h	Postlab:	Incubate cultures.
III. Seed NGM-lite and NGM-lite/amp + IPTG plates with *E. coli*	4	30 min	Lab:	Seed NGM-lite and NGM-lite/amp + IPTG plates.
		24–36 h	Postlab:	Incubate plates.[a]
Stage B: Culturing *C. elegans*				
I. Chunk wild-type *C. elegans*	6	30 min	Prelab:	Chunk wild-type worms to OP50-seeded NGM-lite plates.[b]
	8	20 min	Prelab:	Set up student stations.
		15 min	Lab:	Chunk worms to OP50-seeded NGM-lite plates. Incubate for 48 h.
II. Pick individual *C. elegans*	10	20 min	Prelab:	Make worm picks.
				Set up student stations.
		30 min	Lab:	Pick L4 worms to fresh OP50-seeded NGM-lite plates.
Stage C: Observing Wild-Type and Mutant *C. elegans*[c]				
I. Observe the *C. elegans* life cycle	12–13	20 min	Prelab:	Set up student stations.
		45 min	Lab:	Study the morphology, behavior, and life cycle of wild-type worms under a dissecting microscope.
II. Observe *C. elegans* mutants	9	30 min	Prelab:	Chunk wild-type and mutant *C. elegans* strains (*rol-6*, *bli-1*, *unc-22*, *dpy-11*) to OP50-seeded NGM-lite plates.
				Set up student stations.
	12–13	45 min	Lab:	Examine wild-type and mutant worms; identify differences in development rate, morphology, or movement.

[a]After incubation, seeded plates from Part III of Stage A can be stored in sealed containers for several weeks at 4°C.

[b]If enough plates for each student are available, this extra round of chunking may be skipped. (It is a good idea to prepare extra plates, because some may become contaminated during preparation or student work.)

[c]Both procedures in Stage C require wild-type worms from Parts I or II of Stage B. If you are not completing Parts I or II of Stage C immediately after Stage B, transfer small chunks of wild-type worms to OP50-seeded NGM-lite plates 2 d before completing Stage C.

OVERVIEW OF EXPERIMENTAL METHODS IN STAGE A: CULTURING *E. COLI*

I. STREAK *E. COLI* TO OBTAIN SINGLE COLONIES

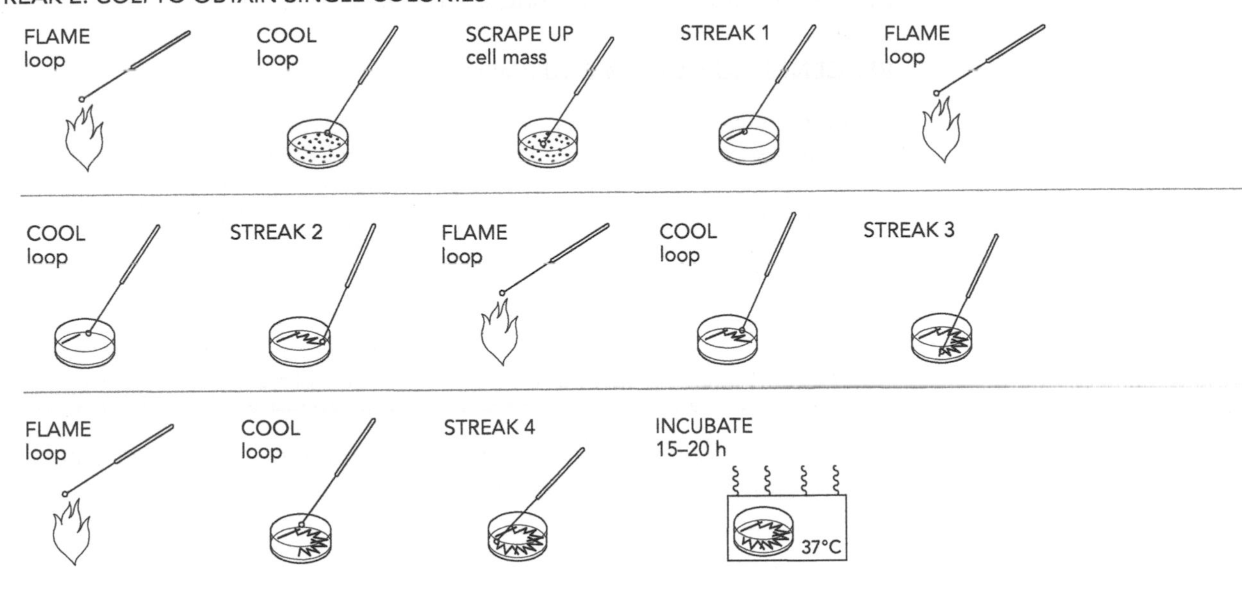

II. GROW *E. COLI* OVERNIGHT CULTURES

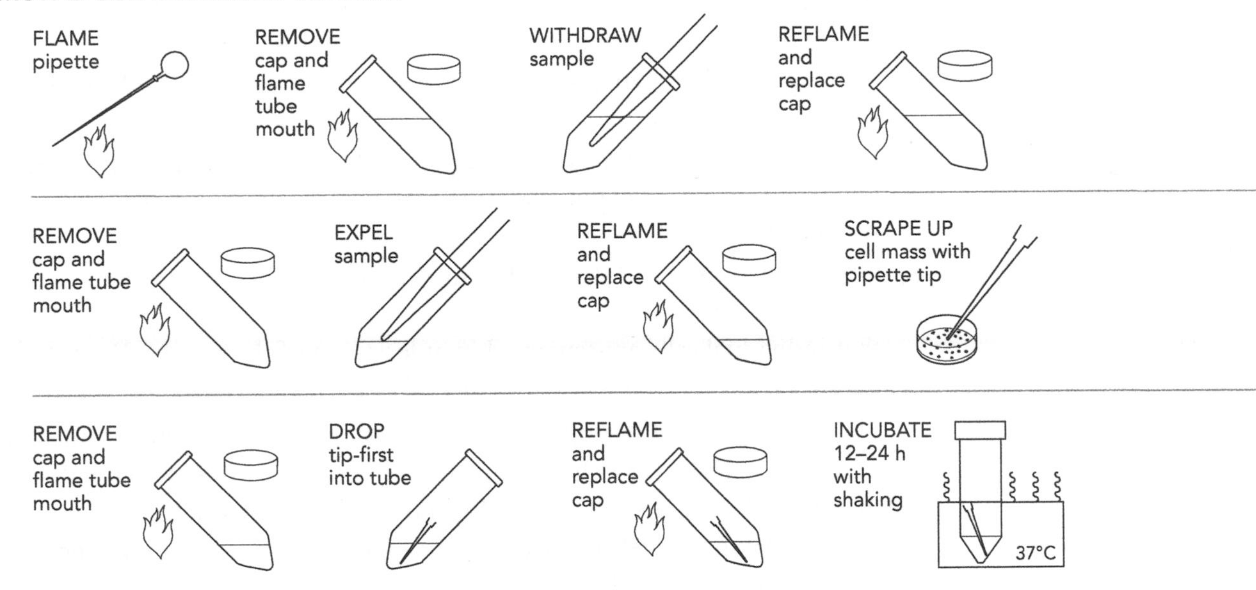

III. SEED NGM-LITE AND NGM-LITE/AMP + IPTG PLATES WITH *E. COLI*

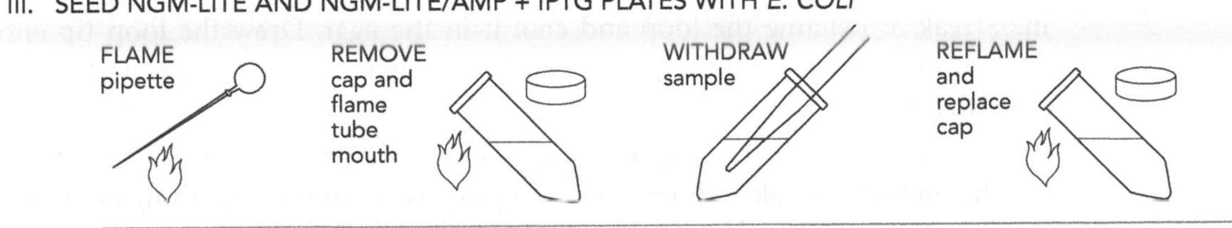

INCUBATE
24–36 h at room temperature

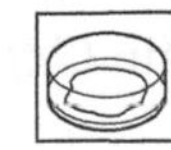

STAGE A: CULTURING *E. COLI*

I. Streak *E. coli* to Obtain Single Colonies

REAGENTS, SUPPLIES, & EQUIPMENT

For each group

Bunsen burner
Inoculating loop
LB agar plate
3 LB + ampicillin <!> (LB/amp) plates
Permanent marker (red)
Bleach (10%) <!> or disinfectant (e.g., Lysol)
E. coli OP50 culture
E. coli RNAi feeding strain cultures (*dpy-11*, *bli-1*, and *unc-22*)
Incubator set at 37°C

To share

"Bio bag" or heavy-duty trash bag

See Cautions Appendix for appropriate handling of materials marked with <!>.

1. Use a red permanent marker to label the *bottom* of the LB agar plate with your group number, the date, and "OP50."

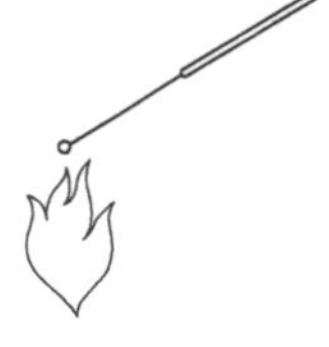

CAUTION! When using an open flame, take appropriate precautions. Make sure that any loose clothing is secured and tie long hair back. Do not lean over the flame.

2. Hold the inoculating loop like a pencil and sterilize the loop in the Bunsen burner flame until it glows red hot.
3. Remove the lid from the *E. coli* OP50 culture plate with your free hand. Do not place the lid on the lab bench; hold the lid face down just above the culture plate to help to prevent contaminants from falling on the plate or lid.
4. Stab the inoculating loop into a clear area of the *E. coli* OP50 culture plate several times to cool it.
5. Use the loop tip to scrape a visible cell mass from a bacterial colony on the *E. coli* OP50 culture plate. Do not gouge the agar. Replace the lid on the *E. coli* OP50 culture plate.

6. Lift the lid of the new LB agar plate just enough to perform streaking as described below. The object is to serially dilute the bacteria with each successive streak, so that individual cells are separated in at least one of the streaks. Do not place the lid on the lab bench; replace the plate lid after each streak.
   i. Streak 1: Glide the loop tip back and forth across the surface of the LB agar to make a streak across the top quarter of the plate. Avoid gouging the agar.

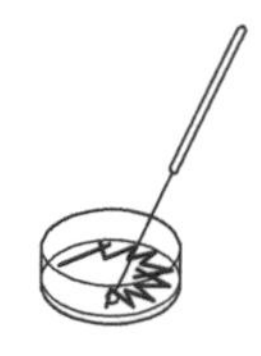

   ii. Streak 2: Reflame the inoculating loop and cool it by stabbing it into the agar away from the first (primary) streak. Draw the loop tip through the end of the primary streak and, without lifting the loop, make a zigzag streak across one quarter of the agar surface.
   iii. Streak 3: Reflame the loop and cool it in the agar. Draw the loop tip *once* through the end of the previous streak and make another zigzag streak in the adjacent quarter.
   iv. Streak 4: Reflame the loop and cool it. Draw the tip *once* through the end of the previous streak and make a final zigzag streak in the remaining quarter of the plate.
   v. Reflame the loop and allow it to cool before placing it on the lab bench.

Make it a habit to always flame the loop one last time.

7. Label the *bottom* of three LB/amp plates with your group number, the date, and the appropriate *E. coli* RNAi feeding strain ("*dpy-11* RNAi," "*bli-1* RNAi," or "*unc-22* RNAi").

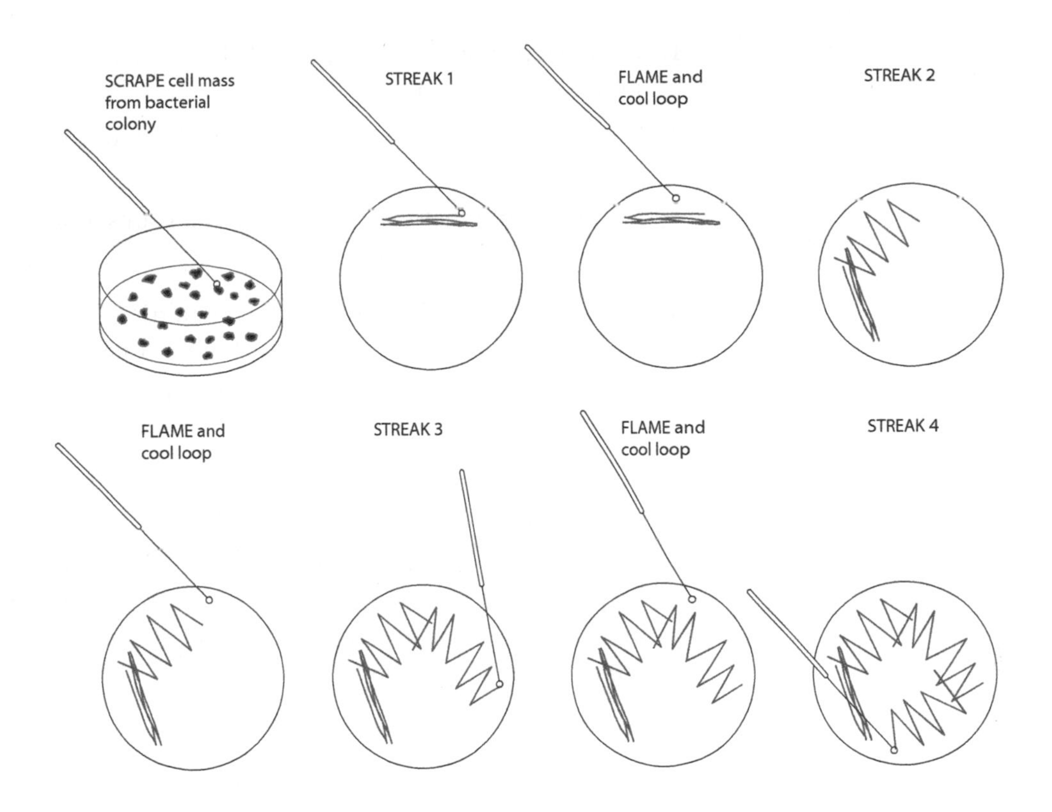

8. Follow the procedures outlined in Steps 2–6 to streak each feeding strain onto a separate LB/amp plate.
9. Place all four plates upside down in a 37°C incubator. Incubate them for 15–20 h, until optimal growth of well-formed colonies is achieved. At this point, colonies should range from 0.5 to 3 mm in diameter.
10. Take time for responsible cleanup.
    i. Segregate unwanted bacterial cultures into a "bio bag" or heavy-duty trash bag for proper disposal.
    ii. Wipe the lab bench with soapy water and 10% bleach or disinfectant at the end of the lab.
    iii. Wash your hands before leaving the lab.

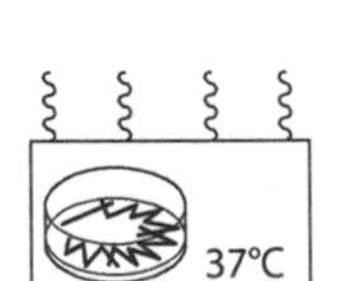

Plates are inverted to prevent condensation that collects on the lids from falling onto the agar, causing the colonies to run together.

II. Grow *E. coli* Overnight Cultures

REAGENTS, SUPPLIES, & EQUIPMENT

For each group
Bunsen burner
4 Culture tubes (15 mL) (sterile)
E. coli OP50 culture plate from Part I
E. coli RNAi feeding strain cultures (*dpy-11*, *bli-1*, and *unc-22*) on LB/amp plates from Part I
Inoculating loop (optional)
LB/amp broth (10 mL)
LB broth (5 mL)
Micropipette tips (10–100 µL, sterile)
Permanent marker (red)
Pipette aid or bulb
2 Pipettes (5 mL) (sterile)
Test tube rack

To share
"Bio bag" or heavy-duty trash bag
Bleach (10%) <!> or disinfectant (e.g., Lysol)
Shaking water bath or incubator set at 37°C

See Cautions Appendix for appropriate handling of materials marked with <!>.

If working in a team, one partner should handle the pipette and the other should handle the tubes and caps.

1. Use a red permanent marker to label a *sterile* 15-mL culture tube with your group number, the date, and "OP50."
2. Use a 5-mL pipette to *sterilely* transfer 2 mL of LB broth into the culture tube as follows:

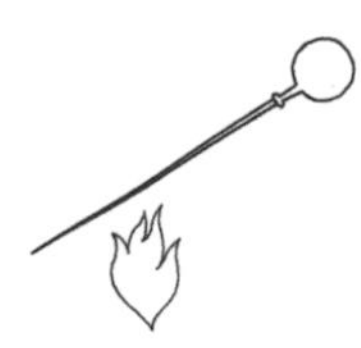

Pipette flaming can be eliminated if individually wrapped pipettes are used.

   i. Make sure that the culture tube cap is unscrewed to the "loose" position.
   ii. Attach a pipette aid or bulb to a 5-mL pipette. Briefly flame the pipette cylinder.
   iii. Remove the cap of the bottle containing LB broth using the little finger of your hand holding the pipette bulb. Flame the mouth of the LB bottle.
   iv. Use the pipette to withdraw 2 mL of LB. Reflame the mouth of the bottle and replace the cap.
   v. Remove the cap of the labeled culture tube from Step 1. Expel the LB into the tube, reflame, and replace the cap.

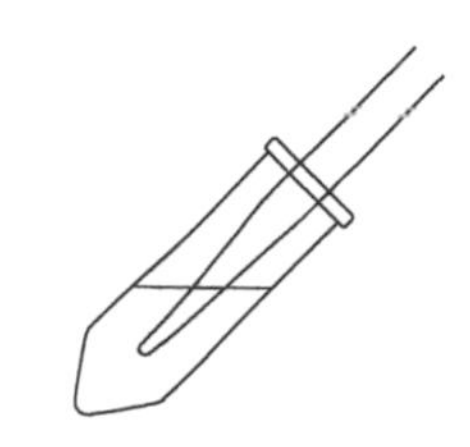

3. Use a *sterile* micropipette tip to scrape a visible cell mass from a selected colony on your *E. coli* OP50 culture plate (from Part I) and drop it tip-first into the culture tube. Reflame and replace the tube cap in the loose position. Alternatively, use an inoculating loop as follows:

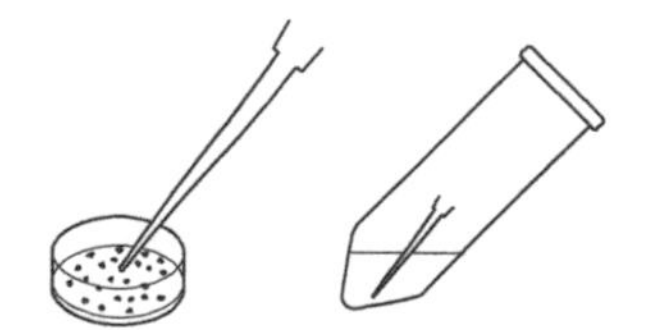

Loop flaming can be eliminated if an individually wrapped, sterile plastic loop is used.

   i. Sterilize the loop in the Bunsen burner flame until it glows red hot. Cool the loop by stabbing it several times into a clear area near the edge of your *E. coli* OP50 culture plate (from Part I).
   ii. Use the loop to scrape a visible cell mass from a selected colony on your *E. coli* OP50 culture plate. Immerse the cell mass in the LB broth and agitate the loop to dislodge the cell mass.
   iii. Replace the tube cap in the loose position. Reflame the loop before setting it on the lab bench.
4. Label three *sterile* 15-mL culture tubes with your group number, the date, and one of the *E. coli* RNAi feeding strains ("*dpy-11* RNAi," "*bli-1* RNAi," or "*unc-22* RNAi").
5. Use a 5-mL pipette to *sterilely* transfer 2 mL of LB/amp broth into each of the three labeled culture tubes as described in Step 2.
6. Use a *sterile* pipette tip (or a flamed and cooled inoculating loop) to transfer a single colony of RNAi feeding bacteria from each of the *dpy-11*, *bli-1*, and *unc-22* LB/amp plates (from Part I) into the appropriate culture tubes as described in Step 3.

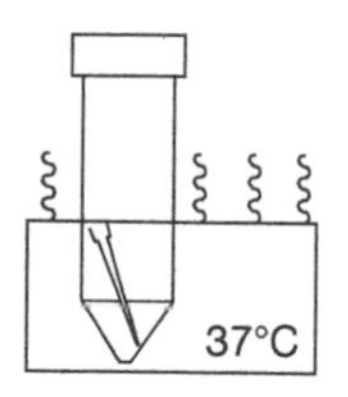

7. Incubate the tubes for 12–24 h in a 37°C shaking water bath (with continuous shaking) or for 24–48 h in a 37°C incubator (without shaking). To allow air flow, do not seal the tube lids during incubation.
8. Take time for responsible cleanup.
   i. Segregate unwanted bacterial cultures and tubes, pipettes, and micropipette tips that have come into contact with cultures into a "bio bag" or heavy-duty trash bag for proper disposal.
   ii. Wipe the lab bench with soapy water and 10% bleach or disinfectant at the end of lab.
   iii. Wash your hands before leaving the lab.

III. Seed NGM-Lite and NGM-Lite/Amp + IPTG Plates with *E. coli*

REAGENTS, SUPPLIES, & EQUIPMENT

For each group

Bunsen burner
E. coli OP50 overnight culture from Part II
E. coli RNAi feeding strain (*dpy-11*, *bli-1*, and *unc-22*) overnight cultures from Part II
3 NGM-lite + ampicillin <!> + isopropyl-β-D-thiogalactopyranoside <!> (NGM-lite/amp + IPTG) plates
4 NGM-lite plates
Permanent markers (black and red)
Pipette aid or bulb
4 Pipettes (5 mL)
Test tube rack

To share

"Bio bag" or heavy-duty trash bag
Bleach (10%) <!> or disinfectant (e.g., Lysol)

See Cautions Appendix for appropriate handling of materials marked with <!>.

1. Label the *bottom* of four NGM-lite plates with your group number and the date with a *black* marker. Use a *red* marker to label the *bottom* of the plates "OP50."

Seed the plates in a clean area to avoid contamination.

2. Use a 5-mL pipette to *sterilely* seed each of the NGM-lite plates with OP50 as follows:
   i. Attach the pipette aid or bulb to the 5-mL pipette. Briefly flame the pipette cylinder.
   ii. Remove the cap from the OP50 overnight culture (from Part II) using the little finger of your hand holding the pipette bulb. Flame the mouth of the OP50 overnight culture.
   iii. Use the pipette to withdraw 1 mL of overnight culture. Reflame and replace cap.
   iv. Add one to two drops of overnight culture to the center of the surface of each NGM-lite plate. The drops should occupy most of the plate surface but *should not touch the edge* of the dish.

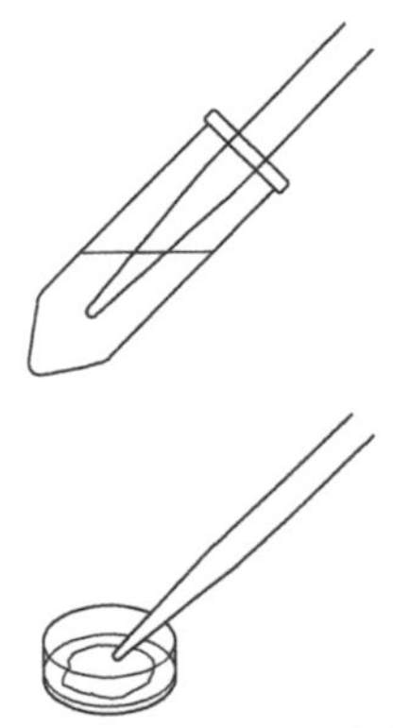

Avoiding the edges of the dish ensures that the worms will remain in the center of the plate.

3. Use a *red* marker to label the *bottom* of three NGM-lite/amp + IPTG plates with your group number, the date, and one of the *E. coli* RNAi feeding strains ("*dpy-11* RNAi," "*bli-1* RNAi," or "*unc-22* RNAi").
4. Use a 5-mL pipette to *sterilely* transfer one to two drops of each feeding strain overnight culture to the appropriate NGM-lite/amp + IPTG plate as described in Step 2.

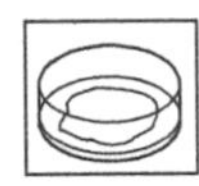

The bacterial lawn should be confluent and dry before any worms are added.

5. Grow the seeded plates *face-up* for 24–36 h at room temperature.
6. Take time for responsible cleanup.
   i. Segregate unwanted bacterial cultures and tubes, pipettes, and micropipette tips that have come into contact with cultures into a "bio bag" or heavy-duty trash bag for proper disposal.
   ii. Wipe the lab bench with soapy water and 10% bleach or disinfectant.
   iii. Wash your hands before leaving the lab.

OVERVIEW OF EXPERIMENTAL METHODS IN STAGE B: CULTURING *C. ELEGANS*

I. CHUNK WILD-TYPE *C. ELEGANS*

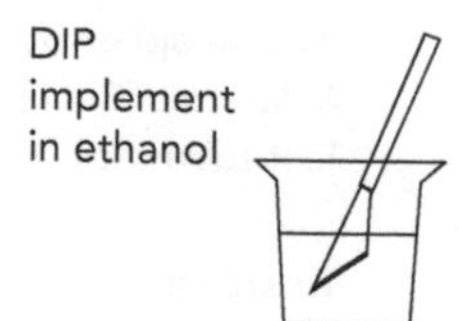

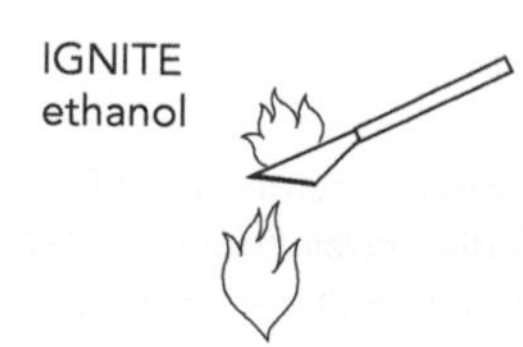

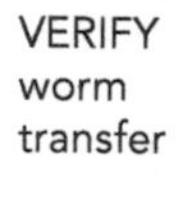
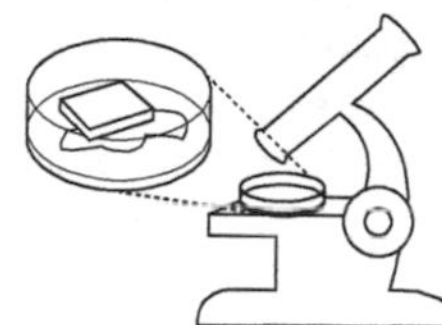

II. PICK INDIVIDUAL *C. ELEGANS*

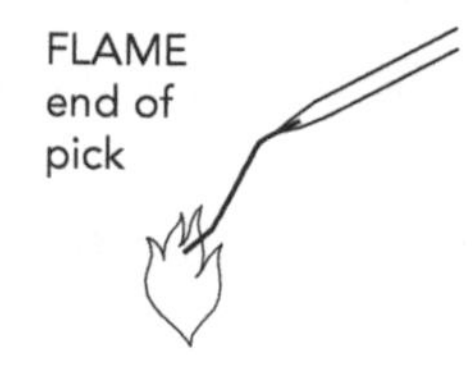

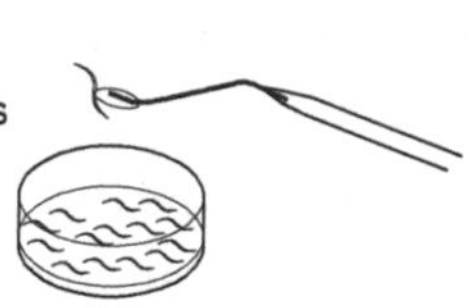

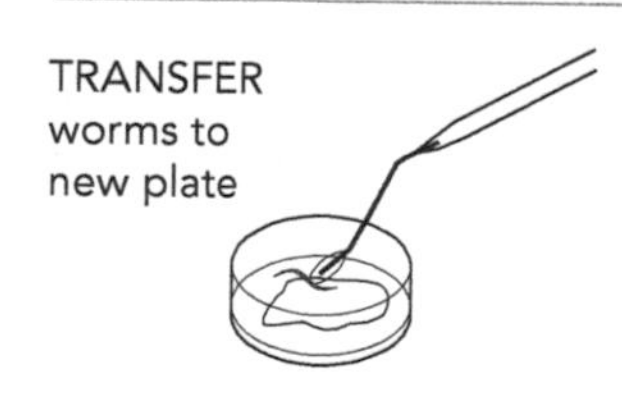

STAGE B: CULTURING *C. ELEGANS*

I. Chunk Wild-Type *C. elegans*

REAGENTS, SUPPLIES, & EQUIPMENT

For each group
Binocular dissecting microscope
Bunsen burner
Ethanol (95%) <!> in a 50- or 100-mL beaker
Metal spatula or forceps
OP50-seeded NGM-lite plate from Part III of Stage A
Permanent marker (black)
Wild-type worms on NGM-lite plate

To share
"Bio bag" or heavy-duty trash bag
Bleach (10%) <!> or disinfectant (e.g., Lysol)

See Cautions Appendix for appropriate handling of materials marked with <!>.

Sterilization prevents cross-contamination with different C. elegans strains and non-OP50 bacteria.

CAUTION! Be extremely careful to avoid igniting the ethanol in the beaker. Do not panic if the ethanol is accidentally ignited. Cover the beaker with a glass Petri dish lid or other non-flammable cover to cut off oxygen and rapidly extinguish the fire.

To feed worm strains, it is easier to transfer a chunk of worm-filled agar from a well-grown plate to a new plate rather than to pick individual worms.

1. Obtain a fresh OP50-seeded NGM-lite plate (from Part III of Stage A) and a plate with wild-type worms.
2. Examine both plates under the dissecting microscope for signs of bacterial or mold contamination—any growth of a different color or morphology (shape) from the OP50 lawn. Obtain a new plate if you detect any contamination.
3. Use a black permanent marker to label the bottom of the fresh OP50-seeded NGM-lite plate with your group number, the date, and "wild type."
4. Use your dissecting microscope to identify a region of the plate of wild-type worms that is densely populated with worms and eggs.
5. Sterilize a metal spatula or forceps by dipping the end of the implement into the beaker of ethanol and then briefly passing it through a Bunsen burner flame to ignite the ethanol. Allow the ethanol to burn off away from the Bunsen flame; the implement will become too hot if left in the flame.
6. Use a sterilized spatula or forceps to cut a 1-cm (~3/8-in) square chunk of agar from the worm- and egg-dense region of the plate identified in Step 4.
7. Carefully remove the piece of agar with worms from the wild-type plate and place it upside down on the lawn of the fresh OP50-seeded NGM-lite plate.

Placing the agar piece upside down makes it easier for the worms to crawl into the new bacterial food source.

8. Examine the new plate under the microscope to verify that you have successfully chunked the worms. Within a few minutes, worms should crawl from the agar chunk and be visible in the bacterial lawn.
9. Store the new plate lid-side down for ~48 h at room temperature before continuing with Part II of Stage B (or Parts I or II of Stage C). Choose a place where the plate will not be disturbed.
10. Take time for responsible cleanup.
    i. Segregate any bacterial cultures that need to be discarded into a "bio bag" or heavy-duty trash bag for proper disposal.
    ii. Wipe the lab bench with soapy water and 10% bleach or disinfectant.
    iii. Wash your hands before leaving the lab.

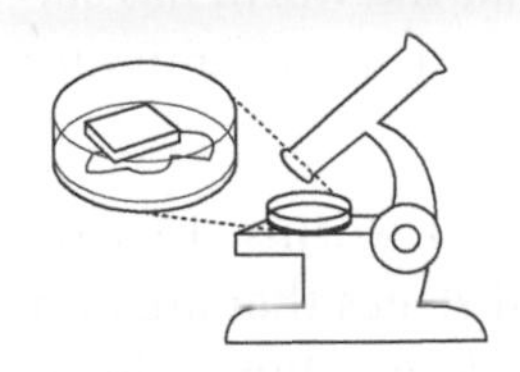

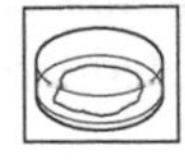

II. Pick Individual *C. elegans*

Repeat this procedure until you can efficiently pick worms to a new plate. Once you feel accomplished, try to pick several worms at once.

REAGENTS, SUPPLIES, & EQUIPMENT

For each group

Binocular dissecting microscope
Bunsen burner
Forceps
OP50-seeded NGM-lite plate from Part III of Stage A
Permanent marker (black)
Wild-type worms on NGM-lite plate from Part I
Worm pick

To share

"Bio bag" or heavy-duty trash bag
Bleach (10%) <!> or disinfectant (e.g., Lysol)

See Cautions Appendix for appropriate handling of materials marked with <!>.

1. Examine the plate of wild-type worms you chunked in Part I under the dissecting microscope. Confirm that no contaminants have grown since chunking. Obtain a new plate of worms if necessary.
2. Use a black permanent marker to label the bottom of the fresh OP50-seeded NGM-lite plate (from Part III of Stage A) with your group number, the date, and "wild type."

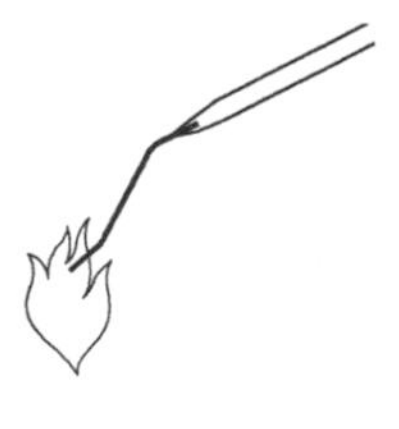

3. Examine a worm pick. The flattened end of the platinum wire should be bent at roughly a 45° angle. Adjust the pick with the forceps if necessary. You may need to adjust your pick from time to time during the course of this exercise.
4. Hold the worm pick like a pencil and sterilize the tip in a Bunsen burner flame until it glows red hot.

Bacteria from a plate that has aged 2–3 wk serve as a better source of sticky bacteria.

5. Attach a glob of bacteria to the worm pick by wiping the flat head across the lawn of bacteria on the fresh OP50-seeded NGM-lite plate, as shown to the left. The bacteria will act like double-stick tape when you pick worms.
6. Open the lid of your plate of worms and identify a large worm.
7. Gently tap the top of the worm with the glob of bacteria on the bottom of the flattened pick. The glob of bacteria will attach the worm to the pick.

8. To transfer the worm, gently wipe the bottom of the pick in the lawn of the fresh OP50-seeded plate. Make sure to avoid using too much force or you may tear the agar surface and possibly crush the worm.
9. Examine the new plate under the microscope to confirm that the worm has survived picking. If it is visibly damaged or fails to move within several minutes, transfer another worm.

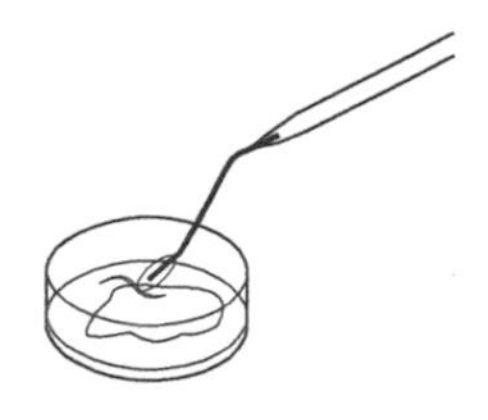

10. For most experiments, it is important to use only one stage of worms. To avoid later confusion, "burn" any embryos or other smaller larval stages that are accidentally transferred to the new plate by heating the worm pick in a Bunsen burner flame until it glows red hot and then immediately touching the flattened end to any unwanted worm. (Alternatively, carefully pick each unwanted worm and flame it in the Bunsen burner.)

11. Store the plate at room temperature. Choose a place where the plate will not be disturbed.
12. Take time for responsible cleanup.
    i. Segregate any bacterial cultures that need to be discarded into a "bio bag" or heavy-duty trash bag for proper disposal.
    ii. Wipe the lab bench with soapy water and 10% bleach or disinfectant.
    iii. Wash your hands before leaving the lab.

STAGE C: OBSERVING WILD-TYPE AND MUTANT *C. ELEGANS*

I. Observe the *C. elegans* Life Cycle

REAGENTS, SUPPLIES, & EQUIPMENT

For each group

Binocular dissecting microscope

Wild-type worms on NGM-lite plate from Part I or II of Stage B

1. Obtain a plate with wild-type worms.
2. Observe the worms under a dissecting microscope. Note any physical (morphological) differences among the worms.
3. Note any differences in behavior, paying particular attention to how they move on the plate.
4. Lift the plate several centimeters (~1 in) above the microscope stage and drop it. Note any changes in worm movement. You may need to tap the plate several times to induce movement.

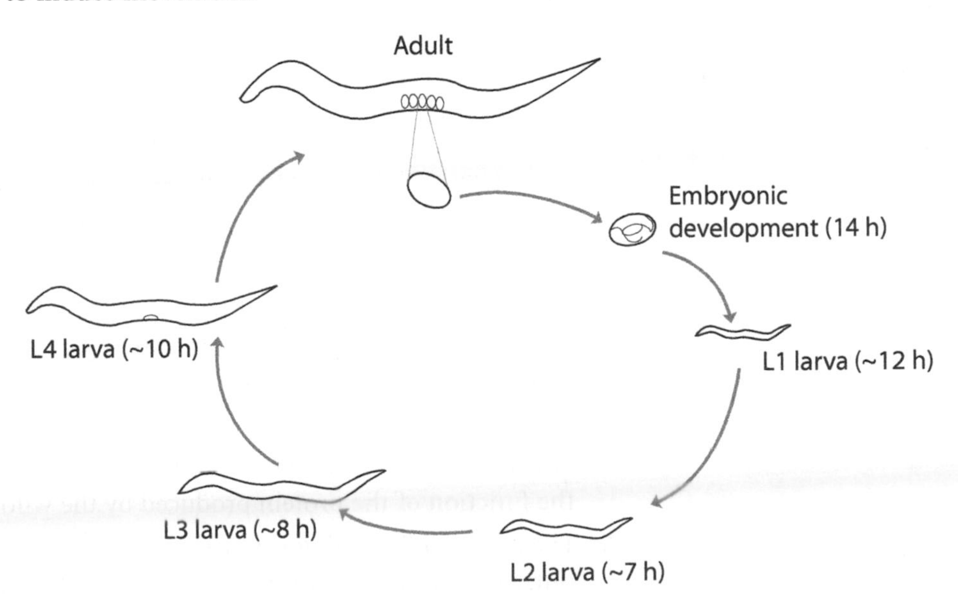

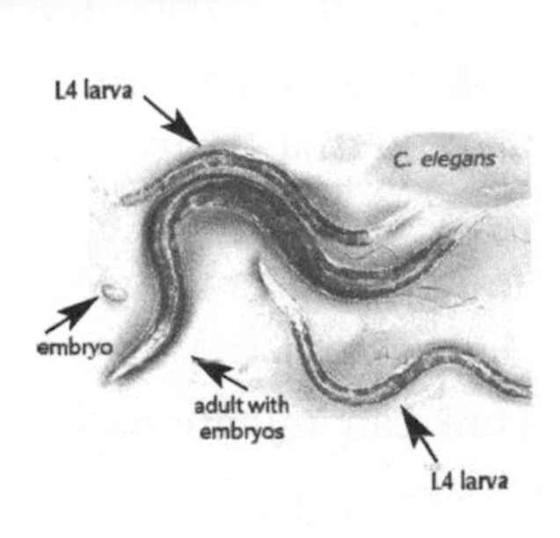

5. Study the diagram of the *C. elegans* life cycle above and attempt to identify an example of each stage of the worm life cycle on the plate.
    i. An adult hermaphrodite is a large worm with embryos inside. (The wild-type strain used in this experiment produces few if any adult males.)
    ii. The embryo is a small, oval object.

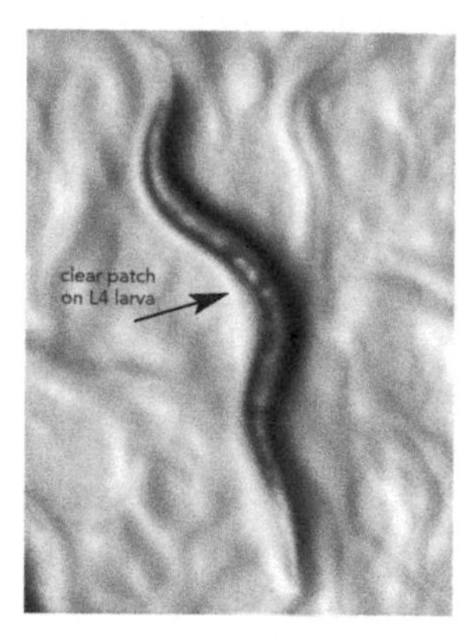

All RNAi experiments in subsequent labs begin with identifying L4 hermaphrodites, so it is important to become proficient at identifying them.

iii. An L1 larva has recently hatched and is the smallest of the four larval stages.

iv. L2 and L3 larvae are larger than L1 worms but not as large as an adult. Examine worms of different sizes to familiarize yourself with these larval stages.

v. The final juvenile stage, an L4 larva, is almost as large as an adult hermaphrodite. The lack of internal embryos is one marker that distinguishes an L4 larva from an adult. A clear, crescent-shaped patch near the center of the body is another characteristic of an L4 larva. The egg-laying structure, called the vulva, will develop in this patch when the L4 molts into an adult.

II. Observe *C. elegans* Mutants

REAGENTS, SUPPLIES, & EQUIPMENT

For each group

Binocular dissecting microscope

Mutant worms on NGM-lite plates (*dpy-11, rol-6, bli-1*, and *unc-22*)

Wild-type worms on NGM-lite plate from Part I or II of Stage B

1. Obtain plates with mutant and wild-type worms.
2. Observe the worms under a dissecting microscope. Note any physical (morphological) differences among the wild-type and mutant worms. Record your observations and make sketches as needed.
3. Note any differences in behavior, paying particular attention to how the wild-type and mutant worms move on the plate. Gently tap the plates on the microscope stage to induce movement. Record your observations and make sketches as needed.

RESULTS AND DISCUSSION

1. How many stages of *C. elegans* development were you able to identify? Describe each stage.
2. Why is it necessary for *C. elegans* to pass through several larval stages and how is this type of development different from humans?
3. How does a hermaphrodite produce offspring without mating?
4. What physical (morphological) differences did you observe in the mutant worms? What differences in behavior or movement did you notice? Did your classmates identify the same characteristics of the mutant *C. elegans*?
5. Based on each mutant phenotype that you observed, what do you think would be the function of the protein produced by the wild-type gene?
6. The mutant *bli-1* and *dpy-11* strains contain mutations that affect the cuticle, the outer layer of the worm that is secreted by the epidermal (skin) cells. These two very different phenotypes show how the nature of the mutation in a strain (the genotype) affects the phenotype.

   i. *bli-1* encodes a collagen. What is a collagen? How can mutations in a collagen affect the cuticle?

   ii. *dpy-11* encodes an enzyme. What do enzymes do? How can mutations in enzymes affect the cuticle?

LABORATORY 4.2

Using *E. coli* Feeding Strains to Induce RNAi and Knock Down Genes

▼ OBJECTIVES

This laboratory demonstrates several important concepts of modern biology. During the course of this laboratory, you will

- Learn about the mechanism of RNA interference (RNAi) and its applications.
- Explore the relationships among genotype, phenotype, and RNAi-induced phenotype.
- Move between in vitro experimentation and in silico computation.

In addition, this laboratory utilizes several experimental and bioinformatics methods in modern biological research. You will

- Use sterile technique to grow *Caenorhabditis elegans*.
- Use dissecting microscopes to observe and analyze cultures of *C. elegans*.
- Disrupt the function of a gene using RNAi.
- Explore gene data online using WormBase.
- Use the Basic Local Alignment Search Tool (BLAST) to identify sequences in databases.

INTRODUCTION

Functional genetic studies typically rely on mutating a particular gene and then looking for physical or behavioral changes in the organism. Mutagenesis is very time-consuming and some techniques do not target specific genes, making it very difficult to study the function of a particular gene if no natural mutation exists. A powerful alternative is to down-regulate genes with RNAi.

RNAi is a mechanism that is thought to have evolved to protect organisms from infection by RNA viruses. When double-stranded RNA (dsRNA) is present in a cell, it is recognized by a protein named Dicer. Dicer is an RNase that cuts these dsRNA molecules into short pieces. These 21–25-base pair (bp) pieces, called small interfering RNAs (siRNAs), are bound by another protein called Argonaute. One strand from the siRNA is then destroyed, leaving only one strand bound to Argonaute. This strand acts as a guide that hybridizes to messenger RNAs (mRNAs) that are complementary to the guide strand. Once these mRNAs are bound, Argonaute cleaves them within the complementary sequence, thereby inhibiting (silencing) gene function.

By deliberately introducing defined sequences of dsRNA that are identical to the sequence of a gene of interest, biologists can observe the physiological consequences of

Craig Mello (*left*) and Andrew Fire (*right*) were awarded a Nobel Prize for their discovery of RNA interference.

(Photograph of C. Mello used with permission of the University of Massachusetts Medical School.)

"silencing" that gene in organisms where RNAi can be induced. Amazingly, this mechanism can be activated in *C. elegans* by simply feeding worms bacteria that express dsRNA corresponding to the part of the gene to be silenced. The dsRNA enters the cells through the intestine and is recognized by the RNAi machinery, leading to silencing. In *C. elegans*, the silencing spreads from cell to cell, turning off the target gene in the entire body.

This laboratory demonstrates how RNAi can be used to deduce the function of genes. Wild-type worms are grown on OP50-seeded nematode growth medium (NGM)-lite plates. These growing wild-type worms are transferred onto OP50-seeded NGM-lite plates as a control, as well as onto plates containing RNAi feeding strains of bacteria (*Escherichia coli*) that target specific genes (*dpy-11, bli-1,* and *unc-22*). Each RNAi feeding strain contains a plasmid with a gene-specific DNA insert. The RNAi feeding strains are grown on NGM-lite plates that contain ampicillin and isopropyl-β-D-thiogalactopyranoside (IPTG); the ampicillin selects for bacteria carrying an RNAi plasmid and the IPTG induces expression of T7 RNA polymerase in the bacteria. T7 RNA polymerase transcribes RNA starting at specific T7 promoters that flank the gene-specific insert in the RNAi plasmid. The polymerase transcribes both strands of the insert, and when transcripts from both strands hybridize, dsRNA with a sequence identical to that of the targeted gene is produced. Worms may then ingest bacteria containing this dsRNA, thereby triggering RNAi. An altered phenotype in the progeny of RNAi-treated worms indicates what happens when the normal function of this gene is lost.

In accompanying bioinformatics exercises, the function of the protein encoded by the *unc-22* gene is examined using the online *C. elegans* database WormBase. WormBase contains the entire *C. elegans* genome sequence and the locations of all genes; an entry for each gene includes a summary of data from genetic, biochemical, and RNAi experi-

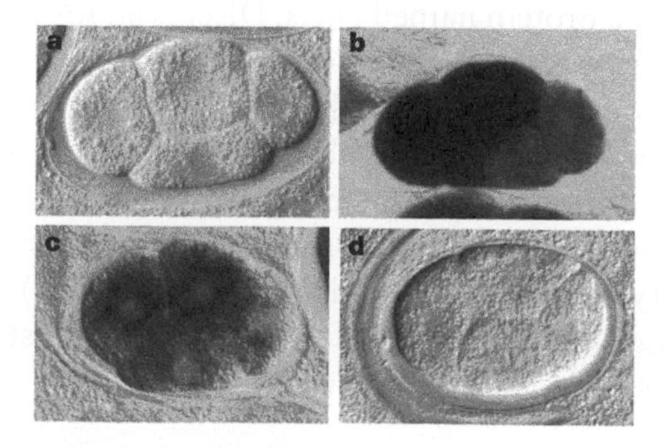

dsRNA treatment reduces mRNA levels. (*a*) Negative control showing lack of staining in the absence of a probe. (*b*) Embryo from untreated parent, showing normal *mex-3* mRNA staining. (*c*) Embryo from a parent treated with antisense RNA, showing reduced *mex-3* expression. (*d*) Embryo from a parent injected with dsRNA, showing loss of *mex-3* mRNA staining.

(Reprinted, with permission, from Fire et al. 1998. *Nature* 391: 806–811; ©Macmillan.)

ments. BLAST is then used to identify the human equivalent of the worm *unc-22* gene. The same computational exercises may be repeated with *dpy-11* and *bli-1*, the other genes examined in the laboratory.

FURTHER READING

De Melo JV, De Souza W, Peixoto CA. 2002. Ultrastructural analyses of the *Caenorhabditis elegans* DR 847 *bli-1(n361)* mutant which produces abnormal cuticle blisters. *J Submicrosc Cytol Pathol* **34:** 291–297.

Fire A, Xu S, Montgomery MK, Kostas SA, Driver SE, Mello CC. 1998. Potent and specific genetic interference by double-stranded RNA in *Caenorhabditis elegans. Nature* **391:** 806–811.

Kaletta T, Hengartner M. 2006. Finding function in novel targets: *C. elegans* as a model organism. *Nat Rev Drug Discov* **5:** 387–399.

Ko FCF, Chow KL. 2002. A novel thioredoxin-like protein encoded by the *C. elegans dpy-11* gene is required for body and sensory organ morphogenesis. *Development* **129:** 1185–1194.

Moerman DG, Benian GM, Waterston RH. 1986. Molecular cloning of the muscle gene *unc-22* in *Caenorhabditis elegans* by Tc1 transposon tagging. *Proc Natl Acad Sci* **83:** 2579–2583.

Simmer F, Tijsterman M, Parrish S, Koushika SP, Nonet ML, Fire A, Ahringer J, Plasterk RH. 2002. Loss of the putative RNA-directed RNA polymerase RRF-3 makes *C. elegans* hypersensitive to RNAi. *Curr Biol* **12:** 1317–1319.

Timmons L, Court DL, Fire A. 2001. Ingestion of bacterially expressed dsRNAs can produce specific and potent genetic interference in *Caenorhabditis elegans. Gene* **263:** 103–112.

PLANNING AND PREPARATION

The following table will help you to plan and integrate the different experimental methods.

Experiment	Day	Time		Activity
I. Transfer wild-type *C. elegans* to OP50-seeded NGM-lite plates	1	20 min	Prelab:	Set up student stations.
		20 min	Lab:	Chunk wild-type worms to OP50-seeded NGM-lite plates. 'Incubate 48 h.
II. Induce RNAi by feeding	3	20 min	Prelab:	Set up student stations.
		60 min	Lab:	Pick L4 worms to the appropriate plates to induce RNAi.
III. Observe and score phenotypes	4	20 min	Prelab:	Set up student stations.
		20 min	Lab:	Examine plates and flame dead worms.
	5	20 min	Prelab:	Set up student stations.
		60 min	Lab:	Observe and score phenotypes.
	6	20 min	Prelab:	Set up student stations.
		60 min	Lab:	Examine plates, score phenotypes, and flame parental worms.
	7	20 min	Prelab:	Set up student stations.
		60 min	Lab:	Observe and score phenotypes.
	8	20 min	Prelab:	Set up student stations.
		60 min	Lab:	Observe and score phenotypes.

OVERVIEW OF EXPERIMENTAL METHODS

I. TRANSFER WILD-TYPE *C. ELEGANS* TO OP50-SEEDED NGM-LITE PLATES

OBSERVE
worms

DIP
implement
in ethanol

IGNITE
ethanol

CUT
chunk with
worms

PICK UP
chunk

TRANSFER
chunk to
new plate

VERIFY
worm
transfer

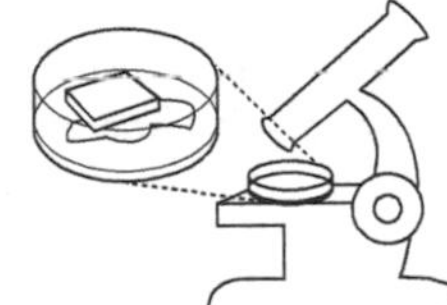

INCUBATE
48 h at room
temperature

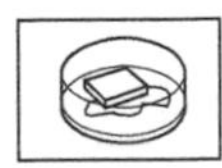

II. INDUCE RNAi BY FEEDING

FLAME
end of
pick

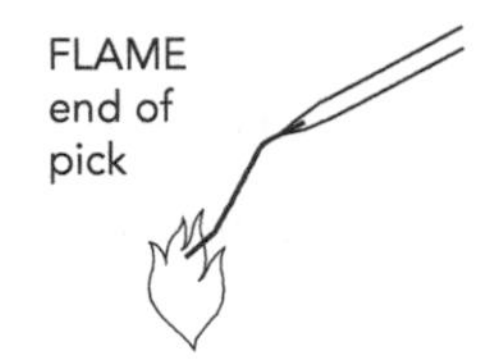

ATTACH
glob of
bacteria

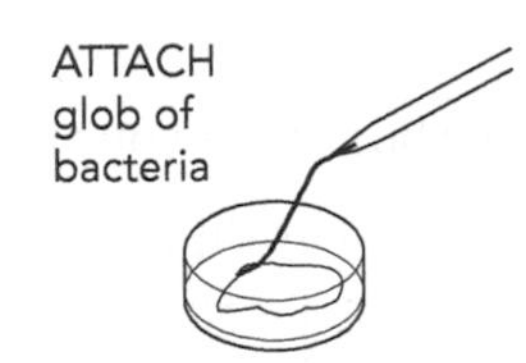

PICK
L4 worms

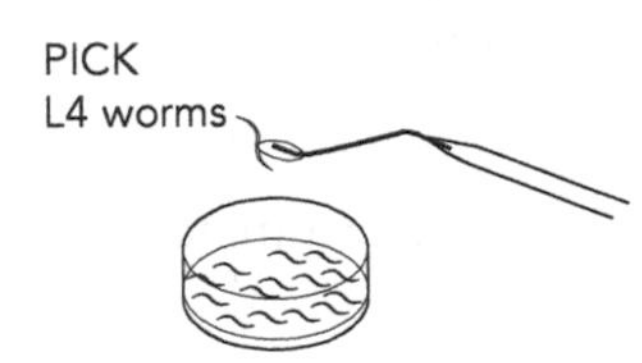

INCUBATE
48 h at room
temperature

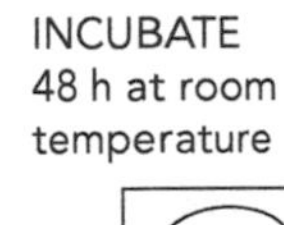
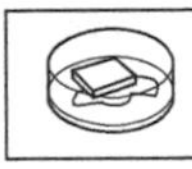

TRANSFER
worms to
new plate

VERIFY
worm transfer
and health

III. OBSERVE AND SCORE PHENOTYPES

OBSERVE
worms

EXPERIMENTAL METHODS

I. Transfer Wild-Type *C. elegans* to OP50-Seeded NGM-Lite Plates

REAGENTS, SUPPLIES, & EQUIPMENT

For each group
Binocular dissecting microscope
Bunsen burner
Ethanol (95%) <!> in a 50- or 100-mL beaker
Metal spatula or forceps
OP50-seeded NGM-lite plate from Part III of Stage A in Laboratory 4.1
Permanent marker (black)
Starter plate of wild-type worms

See Cautions Appendix for appropriate handling of materials marked with <!>.

1. Examine your OP50-seeded plate under the dissecting microscope for signs of bacterial or mold contamination—any growth of a different color or morphology (shape) from the OP50 lawn. Obtain a new plate if you detect any contamination.
2. Use a black marker to label the bottom of your plate with the date and "wild type."
3. Sterilize a metal spatula or forceps by dipping the end of the implement into the beaker of ethanol and then briefly passing it through a Bunsen burner flame to ignite the ethanol. Allow the ethanol to burn off away from the Bunsen flame; the implement will become too hot if left in the flame.
4. Use the sterilized spatula or forceps to cut a 1-cm (~3/8-in) square chunk of agar with worms from the wild-type starter plate.
5. Carefully remove the piece of agar with worms from the wild-type starter plate and place it upside down on the lawn of the fresh OP50-seeded NGM-lite plate. Verify worm transfer.
6. Incubate the plate upside down for 48 h at room temperature. Choose a place where the plate will not be disturbed. This will give the newly transferred worms time to grow.

Sterilization prevents cross-contamination with different C. elegans strains and non-OP50 bacteria.

CAUTION! Be extremely careful to avoid igniting the ethanol in the beaker. Do not panic if the ethanol is accidentally ignited. Cover the beaker with a glass Petri dish lid or other nonflammable cover to cut off oxygen and rapidly extinguish the fire.

To feed worm strains, it is easier to transfer a chunk of worm-filled agar from a well-grown plate to a new plate rather than to pick individual worms.

II. Induce RNAi by Feeding

REAGENTS, SUPPLIES, & EQUIPMENT

For each group
Binocular dissecting microscope
Bunsen burner
E. coli RNAi feeding strains (*dpy-11, bli-1, unc-22*) on NGM lite/amp + IPTG plates from Part III of Stage A in Laboratory 4.1
OP50-seeded NGM-lite plate from Part III of Stage A in Laboratory 4.1
Permanent marker (black)
Wild-type worms on NGM-lite plate from Part I
Worm pick

1. Use a black marker to label the bottom of your OP50-seeded plate with the date and "wild type."
2. Use a black marker to label the bottom of each plate seeded with the *E. coli* RNAi feeding strains (*dpy-11, bli-1,* and *unc-22*) with the date and "wild type."

Placing the agar piece upside down makes it easier for the worms to crawl into the new bacterial food source.

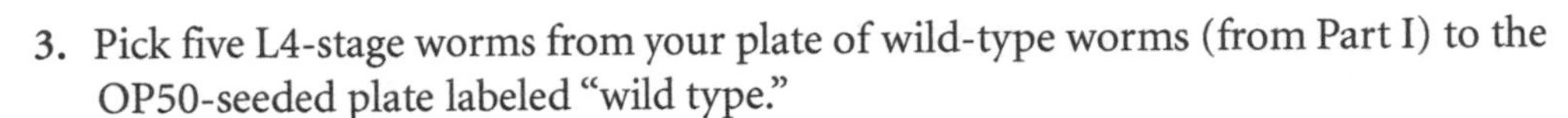

3. Pick five L4-stage worms from your plate of wild-type worms (from Part I) to the OP50-seeded plate labeled "wild type."
4. Examine the transferred worms under the dissecting microscope to confirm that they are the correct stage and were not injured or killed during the picking process.
5. Identify any eggs or young larvae that may have been accidentally transferred. Pick them off the plate and flame them in a Bunsen burner.
6. Repeat the procedure in Steps 3–5 to move five L4-stage wild-type worms to each of the plates seeded with the RNAi feeding strains (*dpy-11*, *bli-1*, and *unc-22*).
7. Incubate the plates upside down at room temperature. Choose a place where the plates will not be disturbed.

Refer to Laboratory 4.1 for Instructions on how to identify L4-stage larvae and pick worms.

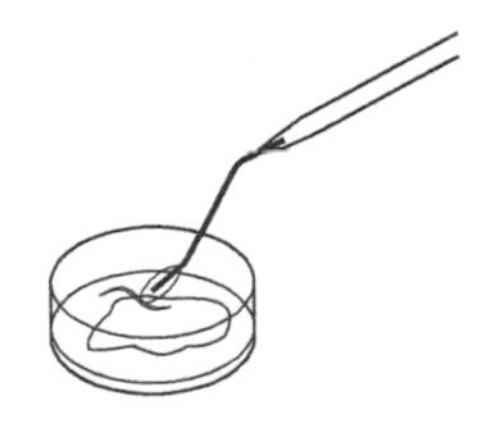

III. Observe and Score Phenotypes

REAGENTS, SUPPLIES, & EQUIPMENT

For each group
Binocular dissecting microscope
Bunsen burner
Wild-type worms on NGM-lite/amp + IPTG plates containing *E. coli* RNAi feeding strains (*dpy-11*, *bli-1*, and *unc-22*) from Part II
Wild-type worms on OP50-seeded NGM-lite plate from Part II
Worm pick

1. On the day after transferring, check that your worms are still healthy. Pick any dead worms off the plate and flame them in a Bunsen burner.
2. Starting on the second day after transferring, carefully examine each plate as follows:
   i. Have any eggs been laid? Have any eggs hatched? If so, are the worms at a larval or adult stage?
   ii. Compare the progeny of the RNAi-treated worms to the progeny of the untreated wild-type worms. Note any differences in morphology or behavior. At which developmental stage are the worms in which you observe these differences (L1, L2–L3, L4, or adult)?
   iii. Make a quantitative assessment of RNAi knockdown on each plate, looking only at the progeny of the original worms that you transferred. Once some of the first generation (F_1) progeny of the RNAi-treated worms reach adulthood, count the number of adult F_1 worms that appear to be normal (wild type) and the number with a different (RNAi-induced) phenotype. As you count each worm, remove it from the plate with a pick and flame it in a Bunsen burner. Do this before the second generation (F_2) reaches adulthood.
3. Repeat Step 2 on a daily basis for a total of 4 d, observing and scoring additional F_1 worms that reach adulthood. On the third day after transferring, pick any remaining adult worms off the plate and flame them in a Bunsen burner. This will avoid confusion when scoring the progeny.
4. Calculate the percentage of F_1 adult worms that were affected on each RNAi plate on each day. How effective was the RNAi treatment? Did the RNAi effect vary over time?
5. Given the phenotypes that you observed, what can you deduce about the function of each gene that was knocked down?

BIOINFORMATICS METHODS

I. Use WormBase to Find Basic Information About the *unc-22* Gene

1. Open the Internet site for WormBase (http://www.wormbase.org/).
2. Enter "*unc-22*" in the text box adjacent to "Find: Any Gene" and then click "Search."
   i. Look at the description of the gene. What does "unc" stand for?
   ii. What sort of protein does the gene encode? What is its function? Follow any blue links to find additional information.
3. Scroll down to the "Location" section to examine a graphical representation of the structure of the *unc-22* gene, as well as a neighboring glucose transporter gene, on *C. elegans* chromosome IV. Several *unc-22* transcripts are depicted, each running from right to left, or 5′ to 3′. Exons are colored boxes, introns are thin black lines, and the 3′-untranslated region is a black arrow. How can there be more than one transcript for this gene?
4. Scroll down to the "Function" section, which describes different alleles (mutations) of the *unc-22* gene. Each allele is given a different code (combination of letters and numbers). Follow the blue link for a complete description of the *unc-22*(*e66*) allele.
5. Return to the previous page ("Gene Summary for unc-22"). In the "Identification" section at the top of the page, find the "Gene model(s)" table. Click on the first entry in the "Nucleotides (coding/transcript)" column to retrieve the nucleotide sequence for the transcript "ZK617.1a.1".
   i. The first entry is the spliced coding region (transcript) as it would be translated by the ribosome. (For simplicity's sake, the transcript is presented as DNA code, rather than RNA.) Exons are shaded in alternating yellow and orange (exon 1 is yellow, exon 2 is orange, exon 3 is yellow, etc.).
   ii. The second entry is the unspliced gene. Exons are colored; introns and untranslated regions are white.
   iii. Copy the entire spliced sequence and save it as a text document for use in Step 1.iv of Part II.

II. Use BLAST to Identify the Human Homolog of the *unc-22* Gene

1. Perform a BLAST search as follows:
   i. Do an Internet search for "ncbi blast."
   ii. Click on the link for the result "BLAST: Basic Local Alignment Search Tool." This will take you to the Internet site of the National Center for Biotechnology Information (NCBI).
   iii. Click on the link "blastx" under the heading "Basic BLAST" to search a protein database using a translated nucleotide query. Why would we do a BLAST search with amino acids rather than with DNA sequence?
   iv. Enter the *unc-22* coding sequence from Step 5.iii of Part I into the search window. This is the query sequence.
   v. Omit any nonnucleotide characters from the window because they will not be recognized by the BLAST algorithm.

vi. Under the "Choose Search Set" section, type "*Homo sapiens*" in the "Organism" window to narrow the search to humans.

vii. Click on "BLAST." This sends your query sequence to a server at NCBI in Bethesda, Maryland. There, the BLAST algorithm will translate the nucleotide sequence and attempt to match the amino acid sequence to the millions of protein sequences stored in its database. While searching, a page showing the status of your search will be displayed until your results are available. This may take only a few seconds or more than 1 min if many other searches are queued at the server.

2. Analyze the results of the BLAST search, which are displayed in three ways as you scroll down the page:

   i. First, a graphical overview illustrates how significant matches (hits) align with the query sequence. Matches of differing lengths are indicated by color-coded bars. What do you notice about the lengths (and colors) of the matches (bars) as you look from the top to the bottom?

   ii. This is followed by a list of significant alignments (hits) with links to the corresponding accession numbers. (An accession number is a unique identifier given to a sequence when it is submitted to a database such as GenBank.) Note the scores in the "E value" column on the right. The Expectation or E value is the number of alignments with the query sequence that would be expected to occur by chance in the database. The lower the E value, the higher the probability that the hit is related to the query. For example, an E value of 1 means that a search with your sequence would be expected to turn up one match by chance. Longer query sequences generally yield lower E values. An alignment is considered significant if it has an E value of less than 0.1. The 0.0 values at the top of the list are obviously very significant, but consider the values further down the list: For example, an E value of 2E-157 denotes 2×10^{-157}, or a 2 preceded by 156 decimal places! Hits with similar E values are usually variants of the same sequence. What human protein is most related to *C. elegans* UNC-22? What is its E value and what does that E value mean?

   iii. Third is a detailed view of the translated *unc-22* sequence (query) aligned to the amino acid sequence of each search hit (subject, abbreviated "Sbjct"). Amino acids that are identical to both sequences are listed in the center row; biochemically similar amino acids are indicated by "+" signs.

3. Once again, identify the first hit in the list of significant alignments. Find the boxed "G" next to the entry and click on the "G." This link will open an Entrez Gene record for this human *unc 22* relative. From the "Display" drop-down menu at the top left, select "Full Report." Scroll down and note that the report includes a graphical representation of transcripts for the gene, its chromosomal location, and a list of references to papers in which the gene is discussed. Follow any of the links to obtain more detailed information about the gene.

   i. Does this gene have a similar function to that of *unc-22*? If so, in what ways are they similar?

   ii. Is this gene important in human health? If so, how?

LABORATORY 4.3

Examining the RNAi Mechanism

▼ OBJECTIVES

This laboratory demonstrates several important concepts of modern biology. During the course of this laboratory, you will

- Learn about the mechanism of RNA interference (RNAi) and its applications.
- Explore the relationships among genotype, phenotype, and RNAi-induced phenotype.
- Move between in vitro experimentation and in silico computation.

In addition, this laboratory utilizes several experimental and bioinformatics methods in modern biological research. You will

- Use sterile technique to grow *Caenorhabditis elegans.*
- Use dissecting microscopes to observe and analyze cultures of *C. elegans.*
- Disrupt the function of a gene using RNAi.
- Extract and purify DNA from *C. elegans.*
- Amplify a specific region of the genome by polymerase chain reaction (PCR) and analyze PCR products by gel electrophoresis.
- Explore gene data online using GenBank, Entrez Gene, and WormBase.
- Use the Basic Local Alignment Search Tool (BLAST), MegaBLAST, and Discontiguous MegaBLAST to identify sequences in databases.

INTRODUCTION

In classical genetics, a mutant phenotype is attributed to one or more mutations in the DNA sequence of a gene. RNAi, a more recently discovered mechanism, silences gene function when double-stranded RNA (dsRNA) corresponding to the gene sequence is present in a cell. This laboratory compares DNA mutation and dsRNA silencing to explore how RNAi disrupts the function of a gene.

Consider the steps that are necessary to change the DNA "blueprint" into a functional protein. First, the DNA sequence of a gene is transcribed into messenger RNA (mRNA); this, in turn, is translated into proteins. How could RNAi affect the function of the gene? One possibility is that the dsRNA is somehow used by machinery in the cells as a way to alter the sequence of the DNA, thereby creating a mutation in the gene. Another possibility is that the dsRNA somehow disrupts one of the subsequent steps in protein production, perhaps by acting at the level of the mRNA. This laboratory is designed to test whether RNAi acts at the level of the DNA sequence, by altering the size of the gene. DNA from a wild-type phenotypically normal worm is com-

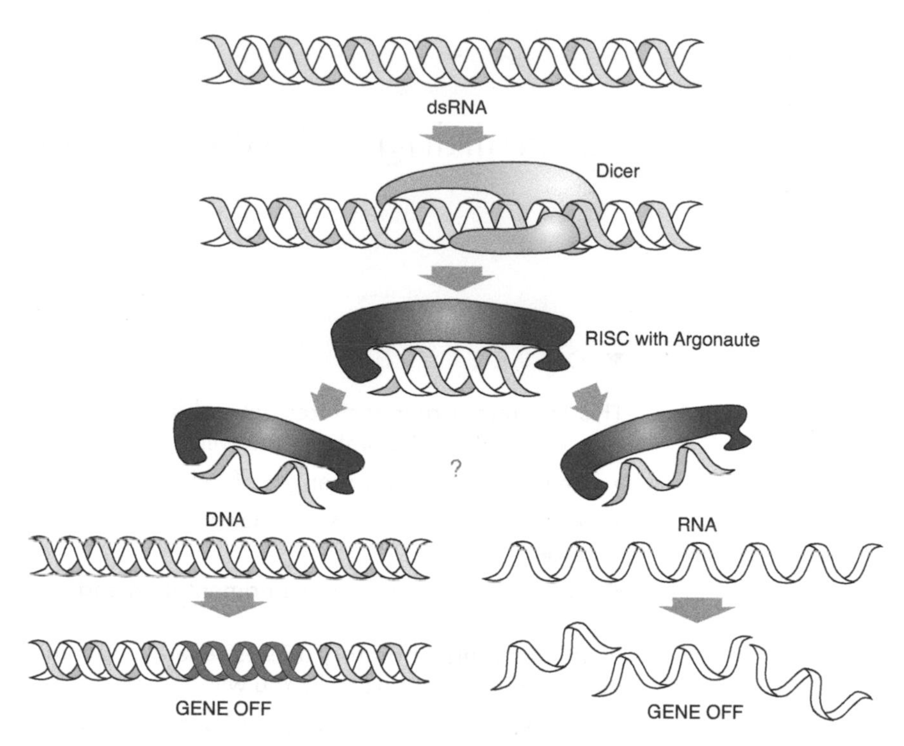

The RNAi machinery recognizes long dsRNA. First, Dicer protein binds to dsRNA and cleaves it. The resulting short interfering RNA (siRNA) is bound by Argonaute protein, and one strand of the siRNA is released. Argonaute turns off the gene with sequence complementary to the siRNA. This laboratory tests whether RNAi alters gene expression by affecting the chromosomal DNA (*left*) or the RNA transcript (*right*).

pared to DNA from two *C. elegans* strains with an identical "Dumpy" trait: one induced by RNAi and one caused by a chromosomal deletion in the *dpy-13* gene.

Wild-type and short (Dumpy) *dpy-13* mutant worms are chunked and fed *Escherichia coli* OP50 on nematode growth medium (NGM)-lite plates. A separate group of wild-type worms is fed a strain of *E. coli* that expresses dsRNA corresponding to the *dpy-13* gene on an NGM-lite plate containing ampicillin and isopropyl-β-D-thiogalactopyranoside (IPTG). On this plate, RNAi will "silence" the *dpy-13* gene when ingested by the wild-type worms, causing the same Dumpy phenotype as seen in the *dpy-13* mutant worms. These phenotypes will be compared to that of the wild-type worms.

After observation, genomic DNA from each type of worm is isolated by "freeze-cracking" and proteinase K extraction. DNA from the *dpy-13* locus is amplified by PCR, and the size of each PCR product is determined by agarose gel electrophoresis. The *dpy-13* mutation used in this experiment affects the size of this PCR product. By analyzing the size of the PCR product from the RNAi-treated worms, students will determine whether RNAi also affects the size of the *dpy-13* gene.

In accompanying bioinformatics exercises, BLAST is used to identify the size and sequence of the wild-type *dpy-13* PCR product (amplicon). The function and structure of the *dpy-13* gene are examined in WormBase, an online *C. elegans* database, as well as in GenBank and Entrez Gene. Finally, relatives of the *dpy-13* gene in *C. elegans* and humans are identified using BLAST-related tools (MegaBLAST and Discontiguous MegaBLAST).

FURTHER READING

Fire A, Xu S, Montgomery MK, Kostas SA, Driver SE, Mello CC. 1998. Potent and specific genetic interference by double-stranded RNA in *Caenorhabditis elegans*. *Nature* **391:** 806–811.

von Mende N, Bird DM, Albert PS, Riddle DL. 1988. *dpy-13:* A nematode collagen gene that affects body shape. *Cell* **55:** 567–576.

PLANNING AND PREPARATION

The following table will help you to plan and integrate the different experimental methods.

Experiment	Day	Time		Activity
	5 d before Part I	60 min	Prelab:	Prepare NGM-lite and NGM-lite/amp + IPTG plates.
	4 d before Part I	15 min	Prelab:	Prepare overnight cultures of *dpy-13* RNAi *E. coli* in LB/amp and OP50 *E. coli* in LB.
	3 d before Part I	30 min	Prelab:	Seed NGM-lite/amp + IPTG plates with *dpy-13* RNAi *E. coli*. Seed NGM-lite plates with OP50 *E. coli*.
I. Transfer wild-type and *dpy-13 C. elegans* to OP50-seeded NGM-lite plates	1	20 min	Prelab:	Set up student stations.
		30 min	Lab:	Chunk wild-type and *dpy-13* worms to OP50-seeded NGM-lite plates. Incubate for 48 hr between Parts I and II.
II. Induce RNAi by feeding	3	20 min	Prelab:	Set up student stations.
		60 min	Lab:	Pick L4 worms to the appropriate plates to induce RNAi.
	4	60 min	Postlab:	Examine plates and flame dead worms.
	5	60 min	Postlab:	Observe and store phenotypes.
	6	60 min	Postlab:	Examine plates, score phenotypes, and flame parental worms.
III. Isolate DNA from *C. elegans*	7	60 min	Prelab:	Prepare lysis buffer. Set up student stations.
		20 min	Lab:	Collect worms for lysis.
		120 min	Postlab:	Incubate to lyse worms.
IV. Amplify DNA by PCR	8	30 min	Prelab:	Aliquot primer/loading dye mix. Set up student stations.
		15 min	Lab:	Set up PCRs.
		70+ min	Postlab:	Amplify DNA in thermal cycler.
V. Analyze PCR products by gel electrophoresis	8	30 min	Prelab:	Prepare agarose gel solution. Dilute TBE electrophoresis buffer. Set up student stations.
		30 min	Lab:	Cast gels.
	9	45+ min	Postlab:	Load DNA samples into gel. Electrophorese samples. Photograph gels.

OVERVIEW OF EXPERIMENTAL METHODS

I. TRANSFER WILD-TYPE AND *dpy-13 C. ELEGANS* TO OP50-SEEDED NGM-LITE PLATES

OBSERVE worms

DIP implement in ethanol

IGNITE ethanol

CUT chunk with worms

PICK UP chunk

TRANSFER chunk to new plate

VERIFY worm transfer

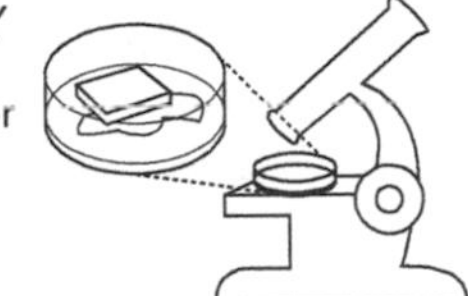

INCUBATE 48 h at room temperature

II. INDUCE RNAi BY FEEDING

FLAME end of pick

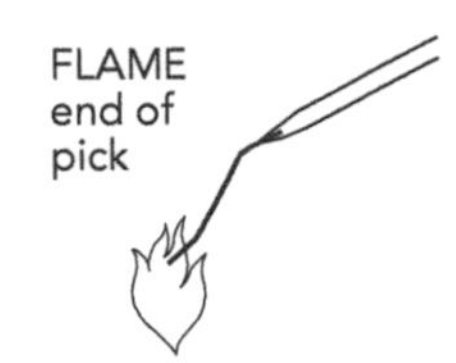

ATTACH glob of bacteria

PICK L4 worms

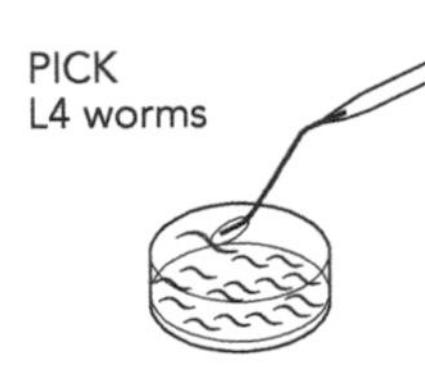

TRANSFER worms to new plate

VERIFY worm transfer and health

III. ISOLATE DNA FROM *C. ELEGANS*

FLAME end of pick

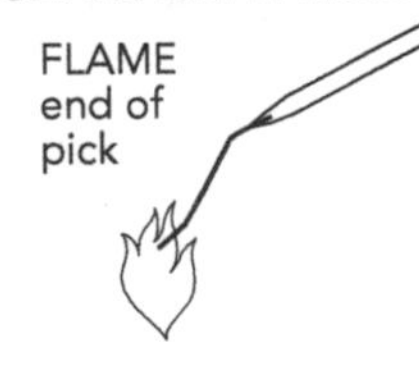

ATTACH glob of bacteria

PICK worms

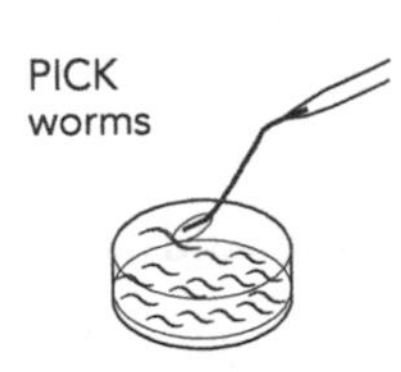

TRANSFER worms to tube

VERIFY worm transfer and health

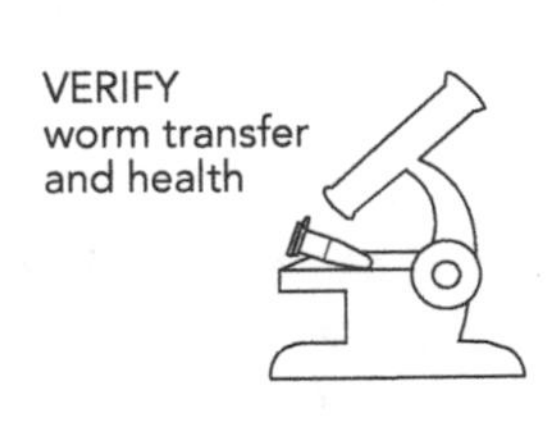

CENTRIFUGE to pellet worms

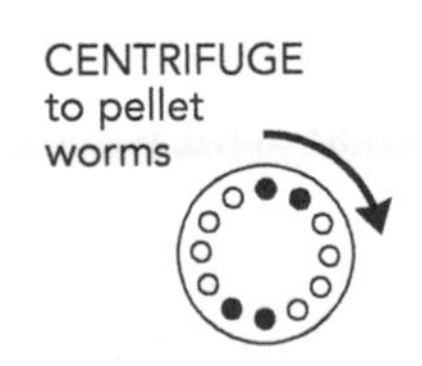

FREEZE 10 min

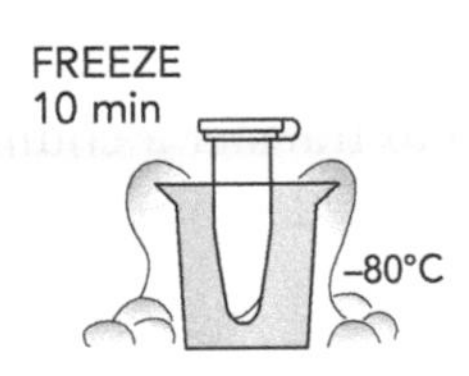

INCUBATE 90 min

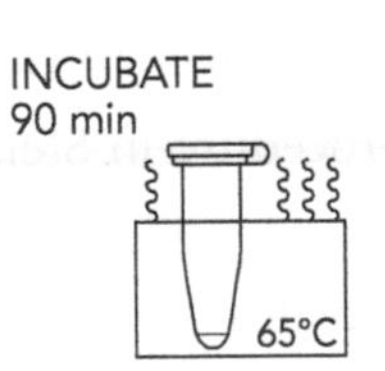

BOIL 15 min

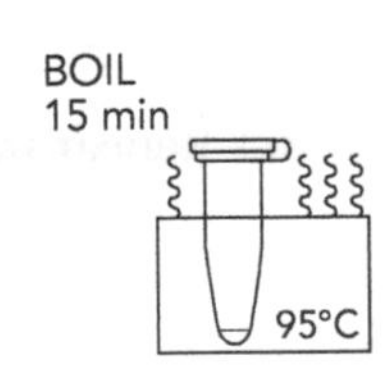

STORE

IV. AMPLIFY DNA BY PCR

ADD primer/loading dye

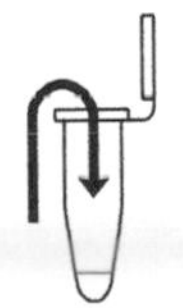

ADD worm DNA

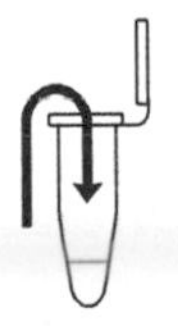

AMPLIFY in thermal cycler

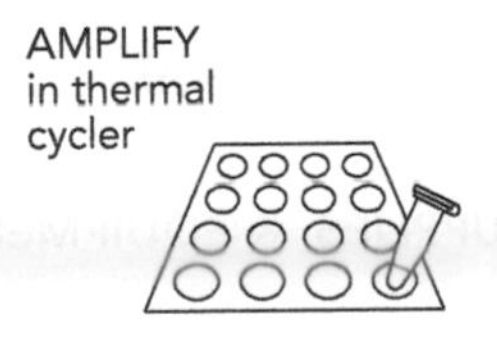

V. ANALYZE PCR PRODUCTS BY GEL ELECTROPHORESIS

POUR gel

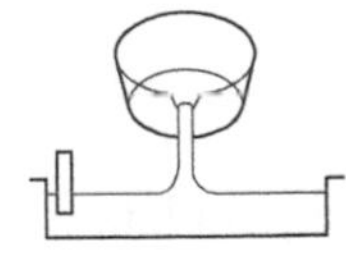

SET 20 min

LOAD gel

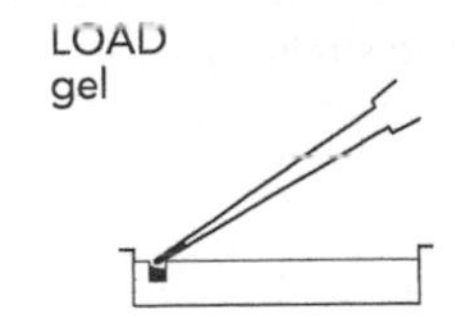

ELECTROPHORESE 130 V

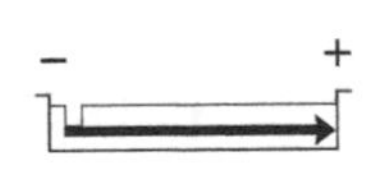

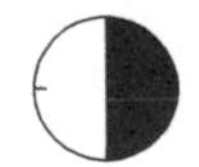

EXPERIMENTAL METHODS

I. Transfer Wild-Type and *dpy-13 C. elegans* to OP50-Seeded NGM-Lite Plates

The dpy-13 strain is a control.

Sterilization prevents cross-contamination with different C. elegans strains and non-OP50 bacteria.

CAUTION! Be extremely careful to avoid igniting the ethanol in the beaker. Do not panic if the ethanol is accidentally ignited. Cover the beaker with a glass Petri dish lid or other nonflammable cover to cut off oxygen and rapidly extinguish the fire.

To feed worm strains, it is easier to transfer a chunk of worm-filled agar from a well-grown plate to a new plate rather than to pick individual worms.

Placing the agar piece upside down makes it easier for the worms to crawl into the new bacterial food source.

REAGENTS, SUPPLIES, & EQUIPMENT

For each student

Binocular dissecting microscope
Bunsen burner
Ethanol (95%) <!> in a 50- or 100-mL beaker
Metal spatula or forceps
2 OP50-seeded NGM-lite plates
Permanent marker (black)
Starter plate of *dpy-13* worms
Starter plate of wild-type worms

See Cautions Appendix for appropriate handling of materials marked with <!>.

1. Examine your OP50-seeded plate under the dissecting microscope for signs of bacterial or mold contamination—any growth of a different color or morphology (shape) from the OP50 lawn. Obtain a new plate if you detect any contamination.
2. Use a black marker to label the bottom of each OP50-seeded plate with the date. Label one plate "wild type" and the other "*dpy-13*."
3. Sterilize a metal spatula or forceps by dipping the end of the implement into the beaker of ethanol and then briefly passing it through a Bunsen burner flame to ignite the ethanol. Allow alcohol to burn off away from the Bunsen flame; the implement will become too hot if left in the flame.
4. Use the sterilized spatula or forceps to cut a 1-cm (~3/8-in) square chunk of agar with worms from the wild-type starter plate.
5. Carefully remove the piece of agar with worms from the wild-type starter plate and place it upside down on the lawn of the fresh OP50-seeded plate labeled "wild type." Verify worm transfer.
6. Repeat the procedure in Steps 3–5 to transfer a chunk of agar with worms from the *dpy-13* starter plate to the OP50-seeded plate labeled "*dpy-13*."
7. Incubate the plates upside down for 48 h at room temperature. Choose a place where the plate will not be disturbed. This will give the newly transferred worms time to grow.

II. Induce RNAi by Feeding

REAGENTS, SUPPLIES, & EQUIPMENT

For each student

Binocular dissecting microscope
Bunsen burner
dpy-13 worms on NGM-lite plate from Part I
E. coli RNAi feeding strain (*dpy-13*) on NGM-lite/amp + IPTG plate
2 OP50-seeded NGM-lite plates
Permanent marker (black)
Wild-type worms on NGM-lite plate from Part I
Worm pick

1. Use a black marker to label the bottom of each OP50-seeded plate with the date. Label one plate "wild type" and the other "*dpy-13*."

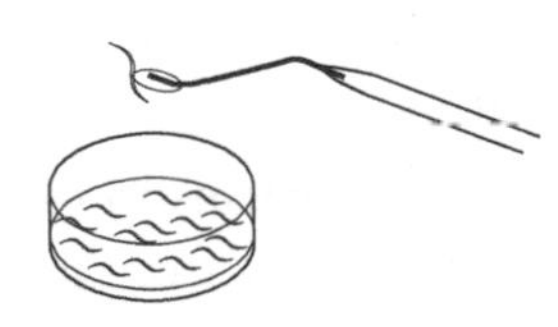

Refer to Laboratory 4.1 for instructions on how to identify L4-stage larvae and pick worms.

2. Use a black marker to label the bottom of the plate seeded with the *dpy-13* RNAi feeding strain with the date and "*dpy-13*."
3. Pick five L4-stage worms from your plate of wild-type worms to the OP50-seeded plate labeled "wild type."
4. Examine the transferred worms under the dissecting microscope to confirm that they are at the correct stage and were not injured or killed during the picking process.
5. Identify any eggs or young larvae that may have been accidentally transferred. Pick them off the plate and flame them in a Bunsen burner.
6. Repeat the procedure in Steps 3–5 to move five L4-stage wild-type worms to the plate seeded with the *dpy-13* RNAi feeding strain.
7. Repeat the procedure in Steps 3–5 to move five L4-stage *dpy-13* worms to the OP50-seeded plate labeled "*dpy-13*."
8. Incubate the plates upside down at room temperature. Choose a place where the plates will not be disturbed.
9. On the day after transferring, check that your worms are still healthy. Note any dead worms, pick them off the plate, and flame them in a Bunsen burner.
10. Starting on the second day after transferring, carefully examine each plate, noting any differences in morphology and behavior. Record your observations in Step 1 of Part III in Results and Discussion. Repeat your observations on a daily basis until you are ready to isolate DNA in Part III. On the third day after transferring, pick the remaining adult worms off the plate and flame them in a Bunsen burner. This will avoid confusion when scoring the progeny.

III. Isolate DNA from *C. elegans*

Assuming that the worms are healthy and exhibit a strong RNAi phenotype, allow 4 d between inducing RNAi in Part II and isolating DNA as described here.

One of the main ingredients in lysis buffer, an enzyme called proteinase K, is a nonspecific protease, an enzyme that cleaves the peptide bonds of proteins. During the incubating step, proteinase K digests protein in the cuticle of the worm and helps to liberate individual cells by destroying the protein fibers of the extracellular matrix that bind cells together. Proteinase K also inactivates cellular proteins, including DNases that interfere with PCR amplification. During the boiling step, the near-boiling temperature lyses the cell and nuclear membranes, releasing DNA and other cell contents. It also inactivates the proteinase K.

REAGENTS, SUPPLIES, & EQUIPMENT

For each group

- Binocular dissecting microscope
- Bunsen burner
- Container with cracked or crushed ice
- dpy-13 and wild-type worms on NGM-lite plates from Part II
- dpy-13 RNAi-treated wild-type worms on an NGM-lite/amp + IPTG plate from Part II
- Lysis buffer (35 μL)*
- Microcentrifuge tube rack
- Micropipettes and tips (1–20 μL)
- 3 PCR tubes (0.2 or 0.5 mL)
- Permanent marker
- Worm pick

To share

- Container with liquid nitrogen <!> or dry ice <!> (or access to a –80°C freezer)
- Microcentrifuge
- Microcentrifuge adaptors for 0.2 or 0.5 mL PCR tubes
- Thermal cycler or water baths at 65°C and 95°C

*Store on ice.
See Cautions Appendix for appropriate handling of materials marked with <!>.

1. Use a permanent marker to label three PCR tubes with your group number. Label one tube "W" (for "wild type"), one "R" (for "RNAi"), and one "D" (for "deletion mutant").

Centrifugation will pellet the worms. Freeze-thawing cracks the tough outer cuticle of the worm.

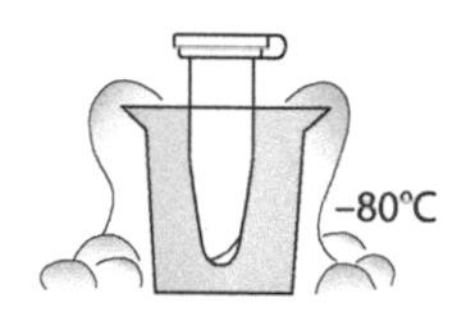

CAUTION! Liquid nitrogen, dry ice, and –80°C freezers can all cause burns and serious damage. Wear appropriate protective eyewear, clothing, and gloves.

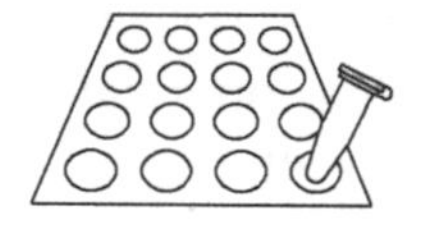

2. Use a micropipette with a fresh tip to add 10 µL of lysis buffer to each tube.
3. Observe the wild-type, *dpy-13*, and *dpy-13* RNAi-treated wild-type worms under a dissecting microscope. Describe the appearance of each of the three types of worms, noting differences in the length and width of the worms.
4. Use a sterilized worm pick to transfer four or five adult wild-type worms to the "W" tube. Swish the tip directly in the lysis buffer to dislodge the worms from the pick.
5. Observe the PCR tube under a dissecting microscope. Verify that at least three worms are alive and writhing in the buffer.
6. Repeat the procedure in Steps 4 and 5 to transfer four or five *dpy-13* RNAi-treated wild-type worms to the "R" tube and verify that at least three live worms are in the lysis buffer.
7. Repeat the procedure in Steps 4 and 5 to transfer four or five *dpy-13* worms to the "D" tube and verify that at least three live worms are in the lysis buffer.
8. To prepare your PCR tubes for centrifugation, "nest" each within adapter tubes as follows:
   i. If your sample is in a 0.5-mL PCR tube, "nest" it within a capless 1.5-mL tube.
   ii. If your sample is in a 0.2-mL PCR tube, "nest" it within a capless 0.5-mL tube and then place both tubes into a capless 1.5-mL tube.
9. Place your tubes, along with those from other groups, in a balanced configuration in a microcentrifuge. Centrifuge the tubes for 5–10 sec at full speed.
10. Freeze your tubes for 10 min in liquid nitrogen, on dry ice, or in a –80°C freezer. (If necessary, freeze overnight in a –20°C freezer.)
11. Place your PCR tubes, along with those from other groups, in a thermal cycler that has been programmed for one cycle of the following profile:

Incubating step:	90 min	65°C
Boiling step:	15 min	95°C

    The profile may be linked to a 4°C hold program.

Worm DNA can be stored for several weeks at 4°C or indefinitely at –20°C.

12. Store your samples on ice or at –20°C until you are ready to proceed to Part IV.

IV. Amplify DNA by PCR

REAGENTS, SUPPLIES, & EQUIPMENT

For each group

C. elegans DNA* from Part III
Container with cracked or crushed ice
dpy-13 primer/loading dye mix (75 µL)*
Microcentrifuge tube rack
Micropipettes and tips (1–100 µL)
Permanent marker
3 Ready-To-Go PCR Beads in 0.2- or 0.5-mL PCR tubes

To Share

Thermal cycler

*Store on ice.

1. Obtain three PCR tubes containing Ready-To-Go PCR Beads. Label each tube with your group number. Label one tube "W" (for "wild type"), one "R" (for "RNAi"), and one "D" (for "deletion mutant").

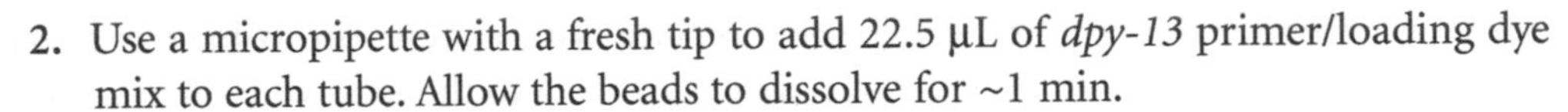

The primer/loading dye mix will turn purple as the PCR bead dissolves.

2. Use a micropipette with a fresh tip to add 22.5 µL of *dpy-13* primer/loading dye mix to each tube. Allow the beads to dissolve for ~1 min.
3. Use a micropipette with a fresh tip to add 2.5 µL of worm DNA (from Part III) directly into the primer/loading dye mix of the appropriate tube. Ensure that no DNA remains in the tip after pipetting.

If the reagents become splattered on the wall of the tube, pool them by pulsing the sample in a microcentrifuge or by sharply tapping the tube bottom on the lab bench.

4. Store your samples on ice until your class is ready to begin thermal cycling.
5. Place your PCR tubes, along with those of other students, in a thermal cycler that has been programmed for 30 cycles of the following profile:

Denaturing step:	30 sec	94°C
Annealing step:	30 sec	55°C
Extension step:	60 sec	72°C

The profile may be linked to a 4°C hold program after the 30 cycles are completed.

6. After thermal cycling, store the amplified DNA on ice or at –20°C until you are ready to proceed to Part V.

V. Analyze PCR Products by Gel Electrophoresis

REAGENTS, SUPPLIES, & EQUIPMENT

For each group

1% Agarose in 1x TBE (hold at 60°C) (50 mL per gel)
Container with cracked or crushed ice
Gel-casting tray and comb
Gel electrophoresis chamber and power supply
Latex gloves
Masking tape
Microcentrifuge tube rack
Micropipette and tips (1–100 µL)
pBR322/BstNI marker (20 µL per gel)*
PCR product* from Part IV
SYBR Green <!> DNA stain (10 µL)
1x TBE buffer (300 mL per gel)

To share

Digital camera or photodocumentary system
UV transilluminator <!> and eye protection
Water bath for agarose solution (60°C)

*Store on ice.
See Cautions Appendix for appropriate handling of materials marked with <!>.

1. Seal the ends of the gel-casting tray with masking tape and insert a well-forming comb.

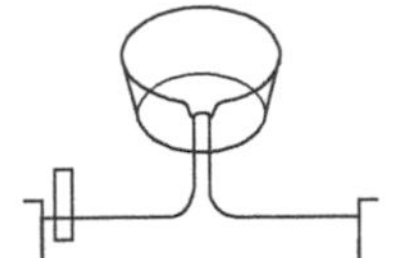

Avoid pouring an overly thick gel, which will be more difficult to visualize.

2. Pour the 1% agarose solution into the tray to a depth that covers about one-third the height of the open teeth of the comb.

The gel will become cloudy as it solidifies.

3. Allow the gel to completely solidify; this takes ~20 min.
4. Remove the masking tape, place the gel into the electrophoresis chamber, and add enough 1x TBE buffer to cover the surface of the gel.

Do not add more buffer than necessary. Too much buffer above the gel channels electrical current over the gel, increasing the running time.

5. Carefully remove the comb and add additional 1x TBE buffer to fill in the wells and just cover the gel, creating a smooth buffer surface.
6. Orient the gel according to the diagram in Step 8, so that the wells are along the top of the gel.
7. Use a micropipette with a fresh tip to load 2 µL of SYBR Green DNA stain to each PCR product (from Part IV). In addition, add 2 µL of SYBR Green DNA stain to 20 µL of pBR322/BstNI marker.

A 100-bp ladder may also be used as a marker.

8. Use a micropipette with a fresh tip to load each sample from Step 7 into the assigned well, according to the following diagram:

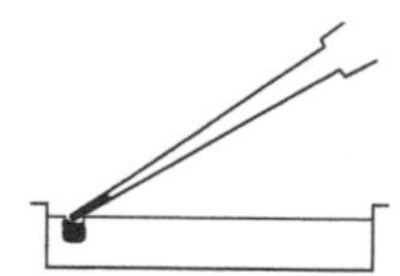

Expel any air from the tip before loading. Be careful not to push the tip of the pipette through the bottom of the sample well.

MARKER	WILD TYPE	RNAi TREATED	DELETION
pBR322/ BstNI	wild type	wild type (*dpy-13* RNAi)	*dpy-13* mutant

9. Run the gel for ~30 min at 130 V. Adequate separation will have occurred when the cresol red dye front has moved at least 50 mm from the wells.
10. View the gel using UV transillumination. Photograph the gel using a digital camera or photodocumentary system.

Transillumination, where the light source is below the gel, increases brightness and contrast.

BIOINFORMATICS METHODS

For a better understanding of the experiment, perform the following bioinformatics exercises before you analyze your results.

I. Use BLAST to Find DNA Sequences in Databases (Electronic PCR)

1. Perform a BLAST search as follows:
   i. Do an Internet search for "ncbi blast."
   ii. Click on the link for the result "BLAST: Basic Local Alignment Search Tool." This will take you to the Internet site of the National Center for Biotechnology Information (NCBI).
   iii. Click on the link "nucleotide blast" (blastn) under the heading "Basic BLAST."
   iv. Enter both primer sequences into the search window. These are the query sequences.

   > The following primer set was used in the experiment:
   >
   > 5′-AGTCGTCTTCTCCGTTATCG-3′ (forward primer)
   > 5′-GAGCAACGCATAAGGCAAAG-3′ (reverse primer)

   v. Omit any nonnucleotide characters from the window because they will not be recognized by the BLAST algorithm.
   vi. Under "Choose Search Set," select the "Nucleotide collection (nr/nt)" database from the drop-down menu.
   vii. Under "Program Selection," optimize for "Somewhat similar sequences (blastn)."
   viii. Click on "BLAST." This sends your query sequences to a server at NCBI in Bethesda, Maryland. There, the BLAST algorithm will attempt to match the primer sequences to the millions of DNA sequences stored in its database. While searching, a page showing the status of your search will be displayed

until your results are available. This may take only a few seconds or more than 1 min if many other searches are queued at the server.

2. Analyze the results of the BLAST search, which are displayed in three ways as you scroll down the page:

   i. First, a graphical overview illustrates how significant matches (hits) align with the query sequence. Matches of differing lengths are indicated by color-coded bars. What do you notice about the lengths (and colors) of the matches (bars) as you look from the top to the bottom?

   ii. This is followed by a list of significant alignments (hits) with links to the corresponding accession numbers. (An accession number is a unique identifier given to a sequence when it is submitted to a database such as GenBank.) Note the scores in the "E-value" column on the right. The Expectation or E value is the number of alignments with the query sequence that would be expected to occur by chance in the database. The lower the E value, the higher the probability that the hit is related to the query. For example, an E value of 1 means that a search with your sequence would be expected to turn up one match by chance. What is the E value of the most significant hit and what does it mean? Note the names of any significant alignments that have E values of less than 1. Do they make sense? What do they have in common?

   iii. Third is a detailed view of each primer (query) sequence aligned to the nucleotide sequence of the search hit (subject, abbreviated "Sbjct"). Note that the first match to the forward primer (nucleotides 1–20) and to the reverse primer (nucleotides 21–40) are within the same subject (accession number). Do the rest of the alignments produce good matches to both primers? Which of the hits would be amplified in vitro in a PCR using the two primers? Why?

3. Click on the accession number link to open the data sheet for the first hit (the first subject sequence).

   i. At the top of the report, note basic information about the sequence, including its length (in base pairs, or bp), database accession number, source, and references to papers in which the sequence is published. What is the source and size of the sequence in which your BLAST hit is located?

   ii. In the middle section of the report, the sequence features are annotated, with their beginning and ending nucleotide positions ("xx..xx"). These features may include genes, coding sequences (cds), regulatory regions, ribosomal RNA (rRNA), or transfer RNA (tRNA). You examine these features more closely in Part II.

   iii. Scroll to the bottom of the data sheet. This is the nucleotide sequence to which the term "Sbjct" refers.

   iv. Use your browser's "Back" button to return to the BLAST results page.

4. Predict the length of the product that the primer set would amplify in a PCR (in vitro) as follows:

   i. Scroll down to the alignments section (third section) of the BLAST results page. Locate the alignment for M23559.1, "*C. elegans* collagen (*dpy-13*) gene, complete cds."

   ii. To which positions do the primers match in the subject sequence?

iii. The lowest and highest nucleotide positions in the subject sequence indicate the borders of the amplified sequence. Subtract the lowest nucleotide position in the subject sequence from the highest nucleotide position in the subject sequence. What is the difference between the coordinates?

iv. Note that the actual length of the amplified fragment includes both ends, so add 1 nucleotide to the result that you obtained in Step 4.iii to obtain the exact length of the PCR product amplified by the two primers.

v. Repeat Steps i–iv for at least two more hits. Do you notice any discrepancies among the calculations? If so, why?

5. Obtain the nucleotide sequence of the amplicon that the primer set would amplify in a PCR (in vitro) as follows:

   i. Open the sequence data sheet for the hit that you identified in Step 4.i by clicking on the accession number link. Scroll to the bottom of the data sheet.

   ii. The bottom section of the data sheet lists the entire nucleotide sequence that contains the PCR product. Highlight all of the nucleotides between the coordinates that you identified in Step 4.iii, from the beginning of the forward primer to the end of the reverse primer.

   iii. Copy and paste the highlighted sequence into a text document. Then, delete all nonnucleotide characters and spaces. This is the amplicon, or amplified PCR product. Save this text document for use in Step 2 of Part III.

In Step 5.iii, you can retain the nucleotide coordinates and spacers to ease readability. Reduce the point size of the font so that each row of 60 nucleotides sits on one line. Then, set the font to Courier or another nonproportional font to align the blocks of sequence.

II. Use GenBank, Entrez Gene, and WormBase to Learn About the Function and Structure of the *dpy-13* Gene

1. Open the data sheet for GenBank accession number M23559.1. The middle of the report, under "FEATURES," contains annotations of sequence features such as promoters, genes, mRNA sequences, and coding sequences (cds), with their beginning and ending nucleotide positions ("xx..xx"). For each mRNA and cds feature, "join" shows the coordinates of coding exons that are spliced together for translation into a polypeptide chain. The use of angle brackets ("<" or ">") indicates that the feature extends beyond the nucleotide position indicated.

   i. How many exons and introns are in the *dpy-13* gene?

   ii. Identify the feature(s) that the two primers span, using the nucleotide coordinates that you identified in Step 4.iii of Part I. Where does the forward primer begin? Where does the reverse primer end? Which features are found between these nucleotide positions?

2. Open the Entrez Gene record for the gene. Find the links to the right of the datasheet header and click "Gene."

3. If the default view does not show the complete report, select "Full Report" from the "Display" pull-down menu. Briefly examine the Entrez Gene record for information about the *dpy-13* gene. Note that the report includes graphic representations of the gene's transcripts and chromosomal location.

4. Follow the "Primary source" link ("WormBase:WBGene00001074") to access more information about the gene in WormBase, the central database for information about *C. elegans* research.

5. Examine the WormBase entry to learn more about the function of the gene. What does "dpy" stand for? What sort of protein does the *dpy-13* gene encode and what is its function?
6. In the "Location" section, examine the visual representation of the gene model.
   i. The top line indicates the chromosomal location of the gene. The chromosome number is indicated with a roman numeral, and the location in kilobase pairs is marked at regular intervals.
   ii. In the gene model, at the bottom is a representation of the transcript for the gene. Exons are colored rectangles and introns are shown with thin black lines.
   iii. How many exons and introns are shown? Does this agree with your findings from Step 1.i?
7. Click on the link to the "Genetics" section, which lists different mutant alleles of the *dpy-13* gene and their corresponding phenotypes. Each allele is given a different code (combination of letters and numbers) that indicate where the allele was isolated (e.g., "e" stands for England and indicates that the allele was isolated in Cambridge). In this laboratory, you used the *e458* allele. Follow the blue link for a complete description of this allele.

III. Identify *dpy-13* Relatives in *C. elegans* and Humans

A. Use MegaBLAST to Identify dpy-13 *Relatives in* C. elegans

1. Open the nucleotide BLAST page by following Steps 1.i–iii in Part I.
2. Paste the *dpy-13* amplicon from Step 5.iii of Part I into the search window.
3. Under "Choose Search Set," select the "Nucleotide collection (nr/nt)" database from the drop-down menu.
4. Under "Program Selection," optimize for "Highly similar sequences (megablast)" and then click on "BLAST."
5. Compare the E values obtained by this search to those values that you obtained in Step 2.ii of Part I. What do you notice? Why is this so?
6. Why do the first several hits have E values of 0?
7. Carefully read the descriptions of the hits with the lowest E values. What do they have in common?

B. Use Discontiguous MegaBLAST to Identify dpy-13 *Relatives in Humans*

1. Open the nucleotide BLAST page by following Steps 1.i–iii in Part I.
2. Paste the *dpy-13* amplicon from Step 5.iii of Part I into the search window.
3. Under "Choose Search Set," select the "Human genomic plus transcript (Human G + T)" database from the drop-down menu.
4. Under "Program Selection," optimize for "More dissimilar sequences (discontiguous megablast)" and then click on "BLAST."
5. Compare the E values obtained by this search to those values that you obtained in Step 5 of Part IIIA. What do you notice? Why is this so?
6. Carefully read the descriptions of the hits with the lowest E values. What do they have in common?

The Discontiguous MegaBLAST algorithm is optimized to search for relatively short blocks of conserved sequence in related genes that have diverged among different organisms.

RESULTS AND DISCUSSION

I. Think About the Experimental Methods

1. Describe the purpose of each of the following steps or reagents used during DNA isolation (Part III of Experimental Methods):
   i. Freezing at –80°C.
   ii. Incubating with proteinase K.
   iii. Boiling at 95°C.
2. What is the purpose of performing each of the following PCRs?
   i. Wild-type worms.
   ii. *dpy-13* mutant worms.
   iii. *dpy-13* RNAi-treated wild-type worms.

II. Interpret Your Gel

1. Observe the photograph of the stained gel containing your PCR samples and those from other students. Orient the photograph with the sample wells at the top. Use the sample gel shown below to help to interpret the band(s) in each lane of the gel.

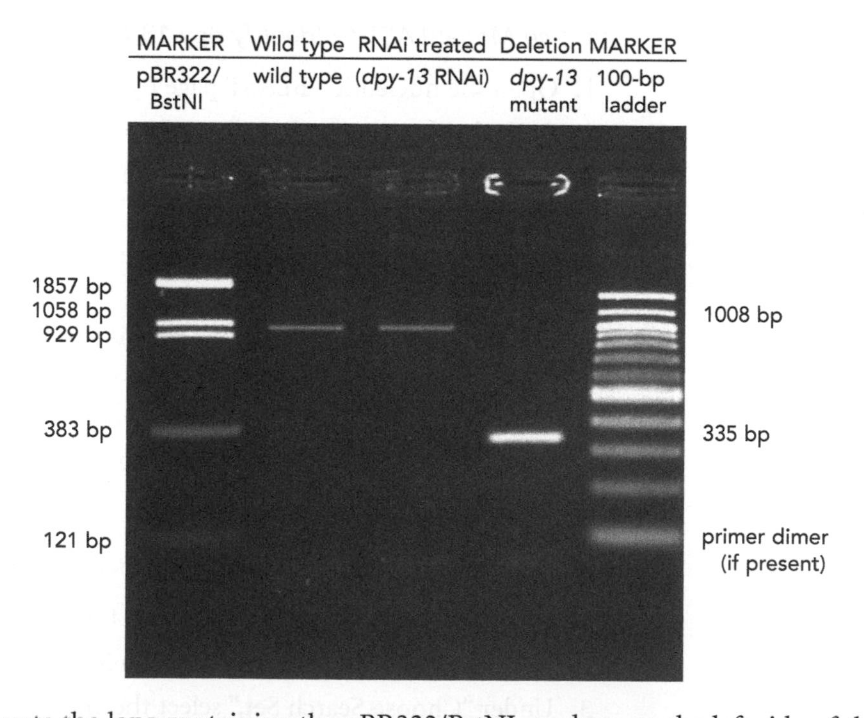

If a 100-bp ladder was used, as shown on the right-hand side of the sample gel, the bands increase in size in 100-bp increments starting with the fastest-migrating band of 100 bp.

2. Locate the lane containing the pBR322/BstNI markers on the left side of the gel. Working from the well, locate the bands corresponding to each restriction fragment: 1857, 1058, 929, 383, and 121 bp. The 1058- and 929-bp fragments will be very close together or may appear as a single large band. The 121-bp band may be very faint or not visible.
3. Scan across the row of results. Notice that virtually all lanes contain one prominent band.

i. The PCR product of wild type should align between the 1058- and 929-bp fragments of the pBR322/BstNI marker.

ii. The PCR product of the *dpy-13* mutant worms should run slightly ahead of the 383-bp marker fragment.

4. To "score" the genotype of the *dpy-13* RNAi-treated wild-type worms, compare the PCR product with the markers and with the bands for the wild-type and *dpy-13* mutant worms. What is the genotype of the *dpy-13* RNAi-treated wild-type worms?

Additional faint bands at other positions occur when the primers bind to chromosome loci other than the exact locus and give rise to "nonspecific" amplification products.

5. It is common to see a diffuse (fuzzy) band that runs ahead of the 121-bp marker. This is "primer dimer," an artifact of the PCR that results from the primers overlapping one another and amplifying themselves. How would you interpret a lane in which you observe primer dimer but no bands as described in Step 3?

III. Correlate *dpy-13* Genotype and Phenotype

1. Describe the appearance of each of the three types of worms from which DNA was isolated in Part III of Experimental Methods:

   i. Wild-type worms on the NGM-lite plate.

   ii. *dpy-13* RNAi-treated wild-type worms on the NGM-lite/amp + IPTG plate.

   iii. *dpy-13* mutant worms on the NGM-lite plate.

2. According to your results, how well does the *dpy-13* genotype predict the phenotype? If the genotypes were not predictive of the phenotypes, explain why this might be the case. What do these results suggest about RNAi?

3. What do you conclude from the experiment? What are the limitations of this experiment? What additional experiments could you do to strengthen your conclusions?

LABORATORY 4.4

Constructing an RNAi Feeding Vector

▼ OBJECTIVES

This laboratory demonstrates several important concepts of modern biology. During the course of this laboratory, you will

- Learn about the mechanism of RNA interference (RNAi) and its applications.
- Explore the relationships among genotype, phenotype, and RNAi-induced phenotype.
- Move between in vitro experimentation and in silico computation.

In addition, this laboratory utilizes several experimental and bioinformatics methods in modern biological research. You will

- Use the Online Mendelian Inheritance in Man (OMIM) database to identify human genes associated with disease.
- Explore gene data online using GenBank, Entrez Gene, and WormBase.
- Use the Basic Local Alignment Search Tool (BLAST) to identify sequences in databases.
- Use sterile technique to isolate and grow pure cultures of bacteria and *Caenorhabditis elegans*.
- Use dissecting microscopes to observe and analyze cultures of bacteria and *C. elegans*.
- Extract and purify DNA from *C. elegans*.
- Amplify a specific region of the genome by polymerase chain reaction (PCR) and analyze PCR products by gel electrophoresis.
- Create recombinant DNA.
- Introduce recombinant DNA into bacteria using transformation and use antibiotic selection to maintain a recombinant bacterial culture.
- Extract and purify DNA from recombinant bacteria.

INTRODUCTION

Scientists use model organisms to study biological processes that would be difficult or impossible to study in humans. *C. elegans* is a model organism that is used to study gene function using RNAi. To induce RNAi and silence a gene of interest, *C. elegans* are fed bacteria that express double-stranded RNA (dsRNA) corresponding to the sequence of the gene. Bacteria can be genetically engineered to produce dsRNA by inserting a gene-specific DNA sequence into a specialized RNAi feeding vector (plasmid). This is a very powerful method to deduce the function of a gene.

This laboratory includes all of the steps that go from identifying the sequence of a gene of interest in humans to assaying the function of a related gene in *C. elegans*. In the first step of this process, a human disease-related gene is identified by searching OMIM, a database of human genes and genetic disorders. The sequence of this gene is then extracted, and BLAST is used to identify sequences in *C. elegans* that are related to the human gene. Information about this gene is obtained from WormBase, a central database for information about *C. elegans*. Finally, primers that specifically amplify part of this gene by PCR are designed using bioinformatics methods; these primers include specific sites (*attB* sites) that will eventually allow cloning of the amplified DNA by recombination.

In the next stage of the experiment, DNA is isolated from wild-type worms to provide a template for PCR amplification, and the primers are used to amplify the gene-specific region by PCR. During PCR, *attB* sites are added to the ends of the amplicons. After the expected size of the PCR product is confirmed by agarose gel electrophoresis, the PCR product is inserted into the RNAi feeding vector between two T7 RNA polymerase promoters using Gateway BP Clonase. The Clonase enzyme contains integrase (Int) and integration host factor (IHF), proteins that catalyze the in vitro recombination between PCR products (containing *attB* sites) and a vector (containing *attP* sites). Int is from the λ virus; it allows the virus to integrate its genome into a host bacterial genome by site-specific recombination. Int recognizes the *attB* sites at the ends of the amplicon and the *attP* sites in the vector, and recombination occurs between the *attB* and *attP* sites. IHF is a protein from *Eschericha coli* that binds to DNA and bends it. IHF increases the efficiency of integration by helping to bring recombination sites close together. This recombination reaction produces a vector that contains the gene-specific PCR product. It is far more efficient than traditional cloning methods.

The products of the recombination reaction are then inserted into *E. coli* bacteria using a transformation procedure. Recombined DNA is added to the bacterial cells (called "competent" bacterial cells), and the cells are "heat-shocked," causing the recombined plasmid to enter the cells. The recombination reaction inserts the gene of interest into the plasmid and excises a fragment of DNA that includes the *ccdB* gene. *ccdB* encodes a protein that poisons *E. coli* DNA gyrase, which unwinds supercoils in DNA and allows *E. coli* to replicate its genome. If no recombination occurs, the protein product encoded by *ccdB* inhibits DNA replication in bacteria that take up the plasmid during transformation.

The vector also contains a gene for antibiotic resistance, which selects for bacteria that have taken up the plasmid. After transformation, samples of transformed cells are spread onto two types of plates: LB plates with an antibiotic (kanamycin) and LB plates without an antibiotic. Individual cells will grow and divide to form colonies on the plates. Transformed cells that form colonies on the LB plates containing kanamycin must contain the plasmid with the kanamycin resistance gene. Furthermore, only cells containing recombined plasmid (i.e., plasmid with the gene of interest) will grow because the product of the *ccdB* gene will inhibit DNA replication in cells containing nonrecombined plasmid. This combination of selection ensures that any bacteria that grow after transformation have incorporated a plasmid that has undergone recombination.

Although the recombination and transformation reactions should be efficient and specific, it is possible that some transformants will not possess the intended plasmid.

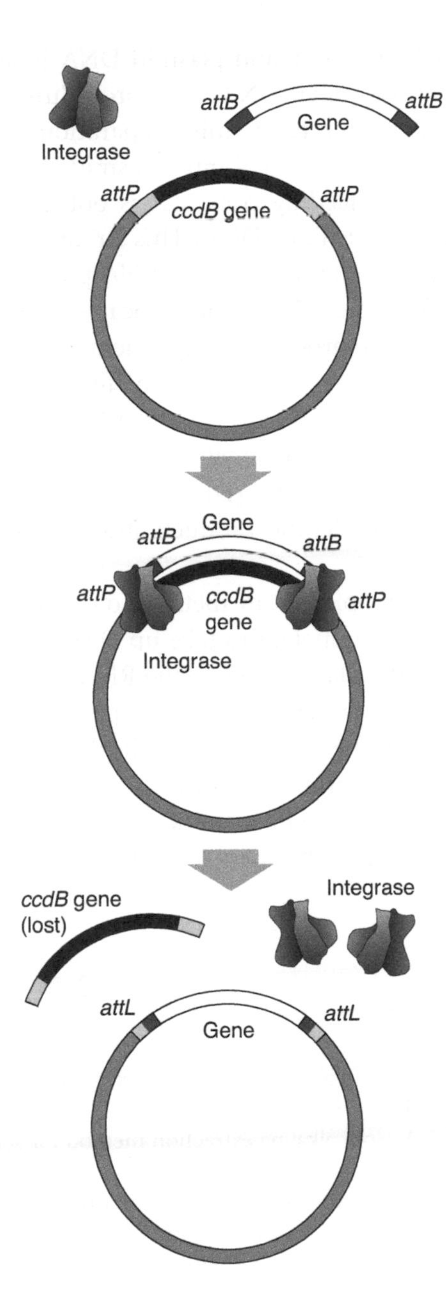

Cloning by recombination. PCR products with *attB* sites can be introduced into a vector with *attP* sites by recombination using integrase. This replaces the *ccdB* gene in the vector with the PCR product.

To ensure that the desired product has been obtained, PCR is used to verify the size of the insert in the colonies. T7 primers, which bind to the regions flanking the insertion site in the plasmid, are used to amplify the insert. The size of the amplified DNA fragment (amplicon) is determined by comparison with a size standard following agarose gel electrophoresis. Finding a band corresponding to the predicted size of the amplicon strongly suggests that the desired plasmid has been obtained.

Overnight cultures are inoculated with cells from the same colonies that are tested by PCR, so that DNA can be isolated from the colonies that possess the correct plasmid. This DNA, which is the newly constructed gene-specific RNAi feeding vector DNA, is isolated from overnight cultures by plasmid minipreparation. To isolate DNA, cells are grown to stationary phase in suspension culture, the cells from 1 mL of culture are harvested and lysed, and plasmid DNA is separated from the cellular proteins, lipids, and chromosomal DNA. This procedure yields 2–5 µg of relatively crude plasmid DNA (hence the term "minipreparation"), in contrast to large-scale preparations that yield 1 mg or more of pure plasmid DNA from a 1-L culture.

Once the desired RNAi feeding construct is obtained and isolated, it must be introduced into *E. coli* strain HT115(DE3). This strain of *E. coli* has been engineered to include an inducible gene that encodes T7 RNA polymerase. T7 RNA polymerase initiates transcription from the T7 promoters located on either side of the insert site in the vector. T7 transcription produces complementary RNA molecules that can bind to form dsRNA with the equivalent sequence of the DNA insert. HT115(DE3) also lacks an RNase that would normally digest the dsRNA produced by the vector. (Note: HT115[DE3] bacteria do not transform efficiently; it is difficult to transform recombination products directly into HT115[DE3] cells. Therefore, the initial transformation to introduce the recombination products into competent *E. coli* is a necessary "extra" step in this process.)

This experiment uses a rapid colony method to transform HT115(DE3) cells. The bacterial cells are rendered competent to take up DNA by treatment with $CaCl_2$, and a brief heat shock causes the plasmid DNA (the RNAi feeding construct) to enter the cells. As before, bacteria that successfully take up plasmid DNA and express the gene for antibiotic resistance can be detected by their ability to grow in the presence of kanamycin. HT115(DE3) bacteria transformed with the gene-specific RNAi plasmid represent a newly created RNAi feeding strain, ready to use to silence the gene of interest in *C. elegans* by following the protocol for Laboratory 4.2.

FURTHER READING

Altschul SF, Gish W, Miller W, Myers EW, Lipman DJ. 1990. Basic local alignment search tool. *J Mol Biol* **215:** 403–410.

Arziman Z, Horn T, Boutros M. 2005. E-RNAi: A web application to design optimized RNAi constructs. *Nucleic Acids Res* **33:** W582–W588.

Birnboim HC, Doly J. 1979. A rapid alkaline extraction method for screening recombinant plasmid DNA. *Nucleic Acids Res* **7:** 1513–1523.

Cohen SN, Chang AC, Hsu L. 1972. Nonchromosomal antibiotic resistance in bacteria: Genetic transformation of *Escherichia coli* by R-factor DNA. *Proc Natl Acad Sci* **69:** 2110–2114.

Dagert M, Ehrlich SD. 1979. Prolonged incubation in calcium chloride improves the competence of *Escherichia coli* cells. *Gene* **6:** 23–28.

Fire A, Xu S, Montgomery MK, Kostas SA, Driver SE, Mello CC. 1998. Potent and specific genetic interference by double-stranded RNA in *Caenorhabditis elegans*. *Nature* **391:** 806–811.

Hanahan D. 1983. Studies on transformation of *Escherichia coli* with plasmids. *J Mol Biol* **166:** 557–580.

Hartley JL, Temple GF, Brasch MA. 2000. DNA cloning using in vitro site-specific recombination. *Genome Res* **10:** 1788–1795.

Ish-Horowicz D, Burke JF. 1981. Rapid and efficient cosmid cloning. *Nucleic Acids Res* **9:** 2989–2998.

Levy AD, Yang J, Kramer JM. 1993. Molecular and genetic analyses of the *Caenorhabditis elegans dpy-2* and *dpy-10* collagen genes: A variety of molecular alterations affect organismal morphology. *Mol Biol Cell* **4:** 803–817.

Mandel M, Higa A. 1970. Calcium-dependent bacteriophage DNA infection. *J Mol Biol* **53:** 159–162.

Marchuk D, Drumm M, Saulino A, Collins FS. 1991. Construction of T-vectors, a rapid and general system of direct cloning of unmodified PCR products. *Nucleic Acids Res* **19:** 1154.

Sambrook J, Russell DW. 2001. *Molecular cloning: A laboratory manual*, 3rd ed. Cold Spring Harbor Laboratory Press, Cold Spring Harbor, NY.

PLANNING AND PREPARATION

The following table will help you to plan and integrate the different experimental methods.

Experiment	Day	Time	Activity
Stage A: Identify a Disease-Causing Gene and Design PCR Primers In Silico			
I. Use OMIM to find human disease genes and identify sequence for the gene of interest	1 d before Stage B	30 min	Lab: Identify a human disease gene.
II. Use BLAST and WormBase to identify a related sequence in *C. elegans*	1 d before Stage B	30 min	Lab: Obtain sequence of related gene from *C. elegans*.
III. Design primers for PCR amplification of the *C. elegans* gene	1 d before Stage B	30 min	Lab: Design primers for gene of interest.
Stage B: Clone Gene-Specific DNA			
	6 d before Part I	60 min	Prelab: Prepare NGM-lite plates.
	5 d before Part I	15 min	Prelab: Prepare overnight culture of OP50 *E. coli* in LB.
	4 d before Part I	30 min	Prelab: Seed NGM-lite plates with OP50 *E. coli*.
	2 d before Part I	30 min	Chunk wild-type worms to OP50-seeded NGM-lite plates.
I. Isolate DNA from *C. elegans*	1	60 min	Prelab: Prepare lysis buffer. Set up student stations.
		20 min	Lab: Collect worms for lysis.
		120 min	Postlab: Incubate to lyse worms.
II. Amplify DNA by PCR	2	30 min	Prelab: Aliquot primer/loading dye mix. Set up student stations.
		15 min	Lab: Set up PCRs.
		70+ min	Postlab: Amplify DNA in thermal cycler.
III. Analyze PCR products by gel electrophoresis	3	30 min	Prelab: Prepare agarose gel solution. Dilute TBE electrophoresis buffer. Set up student stations.
		75+ min	Lab: Cast gels. Load DNA samples into gels. Electrophorese samples. Photograph gels.
V. Recombine PCR product with vector	4	20 min	Prelab: Thaw and aliquot vector DNA and Clonase mix. Set up student stations.
		20 min	Lab: Prepare recombination reactions.
		60 min	Postlab: Incubate recombination reactions.

(Continued on following page.)

Experiment	Day	Time		Activity
Stage C: Transformation of *E. coli*				
	4	60 min	Prelab:	Prepare competent *E. coli* cells.
				Prepare LB and LB/kan plates.
				Prepare LB broth.
I. Transform competent *E. coli* cells with recombined DNA	5	30 min	Prelab:	Adjust water bath to 42°C.
				Aliquot LB.
				Prepare glass spreading beads (if necessary).
				Set up student stations.
		90 min	Lab:	Transform and plate cells.
Stage D: Vector Analysis				
I. Perform single-colony PCR and inoculate overnight cultures		30 min	Prelab:	Aliquot T7 primer/loading dye mix.
				Prepare and aliquot LB plus kanamycin.
				Set up student stations.
		60 min	Lab:	Set up PCRs and overnight cultures.
		120 min	Postlab:	Amplify DNA in thermal cycler.
II. Analyze PCR products by gel electrophoresis	7	30 min	Prelab:	Prepare agarose gel solution.
				Dilute TBE electrophoresis buffer.
				Set up student stations.
		75+ min	Lab:	Cast gels.
				Load DNA samples into gels.
				Electrophorese samples.
				Photograph gels.
Stage E: Isolate Plasmid DNA				
I. Perform plasmid miniprep	7	40 min	Prelab:	Prepare fresh SDS/NaOH solution.
				Prepare aliquots of GTE, potassium acetate/acetic acid, isopropanol, 95% ethanol, and TE.
				Set up student stations.
	8	60 min	Lab:	Purify plasmid DNA.
Stage F: Create RNAi Feeding Strain				
	8	60 min	Prelab:	Streak HT115(DE3) *E. coli* to LB plates.
				Prepare LB and LB/kan plates.
				Prepare LB broth and sterile $CaCl_2$.
I. Transform vector into the RNAi feeding *E. coli* strain	9	30 min	Prelab:	Adjust water bath to 42°C.
				Aliquot LB and $CaCl_2$.
				Prepare glass spreading beads (if necessary).
				Set up student stations.
		90 min	Lab:	Transform and plate cells.

STAGE A: IDENTIFY A DISEASE-CAUSING GENE AND DESIGN PCR PRIMERS IN SILICO

I. Use OMIM to Find Human Disease Genes and Identify Sequence for the Gene of Interest

1. Open the Internet site of the National Center for Biotechnology Information (NCBI) (www.ncbi.nlm.nih.gov).
2. Select the "OMIM" database in the drop-down menu.
3. Enter the disease of your choice in the search window and press "Search" to search OMIM for any genes associated with the disease. Note that for some diseases, no gene-specific data are available.
4. Follow the numbered links to any entries to learn how the genes relate to the disease. Record your findings. Use this information to identify one gene related to the disease of interest.
5. Open the Gene record for the gene: Click on "Gene info" to the right of the OMIM gene entry on the results page and select "Gene."
6. Briefly examine the Entrez Gene record for information about the gene. Information about its location, structure, and function may be provided.
7. To identify the protein sequence encoded by this gene, scroll down the page to the section entitled "NCBI Reference Sequences (RefSeq)." Proteins in the database have accession numbers with the format "NP_XXXXXX.X." (An accession number is a unique identifier given to a sequence.) Click on the accession number for the protein encoded by the gene and select "FASTA" from the drop-down menu. (FASTA is a standard format for displaying sequence.) If more than one protein is encoded by the gene, it is generally best to choose the one with the most amino acids.
8. Highlight the entire sequence, including the ">" at the beginning of the first line. Although the first line includes characters that would normally be excluded from BLAST searches, the ">" tells search programs to ignore the contents of this line.
9. Copy and paste the highlighted sequence into a text document. Save this text document for use in Step 1.iv of Part II.

II. Use BLAST and WormBase to Identify a Related Sequence in *C. elegans*

1. Perform a BLAST search as follows:
   i. Do an Internet search for "ncbi blast."
   ii. Click on the link for the result "BLAST: Basic Local Alignment Search Tool." This will take you to the Internet site of the National Center for Biotechnology Information (NCBI).
   iii. Click on the link to "tblastn" under the heading "Basic BLAST" to search a translated nucleotide database using a protein query. Why would we do a BLAST search with amino acids rather than with DNA sequence?
   iv. Enter the protein sequence from Step 9 of Part I into the search window. This is the query sequence.

v. Under "Choose Search Set," select the "Nucleotide collection (nr/nt)" database from the drop-down menu.

vi. Type "Caenorhabditis elegans" in the "Organism" window to narrow the search to *C. elegans.*

vii. Click on "BLAST" to send the query sequences to a server at NCBI in Bethesda, Maryland. There, the BLAST algorithm will attempt to match the protein sequence to translations of the millions of nucleotide sequences stored in its database. While searching, a page showing the status of your search will be displayed until your results are available. This may take only a few seconds or more than 1 min if many other searches are queued at the server.

2. Analyze the results of the BLAST search, which are displayed in three ways as you scroll down the page:

   i. First, a graphical overview illustrates how significant matches (hits) align with the query sequence. Matches of differing lengths are indicated by color-coded bars. What do you notice about the lengths (and colors) of the matches (bars) as you look from the top to the bottom?

   ii. This is followed by a list of significant alignments (hits) with links to the corresponding accession numbers. Note the scores in the "E value" column on the right. The Expectation or E value is the number of alignments with the query sequence that would be expected to occur by chance in the database. The lower the E value, the higher the probability that the hit is related to the query. For example, an E value of 1 means that a search with your sequence would be expected to turn up one match by chance. Longer query sequences generally yield lower E values. An alignment is considered significant if it has an E value of less than 0.1. Identify any *C. elegans* genes related to the human gene of interest. Find the hits with lowest E values; these are most likely to be related to the gene of interest. If no significant hits are found, the human gene may have no related gene in *C. elegans*. In that case, choose a different human gene to study.

   iii. Third is a detailed view of the protein sequence (query) aligned to the amino acid sequence of each search hit (subject, abbreviated "Sbjct"). Amino acids that are identical to both sequences are listed in the center row; biochemically similar amino acids are indicated by "+" signs.

Sometimes, related genes have diverged too far to be identified by a BLAST search. If you do not find a significant hit by BLAST, searching an ortholog database may work. OrthoDisease (http://orthodisease.sbc.su.se.cgi-bin/index.cgi) is one such database. Some human genes have no related genes in C. elegans. *If no related genes are identified, you will need to identify another human gene to study.*

3. Click on the accession number link to open the data sheet for the first hit (the first subject sequence). To keep a record of information about this gene for future reference, print out the data sheet.

   i. At the top of the report, note basic information about the sequence, including its length (in base pairs, or bp), database accession number, source, and references to papers in which the sequence has been published.

   ii. In the middle section of the report, the sequence features are annotated, with their beginning and ending nucleotide positions ("xx..xx"). These features may include genes, coding sequences (cds), regulatory regions, ribosomal RNA (rRNA), or transfer RNA (tRNA). If this data sheet does not list any "gene" features, return to the previous page and select another hit from the list of significant alignments.

iii. Scroll to the bottom of the data sheet. This is the nucleotide sequence to which the term "Sbjct" refers.

4. Scroll back up to the middle of the data sheet. In the "gene" feature, click the link titled "WormBase:WBGeneXXXXXXXX." This will provide access to more information about the gene in WormBase, the central database for information about *C. elegans* research.
5. Examine the WormBase entry to learn more about the function of the gene. Print the page for future reference.
6. In the "Sequences" section, any transcripts for the gene are summarized. Identify the transcript with the longest length from the table. To download the sequence for the longest transcript, click on the download icon in the "Transcript" column. From the pop-up window, click "download" for the unspliced + UTR sequence and save the sequence to a file.

III. Design Primers for PCR Amplification of the *C. elegans* Gene

1. Conduct a web search for "e-rnai webservice" and follow the link to the E-RNAi website. E-RNAi is a sophisticated web application that designs and evaluates primers for optimized dsRNA constructs.
2. Select "ID or sequence input." From the drop-down menus, select "long dsRNA for RNAi reagent type and "C. elegans" as the organism. Paste the sequence for the transcript from Part I, Step 6 into the target nucleotide sequence window.
3. Click "Submit" and wait for the results to be displayed. On the following page, ensure that the target is selected and click "Submit Selection."
4. To design primers, use the default settings, click "Design," and follow the link to HTML results.
5. Cut and paste the output and save it into the same document that you created in Part I, Step 6.
6. Choose the first suggested primer sequences from the output. Add the *attB1* sequence (GGGGACAAGTTTGTACAAAAAAGCAGGCT) to the 5′ end of the sequence of one of the primers and the *attB2* sequence (GGGGACCACTTTG-TACAAGAAAGCTGGGT) to the 5′ end of the sequence of the other primer. Purchase these primers for use in Part II of Stage B.

OVERVIEW OF STAGE B: CLONE GENE-SPECIFIC DNA

I. ISOLATE DNA FROM *C. ELEGANS*

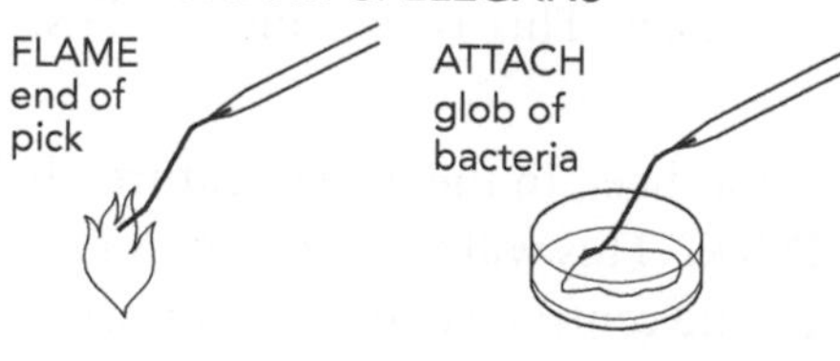

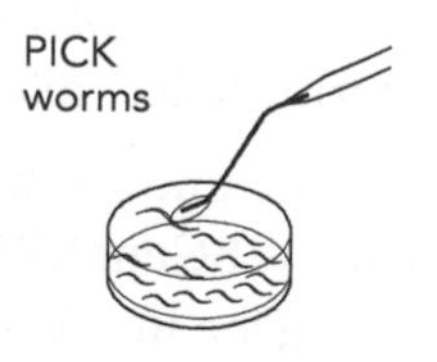

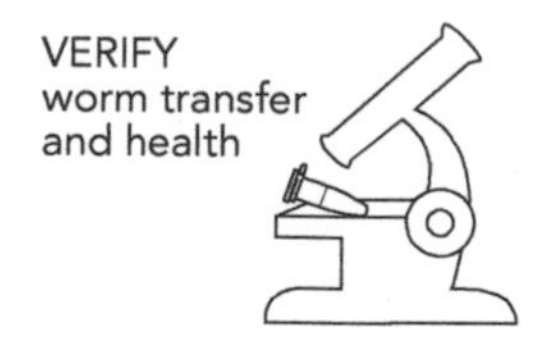

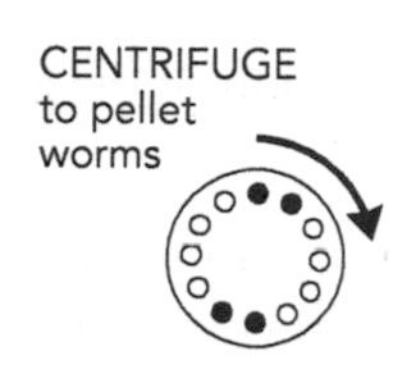

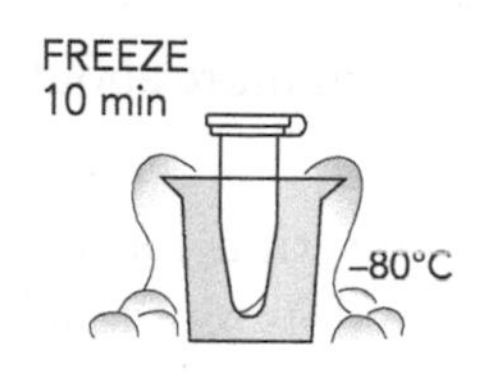

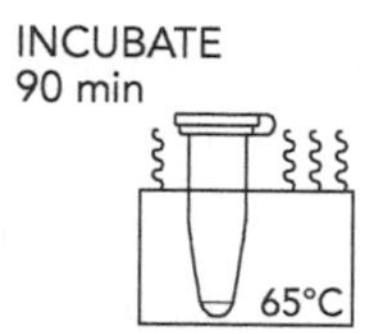

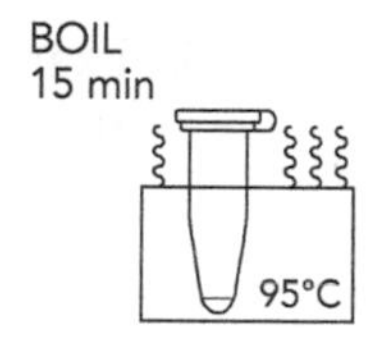

II. AMPLIFY DNA BY PCR

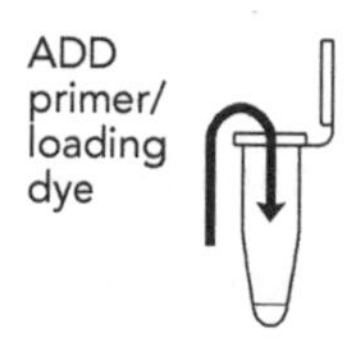

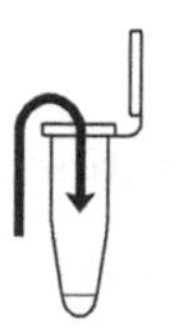

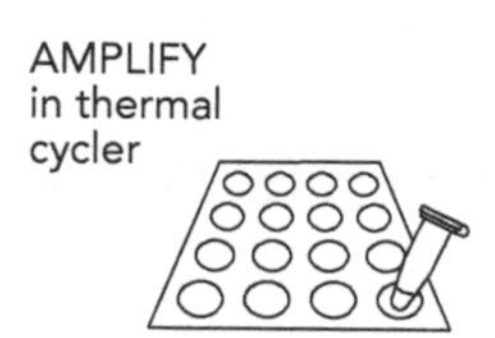

III. ANALYZE PCR PRODUCTS BY GEL ELECTROPHORESIS

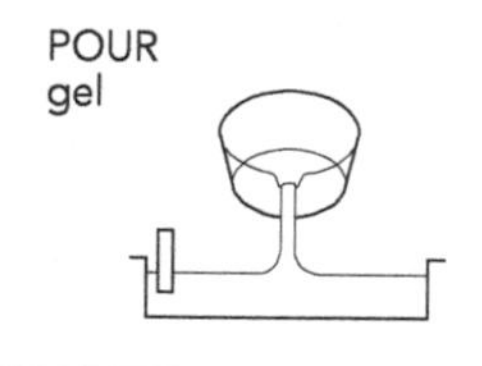

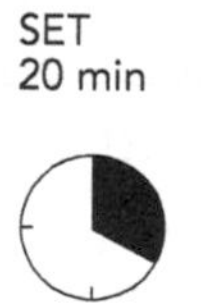

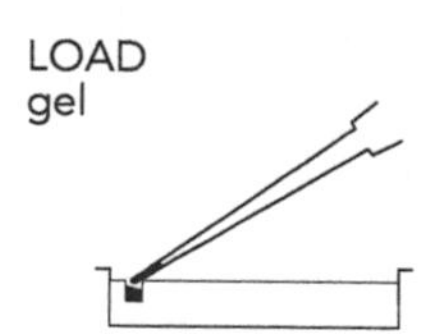

V. RECOMBINE PCR PRODUCT WITH VECTOR

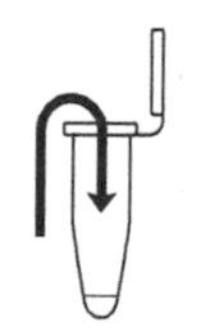

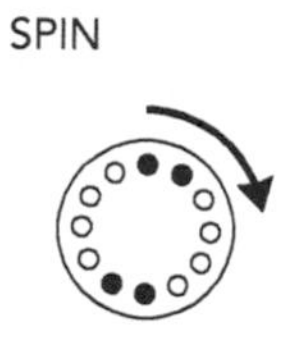

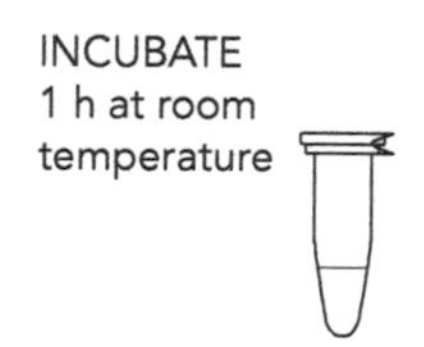

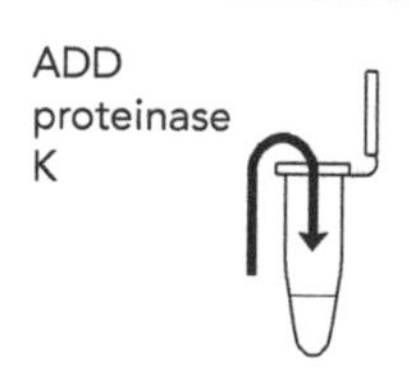

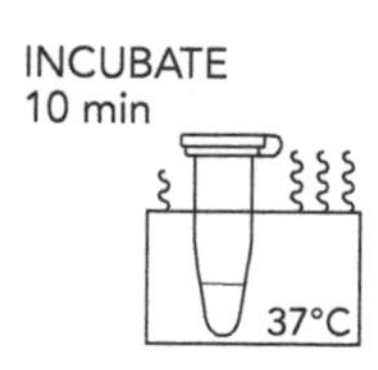

STAGE B: CLONE GENE-SPECIFIC DNA

I. Isolate DNA from *C. elegans*

One of the main ingredients in lysis buffer is an enzyme called proteinase K. Proteinase K is a nonspecific protease, an enzyme that cleaves the peptide bonds of proteins. During the incubating step, proteinase K digests protein in the cuticle of the worm and helps to liberate individual cells by destroying the protein fibers of the extracellular matrix that bind cells together. Proteinase K also inactivates cellular proteins, including DNases that interfere with PCR amplification. During the boiling step, the near-boiling temperature lyses the cell and nuclear membranes, releasing DNA and other cell contents. It also inactivates the proteinase K.

REAGENTS, SUPPLIES, & EQUIPMENT

For each group

Binocular dissecting microscope
Bunsen burner
Container with cracked or crushed ice
Lysis buffer (15 µL)*
Microcentrifuge tube rack
Micropipettes and tips (1–20 µL)
PCR tube (0.2 or 0.5 mL)
Permanent marker
Wild-type worms on NGM-lite plate
Worm pick

To share

Container with liquid nitrogen <!> or dry ice <!> (or access to a –80°C freezer)
Microcentrifuge
Microcentrifuge adapters for 0.2- or 0.5-mL PCR tubes
Thermal cycler or water baths at 65°C and 95°C

*Store on ice.
See Cautions Appendix for appropriate handling of materials marked with <!>.

1. Use a permanent marker to label one PCR tube with your group number and "W" (for "wild type").
2. Use a micropipette with a fresh tip to add 10 µL of lysis buffer to the tube.
3. Place the wild-type worms under a dissecting microscope.
4. Use a sterilized worm pick to transfer four or five adult wild-type worms to the "W" tube. Swish the tip directly in the lysis buffer to dislodge the worms from the pick.

5. Observe the PCR tube under a dissecting microscope. Verify that at least three worms are alive and writhing in the buffer.
6. To prepare your PCR tube for centrifugation, "nest" it within adapter tubes as follows:
   i. If your sample is in a 0.5-mL PCR tube, "nest" it within a capless 1.5-mL tube.
   ii. If your sample is in a 0.2-mL PCR tube, "nest" it within a capless 0.5-mL tube and then place both tubes into a capless 1.5-mL tube.

Centrifugation will pellet the worms. Freeze-thawing cracks the tough outer cuticle of the worm.

7. Place your tube, along with those from other groups, in a balanced configuration in a microcentrifuge. Centrifuge the tubes at full speed for 5–10 sec.

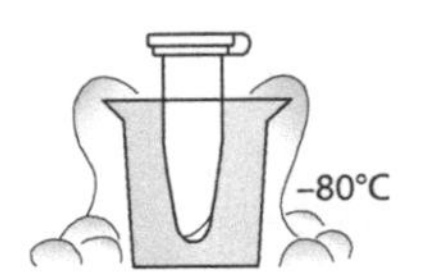

CAUTION! Liquid nitrogen, dry ice, and –80°C freezers can all cause burns and serious damage. Wear appropriate protective eyewear, clothing, and gloves.

8. Freeze your tube for 10 min in liquid nitrogen, on dry ice, or in a –80°C freezer. (If necessary, freeze overnight in a –20° freezer.)
9. Place your PCR tube, along with those from other groups, in a thermal cycler that has been programmed for one cycle of the following profile:

Incubating step:	90 min	65°C
Boiling step:	15 min	95°C

   The profile may be linked to a 4°C hold program.

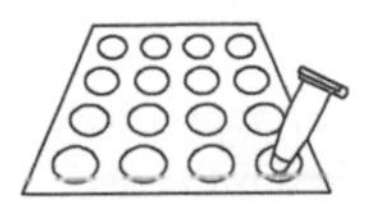

Worm DNA can be stored for several weeks at 4°C or indefinitely at –20°C.

10. Store your sample on ice or at –20°C until you are ready to proceed to Part II.

II. Amplify DNA by PCR

REAGENTS, SUPPLIES, & EQUIPMENT

For each group
C. elegans DNA* from Part I
Container with cracked or crushed ice
Microcentrifuge tube rack
Micropipettes and tips (1–100 μL)
Permanent marker
Primer/loading dye mix (25 μL)*
Ready-To-Go PCR Bead in a 0.2- or 0.5-mL PCR tube

To share
Thermal cycler

*Store on ice.

1. Obtain a PCR tube containing a Ready-To-Go PCR Bead. Label the tube with your group number.

The primer/loading dye mix will turn purple as the PCR bead dissolves.

2. Use a micropipette with a fresh tip to add 22.5 μL of primer/loading dye mix to the tube. Allow the bead to dissolve for ~1 min.
3. Use a micropipette with a fresh tip to add 2.5 μL of worm DNA (from Part I) directly into the primer/loading dye mix. Ensure that no DNA remains in the tip after pipetting.
4. Store your sample on ice until your class is ready to begin thermal cycling.

If the reagents become splattered on the wall of the tube, pool them by pulsing the sample in a microcentrifuge or by sharply tapping the tube bottom on the lab bench.

5. Place your PCR tube, along with those of other groups, in a thermal cycler that has been programmed for 30 cycles of the following profile:

Denaturing step:	30 sec	94°C
Annealing step:	30 sec	60°C
Extending step:	60 sec	72°C

The profile may be linked to a 4°C hold program after the 30 cycles are completed.

6. After thermal cycling, store the amplified DNA on ice or at –20°C until you are ready to proceed to Part III.

III. Analyze PCR Products by Gel Electrophoresis

REAGENTS, SUPPLIES, & EQUIPMENT

For each group
1% Agarose in 1x TBE (hold at 60°C) (50 mL per gel)
Container with cracked or crushed ice
Gel-casting tray and comb
Gel electrophoresis chamber and power supply
Latex gloves
Masking tape
Microcentrifuge tube (1.5 mL)
Microcentrifuge tube rack
Micropipette and tips (1–100 μL)
pBR322/BstNI marker (20 μL per gel)*
PCR product* from Part II
SYBR Green <!> DNA stain (5 μL)
1x TBE buffer (300 mL per gel)

To share
Digital camera or photodocumentary system
UV transilluminator <!> and eye protection
Water bath for agarose solution (60°C)

*Store on ice.
See Cautions Appendix for appropriate handling of materials marked with <!>.

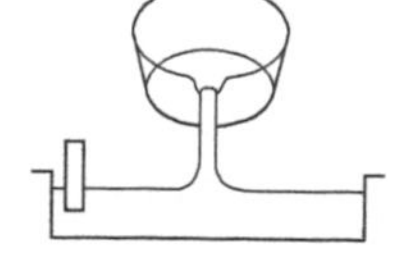

Avoid pouring an overly thick gel, which would be more difficult to visualize.

1. Seal the ends of the gel-casting tray with masking tape and insert a well-forming comb.
2. Pour the 1% agarose gel solution into the tray to a depth that covers about one-third the height of the open teeth of the comb.

The gel will become cloudy as it solidifies.

Do not add more buffer than necessary. Too much buffer above the gel channels electrical current over the gel, increasing the running time.

A 100-bp ladder may also be used as a marker.

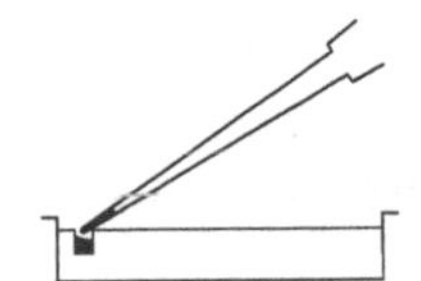

Expel any air from the tip before loading. Be careful not to push the tip of the pipette through the bottom of the sample well.

Transillumination, where the light source is below the gel, increases brightness and contrast.

The 670-bp amplicon is shown as an example.

3. Allow the gel to completely solidify; this takes ~20 min.
4. Remove the masking tape, place the gel into the electrophoresis chamber, and add enough 1X TBE buffer to cover the surface of the gel.
5. Carefully remove the comb and add additional 1X TBE buffer to fill in the wells and just cover the gel, creating a smooth buffer surface.
6. Orient the gel according to the diagram in Step 8, so that the wells are along the top of the gel.
7. Use a micropipette with a fresh tip to transfer 10 μL of your PCR product (from Part II) to a fresh 1.5-mL microcentrifuge tube. Add 1 μL of SYBR Green DNA stain to the 10 μL of PCR product. In addition, add 2 μL of SYBR Green DNA stain to 20 μL of pBR322/BstNI marker.
8. Use a micropipette with a fresh tip to load the marker from Step 7 into the far left well. Then, use another fresh tip to load the sample from Step 7 into your group's assigned well, according to the following diagram:

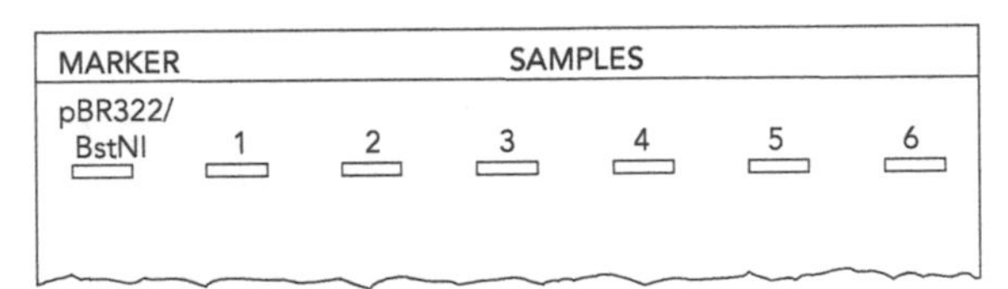

9. Run the gel for ~30 min at 130 V. Adequate separation will have occurred when the cresol red dye front has moved at least 50 mm from the wells.
10. View the gel using UV transillumination. Photograph the gel using a digital camera or photodocumentary system.

IV. Interpret Your Gel

1. Observe the photograph of the stained gel containing your PCR sample and those from other students. Orient the photograph with the sample wells at the top. Use the sample gel shown below to help to interpret the band(s) in each lane of the gel.

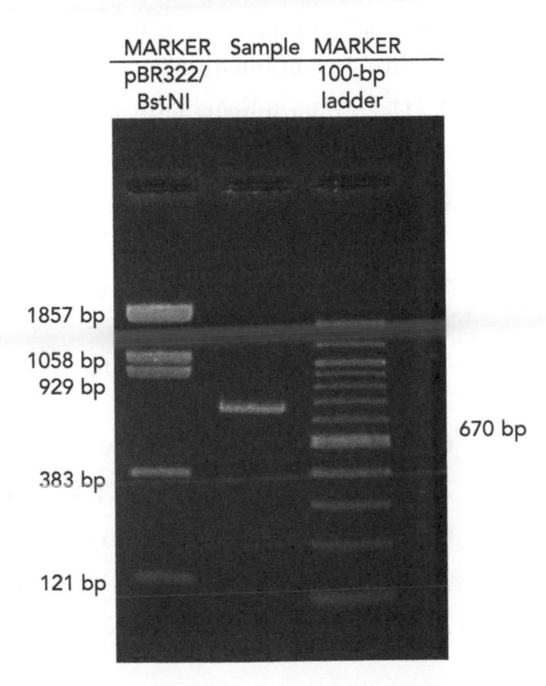

If a 100-bp ladder was used, as shown on the right-hand side of the sample gel on page 603, the bands increase in size in 100-bp increments starting with the fastest-migrating band of 100 bp.

2. Locate the lane containing the pBR322/BstNI markers on the left side of the gel. Working from the well, locate the bands corresponding to each restriction fragment: 1857, 1058, 929, 383, and 121 bp. The 1058- and 929-bp fragments will be very close together or may appear as a single large band. The 121-bp band may be very faint or not visible.
3. Scan across the row of student results. Notice that virtually all student lanes contain one prominent band.

Additional faint bands at other positions occur when the primers bind to chromosome loci other than the exact locus and give rise to "nonspecific" amplification products.

4. It is common to see a diffuse (fuzzy) band that runs ahead of the 121-bp marker. This is "primer dimer," an artifact of the PCR that results from the primers overlapping one another and amplifying themselves. How would you interpret a lane in which you observe primer dimer but no bands as described in Step 3?
5. Before proceeding to Part V, confirm that the PCR gives the expected band size with no nonspecific contaminants. If there are contaminants, how can you modify the PCR conditions to attempt to eliminate these products?

V. Recombine PCR Product with Vector

REAGENTS, SUPPLIES, & EQUIPMENT

For each group

BP Clonase enzyme mix (2 μL)*
Container with cracked or crushed ice
Microcentrifuge tube (1.5 mL)
Microcentrifuge tube rack
Micropipette and tips (1–10 μL)
PCR product* from Part III
Permanent marker
Proteinase K <!> solution (2 μg/μL)*
Vector (4 μL) (37.5 ng/μL)*

To share

Incubator (37°C)
Microcentrifuge
Microcentrifuge adapters for 0.2- or 0.5-mL PCR tubes

*Store on ice.
See Cautions Appendix for appropriate handling of materials marked with <!>.

1. Label an empty 1.5-mL microcentrifuge tube with your group number and "R" (for "recombination").
2. Use a micropipette with a fresh tip to add 4 μL of PCR product (from Part III) to the tube marked "R."
3. Use a micropipette with a fresh tip to add 4 μL of 37.5 ng/μL vector to the tube marked "R."
4. Use a micropipette with a fresh tip to add 2 μL of BP Clonase to the tube marked "R."
5. Tap the tube with your finger to mix and then pulse-spin in a microcentrifuge using an adapter.
6. Incubate the tube for at least 1 h at room temperature.
7. To terminate the reaction, add 1 μL of proteinase K and mix.
8. Incubate the sample for 10 min at 37°C.
9. Store the sample at –20°C until you are ready to proceed to Stage C.

RESULTS AND DISCUSSION

1. Describe the purpose of each of the following steps or reagents used in DNA isolation (Part I):
   i. Freezing at –80°C.
   ii. Incubating with proteinase K.
   iii. Boiling at 95°C.
2. The PCR products used in this experiment include *attB* sites. Why?

OVERVIEW OF STAGE C: TRANSFORMATION OF *E. COLI*

I. TRANSFORM COMPETENT *E. COLI* CELLS WITH RECOMBINED DNA

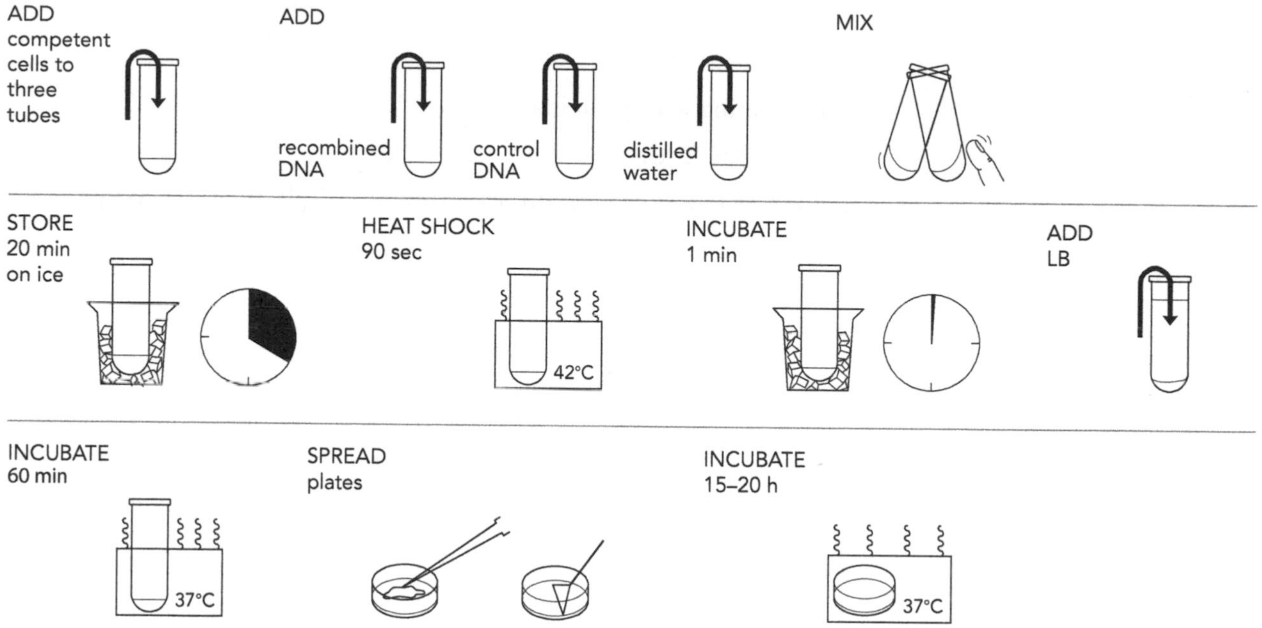

ADD competent cells to three tubes

ADD

recombined DNA

control DNA

distilled water

MIX

STORE 20 min on ice

HEAT SHOCK 90 sec

42°C

INCUBATE 1 min

ADD LB

INCUBATE 60 min

37°C

SPREAD plates

INCUBATE 15–20 h

37°C

STAGE C: TRANSFORMATION OF *E. COLI*

I. Transform Competent *E. coli* Cells with Recombined DNA

The recombination procedure in Stage B created recombinant DNA molecules that confer kanamycin resistance to bacteria after transformation. The following procedure is used to move the recombinant DNA molecules inside the bacterial cells. This entire experiment must be performed under sterile conditions.

REAGENTS, SUPPLIES, & EQUIPMENT

For each group

Bunsen burner (for spreading with a rod only)
Competent DH5α *E. coli* cells* (600 μL)
Container with cracked or crushed ice
3 Culture tubes (15 mL)
Ethanol (95%) <!> in a 50- or 100-mL beaker (for spreading with a rod only)
LB broth (2.5 mL)
3 LB/kan plates
3 LB plates
Micropipettes and tips (1–1000 μL)
Permanent marker
pKAN control DNA (12 μL) (0.005 μg/μL)*
Recombined DNA* from Stage B
Spreading beads (in a 1.5-mL tube) or a spreading rod
Sterile distilled or deionized H_2O (12 μL)
Test tube rack

To share

"Bio bag" or heavy-duty trash bag
Bleach (10%) <!> or disinfectant (e.g., Lysol)
Incubator (37°C)
Shaking water bath (37°C)
Storage container (for spreading beads only)
Water bath (42°C)

*Store on ice.
See Cautions Appendix for appropriate handling of materials marked with <!>.

A. Transform E. coli *Cells*

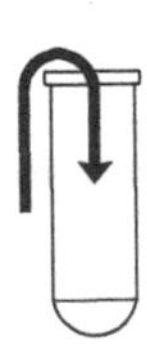

1. Use a permanent marker to label three 15-mL culture tubes with your group number. Label one tube "R" (for "recombined DNA"), one tube "–" (for "negative control"), and one tube "+" (for "positive control").
2. Use a micropipette with a sterile tip to add 200 μL of competent cells to each tube.
3. Use a micropipette with a sterile tip to add 10 μL of recombined DNA (from Stage B) directly into the cell suspension in the tube labeled "R."
4. Use a sterile tip to add 10 μL of 0.005 μg/μL pKAN control DNA directly into the cell suspension in the tube labeled "+."
5. Use a sterile tip to add 10 μL of sterile distilled or deionized H_2O directly into the cell suspension in the tube labeled "–."
6. Close the caps and tap the tubes with your finger to mix. Avoid making bubbles or splashing the suspension on the sides of the tubes.
7. Return all three tubes to ice for 20 min.
8. While the cells are incubating on ice, use a permanent marker to label the bottom of all six plates with your group number and the date. Divide the plates into three sets (with one LB/kan plate and one LB plate per set) and label as follows:
   i. Mark "R" on one LB/kan plate and one LB plate.
   ii. Mark "+" on one LB/kan plate and one LB plate.
   iii. Mark "–" on one LB/kan plate and one LB plate.

It is critical that the cells receive a sharp and distinct shock. If the tubes are not immediately moved from the ice to the water bath, heat shock will not occur. Removing the tubes from the ice and then walking them to the water bath will cause this step to fail.

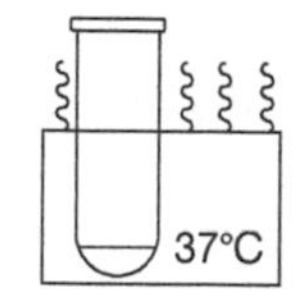

If a shaking water bath is not available, incubate the tubes in a regular 37°C water bath but gently mix the tubes periodically. Alternatively, warm the cells for several minutes in a 37°C water bath and then transfer to a dry shaker inside a 37°C incubator. The cells may be allowed to recover for up to 2 h. A recovery period assures the growth of as many kanamycin-resistant recombinants as possible and can help to compensate for poor recombination or cells of low competence.

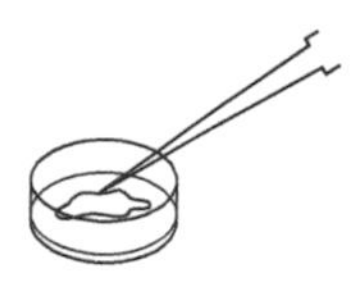

9. Following the 20-min incubation on ice, heat shock the cells in all three tubes as follows:
   i. Carry the ice container with the tubes to the 42°C water bath. Remove the tubes from the ice and immediately immerse them for 90 sec in the 42°C water bath.
   ii. Immediately return all three tubes to ice for at least 1 min more.
10. Use a micropipette with a sterile tip to add 800 μL of LB broth to each tube. Gently tap the tubes with your finger to mix.
11. Allow the cells to recover by incubating all three tubes for 1 h at 37°C in a shaking water bath (with moderate agitation).
12. After cell recovery, use a micropipette with a sterile tip to add either 100 μL or 500 μL of the appropriate cell suspension to each of the labeled plates from Step 8 (use the matrix below as a guide). Make sure to use a fresh sterile tip for each transfer. Then, immediately spread the cells using either a spreading rod or beads as described below in Parts IB and IC, respectively. Do not allow the suspensions to sit on the plates for too long before spreading.

Plate	Recombined DNA ("R")	Positive control ("+")	Negative control ("–")
LB/kan	500 mL	100 mL	100 mL
LB	500 mL	100 mL	100 mL

B. Use a Spreading Rod to Spread Transformed Cells on Plates

1. Sterilize the spreading rod by dipping it into the beaker of ethanol and then briefly passing it through a Bunsen burner flame to ignite the ethanol. Allow the ethanol to burn off away from the Bunsen flame; the spreading rod will become too hot if left in the flame.

CAUTION! Be extremely careful to avoid igniting the ethanol in the beaker. Do not panic if the ethanol is accidentally ignited. Cover the beaker with a glass Petri dish lid or other non-flammable cover to cut off oxygen and rapidly extinguish the fire.

Sterile spreading technique.

2. Lift the lid of the plate only enough to allow spreading. Do not place the lid on the lab bench.
3. Cool the spreader by gently rubbing it on the surface of the agar away from the cell suspension or by touching it to the condensation on the plate lid.
4. Touch the spreader to the cell suspension and gently drag it back and forth several times across the surface of the agar. Then, rotate the plate one-quarter turn and repeat the spreading motion. Try to spread the suspension evenly across the surface of the agar and be careful not to gouge the agar.
5. Replace the plate lid. Return the spreading rod to ethanol without flaming. Let the plate sit for several minutes to allow the suspension to become absorbed into the agar.
6. Repeat Steps 1–5 for the remaining plate(s). After spreading the last plate, reflame the spreader one last time before placing it on lab bench. Proceed to Part *ID* below.

C. Use Beads to Spread Transformed Cells on Plates

1. Lift the lid of the plate enough to add beads. Do not place the lid on the lab bench.
2. Carefully pour five to seven sterile glass spreading beads onto the agar surface.
3. Close the plate lid and use a swirling motion to move the glass beads around the entire surface of the plate. This evenly spreads the cell suspension on the agar surface. Continue swirling until the cell suspension is absorbed into the agar.
4. Repeat Steps 1–3 for the remaining plate(s).
5. After spreading the last plate, invert all plates and gently tap the plate bottoms so that the spreading beads fall onto the plate lids. Carefully pour the beads from each lid into a storage container for reuse. Proceed to Part *ID* below.

D. Incubate the Plates and Clean Up

1. Stack the plates and tape them in a bundle to keep them together.
2. Place the plates upside down in a 37°C incubator and incubate for 15–20 h.
3. Take time for responsible cleanup:

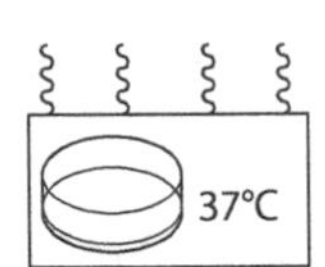

   i. Segregate bacterial cultures and tubes, pipettes, and micropipette tips that have come into contact with cultures into a "bio bag" or heavy-duty trash bag for proper disposal.
   ii. Wipe the lab bench with soapy water and 10% bleach or disinfectant.
   iii. Wash your hands before leaving lab.
4. After the initial 15–20-h incubation, store the plates at 4°C to arrest *E. coli* growth and to slow the growth of any contaminating microbes.

II. Calculate Transformation Efficiencies

1. While observing the colonies through the bottom of the culture plate, use a permanent marker to mark each colony as it is counted. If the experiment worked well, 5–50 colonies should be counted on the "R" LB/kan experimental plate,

An extended recovery period would inflate these numbers. In addition, if plates have been overincubated or left at room temperature for several days, "satellite" colonies may be observed on the LB/kan plates. These satellite colonies radiate from the edges of large, well-established colonies. Nonresistant satellite colonies grow in an "antibiotic shadow" where ampicillin has been broken down by the large resistant colony. Do not include satellite colonies in the count of transformants.

500–5000 colonies on the "+" LB/kan control plate, and 0 colonies on the "–" LB/kan control plate. (If the colonies are densely packed, draw lines on the bottom of the plate to divide it into equal-sized sections. Count one sector that is representative of the whole plate. After counting, multiply the result by the number of sectors to obtain the number of colonies on the entire plate.) In addition, examine the colonies carefully to detect any possible contamination; con-taminating organisms usually look different in color, shape, or size.

2. Record your observation of each plate in the matrix below. If the cell growth is too dense to count individual colonies, record "lawn."

Plate	Recombined DNA ("R")	Positive control ("+")	Negative control ("–")
LB/kan			
LB			

Were the results as expected? Explain possible reasons for variations from the expected results.

3. Compare and contrast the growth on each of the following pairs of plates. What does each pair of results tell you about transformation and/or antibiotic selection?

   i. "–" LB and "–" LB/kan.

   ii. "–" LB/kan and "+" LB/kan.

   iii. "R" LB/kan and "+" LB/kan.

4. Calculate the transformation efficiency of the LB/kan positive ("+") control as follows. Remember that transformation efficiency is expressed as the number of antibiotic-resistant colonies per microgram of intact plasmid DNA. The object is to determine the mass of plasmid that was spread on each plate and was therefore responsible for the transformants observed.

   i. Determine the total mass of control DNA used in Step 4 of Part IA:

   concentration (in μg/μL) x volume (in μL) = mass (in μg).

   ii. Determine the fraction of the cell suspension that was spread onto the "+" LB/kan plate in Step 12 of Part IA:

   $$\frac{\text{volume suspension spread (see Step 12 of Part IA)}}{\text{total volume suspension (see Steps 2 and 10 of Part IA)}} = \text{fraction spread.}$$

   iii. Determine the mass of the plasmid in the cell suspension that was spread onto the "+" LB/kan plate:

   total mass plasmid (Step 4.i) x fraction spread (Step 4.ii) = mass plasmid spread (in μg).

   iv. Determine the number of colonies per microgram of plasmid. Express the answer in scientific notation:

   $$\frac{\text{colonies observed (Step 2)}}{\text{mass plasmid spread (Step 4.iii)}} = \text{transformation efficiency (in colonies/μg).}$$

5. Calculate the transformation efficiency of the LB/kan recombined DNA ("R") as follows.

i. Calculate the mass of vector used in Step 3 of Part V in Stage B. Your result is analogous to the mass that was calculated in Step 4.i for the "+" control.

ii. Use the result from Step 5.i and the procedure outlined in Steps 4.ii–iv to calculate the transformation efficiency of the LB/kan recombined DNA ("R").

6. Compare the transformation efficiencies that you calculated for the control ("+") DNA in Step 4 and the recombined DNA ("R") in Step 5. How can you account for the differences in efficiency? Take into account the formal definition of transformation efficiency.

OVERVIEW OF STAGE D: VECTOR ANALYSIS

I. PERFORM SINGLE-COLONY PCR AND INOCULATE OVERNIGHT CULTURES

ADD
primer/
loading
dye

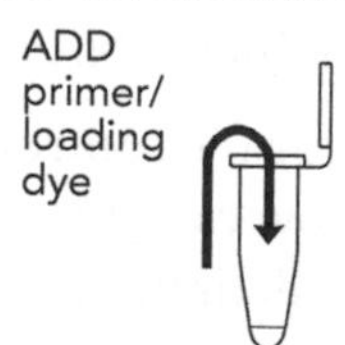

SCRAPE
cell mass with
pipette tip

TRANSFER
part of the
cell mass to
PCR tube

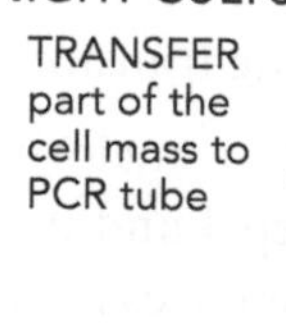

EJECT
tip into tube
with LB

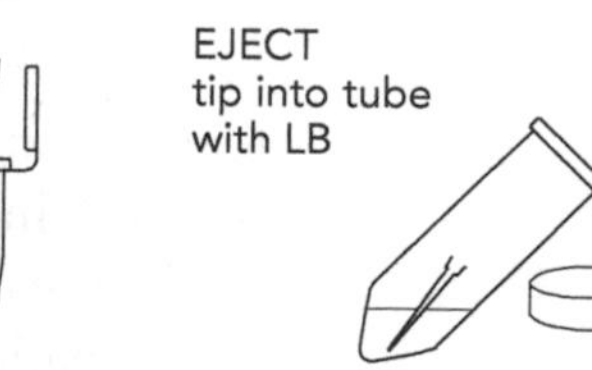

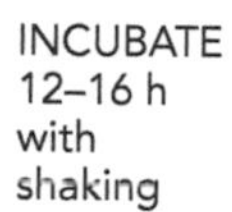

INCUBATE
12–16 h
with
shaking

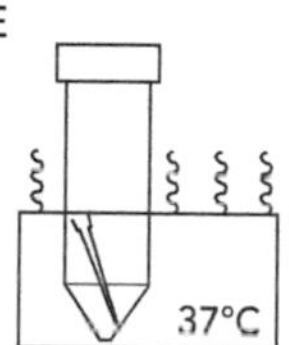

AMPLIFY
in thermal
cycler

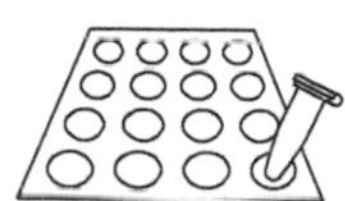

STORE

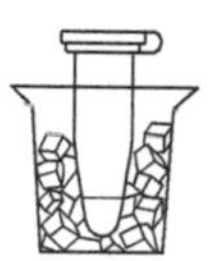

II. ANALYZE PCR PRODUCTS BY GEL ELECTROPHORESIS

POUR
gel

SET
20 min

LOAD
gel

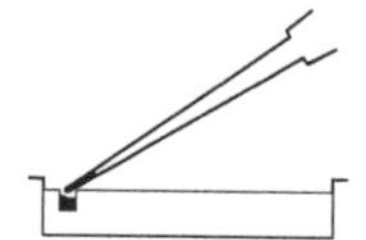

ELECTROPHORESE
130 V

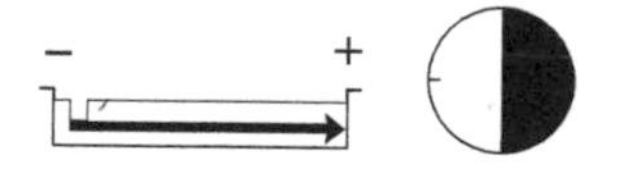

STAGE D: VECTOR ANALYSIS

I. Perform Single-Colony PCR and Inoculate Overnight Cultures

REAGENTS, SUPPLIES, & EQUIPMENT

For each group

Container with cracked or crushed ice
LB/kan plate with transformants marked "R" from Stage C
Microcentrifuge tube rack
Micropipette and tips (1–100 µL)
Permanent marker
4 Ready-To-Go PCR Beads in 0.2- or 0.5-mL PCR tubes
4 Snap-cap tubes (15 mL), each with 2 mL of LB plus kanamycin <!>
T7 primer/loading dye mix (120 µL)*
Test tube rack

To share

Shaking incubator (37°C)
Thermal cycler

*Store on ice.
See Cautions Appendix for appropriate handling of materials marked with <!>.

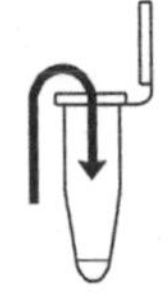

The primer/loading dye mix will turn purple as the PCR bead dissolves.

1. Obtain four PCR tubes containing Ready-To-Go PCR Beads and four 15-mL snap-cap tubes containing 2 mL of LB plus kanamycin. Number each set of tubes "1" through "4" and label them with your group number.
2. Use a micropipette with a fresh tip to add 25 µL of the T7 primer/loading dye mix to each tube. Allow the bead to dissolve for ~1 min.

If the transformation in Stage C worked well, the LB/kan plate marked "R" should possess many bacterial colonies.

3. Circle the location of four colonies on the bottom of the LB/kan plate marked "R" (from Stage C). Number the circled colonies "1" through "4." Then, pick a bit of the cell mass from each colony for PCR and overnight culture as follows:
   i. Use a micropipette with a fresh, sterile tip to pick up a part of the cell mass from colony 1. Aim to have the cell mass at the end of the tip.
   ii. Immediately dip the end of the tip into the PCR tube labeled "1." Twirl the micropipette so that part (but not all) of the cell mass is released into the primer/loading dye mix.
   iii. Immediately after dipping the tip into the PCR tube, inoculate an overnight culture with the rest of the cell mass by ejecting the tip into the 15-mL tube labeled "1." Make sure to use sterile technique.
   iv. Repeat Steps 3.i–iii for the remaining colonies.
   v. Store the LB/kan plate at 4°C, because it may be needed to repeat either the PCR or the culture growth.

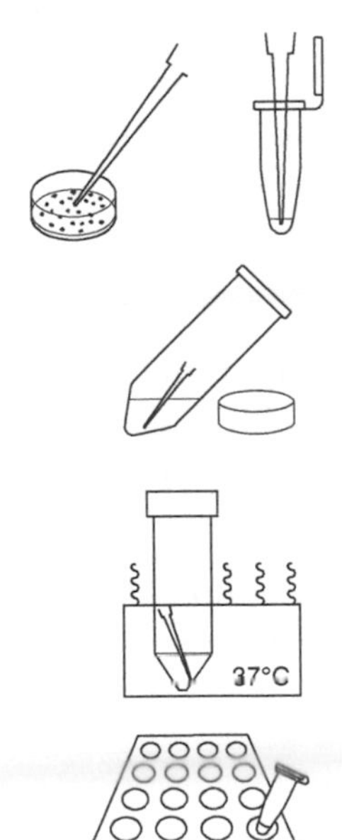

4. Place the culture tubes in a 37°C incubator and shake for 12–16 h (a typical incubation would last overnight). Store your PCR samples on ice until your class is ready to begin thermal cycling.
5. Place your PCR tubes, along with those of other groups, in a thermal cycler that has been programmed for 30 cycles of the following profile:

Denaturing step:	30 sec	94°C
Annealing step:	30 sec	52°C
Extension step:	90 sec	72°C

The profile may be linked to a 4°C hold program after the 30 cycles are completed.

If the reagents become splattered on the wall of the tube, pool them by pulsing the sample in a microcentrifuge or by sharply tapping the tube bottom on the lab bench.

6. After thermal cycling, store the amplified DNA at –20°C until you are ready to proceed to Part II. The overnight cultures can be stored for 2–3 d at 4°C before proceeding to Part II.

II. Analyze PCR Products by Gel Electrophoresis

REAGENTS, SUPPLIES, & EQUIPMENT

For each group

1% Agarose in 1x TBE (hold at 60°C) (50 mL per gel)
Container with cracked or crushed ice
Gel-casting tray and comb
Gel electrophoresis chamber and power supply
Latex gloves
Masking tape
Microcentrifuge tube rack
4 Microcentrifuge tubes (1.5 mL)
Micropipette and tips (1–100 μL)
pBR322/BstNI marker (20 μL per gel)*
PCR products* from Part I
SYBR Green <!> DNA stain (12 μL)
1x TBE buffer (300 mL per gel)

To share

Digital camera or photodocumentary system
UV transilluminator <!> and eye protection
Water bath for agarose solution (60°C)

*Store on ice.

See Cautions Appendix for appropriate handling of materials marked with <!>.

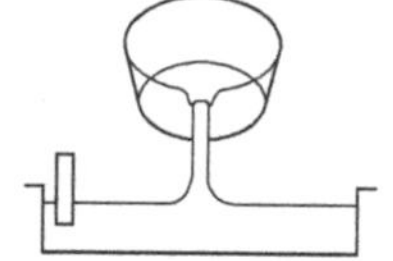

1. Seal the ends of the gel-casting tray with masking tape and insert a well-forming comb.
2. Pour the 1% agarose solution into the tray to a depth that covers about one-third the height of the open teeth of the comb.

Avoid pouring an overly thick gel, which would be more difficult to visualize.

3. Allow the gel to completely solidify; this takes ~20 min.

The gel will become cloudy as it solidifies.

4. Remove the masking tape, place the gel into the electrophoresis chamber, and add enough 1x TBE to cover the surface of the gel.
5. Carefully remove the comb and add additional 1x TBE buffer to fill in the wells and just cover the gel, creating a smooth buffer surface.

Do not add more buffer than necessary. Too much buffer above the gel channels electrical current over the gel, increasing the running time.

6. Orient the gel according to the diagram in Step 8, so that the wells are along the top of the gel.
7. Use a micropipette with a fresh tip to transfer 15 μL of each PCR product (from Part I) to a fresh 1.5-mL microcentrifuge tube. Add 2 μL of SYBR Green DNA stain to each 15-μL sample of PCR product. In addition, add 2 μL of SYBR Green DNA stain to 20 μL of pBR322/BstNI marker.

A 100-bp ladder may also be used as a marker.

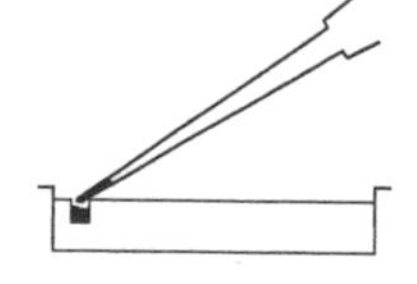

8. Use a micropipette with a fresh tip to load the marker from Step 7 into the far left well. Then, use another fresh tip to load each sample from Step 7 into the assigned well, according to the following diagram:

Expel any air from the tip before loading. Be careful not to push the tip of the pipette through the bottom of the sample well.

MARKER SAMPLES
pBR322/ BstNI 1 2 3 4

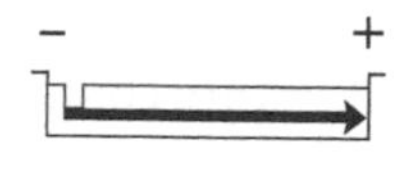

9. Run the gel for ~30 min at 130 V. Adequate separation will have occurred when the cresol red dye front has moved at least 50 mm from the wells.

Transillumination, where the light source is below the gel, increases brightness and contrast.

10. View the gel using UV transillumination. Photograph the gel using a digital camera or photodocumentary system.

RESULTS AND DISCUSSION

1. Why were colonies picked only from the LB/kan plate marked "R"? What event does each colony represent?
2. Observe the photograph of the stained gel containing your PCR samples. Orient the photograph with the sample wells at the top. Use the sample gel shown below to help to interpret the band(s) in each lane of the gel.

The 771-bp amplicon is shown as an example.

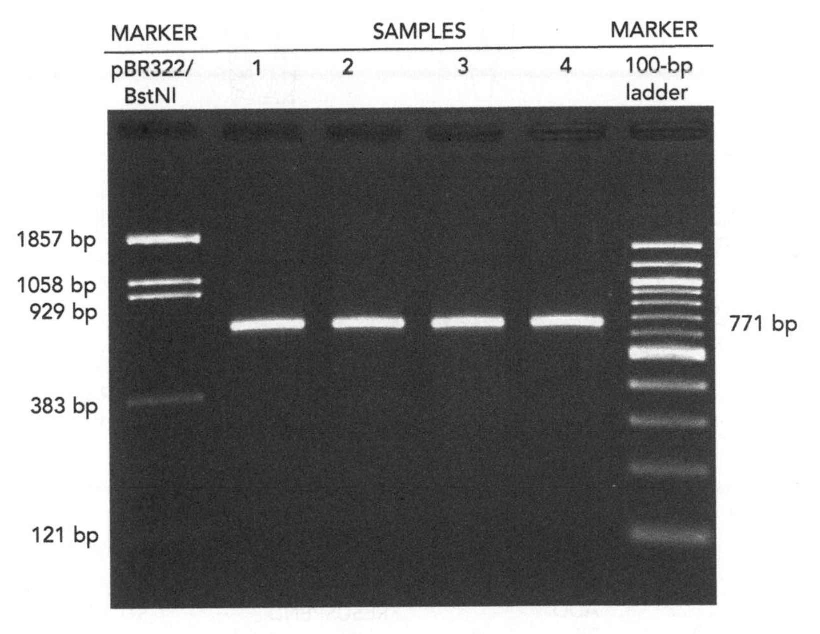

3. Determine the size of the PCR fragments by comparing them to the markers (the DNA bands of known size). Then, compare the size of the PCR fragments in this gel to the size of the PCR fragments that you obtained in Part IV of Stage B. Do any of the colonies (transformants) have the expected insert? Explain how you determined this.

OVERVIEW OF STAGE E: ISOLATE PLASMID DNA

I. PERFORM PLASMID MINIPREP

ADD
overnight
cultures of
transformants

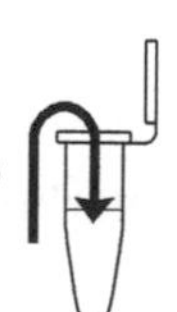

CENTRIFUGE
1 min

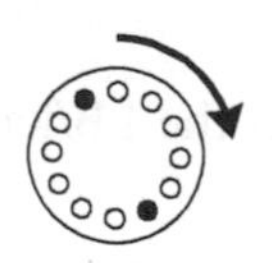

POUR OFF
supernatant

DRAIN

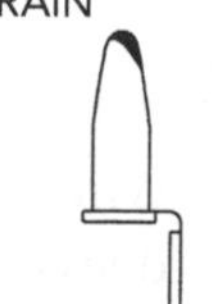

ADD
GTE

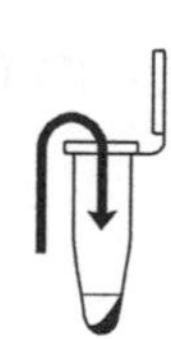

RESUSPEND

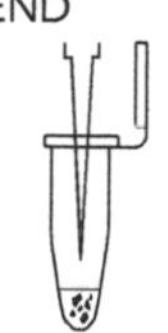

ADD
SDS/NaOH

INCUBATE
5 min

ADD
potassium
acetate/
acetic
acid

INCUBATE
5 min

CENTRIFUGE
5 min

TRANSFER
supernatant

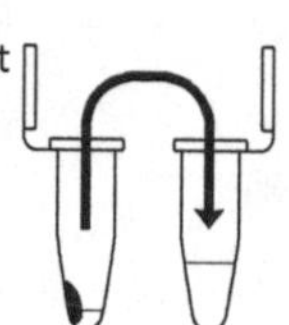

ADD
isopropanol

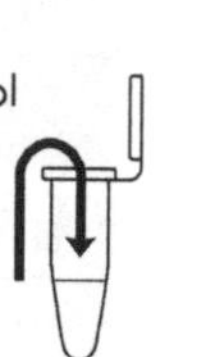

CENTRIFUGE
5 min

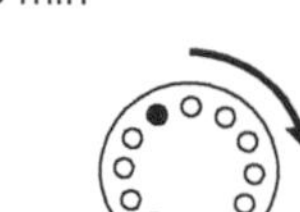

POUR OFF
supernatant

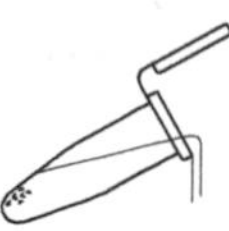

DRAIN

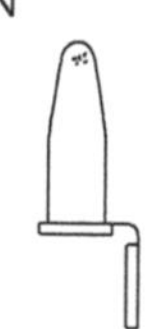

ADD
ethanol

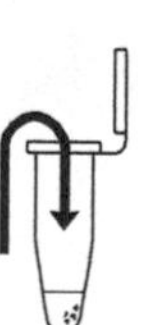

CENTRIFUGE
2 min

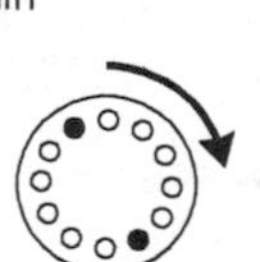

POUR OFF
supernatant

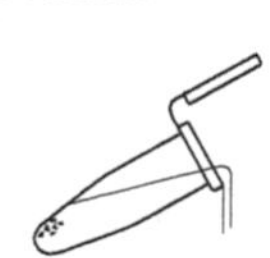

DRAIN

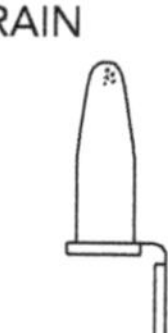

DRY

ADD
TE

RESUSPEND

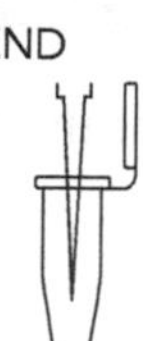

STORE

STAGE E: ISOLATE PLASMID DNA

I. Perform Plasmid Miniprep

REAGENTS, SUPPLIES, & EQUIPMENT

For each group

2 Beakers for waste
Container with cracked or crushed ice
E. coli cultures of transformants from Part I of Stage D
Ethanol (95%) <!> (500 μL)
Glucose/Tris/EDTA <!> (300 μL) (GTE)*
Isopropanol <!> (1000 μL)
Microcentrifuge tube rack
4 Microcentrifuge tubes (1.5 mL)
Micropipettes and tips (10–1000 μL)
Paper towel
Permanent marker
Potassium acetate/acetic acid <!> (600 μL)*
SDS <!>/sodium hydroxide <!> (SDS/NaOH) (500 μL)
Tris <!>/EDTA <!> (TE) (50 μL)

To share

"Bio bag" or heavy-duty trash bag
Bleach (10%) <!> or disinfectant
Microcentrifuge

*Store on ice.
See Cautions Appendix for appropriate handling of materials marked with <!>

1. Obtain the *E. coli* overnight cultures of two transformants (from Part I of Stage D) that contain the correct insert size (as determined in Part III of Stage D).
2. Label two 1.5-mL microcentrifuge tubes with your group number and the number corresponding to the culture tubes. For example, if the culture tube is labeled "2," label the microcentrifuge tube "2" as well.
3. Shake the culture tubes to resuspend the *E. coli* cells that have settled to the bottom of the tube.
4. Use a micropipette to transfer 1000 μL of each overnight suspension into the appropriate tubes.

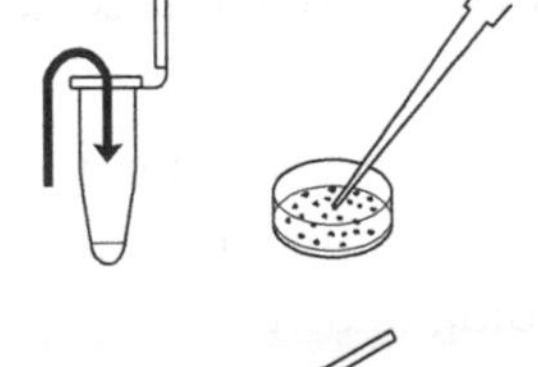

5. Close the caps and place the tubes in a balanced configuration in a microcentrifuge. Centrifuge the tubes at maximum speed for 1 min to pellet the cells.
6. Pour off the supernatant from each tube into a waste beaker for later disinfection.

7. Invert the tubes and tap them gently on the surface of a clean paper towel to drain thoroughly.

As an alternative to Step 6, use a micropipette to remove the supernatant from each tube and discard it in the waste beaker. Do not disturb the cell pellet.

8. Add 100 μL of ice-cold GTE solution to each tube. Resuspend each pellet by pipetting the solution in and out several times. Use a fresh tip for each sample. Hold the tubes up to the light to check that the suspensions are homogeneous and that no visible clumps of cells remain.
9. Add 200 μL of SDS/NaOH solution to each tube. Close the caps and mix the solutions by rapidly inverting the tubes five times.

SDS/NaOH helps to solubilize proteins and dissolves lipids of the cell membrane.

10. Incubate the tubes for 5 min on ice. The suspension will become relatively clear.
11. Add 150 μL of ice-cold potassium acetate/acetic acid to each tube. Close the caps and mix the solutions by rapidly inverting the tubes five times. A white precipitate, composed of proteins and lipids, will immediately appear.

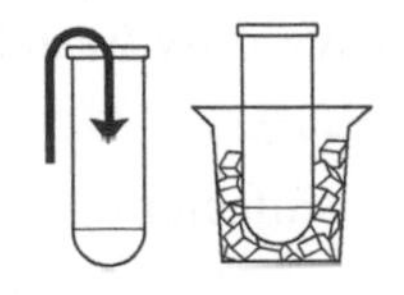

Potassium acetate precipitates proteins and lipids.

12. Incubate the tubes for 5 min on ice.
13. Place the tubes in a balanced configuration in a microcentrifuge. Centrifuge the tubes at maximum speed for 5 min to pellet the precipitate along the side of the tube.

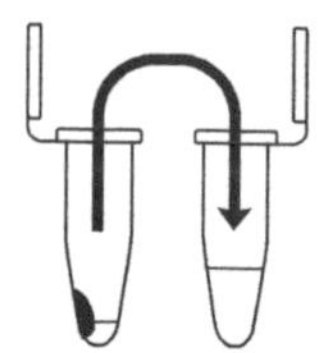

Isopropanol preferentially and rapidly precipitates nucleic acids; however, proteins begin to precipitate with time. Leaving the supernatant in isopropanol for longer than 2 min before proceeding to Step 16 will increase salt precipitation.

14. Label two fresh 1.5-mL microcentrifuge tubes with your group number and the number corresponding to the original culture tubes. Transfer 400 μL of supernatant into the appropriately labeled tubes. Avoid pipetting the precipitate and wipe off any precipitate that clings to the outside of the tip before expelling the supernatant. Discard the old tubes that contain the precipitate.
15. Add 400 μL of isopropanol to each tube of supernatant. Close the caps and mix the supernatant vigorously by rapidly inverting the tubes five times. Incubate the tubes for only 2 min at room temperature.
16. Place the tubes in a balanced configuration in the microcentrifuge, with the hinge of the cap pointing outward. (During centrifugation, nucleic acids, visible or not, will gather on the side of the tube underneath the hinge.) Centrifuge the tubes at maximum speed for 5 min to pellet the nucleic acids.

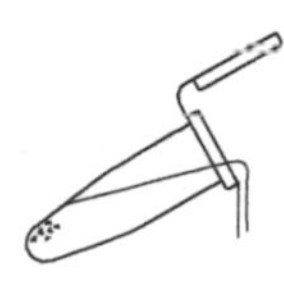

17. Pour off the supernatant from each tube into the second waste beaker. Do not disturb the nucleic acid pellets. Invert the tubes and tap them gently on the surface of a clean paper towel to drain thoroughly.
18. Add 200 μL of 95% ethanol to each tube and close the caps. Flick the tubes several times to wash the pellets.
19. Place the tubes in a balanced configuration in the microcentrifuge and centrifuge them at maximum speed for 2 min.

As an alternative to Step 17 or 20, use a micropipette to remove the supernatant and discard in a waste beaker. Do not disturb the pellet. If the pellet is drawn into the tip, transfer the supernatant to a fresh 1.5-mL tube, recentrifuge, and remove the supernatant again.

In Step 22, the pellet may appear as a tiny teardrop-shaped smear or particles on the bottom side of the tube underneath the hinge. Do not be concerned if you do not see a pellet. A pellet of DNA with little salt contamination is sometimes difficult to see.

20. Pour off the supernatant from each tube into the second waste beaker. Do not disturb the nucleic acid pellets. Invert the tubes and tap them gently on the surface of a clean paper towel to drain thoroughly.
21. Close the caps and pulse the tubes in a microcentrifuge to pool any remaining ethanol. Then, *completely* remove the remaining liquid with a micropipette set at 100 μL. Air-dry the pellets by letting the tubes sit open for 10 min.
22. Hold each tube up to the light to check that no ethanol droplets remain. If ethanol is still evaporating, an alcohol odor can be detected by sniffing the mouth of the tube. All ethanol must be evaporated before proceeding to Step 23.
23. Add 15 μL of TE to each tube. Dissolve the pelleted DNA by pipetting in and out, using a fresh tip for each tube. Make sure to wash the TE down the side of the tube underneath the hinge, where the pellet formed during centrifugation. Check that all DNA is dissolved and that no particles remain in the tip or on the side of the tube.

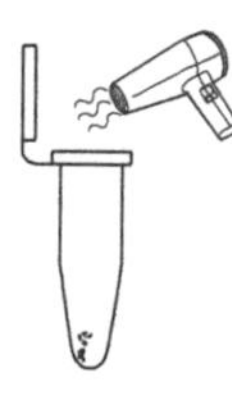

Dry the pellets quickly with a hair dryer. To prevent blowing the pellets away, direct the air across the tube mouths, not into the tubes, for ~3 min.

24. Keep the two DNA/TE solutions separate. DO NOT pool into one tube. The DNA may be used immediately or stored at −20°C until you are ready to proceed to Stage F.
25. Take time for responsible cleanup:
    1. Segregate bacterial cultures and tubes, pipettes, and micropipette tips that have come into contact with cultures into a "bio bag" or heavy-duty trash bag for proper disposal.
    2. Wipe the lab bench with soapy water and 10% bleach or disinfectant.
    3. Wash your hands before leaving lab.

RESULTS AND DISCUSSION

1. Consider three major classes of biologically important molecules: proteins, lipids, and nucleic acids. Which steps of the miniprep procedure act on proteins? On lipids? On nucleic acids?
2. What aspect of plasmid DNA structure allows it to renature efficiently in Step 11?
3. What other kinds of molecules, in addition to plasmid DNA, would you expect to be present in the final miniprep sample? How could you find out?

OVERVIEW OF STAGE F: CREATE RNAi FEEDING STRAIN

I. TRANSFORM VECTOR INTO THE RNAi FEEDING *E. COLI* STRAIN

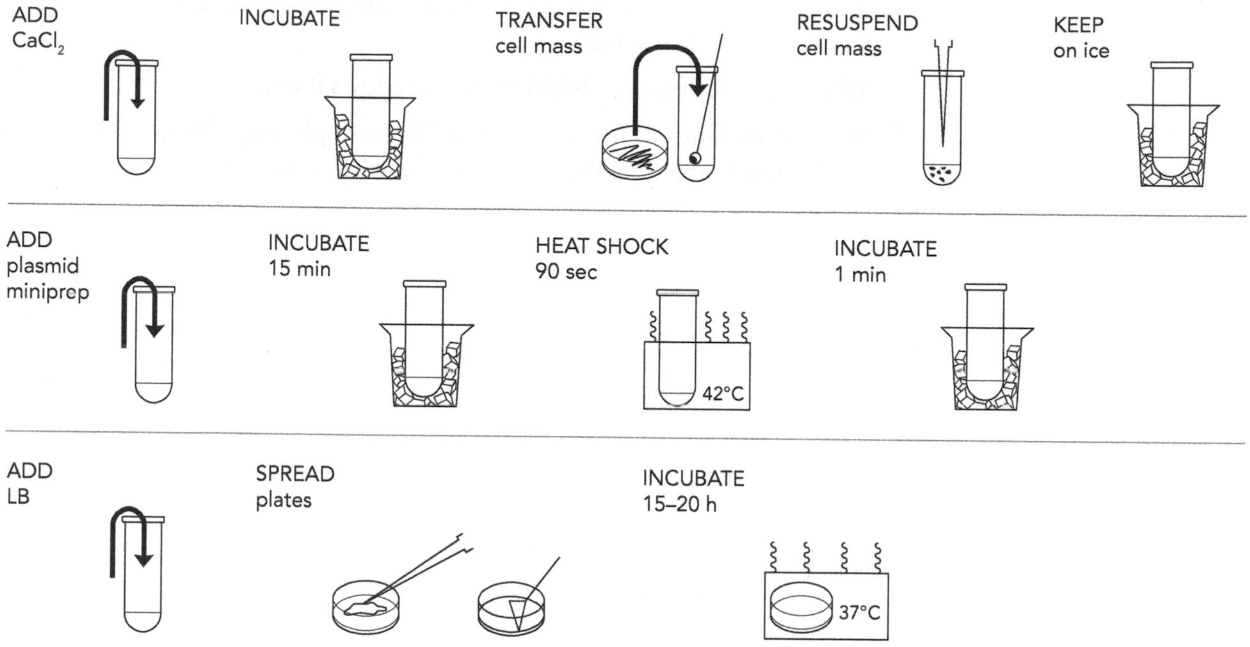

ADD $CaCl_2$

INCUBATE

TRANSFER cell mass

RESUSPEND cell mass

KEEP on ice

ADD plasmid miniprep

INCUBATE 15 min

HEAT SHOCK 90 sec

42°C

INCUBATE 1 min

ADD LB

SPREAD plates

INCUBATE 15–20 h

37°C

STAGE F: CREATE RNAi FEEDING STRAIN

I. Transform Vector into the RNAi Feeding *E. coli* Strain

This entire experiment must be performed under sterile conditions.

REAGENTS, SUPPLIES, & EQUIPMENT

For each group

Bunsen burner
$CaCl_2$ (50 mM) <!> (600 μL)
Container with cracked or crushed ice
2 Culture tubes (15 mL)
Ethanol (95%) <!> in a 50- or 100-mL beaker (for spreading with a rod only)
Innoculating loop
LB broth (600 μL)
2 LB/kan plates
2 LB plates
Microcentrifuge tube rack
Micropipette and tips (1–1000 μL)
Permanent marker
Plasmid miniprep DNA* from Stage E
Spreading beads (in a 1.5-mL tube) or a spreading rod
Starter plate of HT115(DE3) *E. coli* grown overnight on LB plates
Test tube rack

To share

"Bio bag" or heavy-duty trash bag
Bleach (10%) <!> or disinfectant
Incubator (37°C)
Shaking water bath (37°C)
Storage container (for spreading beads only)
Water baths (42°C)

*Store on ice.
See Cautions Appendix for appropriate handling of materials marked with <!>.

Plasmid DNA will be added to the "+plasmid" tube; none will be added to the "–plasmid" tube.

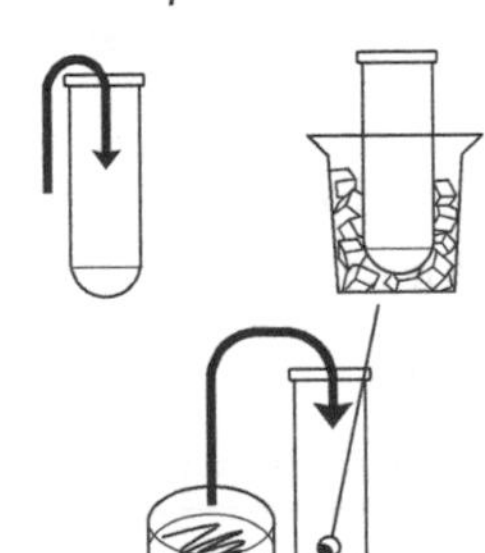

A. Transform HT115(DE3) E. coli *Cells*

1. Use a permanent marker to label one sterile 15-mL tube "+plasmid." Label another 15-mL tube "–plasmid."
2. Use a micropipette and sterile tip to add 250 μL of $CaCl_2$ solution to each tube.
3. Place both tubes on ice.
4. Use a sterile inoculating loop to transfer one or two large colonies from the starter plate of HT115(DE3) *E. coli* to the "+plasmid" tube as follows:
   i. Sterilize the loop in a Bunsen burner flame until it glows red hot. Then, pass the lower half of the shaft through the flame. Cool the loop by stabbing it several times into a clear area near the edge of the starter plate.
   ii. Pick one or two large (3-mm diameter) colonies and scrape them up (the cell mass should be visible on the loop). Be careful not to transfer any agar because impurities in the agar can inhibit transformation.
   iii. Immerse the cell mass in the $CaCl_2$ solution and vigorously tap it against the wall of the tube to dislodge the cell mass. Hold the tube up to the light to make sure that the cell mass is not left on the loop or the side of the tube.
   iv. Reflame the loop before placing it on the lab bench.

If there are no separate colonies on the starter plate, scrape up a small cell mass from a streak. Transformation efficiency decreases if too many cells are added to the $CaCl_2$ solution.

Optimally, flame the mouth of the 15-mL tube after removing and before replacing the cap.

Cells become difficult to resuspend if allowed to clump together in the $CaCl_2$ solution for several minutes.

5. Immediately resuspend the cells in the "+plasmid" tube by repeatedly pipetting in and out using a micropipette with a sterile tip. Pipette carefully to avoid making bubbles in the suspension or splashing the suspension far up the sides of the tube. Hold the tube up to the light to check that the suspension is homogeneous. No visible clumps of cells should remain.
6. Return the "+plasmid" tube to ice.

CAUTION! Keep your nose and mouth away from the tip end when pipetting suspension culture to avoid inhaling any aerosol that might be created.

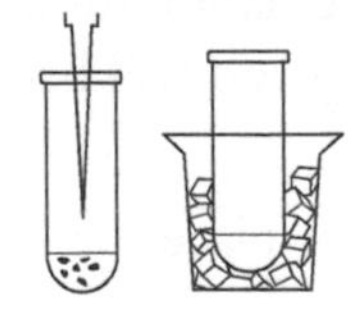

Resuspending the cells in the "+plasmid" tube first allows the cells to preincubate for several minutes at 0°C while the "–plasmid" tube is being prepared. If time permits, both tubes can be preincubated on ice for 5–15 min.

Resuspension is probably the most important variable in obtaining good results.

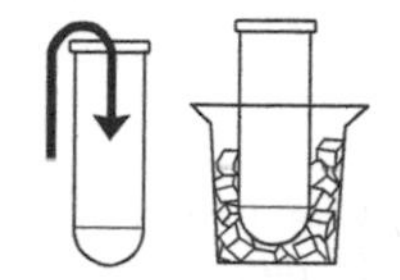

To save plates, the "+LB" or "–LB" plates may be omitted.

It is critical that cells receive a sharp and distinct shock.

If the tubes are not immediately moved from the ice to the water bath, heat shock will not occur. Removing the tubes from the ice and then walking them to the water bath will cause this step to fail.

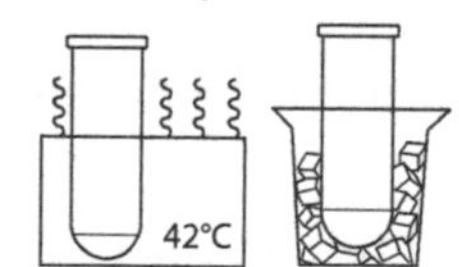

An extended period on ice following the heat shock will not affect the transformation. If necessary, store the "+plasmid" and "–plasmid" tubes on ice in the refrigerator (0°C) for up to 24 h, until there is time to plate the cells. Do not put cell suspensions in the freezer.

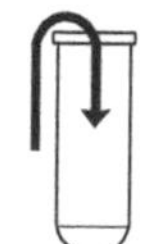

If a shaking water bath is not available, incubate the tubes in a regular 37°C water bath but gently mix the tubes periodically. Alternatively, warm the cells for several minutes in a 37°C water bath and then transfer to a dry shaker inside a 37°C incubator. The cells may be allowed to recover for up to 2 h.

7. Transfer a second mass of cells to the "–plasmid" tube as described in Steps 4 and 5.
8. Return the "–plasmid" tube to ice. Both tubes should be on ice. Double-check that the cells have been completely resuspended in both tubes.
9. Use a micropipette to add 1 µl of plasmid miniprep DNA (from Stage E) directly into the cell suspension in the "+plasmid" tube. Tap the tube with a finger to mix. Avoid making bubbles in the suspension or splashing the suspension up the sides of the tube.
10. Return the "+plasmid" tube to ice. Incubate both tubes for 15 min on ice.
11. While the cells are incubating on ice, use a permanent marker to label two LB plates and two LB/kan plates with your group number, the date, and the following identifiers:
    i. Label one LB/kan plate "+." This is the experimental plate.
    ii. Label the other LB/kan plate "–." This is the negative control.
    iii. Label one LB plate "+." This is a positive control.
    iv. Label one LB plate "–." This is a negative control.
12. Following the 15-min incubation on ice, heat shock the cells in the "+plasmid" and "–plasmid" tubes as follows:
    i. Carry the ice container to the 42°C water bath. Remove the tubes from the ice and immediately immerse them in the 42°C water bath for 90 sec.
    ii. Immediately return both tubes to ice for at least 1 min more.
13. Use a micropipette with a sterile tip to add 250 µL of LB broth to each tube. Gently tap the tubes with your finger to mix.
14. Allow the cells to recover by incubating both tubes in a shaking water bath (with moderate agitation) for 1 h at 37°C.
15. After cell recovery, use a micropipette with a sterile tip to add 100 µL of the appropriate cell suspension to each of the labeled plates from Step 11 (use the matrix below as a guide). Make sure to use a fresh sterile tip for each transfer. Then, immediately spread the cells using either a spreading rod or beads as described below in Parts IB and IC, respectively. Do not allow the suspensions to sit on the plates for too long before spreading.

Plate	Transformed cells ("+plasmid")	Nontransformed cells ("–plasmid")
LB/kan	100 mL	100 mL
LB	100 mL	100 mL

B. Use a Spreading Rod to Spread Transformed Cells on Plates

1. Sterilize the spreading rod by dipping it into the beaker of ethanol and then briefly passing it through a Bunsen burner flame to ignite the ethanol. Allow the ethanol to burn off away from the Bunsen flame; the spreading rod will become too hot if left in the flame.
2. Lift the lid of the plate only enough to allow spreading. Do not place the lid on the lab bench.

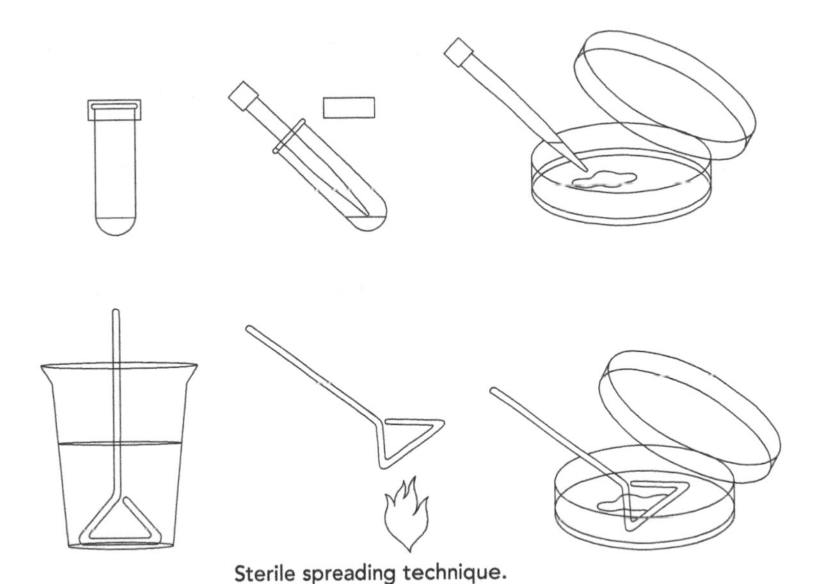
Sterile spreading technique.

CAUTION! Be extremely careful to avoid igniting the ethanol in the beaker. Do not panic if the ethanol is accidentally ignited. Cover the beaker with a glass Petri dish lid or other nonflammable cover to cut off oxygen and rapidly extinguish the fire.

3. Cool the spreader by gently rubbing it on the surface of the agar away from the cell suspension or by touching it to the condensation on the plate lid.
4. Touch the spreader to the cell suspension and gently drag it back and forth several times across the surface of the agar. Then, rotate the plate one-quarter turn and repeat the spreading motion. Try to spread the suspension evenly across the surface of the agar and be careful not to gouge the agar.
5. Replace the plate lid. Return the spreading rod to ethanol without flaming. Let the plate sit for several minutes to allow the suspension to become absorbed into the agar.
6. Repeat Steps 1–5 for the remaining plate(s). After spreading the last plate, reflame the spreader one last time before placing it on lab bench. Proceed to Part I*D* below.

C. Use Beads to Spread Transformed Cells on Plates

1. Lift the lid of the plate enough to add beads. Do not place the lid on the lab bench.
2. Carefully pour five to seven sterile glass spreading beads onto the agar surface.
3. Close the plate lid and use a swirling motion to move the glass beads around the entire surface of the plate. This evenly spreads the cell suspension on the agar surface. Continue swirling until the cell suspension is absorbed into the agar.
4. Repeat Steps 1–3 for the remaining plate(s).
5. After spreading the last plate, invert all plates and gently tap the plate bottoms so that the spreading beads fall onto the plate lids. Carefully pour the beads from each lid into a storage container for reuse. Proceed to Part I*D* below.

D. Incubate the Plates and Clean Up

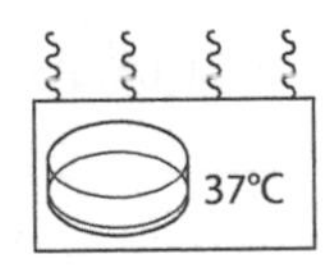

1. Stack the plates and tape them in a bundle to keep them together.
2. Place the plates upside down in a 37°C incubator and incubate for 15–20 h.
3. Take time for responsible cleanup:

i. Segregate bacterial cultures and tubes, pipettes, and micropipette tips that have come into contact with cultures into a "bio bag" or heavy-duty trash bag for proper disposal.

ii. Wipe the lab bench with soapy water and 10% bleach or disinfectant.

iii. Wash your hands before leaving lab.

4. After the initial 15–20-h incubation, store the plates at 4°C to arrest *E. coli* growth and to slow the growth of any contaminating microbes. Once transformants are obtained on the +LB/kan plate, a new RNAi feeding strain has been created. Follow the instructions in Laboratories 4.1 and 4.2 to use this newly created RNAi feeding strain to assess the function of the targeted gene. Note, however, that kanamycin must be used in place of ampicillin when following these procedures.

II. Calculate Transformation Efficiencies

1. While observing the colonies through the bottom of the culture plate, use a permanent marker to mark each colony as it is counted. If the experiment worked well, 50–500 colonies should be observed on the "+" LB/kan control plate and 0 colonies on the "–" LB/kan control plate. (If the colonies are densely packed, draw lines on the bottom of the plate to divide it into equal-sized sections. Count one sector that is representative of the whole plate. After counting, multiply the result by the number of sectors to obtain the number of colonies on the entire plate.) In addition, examine the colonies carefully to detect any possible contamination; contaminating organisms usually look different in color, shape, or size.

An extended recovery period would inflate these numbers. In addition, if plates have been overincubated or left at room temperature for several days, "satellite" colonies may be observed on the LB/kan plates. These satellite colonies radiate from the edges of large, well-established colonies. Nonresistant satellite colonies grow in an "antibiotic shadow" where kanamycin has been broken down by the large resistant colony. Do not include satellite colonies in the count of transformants.

2. Record your observation of each plate in the matrix below. If the cell growth is too dense to count individual colonies, record "lawn."

Plate	Transformed cells ("+plasmid")	Nontransformed cells ("–plasmid")
LB/kan		
LB		

Were the results as expected? Explain possible reasons for variations from the expected results.

3. Compare and contrast the growth on each of the following pairs of plates. What does each pair of results tell you about transformation and/or antibiotic selection?

i. "+" LB and "–" LB.

ii. "–" LB/kan and "–" LB.

iii. "+" LB/kan and "–" LB/kan.

iv. "+" LB/kan and "+" LB.

4. Calculate the transformation efficiency of the LB/kan positive ("+") control as follows. Remember that transformation efficiency is expressed as the number of antibiotic-resistant colonies per microgram of intact plasmid DNA. The object is to determine the mass of plasmid that was spread on each plate and was therefore responsible for the transformants observed.

i. Determine the total mass of plasmid DNA used in Step 9 of Part IA. Assume that the concentration of miniprep DNA is 0.05 μg/μL.

concentration (in μg/μL) x volume (in μL) = mass (in μg).

ii. Determine the fraction of the cell suspension that was spread onto the "+" LB/kan plate in Step 15 of Part IA:

$$\frac{\text{volume suspension spread (see Step 15 of Part I)}}{\text{total volume suspension (see Steps 2 and 13 of Part IA)}} = \text{fraction spread.}$$

iii. Determine the mass of the plasmid in the cell suspension that was spread onto the "+" LB/kan plate:

total mass plasmid (Step 4.i) × fraction spread (Step 4.ii) = mass plasmid spread (in μg).

iv. Determine the number of colonies per microgram of plasmid. Express the answer in scientific notation:

$$\frac{\text{colonies observed (Step 2)}}{\text{mass plasmid spread (Step 4.iii)}} = \text{transformation efficiency (in colonies/μg).}$$

5. What factors might influence transformation efficiency?

6. Let's say that your favorite gene (*YFG*) is cloned into a plasmid and used to transform *E. coli* according to the protocol described in this laboratory. Follow the steps below to calculate the number of molecules (copies) of plasmid that are present in a culture 200 min after transformation.

   i. Using 0.2 μg of plasmid/*YFG*, you achieve a transformation efficiency equal to 10^6 colonies per microgram of intact plasmid/*YFG*. How many transformants are in the culture?

   ii. Plasmid/*YFG* grows at an average copy number of 100 molecules/transformed cell. How many copies of plasmid/*YFG* are in the culture?

   iii. Following heat shock, the entire 250 μL of cell suspension is used to inoculate 25 mL of fresh LB broth. The culture is incubated with shaking at 37°C. The transformed cells enter log phase 60 min after inoculation and then begin to double an average of once every 20 min. After 200 min, the culture enters a stationary phase. How many doublings have occurred? How many molecules (copies) of plasmid/*YFG* are present in the culture?

7. Let's say the transformation protocol described in Part I is used with 10 μL of intact plasmid DNA at different concentrations. The chart below lists the numbers of colonies that are obtained when 100 μL of transformed cells are plated on selective medium. Use the chart to record your answers to the questions that follow.

Concentration (μg/μg)	Mass of DNA (μg)	No. of colonies	Transformation efficiency (colonies/μg)
0.00001		4	
0.00005		12	
0.0001		32	
0.0005		125	
0.001		442	
0.005		542	
0.01		507	
0.05		475	
0.1		516	

i. Use the procedure in Step 4 to calculate the mass of DNA and transformation efficiency at each concentration.

ii. Plot a graph of DNA mass versus colony number.

iii. Plot a graph of DNA mass versus transformation efficiency.

iv. What is the relationship between DNA mass and transformation efficiency?

v. At what point does the transformation reaction appear to be saturated?

vi. What is the true transformation efficiency?

Laboratory Planning and Preparation

Laboratories 4.1–4.4

AT THE OUTSET OF THE LABORATORY EXERCISE, divide the class into teams and assign each team a number. This will make it easier to mark and identify the tubes and plates used in each experiment. It will be important for students to keep track of both the type of bacteria and the type of worm on the plates. Color coding is recommended to make this easier. Label the type of bacteria on the plate (e.g., "*dpy-10* RNAi") in red and the type of worm on the plate (e.g., "*dpy-11*") in black. Pen markings can make it difficult to view the worms; therefore, place all writing on the bottom of the plate along the outside edge.

CULTURING BACTERIA

Most *Caenorhabditis elegans* researchers feed their worms *Escherichia coli* OP50, primarily because it grows slowly and forms low-density lawns that do not overwhelm the worms. However, DH5α, HB101, and other slow-growing strains can also be used. Each of the *E. coli* RNA interference (RNAi) feeding strains used in the laboratory exercises (*dpy-11, dpy-13, bli-1,* and *unc-22*) contains a plasmid that expresses double-stranded RNA (dsRNA) corresponding to a gene targeted for silencing.

E. coli strains can be obtained from the DNA Learning Center (DNALC) through the Silencing Genomes website (www.silencinggenomes.org). Bacterial strains not used for transformation should be stored at 4°C and used within 2 wk of receipt. To maintain strains for longer periods, follow the steps in Part I of Stage A in Laboratory 4.1 to streak the bacterial strains to new plates and incubate them; after incubation, seal the plates and store them at 4°C. Repeat streaking every 4 wk. Bacterial strains used for transformation should be stored sealed at room temperature and maintained by streaking every 2 wk.

E. coli is grown on Luria-Bertani (LB) broth, a nutrient-rich medium composed of salt (NaCl), B vitamins (from yeast extract), and amino acids (from tryptone, which is milk protein digested with trypsin). The addition of agar stiffens LB for culture plates. Presterilized ready-to-pour LB agar is a great convenience. It needs only to be melted in a microwave oven or boiling water bath, cooled to approximately 60°C, and poured into sterile culture plates.

LB plates incorporating ampicillin, tetracycline, or kanamycin are used to maintain selection for antibiotic-resistance genes carried by the *E. coli* RNAi feeding strains. Ampicillin and kanamycin select for bacteria carrying the RNAi feeding vector in Laboratories 4.1–4.3 and 4.4, respectively. The bacterial strain used with the feeding vectors is HT115(DE3); this strain contains a tetracycline-resistance gene that disrupts an RNase gene. However, tetracycline inhibits RNAi and should not be used to culture these strains.

Ampicillin and kanamycin are stable in agar plates, the thresholds for selection are relatively broad, and contaminants are infrequent. These antibiotics, like most, are inactivated by heat. Therefore, it is important to allow the agar solution to cool until the container can be handheld comfortably (~60°C) before adding antibiotics.

C. elegans is grown on plates containing nematode growth medium (NGM) lite, and these plates must be inoculated, or seeded, with the bacterial strain. Instructions for seeding four NGM-lite plates per group with OP50 are provided in Part III of Stage A in Laboratory 4.1. These four plates are sufficient for the exercises in both Laboratories 4.1 and 4.2 (two OP50-seeded NGM-lite plates are used in Stage B of Laboratory 4.1, and the other two OP50-seeded NGM-lite plates are used in Laboratory 4.2).

E. coli RNAi feeding strains *dpy-11*, *bli-1*, and *unc-22* are used in Laboratory 4.2; NGM-lite/amp + IPTG plates seeded with each of these feeding strains are prepared in Stage A of Laboratory 4.1. *E. coli* RNAi feeding strain *dpy-13* is used in Laboratory 4.3. When performing Laboratory 4.3, each group should follow the same procedure as in Stage A of Laboratory 4.1: Prepare a total of four OP50-seeded NGM-lite plates, but instead of preparing three NGM-lite/amp + IPTG plates seeded with *dpy-11*, *bli-1*, and *unc-22*, prepare just one NGM-lite/amp + IPTG plate seeded with *dpy-13*.

Laboratory 4.4 describes how to construct a feeding vector; this vector can then be used to silence the gene of interest by growing bacteria, seeding plates, and feeding worms as described in Laboratories 4.1 and 4.2. Note, however, that Laboratories 4.1 and 4.2 use plates prepared with ampicillin; the ampicillin must be substituted with kanamycin if using a vector prepared in Laboratory 4.4. In addition, make sure to plan for the correct number of plates for the class, depending on the amount of different bacterial strains used.

Selection and Use of Supplies and Equipment

Presterilized, disposable, 5-mL plastic pipettes are convenient and are supplied in a bulk pack or individually wrapped. Open bulk-packed pipettes immediately before use. To dispense, cut one corner of the plastic wrapper at the end opposite to the pipette tips. Avoid touching and contaminating the wrapper opening: Tap the bag to push the pipette end through the cut opening and reclose the bag with tape to keep its contents sterile for future use. To use individually wrapped pipettes properly, peel back only enough of the wrapper to expose the wide end of the pipette and affix the end into the pipette aid or bulb. Completely peel back the wrapper immediately before use.

A Bunsen burner is indispensable to a microbiologist. There are several good reasons for working near a Bunsen burner and flaming a tube mouth after removing a cap and before replacing it. A desktop flame warms the air around it, creating an updraft that can prevent contaminating microbes from falling (by gravity) into open tubes and culture plates. Similarly, flaming the mouth of a tube or container creates an outward convection current that prevents microorganisms from falling in. The direct heat may also kill microbes that can collect on the exposed tube lip. Even so, the effect of flaming may be primarily psychological when fresh sterile supplies are used and manipulations are done quickly. Especially when using sterile plasticware and individually wrapped supplies, flaming can be omitted without compromising sterility. If you feel that you must flame plasticware, do so briefly to avoid melting the plastic.

CULTURING *C. ELEGANS*

All worms can be obtained from the DNA Learning Center (DNALC) through the Silencing Genomes website (www.silencinggenomes.org). Upon receipt, store the worms at temperatures ranging from 11°C to 23°C. Outside of this temperature range, *C. elegans* begins to suffer from temperature shock, which can lead to sterility, chromosome segregation defects, and death. Likewise, when using the worms for experiments, make sure that the temperature stays below 23°C for all incubations at "room temperature."

It is best to store cultures in a covered, clean, plastic or aluminum box. Ensure that the box does not seal completely to avoid mold and asphyxia. If possible, avoid storing materials in cardboard boxes, because doing so increases the chance of mite infestation. Worms should be transferred to new plates as described in Part I of Stage B in Laboratory 4.1 within 1 wk of their receipt.

C. elegans is grown on NGM lite, which contains salt, tryptone, agar, potassium phosphate, potassium dihydrogen phosphate, and cholesterol. "RNAi plates" (NGM-lite/amp + IPTG or NGM-lite/kan + IPTG) include additional components related to the vector in the RNAi feeding strains. The feeding vector—the plasmid that expresses dsRNA in the RNAi feeding strains—possesses an ampicillin- or kanamycin-resistance gene that maintains selection for the vector. IPTG induces expression of T7 RNA polymerase in *E. coli* strain HT115(DE3), which transcribes both strands of the vector target gene insert to form dsRNA. OP50 is a "wild-type" strain of *E. coli* that has no plasmid vector and, thus, should be grown on plain NGM-lite plates.

Chunking and Picking Worms

To expand the number of plates with growing worms, worms can be "chunked" as described in Part I of Stage B in Laboratory 4.1. In Laboratory 4.2, for example, you may wish to chunk mutant strains (*bli-1*, *unc-22*, and *dpy-11*) to OP50-seeded NGM-lite plates for phenotypic comparison. Although this is optional, having mutant worm strains available as references makes it much easier to identify the RNAi phenotypes. Because each strain is homozygous for the mutation, every worm on the plate will exhibit the mutant phenotype. Penetrance—the percentage of worms expressing the knockdown phenotype—will vary among RNAi-treated worms. To minimize the number of plates used and the time needed to pick worms, each student group can perform RNAi with one of the three RNAi feeding strains (*bli-1*, *unc-22*, and *dpy-11*). Groups can then share plates to observe the results of RNAi for each gene.

If you need a worm strain from a plate that is so old that the agar has dehydrated into a chip, it is possible to recover such strains by rehydrating the chip. If the chip is completely dried, it rarely yields viable worms. However, if any part of it still adheres to the plate (meaning that there is still some moisture in it), you may still be able to recover viable worms. Simply add distilled water to the agar chip and let the chip soak for a few hours. Pour the water that has not been absorbed by the agar chip onto a seeded plate to recover any worms that may have floated off the chip. Cut the rehydrated chip into three or four chunks and flip the chunks onto freshly seeded plates. In a couple of days, a few worms might crawl out of the rehydrated chips. Successful recovery of viable worms by this method varies markedly and should not be relied on for general care of worms.

"Picking" individual worms is the starting point for genetic crosses and RNAi experiments. Students should be proficient at picking before starting the RNAi experiments that begin in Laboratory 4.2. Picking worms requires good hand-eye coordination and can be difficult for students. Each student should be encouraged. Emphasize the importance of practicing enough to learn this key skill during Part II of Stage B in Laboratory 4.1. Before Laboratory 4.1, construct one worm pick for each student team using the procedure below. Sterilized toothpicks can be used in place of worm picks; use each toothpick once and then treat as bacterial waste.

Constructing a Worm Pick

REAGENTS, SUPPLIES, & EQUIPMENT

Binocular dissecting microscope
Bunsen burner
File
Pasteur pipette, glass
Platinum wire (28–32 gauge)
Pliers (fine, smooth)
Scissors
Tweezers (optional)

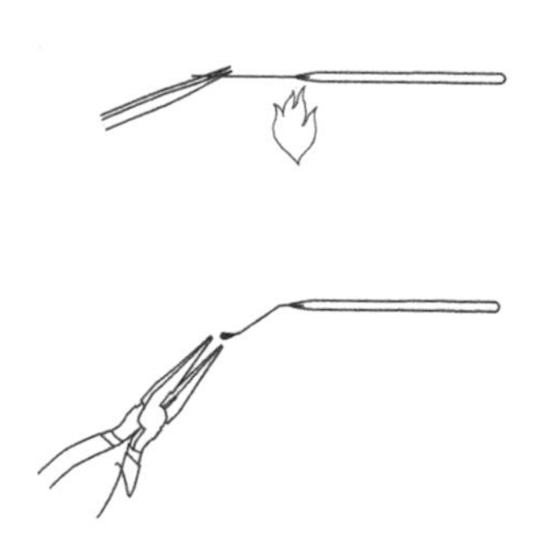

1. Cut a 2–3-cm (1-in) piece of platinum wire. Use 28–32-gauge wire; thicker platinum wire is too stiff and will gouge the agar.
2. Insert the wire into a short-nosed glass Pasteur pipette.
3. Heat the pipette tip over a Bunsen flame. Melt the glass around the wire while holding the wire in a horizontal position with pliers or tweezers (see the drawing at left). When the glass cools, the wire will be held in position.
4. Flatten the tip of the wire into a disc (~2 mm or 1/16 in). Pinch the tip with a pair of fine, smooth pliers; any roughness will make a bad pick. Alternatively, use a smooth metal object to flatten the pick on a hard surface.
5. Examine the tip under a microscope and smooth any rough edges with a file.
6. Bend the tip and shaft at a 45° angle as shown in the drawing at left.

Preventing and Treating Contamination

Worm strains can become contaminated quite easily, especially in a classroom setting. It is advisable to keep more than one plate of commonly used strains and simply throw away any plates that become contaminated. It is important to isolate and dispose of any contaminated strains as soon as possible. To prevent the spread of contamination, tape contaminated plates shut before disposing of them.

Routinely clean the plate storage boxes: Aluminum boxes can be autoclaved, plastic boxes can be bleached, and cardboard boxes can be baked for 1 h at 80°C. In addition, check newly seeded plates for contaminants before using them. It is a good idea to allow seeded plates to grow for a few days at room temperature before using them; this allows time for any contaminants to grow large enough to be identifiable. Sealing plates with Parafilm can help to minimize contamination.

The following is a list of contaminants that can be found on worm plates and suggestions for dealing with them if you cannot simply discard the plate.

1. *Fungi.* Transfer the worms to a fresh plate every 30 min for 2 h. The fungi carried on the worms will become stuck in the bacteria. After a few transfers, the worms should be free of fungal contaminants. If plate containers become contaminated

with fungus, they must be decontaminated or thrown out (if they are made of cardboard).

2. *Non-OP50 bacteria.* Although serial transfer is usually effective for cleaning fungal infections, it is very difficult to remove bacterial contamination using this technique. Instead, it is usually more effective to decontaminate worms in an alkaline hypochlorite solution (see recipe on p. 646). To clean the worms, add 30 µL of hypochlorite solution to the unseeded portion of a fresh plate and then pick ~10 gravid adults (worms with embryos inside) from the contaminated plate into the drop. The adults and most of the bacteria will quickly dissolve, but the eggs will survive. Leave the plate right-side up overnight. The drop of liquid will be absorbed, but the eggs that survived will hatch and the resulting L1s will crawl onto the bacterial lawn. (Note: If dealing with a mutant worm that has a movement defect, it is often necessary to move the hatched L1s to the bacterial lawn. This is generally the case for the more severe *unc* mutations.) On the day after the hypochlorite treatment, move the hatched larvae to a new plate; otherwise, bacteria that survived the hypochlorite treatment will eventually recontaminate the worms.

   Some labs use antibiotics in their agar. Streptomycin or nystatin can be added directly to the cooled NGM-lite solution before pouring. For 1 L of NGM-lite, add 0.14 g of streptomycin sulfate (for a final concentration of 140 µg/mL) or 0.007 g of nystatin (for a final concentration of 7 µg/mL). The streptomycin will reduce bacterial contamination and the nystatin will reduce yeast contaminants. However, neither will reduce mold contamination. (Note: To use this approach, you must have an OP50 strain that can survive the antibiotic treatment. OP50-1, a streptomycin-resistant strain, is available from the Caenorhabditis elegans Genetics Center at http://www.cbs.umn.edu/CGC.)

3. *Phage.* Bacterial phage contamination appears as a clear spot in the bacterial lawn. The bacterial lawn will also become thick and sticky. Perform a hypochlorite treatment as described above to clean phage contamination. It is very difficult to get rid of phage, and they greatly reduce worm strain health.

4. *Mites.* Mites eat nematodes, import fungus, and can cross-contaminate worm strains by carrying larvae or eggs between plates. Mite infestation is a sign that you have (1) too many old plates or (2) a source of mites nearby (fruit fly stocks are a common source of mites). To combat mite infestation, discard any old plates (ones that appear cracked and dried out). Wipe storage containers with ethanol and clean them as described above. In addition, wipe all benchtops with ethanol and bleach and, if possible, set the worm incubators to the highest temperature for a few hours (without the worms inside).

Storing Worms

Worms are, in general, very robust and easy to keep alive. They can survive for long periods of time (weeks to months) as long as they do not get too hot or cold (above 23°C or below 11°C). For incubation at room temperature, make sure that the temperature stays below 23°C; worms will not survive for long if the temperature is higher.

Short-Term Storage

Worms kept at room temperature (20°C–22.5°C) will eventually starve. Such a plate can be maintained for 1 or 2 wk but will eventually dry out and the worms will die. Worms can be recovered from starved plates by cutting a chunk of agar to a fresh plate (see the section Chunking and Picking Worms on p. 629). Instead of chunking, you can also maintain strains by picking worms to a new plate. Plates started by picking only a few worms will not starve as quickly as those maintained by chunking.

Long-Term Storage

Freshly starved, clean plates can be sealed with Parafilm and stored for many months (sometimes up to 1 yr) at 15°C. Plates must be fully starved before long-term storage begins or bacterial growth can deplete the oxygen in a sealed plate and kill the strain. Bacterial contamination often causes strains to become sick or arrests growth of worms; fungal contamination will usually kill a strain stored at 15°C. Therefore, only store clean plates at 15°C and check stocks once a month for fungal contamination.

If you have access to a –80°C freezer or a liquid nitrogen tank, worms can be frozen indefinitely and thawed when needed. To store worms in this manner, follow the procedure below. The plates should be free from contamination, although sometimes this is unavoidable. Keep in mind that freezing a contaminated strain means that the strain will still be contaminated when it is thawed, so it is important to maintain sterile technique while freezing strains. Avoid cross-contamination of strains; never reuse pipettes or tubes.

Cryogenic Storage

REAGENTS, SUPPLIES, & EQUIPMENT

CryoTubes (Nunc) (1 or 1.5 mL)
Freezer at –80°C
Freezing solution
M9 buffer
NGM-lite plates, seeded as described in Laboratory 4.1, Stage A, Part III
Permanent marker
Pipette aid or bulb
Pipettes (5 mL)
Rubber bands
Strain of worms grown on one large (15 cm) plate or three small (6 cm) plates
Styrofoam racks, in which 15-mL disposable centrifuge tubes are packaged
Test tube

1. Freeze the worms *within 1 d of starvation*. There should be many starved L1s on the plate but no food. Starvation probably causes the worms to enter a metabolically altered state that is resistant to damage by freezing. If you wait too long after starvation to begin the freezing process, the worms will burrow into the agar, but if the plates are not completely starved, the freezing will not work.
2. For each strain that you wish to freeze, arrange four Nunc CryoTubes in a styrofoam rack. Use a permanent marker to label each CryoTube with the strain name and the month/year of freezing.
3. Wash the worms from the starved plate into 2 mL of M9 buffer by pipetting M9 onto the plate and gently swirling the worm plate a few times to wash the larvae into the buffer. Use a pipette to move the M9/worm solution to a test tube. Some of the buffer will be lost by absorption into the agar, so you may need to add more M9 buffer to bring the volume back to 2 mL.
4. Add an equal volume (2 mL) of freezing solution to the test tube. Swirl to mix well.

5. Aliquot 1 mL of the mixture to each labeled CryoTube. Cap the tubes and discard the test tube. (Tip: When freezing multiple strains, only uncap the tubes for one strain at a time to avoid cross-contamination.)
6. Freeze the vials slowly to –80°C. This is accomplished by placing the vials in a Styrofoam rack, placing another inverted rack on top of the first, fastening the two racks together with rubber bands (do not use tape because it will lose adhesion in the freezer), and placing the racks in a –80°C freezer.
7. One to 2 d after the tubes have been placed in the freezer, perform a test thaw with one of the four tubes: Allow the tube to thaw at room temperature and empty the contents of the tube onto two or three seeded plates once the thaw is complete. (Note: Do not leave the worms in the defrosted freezing solution for longer than necessary during the thaw, because prolonged exposure to glycerol can be toxic to worms.)
8. After 1 d, viable worms should be observable on the plate(s). Verify that the phenotype is correct. If the test thaw is successful, move the other vials to a permanent storage location (i.e., a well-labeled freezer box). (Note: It is sometimes necessary to move freshly thawed worms to a new plate, because the glycerol from the freezing solution increases bacterial growth and this can overwhelm the plate surface.)

ISOLATING WORM DNA

If a –80°C freezer is available, worms frozen in lysis buffer are stable indefinitely. In Laboratory 4.3, the worms may be frozen at Step 10 of Part III; in Laboratory 4.4, the worms may be stored at Step 8 of Stage B, Part I. If the worms are frozen for a prolonged period of time, the addition of another aliquot (10 µg/µL) of proteinase K may be advisable before proceeding to the next step.

AMPLIFYING DNA BY PCR

Each Ready-To-Go PCR Bead contains reagents so that a final reaction volume of 25 µL contains 2.5 U *Taq* DNA polymerase, 10 mM Tris-HCl (pH 9.0), 50 mM KCl, 1.5 mM $MgCl_2$, and 200 µM of each dNTP. Each primer/loading dye mix includes the appropriate primer pair (0.26 pmol/µL of each primer), 13.8% sucrose, and 0.0081% cresol red.

The lyophilized *Taq* DNA polymerase in the Ready-To-Go PCR Bead becomes active immediately after adding the primer/loading dye mix. In the absence of thermal cycling, nonspecific priming at room temperature allows the *Taq* DNA polymerase to begin generating erroneous products, which can show up as extra bands in gel analysis. Therefore, work quickly! Make sure that the thermal cycler is set and have all experimenters set up their PCRs as a coordinated effort. Add primer/loading dye mix to all reaction tubes, add each template, and begin thermal cycling as quickly as possible. Hold the reactions on ice until all groups are ready to load into the thermal cycler.

For convenience, these laboratories are designed to be used with Ready-To-Go PCR Beads and primer/loading dye mix. If desired, all PCR components can be prepared from scratch by following the recipes included in Recipes for Reagents and Stock Solutions. Each 25-µL reaction should contain the following:

PCR buffer (10x)	2.5 μL
dNTPs (10 mM)	1.0 μL
Forward primer (20 μM)	1.25 μL
Reverse primer (20 μM)	1.25 μL
Taq polymerase (1–5 U/μL)	1.0 μL
Template DNA	2.5 μL
Distilled or deionized water	15.5 μL

Surprisingly, a large number of failures in this experiment are due to the simple fact that students did not get template DNA into their PCRs! Observe each student adding DNA to the reaction. Students must see the small volume of template DNA in their pipette tip, then eject the DNA directly into the reaction, and finally confirm that the tip is empty before removing it from the PCR tube.

If your thermal cycler does not have a heated lid, add a drop of mineral oil on top of each reaction before thermal cycling. Be careful to avoid touching the dropper tip to the tube or sample; this will contaminate the mineral oil.

PCR amplification from crude cell extracts is biochemically demanding and requires the precision of automated thermal cycling. However, amplification of short DNA fragments such as the ones in these experiments is not technically difficult. Therefore, the recommended amplification times and temperatures should work adequately for all types of thermal cyclers.

ANALYZING DNA BY GEL ELECTROPHORESIS

Preparing and Loading Gels

The cresol red and sucrose in the primer mix function as loading dye, so that amplified samples can be loaded directly into an agarose gel. This is a nice time-saver. However, because the concentrations of sugar and cresol red are low, this mix is more difficult to use than typical loading dyes. Students should be encouraged to load carefully.

If the Ready-To-Go PCR Beads and primer/loading dye mix are not used and, instead, a traditional PCR is run, 10x loading dye should be added to each reaction before it is run out on the gel.

Plasmid pBR322 digested with the restriction endonuclease BstNI is an inexpensive marker and produces fragments that are useful as size markers in the experiment. Use 20 μL of a 0.1 μg/μL stock solution of this DNA ladder per gel. Other markers or a 100-bp ladder may be substituted.

If mineral oil was used during PCR, pierce the pipette tip through the layer of mineral oil to withdraw the sample. Do not pipette any mineral oil; leave the mineral oil behind in the original tube.

Staining Gels

In Part V of Laboratory 4.3 and Stages B (Part III) and D (Part II) of Laboratory 4.4, SYBR Green, a fluorescent staining dye, is added to each sample before electrophoresis. According to the Ames test (a method used to estimate the mutagenic properties of a chemical), SYBR Green is much less mutagenic than the classical stain ethidium bromide. Furthermore, bands are immediately visible after electrophoresis and no poststaining steps are necessary. SYBR Green can be imaged with the same filter set and UV transillumination that are used with ethidium bromide, but SYBR Green is more sensitive than ethidium bromide and produces much less background on a stained agarose gel.

As an alternative to SYBR Green, the gel can be stained with ethidium bromide or *Carolina*BLU after electrophoresis but before viewing and photographing. To stain the gel using ethidium bromide or *Carolina*BLU, omit Step 7 from the procedure (Part V of Laboratory 4.3; Part III, Stage B of Laboratory 4.4; or Part II, Stage D of Laboratory 4.4) and then, after Step 9, follow one of the methods described under "Staining a Gel with Ethidium Bromide or *Carolina*BLU" below.

Always view and photograph gels as soon as possible after electrophoresis or staining/destaining. Over time, the small-sized PCR products will diffuse through the gel, lose sharpness, and disappear. Refrigeration will slow diffusion somewhat, so if absolutely necessary, gels can be wrapped in plastic wrap and stored for up to 24 h at 4°C. For best results, view and photograph gels immediately after staining/destaining is complete.

Staining a Gel with Ethidium Bromide or Carolina*BLU*

REAGENTS, SUPPLIES, & EQUIPMENT

*Carolina*BLU Final Stain (for *Carolina*BLU staining only)
Digital camera or photodocumentary system
Ethidium bromide (1 µg/mL) <!> (for ethidium bromide staining only)
Gel containing DNA fragments that have been separated by electrophoresis
Staining trays
UV transilluminator <!> and eye protection (for ethidium bromide staining only)
White light transilluminator (for *Carolina*BLU staining only)

See Cautions Appendix for appropriate handling of materials marked with <!>.

A. Staining with Ethidium Bromide

1. Place your gel in a staining tray and cover it with ethidium bromide solution. Allow it to stain for 10–15 min.
2. Decant the stain back into the storage container for reuse and rinse the gel in tap water.
3. View the gel using UV transillumination. Photograph the gel using a digital camera or photodocumentary system.

B. Staining with Carolina*BLU*

1. Place your gel in a staining tray and cover it with *Carolina*BLU Final Stain. Allow it to sit for 20–30 min with (optional) gentle agitation.
2. After staining, pour the stain back into the bottle for future use. (The stain can be used six to eight times.)
3. To destain the gel, cover it with deionized or distilled water.
4. Change the water three to four times during the course of 30–40 min. Agitate the gel occasionally.
5. View the gel using white light transillumination. Photograph the gel using a digital camera or photodocumentary system.

If desired, *Carolina*BLU Gel and Buffer Stain can be used to stain the DNA while it is being electrophoresed. This prestaining procedure will allow you to visualize your results before the end of the gel run. However, the standard staining procedure outlined in "Staining with *Carolina*BLU" above is still required for optimum viewing.

To prestain the gel during electrophoresis, add *Carolina*BLU Gel and Buffer Stain in the amounts indicated in the tables below. Note that the amount of stain added is dependent on the voltage used for electrophoresis. Do not use more stain than recommended. This may precipitate the DNA in the wells and create artifact bands.

Volume of *Carolina*BLU stain to add to the agarose gel:

Voltage	Agarose gel volume	Stain volume
<50 Volts	30 mL	40 µL (1 drop)
	200 mL	240 µL (6 drops)
	400 mL	520 µL (13 drops)
>50 Volts	50 mL	80 µL (2 drops)
	300 mL	480 µL (12 drops)
	400 mL	640 µL (16 drops)

Volume of CarolinaBLU stain to add to the 1x TBE buffer:

Voltage	1x TBE buffer volume	Stain volume
<50 Volts	500 mL	480 µL (12 drops)
	3000 mL	3 mL (72 drops)
>50 Volts	500 mL	960 µL (24 drops)
	2600 mL	5 µL (125 drops)

Gels containing *Carolina*BLU may be prepared 1 d ahead of the lab day, if necessary. However, gels stored longer tend to fade and lose their ability to stain DNA bands during electrophoresis.

RECOMBINATION

The vector used in Laboratory 4.4 (Part V of Stage B) is pPR244. It can be obtained from the DNA Learning Center (DNALC) through the Silencing Genomes website (www.silencinggenomes.org).

Obtain fresh BP Clonase mix. BP Clonase comes with a 2-µg/µL proteinase K solution that is also required to complete the procedure. BP Clonase should be stored carefully on dry ice or in a –80°C freezer when not in use. If neither is available, a –20°C freezer will work. BP Clonase is very sensitive to heat, so minimize the amount of time that the enzyme mix is thawed. The mix may be dispensed directly into each experimenter's reaction tube by the instructor if there is concern that the students will waste or mishandle it. Make sure to return any remaining enzyme mix to the freezer as soon as possible.

TRANSFORMING BACTERIA

Classic Transformation

Most transformation protocols can be conceptualized as four major steps.

1. *Preincubation.* Cells are suspended in a solution of cations and incubated at 0°C. The cations are thought to complex with exposed phosphates of lipids in the *E. coli* cell membrane. The low temperature freezes the cell membrane, stabilizing the distribution of charged phosphates.
2. *Incubation.* DNA is added, and the cell suspension is incubated again at 0°C. The cations are thought to neutralize negatively charged phosphates in the DNA and cell membrane. With these charges neutralized, the DNA molecule is free to pass through the cell membrane.
3. *Heat shock.* The cell/DNA suspension is briefly incubated at 42°C and then returned to 0°C. The rapid temperature change creates a thermal imbalance on either side of the *E. coli* membrane, which is thought to create a draft that sweeps plasmids into the cell.
4. *Recovery.* LB broth is added to the cell/DNA suspension and incubated at 37°C (ideally with shaking) before plating on selective media. Transformed cells recover from the treatment, amplify the transformed plasmid, and begin to express the antibiotic-resistance protein.

It is important to follow the transformation procedures in Laboratory 4.4 very carefully to ensure success. Emphasize the importance of keeping the cells on ice and heat-shocking the cells abruptly, because these steps are critical.

See pages 607–608 for instructions on how to prepare competent cells for Stage C of Laboratory 4.4. If possible, schedule experiments so that competent cells are prepared 1 d before transformation with the recombinant molecules produced in Part V of Stage B. "Seasoning" cells for 12–24 h at 0°C (an ice bath inside the refrigerator) generally increases transformation efficiency fivefold to 10-fold. This enhanced efficiency will help to ensure successful cloning of recombinant molecules. Alternatively, purchase highly competent cells and store as directed until needed.

pKAN (Carolina Biological), the control plasmid used in Stage C of Laboratory 4.4, is resistant to kanamycin. Kanamycin interferes with translation, killing nonresistant bacteria. It is therefore critical that transformed cells be incubated for at least 1 h with shaking after recovery to allow time for the resistance gene to be expressed. If cells are plated prematurely, even cells containing the plasmid will be killed.

Colony Transformation

Colony transformation is used in Stage F of Laboratory 4.4; it is a simplification of the classic transformation protocol used in Stage C, which requires midlog phase cells grown in liquid culture. This abbreviated protocol begins with *E. coli* colonies scraped from an agar plate. As in classic transformation, it is critical to incubate cells for at least 1 hr after colony transformation to allow cells to express kanamycin resistance before plating the cells. The procedure entails minimal preparation time and is virtually foolproof. However, what is gained in simplicity and time is lost in efficiency.

The transformation efficiencies achieved with the colony protocol (5×10^3 to 5×10^4 colonies per microgram of plasmid) are 2–200 times less than those of the classic protocol (5×10^4 to 5×10^6 colonies per microgram). Colony transformation is perfectly suitable for transforming *E. coli* with purified intact plasmid DNA. However, it will give poor results with ligated DNA, which is composed of relaxed, circular plasmid and linear plasmid DNA. These forms yield 5–100 times fewer transformants than an equivalent mass of intact supercoiled plasmid.

Prolonged reculturing (passaging) of *E. coli* can result in a loss of competence that makes the bacteria virtually impossible to transform using the colony method. There is some evidence that loss of transforming ability may also result from exposure of cells to temperatures below 4°C. Therefore, take care to store stab/slant cultures and streaked plates at room temperature. If there is a severe drop in the number of transformants—from the expected 50–500 colonies per plate to essentially zero—discard the culture and obtain a fresh one.

Plan ahead. Be sure to have a streaked plate or stab/slant culture of viable *E. coli* HT115(DE3) cells from which to streak starter plates. One day before performing the experiment described in Stage F of Laboratory 4.4, streak out several fresh starter plates of *E. coli* HT115(DE3) on LB plates. In addition, streak the *E. coli* strain on an LB/kan plate to make sure that a kanamycin-resistant strain has not been used in error.

Selection and Use of Reagents, Supplies, and Equipment

Presterilized Supplies

Presterilized supplies can be used to good effect in transformations; 15-mL culture tubes and individually packaged 100–1000-μL micropipette tips are handy. A 3-mL transfer pipette, marked in 250-μL gradations, can be substituted for a 100–1000-μL micropipette with no loss of speed or accuracy.

Technically, everything used in this experiment should be sterilized. However, it is acceptable to use clean but nonsterile 1.5-mL tubes for aliquots of calcium chloride, LB broth, and plasmid DNA, provided that they will be used within 1 or 2 d. Clean, nonsterile 1–10-μL micropipette tips can be used for adding DNA to cells. Plastic supplies, if not handled before use, are rarely contaminated. Antibiotic selection covers such minor lapses of sterile technique.

Test Tubes

The type of test tube used is a critical factor in achieving high-efficiency transformation and may also be important in the colony transformation protocol. Therefore, we recommend using a presterilized 15-mL (17 x 100-mm) polypropylene culture tube. The critical heat-shock step has been optimized for the thermal properties of a 15-mL polypropylene tube. Tubes of different materials (such as polycarbonate) or thicknesses conduct heat differently. In addition, the small volume of cell suspension forms a thin layer across the bottom of a 15-mL tube, allowing heat to be quickly transferred to all cells. A smaller tube (such as a 1.5-mL tube) increases the depth of the cell suspension through which heat must be conducted. Thus, any change in tube specifications requires recalibrating the duration of the heat shock to obtain optimal transformation efficiency. The Becton Dickinson Falcon 2059 is the standard for transformation experiments.

Glass Spreading Beads

An alternative to using a traditional cell spreader is to use sterile glass spreading beads. Place five to seven 3-mm silica beads on each agar plate after adding the cell suspension. Swirl the beads around the plate until the cells have been evenly spread. No flame/ethanol is required for this method, thus lowering potential fire hazards. If using spreading beads, prepare four tubes of beads for each group before beginning the experiment. Carefully place five to seven beads into each sterile 1.5-mL tube; the tube can be used as a scooper. The beads can be used directly from the package or autoclaved before use.

Water Bath

A constant-temperature water bath can be made by maintaining a trickle flow of 42°C tap water into a Styrofoam box. Monitor the temperature with a thermometer. An aquarium heater can be used to maintain the temperature.

Ethanol

Maintain a beaker of ethanol exclusively for flaming. To retard evaporation, keep the ethanol in a beaker covered with Parafilm, plastic wrap, or, if using a small beaker, the lid from a Petri dish.

Purified Water

Extraneous salts and minerals in the transformation buffer can affect the results. Use the most highly purified water available; pharmacy-grade distilled water is recommended. It might pay to obtain from a local research center or hospital several liters of water purified through a multistage ion-exchange system, such as Milli-Q.

Recipes for Reagents and Stock Solutions

Laboratories 4.1–4.4

THE SUCCESS OF THE LABORATORY EXERCISES DEPENDS on the use of high-quality reagents. Follow the recipes with care and pay attention to cleanliness. Use a clean spatula for each ingredient or carefully pour each ingredient from its bottle.

The recipes are organized alphabetically within seven sections, according to experimental procedure. Stock solutions that are used for more than one procedure are listed once, according to their first use in the laboratories.

CAUTION: See the Cautions Appendix for appropriate handling of materials marked with <!>.

- Culturing Bacteria
  - Ampicillin (10 mg/mL)
  - Ethanol:Water (1:1)
  - Kanamycin (2.5 mg/mL)
  - Luria-Bertani (LB) Agar Plates
  - Luria-Bertani (LB) Broth
  - Sodium Hydroxide (4 N)
- Culturing *C. elegans*
  - Cholesterol (5 mg/mL)
  - Freezing Solution
  - Hypochlorite Solution
  - Isopropyl-β-D-Thiogalactopyranoside (IPTG) (1 M)
  - M9 Buffer
  - Magnesium Sulfate ($MgSO_4$) (1 M)
  - NGM-Lite Plates
  - Sodium Hydroxide (1 N)
- Isolating Worm DNA
  - Lysis Buffer
  - Magnesium Chloride ($MgCl_2$) (1 M)
  - PCR Buffer (10x)
  - Potassium Chloride (KCl) (5 M)
  - Proteinase K (20 mg/mL)
  - Tris-HCl (1 M, pH 8.0 and 8.3)
- Amplifying DNA by PCR
  - Cresol Red Dye (1%)
  - Cresol Red Loading Dye
  - dNTPs (10 mM)
  - Primer/Loading Dye Mix
- Analyzing DNA by Gel Electrophoresis
  - 100-bp DNA Ladder (0.125 μg/μL)
  - Agarose (1%)

Ethidium Bromide Staining Solution (1 µg/mL)
Loading Dye (10x)
pBR322/BstNI Size Marker (0.1 µg/µL)
SYBR Green DNA Stain
Tris/Borate/EDTA (TBE) Electrophoresis Buffer (1x)
Tris/Borate/EDTA (TBE) Electrophoresis Buffer (10x)

- Transforming Bacteria
  - Calcium Chloride ($CaCl_2$) (1 M)
  - Calcium Chloride ($CaCl_2$) (50 mM)
  - Classic Competent Cells
- Preparing Plasmid Minipreps
  - Ethylenediaminetetraacetic Acid (EDTA) (0.5 M, pH 8.0)
  - Glucose/Tris/EDTA (GTE)
  - Potassium Acetate (5 M)
  - Potassium Acetate/Acetic Acid
  - Sodium Dodecyl Sulfate (SDS) (1%)/Sodium Hydroxide (0.2 N)
  - Sodium Dodecyl Sulfate (SDS) (10%)
  - Tris/EDTA (TE) Buffer

General Notes

- Typically, solid reagents are dissolved in a volume of deionized or distilled water equivalent to 70%–80% of the finished volume of buffer. This leaves room for the addition of acids or base to adjust the pH. Finally, water is added to bring the solution up to the final volume.
- Buffers are typically prepared as 1x or 10x solutions. Solutions at a concentration of 10x are diluted when mixed with other reagents to produce a working concentration of 1x.
- Storage temperatures of 4°C and –20°C refer to normal refrigerator and freezer temperatures, respectively.

CULTURING BACTERIA

Ampicillin (10 mg/mL)

Makes 100 mL.
Store for 1 yr at –20°C or for 3 mo at 4°C.

1. Add 1 g of ampicillin sodium salt <!> (MW 371.40) to 100 mL of deionized or distilled water in a clean 250-mL flask. Stir to dissolve.
2. Prewash a 0.45- or 0.22-µm sterile filter (Nalgene or Corning) by drawing 50–100 mL of deionized or distilled water through the filter.
3. Pass the ampicillin solution through the washed filter.
4. Dispense 10-mL aliquots into sterile 15-mL tubes (Falcon 2059 or equivalent) and freeze at –20°C.

Note: Use ampicillin sodium salt, which is very soluble in water; the free acid form is difficult to dissolve.

Ethanol:Water (1:1)

Makes 100 mL.
Store indefinitely at room temperature.

1. Add 50 mL of 95%–100% ethanol <!> to 50 mL of distilled or deionized water in a clean 200-mL bottle.
2. Mix well and store in a sealed container.

Kanamycin (2.5 mg/mL)

Makes 100 mL.
Store for 1 yr at –20°C or for 3 mo at 4°C.

1. Add 0.25 g of kanamycin sulfate <!> (MW 582.6) to 100 mL of deionized or distilled water in a clean 250-mL flask. Stir to dissolve.
2. Prewash a 0.45- or 0.22-µm sterile filter (Nalgene or Corning) by drawing 50–100 mL of deionized or distilled water through the filter.
3. Pass the kanamycin solution through the washed filter.
4. Dispense 10-mL aliquots in sterile 15-mL tubes (Falcon 2059 or equivalent) and freeze at –20°C.

Luria-Bertani (LB) Agar Plates

Makes 35–40 plates.
Store for 3 mo at 4°C or room temperature without antibiotics. If antibiotics are added, store for 3 mo at 4°C.

1. Weigh out the following:

   Tryptone 10 g
   Yeast extract 5 g
   NaCl (MW 58.44) 10 g
   Agar 15 g

   Alternatively, use 40 g of premix containing all of these ingredients.
2. Add all ingredients to a clean 2-L flask that has been rinsed with deionized or distilled water.
3. Add 1 L of deionized or distilled water to the flask.
4. Add 0.5 mL of 4 N NaOH <!>.
5. Stir the solution to dissolve the dry ingredients, preferably using a magnetic stir bar. Any undissolved material will dissolve during autoclaving.
6. Cover the mouth of the flask with aluminum foil and autoclave the solution for 15 min at 121°C.
7. Allow the solution to cool in a water bath until the flask can be held by bare hands (55°C–60°C). If the solution cools too long and the agar begins to solidify, remelt

by briefly autoclaving for 5 min or less or by heating the solution in a microwave oven for a few minutes.

8. While the agar is cooling, carefully cut open the end of the plastic sleeves containing presterilized polystyrene plates and save the sleeves for storing the poured plates. Spread the culture plates out on the lab bench in preparation for pouring. Mark the bottoms with the date and a description of the media (e.g., LB, LB/amp, LB/kan, LB/amp + tet, or LB/kan + tet).
9. If antibiotics are required, sterilely add 10 mL of stock antibiotic (10 mg/mL ampicillin <!> or 2.5 mg/mL kanamycin <!>) when the agar flask is cool enough to hold. Swirl the flask to mix the antibiotic and agar solution.
10. When the agar flask is cool enough to hold and antibiotics have been added (if required), lift the lid of a culture plate only enough to pour the solution. Do not place the lid on the lab bench. Quickly pour just enough agar to cover the bottom of the plate (~25–30 mL). Tilt the plate to spread the agar evenly and immediately replace the lid.
11. Continue pouring agar into plates. Occasionally flame the mouth of the flask to maintain sterility.
12. To remove bubbles in the surface of the poured agar, touch the plate surface with the flame from a Bunsen burner while the agar is still in a liquid state.
13. Allow the agar to solidify undisturbed.
14. If possible, incubate the plates lid-side down for several hours at 37°C (overnight if convenient). This dries the agar, limiting condensation when plates are stored under refrigeration. It also allows the ready detection of any contaminated plates.
15. Stack the plates in their original sleeves for storage.

Notes: Ampicillin and kanamycin are destroyed by heat; therefore, it is essential to cool the agar before adding the antibiotic. In a pinch, antibiotic-containing plates can be made quickly by evenly spreading 200 μL of stock antibiotic (10 mg/mL ampicillin or 2.5 mg/mL kanamycin) on the surface of an LB agar plate. Allow the agar to absorb the antibiotic for 10–20 min before use. Outdated antibiotic plates can also be refurbished in this manner.

Luria-Bertani (LB) Broth

Makes 1 L.
Store indefinitely at room temperature without antibiotics or for 3 mo at 4°C with antibiotics.

1. Weigh out the following:

Tryptone	10 g
Yeast extract	5 g
NaCl (MW 58.44)	10 g

   Alternatively, use 25 g of premix containing all of these ingredients.
2. Add all ingredients to a clean 2-L flask that has been rinsed with deionized or distilled water.

3. Add 1 L of deionized or distilled water to the flask.
4. Add 0.5 mL of 4 N NaOH <!>.
5. Stir the solution to dissolve the dry ingredients, preferably using a magnetic stir bar.
6. Aliquot and sterilize the solution as follows:
   - If preparing LB broth for midlog cultures, split the LB broth into two 500-mL aliquots, each in a 2-L flask. Plug the top of the flask with cotton or foam and cover with aluminum foil. (Alternatively, cover with aluminum foil only.) Autoclave for 15–20 min at 121°C.
   - If preparing LB broth for general use in transformations, dispense 100-mL aliquots into each of 10 150–250-mL bottles. Loosely put on the caps and autoclave for 15–20 min at 121°C. To help guard against breakage, autoclave the bottles in a shallow pan with a small amount of water. (Alternatively, prewash a 0.45- or 0.22-µm sterile filter [Nalgene or Corning] by drawing 50–100 mL of deionized or distilled water through the filter. Then, pass the LB broth through the filter and dispense 100-mL aliquots into each of 10 150–250-mL bottles.)
7. If antibiotics are required, sterilely add 1 mL of stock antibiotic (10 mg/mL ampicillin <!> or 2.5 mg/mL kanamycin <!>) to 100 mL of cool LB broth and swirl to mix.

Note: LB broth can be considered sterile as long as the solution remains clear. Cloudiness is a sign of contamination by microbes. Always swirl the solution to check for bacterial or fungal cells that may have settled at the bottom of the flask or bottle.

Sodium Hydroxide (4 N)

Makes 100 mL.
Store at room temperature (indefinitely).

1. Slowly add 16 g of NaOH <!> pellets (MW 40.00) to 80 mL of deionized or distilled water, with stirring. The solution will get very hot.
2. When the NaOH pellets are completely dissolved, add water to a final volume of 100 mL.

CULTURING *C. ELEGANS*

Cholesterol (5 mg/mL)

Makes 10 mL.
Store for 1 yr at –20°C or for 3 mo at 4°C.

1. Add 50 mg of cholesterol to 10 mL of 100% ethanol <!> in a sterile 15-mL tube (Falcon 2059 or equivalent).
2. Stir to dissolve.

Freezing Solution

Makes 1 L.
Store indefinitely at room temperature.

1. Add the following ingredients to 500 mL of deionized or distilled water in a 2-L flask:

NaCl	5.85 g
KH_2PO_4	6.8 g
Glycerol	300 g
NaOH (1 N) <!>	5.6 mL
$MgSO_4$ (1 M) <!>	300 µL

2. Stir to dissolve, preferably using a magnetic stir bar.
3. Add deionized or distilled water to bring the total volume to 1 L. Transfer to a 1-L bottle.
4. Make sure that the bottle cap is loose and autoclave for 15 min at 121°C.
5. After autoclaving, cool the solution to room temperature and tighten the lid for storage.

Hypochlorite Solution

Makes 500 µL.
Store for 1 wk at room temperature.

1. Add the following to a 1.5-mL tube:

NaOH (4 N) <!>	200 µL
Bleach with 10%–20% hypochlorite (NaOCl) <!>	300 µL

2. Mix by pipetting in and out and keep at room temperature until use.

Note: Commercial bleach brands vary in hypochlorite concentration, so the concentration of NaOH and relative volumes used may need to be adjusted.

Isopropyl-β-D-Thiogalactopyranoside (IPTG) (1 M)

Makes 5 mL.
Store indefinitely at –20°C.

1. Add 1.19 g of IPTG <!> (MW 238.3) to 5 mL of deionized or distilled water in a clean 15-mL tube.
2. Mix or rock gently to dissolve.
3. Pass the IPTG solution through a 0.45- or 0.22-µm sterile filter (Nalgene or Corning).
4. Dispense the solution into a sterile 15-mL tube (Falcon 2059 or equivalent) and freeze at –20°C.

M9 Buffer

Makes 1 L.
Store indefinitely at room temperature.

1. Add the following ingredients to 500 mL of deionized or distilled water in a 1-L bottle:

KH_2PO_4	3 g
Na_2HPO_4	6 g
NaCl	5 g
$MgSO_4$ (1 M) <!>	1 mL

2. Stir to dissolve, preferably using a magnetic stir bar.
3. Add deionized or distilled water to bring the total volume to 1 L.
4. Make sure that the bottle cap is loose and autoclave the solution for 15 min at 121°C.
5. After autoclaving, cool the solution to room temperature and tighten the cap for storage.

Magnesium Sulfate ($MgSO_4$) (1 M)

Makes 100 mL.
Store indefinitely at room temperature.

1. Dissolve 24.6 g of $MgSO_4 \cdot 7H_2O$ <!> (MW 246.47) in 80 mL of deionized or distilled water.
2. Add deionized or distilled water to make a total volume of 100 mL of solution.
3. Make sure that the bottle cap is loose and autoclave for 15 min at 121°C.
4. After autoclaving, cool the solution to room temperature and tighten the lid for storage.

NGM-Lite Plates

Makes 35–40 plates.
Store for 2 mo at 4°C.

1. Weigh out the following:

NaCl	2 g
Tryptone	4 g
KH_2PO_4	3 g
K_2HPO_4	0.5 g
Agar	20 g

2. Add all ingredients to a clean 2-L flask that has been rinsed with deionized or distilled water.
3. Add 1 L of deionized or distilled water to the flask.
4. Add 1 mL of 5 mg/mL cholesterol (prepared in 100% ethanol<!>).
5. Stir the solution to dissolve the dry ingredients, preferably using a magnetic stir bar. Any undissolved material will dissolve during autoclaving.

6. Cover the mouth of the flask with aluminum foil and autoclave the solution for 15 min at 121°C.
7. Allow the solution to cool in a water bath until the flask can be held by bare hands (55°C–60°C). (If the solution cools too long and the agar begins to solidify, remelt by briefly autoclaving for 5 min or less or heat the solution in a microwave oven for a few minutes.)
8. While the agar is cooling, carefully cut open the end of the plastic sleeves containing presterilized 6-cm plastic Petri dishes and save the sleeves for storing the poured plates. Spread the dishes out on the lab bench in preparation for pouring. Mark the bottoms with the date and a description of the media (e.g., NGM-lite, NGM-lite/amp + IPTG, or NGM-lite/kan + IPTG).
9. If plates containing antibiotics and IPTG are required, sterilely add 10 mL of 10 mg/mL ampicillin <!> or 2.5 mg/mL kanamycin <!> and 0.4 mL of 1 M IPTG <!> when the agar flask is cool enough to hold. Swirl the flask to mix the antibiotics and IPTG with the agar solution.
10. When the agar flask is cool enough to hold and antibiotics and IPTG have been added (if required), lift the lid of a Petri dish only enough to pour the solution. Do not place the lid on the lab bench. Add enough to fill the dishes about half full. Immediately replace the lid.
11. Continue pouring agar into the dishes. Occasionally flame the mouth of the flask to maintain sterility.
12. To remove bubbles from the surface of the poured agar, touch the plate surface with the flame from a Bunsen burner while the agar is still liquid. (This is necessary if the plate has bubbles that are large enough for worms to crawl inside.)
13. Allow the agar to solidify undisturbed.
14. Allow the plates to cool on the bench for at least 1 d at room temperature before seeding them.
15. Stack the dishes in their original sleeves for storage.

Notes: Eric Lambie originally described NGM-lite plates in the February 1995 issue of *Worm Breeder's Gazette.* These plates contain a simple, complete medium for culturing *C. elegans.* Antibiotics and IPTG are destroyed by heat; therefore, it is essential to cool the agar before adding them.

Sodium Hydroxide (1 N)

Makes 100 mL.
Store indefinitely at room temperature.

1. Slowly add 4 g of NaOH <!> pellets (MW 40.00) to 80 mL of deionized or distilled water, with stirring. The solution will get very hot.
2. When the NaOH pellets are completely dissolved, add water to a final volume of 100 mL.

ISOLATING WORM DNA

Lysis Buffer

Makes 100 µL.
Prepare fresh on day of use.

1. Combine the following in a 1.5-mL tube:

Distilled or deionized water	85 µL
PCR buffer (10x)	10 µL
Proteinase K (20 mg/mL) <!>	5 µL

2. Mix well, aliquot 35 µL per group, and store aliquots on ice.

Magnesium Chloride ($MgCl_2$) (1 M)

Makes 100 mL.
Store indefinitely at room temperature.

1. Dissolve 20.3 g of $MgCl_2 \cdot 6H_2O$ <!> (MW 203.30) in 80 mL of deionized or distilled water.
2. Add deionized or distilled water to make a total volume of 100 mL of solution.
3. Make sure that the bottle cap is loose and autoclave for 15 min at 121°C.
4. After autoclaving, cool the solution to room temperature and tighten the lid for storage.

PCR Buffer (10x)

Makes 10 mL.
Store indefinitely at –20°C.

1. Combine the following in a 15-mL tube:

Deionized or distilled water	7.85 mL
Tris-HCl (1 M, pH 8.3)	1 mL
KCl (5 M) <!>	1 mL
$MgCl_2$ (1 M)	0.15 mL

2. Mix well.

Potassium Chloride (KCl) (5 M)

Makes 100 mL.
Store indefinitely at room temperature.

1. Dissolve 37.3 g of KCl <!> (MW 74.55) in 70 mL of deionized or distilled water.
2. Add deionized or distilled water to make a total volume of 100 mL of solution.
3. Make sure that the bottle cap is loose and autoclave for 15 min at 121°C.
4. After autoclaving, cool the solution to room temperature and tighten the lid for storage.

Proteinase K (20 mg/mL)

Makes 10 mL.
Store indefinitely at –20°C.

1. Dissolve 200 mg of proteinase K <!> in deionized or distilled water to a final volume of 10 mL.
2. Make 1-mL aliquots in sterile 1.5-mL tubes.

Tris-HCl (1 M, pH 8.0 and 8.3)

Makes 100 mL.
Store indefinitely at room temperature.

1. Dissolve 12.1 g of Tris base <!> (MW 121.10) in 70 mL of deionized or distilled water.
2. Adjust the pH by slowly adding concentrated hydrochloric acid (HCl) <!> for the desired pH listed below.

pH 8.0	5.0 mL
pH 8.3	4.5 mL

   Monitor with a pH meter or strips of pH paper. (If neither is available, adding the volumes of concentrated HCl listed here will yield a solution with approximately the desired pH.)
3. Add deionized or distilled water to make a total volume of 100 mL of solution.
4. Make sure that the bottle cap is loose and autoclave for 15 min at 121°C.
5. After autoclaving, cool the solution to room temperature and tighten the lid for storage.

Notes: A yellow-colored solution indicates poor-quality Tris. If your solution is yellow, discard it and obtain a Tris solution from a different source. The pH of Tris solutions is temperature dependent, so make sure to measure the pH at room temperature. Many types of electrodes do not accurately measure the pH of Tris solutions; check with the manufacturer to obtain a suitable one.

AMPLIFYING DNA BY PCR

Cresol Red Dye (1%)

Makes 50 mL.
Store indefinitely at room temperature..

1. Weigh out 500 mg of cresol red dye <!>.
2. In a 50-mL tube, mix the cresol red dye with 50 mL of distilled water.

Cresol Red Loading Dye

Makes 50 mL.
Store indefinitely at –20°C.

1. In a 50-mL tube, dissolve 17 g of sucrose in 49 mL of distilled water.
2. Add 1 mL of 1% cresol red dye <!> and mix well.

dNTPs (10 mM)

Makes 100 μL.
Store for up to 1 yr at −20°C.

1. Add 10 μL each of 100 mM dTTP, dATP, dGTP, and dCTP.
2. Add 60 μL of deionized or distilled water and mix.

Primer/Loading Dye Mix

Makes enough for 50 PCRs.
Store for up to 1 yr at −20°C.

1. Obtain the primers for the laboratory exercise and dissolve each at a concentration of 15 pmol/μL.
   - The primer sequences for Stage B, Part II of Laboratory 4.3 (*dpy-13*) are as follows:

     5′-AGTCGTCTTCTCCGTTATCG-3′ (forward primer)
     5′-GAGCAACGCATAAGGCAAAG-3′ (reverse primer)

   - Primers for Stage B, Part II of Laboratory 4.4 are designed during Stage A and can be ordered from custom oligonucleotide suppliers (e.g., Sigma-Genosys). Make sure that you provide ample time for delivery of primers; custom primers can usually be delivered 3–5 d after they are ordered, although this will vary by supplier.
   - The universal T7 primer for Stage D, Part I of Laboratory 4.4 is as follows:

     5′-CGTAATACGACTCACTATAG-3′

2. In a 1.5-mL tube, mix the following:

Distilled water	640 μL
Cresol red loading dye <!>	460 μL
Forward primer (15 pmol/μL)	20 μL
Reverse primer (15 pmol/μL)	20 μL

   (For the T7 primer/loading dye mix, there is only one primer, so use 660 μL of distilled water.)

3. Vortex to mix.

Note: This primer/loading dye mix is for use with Ready-To-Go PCR Beads. puRe *Taq* Ready-to-Go PCR Beads (GE Healthcare) are available in individual thin-wall tubes (0.2 or 0.5 mL) or in 96-well plates.

ANALYZING DNA BY GEL ELECTROPHORESIS

100-bp DNA Ladder (0.125 μg/μL)

Makes 100 μL.
Store for up to 1 yr at −20°C.

1. Obtain a stock solution of 100-bp DNA Ladder (0.5 μg/μL) from New England Biolabs (N3231). Store the solution at –20°C.
2. Dilute a small amount of the stock at a time by mixing the following in a 1.5-mL tube:

Deionized or distilled water	50 μL
Cresol red loading dye <!>	25 μL
DNA Ladder	25 μL

Agarose (1%)

Makes 200 mL.
Use fresh or store solidified agarose for several weeks at room temperature.

1. In a 600-mL beaker or Erlenmeyer flask, add 200 mL of 1x TBE electrophoresis buffer and 2 g of agarose (electrophoresis grade).
2. Stir to suspend the agarose.
3. Dissolve the agarose using one of the following methods:
    - Cover the flask with aluminum foil and heat the solution in a boiling water bath (double boiler) or on a hot plate until all of the agarose is dissolved (~10 min).
    - Heat the flask uncovered in a microwave oven at high setting until all of the agarose is dissolved (3–5 min per beaker).
4. Swirl the solution and check the bottom of the beaker to make sure that all agarose has dissolved. (Just before complete dissolution, particles of agarose appear as translucent grains.) Reheat for several minutes if necessary.
5. Cover the agarose solution with aluminum foil and hold in a hot water bath (at ~60°C) until ready for use. Remove any "skin" of solidified agarose from the surface before pouring the gel.

Notes: Samples of agarose powder can be preweighed and stored in capped test tubes until ready for use. Solidified agarose can be stored at room temperature and then remelted over a boiling water bath (15–20 min) or in a microwave oven (3–5 min per beaker) before use. When remelting, evaporation will cause the agarose concentration to increase; if necessary, compensate by adding a small volume of water. Always loosen the cap when remelting agarose in a bottle.

Ethidium Bromide Staining Solution (1 μg/mL)

Makes 500 mL.
Store indefinitely in the dark at room temperature.

1. Add 100 μL of 5 mg/mL ethidium bromide <!> to 500 mL of deionized or distilled water.
2. Store the ethidium bromide in a dark (preferably opaque), unbreakable container or wrap the container in aluminum foil.

3. Label the container **"CAUTION: Ethidium Bromide. Mutagen and cancer-suspect agent. Wear rubber gloves when handling."**

Note: Ethidium bromide is light sensitive.

Loading Dye (10x)

Makes 100 mL.
Store indefinitely at room temperature.

1. Dissolve the following ingredients in 60 mL of deionized or distilled water:

Bromophenol blue (MW 669.96) <!>	0.25 g
Xylene cyanol (MW 538.60) <!>	0.25 g
Sucrose (MW 342.3)	50.0 g
Tris-HCl (1 M, pH 8.0)	1 mL

2. Add deionized or distilled water to make a total volume of 100 mL.

pBR322/BstNI Size Marker (0.1 μg/μL)

Makes 100 μL.
Store for up to 1 yr at –20°C.

1. Add 1 μL of 10 μg/μL pBR322 to 84 μL of deionized or distilled water.
2. Add 10 μL of 10x buffer (provided by the supplier of the BstNI enzyme).
3. Add 5 μL of 10 U/μL BstNI restriction enzyme and incubate for 60 min at 60°C.
4. Electrophorese 5 μL (plus 1 μL cresol red loading dye <!>) in a 1%–2% agarose gel to check for complete digestion. Exactly five bands should be visible, corresponding to 1857, 1058, 929, 383, and 121 bp. Any additional bands indicate incomplete digestion; if this is the case, add additional enzyme and incubate again at 60°C.

Note: pBR322 precut with restriction enzyme BstNI is also available from New England BioLabs (N3031).

SYBR Green DNA Stain

Makes 100 μL.
Store in the dark for 3 mo at 4°C.

1. Obtain a stock solution of 10,000x SYBR Green I <!>, which comes dissolved in DMSO <!> and is stored at –20°C. Allow the dye to thaw for ~10 min at room temperature.
2. Add 1 μL of 10,000x SYBR Green I in DMSO to 100 μL of sucrose solution. Mix thoroughly.

Note: SYBR Green is light sensitive.

Tris/Borate/EDTA (TBE) Electrophoresis Buffer (1x)

Makes 10 L.
Store indefinitely at room temperature.

1. In a spigoted carboy, add 9 L of deionized or distilled water to 1 L of 10x TBE electrophoresis buffer.
2. Stir to mix.

Tris/Borate/EDTA (TBE) Electrophoresis Buffer (10x)

Makes 1 L.
Store indefinitely at room temperature.

1. Add the following dry ingredients to 700 mL of deionized or distilled water in a 2-L flask:

NaOH <!> (MW 40.00)	1 g
Tris base <!> (MW 121.10)	108 g
Boric acid <!> (MW 61.83)	55 g
EDTA <!> (disodium salt, MW 372.24)	7.4 g

2. Stir to dissolve, preferably using a magnetic stir bar.
3. Add deionized or distilled water to bring the total volume to 1 L.

Note: If the stored 10x TBE comes out of solution, place the flask in a water bath (37°C–42°C) and stir occasionally until all solid matter goes back into solution.

TRANSFORMING BACTERIA

Calcium Chloride ($CaCl_2$) (1 M)

Makes 100 mL.
Store indefinitely at room temperature.

1. Dissolve 11.1 g of anhydrous $CaCl_2$ <!> (MW 110.99) or 14.7 g of $CaCl_2 \cdot 2H_2O$ <!> (MW 146.99) in 80 mL of deionized or distilled water.
2. Add deionized or distilled water to make a total volume of 100 mL of solution.

Calcium Chloride ($CaCl_2$) (50 mM)

Makes 1 L.
Store indefinitely at 4°C or room temperature.

1. Mix 50 mL of 1 M $CaCl_2$ <!> with 950 mL of deionized or distilled water.
2. Prerinse a 0.45- or 0.22-µm sterile filter by drawing 50–100 mL of deionized or distilled water through it.

3. Pass the $CaCl_2$ solution through the prerinsed filter.
4. Dispense aliquots into presterilized 50-mL conical tubes or autoclaved 150–250-mL bottles.

Classic Competent Cells

Makes 2 mL. To scale up, use more tubes starting with Step 6.
Prepare fresh. Store the cells in a beaker of ice in the refrigerator (~0°C) until ready for use. "Seasoning" for up to 24 h at 0°C increases the competency of the cells five-fold to 10-fold.

This entire procedure must be performed under sterile conditions.

1. On the day before preparing competent cells, start an overnight culture of DH5α or other *E. coli* strain.
2. Approximately 2–4 h before preparation, begin a midlog suspension. Add 1 mL of overnight culture to 100 mL of liquid LB at 37°C. Allow the culture to incubate at 37°C with continuous shaking.
3. Grow the culture to an OD_{550} of 0.3–0.5.
   - If a spectrophotometer is available, sterilely withdraw a 1-mL sample of the culture ~1 h after inoculation and measure the absorbance (optical density) at 550 nm. Repeat the procedure in 20-min intervals.
   - If a spectrophotometer is not available, it can be safely assumed that the culture has reached an OD_{550} of 0.3–0.5 after 2 h and 15 min of incubation with continuous shaking.
4. Store the midlog culture on ice until you are ready to begin preparing competent cells.
5. Place a sterile tube of 50 mM $CaCl_2$ <!> on ice.
6. Obtain two 15-mL tubes, transfer 10 mL of midlog cells to each tube, and securely close the caps.
7. Place both tubes of cells in a balanced configuration in the rotor of a clinical centrifuge. Centrifuge at 1500*g* for 10 min at 4°C to pellet the cells.
8. Sterilely pour off the supernatant from each tube into a waste beaker for later disinfection as follows. Do not disturb the cell pellet.
   i. Remove the cap from the culture tube and briefly flame the mouth. Do not place the cap on the lab bench.
   ii. Carefully pour off the supernatant. Invert the culture tube and tap the mouth gently on the surface of a clean paper towel to drain thoroughly.
   iii. Reflame mouth of the culture tube and replace the cap.
9. Use a 5- or 10-mL pipette to sterilely add 5 mL of ice-cold 50 mM $CaCl_2$ solution to each culture tube as follows:
   i. Remove the cap from the $CaCl_2$ tube. Do not place the cap on the lab bench.
   ii. Withdraw 5 mL of 50 mM $CaCl_2$ and replace the cap.

iii. Remove the cap of the culture tube. Do not place the cap on the lab bench.

iv. Expel the $CaCl_2$ into the culture tube and replace the cap.

10. Immediately finger-vortex as follows to resuspend the pelleted cells in each tube:

    i. Tightly close the cap of the culture tube.

    ii. Hold the upper part of tube securely with your thumb and index finger.

    iii. With the other hand, vigorously hit the bottom end of the tube with your index finger or thumb to create a vortex that lifts the cell pellet off the bottom of the tube. Continue "finger-vortexing" until all traces of the cell mass are completely resuspended. This may take a couple of minutes, depending on technique.

    iv. Hold the tube up to the light to check that the suspension is homogeneous. No visible clumps of cells should remain.

11. Return both tubes to ice and incubate for 20 min.

12. Following incubation, respin the cells in a clinical centrifuge at 1000*g*–2000*g* for 5 min at 4°C. This time, the cell pellet will be more spread out on the bottom of the tube due to the $CaCl_2$ treatment.

13. Sterilely pour off the supernatant from each tube into a waste beaker as described in Step 8. Do not disturb the cell pellet.

14. Use a 100–1000-µL micropipettor (or a 1-mL pipette) to sterilely add 1000 µL (1 mL) of ice-cold 50 mM $CaCl_2$ to each tube using the same technique as described in Step 9.

15. Close the caps tightly and immediately finger-vortex to resuspend the pelleted cells in each tube. Hold the tube up to the light to check that the suspension is homogeneous. No visible clumps of cells should remain.

PREPARING PLASMID MINIPREPS

Ethylenediaminetetraacetic Acid (EDTA) (0.5 M, pH 8.0)

Makes 100 mL.
Store indefinitely at room temperature.

1. Add 18.6 g of EDTA <!> (disodium salt, MW 372.24) to 80 mL of deionized or distilled water.
2. Adjust the pH by slowly adding ~2.2 g of sodium hydroxide (NaOH) <!> pellets (MW 40.00); monitor with a pH meter or strips of pH paper. (If neither is available, adding 2.2 g of NaOH pellets will make a solution of ~pH 8.0.)
3. Mix vigorously with a magnetic stirrer or by hand.
4. Add deionized or distilled water to make a total volume of 100 mL of solution.
5. Make sure that the bottle cap is loose and autoclave for 15 min at 121°C.
6. After autoclaving, cool the solution to room temperature and tighten the lid for storage.

Note: Use only the disodium salt of EDTA. EDTA will only dissolve after the pH has reached 8.0 or higher.

Glucose/Tris/EDTA (GTE)

Makes 100 mL.
Store indefinitely at 4°C or room temperature.

1. Combine the following:

Glucose (MW 180.16)	0.9 g
Tris-HCl (1 M, pH 8.0)	2.5 mL
EDTA <!> (0.5 M, pH 8.0)	2 mL
Deionized or distilled water	94.5 mL

2. Mix well.

Potassium Acetate (5 M)

Makes 200 mL.
Store indefinitely at room temperature.

1. Add 98.1 g of potassium acetate (MW 98.14) to 160 mL of deionized water.
2. Add deionized or distilled water to make a total volume of 200 mL of solution.

Potassium Acetate/Acetic Acid

Makes 100 mL.
Store indefinitely at 4°C or room temperature.

1. Add 60 mL of 5 M potassium acetate and 11.5 mL of glacial acetic acid <!> to 28.5 mL of deionized or distilled water.
2. Mix thoroughly.

Sodium Dodecyl Sulfate (SDS) (1%)/Sodium Hydroxide (0.2 N)

Makes 10 mL.
Store for several days at room temperature.

1. Add 1 mL of 10% SDS <!> and 0.5 mL of 4 N NaOH <!> to 8.5 mL of distilled water.
2. Mix thoroughly.

Notes: If a soapy precipitate forms, warm the solution by placing the tube in a beaker of hot tap water and shake gently to dissolve. SDS is the same as sodium lauryl sulfate.

Sodium Dodecyl Sulfate (SDS) (10%)

Makes 100 mL.
Store indefinitely at room temperature.

1. Dissolve 10 g of electrophoresis-grade SDS <!> (MW 288.37) in 80 mL of deionized or distilled water.

2. Add deionized or distilled water to make a total volume of 100 mL of solution.

Note: SDS is the same as sodium lauryl sulfate.

Tris/EDTA (TE) Buffer

Makes 100 mL.
Store indefinitely at room temperature.

1. Combine the following:

Tris-HCl (1 M, pH 8.0)	1 mL
EDTA <!> (0.5 M, pH 8.0)	200 μL
Deionized or distilled water	99 mL

2. Mix thoroughly.

Answers to Questions

Laboratories 4.1–4.4

LABORATORY 4.1: CULTURING AND OBSERVING *C. ELEGANS*

Answers to Questions in Stage C: Observing Wild-type and Mutant *C. elegans:* Results and Discussion

1. **How many stages of *C. elegans* development were you able to identify? Describe each stage.**

   The *C. elegans* life cycle consists of six different stages: the embryonic stage, four larval stages (L1–L4), and the adult stage. Nematodes continue to grow between molts, so each larval stage consists of worms of varying sizes. Although this makes it difficult to clearly distinguish most larval stages, L4s are larger than the other larvae but lack the embryos and vulva of adults. Students should be able to distinguish five stages: embryo, L1, L2–L3, L4, and adult.

2. **Why is it necessary for *C. elegans* to pass through several larval stages and how is this type of development different from humans?**

   Most organisms molt because there is a physical limit to growth in each larval phase. However, it is not clear that this is important in *C. elegans* development. Some scientists believe that the larval stages were once important for the ancestors of *C. elegans* and that the molts are remnants from that time. Like many higher organisms, humans do not develop in distinct phases, where each stage has a physical limit to growth. Rather, we grow more or less continuously by adding new cells to existing tissues and organs.

3. **How does a hermaphrodite produce offspring without mating?**

   A hermaphrodite worm produces both spermatocytes and oocytes, enabling a single worm to self-fertilize within its body to produce embryos.

4. **What physical (morphological) differences did you observe in the mutant worms? What differences in behavior or movement did you notice? Did your classmates identify the same characteristics of the mutant *C. elegans*?**

   Identifying *C. elegans* phenotypes is subjective, and different students may focus on different aspects of the mutant phenotypes. Comparing different student observations of the same mutant is an interesting way to examine subjectivity in science. The panel on the following page will aid you in leading student discussion.

5. **Based on each mutant phenotype that you observed, what do you think would be the function of the protein produced by the wild-type gene?**

   The shortened body in *dpy-11* mutants suggests that the protein product of the *dpy-11* gene is required to develop full body length. Because the *unc-22* mutation

C. elegans phenotypes

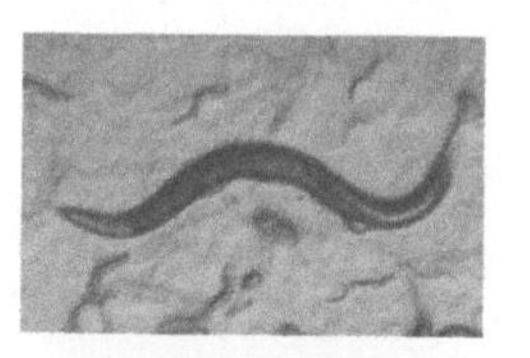

Wild type, very active. Exhibits graceful, serpentine movements and produces tracks in the agar.

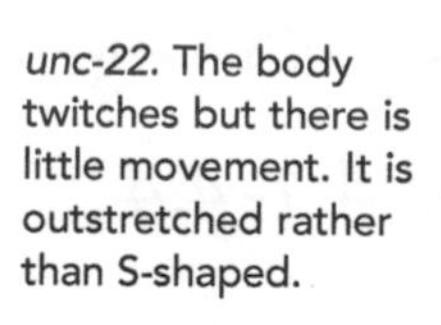

unc-22. The body twitches but there is little movement. It is outstretched rather than S-shaped.

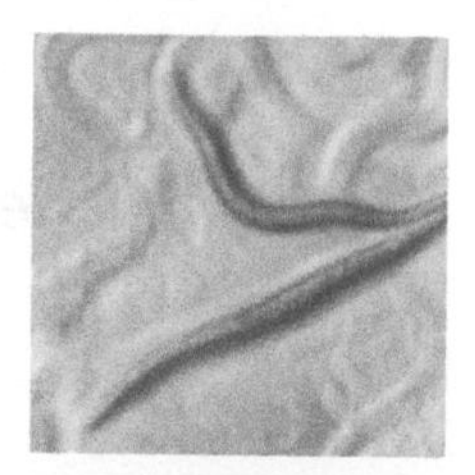

bli-1. The clear area on the side of the worm is a blister in the cuticle. Blisters are most obvious in old adults and may inhibit movement.

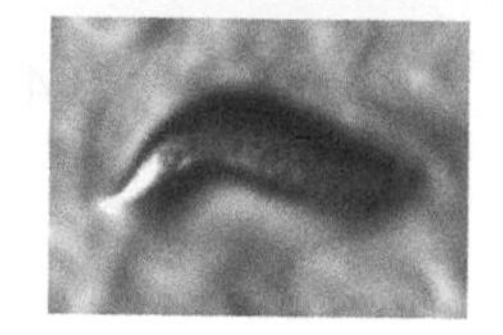

dpy-11. Short and plump.

causes movement disorders, this gene may encode a protein that is a component of the nervous or muscle system or is required for the development of those systems. *bli-1* seems to encode a protein that is needed to form the interface between layers of the body wall or to prevent extracellular fluid buildup.

6.i. *bli-1* encodes a collagen. What is a collagen? How can mutations in a collagen affect the cuticle?

Collagens are important for connective tissue. There are many collagens, and they can have different roles. BLI-1 is important for connecting different layers of the adult cuticle. When it is absent, these different layers can separate and the gap between the layers fills with fluid.

6.ii. *dpy-11* encodes an enzyme. What do enzymes do? How can mutations in enzymes affect the cuticle?

Enzymes are proteins that catalyze reactions. DPY-11 is a thioredoxin-like protein, an enzyme that modifies other proteins. The protein is expressed in the cells of the epidermis, where it is predicted to be located in the membrane of the endoplasmic reticulum. It is important for maintaining the correct geometry of the cuticle in *C. elegans*. Although it is not known how DPY-11 performs this function, it is probably required to modify proteins such as collagens that are secreted from epidermal cells to form the cuticle.

LABORATORY 4.2: USING *E. COLI* FEEDING STRAINS TO INDUCE RNAi AND KNOCK DOWN GENES

Answers to Experimental Methods Questions

III. Observe and Score Phenotypes

2.i. Have any eggs been laid?

By the second day after transferring, the worms should have laid embryos. If no embryos have been laid, the worms may be unhealthy or dead.

Have any eggs hatched? If so, are the worms at a larval or adult stage?

The eggs should hatch 1 d after being laid, and larval worms should be present. These larvae will grow to adulthood in 3–4 d.

2.ii. At which developmental stage are the worms in which you observe these differences (L1, L2–L3, L4, or adult)?

The RNAi phenotypes are typically noticeable only in adult worms. The phenotypes of RNAi-treated worms are similar to those of the worm strain carrying a mutation in the same gene: *bli-1* adults develop fluid-filled blisters in the cuticle, *unc-22* adults move slowly and twitch, and *dpy-11* adults have short bodies.

4. How effective was the RNAi treatment? Did the RNAi effect vary over time?

The RNAi treatment will vary both in the percentage of worms affected (penetrance) and in the severity of the phenotype (expressivity). Penetrance and expressivity may be higher in F_1 progeny that mature later. Because they hatch later, their mother has a longer exposure to the RNAi feeding bacteria and more time to develop a stronger RNAi response. The RNAi treatment should yield 70%–100% blistered worms for *bli-1*, 60%–100% uncoordinated worms for *unc-22*, and 55%–100% Dumpy worms for *dpy-11*.

5. Given the phenotypes that you observed, what can you deduce about the function of each gene that was knocked down?

bli-1 is required for development of the outer cuticle, *unc-22* is required for normal muscle development and function, and *dpy-11* is required for normal body length.

Answers to Bioinformatics Questions

I. Use WormBase to Find Basic Information About the *unc-22* Gene

2.i. What does "unc" stand for?

C. elegans genes that were first identified by a mutant phenotype are named with a three-letter abbreviation of that phenotype. "unc" is an abbreviation for "uncoordinated."

2.ii. What sort of protein does the gene encode? What is its function?

unc-22 encodes the UNC-22 protein, also known as twitchin. It is a giant muscle protein that is required for normal muscle development (morphology) and for regulating the actomyosin contraction–relaxation cycle.

3. How can there be more than one transcript for this gene?

Different transcripts are the result of alternative splicing, in which gene exons are spliced together in different arrangements.

II. Use BLAST to Identify the Human Homolog of the *unc-22* Gene

1.iii. Why would we do a BLAST search with amino acids rather than with DNA sequence?

To conserve protein function through evolutionary time, key amino acid sequences must be maintained. The degenerate nature of the genetic code—several different codons specifying the same amino acid—permits some DNA mutations to occur during evolution, provided that the protein sequence is maintained. Because identical protein sequences could be encoded by different DNA sequences, a protein BLAST search provides a better assessment of conservation of gene function among different species.

2.i. What do you notice about the lengths (and colors) of the matches (bars) as you look from the top to the bottom?

At the top of the table, there are several long matches with sections of different colors. Red sections match *unc-22* very well. The lengths of the matches and the amount of red decrease from top to bottom.

2.ii. What human protein is most related to *C. elegans* UNC-22? What is its E value and what does that E value mean?

UNC-22 is most related to titin (TTN). An E value of 0.0 is highly significant, meaning that the two proteins are closely related—with little chance that the match is a chance occurrence.

3.i. Does this gene have a function similar to that of *unc-22*? If so, in what ways are they similar?

Yes. They are both large muscle proteins required for muscle structure and contraction.

3.ii. Is this gene important in human health? If so, how?

Yes. Mutations in the gene for titin have been identified in some patients with hypertrophic cardiomyopathy, and antibodies against titin are produced in patients with the autoimmune disease scleroderma.

LABORATORY 4.3: EXAMINING THE RNAi MECHANISM

Answers to Bioinformatics Questions

I. Use BLAST to Find DNA Sequences in Databases (Electronic PCR)

2.i. What do you notice about the lengths (and colors) of the matches (bars) as you look from the top to the bottom?

Typically, most of the significant alignments will have complete matches to the forward and reverse primers. Partial alignments at the bottom of the list may correspond to only one primer or parts of both primers.

2.ii. What is the E value of the most significant hit and what does it mean?

The lowest E value obtained for a match to both primers is typically about 0.49. This denotes a probability of 0.49 (or 49 times out of 100) to match the hit sequence just by chance.

Note the names of any significant alignments that have E values of less than 1. Do they make sense? What do they have in common?

Several hits are from *C. elegans* and refer to *DumPY* or *dpy-13*. The cosmid sequence is a very large segment of chromosomal DNA cloned into a cosmid vector, which was the basic unit of sequence information used to assemble the *C. elegans* genome. Thus, they make sense. Hits from *Oryza sativa* (rice) and other species are perplexing, but remember that a BLAST match does not necessarily prove a meaningful (functional) relationship. In this case, close examination shows that the hits to rice and other species have poor matches to the primer sequences.

2.iii. Do the rest of the alignments produce good matches to both primers?

A few BLAST hits contain good matches to both primers, but most of the rest of the alignments are shorter partial matches to one primer or parts of both primers.

Which of the hits would be amplified in vitro in a PCR using the two primers? Why?

Only hits that contain complete matches to both primers would be amplified in a PCR. PCR requires a set of primers that bind to complementary sequences on each of the antiparallel DNA strands, flanking the region to be amplified.

3.i. What is the source and size of the sequence in which your BLAST hit is located?

The BLAST hit is to a *C. elegans* mRNA that codes for DPY-13 protein. It is 909 nucleotides long.

4.ii. To which positions do the primers match in the subject sequence?

The forward primer matches positions 167–186 and the reverse primer matches positions 1155–1174.

4.iii. Subtract the lowest nucleotide position in the subject sequence from the highest nucleotide position in the subject sequence. What is the difference between the coordinates?

1174 – 167 = 1007 nucleotides. These are the coordinates in accession M23559.1.

4.iv. Note that the actual length of the amplified fragment includes both ends, so add 1 nucleotide to the result that you obtained in Step 4.iii to obtain the exact length of the PCR product amplified by the two primers.

1007 + 1 = 1008 nucleotides.

4.v. Repeat Steps i–iv for at least two more hits.

For accession NM_068118.4, 967 – 68 + 1 = 900 nucleotides. For accession U42437.2, 23,234 – 22,227 + 1 = 1008 nucleotides.

Do you notice any discrepancies among the calculations? If so, why?

The correct calculation of 1008 nucleotides is predicted from genomic DNA sequences, and genomic DNA is the template used during PCR. The smaller product (900 nucleotides) is predicted from an mRNA sequence, which is missing 108 nucleotides of intron sequence.

II. Use GenBank, Entrez Gene, and WormBase to Learn About the Function and Structure of the *dpy-13* Gene

1.i. How many exons and introns are in the *dpy-13* gene?

The *dpy-13* gene has three exons and two introns.

1.ii. Where does the forward primer begin? Where does the reverse primer end? Which features are found between these nucleotide positions?

The amplicon begins in the first exon; extends through the first intron, second exon, second intron, and third exon; and ends in the 3′-untranslated region, just short of the poly(A) signal.

5. What does "dpy" stand for?

"Dpy" stands for "Dumpy," which describes the short and fat phenotype of this mutant.

What sort of protein does the *dpy-13* gene encode and what is its function?

The gene encodes a protein named DPY-13, which is a member of the collagen family. Collagens are abundant proteins that are a major component of the extracellular matrix secreted by cells.

6.iii. How many exons and introns are shown? Does this agree with your findings from Step 1.i?

There are three exons and two introns. So, yes, the findings agree.

III. Identify *dpy-13* Relatives in *C. elegans* and Humans

A. Use MegaBLAST to Identify dpy-13 Relatives in C. elegans

5. Compare the E values obtained by this search to those values that you obtained in Step 2.ii of Part I. What do you notice? Why is this so?

Many of the hits have extremely low E values (with many decimal places). This is because the query sequence is longer.

6. Why do the first several hits have E values of 0?

These hits completely match the query. They are the same *dpy-13* transcripts identified in Step 2 of Part I.

7. Carefully read the descriptions of the hits with the lowest E values. What do they have in common?

Many are collagen genes. This appears to be a large gene family in *C. elegans*.

B. Use Discontiguous MegaBLAST to Identify dpy-13 Relatives in Humans

5. Compare the E values obtained by this search to those values that you obtained in Step 5 of Part IIIA. What do you notice? Why is this so?

The E values are higher because the hits match only a fraction of the query sequence.

6. Carefully read the descriptions of the hits with the lowest E values. What do they have in common?

They are collagen genes. One of the collagens is implicated in a dominant genetic disorder, Ehlers-Danlos Syndrome. The major symptoms of this disorder include unstable joints that are prone to dislocations, joint pain, and early onset osteoarthritis, as well as fragile skin that tears or bruises easily, slow wound healing, and severe scarring.

Answers to Results and Discussion Questions

I. Think About the Experimental Methods

1. **Describe the purpose of each of the following steps or reagents used during DNA isolation (Part III of Experimental Methods):**

1.i. **Freezing at −80°C.**

Freezing-thawing cracks the tough outer cuticle of the worm, providing the lysis buffer access to internal body tissues and cells.

1.ii. **Incubating with proteinase K.**

This proteolytic enzyme digests the protein in the cuticle. It helps to liberate individual cells by digesting protein fibers of the extracellular matrix—collagen, laminins, fibronectin, and elastin—that bind cells together. It also inactivates cellular proteins, including DNases, that degrade DNA or interfere with PCR amplification.

1.iii. **Boiling at 95°C.**

The near-boiling temperature lyses individual cells and denatures the proteinase K, which could digest the *Taq* polymerase needed during PCR.

2. **What is the purpose of performing each of the following PCRs?**

2.i. **Wild-type worms.**

This reaction detects the presence of the wild-type *dpy-13* gene.

2.ii. ***dpy-13* mutant worms.**

This reaction detects the presence of a chromosomal deletion in the *dpy-13* gene.

2.iii. ***dpy-13* RNAi-treated wild-type worms.**

This reaction detects whether RNAi alters the size of the *dpy-13* gene.

II. Interpret Your Gel

4. **What is the genotype of the *dpy-13* RNAi-treated wild-type worms?**

The PCR products of the *dpy-13* RNAi-treated wild-type worms should align between the 1058- and 929-bp fragments of the pBR322/BstNI marker. Thus, these worms have the same genotype as the wild-type worms that were not treated with RNAi.

5. **How would you interpret a lane in which you observe primer dimer but no bands as described in Step 3?**

The presence of primer dimer confirms that the reaction contained all components necessary for amplification but that there was insufficient template to amplify the target sequence.

III. Correlate *dpy-13* Genotype and Phenotype

1. **Describe the appearance of each of the three types of worms from which DNA was isolated in Part III of Experimental Methods:**

1.i. **Wild-type worms on the NGM-lite plate.**

The wild-type worms are long and slender, with an S-shaped body. They are very active and move gracefully.

1.ii. ***dpy-13* RNAi-treated wild-type worms on the NGM-lite/amp + IPTG plate.**

The *dpy-13* RNAi-treated wild-type worms are short and fat (Dumpy). They often have more trouble moving.

1.iii. ***dpy-13* mutant worms on the NGM-lite plate.**

The *dpy-13* mutant worms are short and fat (Dumpy). They often have more trouble moving.

2. **According to your results, how well does the *dpy-13* genotype predict the phenotype? If the genotypes were not predictive of the phenotypes, explain why this might be the case. What do these results suggest about RNAi?**

The *dpy-13* RNAi-treated worms have a genotype different from that of the *dpy-13* mutant worms. However, the fact that RNAi-treated worms have the same phenotype as the mutant worms suggests that RNAi disrupts the function of the same or a similar gene. The *dpy-13* RNAi "phenocopies" the loss of gene function in the *dpy-13* mutant. Students should note variability in the phenotype of RNAi-treated worms, from apparently normal to obviously Dumpy. Thus, RNAi-induced phenotypes have variable expressivity compared to the phenotypes induced by a homozygous mutation in the gene.

3. **What do you conclude from the experiment?**

This experiment supports the contention that RNAi does not work by altering the size of the gene.

What are the limitations of this experiment?

This experiment cannot rule out the possibility that RNAi induced a point mutation in the *dpy-13* gene that cannot be detected by simple agarose gel electrophoresis.

What additional experiments could you do to strengthen your conclusions?

Sequencing the *dpy-13* gene in wild-type and RNAi-treated worms would provide direct evidence that the gene sequence is not altered in RNAi-treated animals. Moving RNAi-treated worms to OP50-seeded NGM-lite plates and observing successive generations would reveal that the Dumpy trait diminishes or disappears over time in the absence of dsRNA-producing bacteria. This is not consistent with the inheritance of a homozygous mutation, such as *dpy-13(e458)*, in which the trait persists in 100% of the offspring through successive generations.

LABORATORY 4.4: CONSTRUCTING AN RNAi FEEDING VECTOR

Answers to Questions in Stage A: Identify a Disease-Causing Gene and Design PCR Primers In Silico

II. Use BLAST and WormBase to Identify a Related Sequence in *C. elegans*

1.iii. **Why would we do a BLAST search with amino acids rather than with DNA sequence?**

To conserve protein function through evolutionary time, key amino acid sequences must be maintained. The degenerate nature of the genetic code—several different codons specifying the same amino acid—permits some DNA mutations to occur during evolution, provided that the protein sequence is maintained. Because identical protein sequences could be encoded by different DNA sequences, a protein BLAST search provides a better assessment of conservation of gene function among different species.

2.i. What do you notice about the lengths (and colors) of the matches (bars) as you look from the top to the bottom?

Typically, there will be several highly significant matches, followed by a number of partial matches.

Answers to Questions in Stage B: Clone Gene-Specific DNA

IV. Interpret Your Gel

4. How would you interpret a lane in which you observe primer dimer but no bands as described in Step 3?

The presence of primer dimer confirms that the reaction contained all components necessary for amplification but that there was insufficient template to amplify the target sequence.

5. If there are contaminants, how can you modify the PCR conditions to attempt to eliminate these products?

Increasing the annealing temperature makes it more difficult for a primer to hybridize to an imperfectly matched sequence. Decreasing the concentration of primers or reducing the number of cycles would also reduce the accumulation of nonspecific products.

Answers to Results and Discussion Questions in Stage B

1. Describe the purpose of each of the following steps or reagents used in DNA isolation (Part I):

1.i. Freezing at −80°C.

Freezing-thawing cracks the tough outer cuticle of the worm, providing the lysis buffer access to internal body tissues and cells.

1.ii. Incubating with proteinase K.

This proteolytic enzyme digests the protein in the cuticle. It helps to liberate individual cells by digesting protein fibers of the extracellular matrix—collagen, laminins, fibronectin, and elastin—that bind cells together. It also inactivates cellular proteins, including DNases, that degrade DNA or interfere with PCR amplification.

1.iii. Boiling at 95°C.

The near-boiling temperature lyses individual cells and denatures the proteinase K, which could digest the *Taq* polymerase needed during PCR.

2. The PCR products used in this experiment include *attB* sites. Why?

attB sites are 21-bp sequences that are recognized by the enzymes in BP Clonase. *attB1* sites recombine specifically with *attP1* sites, and *attB2* sites recombine with *attP2* sites. This means that the direction in which the PCR fragment is inserted in the vector is determined by the primer that has the *attB1* sequence and the primer that has the *attB2* sequence. This is called "directional" cloning. For RNAi, the direction of the insertion does not matter. The purpose of the vector is to transcribe from both strands using T7 promoters upstream of the insertion site on both strands. This will be accomplished when the PCR product is inserted into the vector in either orientation.

Answers to Questions in Stage C: Transformation of *E. coli*

II. Calculate Transformation Efficiencies

2. Were the results as expected? Explain possible reasons for variations from the expected results.

Transformation with the positive control should produce thousands of colonies on the LB/kan plate. Transformation with the recombined DNA should show growth on LB/kan, with ~100- to 1000-fold fewer colonies than the positive control. No growth should occur for the negative control.

If a few colonies are found where they are unexpected, inadequate sterilization of the spreading rod or poor sterile technique are possible explanations. Alternatively, one of the reagents may be contaminated with a small quantity of kanamycin-resistant bacteria or DNA conferring kanamycin resistance. If large numbers of colonies are seen on plates where no growth is expected, consider that the antibiotic may still be active in the plates. A more trivial explanation is that cells were spread on the wrong plate.

If no growth is observed on a plate where you expect to see colonies, one possibility is that the cells were spread with a spreader that was too hot. If the transformations with recombined DNA show no growth but the controls exhibit the expected number of colonies, the recombination reaction may not have gone to completion. If samples of the ligated DNA are still available, separate them by electrophoresis to gauge the extent of the recombination. If all plates show much less growth than expected, the bacteria may not have been rendered sufficiently "competent" to accept plasmid DNA.

3. Compare and contrast the growth on each of the following pairs of plates. What does each pair of results tell you about transformation and/or antibiotic selection?

3.i. "–" LB and "–" LB/kan.

Growth on the "–" LB plate confirms that the cells used in the experiment are viable. No growth on the "–" LB/kan plate confirms that the kanamycin in the plates is active.

3.ii. "–" LB/kan and "+" LB/kan.

There should be no growth on the "–" LB/kan plate. This indicates that the kanamycin in the LB/kan plates is working correctly and that the strain of bacteria used for the transformation is not kanamycin resistant. In other words, it confirms that the plasmid is responsible for kanamycin resistance. The "+" LB/kan positive control should yield hundreds or thousands of colonies, indicating that the cells are highly competent.

3.iii. "R" LB/kan and "+" LB/kan.

There should be ~100x less growth on the "R" LB/kan plate compared to the "+" LB/kan positive control plate. This occurs because the total mass of ligated DNA is much smaller than the total mass of control DNA.

4.i. Determine the total mass of control DNA used in Step 4 of Part IA: concentration (in µg/µL) × volume (in µL) = mass (in µg).

0.005 µg/µL × 10 µL = 0.05 µg.

4.ii. Determine the fraction of the cell suspension that was spread onto the "+" LB/kan plate in Step 12 of Part IA:

$$\frac{\text{volume suspension spread (see Step 12 of Part IA)}}{\text{total volume suspension (see Steps 2 and 10 of Part IA)}} = \text{fraction spread.}$$

100 µL/1000 µL = 0.1.

4.iii. Determine the mass of the plasmid in the cell suspension that was spread onto the "+" LB/kan plate:
total mass plasmid (Step 4.i) × fraction spread (Step 4.ii) = mass plasmid spread (in µg).

0.05 µg × 0.1 = 0.005 µg.

4.iv. Determine the number of colonies per microgram of plasmid. Express the answer in scientific notation:

$$\frac{\text{colonies observed (Step 2)}}{\text{mass plasmid spread (Step 4.iii)}} = \text{transformation efficiency (in colonies/µg).}$$

2000 colonies/0.005 µg = 4×10^5 transformants/µg.
(Substitute your result for "2000 colonies.")

5.i. Calculate the mass of vector used in Step 3 of Part V in Stage B.

37.5 ng/µL × 4 µL = 150 ng = 0.15 µg.

5.ii. Use the result from Step 5.i and the procedure outlined in Steps 4.ii–iv to calculate the transformation efficiency of the LB/kan recombined DNA ("R").

fraction of cell suspension spread = 500 µL/1000 µL = 0.5,
mass of plasmid spread = 0.15 µg × 0.5 = 0.075 µg,
transformation efficiency = 50 colonies/0.075 µg = 6.7×10^2 transformants/µg.
(Substitute your result for "50 colonies.")

6. Compare the transformation efficiencies that you calculated for the control ("+") DNA in Step 4 and the recombined DNA ("R") in Step 5. How can you account for the differences in efficiency? Take into account the formal definition of transformation efficiency.

Transformation efficiency for the recombined DNA is expected to be lower than that for the control. Calculations of transformation efficiency should be done with the mass of relevant DNA. For the recombined DNA in Step 5, the calculation was done with the total amount of plasmid in the tube. Only a portion of this plasmid DNA will be recombined with insert. Hence, the calculation in Step 5 underestimates the transformation efficiency of the cells. Also, recombined DNA is not supercoiled. Supercoiled DNA, which makes up a large fraction of purified plasmid

DNA, yields more transformants than nonsupercoiled DNA, accounting for much of the difference in expected transformation efficiencies between control DNA and recombined DNA.

Answers to Questions in Stage D: Vector Analysis: Results and Discussion

1. **Why were colonies picked only from the LB/kan plate marked "R"?**

   The kanamycin selects against untransformed cells, only allowing transformed cells to reproduce.

   What event does each colony represent?

   Each colony consists of descendants from a single bacterium transformed with the vector.

3. **Do any of the colonies (transformants) have the expected insert? Explain how you determined this.**

   One T7 primer will anneal 37 bp from the insertion site, and the other T7 primer will anneal 64 bp from the insertion site. Therefore, the PCR fragments observed here should be 101 bp longer than the PCR fragments observed in Part IV of Stage B. Students should compare these results to determine if they have the correct insert. Alternatively, the E-RNAi output includes the size of the insert, and adding 101 bp to this size will give the size of the PCR fragment that is expected here.

Answers to Questions in Stage E: Isolate Plasmid DNA: Results and Discussion

1. **Consider three major classes of biologically important molecules: proteins, lipids, and nucleic acids. Which steps of the miniprep procedure act on proteins? On lipids? On nucleic acids?**

   SDS/NaOH helps to solubilize proteins and lipids, which are precipitated after addition of potassium acetate. Additional proteins are removed into the isopropanol. Both isopropanol and ethanol precipitate nucleic acids.

2. **What aspect of plasmid DNA structure allows it to renature efficiently in Step 11?**

   The two DNA strands are interlinked.

3. **What other kinds of molecules, in addition to plasmid DNA, would you expect to be present in the final miniprep sample? How could you find out?**

   RNA molecules should also be present. They can be visualized on ethidium bromide–stained agarose gels and removed following incubation with RNase.

Answers to Questions in Stage F: Create RNAi Feeding Strain

II. Calculate Transformation Efficiencies

2. **Were the results as expected? Explain possible reasons for variations from the expected results.**

   If growth is observed where it is not expected (e.g., negative control on LB/kan) or if growth is not observed where it is expected (e.g., transformed cells on LB/kan), verify that the proper type of cells was spread on the appropriate plates. If growth appears on all plates, verify that the kanamycin in the plates is still active.

3. **Compare and contrast the growth on each of the following pairs of plates. What does each pair of results tell you about transformation and/or antibiotic selection?**

3.i. **"+" LB and "–" LB.**

Growth on these plates confirms that the cells used in the experiment are viable.

3.ii. **"–" LB/kan and "–" LB.**

No growth on the "–" LB/kan plate confirms that the kanamycin in the plates is active.

3.iii. **"+" LB/kan and "–" LB/kan.**

Growth on "+" LB/kan and no growth on "–" LB/kan confirm that the plasmid is responsible for kanamycin resistance.

3.iv. **"+" LB/kan and "+" LB.**

Much less growth on "+" LB/kan as compared to "+" LB confirms that transformation is a rare event.

4.i. **Determine the total mass of plasmid DNA used in Step 9 of Part IA. Assume that the concentration of miniprep DNA is 0.05 μg/μL.**

concentration (in μg/mL) × volume (in μL) = mass (in μg).

0.05 μg/μL × 1 μL = 0.05 μg.

4.ii. **Determine the fraction of the cell suspension that was spread onto the "+" LB/kan plate in Step 15 of Part IA:**

$$\frac{\text{volume suspension spread (see Step 15 of Part IA)}}{\text{total volume suspension (see Steps 2 and 13 of Part IA)}} = \text{fraction spread.}$$

100 μL/500 μL = 0.2.

4.iii. **Determine the mass of the plasmid in the cell suspension that was spread onto the "+" LB/kan plate:**

total mass plasmid (Step 4.i) × fraction spread (Step 4.ii) = mass plasmid spread (in μg)

0.05 μg × 0.2 = 0.01 μg.

4.iv. **Determine the number of colonies per microgram of plasmid. Express the answer in scientific notation:**

$$\frac{\text{colonies observed (Step 2)}}{\text{mass plasmid spread (Step 4.iii)}} = \text{transformation efficiency (in colonies/μg).}$$

500 colonies/0.01 μg = 5×10^4 transformants/μg.
(Substitute your result for "500 colonies.")

5. **What factors might influence transformation efficiency?**

Factors that can influence transformation efficiency include the health of the *E. coli* cells, how well the cells are resuspended in $CaCl_2$, temperature and duration of heat shock, amount of plasmid DNA used, spreading technique, and length of recovery period.

6.i. **Using 0.2 μg of plasmid/*YFG* you achieve a transformation efficiency equal to 10^6 colonies per microgram of intact plasmid/*YFG*. How many transformants are in the culture?**

Because the transformation efficiency was 1×10^6 colonies per microgram of plasmid and 0.2 µg of plasmid was used in the experiment, there must be 2×10^5 transformants (1×10^6 transformants/µg × 0.2 µg = 2×10^5 transformants).

6.ii. Plasmid/*YFG* grows at an average copy number of 100 molecules/transformed cell. How many copies of plasmid/*YFG* are in the culture?

Because the average copy number of the plasmid is 100, there must be 2×10^7 copies of plasmid/*YFG* (2×10^5 transformants × 100 copies/transformant = 2×10^7 copies).

6.iii. How many doublings have occurred?

The culture spent 140 min (200 – 60 min) in log phase, during which it doubled every 20 min. This represents seven doublings (140 min/20 min = 7).

How many molecules (copies) of plasmid/*YFG* are present in the culture?

If the culture underwent seven doublings, or a factor of 128 (2^7), it would contain 2.56×10^9 copies of plasmid/*YFG*.

7.i. Use the procedure in Step 4 to calculate the mass of DNA and transformation efficiency at each concentration.

Concentration (µg/µL)	Mass of DNA (µg)	No. of colonies	Transformation efficiency (colonies/µg)
0.00001	0.0001	4	2×10^5
0.00005	0.0005	12	1.2×10^5
0.0001	0.001	32	1.6×10^5
0.0005	0.005	125	1.25×10^5
0.001	0.01	442	2.21×10^5
0.005	0.05	542	5.42×10^4
0.01	0.1	507	2.54×10^4
0.05	0.5	475	4.75×10^3
0.1	1	516	2.58×10^3

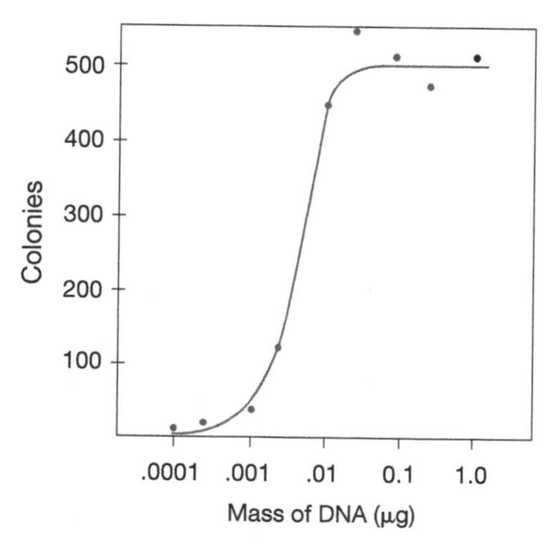

(Adapted, with permission, from Pearson Education Inc.; ©1996 Benjamin/Cummings Publishing Inc.)

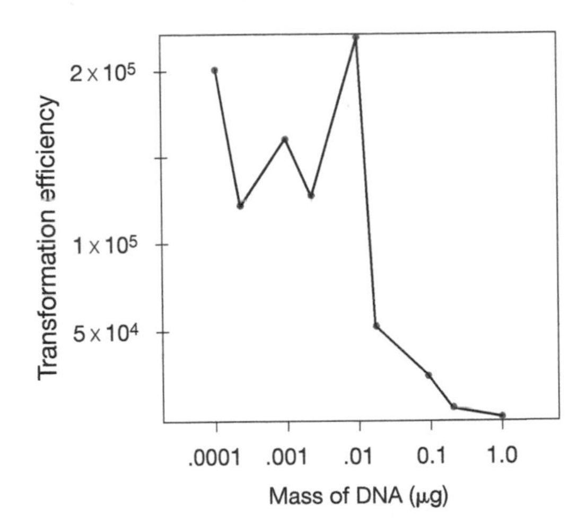

(Reprinted, with permission, from Pearson Education Inc.; ©1996 Benjamin/Cummings Publishing Inc.)

7.ii. Plot a graph of DNA mass versus colony number.

7.iii. Plot a graph of DNA mass versus transformation efficiency.

7.iv. What is the relationship between DNA mass and transformation efficiency?

Transformation efficiency remains relatively constant until the cells become saturated with respect to DNA. Thereafter, transformation efficiency declines with increasing mass of DNA.

7.v. At what point does the transformation reaction appear to be saturated?

The saturation point appears to be at a concentration of about 0.01 µg/µL.

7.vi. What is the true transformation efficiency?

The true transformation efficiency is the one that was calculated for the saturation point.

APPENDIX 1

Products Available from Carolina Biological Supply Company

The following products from Carolina Biological Supply Company (www.carolina.com) support the laboratory activities in *Genome Science.*

Lab 2.1 Using Mitochondrial DNA Polymorphisms in Evolutionary Biology

211236	Extraction and Amplification Kit with 0.5-mL Tubes
211236A	Extraction and Amplification Kit with 0.2-mL Tubes
211237	Extraction, Amplification, and Electrophoresis Kit with Ethidium Bromide and 0.5-mL Tubes
211237A	Extraction, Amplification, and Electrophoresis Kit with Ethidium Bromide and 0.2-mL Tubes
211238	Extraction, Amplification, and Electrophoresis Kit with *Carolina*BLU and 0.5-mL Tubes
211238A	Extraction, Amplification, and Electrophoresis Kit with *Carolina*BLU Stain and 0.2-mL Tubes

Lab 2.2 Using an *Alu* Insertion Polymorphism to Study Human Populations

211230	Extraction and Amplification Kit with 0.5-mL Tubes
211230A	Extraction and Amplification Kit with 0.2-mL Tubes
211231	Extraction, Amplification, and Electrophoresis Kit with Ethidium Bromide and 0.5-mL Tubes
211231A	Extraction, Amplification, and Electrophoresis Kit with Ethidium Bromide and 0.2-mL Tubes
211232	Extraction, Amplification, and Electrophoresis Kit with *Carolina*BLU Stain and 0.5-mL Tubes
211232A	Extraction, Amplification, and Electrophoresis Kit with *Carolina*BLU Stain and 0.2-mL Tubes
211502	*Alu* PV92 Primers

Lab 2.3 Using a Single-Nucleotide Polymorphism to Predict Bitter-Taste Ability

211376	Extraction and Amplification Kit with 0.5-mL Tubes
211377	Extraction and Amplification Kit with 0.2-mL Tubes
211378	Extraction, Amplification, and Electrophoresis Kit with Ethidium Bromide and 0.5-mL Tube
211379	Extraction, Amplification, and Electrophoresis Kit with Ethidium Bromide and 0.2-mL Tubes

211380	Extraction, Amplification, and Electrophoresis Kit with *Carolina*BLU and 0.5-mL Tubes
211381	Extraction, Amplification, and Electrophoresis Kit with *Carolina*BLU and 0.2-mL Tubes
211376C	PTC Perishables (primer/loading dye mix, marker, and HaeIII enzyme)
211712	HaeIII enzyme, 40 µL (400 U)
211508	PTC Tasting Primers Set

Lab 3.4 Detecting Genetically Modified Foods by Polymerase Chain Reaction

211366	Extraction and Amplification Kit with 0.5-mL Tubes
211367	Extraction and Amplification Kit with 0.2-mL Tubes
211368	Extraction, Amplification, and Electrophoresis Kit with Ethidium Bromide and 0.5-mL Tubes
211369	Extraction, Amplification, and Electrophoresis Kit with Ethidium Bromide and 0.2-mL Tubes
211370	Extraction, Amplification, and Electrophoresis Kit with *Carolina*BLU and 0.5-mL Tubes
211371	Extraction, Amplification, and Electrophoresis Kit with *Carolina*BLU and 0.2-mL Tubes

Lab 3.5 Using DNA Barcodes to Identify and Classify Living Things

211385	DNA Barcode Amplification Kit
211386	DNA Barcode Amplification and Electrophoresis Kit with *Carolina*BLU Stain
211387	DNA Barcode Amplification and Electrophoresis Kit with GelGreen Stain

Lab 3.6 Detecting Epigenetic DNA Methylation in *Arabidopsis*

(*To be available soon.*)

Lab 4.1 Culturing and Observing *C. elegans*

211390	Culturing and Observing *C. elegans* Kit

Lab 4.2 Using *E. coli* Feeding Strains to Induce RNAi and Knock Down Genes

211391	Inducing RNAi by Feeding Kit

Lab 4.3 Examining the RNAi Mechanism

211392	RNAi and Amplification Kit
211393	RNAi, Amplification, and Electrophoresis Kit with *Carolina*BLU
211394	RNAi, Amplification, and Electrophoresis Kit with Ethidium Bromide

APPENDIX 2

Cautions

GENERAL CAUTIONS

Please note that the Cautions Appendix in this manual is not exhaustive. Readers should always consult individual manufacturers and other resources for current and specific product information. Chemicals and other materials discussed in text sections are not identified by the icon <!> used to indicate hazardous materials in the protocols. However, without special handling, they may be hazardous to the user. Please consult your local safety office or the manufacturer's safety guidelines for further information. The following general cautions should always be observed.

- **Before beginning the procedure,** become completely familiar with the properties of substances to be used.
- **The absence of a warning** does not necessarily mean that the material is safe, because information may not always be complete or available.
- **If exposed to toxic substances,** contact your local safety office immediately for instructions.
- **Use proper disposal procedures** for all chemical, biological, and radioactive waste.
- **For specific guidelines on appropriate gloves to use,** consult your local safety office.
- **Handle concentrated acids and bases** with great care. Wear goggles and appropriate gloves. A face shield should be worn when handling large quantities.

  Do not mix strong acids with organic solvents because they may react. Sulfuric acid and nitric acid especially may react highly exothermically and cause fires and explosions.

  Do not mix strong bases with halogenated solvent because they may form reactive carbenes that can lead to explosions.
- **Handle and store pressurized gas containers** with caution because they may contain flammable, toxic, or corrosive gases; asphyxiants; or oxidizers. For proper procedures, consult the Material Safety Data Sheet that must be provided by your vendor.
- **Never pipette** solutions using mouth suction. This method is not sterile and can be dangerous. Always use a pipette aid or bulb.
- **Keep halogenated and nonhalogenated solvents separately** (e.g., mixing chloroform and acetone can cause unexpected reactions in the presence of bases). Halogenated solvents are organic solvents such as chloroform, dichloromethane, trichlorotrifluoroethane, and dichloroethane. Nonhalogenated solvents include pentane, heptane, ethanol, methanol, benzene, toluene, *N,N*-dimethylformamide (DMF), dimethyl sulfoxide (DMSO), and acetonitrile.

- **Laser radiation,** visible or invisible, can cause severe damage to the eyes and skin. Take proper precautions to prevent exposure to direct and reflected beams. Always follow the manufacturer's safety guidelines and consult your local safety office. See flash lamps caution below for more detailed information.
- **Flash lamps,** due to their light intensity, can be harmful to the eyes. They can also explode on occasion. Wear appropriate eye protection and follow the manufacturer's guidelines.
- **Photographic fixatives, developers, and photoresists** also contain chemicals that can be harmful. Handle them with care and follow the manufacturer's directions.
- **Power supplies and electrophoresis equipment** pose serious fire hazard and electrical shock hazards if not used properly.
- **Use of microwave ovens and autoclaves** in the lab require certain precautions. Accidents have occurred involving their use (e.g., when melting agar or bacto-agar stored in bottles or when sterilizing). If the screw top is not completely removed and there is inadequate space for the steam to vent, the bottles can explode and cause severe injury when the containers are removed from the microwave or autoclave. Always completely remove bottle caps before microwaving or autoclaving. An alternative method for routine agarose gels that do not require sterile agar is to weigh out the agar and place the solution in a flask.
- **Ultrasonicators** use high-frequency sound waves (16–100 kHz) for cell disruption and other purposes. This "ultrasound," conducted through air, does not pose a direct hazard to humans, but the associated high volumes of audible sound can cause a variety of effects, including headache, nausea, and tinnitus. Direct contact of the body with high-intensity ultrasound (not medical imaging equipment) should be avoided. Use appropriate ear protection and display signs on the door(s) of laboratories where the units are used.
- **Use extreme caution when handling cutting devices,** such as microtome blades, scalpels, razor blades, or needles. Microtome blades are extremely sharp! Use care when sectioning. If unfamiliar with their use, have an experienced user demonstrate proper procedures. For proper disposal, use the "sharps" disposal container in your lab. Discard used needles *unshielded*, with the syringe still attached. This prevents injuries and possible infections when manipulating used needles because many accidents occur while trying to replace the needle shield. Injuries may also be caused by broken Pasteur pipettes, coverslips, or slides.
- **Procedures for the humane treatment of animals** must be observed at all times. Consult your local animal facility for guidelines. Animals, such as rats, are known to induce allergies that can increase in intensity with repeated exposure. Always wear a lab coat and gloves when handling these animals. If allergies to dander or saliva are known, wear a mask.

GENERAL PROPERTIES OF COMMON CHEMICALS

The hazardous materials list can be summarized in the following categories.

- Inorganic acids, such as hydrochloric, sulfuric, nitric, or phosphoric, are colorless liquids with stinging vapors. Avoid spills on skin or clothing. Spills should be diluted with large amounts of water. The concentrated forms of these acids can destroy paper, textiles, and skin and cause serious injury to the eyes.

- Inorganic bases, such as sodium hydroxide, are white solids that dissolve in water and under heat development. Concentrated solutions will slowly dissolve skin and even fingernails.
- Salts of heavy metals are usually colored, powdered solids that dissolve in water. Many of them are potent enzyme inhibitors and therefore toxic to humans and the environment (e.g., fish and algae).
- Most organic solvents are flammable volatile liquids. Avoid breathing the vapors, which can cause nausea or dizziness. Also avoid skin contact.
- Other organic compounds including organosulphur compounds, such as mercaptoethanol or organic amines, can have very unpleasant odors. Others are highly reactive and should be handled with appropriate care.
- If improperly handled, dyes and their solutions can stain not only your sample, but also your skin and clothing. Some are also mutagenic (e.g., ethidium bromide), carcinogenic, and toxic.
- Nearly all names ending with "ase" (e.g., catalase, β-glucuronidase, or zymolyase) refer to enzymes. There are also other enzymes with nonsystematic names such as pepsin. Many of them are provided by manufacturers in preparations containing buffering substances, etc. Be aware of the individual properties of materials contained in these substances.
- Toxic compounds are often used to manipulate cells. They can be dangerous and should be handled appropriately.
- Be aware that several of the compounds listed have not been thoroughly studied with respect to their toxicological properties. Handle each chemical with appropriate respect. Although the toxic effects of a compound can be quantified (e.g., LD_{50} values), this is not possible for carcinogens or mutagens where one single exposure can have an effect. Also realize that dangers related to a given compound may also depend on its physical state (fine powder vs. large crystals/diethylether vs. glycerol/dry ice vs. carbon dioxide under pressure in a gas bomb). Anticipate under which circumstances during an experiment exposure is most likely to occur and how best to protect yourself and your environment.

HAZARDOUS MATERIALS

In general, proprietary materials are not listed here. Kits and other commercial items as well as most anesthetics, sedatives, dyes, fixatives, embedding media, stains, herbicides, and fungicides are also not included. Anesthetics and antibiotics also require special care. Follow the manufacturer's safety guidelines that accompany these products.

Acetic acid (concentrated) must be handled with great care. It may be harmful by inhalation, ingestion, or skin absorption. Wear appropriate gloves and goggles. Use in a chemical fume hood.

Acetic acid (glacial) is highly corrosive and must be handled with great care. It may be a carcinogen. Liquid and mist cause severe burns to all body tissues. It may be harmful by inhalation, ingestion, or skin absorption. Wear appropriate gloves and goggles and use in a chemical fume hood. Keep away from heat, sparks, and open flame.

Ampicillin may be harmful by inhalation, ingestion, or skin absorption. Wear appropriate gloves and safety glasses and use in a chemical fume hood.

Bleach (Sodium hypochlorite), NaOCl, is poisonous, can be explosive, and may react with organic solvents. It may be fatal by inhalation and is also harmful by ingestion and destructive to the skin. Wear appropriate gloves and safety glasses and use in a chemical fume hood to minimize exposure and odor.

Boric acid, H_3BO_3, may be harmful by inhalation, ingestion, or skin absorption. Wear appropriate gloves and goggles.

$CaCl_2$, *see* **Calcium chloride**

Calcium chloride, $CaCl_2$, is hygroscopic and may cause cardiac disturbances. It may be harmful by inhalation, ingestion, or skin absorption. Do not breathe the dust. Wear appropriate gloves and safety goggles.

Carbon dioxide, CO_2, in all forms may be fatal by inhalation, ingestion, or skin absorption. In high concentrations, it can paralyze the respiratory center and cause suffocation. Use only in well-ventilated areas. In the form of dry ice, contact with carbon dioxide can also cause frostbite. Do not place large quantities of dry ice in enclosed areas such as cold rooms. Wear appropriate gloves and safety goggles.

Cresol red may be harmful by inhalation, ingestion, and skin absorption. Wear appropriate gloves and safety goggles and use in a chemical fume hood, Do not breathe the dust.

Dry ice, *see* **Carbon dioxide**

EDTA, *see* **Ethylenediamenetetraacetic acid**

Ethanol (EtOH), CH_3CH_2OH, is highly flammable and may be harmful by inhalation, ingestion, or skin absorption. Wear appropriate gloves and safety glasses. Keep away from heat, sparks, and open flame.

Ethidium bromide is a powerful mutagen, is toxic and may cause irreversible effects. Consult the local institutional safety officer for specific handling and disposal procedures. Avoid breathing the dust. Wear appropriate gloves when working with solutions that contain this dye.

Ethylenediaminetetraacetic acid (EDTA) may be harmful by inhalation, ingestion, or skin absorption. Wear appropriate gloves and safety glasses. Severe over-exposure can result in death.

Hydrochloric acid, HCl, is volatile and may be fatal if inhaled, ingested, or absorbed through the skin. It is extremely destructive to mucous membranes, upper respiratory tract, eyes, and skin. Wear appropriate gloves and safety glasses and use with great care in a chemical fume hood. Wear goggles when handling large quantities.

IPTG, *see* **Isopropyl-β-D-thiogalactopyranoside**

Isopropanol is flammable and irritating. It may be harmful by inhalation, ingestion, or skin absorption. Wear appropriate gloves and safety glasses. Do not breathe the vapor. Keep away from heat, sparks, and open flame.

Isopropyl-β-D-thiogalactopyranoside (IPTG) may be harmful by inhalation, ingestion, or skin absorption. Wear appropriate gloves and safety glasses.

Kanamycin may be harmful by inhalation, ingestion, or skin absorption. Wear appropriate gloves and safety glasses. Use only in a well-ventilated area.

KCl, *see* **Potassium chloride**

Liquid nitrogen (LN_2) can cause severe damage due to extreme temperature. Handle frozen samples with extreme caution. Do not breathe the vapors. Seepage of liquid nitrogen into frozen vials can result in an exploding tube upon removal from liquid nitrogen. Use vials with O-rings when possible. Wear cryo-mitts and a face mask. Do not allow the liquid nitrogen to spill onto your clothes. Do not breathe the vapors.

Magnesium chloride, $MgCl_2$**,** may be harmful by inhalation, ingestion, or skin absorption. Wear appropriate gloves and safety glasses and use in a chemical fume hood.

Magnesium sulfate, $MgSO_4$**,** presents chronic health hazards and affects the central nervous system and the gastrointestinal tract. It may be harmful by inhalation, ingestion, or skin absorption. Wear appropriate gloves and safety glasses and use in a chemical fume hood.

Methylene blue is irritating to the eyes and skin. It may be harmful by inhalation, ingestion, or skin absorption. Wear appropriate gloves and safety glasses.

$MgCl_2$, *see* **Magnesium chloride**

$MgSO_4$, *see* **Magnesium sulfate**

NaOH, *see* **Sodium hydroxide**

Potassium chloride, KCl, may be harmful by inhalation, ingestion, or skin absorption. Wear appropriate gloves and safety glasses.

Proteinase K is an irritant and may be harmful by inhalation, ingestion, or skin absorption. Wear appropriate gloves and safety glasses.

SDS, *see* **Sodium dodecyl sulfate**

Sodium dodecyl sulfate (**SDS**) is toxic, an irritant, and poses a risk of severe damage to the eyes. It may be harmful by inhalation, ingestion, or skin absorption. Wear appropriate gloves and safety goggles. Do not breathe the dust.

Sodium hydroxide, NaOH, and **solutions containing NaOH,** are highly toxic and caustic and should be handled with great care. Wear appropriate gloves and a face mask. All other concentrated bases should be handled in a similar manner.

SYBR GREEN/GOLD is supplied by the manufacturer as a 10,000-fold concentrate in DMSO which transports chemicals across the skin and other tissues. Wear appropriate gloves and safety glasses and decontaminate according to your safety office guidelines. *See* **DMSO**.

Tetracycline may be harmful by inhalation, ingestion, or skin absorption. Wear appropriate gloves and safety glasses and use in a chemical fume hood.

Transilluminator, *see* **UV light**

Tris is an irritant and may be harmful by inhalation, ingestion, or skin absorption. Wear appropriate gloves and safety glasses.

UV light and/or **UV radiation** is dangerous and can damage the retina. Never look at an unshielded UV light source with naked eyes. Examples of UV light sources that are common in the laboratory include hand-held lamps and transilluminators. View only through a filter or safety glasses that absorb harmful wavelengths. UV radiation is also mutagenic and carcinogenic. To minimize exposure, make sure that the UV light source is adequately shielded. Wear protective appropriate gloves when holding materials under the UV light source.

Index

H

I